Mobile Intelligence

**WILEY SERIES ON PARALLEL
AND DISTRIBUTED COMPUTING**

Editor: Albert Y. Zomaya

A complete list of titles in this series appears at the end of this volume.

Mobile Intelligence

Edited by

Laurence T. Yang
Agustinus Borgy Waluyo
Jianhua Ma
Ling Tan
Bala Srinivasan

WILEY

A JOHN WILEY & SONS, INC., PUBLICATION

Library of Congress Cataloging-in-Publication Data:

Yang, Laurence T.
  Research in mobile intelligence / Laurence T. Yang.
      p. cm.
  Includes bibliographical references and index.
  ISBN 978-0-470-19555-0 (cloth)
  1. Mobile computing–Research. 2. Computational intelligence–Research. 3. Wireless communication systems–Research. I. Waluyo, Agustinus Borgy II. Ma, Jianhua III. Tan, Ling IV. Srinivasan, Bala V. Title.
  QA76.59.Y35 2010
  004.16–dc22
                                                                      2009034629

10 9 8 7 6 5 4 3 2 1

Contents

Preface   ix

Contributors   xv

Part I   Mobile Data and Intelligence

1. **A Survey of State-of-the-Art Routing Protocols for Mobile
   Ad Hoc Networks**   3
2. **Connected Dominating Set for Topology Control in Ad Hoc
   Networks**   26
3. **An Intelligent Way to Reduce Channel Under-utilization in Mobile
   Ad Hoc Networks**   43
4. **Mobility in Publish/Subscribe Systems**   62
5. **Cross-Layer Design Framework for Adaptive Cooperative Caching
   in Mobile Ad Hoc Networks**   87
6. **Recent Advances in Mobile Agent-Oriented Applications**   106

Part II   Location-Based Mobile Information Services

7. **KCLS: A Cluster-Based Location Service Protocol and Its
   Applications in Multihop Mobile Networks**   143
8. **Predictive Location Tracking in Cellular and in Ad Hoc Wireless
   Networks**   163
9. **An Efficient Air Index Scheme for Spatial Data Dissemination
   in Mobile Computing Environments**   191
10. **Next Generation Location-based Services: Merging Positioning
    and Web 2.0**   213

Part III   Mobile Mining

11. **Data Mining for Moving Object Databases**   239
12. **Mobile Data Mining on Small Devices Through Web Services**   264

Part IV   Mobile Context-Aware and Applications

13. **Context Awareness: A Formal Foundation**   279
14. **Experiences with a Smart Office Project**   294
15. **An Agent-Based Architecture for Providing Enhanced
    Communication Services**   320

Part V   Mobile Intelligence Security

16. **MANET Routing Security**   345
17. **An Online Scheme for Threat Detection Within Mobile
    Ad Hoc Networks**   380
18. **SMRTI: Secure Mobile Ad Hoc Network Routing with Trust Intrigue** 412
19. **Managing Privacy in Location-based Access Control Systems**   437

Part VI   Mobile Multimedia

20. **VoiceXML-Enabled Intelligent Mobile Services**   471
21. **User Adaptive Video Retrieval on Mobile Devices**   488
22. **A Ubiquitous Fashionable Computer with an i-Throw Device
    on a Location-based Service Environment**   510
23. **Energy Efficiency for Mobile Multimedia Replay**   530

Part VII   Intelligent Network

24. **Efficient Data-Centric Storage Mechanisms in Wireless
    Sensor Networks**   549

25. Tracking in Wireless Sensor Networks **573**
26. DDoS Attack Modeling and Detection in Wireless Sensor Networks **595**
27. Energy-Efficient Pattern Recognition for Wireless Sensor Networks **627**

Index **661**

Preface

Nowadays, it is very common to see people everywhere and at anytime use mobile telecommunications devices, such as cellular/mobile phones, Personal Digital Assistants (PDAs), or laptop equipped with wireless connectivity to communicate with the network and access information. This phenomenon, which is generally known as mobile computing is increasing rapidly world wide. Mobile phones companies have been racing to release the latest version of their products to the market and just recently Apple Inc. introduced its first ever and revolutionary mobile phone, iPhone, which was long awaited by most users. It is a challenge for the mobile service providers to keep up with the information services provided to mobile users in order to gain commercial advantage.

Thanks to the proliferation of wireless technology that has led to the emergence of mobile computing paradigm. Mobile computing introduces anytime and anywhere computing and hence offers many promising applications that have never existed before. However, there are also important issues such as limited storage restriction, frequency of disconnection, narrow bandwidth capacity, security, asymmetric communications costs and bandwidth, and small screen size that need to be addressed before the true potential of mobile computing can be realized. Computational intelligence has emerged as a powerful and indispensable method of analyzing a variety of problems in research. It uses learning, adaptive, or evolutionary computations to create programs that are, in some sense, intelligent. It has been adopted to resolve issues in many different areas including security, marketing, image-quality prediction, and risk assessment. As such, it shows a clear potential in addressing essential issues in mobile computing.

The use of computational intelligence approaches to mobile paradigm is very promising. The combination of these two disciplines creates a novel theme of what we called mobile intelligence. Mobile intelligence can benefit a wide variety of applications including database query processing, multimedia, network security, information retrieval, commerce systems, web traffic and applications, and search engines to name a few.

This book covers a comprehensive state-of-the-art in various applications of computational intelligence to mobile paradigm including mobile data intelligence, mobile mining, mobile intelligence security, mobile agent, location-based mobile information services, mobile context-aware and applications, intelligent networks, and mobile multimedia. This book attempts to dedicate to the research on this theme with the purpose to advance and disseminate the most recent research under this theme.

The chapters presented in this book are invited only and selected from prominent researchers in the area. The chapters are divided into seven research domains namely: mobile data and intelligence (chapters 1–6), location-based mobile information services (chapters 7–10), mobile mining (chapters 11 and 12), mobile context-aware and applications (chapters 13–15), mobile intelligence security (chapters 16–19), mobile multimedia (chapters 20–23), and intelligent network (chapters 24–27).

Part I (Mobile Data and Intelligence) comprises of six chapters, each of which concerns with efficient and effective mobile data retrieval by addressing different aspects in wireless environment. Chapter 1 presents a survey of existing intelligent routing protocols for mobile *ad hoc* networks and provides reviews on several key elements. It also discusses how these protocols are evaluated considering the packet throughput and latency performance. This survey is an excellent starting point to learn about wireless routing protocol in mobile *ad hoc* networks (MANET). In chapter 2, topology management for connected dominating set (CDS) formation in MANET is proposed. An area algorithm is deployed to reduce the number of nodes in the CDS, and thus it helps to efficiently manage the networks. Chapter 3 presents an intelligent scheme to optimize the medium access control (MAC) mechanism in MANET by releasing the unused reserved channel. As such, channel efficiency can be improved while increasing the channel throughput. Chapter 4 introduces publish/subscribe messaging paradigm in mobile environment. Such approach decouples the communication between providers and user's data, and hence benefits mobile applications by its ability to hide intermittent disconnections. Further, it is incorporated with filtering scheme to select only messages of interest to the users, which in turns optimizes the bandwidth utilization over limited bandwidth available in the wireless network. Chapter 5 defines cluster-based cooperation caching and pre-fetching scheme in MANET to improve query performance. Some open issues have been highlighted and described in details. The deployment of intelligent mobile agent as a means to support transparent provision of data and services in mobile environment is discussed in chapter 6. It also describes a meaningful background on how mobile agent can help provide users with anytime and anywhere access of information.

Part II (Location-Based Mobile Information Services) is dedicated to intelligent approaches to resolve location-based information retrieval in mobile environment. There are four chapters in this part. Chapter 7 presents an intelligent location service protocol in MANET based on the clustering architecture. This protocol, which is designed to support multi-hop wireless networks, offers various advantages including overheads reduction, scalability, fault-tolerance, and better location information accuracy. Chapter 8 reports a study on predictive location tracking in cellular as well as *ad hoc* wireless networks. This chapter identifies the benefits of being able to forecast the future locations on mobile users, investigates the key factors in location predictive, and defines a feasible model for location tracking. Such model helps to reduce the latency and resource consumption in any wireless network as well as prolong the lifetime of the network. Although significant progresses have made in location predicting, the chapter defines further important issues that are left open for investigations.

Chapter 9 concerns with efficient wireless indexing schemes for mobile data broadcasting. In this chapter, several indexing schemes from the literatures have been classified and discussed in details. Further, the authors propose cell-based distributed spatial index (CEDI), which instead of incorporating pointer for each broadcast item intelligently incorporates distributed structure for a group of data items. Thus, it eliminates listening to redundant data items when processing window query, which results in better power efficiency and query access time. Chapter 10 envisions the deployment of Web 2.0 for location-based services (LBS) in mobile environment. It provides a comprehensive discussion on how Web 2.0 may benefit LBS, how it works, and how it can be implemented. Some open issues have been highlighted, including the need for privacy preservation schemes.

Part III (Mobile Mining) contains two chapters on learning interesting pattern and knowledge from data generated by mobile users. Chapter 11 presents a comprehensive review on mobile mining. This is an excellent reference to learn about applications of data mining for moving objects, current trend, issues, and methods. Chapter 12 provides an interesting discussion on pervasive data mining from mobile devices through the use of Web services. This chapter conveys the possibility of implementing mobile Web services that allow execution of data mining tasks remotely from mobile devices and receive the results of such service accordingly.

Part IV (Mobile Context-Aware and Applications) consists of three chapters, which investigate the importance of intelligent schemes as a means to support context-aware mobile applications. A formal foundation of context-aware computing is first introduced in chapter 13. In this chapter, a framework to understand basic principles of context-aware computing is presented. This may form a basis for a formal design and evaluations of context-aware technologies. Chapter 14 describes the authors' experience in developing ubiquitous middleware, mobile-agent system for handling user's context, location tracking, and prediction techniques as a means to support dynamic office room allocation of the employees in order to reduce operational costs. The use of mobile agent for context-aware computing has been of much interest, and its potential to resolve issues in personal communication between users is explored in chapter 15. In this chapter, the authors propose agent-based context-aware communication system for personal communication services based on the user's needs and preferences. The proposed system also incorporates a mechanism for handling conflicting users' policies.

Part V (Mobile Intelligence Security) is comprised of four chapters, which attempt to solve security aspects in mobile environment by deploying various intelligent schemes. Chapter 16 reviews the existing routing approaches in MANET and pinpoints their limitations especially in addressing selfish or malicious entities that selectively drop packets rather than forward them. The chapter defines a new scheme called robust source routing (RSR), which adopts the concept of forerunner packets. This scheme informs all nodes along the path about the flow of the packets and the time frame to receive each packet. As such, malicious agent can be identified and isolated. Chapter 17 introduces an online scheme for threat detection within MANET. The proposed scheme is designed using a special form of neural network, known as the graph neuron (GN). The scheme implements a pattern recognition approach, where the states of the network are considered as patterns. These patterns are collected and analyzed in real-time for discovering network intrusions and threat detection. An extension of GN called distributed hierarchical graph neuron (DHGN) is presented and its superior performance over standard GN is shown. Further investigation on secure routing protocols in MANET is discussed in Chapter 18. This chapter emphasizes the importance to evaluate trustworthiness between intermediate nodes in order to enhance the security level in MANET. A repudiation-based trust model called SMRTI or secure mobile *ad hoc* network routing with trust intrigue is introduced. This model is able to capture and evaluate the evidence of trustworthiness through several means including recommendations approach. It is then applied to predict the potential of a node being malicious. Chapter 19 conveys a detailed discussion on the importance, design, and requirements of location-based access control system. Privacy issue is also highlighted and a matric called relevance has been defined to use for both measuring the degree of location privacy and accuracy. It is complimented with some examples and case studies.

Part VI (Mobile Multimedia) focuses on the application of intelligent methods for multimedia retrieval in mobile environment. There are four chapters in this section. Chapter 20 presents the development of intelligent mobile context-aware recommender system based on VoiceXML called VOICE. Specifically, restaurant recommendation system has been showcased and its effective uses to provide user with a selection of restaurants following the user's preferences and the direction to the selected restaurant, has been reported. Chapter 21 addresses the issue of efficient and effective multimedia content management and retrieval in mobile computing. A system called MoVR is proposed and presented in detail. The system deploys intelligent schemes for personal video retrieval including hierarchical markov model mediator (HMMM) to model and organize the videos, HMMM-based profiles to capture and store individual user's access histories and preferences such that the system can provide the "personalized recommendation," and the fuzzy association concept to empower the framework so that the users can make their choice. To enhance efficient processing, computationally intensive operations are executed in the server side, while mobile clients mainly target to manage the retrieved media and user feedbacks for the current query. The system also supports information caching for the mobile devices. A prototype of the system on mobile-based soccer video is also showcased. In chapter 22, the ubiquitous fashionable computer (UFC) is introduced. UFC is a wearable computer that exploits ubiquitous computing environment. The system supports the interoperability of different communication links (i.e., Bluetooth, Zigbee) and intuitive use between user interfaces and UFC. The user interface deployed is called i-Throw or intuitive input devices, which recognizes humans' hand gestures, direction and control ubiquitous devices with the gestures. The prototype system has been developed and its effective uses are shown. Chapter 23 presents in-depth experiments on a variety of video codecs deployed on mobile devices with focus on energy consumptions. A review based on the experimental data is presented. This review is much useful to determine the most optimized codecs and their parameters for mobile video.

Part VII (Intelligent Network) is going one step ahead from mobile computing to pervasive computing whereby the notion of computing is seamlessly integrated with the environment and mostly involved wireless sensor devices. It consists of four chapters on the advanced techniques in solving various issues in wireless sensor network (WSN). Chapter 24 addresses the issue of data storage in WSN. Existing solutions on this issue is discussed and compared comprehensively. At the end, some open issues and future directions concerning data-centric storage in WSN are highlighted. Due to numerous deployments of sensor devices in a typical WSN set up, it is of importance to incorporate tracking mechanism especially when the sensors are located in harsh environment and far from human interventions. Chapter 25 provides a solution with this tracking issue by introducing CollECT. CollECT is designed to incorporate vicinity triangulation, event determination, and border sensor selection procedures. It is able to achieve event detection and tracking promptly. Other approaches in the literatures are also discussed and some open issues for future studies are highlighted. Attack on WSN is another essential challenge to overcome. Such an attack may corrupt the network and cause instability. Chapter 26 deals with this issue and presents the model of a distributed denial of service (DDoS) attack. DDoS is considered as flooding attacks that exploit the asymmetry between the network line rate and the node's processing capabilities. Its aim is to manipulate the intensity of the traffic in order to incapacitate the victim or its network. A centralized approach based on self-organising maps (SOMs) to detect such an attack is proposed. SOM neural network is trained with patterns of network traffic, consisting of both attack and normal condition. The proposed SOM-based approach is showing promise in detecting attacks in constant network traffic environment. Last but not least, chapter 27 addresses another critical issue in WSN, that is energy efficiency. In this chapter, a pattern recognition approach called voting graph neuron (VGN) is proposed. This approach employs collaboration among sensor nodes to detect events and introduces a novel energy-efficient template-matching approach. This approach is able to alternate sensor nodes between active and sleep modes to dynamically control their collaboration and conserve energy.

As a very new area of research, mobile intelligence has created a wide range of opportunities for researchers, engineers, and developers to create new interesting applications for both the end users and businesses. As more knowledge about mobile users is gained in the near future, it is, therefore, expected that the upcoming applications and systems will be able to adapt more seamlessly our daily lives.

Readers of this book will not only learn different areas of mobile computing and its issues but also various intelligent approaches that contribute in addressing these issues as well as to discover other potential elements in mobile paradigm.

We hope that you enjoy the book. Comments and feedbacks from readers would be greatly appreciated.

Laurence T. Yang
Agustinus Borgy Waluyo
Jianhua Ma
Ling Tan
Bala Srinivasan

Contributors

A. H. Muhamad Amin, *Clayton School of Information Technology, Monash University, Clayton, Victoria, Australia*, An Online Scheme for Threat Detection Within Mobile Ad Hoc Networks

Claudio Agostino Ardagna, *Dipartimento di Tecnologie dell'Informazione, Universit'a di Milano, Milan, Italy*, Managing Privacy in Location-based Access Control Systems

F. Bagci, *Institute of Computer Science, University of Augsburg, Augsburg, Germany*, Experiences with a Smart Office Project

Zubair A. Baig, *Faculty of Information Technology, Monash University, Clayton, Victoria, Australia*, DDoS Attack Modeling and Detection in Wireless Sensor Networks

Venkat Balakrishnan, *Information and Networked Systems Security (INSS) Research Group, Department of Computing, Macquarie University, Sydney, Australia*, SMRTI: Secure Mobile Ad Hoc Network Routing with Trust Intrigue

M. Baqer, *Clayton School of Information Technology, Monash University, Clayton, Victoria, Australia*, Pattern Recognition Approach for Energy-Constraint Wireless Networks

Panayiotis Bozanis, *Department of Computer and Communication Engineering, University of Thessaly, Thessaly, Greece*, Predictive Location Tracking in Cellular and in Ad Hoc Wireless Networks

Debasish Chakraborty, *Research Institute of Electrical Communication, Tohoku University, Sendai, Japan*, An Intelligent Way to Reduce Channel Under-utilization in Mobile Ad hoc Networks

Chih-Yung Chang, *Department of Computer Science and Information Engineering, Tamkang University*, Efficient Data-Centric Storage Mechanisms in Wireless Sensor Networks

Chih-Yung Chang, *Department of Computer Science and Information Engineering, Tamkang University*, Tracking in Wireless Sensor Networks

Min Chen, *Department of Computer Science, University of Montana, Missoula, MT*, User Adaptive Video Retrieval on Mobile Devices

Shu-Ching Chen, *School of Computing and Information Sciences, Florida International University, Miami, FL*, User Adaptive Video Retrieval on Mobile Devices

Claude Crépeau, *School of Computer Science, McGill University, Montréal, Quebec, Canada*, MANET Routing Security

Marco Cremonini, *Dipartimento di Tecnologie dell'Informazione, Universit'a di Milano, Milan, Italy*, Managing Privacy in Location-based Access Control Systems

Amitava Datta, *School of Computer Science and Software Engineering, University of Western Australia, Perth, WA, Australia*, A Survey of State-of-the-Art Routing Protocols for Mobile Ad Hoc Networks

Carlton R. Davis, *School of Computer Science, McGill University, Montréal, Quebec, Canada*, MANET Routing Security

Mieso K. Denko, *Department of Computing and Information Science, University of Guelph, Guelph, Canada*, Cross-Layer Design Framework for Adaptive Cooperative Caching in Mobile Ad Hoc Networks

Nikos Dimokas, *Department of Informatics, Aristotle University of Thessaloniki, Thessaloniki, Greece*, Predictive Location Tracking in Cellular and in Ad Hoc Wireless Networks

Xing Gao, *Department of Computer Science and Engineering, The Pennsylvania State University, University Park, PA*, Recent Advances in Mobile Agent-Oriented Applications

Bo Han, *Department of Computer Science, University of Maryland, College Park, MD*, Connected Dominating Set for Topology Control in Ad Hoc Networks

Ali R. Hurson, *Department of Computer Science and Engineering, The Pennsylvania State University, University Park, PA*, Recent Advances in Mobile Agent-Oriented Applications

Chong-Sun Hwang, *Department of Computer Science, Korea University, Seoul, Korea*, An Efficient Air Index Scheme for Spatial Data Dissemination in Mobile Computing Environments

SeokJin Im, *Department of Computer Science, Korea University, Seoul, Korea*, An Efficient Air Index Scheme for Spatial Data Dissemination in Mobile Computing Environments

Yoshiharu Ishikawa, *Information Technology Center, Nagoya University, Furo-cho, Chikusa-ku, Nagoya, Japan*, Data Mining for Moving Object Databases

Hans-Arno Jacobsen, *Department of Electrical and Computer Engineering, University of Toronto, Toronto, Canada*, Mobility in Publish/Subscribe Systems

Evens Jean, *Department of Computer Science and Engineering, The Pennsylvania State University, University Park, PA*, Recent Advances in Mobile Agent-Oriented Applications

Weijia Jia, *Department of Computer Science, City University of Hong Kong, Kowloon, Hong Kong*, Connected Dominating Set for Topology Control in Ad Hoc Networks

Yu Jiao, *Applied Software Engineering Research Group, Computational Sciences and Eng. Division, Oak Ridge National Laboratory, Oak Ridge, TN*, Recent Advances in Mobile Agent-Oriented Applications

Axel Küpper, *Mobile and Distributed Systems Group, Ludwig Maximilian University, Munich, Germany*, Next Generation Location-based Services: Merging Positioning and Web 2.0

Dimitrios Katsaros, *Department of Computer & Communication Engineering, University of Thessaly, Thessaly, Greece*, Predictive Location Tracking in Cellular and in Ad Hoc Wireless Networks

Dimitrios Katsaros, *Department of Informatics, Aristotle University of Thessaloniki, Thessaloniki, Greece*, Predictive Location Tracking in Cellular and in Ad Hoc Wireless Networks

A. I. Khan, *Clayton School of Information Technology, Monash University, Clayton, Victoria, Australia*, An Online Scheme for Threat Detection Within Mobile Ad Hoc Networks

A. I. Khan, *Clayton School of Information Technology, Monash University, Clayton, Victoria, Australia*, Energy-Efficient Pattern Recognition for Wireless Sensor Networks

Asad I. Khan, *Faculty of Information Technology, Monash University, Clayton, Victoria, Australia*, DDoS Attack Modeling and Detection in Wireless Sensor Networks

F. Kluge, *Institute of Computer Science, University of Augsburg, Augsburg, Germany*, Experiences with a Smart Office Project

Stan Kurkovsky, *Central Connecticut State University, New Britain, LT*, VoiceXML-Enabled Intelligent Mobile Services

Jupyung Lee, *Korea Advanced Institute of Science and Technology, Daejeon, Korea*, A Ubiquitous Fashionable Computer with an i-Throw Device on a Location-based Service Environment

Supeng Leng, *University of Electronic Science and Technology of China, China*, KCLS: A Cluster-Based Location Service Protocol and Its Applications in Multihop Mobile Networks

Seung-Ho Lim, *Korea Advanced Institute of Science and Technology, Daejeon, Korea*, A Ubiquitous Fashionable Computer with an i-Throw Device on a Location-based Service Environment

Chu-Hsing Lin, *Department of Computer Science and Information Engineering, Tunghai University, Taichung, Taiwan*, Energy Efficiency for Mobile Multimedia Replay

Jung-Chun Liu, *Department of Computer Science and Information Engineering, Tunghai University, Taichung, Taiwan*, Energy Efficiency for Mobile Multimedia Replay

Luigi Logrippo, *Département d'informatique et ingénierie, Université du Québec en Outaouais, Québec, Canada*, An Agent-Based Architecture for Providing Enhanced Communication Services

Luigi Logrippo, *School of Information Technology and Engineering, University of Ottawa, Ottawa, Canada*, An Agent-Based Architecture for Providing Enhanced Communication Services

Phillip Lucs, *Information and Networked Systems Security (INSS) Research Group, Department of Computing, Macquarie University, Sydney, Australia*, SMRTI: Secure Mobile Ad Hoc Network Routing with Trust Intrigue

Muthucumaru Maheswaran, *School of Computer Science, McGill University, Montréal, Quebec, Canada*, MANET Routing Security

Yannis Manolopoulos, *Department of Informatics, Aristotle University of Thessaloniki, Thessaloniki, Greece*, Predictive Location Tracking in Cellular and in Ad Hoc Wireless Networks

Vinod Muthusamy, *Department of Electrical and Computer Engineering, University of Toronto, Toronto, Canada*, Mobility in Publish/Subscribe Systems

Naoki Nakamura, *Tohoku University School of Medicine, Tohoku University, Sendai, Japan*, An Intelligent Way to Reduce Channel Under-utilization in Mobile Ad hoc Networks

Mats Neovius, *Åbo Akademi University, Åbo, Finland*, Context Awareness: A Formal Foundation

Machigar Ongtang, *Department of Computer Science and Engineering, The Pennsylvania State University, University Park, PA*, Recent Advances in Mobile Agent-Oriented Applications

Ki-Woong Park, *Korea Advanced Institute of Science and Technology, Daejeon, Korea*, A Ubiquitous Fashionable Computer with an i-Throw Device on a Location-based Service Environment

Kyu Ho Park, *Korea Advanced Institute of Science and Technology, Daejeon, Korea*, A Ubiquitous Fashionable Computer with an i-Throw Device on a Location-based Service Environment

A. Pietzowski, *Institute of Computer Science, University of Augsburg, Augsburg, Germany*, Experiences with a Smart Office Project

Romelia Plesa, *School of Information Technology and Engineering, University of Ottawa, Ottawa, Canada*, An Agent-Based Architecture for Providing Enhanced Communication Services

Thomas E. Potok, *Applied Software Engineering Research Group, Computational Sciences and Eng. Division, Oak Ridge National Laboratory, Oak Ridge, TN*, Recent Advances in Mobile Agent-Oriented Applications

Pierangela Samarati, *Dipartimento di Tecnologie dell'Informazione, Universit'a di Milano, Milan, Italy*, Managing Privacy in Location-based Access Control Systems

B. Satzger, *Institute of Computer Science, University of Augsburg, Augsburg, Germany*, Experiences with a Smart Office Project

Adam Sharp, *Central Connecticut State University, New Britain, LT*, VoiceXML-Enabled Intelligent Mobile Services

Kuei-Ping Shih, *Department of Computer Science and Information Engineering, Tamkang University*, Efficient Data-Centric Storage Mechanisms in Wireless Sensor Networks

Kuei-Ping Shih, *Department of Computer Science and Information Engineering, Tamkang University*, Tracking in Wireless Sensor Networks

Norio Shiratori, *Research Institute of Electrical Communication, Tohoku University, Sendai, Japan*, An Intelligent Way to Reduce Channel Under-utilization in Mobile Ad hoc Networks

Mei-Ling Shyu, *Department of Electrical and Computer Engineering, University of Miami, Coral Gables, FL*, User Adaptive Video Retrieval on Mobile Devices

MoonBae Song, *Graduate School of Systems and Information Engineering, University of Tsukuba, Tsukuba, Japan*, An Efficient Air Index Scheme for Spatial Data Dissemination in Mobile Computing Environments

Subbiah Soundaralakshmi, *School of Computer Science and Software Engineering, University of Western Australia, Perth, WA, Australia*, A Survey of State-of-the-Art Routing Protocols for Mobile Ad Hoc Networks

Domenico Talia, *DEIS, University of Calabria, Rende (CS), Italy*, Mobile Data Mining on Small Devices Through Web Services

Jun Tian, *Department of Computing and Information Science, University of Guelph, Guelph, Canada*, Cross-Layer Design Framework for Adaptive Cooperative Caching in Mobile Ad Hoc Networks

Georg Treu, *Mobile and Distributed Systems Group, Ludwig Maximilian University, Munich, Germany*, Next Generation Location-based Services: Merging Positioning and Web 2.0

W. Trumler, *Institute of Computer Science, University of Augsburg, Augsburg, Germany*, Experiences with a Smart Office Project

Paolo Trunfio, *DEIS, University of Calabria, Rende (CS), Italy*, Mobile Data Mining on Small Devices Through Web Services

Uday Tupakula, *Information and Networked Systems Security (INSS) Research Group, Department of Computing, Macquarie University, Sydney, Australia*, SMRTI: Secure Mobile Ad Hoc Network Routing with Trust Intrigue

T. Ungerer, *Institute of Computer Science, University of Augsburg, Augsburg, Germany*, Experiences with a Smart Office Project

Vijay Varadharajan, *Information and Networked Systems Security (INSS) Research Group, Department of Computing, Macquarie University, Sydney, Australia*, SMRTI: Secure Mobile Ad Hoc Network Routing with Trust Intrigue

Sabrina De Capitani di Vimercati, *Dipartimento di Tecnologie dell'Informazione, Universit'a di Milano, Milan, Italy*, Managing Privacy in Location-based Access Control Systems

Sheng-Shih Wang, *Department of Information Management, Minghsin University of Science and Technology*, Tracking in Wireless Sensor Networks

Sheng-Shih Wang, *Department of Information Network Technology, Chihlee Institute of Technology*, Efficient Data-Centric Storage Mechanisms in Wireless Sensor Networks

David Wei, *CIS Department, Fordham University, Bronx, NY*, An Intelligent Way to Reduce Channel Under-utilization in Mobile Ad hoc Networks

Lu Yan, *University College London, London, UK*, Context Awareness: A Formal Foundation

Jong-Woon Yoo, *Korea Advanced Institute of Science and Technology, Daejeon, Korea*, A Ubiquitous Fashionable Computer with an i-Throw Device on a Location-based Service Environment

Gwo-Jong Yu, *Department of Computer and Information Science, Aletheia University*, Tracking in Wireless Sensor Networks

Gwo-Jong Yu, *Department of Computer Information Science, Aletheia University*, Efficient Data-Centric Storage Mechanisms in Wireless Sensor Networks

Liren Zhang, *University of South Australia, Australia*, KCLS: A Cluster-Based Location Service Protocol and Its Applications in Multihop Mobile Networks
Lizhuo Zhang, *Department of Computer Science, City University of Hong Kong, Kowloon, Hong Kong*, Connected Dominating Set for Topology Control in Ad Hoc Networks
Yan Zhang, *Simula Research Laboratory, Norway*, KCLS: A Cluster-Based Location Service Protocol and Its Applications in Multihop Mobile Networks
Na Zhao, *School of Computing and Information Sciences, Florida International University, Miami, FL*, User Adaptive Video Retrieval on Mobile Devices

Part I

Mobile Data and Intelligence

Chapter 1

A Survey of State-of-the-Art Routing Protocols for Mobile Ad Hoc Networks

Amitava Datta and Subbiah Soundaralakshmi

School of Computer Science and Software Engineering, University of Western Australia, Perth, WA, Australia

1.1	Introduction	3
1.2	A Taxonomy of MANET Routing Protocols	5
1.3	Proactive Routing Protocols	6
1.4	Reactive Routing Protocols	10
1.5	Other Routing Protocols	17
1.6	Conclusion	23
	References	24

1.1 INTRODUCTION

Mobile ad hoc networks (MANETs) have opened up many new possibilities in using computer networks in improvised scenarios where traditional networking infrastructure is unavailable. Mobile networks are usually based on wireless communication and hence many of the established technologies developed for wired networks are not directly applicable in mobile networks. In particular, nodes forming a mobile network are not tied to any infrastructure and can form a network on the fly and for a short

Mobile Intelligence. Edited by Laurence T. Yang, Agustinus Borgy Waluyo, Jianhua Ma, Ling Tan, and Bala Srinivasan
Copyright © 2010 John Wiley & Sons, Inc.

period of time. The main difficulties in forming a mobile network are the mobility of the nodes, the nature of the wireless medium, the energy constraints of small mobile nodes, and the possibility that nodes may join or leave a network anytime during the lifetime of the network.

Since nodes are mobile, the routes in the network usually have a short life span. A route may or may not exist for the entire duration of a data communication session, unlike in wired networks, where nodes are usually present in fixed geographical positions. The wireless medium has the constraint that any communication by a node is done through broadcasting of a packet. Since the bandwidth in the wireless medium is much less compared to wired networks, this poses the problem that almost all communications take up large amount of bandwidth due to the flooding of packets. Moreover, the contention for limited bandwidth results in packet collisions and retransmissions, further wasting the available bandwidth. The nodes in a mobile ad hoc network are usually powered by batteries that may not be rechargeable during a network session. This imposes the added constraint that nodes should not waste their energy in retransmitting packets that have been lost due to collision. Since nodes spend almost equal amount of energy in transmitting and receiving packets, even overhearing packets destined for other nodes results in wastage of energy.

Routing of packets is one of the most basic activities in any computer network, wired or wireless. All applications in a MANET depend on reliable and efficient routing of packets. Hence, it is extremely important to design routing protocols that can work within the constraints of a mobile ad hoc network and provide support for all higher level applications. It is not surprising that tremendous amount of research effort has been invested in designing efficient routing protocols for MANETs during the past 10 years. The MANET working group within the Internet Engineering Task Force (IETF) [8] is considering several of these protocols for standardization. However, the task of the MANET working group is considerably difficult due to the existence of many different routing protocols in the literature. Currently, there are four protocols under consideration by the working group and the Internet drafts for these protocols are available for comments.

Our aim in this chapter is to provide a concise yet comprehensive view of many different routing protocols proposed for MANETs. We will pay special attention to the protocols under consideration by the MANET working group as naturally these protocols are some of the most efficient and reliable routing protocols proposed until now. However, we will also discuss the evolution of routing protocols from a historical point of view and discuss other protocols that have significantly different yet interesting ideas. More details about many of the protocols discussed in this chapter can be found in the paper by Belding-Royer and Toh [19] and the book edited by Perkins [15].

The rest of the chapter is organized as follows. We discuss a taxonomy of routing protocols in Section 1.2. We discuss the class of proactive routing protocols in Section 1.3 and reactive routing protocols in Section 1.4. We discuss some other classes of protocols in Section 1.5. Finally, we conclude with some comments in Section 1.6.

1.2 A TAXONOMY OF MANET ROUTING PROTOCOLS

The main aims of all routing protocols designed for MANETs are to achieve a high level of performance, in terms of high throughput, low latency, and low energy expenditure by individual nodes. However, these aims are quite often contradictory in the sense that a routing protocol might have to sacrifice one of them in order to satisfy another. For example, assume that we are trying to design a routing protocol that aims for low latency in packet delivery. This may be a quality of service (QoS) requirement in a network that delivers multimedia content from one node to another. If individual nodes want to deliver or forward packets very fast toward a destination node, they must have a very clear idea about the network topology and the routes to the destination should be as accurate as possible. However, the collection of accurate topology information requires exchange of local views of topology among the nodes. In other words, each node should inform other nodes about its neighbors frequently so that all nodes have up-to-date information about the network topology. This type of information exchange is done through sending *control messages* (this name is used to differentiate from the actual data packets) and requires the nodes to spend substantial amount of energy. Hence, a protocol may have to sacrifice battery power in order to achieve low latency.

All routing protocols implicitly assume that nodes in a MANET cooperate with each other in delivering packets. Nodes in a MANET can be classified into three categories from the point of view of a packet. A node may be a sender, receiver, or a forwarding node for the packet. A forwarding node tries its best to send a packet toward its destination. The question of security in MANETs is beyond the scope of this chapter and we will assume that all nodes in the network are trusted. We refer the reader to the paper by Pirzada et al. [18] and the references therein for more discussion on security in MANETs.

The main routing protocols for MANET can be classified into two categories, *proactive* and *reactive*, depending on how a protocol collects information about the topology of the network. Proactive protocols try to reduce latency in packet delivery by aggressively disseminating topology information throughout the network. This, however, has a detrimental effect that much of the available bandwidth in the wireless medium is used for sending control messages. Hence, a challenging problem in designing a proactive protocol is to reduce the effect of control messages in the network while still achieving an acceptable level of latency. We will discuss this issue in more detail in the next section.

Reactive protocols try to minimize the wastage of bandwidth by reducing the amount of control messages in the network. They try to find routes on-demand and do not depend on proactive collection of topology information for finding routes. However, this approach quite often increases latency in packet delivery. A challenging problem in designing reactive protocols is to reduce latency while maintaining the low volume of control messages. Mobile nodes executing reactive protocols quite often resort to indirect means such as overhearing passing wireless traffic to improve their knowledge of network topology. We will discuss reactive protocols in depth in Section 1.4.

There are other protocols that combine the advantages of both proactive and reactive protocols while eliminating their disadvantages. These protocols use a proactive protocol within a small neighborhood of each node so that the volume of control messages remain manageable and use a reactive protocol over the entire network. The zone routing protocol (ZRP) is the most notable among these protocols, and we will discuss this protocol briefly in Section 1.5.

We will also discuss an important class of protocols in Section 1.5, called link reversal protocols. These protocols are based on an elegant idea of maintaining a rooted directed acyclic graph (DAG) in a MANET. The main thrust in these protocols is to maintain the DAG (and hence routes); however, usually the overhead in these protocols is quite high.

1.3   PROACTIVE ROUTING PROTOCOLS

Proactive routing protocols try to collect as much information about the MANET as possible through proactive exchange of messages about their local topology. One of the earliest protocols for MANETs was the destination sequenced distance vector (DSDV) protocol [16], which is one of the best known proactive routing protocols. The DSDV protocol has a large overhead of control messages and hence it fell out of favor due to the emergence of more efficient reactive protocols such as the dynamic source routing (DSR) and the ad hoc on-demand distance vector (AODV) protocol. However, the low latency in packet delivery is one of the most attractive aspects of the proactive protocols. Considerable work has been done in recent years to reduce the overhead of the DSDV protocol and one of the most promising proactive protocols called the optimal link state routing (OLSR) protocol [13] is currently under consideration by the MANET working group. In this section, we will first discuss the DSDV protocol and then the OLSR protocol.

1.3.1   The Destination Sequenced Distance Vector Protocol

We should first list a few points about MANET routing protocols that are applicable for all the protocols discussed in this chapter. A routing protocol is a distributed algorithm executed by each node in a MANET. In other words, each node executes a local copy of the protocol on the data that they collect locally. Moreover, this distributed execution of a protocol aims to achieve some global performance goals such as high throughput, low latency, and minimizing the overall expenditure of energy. In the following, we will use the terms *packets* and *messages* to mean packets sent by nodes in the wireless medium.

DSDV is a table-driven protocol, in the sense that all routing decisions are taken by individual nodes based on their local routing tables. There are two parts in the DSDV protocol, namely, keeping the local routing tables as up-to-date as possible and computing routes with the help of the local routing tables. We will first discuss the second part as it will be clear how nodes find the best routes by executing the DSDV protocol. We will then discuss how nodes update their routing tables.

First, we assume that each node has collected up-to-date information about the topology of the network in its routing table. Given a source node S and a destination node D, the purpose of a routing table is to find a best path according to some metric between S and D. In a MANET, intermediate nodes (i.e., nodes other than S and D) forward the packets that they receive from S toward D. Hence, the best path from the point of view of an intermediate node I is a path starting at I with D as the destination.

We will take an abstract graph theoretical view of the routing table, but this view can be applied to any other representation of the table without much modification. We consider the nodes in the MANET as nodes of our graph. There is a link or edge between two nodes if they are within the transmission range of each other. In practice, links between neighboring nodes in a MANET may not be bidirectional as the transmission ranges of both the neighbors may not be equal. We will take the simplified view that links are always bidirectional to keep the descriptions of the protocols simpler. However, our descriptions can be modified for the case when the links are unidirectional. We can now view the routing table as an adjacency matrix of the graph representing the underlying MANET. We can indicate an edge between nodes i and j by a 1 in the entry at the intersection of the ith row and the jth column. Similarly, a 0 indicates the absence of an edge.

It is now possible to find shortest paths from this adjacency matrix by running Dijkstra's shortest path algorithm. Suppose the source node (every node is potentially a source node for packets) or an intermediate node I has to find a best path for a packet to the destination D. The node runs Dijkstra's shortest path algorithm on its routing table and finds the shortest path to D. This shortest path must pass through one of its neighbors. The task is then to forward the packet to this neighbor. The neighbor in turn runs Dijkstra's shortest path algorithm on its own routing table to find a best path to the destination and repeats the process of forwarding the packet. We have assumed till now that hop count is the only metric for finding best paths using Dijkstra's algorithm; however, we can use several other metrics that are relevant for the wireless medium. Some other metrics could be the bandwidth as a cost on a link, the possible delay of a link, and so on. The choice of shortest path is illustrated in Figure 1.1.

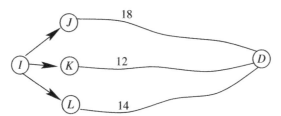

Figure 1.1   Illustration for selection of a forwarding node in the DSDV protocol. Node I finds three different paths to the destination D by running Dijkstra's shortest path algorithm on its routing table. I chooses neighbor K as the next hop of the packet since the cost of the path to D through K is least. The node IDs are shown inside the nodes and cost of the paths from J, K, and L are shown on the paths.

We now turn our attention to the other part of the DSDV protocol, namely, collection of topology information. If we expect nodes to find best paths by executing Dijkstra's shortest path algorithm, the routing table at each node should be as up-to-date as possible. Note that we can possibly never have the situation that all the routing tables in all the nodes of the MANET contain correct information. In other words, all the links that are recorded as links between neighbors may not exist at any time during the execution of the protocol due to the mobility of the nodes. Consider three nodes i, j, and k in a MANET. The node i may have noted the information that there is a link between j and k. However, this information is usually relatively old and by the time i uses this information to compute shortest paths, the link between j and k might have already broken due to the mobility of these two nodes.

This is a very important point to note as it differentiates a centralized algorithm from a distributed one; in particular, centralized algorithms usually have complete input before the execution starts. A distributed algorithm has only a partial view of the input at each node and the node executing the algorithm has to take decisions using this partial input. Moreover, routing protocols in a MANET do not even have a correct partial input due to the mobility of the nodes. Hence, any routing protocol in a MANET is not an exact algorithm in the traditional sense. It is in a sense an algorithm that tries to achieve best results using incomplete and partially incorrect inputs.

The nodes executing the DSDV protocol exchange their routing tables to update their knowledge of the network topology. Consider a node i and its routing table. Suppose i currently has three neighbors j, k, and l. If any of these three neighbors, say j, moves out of the transmission range of i, there is a change in the routing table of i. This type of change triggers a broadcast of the routing table from i to all the other nodes in the network, so that all the other nodes can update the routing tables with the changed topology. The broadcast is done by i sending its routing table to all its current neighbors and the neighbors sending it to their neighbors, and so on. If a node m receives such a broadcast from another node i, that may change the routing table of m, triggering a broadcast from m.

This is clearly an expensive process as all nodes in a MANET need to broadcast their routing tables due to changes in their local topology triggered by the mobility of the nodes. A large part of the bandwidth may be consumed by these update packets, especially in high-mobility scenarios. Several suggestions were made in the original proposal of DSDV to alleviate this overhead by sending incremental updates of routing tables instead of full updates. The idea of an incremental update is to broadcast a part of the routing table that has changed instead of sending the whole routing table. However, the fact remains that each update floods the entire network and the protocol becomes too inefficient in terms of throughput of data packets even in moderate-mobility scenarios.

We conclude the discussion of the DSDV protocol by mentioning another of its important features, assignment of sequence numbers to packets. One of the effects of broadcasting a packet, say p, from one node i to all other nodes in a MANET is that another node j gets multiple copies of p. Moreover, if i has sent two different updates of its routing table at two different instances, there is no guarantee that these two packets will arrive at j in the correct order, that is, the packet sent earlier may reach later. Since DSDV depends on the correct topology information for routing, it is very important that nodes use most recent information for updating their routing tables. One way of ensuring that nodes use most recent information is to time-stamp each packet with the current time. That way if j receives two packets from i, it can decide which one is more recent. However, this scheme works only when all nodes maintain synchronized clocks. Clock synchronization is a difficult task in a distributed system, and in particular in a MANET, since we do not expect the nodes have access to any infrastructure. Another alternative is to use sequence numbers that work as logical clocks. Each node stamps the packets that it broadcasts with an increasing integer called a sequence number. Any node j receiving two packets from a node i can decide which packet is more recent by comparing the sequence numbers in the packets.

1.3.2   The Optimized Link State Routing Protocol

Although the DSDV protocol is attractive due its low latency in finding routes, it has a very high overhead of control packets due to the broadcasting of the routing tables. Any broadcast floods the entire network and takes up a large portion of the available bandwidth in the wireless medium. The OLSR protocol is a relatively recent attempt to reduce the control overhead of the DSDV protocol in order to increase the throughput of packet delivery. The OLSR protocol tries to improve the DSDV protocol in two ways, by reducing the size of the updates and by reducing the effect of the broadcasts. We discuss these two improvements below.

Nodes executing the OLSR protocol broadcast link states rather than routing tables. Suppose a node i currently has three neighbors j, k, and l. If at least one of the neighbors, say k, moves out of the transmission range of i, there is a change in topology and i should inform other nodes about it through a broadcast. Note that it is sufficient for i to inform that the link $i–k$ has broken and other nodes can update their routing tables with this information. Hence, broadcasting link state information instead of routing tables is better to keep the volume of control messages low.

The main improvement in the OLSR protocol, however, comes from reducing the effect of broadcasting of packets. Note that, for a node i, all its neighbors receive any packet broadcast by i due to the fact that wireless transmission is omnidirectional. However, there is no need for all the neighbors of i to rebroadcast the packet again. It is desirable to choose only a subset of neighbors of i to rebroadcast the packet. The OLSR protocol uses a subset of neighbors called *multipoint relays* to broadcast its packets further. Strictly speaking, the concept of multipoint relays is not a part of the OLSR protocol. Rather, it is a concept used for reducing the volume of broadcast packets in wireless networks. We can illustrate the use of multipoint relays through an example. Consider again a node i and its three neighbors j, k, and l. These three neighbors are called one-hop neighbors of i. A two-hop neighbor can be reached from i in two transmissions. Suppose i has six two-hop neighbors a, b, c, d, e, and f. Clearly, each of these two-hop neighbors has at least one of the one-hop neighbors

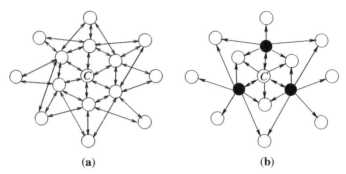

(a)   (b)

Figure 1.2   Illustration for simple flooding and flooding through multipoint relays. (a) This figure shows simple flooding. Every node that receives the packet initially sent by the central node C broadcasts it again. A double arrow between a pair of nodes indicates the receipt of broadcasts from each other (we have assumed symmetric communication links). (b) The number of packets has been reduced considerably by using the multipoint relay nodes (dark nodes). These nodes are one-hop neighbors of C and collectively neighbors of all the two-hop neighbors of C. The total number of packets is reduced considerably as the one-hop neighbors that are not multipoint relays do not broadcast.

of i as a neighbor (otherwise, they cannot be two-hop neighbors of i). As we have mentioned above, the multipoint relays for node i, denoted by $MPR(i)$, are a subset of one-hop neighbors of i such that all the two-hop neighbors of i are neighbors of the nodes in the subset $MPR(i)$. To illustrate this, suppose k and l collectively are neighbors of a, b, c, d, e, and f. In that case, we can choose k and l as the members of $MPR(i)$ so that only these two nodes forward the packets broadcast by i. There is no need for j to broadcast the packets from i. These MPR subsets are chosen by all the nodes in the MANET by keeping track of their one-hop and two-hop neighbors. It has been shown that the broadcast overhead can be significantly reduced by using the MPR sets. The process of broadcasting through multipoint relays is illustrated in Figure 1.2.

The OLSR protocol is currently under consideration by the MANET working group of IETF and it has the desirable properties of high throughput and low latency.

1.4 REACTIVE ROUTING PROTOCOLS

The main design aim of reactive routing protocols is to reduce the control packet over-head of proactive protocols. These protocols do not maintain routing tables proactively and, as a result, cannot find routes as soon as they are required. There is usually a delay or latency in finding routes. This results in high throughput in packet delivery as the available bandwidth is utilized for delivery of data packets rather than regular flooding of control packets as in the proactive protocols. However, reactive protocols also need flooding or broadcasting of packets, which occurs on-demand. We will dis-cuss below two of the most important reactive protocols, namely, the DSR and the AODV protocol.

1.4.1 The Dynamic Source Routing Protocol

The DSR protocol [9, 10] has two distinct phases, *route discovery* and *route maintenance*. The route discovery phase starts when a source node, say S, wants to find a route to a destination node D. Once a route to D has been found, the route maintenance phase starts while S transfers its data packets to D using the discovered route. We discuss these two phases in detail below.

A source node S starts route discovery by sending a *route request* (RREQ) packet to its neighbors. A RREQ packet has an identifier that includes a source node, a destination node, and a sequential ID that is an integer. If an intermediate node I receives a RREQ packet and it does not know a route to the destination node D, it takes one of the two actions. If it is a new RREQ packet (i.e., I has not seen this RREQ before), I broadcasts the packet to its neighbors after attaching its ID on the header of the RREQ packet. I also stores this packet in a list so that it can compare the identifier of this packet with future RREQ packets. If it is an old RREQ packet, that is, I has already received this packet in the past, I simply drops the packet.

If I knows a route to the destination D (we will discuss below how I may know a route), it initiates a *route reply* (RREP) packet by attaching the route I to D and the route from S to I in the header of the RREP packet and sending the packet back to the neighbor from whom it received the corresponding RREQ packet. Note that the RREP packet has the accumulated route from S to I in its header now. Hence, any intermediate node that receives such a RREP packet knows exactly the neighbor to whom it should send the RREP packet back. Eventually, the source node S receives the RREP packet and the route discovery phase ends. If no intermediate node knows a route to D, the RREQ packet reaches the destination D (provided the destination is in the same connected part of the network as S) and D sends the RREP packet. If the source node S does not receive a route reply within a specified period of time, it can initiate a new RREQ packet after assigning a new ID to the packet. The route discovery process in the DSR protocol is illustrated in Figure 1.3.

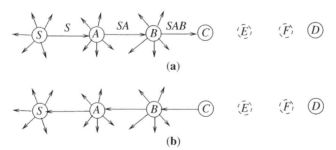

(a)

(b)

Figure 1.3 Illustration for the route discovery phase in the DSR protocol. (a) S starts the route discovery by broadcasting a RREQ packet. The arrows surrounding a node indicate omnidirectional broadcast. Every node appends its ID to the source route in the header of the RREQ packet. Node C has a route to destination D through the nodes E and F. (b) C sends a RREP packet back to S. Each node determines the next hop of the RREP packet by examining the source route.

One of the most important aspects of the route discovery process is the attaching of the IDs of all intermediate nodes in the header of the RREQ packet as well as the RREP packet. This list of IDs is sometimes called the *source route*. Recall that nodes executing the DSR protocol do not maintain routing tables, and hence a mechanism is needed through which a node can decide where to forward a packet when it receives a RREQ or RREP packet. The case for a RREQ packet is easier as it needs to be broadcast to all the neighbors. However, the purpose of a RREP packet is to deliver the packet to the source node S that initiated the route discovery. Hence, every intermediate node that receives a RREP packet should know the previous node (toward S) that originally sent it the corresponding RREQ packet. The purpose of attaching the source route to the header of a RREP packet is to provide this information. Moreover, RREQ packets also need to carry the source route as the node that initiates the RREP packet needs to copy this information from the corresponding RREQ packet.

Once S has received a RREP packet, it has a route to D and also knows the IDs of all the nodes along this route. S can now start the transfer of data packets using this route. S attaches the entire source route in the header of each data packet so that any intermediate node can correctly forward the packet to one of its neighbors along the source route. It is very important to maintain a route discovered in the route discovery phase. A route may break in the middle of data transmission due to the mobility of the nodes. For example, suppose a link in the route is G–H, where G and H are the end nodes of the link. If now H moves out of the transmission range of G, the entire route from S to D will be broken. The node G in this case sends a *route error* (RERR) packet to S by using the source route in the header. A RERR packet is almost similar to a RREP packet except that the purpose now is to inform S that the route it was using has broken. S has now two choices, either to start using an alternate route if it has one or start the route discovery phase again to find a new route.

Till now we have mentioned that nodes executing the DSR protocol do not use a routing table. However, in practice, running route discovery phase again and again is too expensive in a MANET as each route discovery is essentially a flooding of the entire network using RREQ packets. Hence, DSR uses a data structure called a *route cache* to reduce the effect of flooding. The purpose of a route cache is to store any information that a node can gather either from the packets that it receives or forwards or from the passing traffic. In particular, nodes may utilize the promiscuous mode operation allowed in IEEE 802.11 standards, where a node can overhear traffic that is not intended for it. The working of a route cache can be explained through a simple example. Consider a node I and a destination D. Suppose I worked as an intermediate node for forwarding data packets to D for a source node P. Assume also that the path from I to D is through some other intermediate nodes $E, F,$ and G. Now I stores this route fragment I–E–F–G–D in its route cache. If I receives a RREQ packet for the destination D from some other source S in future, it can use this cached route for sending a RREP packet to S.

However, cached route may not always provide correct information due to the mobility of the nodes in a MANET. For example, in the example above, suppose node G has moved out of the transmission range of F and the link F–G has broken as a result. However, I is not aware of this link breakage. If I now sends a RREP in reply to the RREQ from S, it will report a route that has already broken. This causes a problem since this will result in further route requests from S. A better policy is to time out old entries from the cache as most probably old entries are invalid due to the mobility in the network. The ns-2 simulator [12] adopts this policy and times out all cache entries that are older than 20 s.

The DSR protocol employs several other optimizations in order to reduce the control packet overhead and improve throughput. We discuss here only two of these optimizations due to space constraint. The first optimization is called *packet salvaging*. As we have mentioned, an intermediate node that detects link breakage in a route sends a RERR packet to the source node of that route. In addition to that, such an intermediate node tries to deliver the packet by finding an alternate route to the destination from its route cache. In other words, it tries to salvage the data packet by sending it to the destination through an alternate route. Packet salvaging reduces the possibility of loss of data packets in nodes that have experienced link failures with their neighbors while forwarding packets for other nodes. Since the source node continues to send packets until it receives a RERR packet informing the breakage of a route, the buffers in the intermediate nodes may overflow when there is a large amount of traffic in the network.

Another important optimization is related to sending RREP packets in reply to route requests. In certain situations, the route caches in the nodes may be quite up-to-date and many different nodes may be in a situation to send route replies. However, the network may get flooded due to many route replies. Hence, DSR uses a strategy where a node waits for a random period of time before sending a route reply. If it overhears that another node has already sent a route reply for the same route request, or the source node has already started using an alternate route, it does not send a route reply. This optimization reduces the overhead in the network considerably.

DSR is one of the most efficient protocols in terms of throughput even in high-mobility scenarios. Moreover, DSR almost always finds shortest paths through its route discovery mechanism. However, one of the drawbacks of the DSR protocol is the use of source routes in every packet. The number of IDs in a source route increases as the lengths of the routes increase. Since the wireless medium usually supports relatively small packet size, it is not possible to keep an entire source route in a single packet if a route is long. On the other hand, there is no guarantee of delivery of packets in the correct sequence in the wireless medium. Hence, the problem cannot be solved by splitting source routes in multiple packets. Hence, DSR is a very efficient protocol for MANETs that are relatively small and when the routes are up to about 10 hops long.

1.4.2 The Ad Hoc On-Demand Distance Vector Protocol

The AODV protocol [1, 17] tries to remove the main drawback of the DSR protocol by eliminating source routes from its control and data packets. Moreover, AODV was the first protocol to support multicasting in a MANET. Till now we have discussed routing from the perspective of unicasting, that is, one source node sending packets to one destination node. However, this is only one of several types of communication used in a typical network. In multicasting, a source node may want to send the same packet to several destination nodes (called a *multicast group*). Broadcasting is a special case of multicasting, when a source node wants to send the same packet to all the nodes in the network. It is possible to support multicasting through finding unicast (or one-to-one) routes from the source to all the nodes in a multicast group. However, discovering and maintaining separate routes to multiple nodes is usually more expensive. It is desirable to have a common routing mechanism for all members in a multicast group. AODV solves this problem in an elegant way, as we will discuss later in this section.

AODV is in a sense a table-driven as well as a reactive protocol. Each node executing AODV maintains a routing table; however, this routing table is local and contains information only about the neighboring nodes. In contrast to DSDV, there is no need to update the routing tables through global broadcasting. AODV also has two phases, route discovery and route maintenance, like DSR. The route discovery phase is almost similar to that of DSR. The only two differences are that AODV does not use source routes and each RREQ packet is stamped with a sequence number by the source node of the packet. A source node S uses four different fields in the RREQ packet, namely, its own IP address, the IP address of the destination, its own sequence number, and the last known sequence number of the destination D. S may have obtained the sequence number of D from a packet that it had forwarded in the past; however, S does not have a route to D at present. The purpose of including the last known sequence number of D in the packet is to inform other nodes about the quality of the information that S has about D. Suppose the sequence number of D that S has is seq_1 and another node I has a routing table entry for D with a sequence number seq_2. Suppose I now receives a RREQ packet from S for destination D. Recall that sequence numbers are integers and I can compare these two sequence numbers. If $seq_1 > seq_2$, it is clear that S has more recent information about D compared to I and I need not reply to this RREQ packet as its information about D is old. On the other hand, if $seq_1 < seq_2$, I has more recent information about D compared to S. Hence, the sequence numbers are used as a logical clock in a way similar to the DSDV protocol.

AODV uses a mechanism where every node along a route sets up forward and reverse paths during the route discovery phase. We will illustrate this mechanism through an example. Suppose an intermediate node J has received a RREQ packet from another node I and the source for this RREQ packet is a node S. J sets a *reverse route* entry in its routing table with I as the destination of this entry. If J receives a RREP in reply to this RREQ in future, this reverse route entry helps J to route the packet to S through I. J also stores the current time with this reverse route entry. The reverse route entry is deleted if it is not used within a specified lifetime.

Suppose a node M receives a RREP packet from one of its neighbors N. M creates *forward path* entry in its routing table with N as the destination of the forward path. If the source node S sends data packets in future, M can send these data packets to D through N. A node J can reply to a RREQ packet with a RREP only if it has an

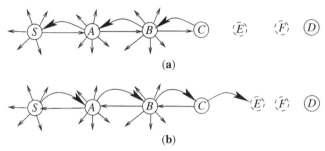

(a)

(b)

Figure 1.4 Illustration for the establishment of forward and reverse paths in the route discovery phase of the AODV protocol. (a) The propagation of the RREQ packet from the source node S is shown by the straight arrows. Each node makes a reverse path entry in its routing table by noting the ID of the neighbor from which it received the RREQ packet. These reverse path entries are illustrated by the bold and curved arrows. (b) The node C initiates a route reply by sending a RREP packet since it has a route to the destination D. When a node receives a RREP packet from one of its neighbors, it makes a forward path entry in its routing table by noting the ID of the neighbor from which the RREP has arrived. The propagation of the RREP packet is shown by the straight arrows and the forward path entries by the bold and curved arrows. Note that nodes such as E and F already have forward path entries as they are parts of a path to D.

unexpired route for D with a higher sequence number (compared to the sequence number in the RREQ packet) from D. The RREP packet follows the reverse path entries along the path through which the RREQ packet traveled to J. Figure 1.4 illustrates the forward and reverse path entries in the AODV protocol.

The route maintenance phase starts if there is a route breakage due to link failure during the data transfer phase along an established route. The sending of RERR packets is exactly similar to the DSR protocol, with the exception that nodes executing the AODV protocol do not try to salvage packets when a link breaks. When the source receives a RERR packet, it initiates a new route discovery phase.

Every node keeps only a single forward path entry for each route in the original AODV protocol. However, this may be inefficient because a new route discovery phase begins every time there is a route breakage, incurring large control packet overheads. There is a variant of the AODV protocol called the *multipath AODV* (AOMDV) [11] in which each node keeps multiple forward path entries for each route. For example, suppose a source node S has requested a route for a destination D and the resulting RREQ packet has passed through an intermediate node J. J may receive multiple RREP packets in future in reply to this RREQ. Assume that J receives RREP packets from three of its neighbors U, V, and W. In the original AODV protocol, J chooses only one of these three neighbors for keeping a forward path entry in its routing table and rejects the other two. In the AOMDV protocol, J keeps forward path entries to all the three neighbors. However, J marks these forward path entries with the times when the corresponding RREP packet was received and times out these entries if they become too old before they are used. Assume now that J chooses one of these entries, say to U, for forwarding data packets when S starts transmitting data. In case U moves out of the transmission range of J in future by breaking the J–U link and also the path

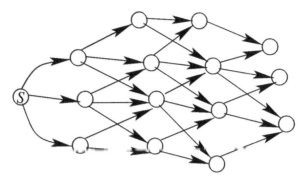

Figure 1.5   An illustration for the AOMDV protocol. In general, each node maintains multiple forward path entries in its routing table corresponding to each RREP it receives. These forward path entries are shown by bold arrows.

to the destination *D*, *J* can start using another of its forward path entries, say the link to *V*, for forwarding packets to *D*, provided the link to *V* is not too old. Hence, the need for sending a RERR packet to *S* due to the failure of the *J–U* link is eliminated in the AOMDV protocol. However, a RERR packet needs to be sent if the last forward path entry in *J*'s routing table breaks. The AOMDV protocol has better performance in terms of lower packet overhead compared to the AODV protocol. We illustrate the forward path entries in the AOMDV protocol in Figure 1.5.

We now discuss how AODV supports multicasting in a MANET. The main aim is to maintain a *multicasting tree* for each *multicasting group* of nodes. A multicasting tree consists of two kinds of nodes, *multicast group members* and *multicast tree members*. All other nodes in the MANET are *non-tree nodes* from the point of view of a multicasting group. Multicast tree members are not members of the multicast group; however, they are required for maintaining connectivity of the multicasting tree. It is important to keep a multicasting tree connected as a packet received by any member of a tree can send the packet to any other member. Also, it is important to have a tree instead of a graph to avoid routing loops.

A multicasting tree is considered as a single entity for the purpose of route discovery in AODV. Recall that any RREQ packet should contain a destination node ID. A node *I* can send a RREQ packet to a multicasting group *M* in two different cases, either *I* wants to send packets to the nodes in *M* (a data request) or *I* wants to join *M* (a join request). In case of a data request, any node that has a current path to the multicasting group can send a RREP packet in reply to the RREQ. The meaning of a current path is same as in the original or unicasting version of AODV that we have discussed before. Also, a path to the multicasting group means a path to any member of the multicasting group. In case of a join request, only a member of the multicasting tree can send the RREP. The setting of reverse path and forward path entries is similar to the unicast version of AODV, with one exception. The sender of a RREQ packet with a join request may receive multiple potential branches that connect to the multicasting tree, since the RREP packets travel through different routes. The sender should select only one of these branches as a path to the multicasting group by sending a *multicast activation* message.

It is important to maintain a multicast tree so that the tree does not get disconnected due to the mobility of the group members. Usually a special node is selected as the leader of the multicast tree and each node knows the ID of this leader node and also the path to the leader node. When a link breaks, the node that is closer to the leader initiates a route repair by sending a RREQ as a join request. The link is repaired when the corresponding RREP comes back and the node has found a new path to the multicasting group.

1.5 OTHER ROUTING PROTOCOLS

Till now we have discussed the two main categories of routing protocols for MANETs, namely, proactive and reactive protocols. However, there are other protocols that cannot be classified as purely reactive or purely proactive. We discuss two such important classes of protocols in this section, namely, the zone routing protocol (ZRP) and link reversal routing protocols.

1.5.1 The Zone Routing Protocol

The zone routing protocol [7] takes advantage of both proactive and reactive routing strategies in a MANET. Recall that the main drawback of proactive protocols is their large overhead due to proactive exchange of routing tables or link state information and that of reactive protocols is their high latency. ZRP tries to overcome both of these drawbacks while preserving the advantages of both proactive and reactive protocols.

Each node N executing the ZRP in a MANET maintains its routing zone. A routing zone is a k-hop neighborhood of N, where k is usually a small number from 2 to 4. k is also called the *radius* of a routing zone and is same for all nodes in the MANET. Each node routes proactively within its routing zone, that is, it tries to maintain a complete routing table for all the nodes in its routing zone, like the DSDV protocol. The routing outside this routing zone is done reactively, like the DSR or AODV protocol.

Suppose node S wants to start a data communication with node D. If D is within the routing zone of S, then there is no need for route discovery, as S can find a route to D by consulting its routing table. If D is outside the routing zone of S, there is a need for route discovery; however, the route discovery process is not as expensive as that of a reactive protocol. S initiates route discovery by sending a RREQ packet as in case of reactive protocols, but this RREQ packet is sent more efficiently by using *border nodes* in each routing zone. A border node is a node on the periphery of a routing zone. For example, suppose the radius of the routing zone is 2. For node S, all the nodes that are two-hop neighbors of S are the border nodes of the routing zone of S. Since S has a routing table for its routing zone, it can quickly route the RREQ packet to all these border nodes by consulting its routing table. Suppose B is one of

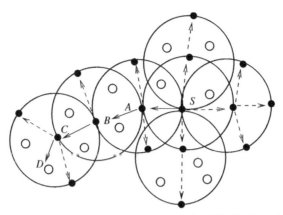

Figure 1.6   An illustration for the propagation of RREQ in the ZRP. The large circles represent the routing zones of individual nodes. The small solid circles represent border nodes and the empty circles represent other nodes. The solid arrows represent the propagation of the RREQ that succeeds in finding a route. *C* can send a RREP since the destination *D* is within its routing zone. The RREQ is propagated from one node to another on its zone boundary. The dashed arrows indicate the propagation of RREQs that are not successful.

these border nodes in the routing zone of *S*. There are two possibilities, either *D* is within the routing zone of *B* (recall that each node proactively maintains its routing zone) or *D* is not within the routing zone of *B*. In the first case, *B* can send a RREP packet back to *S* as it has a path to *D* from its routing table. *B* needs to forward the RREQ further in the second case and this is again done by using the border nodes of *B*'s routing zone. The propagation of the RREQ packet continues until either it reaches *D* or it reaches a node *C* such that *D* is within the routing zone of *C*. This process is illustrated in Figure 1.6.

It is easy to see that the propagation of a RREQ packet in the ZRP is more efficient compared to that in a pure reactive protocol such as DSR or AODV. Every node that receives the packet needs to send the packet to its border nodes and the routes to the border nodes can be found consulting the corresponding routing tables. A RREP packet is also propagated back in exactly the same way, that is, using border nodes. Hence, overall ZRP achieves a better performance in terms of lower latency compared to pure reactive protocols.

We need to ask the question whether the proactive maintenance of routing tables within routing zones incurs considerable overhead in the ZRP. The overhead is indeed high if the routing zones are large, and Haas and Perlman [7] report that the overhead is significantly lower if the radius of a routing zone is kept small, typically 2 or 3. This choice of the radius maintains both low latency and high throughput even under high-mobility scenarios.

We close our discussion on ZRP by mentioning an important optimization that needs to be done to make the protocol efficient. There is usually a significant overlap among the routing zones of different nodes in a MANET. For example, consider two neighboring nodes *I* and *J*. The routing zones of *I* and *J* will contain almost the same nodes, except for few exceptions. Moreover, if a node *M* is a border node for the routing zone of another node *N*, *N* is also a border node for the routing zone of *M*. If the RREQ packets are sent always to the border nodes of routing zones, as we have discussed above, the RREQ packets will flood the same part of the network again and again. This is clearly undesirable. ZRP tries to remove this type of flooding by directing a RREQ packet to the parts of MANET where the packet has not yet arrived and avoids flooding the parts where the packet has already arrived. A node that has broadcast a RREQ suppresses the same RREQ in future if it gets the RREQ again. Similarly, nodes that have overheard the RREQ in the promiscuous mode know that the RREQ has arrived in their part of the network and avoid forwarding the RREQ in future. This simple optimization reduces the overhead of route discovery considerably while using the ZRP.

1.5.2 Link Reversal Routing Protocols

The main idea behind link reversal routing is to treat a MANET as a directed graph. The first such protocol was designed for static packet radio networks by Gafni and Bertsekas [6] and the MANET community later extended the idea to networks of mobile nodes. We first discuss the protocol in Ref. [6].

1.5.2.1 Gafni–Bertsekas Protocol

Consider a network of static wireless nodes with a single destination node *D* and possibly many source nodes. The source nodes want to send packets to the destination node. The link between a pair of nodes is determined by their transmission radii. We will assume for simplicity that all links between neighboring nodes are bidirectional, that is, they are within the transmission range of each other. However, the protocol works equally well for the case when the links are asymmetric.

The problem of routing to a single destination *D* can be naturally framed as a problem on a directed acyclic graph such that the nodes in the wireless network are nodes in this graph and the edges between neighboring nodes in the network are edges of the graph. Every node other than *D* has at least one outgoing edge and *D* is the only node without any outgoing edge. Hence, *D* acts as a sink for all the packets transmitted in the network. A packet transmitted over a directed edge is forwarded by the recipient of the packet over an outgoing edge and this process continues until the packet reaches the destination *D*. It cannot be forwarded again as *D* does not have any outgoing edges. Hence, the routing scheme becomes very simple. A node (other than the destination) does not need to know its position in the network or the position of the destination. If it has to route a packet to the destination, it simply needs to forward the packet over one of its outgoing edges and the packet is guaranteed to reach the destination since there is no loop in the DAG and also *D* is the only sink, or node without any outgoing edges, in the network. Hence, a node executing the Gafni–Bertsekas protocol does not need to either maintain routing tables like proactive protocols or run a route discovery

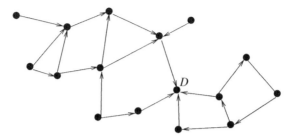

Figure 1.7   The view of a wireless network as a directed acyclic graph. The destination D is the only node that does not have any outgoing edge, every other node has at least one outgoing edge. It is easy to see that a node needs to forward a packet over one of its outgoing edges and the packet will eventually reach D.

phase like reactive protocols. A sample wireless network as a DAG is shown in Figure 1.7.

The success of the Gafni–Bertsekas protocol of course depends on route maintenance; that is, we have to ensure that always the graph is maintained as a DAG and also D is the only sink in the graph. Let us first examine how the DAG is established in the first place. Initially, the network is just a graph with links between neighbors when they are within the transmission range of each other. Only the neighbors of D have directed links to D. The initialization of the network as a DAG starts when a node S other than D needs a path to D for routing packets. S floods a QRY packet in the network in the usual manner. Any node that has a path to D replies to this QRY by sending a RPY packet. The RPY packets travel exactly in the opposite direction compared to the QRY packets. We need to explain at this point the meaning of a *path to the destination* in link reversal routing protocols. Such a path does not have any global meaning from a node's point of view in the sense that we have seen in proactive and reactive routing protocols. For a node N, a path to the destination D exists if N has at least one outgoing directed edge.

A node can send a RPY packet if it is either the destination D or one of the neighbors of D, as only the neighbors of D have a path to D due to their outgoing links to D. Suppose a node I has received a RPY packet from a node J and I and J are neighbors. I can now mark its link to J as an outgoing link as J has a route to D. After this, I also has a route to D and I can now send RPY packets to its neighbors. The initialization proceeds in this way until all the nodes in the network receive RPY packets and set their outgoing directed links. The network now is a DAG with only one sink D as we desired.

A node can lose a link in a static wireless network if nodes are allowed to switch off, that is, they stop participating in the network. Recall that the Gafni–Bertsekas protocol was designed for a packet radio network, where participation is voluntary. In general, there is no harm if a node withdraws from the network, until the situation arises that another node, say T, other than the destination becomes a sink. In other words, T loses all its outgoing links and becomes another sink in the network. This is clearly undesirable, as any packet reaching T will be stuck there and will not be delivered to *D*. The Gafni–Bertsekas protocol rectifies this situation by two mechanisms called *full reversal* and *partial reversal*.

The main activity in a link reversal routing protocol is route maintenance. The two reversal mechanisms maintain the directed acyclicity of the network graph and hence maintain routes in the graph for a destination *D*. In the full reversal mechanism, node *I* that has lost all of its outgoing links reverses all its incoming links to make them outgoing. As a result, one or more of *I*'s neighbors may lose their outgoing links and the full reversal process continues until all the nodes have at least one outgoing link each. It can be shown that the full reversal process terminates within a finite time (usually a short time in a wireless network) if the network is connected, that is, the destination can be reached from each node. The partial reversal mechanism is more selective. Consider two neighboring nodes *I* and *J*. Suppose *I* has just reversed all its links and *J* has lost its last outgoing link as a result. *J* does not reverse its link to *I* and instead reverses its other incoming links in the partial reversal mechanism. The partial reversal mechanism is illustrated in Figure 1.8.

The Gafni–Bertsekas protocol guarantees a directed acyclic graph once the network stabilizes. However, the network may have loops and hence the acyclicity property may not hold temporarily when the partial or full reversal mechanisms are in action. However, the reversal mechanisms usually terminate within a short time in a wireless network. Although this protocol is quite elegant and efficient in static wireless networks, one of its main drawbacks is that it does not work in partitioned networks. It can be shown easily that the full or partial reversal mechanism may go into an indefinite cycle (the nodes continue reversing their links for an indefinite period of

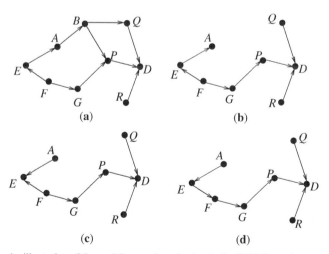

Figure 1.8 An illustration of the partial reversal mechanism in the Gafni–Bertsekas protocol. (a) Each node except the destination *D* has at least one outgoing link. (b) *B* moves away and *A* loses its last outgoing link. (c) *A* reverses its incoming links, the link to *E* in this case. *E* now loses its last outgoing link. (d) *E* does not reverse its incoming link to *A* as *A* has just reversed it. *E* instead reverses its other incoming link to *F*. The network is now again a DAG with only one sink (*D*).

time) for nodes that are partitioned from the destination D. This wastes bandwidth of the wireless network. Since mobile ad hoc networks may get partitioned from time to time due to the mobility of nodes, this protocol cannot be directly used in a MANET.

We need to mention another important point about the Gafni–Bertsekas protocol. Throughout our discussion we have assumed only one destination in the network. However, this is not realistic as usually any node in a wireless network may act as a source or destination of packets. However, the protocol can be extended easily for multiple destinations. Conceptually, we need to maintain k different DAGs for k destinations. Note that there is no global maintenance of DAG in the Gafni Bertsekas protocol, rather individual nodes store local information to get a global DAG in a distributed fashion. A node can keep track of its links with its neighbors in a local routing table and assign status to these links as incoming and outgoing. Also, nodes need to change the status of these links during the link reversal process. It is possible to support k destinations by keeping k copies of each link with a neighbor and assigning a status (incoming or outgoing) for each copy according to the status of the DAG for a particular destination. A link between neighbors I and J could be incoming for node I for a destination D_p and outgoing for I for another destination D_q. When I receives a packet for destination D_p, it cannot forward that packet to J because the direction of the link is from J to I. On the other hand, when it receives a packet for destination D_q, it can forward the packet to J since the direction of the link is from I to J for destination D_q.

1.5.2.2 Temporally Ordered Routing Algorithm

The TORA protocol [4, 14] was designed as a modification of the Gafni–Bertsekas algorithm for mobile ad hoc networks. We cannot discuss this protocol in detail due to limitation of space and a detailed description can be found in the review article by Corson and Park [5]. We discuss only the essential components of this protocol that make it suitable for using in MANETs.

The main aim of the TORA protocol is to make it suitable for networks that may become partitioned, as is the case for MANETs. It retains the core ideas behind the Gafni–Bertsekas protocol, such as using the network as a DAG and maintaining the DAG through full and partial reversal of links. However, it has some extra mechanisms for detecting partitions in a MANET.

The direction of the links between neighbors in the TORA protocol is enforced through a height assignment to each end point of a link. Suppose I and J are two neighboring nodes and we assign integers as heights to I and J to fix the direction of the link between them. For example, if I is assigned a height of 4 and J a height of 7, then J has a greater height compared to I and the direction of the link between them will be from J to I. In other words, the link between I and J will be incoming for I and outgoing for J. If I wants to reverse this link, it has to make its own height higher than J, say 8. Link reversals in TORA are done by changing the heights of the two end nodes of a link. However, the height assignment in TORA is more complex, with each height consisting of five components. Two heights are compared according to the lexicographic ordering of these five components. When a node wants to change its height, it usually tries to borrow the height of one of its neighbors and increases one of the components so that its height becomes higher than that neighbor. This is basically the idea of partial reversal in the Gafni–Bertsekas algorithm. There are several rules in changing heights in TORA and we will not discuss all the rules; however, we will only discuss the way TORA detects partition in the network.

One of the components in a height is the *origin ID* or *oid* and the other is time. There are situations when a node cannot borrow a neighbor's height and increase it to do partial reversal. In such cases, a node (say *I*) initiates a global new height by assigning the first component as the current time. Since current time is the global highest time, the new height becomes lexicographically highest in the entire network. The node *I* initiating a new height also sets the *oid* field of the new height with its own ID. It is then possible for *I* to detect in future if all of its neighbors have the height that it had initiated. This means all of its neighbors have tried to borrow *I*'s height and increase it to establish outgoing links to *I*. This is an indication of a partition in the network as no node is able to force a new route to the destination by increasing others' heights. Nodes executing TORA erase all their heights in such cases as there is no point in trying to find new routes since the destination is unreachable. The destination may become reachable in future again due to the mobility of nodes when there are new nodes between the partitions and these nodes connect the partitions.

Although TORA is an elegant protocol for routing in MANETs, it has quite high overhead in high-mobility scenarios. In particular, the link reversals become excessive when there are frequent link breakages in the network. It has been shown [2] that the performance of TORA becomes worse compared to reactive protocols in high-mobility scenarios. However, the performance of TORA is comparable to protocols such as DSR and AODV in low- and moderate-mobility scenarios.

1.6 CONCLUSION

We have reviewed several key protocols for routing in mobile ad hoc networks. However, this review is not comprehensive as we have left many other protocols from our discussion due to limitation of space. One such class of protocols is the class of position-based routing protocols [20], where each node knows the locations of all the other nodes through a location service such as global positioning system (GPS). It can be shown that the overheads of reactive protocols can be reduced considerably in this class of protocols. We refer the reader to the review article by Stojmenovic [20] for more details.

We conclude this chapter by briefly discussing how the MANET routing protocols are evaluated in terms of their performance, such as throughput and latency in packet delivery. Since a real evaluation of a protocol through deploying mobile nodes is expensive, most such evaluations are done through simulations in a discrete event simulator. The network simulator [12] or ns-2 has become the most popular simulator in the research community for this purpose. ns-2 has been developed for over a decade by volunteers who are researchers and PhD students in different universities worldwide and it is still in the process of development with new protocols and features being added every year. It is written in C++, one of the most popular object-oriented programming languages. The protocols discussed in this chapter are usually evaluated under the *random waypoint* model of mobility of nodes. In this model, a node moves in a random direction with a random speed (bounded above by a maximum speed) and then pauses for a specified time before moving in another random direction again. This process continues throughout the simulation period. The range of mobilities (in terms of maximum speed) usually considered by researchers is in the range of 1 m/s (pedestrian speed) to 20 m/s (speed of a car). Much more details about performance evaluation of specific protocols can be found from the references cited in this chapter.

REFERENCES

1. E. Belding-Royer and C. E. Perkins. Evolution and future directions of the ad hoc on-demand distance-vector routing protocol. *Ad Hoc Networks*, 1(1):125–150, 2003.
2. J. Broch, D. A. Maltz, D. B. Johnson, Y. C. Hu, and J. Jetcheva. A performance comparison of multi-hop wireless ad hoc network routing protocols. *Proceedings of the ACM MOBICOM 1998*, pp. 85–97.
3. M. S. Corson and J. Macker. Mobile ad hoc networking (MANET): routing protocol performance issues and evaluation considerations, *IETF MANET, RFC 2501*, 1999. http://www.ietf.org/html .charters/manet-charter.html.
4. M. S. Corson and A. Ephremides. A distributed routing algorithm for mobile wireless networks. *ACM/Baltzer Wireless Networks J.*, 1(1):61–82, 1995.
5. M. S. Corson and V. Park. In: C. Perkins (Ed.), Link Reversal Routing in Ad Hoc Networking Addison-Wesley, 2001, pp. 255–298, Chapter 8.
6. E. Gafni and D. Bertsekas. Distributed algorithms for generating loop-free routes in networks with frequently changing topology. *IEEE Trans. Commun.*, 29(1):11–18, 1981.
7. Z. J. Haas and M. R. Perlman. ZRP: a hybrid framework for routing in ad hoc networks. In: C. Perkins (Ed.), *Ad Hoc Networking*, Addison-Wesley, 2001, pp. 221–253, Chapter 7.
8. IETF MANET working group. http://www.ietf.org/html.charters/manet-charter.html.
9. D. B. Johnson, D. A. Maltz, and J. Broch. DSR: the dynamic source routing protocol for multi-hop wireless ad hoc networks. In: C. Perkins (Ed.), *Ad Hoc Networking*, Addison-Wesley, 2001, pp. 139–172, Chapter 5.
10. D. B. Johnson, D. A. Maltz, and Y. Hu. The dynamic source routing protocol for mobile ad hoc networks. *IETF MANET, Internet Draft*, 2003. http://www.ietf.org/html.charters/ manet-charter.html.
11. M. K. Marina and S. Das. On-demand multipath distance vector routing in ad hoc networks *Proceedings Ninth International Conference on Network Protocols (ICNP)*, pp. 14–23, 2001.
12. NS : The Network Simulator. http://www.isi.edu/nsnam/ns/.
13. Optimized link state routing protocol. *IETF MANET, RFC 3626*, 2003. http://www.ietf.org/ html.charters/manet-charter.html
14. V. Park and S. Corson. Temporally ordered routing algorithm (TORA) Version 1: functional specification. *IETF MANET, Internet Draft*, 2001. http://www.ietf.org/html.charters/manet-charter.html
15. C. Perkins (Ed.), *Ad Hoc Networking*. Addison Wesley, 2001.
16. C. Perkins and P. Bhagwat. DSDV: routing over a multihop wireless network of mobile computers. In: C. Perkins (Ed.), *Ad Hoc Networking*, Addison-Wesley, 2001, pp. 53–74, Chapter 3.

17. C. Perkins, E. Belding-Royer, and S. Das. Ad hoc on-demand distance vector (AODV) routing, *IETF RFC 3591*, 2003.
18. A. A. Pirzada, C. McDonald, and A. Datta. Performance comparison of trust-based reactive routing protocols. *IEEE Trans. Mobile Comput.*, 5(6):695–710, 2006.
19. E. M. Belding-Royer and C. K. Toh. A review of current routing protocols for ad hoc mobile wireless networks. *IEEE Personal Commun. Mag.*, 6(2):46–55, 1999.
20. I. Stojmenovic. Position-based routing in ad hoc networks. *IEEE Commun. Mag.*, 40(7):128–134, 2002.

Chapter 2

Connected Dominating Set for Topology Control in Ad Hoc Networks

Bo Han,[1] Lizhuo Zhang,[2] and Weijia Jia[2]

[1] *Department of Computer Science, University of Maryland, College Park, MD*
[2] *Department of Computer Science, City University of Hong Kong, Kowloon, Hong Kong*

2.1	Introduction	26
2.2	Related Work	28
2.3	Network Assumptions and Preliminaries	29
2.4	Area-Based CDS Formation Algorithm	30
2.5	Experimental Simulations	36
2.6	Conclusion and Future Work	39
	Acknowledgments	40
	References	41

2.1 INTRODUCTION

In wireless ad hoc networks that are formed by autonomous mobile devices communicating through radio, topology control plays an important role in the performance of the protocols used in the network, such as routing, clustering, and broadcasting. There are two approaches for topology control in ad hoc networks—transmission range control and hierarchical topology organization (clustering). The goal of this technique is to control the topology of the graph representing the communication links between network nodes, with the purpose of maintaining some global graph property (such as

Mobile Intelligence. Edited by Laurence T. Yang, Agustinus Borgy Waluyo, Jianhua Ma, Ling Tan, and Bala Srinivasan
Copyright © 2010 John Wiley & Sons, Inc.

connectivity) while reducing energy consumption. Moreover, topology control has the positive effect of reducing contention when accessing wireless channels. In general, when the nodes' transmission ranges are relatively short, many nodes can transmit simultaneously without interfering with each other, and the network capacity is thus increased. Ideally, the nodes' transmission ranges should be set to the minimum value such that the communication graph is connected.

As mentioned above, transmission range control is a general approach for topology control in ad hoc networks. The construction of hierarchical topology (clustering) is another effective solution. Cluster-based constructions are commonly regarded as a variant of topology control in the sense that energy-consuming tasks can be shared among the members of a cluster. The basic idea of clustering is to group the network nodes that are in physical proximity to provide the network with a logical organization that is smaller in scale, and hence simpler to manage [6]. The notion of cluster organization has been investigated for ad hoc networks since their appearance. Baker et al. [27] introduced a fully distributed linked cluster architecture and demonstrated adaptability of the network to topological changes. With the advent of multimedia communications, the use of the cluster has been revisited by Gerla and Tsai [22] with the emphasis on the allocation of resources to support the multimedia traffic in ad hoc networks. Basagni proposed a distributed clustering algorithm that generalizes these clustering protocols in that the choice of the clusterhead is performed based on a generic "weight" associated with a node [17]. This attribute basically expresses how fit that node is to become a clusterhead. Clustering algorithms have also been proposed explicitly for wireless sensor networks. Among these protocols, one of the first is the low-energy adaptive clustering hierarchy (LEACH) protocol presented in Ref. [28]. LEACH uses randomized rotation of the clusterheads to evenly distribute the energy load among the sensors for network longevity.

Although wireless ad hoc networks have no physical infrastructure, it is natural to construct clusters through connected dominating set (CDS) formation. In general, a dominating set (DS) of a graph $G = (V, E)$ is a subset $V' \subset V$ such that each node in $V - V'$ is adjacent to at least one node in V', and a connected dominating set is a dominating set whose induced subgraph is connected. It has been pointed out that "the most basic clustering that has been studied in the context of ad hoc networks is based on dominating sets" [4]. Moreover, the CDS can also play an important role in message broadcasting in ad hoc networks [3]. Unfortunately, the dominating set and connected dominating set problems have been shown to be NP-complete [5]. Even for a unit disk graph (UDG) [1], the problem of finding a minimum CDS (MCDS) is still NP-complete [7].

This chapter presents a novel distributed algorithm, named the *Area* algorithm, for CDS formation in wireless ad hoc networks. In this algorithm, we partition the nodes into different areas and *selectively* connect two dominators that are two or three hops away. Note that the clusterheads in most clustering algorithms [17, 22] usually form a DS. Since they focused on clusterhead selection, the clusterheads and gateways (selected to connect two clusterheads) construct a CDS with a relatively large size. Thus, our contribution mainly lies in that we introduce the *Area* concept to significantly reduce the number of connectors that connect two neighboring dominators, therefore reduce the size of final CDS.

The rest of the chapter is organized as follows. Section 2.2 introduces the related work. Section 2.3 describes the network assumption and some preliminaries. In Section 2.4, we present our novel distributed CDS formation algorithm and give the performance analysis. Section 2.5 presents the simulation results. We point out future directions and summarize major results in Section 2.6.

2.2   RELATED WORK

In this section, we discuss related work with respect to topology control in two categories: transmission range control and hierarchical topology organization (in the context of connected dominating set formation).

2.2.1   Transmission Range Control

Most of the existing topology control algorithms select a less-than-normal transmission range (also called the actual transmission range) while maintaining network connectivity. Centralized algorithms [26] construct optimized solutions based on global information and, therefore, are not suitable in wireless ad hoc networks. Some probabilistic algorithms [8] adjust transmission range to maintain an optimal number of neighbors. However, probabilistic algorithms do not provide hard guarantees on network connectivity. Most of the localized topology control algorithms use nonuniform actual transmission ranges computed from one-hop information (under the normal transmission range) and take advantage of some original research topics in computational geometry, such as the minimum spanning tree [20], the Delaunay triangulation [19], or the relative neighborhood graph [21]. Most of these contributions mainly considered energy efficiency of paths in the resulting topology. The CBTC algorithm [18] was the first construction to focus on several desired properties. A nice literature review of transmission range control can be found in Ref. [4].

2.2.2   Connected Dominating Set

Das et al. proposed a MCDS-based routing algorithm for wireless ad hoc networks [2]. This algorithm is a distributed version of Guha and Khuller's centralized algorithm to calculate a connected dominating set [10]. The algorithm proposed by Wu and Li first finds a connected dominating set and then prunes certain redundant nodes from the CDS using two rules (Rule 1 and Rule 2) [11]. In the first phase, each node is marked true (dominator) if it has two unconnected neighbors. According to Rule 1, a marked node can unmark itself if its neighbor set is covered by another neighboring marked node. According to Rule 2, a marked node can unmark itself if its neighborhood is covered by two other neighboring directly connected marked nodes. The combination of Rules 1 and 2 is fairly efficient in reducing the size of CDS. This algorithm is fully localized, but does not guarantee a good approximation ratio. Hereafter, this algorithm is referred to as Rule 1&2 (so named for the two pruning rules). Stojmenovic et al. also presented a distributed construction of CDS in the context of clustering and broadcasting [12]. The solution proposed in Ref. [23] relies on all nodes having a common clock and requires two-hop neighbor information. In CEDAR [13], a virtual infrastructure called the *core* is constructed to approximate a *minimum dominating set* (not connected) of the underlying network.

For distributed clustering algorithm, it is undesirable to have neighboring clusterheads [17]. It is also undesirable to have one-hop away neighboring dominators in dominating set formation. This leads to the well-known concept of *maximal independent set* (MIS). An independent set of graph $G = (V, E)$ is a subset $S \subset V$ such that for any pair of vertices in S, there is no edge between them. Obviously, an MIS S is also an independent DS. The two heuristic algorithms proposed by Alzoubi et al. [14] take advantage of the property of MIS, thus may guarantee a constant approximation ratio of 8 and 12, respectively. Although these two algorithms are distributed, they are not localized. To address the problem of nonlocalized computation, Alzoubi et al. also proposed a message-optimal localized algorithm with linear time and message complexity [15]. This algorithm can be briefly described as two phases. In the first phase, an MIS is constructed. As mentioned above, this MIS is also a DS. In the second phase, each dominatee identifies the dominators that are at most two hops away from itself and broadcasts this information. Using such information from all neighbors, each dominator identifies a path to each dominator that is at most three hops away from itself and informs all nodes in this path to become the connectors and join the final CDS. The approximation ratio of this algorithm is bounded by 192. For simplicity, we call this algorithm AWF in the following. Recently, Wang et al. proposed an efficient distributed method to construct a low-cost weighted minimum connected dominating set [24].

2.3 NETWORK ASSUMPTIONS AND PRELIMINARIES

In this chapter, we assume that an ad hoc network comprises a group of nodes communicating with the same transmission range. Scheduling of transmission is the responsibility of the MAC layer. Each node has a unique ID and each node knows the ID and degree of its neighbors, which can be achieved through periodically broadcasting "HELLO" messages by each node. Since the emphasis of this chapter is on the CDS formation, we do not consider the node mobility. We call the nodes in the dominating set *dominators*, the nodes not in the dominating set *dominatees*, and the nodes that connect two or three hops away dominators *connectors*. Especially, we call the connectors that connect dominators two and three hops away *one-hop connector* and *two-hop connector*, respectively. Next, we give some well-known preliminaries.

Preliminary 2.1. *By building a dominating set through MIS construction, for every node u, the number of dominators inside the disk centered at u with radius k-unit is bounded by a constant l_k.*

Proof. Alzoubi et al. gave a proof through calculation of $l_k < (2k+1)^2 - 1$ [15]. When $k = 2, 3$, we have $l_k = 23, 47$. Recently, Li et al. have proved that $l_3 = 42$ [16].

Preliminary 2.2. *Let G be a UDG and opt be the size of a minimum CDS for G, then the size of any MIS for G is at most* $3.8 \times opt + 1.2$.

The proof of this preliminary bounds the size of any MIS in *G* and can be found in Ref. [9].

Preliminary 2.3. *In a DS, the maximum distance to another closest dominator from any dominator is* 3.

Proof. By contradiction. Assume that the maximum distance from a dominator *u* to the closest dominator *v* is 4, and the shortest path between *u* and *v* is $\{u, x, y, z, v\}$. According to the definition of dominating set, node *y* must have a dominator, say *w*, which is one hop closer (three hops) to *u* than *v*. This contradicts the assumption that *v* is the closest dominator to *u*.

2.4 AREA-BASED CDS FORMATION ALGORITHM

2.4.1 Overview

A well-known method for building connected dominating set is to construct an MIS first, which is also a dominating set, and then add some connectors to guarantee the connectivity. This method was utilized by Alzoubi et al. [14, 15]. The algorithms in Ref. [14] were implemented by first electing a leader *r* among the nodes, which was going to be the root of a spanning tree *T*. The approximation ratios of these algorithms are attractive; however, the message complexity $O(n \log n)$, which is bounded by the distributed leader election, is quite high in real practice [6]. Moreover, they are not localized algorithms. The algorithm presented in Ref. [15] has an optimal message complexity $O(n)$, but it connects any pair of dominators (at most three hops away) by adding one or two connectors. Consequently, the resultant CDS has a relative large size with some redundant connectors.

Our main objective of this Area algorithm is to reduce the size of CDS. We use the *most-valued nodes* as the metric to select the nodes among all nodes in the graph for the CDS. The *value* of a node is a performance-related characteristic such as node ID, node degree, or remaining battery life. In this chapter, we define two kinds of *most-valued nodes*, one is the nodes with the minimum ID among all the candidates of dominators or connectors (the resulting Area algorithm is called Min ID) and the other is the nodes with the maximum degree among all the candidates (hence, called Max Degree). In the following description of Area algorithm, we will use node degree as the selection metric. We stress that the proposed algorithm can also be easily extended to support other node values.

2.4.2 Max Degree Algorithm

Define the rank of node u to be an ordered pair of (δ_u, id_u), where δ_u is the node degree and id_u is the node ID of u. We say that a node u with rank (δ_u, id_u) has a higher order than a node v with rank (δ_v, id_v) if $\delta_u > \delta_v$, or $\delta_u = \delta_v$ and $id_u < id_v$. Each node is in one of the four states: *unmarked, dominatee, dominator*, or *connector*. Each node is initially in an unmarked state and subsequently enters either the dominatee or dominator state. The connector state can only be entered from the dominatee state. In this Area algorithm, we partition the nodes into different areas and each area is supposed to have a unique area ID. Thus, each node is also assigned an area ID to indicate which area it belongs to. For simplicity of description, we first give some definitions below.

Definition 2.1 (Seed Dominator). A dominator that has the highest rank among its one-hop neighbors.

Definition 2.2 (Nonseed Dominator). A dominator that has at least one one-hop neighbor with higher rank.

Definition 2.3 (Border Dominator). A dominator that has two or three hops away neighboring dominators with different area IDs.

2.4.2.1 Area Formation

First, an unmarked node u with the highest rank among its unmarked one-hop neighbors becomes a dominator and broadcasts a DOMINATOR message to its neighbors. Note that such a node does exist in the beginning. After receiving a DOMINATOR message, a node, say v, changes its state to be the dominatee if its current state is unmarked. If it is the first time that v receives a DOMINATOR message, v also broadcasts a DOMINATEE message to its neighbors. The same procedure is repeated until each node becomes either a dominator or a dominatee.

In fact, seed dominators are the starting points of the process of these MIS-based algorithms and they are the cores of the areas. The ID of a seed dominator automatically becomes the ID of the corresponding area. During the area formation, we add an item, *Area ID*, into the DOMINATOR message to indicate the area that the dominator belongs to. When an unmarked node receives the first DOMINATOR message, it becomes a dominatee of the area indicated in this message. Each dominatee also inserts its area ID into the DOMINATEE message that it broadcasts to its neighbors. Then every nonseed dominator can know the area it belongs to from its neighboring dominatees. If neighboring dominatees have different area IDs, the nonseed dominator can arbitrarily select one area to join. The nodes with the same area ID form an area eventually.

2.4.2.2 *Area Connection*

After the nodes are partitioned into different areas, the following steps are executed by the related nodes:

1. Each dominatee broadcasts a ONE-HOP-DOMINATOR message that contains all node IDs and area IDs of its one-hop away dominators.

2. After receiving a ONE-HOP-DOMINATOR message, each node knows its two-hop away dominators and the corresponding neighbors to connect these dominators. The neighbor with higher rank has the priority to be chosen as a connector (maybe NOT the connector in the final CDS).

3. Upon reception of the ONE-HOP-DOMINATOR message from all its neighboring dominatees, a dominatee broadcasts a TWO-HOP-DOMINATOR message that contains all node IDs and area IDs of its two-hop away dominators.

4. After receiving a TWO-HOP-DOMINATOR message, each dominator knows its three-hop away neighboring dominators and the relevant neighbors to connect these dominators. Also, the neighbor with higher rank has the priority to become a connector.

After having the knowledge of all the two-hop and three-hop away neighboring dominators, each dominator can know whether it is a border dominator. Dominators inside an area try to connect only their two-hop away neighboring dominators with larger IDs by selecting one connector. Border dominators connect only one two-hop or three-hop away neighboring dominator with larger ID in an adjacent area by selecting one or two connectors. That is, if a border dominator has connected to a two-hop away neighboring dominator in an adjacent area, it will not try to connect the three-hop away neighboring dominator in the same adjacent area. Then the dominating set is constructed through a sweep of the network spreading outward from the seed dominators. To illustrate the algorithm, Figure 2.1 gives an example of the CDS formation using Max Degree algorithm.

2.4.2.3 *Example*

In Figure 2.1, the IDs of nodes are labeled beside the nodes. Black nodes represent the dominators, black nodes with outer circle represent the seed dominators, and gray nodes represent the connectors. A possible execution scenario is shown in Figure 2.1(b)–(d), as explained below.

1. Initially, all nodes are unmarked (Figure 2.1(a)).

2. Nodes 7 and 14 declare themselves as dominators, since they have the highest ranks among their unmarked one-hop neighbors. These two dominators are also seed dominators. After receiving a DOMINATOR message, nodes 5, 9, 13, 15, 16, 20, 22, 23, 24, 25, 26, and 27 declare themselves as the dominatees and broadcast DOMINATEE messages (Figure 2.1(b)).

3. After receiving DOMINATEE messages from their neighbors, nodes 6, 10, 18, 19, and 21 declare themselves as dominators and broadcast DOMINATOR

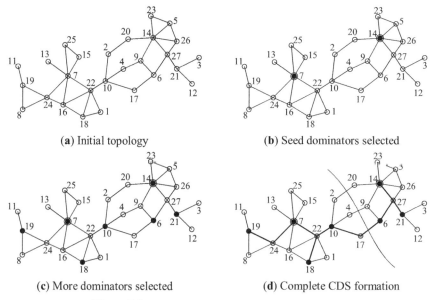

(**a**) Initial topology   (**b**) Seed dominators selected

(**c**) More dominators selected   (**d**) Complete CDS formation

Figure 2.1   CDS construction by Max Degree algorithm.

messages. The reason is that all their neighbors with higher ranks became dominatees, thus their ranks become the highest among their unmarked neighbors. At this time, all the dominators form an MIS and there are two areas centered at dominators 7 and 14, respectively. Suppose dominators 10 and 6 choose to join the areas with ID 7 and 14, respectively (Figure 2.1(c)).

4. After each dominatee broadcasts ONE-HOP-DOMINATOR and TWO-HOP-DOMINATOR messages, every dominator knows its two-hop and three-hop away neighboring dominators. According to the definition, dominators 6, 10, and 14 know that they are border dominators. Finally, dominatees 22 and 24 are selected to become connectors by dominator 7 to connect dominators 10, 18, and 19; dominatee 27 is selected as connector by dominator 6 to connect dominator 14. Dominatee 17 is selected by dominator 6 to connect the two adjacent areas. Obviously, all the black and gray nodes form a connected dominating set of the graph and the induced subgraph is indicated by the thick black lines (Figure 2.1(d)).

Note that compared to dominatee 9 whose rank is $(3, 9)$, dominatee 27 has a higher rank $(4, 27)$, so it is selected by dominator 6 to connect dominator 14. Dominatee 27 is also the only node that connects dominators 14 and 21. Since dominator 10 has a two-hop away neighboring dominator (node 6) in the adjacent area, it will not try to connect its three-hop away neighboring dominator (node 14) in the same area. From the above example, we can see that the benefit of using the area concept is that dominators can selectively connect to their two or three hops away neighboring dominators, thus reduce the size of the final CDS.

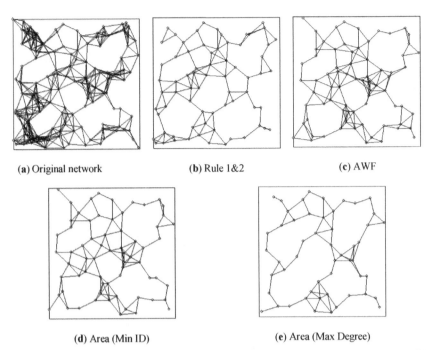

(a) Original network   (b) Rule 1&2   (c) AWF

(d) Area (Min ID)   (e) Area (Max Degree)

Figure 2.2   A sample network with 140 nodes: (a) entire network, (b) CDS by Rule 1&2, (c) CDS by AWF, (d) CDS by Min ID, and (e) CDS by Max Degree.

Figure 2.2 shows a comparison of our two Area algorithms, Min ID and Max Degree with Rule 1&2 [11] and AWF [15]. We use a sample network with 200 nodes. The original topology of the network is depicted in Figure 2.2(a). Figure 2.2(b)–(e) shows the CDS generated by Rule 1&2 (Figure 2.2(b)), AWF (Figure 2.2(c)), Min ID (Figure 2.2(d)), and Max Degree algorithms (Figure 2.2(e)), respectively. In these four figures, only nodes in the CDS and their induced graph are shown. The size of the CDS constructed by these four algorithms is 101 (Rule 1&2), 98 (AWF), 76 (Min ID), and 55 (Max Degree), respectively. We can see that the CDS constructed by Max Degree algorithm contains the least number of nodes, followed by Min ID algorithm.

2.4.3   Performance Analysis

In this subsection, we show the correctness, analyze the time and message complexity of the Area algorithm, and give the approximation ratio of this algorithm.

Theorem 2.1. *The dominators and connectors selected by the Area algorithm form a CDS.*

Proof. In this Area algorithm, it can be easily proved that each dominator has at least one two-hop away neighboring dominator in the area it belongs to if there is any other dominator existing in the same area. To connect each pair of two-hop away dominators, we can guarantee the connectivity inside these areas. Note that we also connect two adjacent areas using at least one path; thus, this theorem is proved.

Theorem 2.2. *The Area algorithm has both time and message complexity of $O(n)$.*

Proof. The time complexity of this algorithm is bounded by MIS construction that has the worst time complexity $O(n)$. The worst case occurs when all nodes are distributed in a line and in either ascending or descending order of their ranks. The rest of the process has time complexity at most $O(n)$. Since each node sends a constant number of messages, the total number of messages is also $O(n)$.

Theorem 2.3. *Let G be a unit disk graph and opt be the size of a minimum CDS for G, then the size of CDS constructed by this Area algorithm is within a constant approximation ratio of opt.*

Proof. From Preliminary 2.2, we know that the size of MIS is at most $3.8 \times opt + 1.2$. Since each pair of nodes in MIS introduces at most two nodes to CDS, from Preliminary 2.1, the number of nodes in the CDS is at most $(42 \times 2/2 + 1) \times (3.8 \times opt + 1.2) = 163.4 \times opt + 51.6$.

The density μ of graph can be calculated as

$$\mu(r) = (n\pi r^2)/A \tag{2.1}$$

where n is the number of nodes in the graph, A is the area of the graph, and r is the transmission range. Let D be the maximum density of packing of n equal circles in a circle. The upper bound of D is given in Ref. [25]:

$$D \leq \frac{n}{\left[1 - \frac{\sqrt{3}}{2} + \sqrt{\frac{3}{4} + \frac{2\sqrt{3}}{\pi}(n-1)}\right]^2} \tag{2.2}$$

From (2.1) and (2.2), we can further refine l_2 in Preliminary 2.1 to be 21. Remember that, in this Area algorithm, the dominators and one-hop connectors form a CDS in the induced subgraph of each area. Let opt' be the size of a minimum CDS for the induced subgraph of an area. Through the similar analysis of Theorem 2.3, the size of CDS for an area is at most $(21/2+1) \times (3.8 \times opt' + 1.2) < 44 \times opt' + 14$.

2.4.4 Discussion of Mobility Issues

Mobility management is an important research topic in wireless ad hoc networks and sometimes node mobility can have good effects on the protocol design. For example, epidemic routing [29] exploits rather than overcomes mobility to achieve a certain degree of connectivity in occasionally partitioned networks; message ferrying is a mobility-assisted approach that can provide efficient data delivery in sparse ad hoc networks where network partitions can last for a significant period [30]. However, node mobility is treated as undesirable for the CDS construction, since it will cause inconsistent local view. Dynamic topology change resulting from node mobility can be handled by the methods proposed in Refs. [11, 15]. Generally, there are two kinds of mechanism: periodical reconstruction and on-demand update. Each method has its own pros and cons. In the former scheme, the period of time elapses before the reconstruction is critical to the system performance. On-demand update is efficient in slight topology change and will lose its effectiveness when facing major topology change. In our future work, we plan to integrate both of these two methods with the proposed Area algorithm to maintain the CDS for mobile ad hoc networks. However, how to update the topology efficiently while preserving the approximation quality is still an open problem.

The next section gives extensive simulation study to verify the efficiency of this Area algorithm in terms of the size of CDS and the communication overhead.

2.5 EXPERIMENTAL SIMULATIONS

We compare the performance of the proposed two Area algorithms, Min ID and Max Degree, with Rule 1&2 [11] and Alzoubi's algorithm [15] in this section. In the simulation scenario, a given number of nodes (ranging from 60 to 200 with an increment step of 20 and from 200 to 1000 with an increment step of 100, respectively) were randomly and uniformly distributed in a square simulation area of size 100 by 100 units. Each node has a fixed transmission range r ($r = 15$ and 30 units, respectively). All the simulation results presented here were obtained by running these algorithms on 300 connected graphs. This allows us to test these algorithms on increasing density of network from $n = 60$, $r = 15$, and $\mu(r) = 4$ (sparse network) to $n = 1000$, $r = 30$, and $\mu(r) = 283$ (very dense network).

When the CDS is used for routing in ad hoc networks, the number of nodes responsible for routing can be reduced to the number of nodes in the CDS. Thus, we prefer smaller size of CDS. Figure 2.3(a) and (b) shows the simulation results when the node's transmission range is 15 units. Figure 2.3(a) shows the trend when the number of nodes in the network ranges from 60 to 200 (the corresponding graph is sparse), whereas Figure 2.3(b) shows the trend when the number of nodes in the network ranges from 200 to 1000 (the corresponding graph is dense). The number of nodes in the CDS increases when more nodes join the network because the number of dominators increases and more nodes may be selected as the connectors, thus the size of CDS increases. From the two figures, we also notice that the size of CDS is more sensitive to the number of nodes in the range from 60 to 200 (sparse network) than to that in the range from 200 to 1000 (dense network). As the number of nodes increases, the gap between the two Area algorithms and the other two becomes significant. And among these four algorithms, the performance of Max Degree is the best. When the number of nodes in the network reaches 1000, the number of nodes in the CDS constructed by Max Degree is only about 60% of that constructed by Rule 1&2 or AWF.

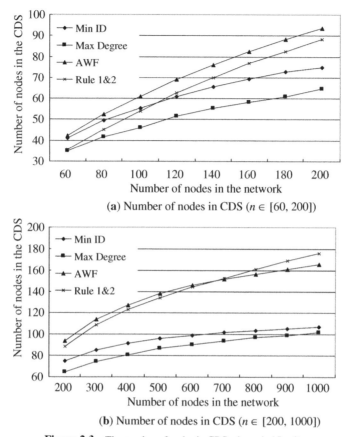

(a) Number of nodes in CDS ($n \in [60, 200]$)

(b) Number of nodes in CDS ($n \in [200, 1000]$)

Figure 2.3   The number of nodes in CDS when r is 15 units.

Figure 2.4(a) and (b) shows the results when the node's transmission range is set as 30 units and the number of nodes in the networks ranges from 60 to 200 and from 200 to 1000, respectively. When the transmission range increases, as more nodes may be connected, the network becomes denser if the number of nodes is fixed. In this case, the size of CDS increases only slightly as the size of the network increases. Based on our simulation results, we find that among these four algorithms, Min ID outperforms the other three, followed by Max Degree, in very dense networks. Comparing Figure 2.3(a) and (b) with Figure 2.4(a) and (b), we find that increasing the node's transmission range can increase the coverage area of each node and, therefore, increase the density of the network, which leads to a smaller size of the CDS. When the number of nodes in the network reaches 1000, the number of nodes in the CDS constructed by Min ID is only about 45% of that constructed by Rule 1&2.

We also compared the number of two-hop connectors in the CDS constructed by Min ID, Max Degree, and AWF and the simulation results are given in Figure 2.5(a)

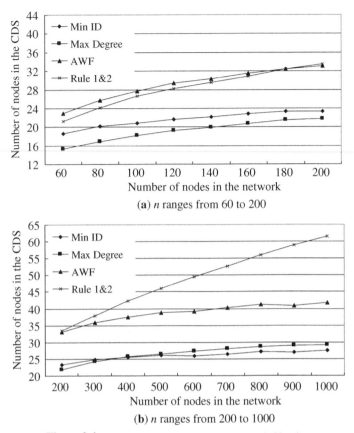

Figure 2.4   The number of nodes in CDS when *r* is 30 units.

and (b), respectively. We can see that both Min ID and Max Degree select much less two-hop connectors than AWF. In very dense networks, the number of two-hop connectors in the CDS constructed by Min ID and Max Degree approximates to zero.

Figure 2.6(a) and (b) relates the message overhead to the number of nodes in the network (ranging from 100 to 1000 with an increment step of 100) when the transmission range *r* is 15 and 30 units, respectively. In both cases, the y-axis denotes the average number of bytes of messages transmitted by the nodes. Among these algorithms, Rule 1&2 consumes the most number of bytes of messages because in this algorithm each node needs to exchange its one-hop neighbor information with its one-hop neighbors. When $n = 1000$ and $r = 15$, this kind of information exchange accounts for about 96% of the total message overhead and demands significant time and energy consumption. AWF, Min ID, and Max Degree show similar performance

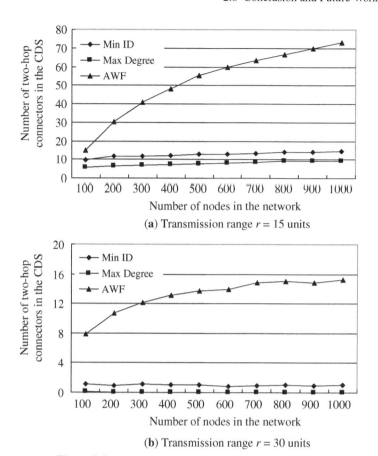

(a) Transmission range $r = 15$ units

(b) Transmission range $r = 30$ units

Figure 2.5   The number of nodes in CDS when r is 30 units.

because in these algorithms each node requires only the knowledge of its one-hop neighbors and a constant number of two-hop and three-hop neighbors. Compared to AWF, Min ID and Max Degree introduce slightly more message overhead in sparse networks because one extra item, area ID, is associated with each node in the message exchange. When $n = 1000$ and $r = 30$, the number of bytes of messages consumed by Rule 1&2 is 439,808, nearly 13 times more than that of the other three. To show the difference between AWF, Min ID, and Max Degree clearly, in Figure 2.6(b) we omit the curve of Rule 1&2.

2.6 CONCLUSION AND FUTURE WORK

In this chapter, we proposed a novel distributed algorithm for connected dominating set formation in wireless ad hoc networks. In this *Area* algorithm, we partition the

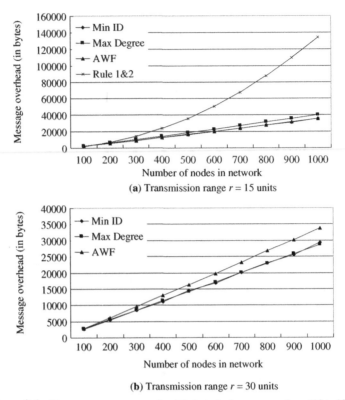

(a) Transmission range $r = 15$ units

(b) Transmission range $r = 30$ units

Figure 2.6   The average message overhead (in bytes) when n ranges from 100 to 1000.

nodes into different areas and selectively connect dominators that are two or three hops away. The time complexity and message complexity of this algorithm are both $O(n)$. Moreover, this algorithm is localized, in which simple local node behavior achieves a desired global objective. From the simulation study, we have observed that this *Area* algorithm always outperforms Rule 1&2 [11] and AWF [15] regardless of the size and density of the networks in terms of the size of CDS. For simplicity, we did not consider the issues of node energy and mobility. Our current work focuses on using the integration of residue energy and node mobility as the selection criteria instead of node ID and node degree.

ACKNOWLEDGMENTS

The work was supported by grants from the Research Grants Council of the Hong Kong SAR, China No. (CityU 114908) and CityU Applied R & D Funding (ARD-(Ctr)) Nos. 9681001 and 9678002.

REFERENCES

1. B. N. Clark, C. J. Colbourn, and D. S. Johnson. Unit disk graphs. *Discrete Math.*, 86(1–3):165–177, 1990.
2. B. Das, R. Sivakumar, and V. Bhargavan. Routing in ad-hoc networks using a spine, *Proceedings of ICCCN'1997*, pp. 1–20.
3. J. Wu and F. Dai. A generic distributed broadcast scheme in ad hoc wireless networks, *Proceedings of ICDCS'2003*, May 2003, pp. 460–468.
4. R. Rajaraman. Topology control and routing in ad hoc networks: a survey, *SIGACT News*, 33(2):60–73, 2002.
5. M. Garey and D. Johnson. *Computers and Intractability: A Guide to the Theory of NP-Completeness*, Freeman, New York, 1979.
6. S. Basagni, M. Mastrogiovanni, and C. Petrioli. A performance comparison of protocols for clustering and backbone formation in large scale ad hoc network, *Proceedings of MASS'2004*, October 2004, pp. 70–79.
7. M. V. Marathe, H. Breu, H. B. Hunt III, S. S. Ravi, and D. J. Rosenkrantz. Simple heuristics for unit disk graphs. *Networks*, 25:59–68, 1995.
8. D. Blough, M. Leoncini, G. Resta, and P. Santi. The K-neigh protocol for symmetric topology control in ad hoc networks. *Proceedings of MobiHoc'2003*, June 2003, pp. 141–152.
9. W. Wu, H. Du, X. Jia, Y. Li, S. C.-H. Huang, and D.-Z. Du. Maximal independent set and minimum connected dominating set in unit disk graphs, submitted.
10. S. Guha and S. Khuller. Approximation algorithms for connected dominating sets. *Algorithmica*, 20(4):374–387, 1998.
11. J. Wu and H. Li. On calculating connected dominating set for efficient routing in ad hoc wireless networks, *Proceedings of DIALM'1999*, pp. 7–14.
12. I. Stojmenovic, S. Seddigh, and J. Zunic. Dominating sets and neighbor elimination based broadcasting algorithms in wireless networks, *IEEE Trans. Parallel Distributed Syst.*, 13(1):14–25, 2002.
13. R. Sivakumar, P. Sinha, and V. Bharghavan. CEDAR: a core-extraction distributed ad hoc routing algorithm, *IEEE J. Selected Areas Commun.*, 17(8):1454–1465, 1999.
14. K. M. Alzoubi, P.-J. Wan, and O. Frieder. Distributed heuristics for connected dominating set in wireless ad hoc networks, *IEEE ComSoc/KICS J. Commun. Networks*, 4(1):22–29, 2002.
15. K. Alzoubi, P.-J. Wan, and O. Frieder. Message-optimal connected dominating sets in mobile ad hoc networks, *Proceedings of MobiHoc'2002*, June 2002, pp. 157–164.
16. Y. Li, S. Zhu, M. T. Thai, and D.-Z. Du. Localized construction of connected dominating set in wireless networks, *Proceedings of TAWN'2004*, June 2004.
17. S. Basagni, Distributed clustering for ad hoc networks, *Proceedings of I-SPAN'1999*, June 1999, pp. 310–315.
18. R. Wattenhofer, L. Li, P. Bahl, and Y.-M. Wang. Distributed topology control for power efficient operation in multihop wireless ad hoc networks, *Proceedings of IEEE INFOCOM'2001*, vol. 3, April 2001, pp. 1388–1397.
19. X.-Y. Li, G. Calinescu, and P.-J. Wan. Distributed construction of a planar spanner and routing for ad hoc wireless networks. *Proceedings of IEEE INFOCOM'2002*, vol. 3, June 2002, pp. 1268–1277.
20. R. Ramanathan and R. Rosales-Hain. Topology control of multihop wireless networks using transmit power adjustment, *Proceedings of IEEE INFOCOM' 2000*, vol. 2, March 2000, pp. 26–30.
21. B. Karp and H. Kung. GPSR: greedy perimeter stateless routing for wireless networks, *Proceedings of MOBICOM'2000*, August 2000, pp. 243–254.
22. M. Gerla and J. T.-C. Tsai. Multicluster, mobile, multimedia radio network, *Wireless Networks*, 1(3):255–265, 1995.
23. L. Bao and J. J. Garcia-Luna-Aceves. Topology management in ad hoc networks, *Proceedings of MobiHoc'2003*, June 2003, pp. 129–140.

24. Y. Wang, W. Wang, and X.-Y. Li, Distributed low-cost backbone formation for wireless ad hoc networks, *Proceedings of MobiHoc'2005*, May 2005, pp. 2–13.
25. Z. Gaspar and T. Tarnai. Upper bound of density for packing of equal circles in special domains in the plane, *Periodica Polytech. Ser. CIV. Eng.*, 44(1):13–32, 2000.
26. E. L. Lloyd, R. Liu, M. V. Marathe, R. Ramanathan, and S. S. Ravi. Algorithmic aspects of topology control problems for ad hoc networks, *Proceedings of MobiHoc'2002*, June 2002, pp. 123–134.
27. D. J. Baker, A. Ephremides, and J. A. Flynn. The design and simulation of a mobile radio network with distributed control. *IEEE J. Selected Areas Commun.*, 2(1):226–237, 1984.
28. W. R. Heinzelman, A. Chandrakasan, and H. Balakrishnan. Energy efficient communication protocol for wireless microsensor networks, *Proceedings of HICSS'2000*, January 2000, pp. 3005–3014.
29. A. Vahdat and D. Becker. Epidemic routing for partially connected ad hoc networks, Technical Report CS-200006, Duke University, April 2000.
30. W. R. Zhao, M. Ammar, and E. Zegura. A message ferrying approach for data delivery in sparse mobile ad hoc networks, *Proceedings of MOBIHOC'2004*, May 2004, pp. 187–198.

Chapter 3

An Intelligent Way to Reduce Channel Under-utilization in Mobile Ad hoc Networks

Naoki Nakamura,[1] Debasish Chakraborty,[2] David Wei,[3] and Norio Shiratori[2]

[1] *Tohoku University School of Medicine, Tohoku University, Sendai, Japan*
[2] *Research Institute of Electrical Communication, Tohoku University, Sendai, Japan*
[3] *CIS Department, Fordham University, Bronx, NY*

3.1 Introduction   43
3.2 Related Works   45
3.3 Background   45
3.4 Enhancements for Efficient Channel Utilization   51
3.5 Performance Evaluation and Discussions   56
3.6 Summary   60
References   60

3.1 INTRODUCTION

In wireless network, performance is dependent on medium access control protocol. Carrier sense multiple access (CSMA) is commonly used for its simplicity. But CSMA is unable to handle the hidden terminal problem, especially in *ad hoc* networks, where multihop communication among nodes is common. To overcome this problem, a frame exchange protocol is used, called request to send/clear to send (RTS/CTS) handshaking. It was first proposed in Ref. [1].

Mobile Intelligence. Edited by Laurence T. Yang, Agustinus Borgy Waluyo, Jianhua Ma, Ling Tan, and Bala Srinivasan

There have been a lot of researches on developing the wireless medium access control (MAC) that efficiently shares limited resources between all stations [1, 2]. At present, IEEE 802.11 MAC is clearly the most accepted and widely used wireless technology. The IEEE 802.11 works based on carrier sense multiple access with collision avoidance (CSMA/CA) and adopts a random access scheme where packets are sent randomly to reduce the number of collisions as much as possible. In addition, IEEE 802.11 introduces a mechanism called RTS/CTS handshaking and virtual carrier sensing to further reduce the chance of collisions that can occur due to hidden terminal problems.

However, it is observed [3] that hidden and exposed terminal problems are exacerbated in mobile ad hoc network (MANET) while using IEEE 802.11. The ultimate result is heavy degradation in throughput and instability of networks. In Ref.[4], it is shown that this problem is more severe in large and dense *ad hoc* networks. So, improvement of performance degradation for IEEE 802.11 over the MANET is an important issue.

"False blocking" problem unnecessarily prohibits nodes from transmitting at a given instant [5]. In worst case, all the neighboring nodes are blocked and can not transmit frames and they are put into the deadlock state. This happens when RTS frame reserves the channel but the channel remains unused. Ray et al. [5] proposed "RTS validation," where a channel is released when each node assumes that CTS is missing, after it receives RTS frame, based on the physical carrier sensing.

In Ref. [6], with the same motivation, we proposed a scheme called "extra frame transmission" to manipulate frame transmission during RTS/CTS handshaking. When no CTS is received for some specific duration after node sends an RTS frame, it will immediately send another small frame to other destination. Both these schemes [5, 6] reuse the channel unnecessarily reserved. The main difference between RTS validation scheme and extra frame transmission scheme is, which node to detect the interruption of RTS/CTS handshaking. In Ref. [6], it is done by sender, whereas in Ref. [5] neighboring nodes are responsible for interruption detection.

In this chapter, we propose another type of extra frame, called "reverse extra frame." We also note that there is scope for improvement when channel release schemes are not applicable. To reuse the channel as much as possible, we modify NAV operations to increase the chance of channel reuse. In addition, focusing on the fact that these schemes can work independently, we combine modified schemes together. Moreover, our proposed mechanisms are free from compatibility problems with standard IEEE 802.11. Results from simulations verify the effectiveness of our schemes. It is observed that our combination of schemes leads to higher gain in throughput compared to IEEE 802.11.

The rest of this chapter is organized as follows. In Section 3.2, we discuss the related works. Basic operations of the IEEE 802.11 are explained and its operation in MANET during RTS/CTS failure is described in Section 3.3. Our proposed scheme of our enhancement of IEEE 802.11 is explained in Section 3.4. The effectiveness of our scheme and evaluation of our proposal are discussed in Section 3.5. Finally, we conclude our work in Section 3.6.

3.2 RELATED WORKS

The CSMA protocol as proposed in Ref. [7] and its susceptibility to the hidden terminal problem was noted in Ref. [8], where the authors proposed a solution called busy-tone multiple access (BTMA) protocol. To reduce the effect of hidden terminal problem, multiple access with collision avoidance (MACA) protocol was proposed in Ref. [1]. MACA uses RTS/CTS mechanism to avoid the hidden terminal problem, but does not include any positive acknowledgement to ensure the integrity of the DATA transmission. A positive acknowledgement scheme as added in the MACAW protocol [2]. The MACAW protocol also requires nodes to send a packet called DATA-send (DS) to indicate that a DATA packet transmission is about to begin; however, this mechanism is not part of the IEEE 802.11 standard, so it may not be compatible with the standard protocol. Ray et al. [5] noted that in a general multihop network, the RTS/CTS mechanism cannot completely eliminate DATA packet collisions due to the masked node problem. They also observed that it holds true even under idealized conditions such as negligible control packet size, negligible propagation delay, and identical interference and packet-sensing ranges. Nevertheless, the RTS/CTS mechanism greatly reduces the hidden terminal problem and is desirable to be deployed in general. Although in Ref. [9] blocking in wireless network has been discussed, but the severity of false blocking was first described in Ref. [5] work. Shigeyasu et al. [10] proposed a new MAC protocol in order to overcome the false blocking problem in wireless LANs.

RTS validation has been proposed to mitigate the false blocking problem where the nodes that have received RTS inhibit themselves from transmitting in the chain [5]. Upon overhearing an RTS frame, nodes listen to the medium whether the corresponding DATA frame transmission has taken place or not. They do this based on physical carrier sensing. If transmission has not taken place, the medium should have remained idle for an expected duration. At this point, nodes start to overhear the DATA frame transmission since they have received an RTS frame. When the medium remains idle for the specified duration since the node received an RTS frame, it will conclude that an interruption of RTS/CTS handshaking has occurred. Then the node will release the NAV registered by that RTS frame and stop deferring. Subsequently, each node releases the channel independently. Harada et al. [11] introduced a new frame, called cancel RTS (CRTS), which left the reserved channel free in order to decrease degradation of channel utilization caused by failure to obtain a channel during RTS/CTS handshaking. In this scheme, when a sender node does not receive CTS correctly for its RTS, it sends a CRTS frame. Then, neighboring nodes, upon overhearing CRTS, cancel the NAV set by the RTS. However, the introduction of a new frame can cause compatibility problems with standard IEEE 802.11.

3.3 BACKGROUND

The IEEE 802.11 MAC layer covers three functional areas: reliable data delivery access, control, and security. We will discuss mainly about first two functions, as they are more closely related to our proposal in this chapter.

3.3.1 Reliable Data Delivery

Like any wireless network, a wireless LAN using IEEE 802.11 physical and MAC layer is not reliable. Noise, interference, and other propagation have adverse effect and cause significant number of frame loses. The reliability mechanism can be handled at the higher level as TCP. But the times used at higher levels are usually in the order of seconds. It is, therefore, more efficient to deal with errors at the MAC layer level. To solve this problem, IEEE 802.11 includes a frame-exchange protocol, called RTS/CTS handshaking. When a station receives a data frame from another station, it returns an acknowledgment (ACK) frame to the source station. This exchange should not be interrupted by a transmission from any other station. If the source station does not receive an ACK within a certain period of time, it will retransmit the frame. To enhance further reliability to this scheme, a four-way-handshaking scheme has been introduced. In this scheme, the source issues a RTS frame to the destination. The destination then responds with a CTS. After receiving the CTS, the source transmits the data frame and the destination responds with a ACK. Both RTS and CTS alert all the neighboring nodes that are within transmission range to refrain from transmitting to avoid collision. The RTS/CTS is an optional function of the MAC that can be disabled.

3.3.2 Access Control

The 802.11 working group considered two types of proposals for a MAC algorithm: distributed access controls, which, like ethernet, distribute the decision to transmit over all the nodes using a carrier-sense mechanism; and centralized access protocols, which involve regulation of transmission by a centralized decision maker. A distributed access protocol makes sense for an ad hoc network of peer workstations and may also be attractive in other wireless LAN configurations that consist primarily of bursty traffic. The IEEE 802.11 describes two medium access functions called point coordination function (PCF) and distributed coordination function (DCF). The lower sublayer of the MAC layer is the distributed coordination function (DCF). DCF uses contention algorithm to provide access to all traffic. The DCF sublayer is based on carrier sense multiple access (CSMA) technique. In this chapter, we focus on IEEE 802.11 DCF that provides a distributed access mechanism scheme over ad hoc networks.

IEEE 802.11 DCF is based on CSMA/CA. It provides basic access mechanism using two-way handshaking and four-way handshaking. We present only the basic functional overview of the IEEE 802.11 standard here. More details can be found in Ref. [13].

If the medium is sensed to be free for a DCF interframe space (DIFS) interval, the transmission may proceed. On the other hand, if the medium is busy, the station must defer its transmission until the end of the current transmission. Then, it will wait for an additional DIFS interval and generate a random back-off timer before transmission. The counter is decreased as long as the medium is sensed as idle and frozen when the medium is busy and resumed when the medium is sensed as idle again for a time longer than a DIFS interval. Only when the backoff counter reaches zero, the station can transmit its packets.

A station desiring to transmit senses the medium whether another station is transmitting before initiating a transmission. If the medium is sensed to be free for a DCF interframe space interval, the transmission may proceed. On the other hand, if the medium is busy, the station must defer it transmission until the end of the current transmission. Then, it will wait for an additional DIFS interval and generate a random backoff timer before transmission. The counter is decreased as long as the medium is sensed as idle and frozen when the medium is busy and resumed when the medium is sensed as idle again for a time longer than a DIFS interval. Only when the backoff counter reaches zero, the station can transmit its packets.

To ensure that backoff maintains stability, a technique known as binary exponential backoff is used. A station will attempt to transmit repeatedly in the face of repeated collisions, but after each collision, the mean value of the random delay is doubled. The binary exponential backoff provides a means of handling a heavy load. Repeated failed attempts to transmit result in longer and longer backoff times, which helps to smooth out the load. Without such a backoff, two or more stations attempt to transmit at the same time, causing a collision. These stations then immediately attempt to retransmit, causing a new collision.

The backoff counter is uniformly chosen between $(0, \omega - 1)$. The value ω, known as contention window (CW), represents the contention level in the channel. At the first transmission attempt, ω is set to CW_{min}. After each transmission failure, ω is doubled up to a maximum value of CW_{max}. Figure 3.1 illustrates the basic access mechanism.

Since stations cannot listen to the channel while transmitting, collision detection is not possible in wireless medium. An ACK is transmitted by the receiving station to confirm the successful reception. Receiving station waits for a short interframe space (SIFS) interval after receiving data frame correctly. After that, it sends back an ACK to the sending station. In case of missing an ACK, sender assumes a transmission loss and schedules retransmission after doubling the CW.

3.3.3 Hidden and Exposed Terminal Problems

Hidden nodes in a wireless network refer to nodes that are out of range of other nodes or a collection of nodes. The hidden node problem occurs when a node is visible

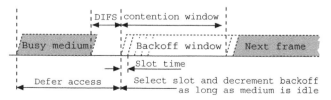

Figure 3.1 The basic access mechanism.

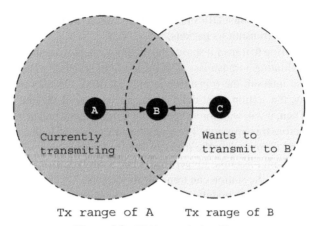

Figure 3.2   Hidden terminal problem.

from a wireless hub, but not from other nodes communicating with the said hub. This leads to difficulties in media access control. In the classic hidden terminal situation, as station "B" can hear both "A" and "C" station, but "A" and "C" cannot hear each other. Therefore, "A" and "C" unable to avoid colliding with each other as shown in Figure 3.2. In wireless networks, the exposed terminal problem occurs when a node is prevented from sending packets to other nodes due to a neighboring transmit ion. In the exposed terminal case, a well-sited station "A" can hear far away station "C." Even though "A" is too far from "C" to interfere with its traffic to other nearby stations, "A" will defer to it unnecessarily, thus wasting an opportunity to reuse the channel locally as illustrated in Figure 3.3. Sometimes there can be so much traffic in the remote area that the well-sited station seldom transmits.

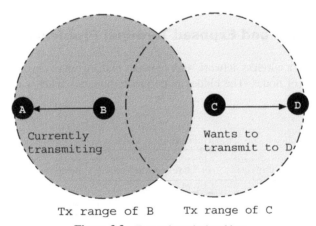

Figure 3.3   Exposed terminal problem.

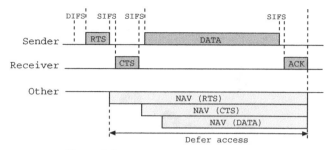

Figure 3.4   The RTS/CTS access mechanism.

3.3.4 RTS/CTS Handshaking Mechanism and Virtual Carrier Sensing

The RTS/CTS access method is provided as an option in IEEE 802.11 to reduce the collisions caused by hidden terminal problem. A station that needs to transmit large data frame (longer than predefined RTS threshold value) follows the backoff procedure as the basic mechanism described before. After that, instead of sending data frame, it sends a special short control frame called RTS. This frame includes information about the source, destination, and duration required by the following transaction (CTS, DATA, and ACK transmission). Upon receiving the RTS, the destination responds with another control frame called CTS, which also contains the same information. The transmitting station is allowed to transmit data only if the CTS frame is received correctly.

All other nodes overhearing either RTS and/or CTS frame adjust their network allocation vector (NAV) to the duration specified in RTS/CTS frames. The NAV contains period of time in which the channel will be unavailable and is used as virtual carrier sensing. Figure 3.4 depicts the RTS/CTS mechanism. Stations defer transmissions if either physical or virtual sensing finds the channel being busy. Nevertheless, if receiver's NAV is set while data frame is received, DCF allows the receiver to send the ACK frame.

The effectiveness of RTS/CTS mechanism is shown in Ref. [12] as it can early detect the collisions by the lack of CTS. Here, it considered that an absence of CTS implies a collision has occurred and thus can effect an early detection. However, the protocol cannot free or reallocate the channel that was already reserved by the previous RTS frame. Stations receiving only the RTS frame but not CTS cannot assume that transmission does not take place. Therefore, they defer for an interval declared in last RTS. This results in wasting of channel capacity around the sender node.

3.3.5 RTS/CTS-Induced False Blocking

In this section, we analyze the situations when CTS is not received at the sender and how to improve the channel utilization in each occasion.

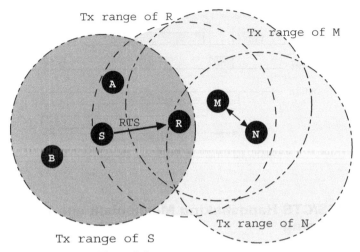

Figure 3.5   Illustration of situation 2.

Situation 1: Backoff timers at two or more stations reach zero at the same time and send the RTS frame simultaneously, so the sender does not receive the CTS frame. This happens more frequently as network traffic increases.

Situation 2: As is illustrated in Figure 3.5, station *S* starts the RTS/CTS sequence while another transmission which interferes with the reception is been carrying on, say, between *N* and *M*. Even if the RTS correctly reaches the receiver, the virtual carrier sensing at station *R* will forbid the CTS response.

Situation 3: This situation occurs when the intended receiver, *R*, moves to a new position, which is out of communication range of *S* as shown in Figure 3.6. Hence, it cannot receive RTS from *S*.

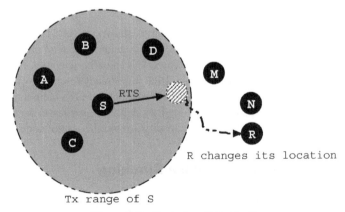

Figure 3.6   Illustration of situation 3.

The above situations are regularly found in MANET where stations route packets through each other in multihop fashion, as stations are free to move arbitrarily. In wireless network, only node is allowed to transmit at a particular time and many nodes around the receiver may be blocked. The neighbors of the blocked node are unaware of this blocking. So, a node may initiate a communication with a node that is presently blocked and consequently the destination does not respond to the RTS packet. However, the sender interprets it as channel contention and enters backoff. The neighboring nodes are prevented from decrementing backoff counter and sending packets because of the NAV set by RTS.

This false blocking is a consequence when all the nodes that receive RTS inhibit themselves from transmitting. This problem can get severe when it is occurred in circular way, which can create pseudodeadlock [5]. This leads to lower channel utilization and route failure. Therefore, releasing unused channel is important for channel stability. RTS validation reduced the above problem, but there is still wasted channel capacity.

By our proposed schemes, we try to reuse the wasted channel capacity as much as possible.

3.4 ENHANCEMENTS FOR EFFICIENT CHANNEL UTILIZATION

In this section, we present three different approaches and their modification for optimal use of otherwise wasted channel capacity in MANET. In our proposed schemes, we tried to unlock the unnecessarily blocked channel by using NAV update and also tried some aggressive methods to recover as much as possible and minimize the channel lose due to false blocking.

We have described each of our proposed schemes (i) modification of NAV operation, (ii) extra frame transmission (EFT), (iii) combination of schemes, and (iv) reverse extra frame transmission (R-EFT) in different subsequent subsections.

3.4.1 Modification of NAV Operation

In case of RTS validation mechanism [5], when the node has already been deferred, it can not set NAV back to the previous value that has already been set by other RTS frames. Besides that, RTS validation may not be always available. Higher the amount of network traffic, more frequent the unavailability of RTS validation. As a result, RTS validation can not fully utilize the unused channel. Thus, the efficiency of channel reuse will be reduced. Improvement is possible if RTS validation works irrespective of NAV set.

With the above considerations, we modify the NAV operation with three new variables as follows:

(i) We divided the original NAV in two parts: one is set of NAV_k indexed with the corresponding node's ID and the other is NAV_{other}.

(ii) NAV used for the operation is calculated by the maximum value in the sets of NAV_k and NAV_{other}.

(iii) NAV_k is adjusted when overhearing RTS/DATA frame from node $node_k$, and NAV_{other} is adjusted by cases, other than RTS/DATA frame, like receiving CTS frame or suffering from collision.

(iv) Allowing NAV_k to override by newer value in the duration field. It means, when node overhears RTS/ DATA frame of node k then NAV_k will be updated by the duration the frame has.

(v) If needed, RTS validation will reset NAV_k.

NAV updating scheme can handle NAV for multiple nodes with record of corresponding senders' ID, related to failure of RTS/CTS handshaking, and gives the flexibility and convenience to cancel the NAV even if NAV has already set. This modification is helpful for RTS validation to cancel the NAV and improve the channel capacity.

In addition, there is benefit of NAV updating in extra frame transmission schemes, discussed in the following subsection.

3.4.2 Extra Frame Transmission

Extra frame transmission works as shown in Figure 3.7. After sender has transmitted RTS for receiver 1 and has waited until it conceives that CTS will not come back, it picks a frame from the sending queue and immediately transmits it to the alternate receiver, if an appropriate one exists. We explain the term "appropriate frame" in the following paragraphs. The extra frame will be removed from the queue if the transmission is completed (confirmed by ACK from receiver) [13] or the transmitted extra frame is broadcasted. Regardless of the success of extra frame transmission, the sender goes back to normal operation by scheduling the retransmission of the original frame with doubled CW.

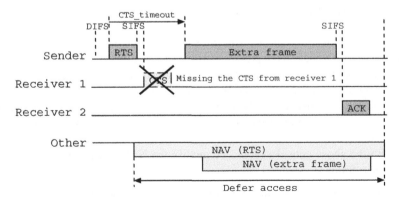

Figure 3.7   Extra frame transmission.

Since the standard protocol does not specify the timeout value for CTS response, sender normally stays idle till the end of the allocated duration. However, the RTS/CTS sequence uses strict timing. So, we introduce a new parameter, handshake timeout S, which accounts for the maximum time that may be required to receive a CTS. Having waited for this handshake timeout S, the sender is assured that the CTS response from receiver will not come at all.

The transmitted extra frame should maintain the following properties:

1. Extra frame should be destined to a station different from the currently attempted one. Since no reply is received from the current destination, any further attempt to the same station will be futile.

2. The selected extra frame should be a broadcast frame (which is irrelevant to RTS threshold), or unicast frame smaller than RTS threshold. So, this frame can be immediately transmitted without following RTS/CTS frame exchange protocol.

3. The chosen extra frame should be first in the queue destined for a particular receiver. For example, if there are two frames for the same destination at the sender, and the first frame is larger than RTS threshold but the second frame is not, then none of the frames will be sent. Even though second frame may satisfy the first two conditions. This constraint is considered to avoid any out-of-order transmission.

It has been observed that with the increase in traffic load and node density, the chance of false blocking increases. In that case, there is a fair chance of a successful transmission of this extra frame because channel around sender has been already reserved by previous RTS frame. In this process, we can deliver a frame that cannot be sent in normal operation.

With NAV updating schemes, introduced earlier, the benefits of this scheme can be double folded. The nodes who received the RTS and blocked their channel will be allowed to cancel the original NAV duration.

Actually, nodes will readjust the previously set NAV with the duration of extra frame. NAV of extra frame should have shorter value in duration field than current NAV for the RTS sender; node can indirectly cancel its NAV. So, if the selected extra frame is a broadcast frame, whose NAV duration is zero, the overhearing nodes can reset the NAV value for RTS and completely cancel the NAV. Even if RTS threshold is set to zero, extra frame can release the channel effectively.

3.4.3   Combination of RTS Validation and Extra Frame Transmission

Extra frame transmission and RTS validation [5] work-independently on sender node and neighboring nodes, respectively. For further performance improvement, we propose an approach to combine RTS validation and extra frame.

Since an appropriate extra frame cannot be always available in the waiting queue of sender node, the channel reuse is not available as frequently as RTS validation.

although it may available, it has the ability to deliver data as an extra frame. To utilize this ability, we entrust mainly extra frame transmission with delivering the extra frame transmission and entrust RTS validation with releasing channel. To work together in parallel, we set two parameters, handshake timeout N and handshake timeout S, as follows:

$$\text{Handshake\_Timeout\_S} : \text{RTS\_Tx\_time} + \text{propagation\_delay}$$

$$+ \text{SIFS} + \text{CTS\_Tx\_time} + \text{propagation\_delay}$$

$$\text{Handshake\_Timeout\_N: propagation\_delay} + \text{SIFS} + \text{CTS\_Tx\_time}$$

$$+ \text{propagation\_delay} + \text{SIFS} + \text{propagation\_delay} + \text{SIFS}$$

where T_x represents the transmission time.

With these parameters, when a node detects the interruption of RTS/CTS handshaking, and if an extra frame is available, the extra frame will deliver a small data as well as release the channel by virtue of NAV updating. Even if there are no extra frames, RTS validation just releases the NAV. Therefore, complementing both mechanisms together leads to improvement in channel efficiency.

3.4.4   Reverse Extra Frame Transmission

To allow the neighboring nodes, who overhear the RTS, we propose an aggressive way of channel reuse by introducing a new type of extra frame called "reverse extra frame." The timing to transmit it is same as that of releasing channel of RTS validation. So, it can be said to be a subset of RTS validation. The algorithm for combining the schemes, including reverse extra frame, is shown in Figure 3.8.

The idea stem from the fact that, generally speaking, once RTS frame has been sent, it is relatively free from collision for the duration specified in the RTS frame. If one of the neighboring nodes can send a frame to the sender, its frame is expected to reach successfully. To exploit this relatively safe period of time to reuse the channel, we allow the neighboring nodes to send an extra frame to the node that originates the RTS. Reverse extra frame transmission works as shown in Figure 3.9.

It is not possible to completely prevent the reverse extra frames from collision because there can be several eligible candidates for sending an extra frame. To reduce the probability of collision, the following constraints are introduced:

- Reverse extra frame should be the first frame in the queue and be destined to the node RTS frame had sent.
- The length of the duration in reverse extra frame should be smaller than that of the duration specified in RTS.
- The node should have a short backoff timer that would have expired if node does not receive RTS frame.

If an appropriate reverse extra frame is found, it will be sent immediately and will be removed from the queue if the transmission is completed (confirmed by ACK

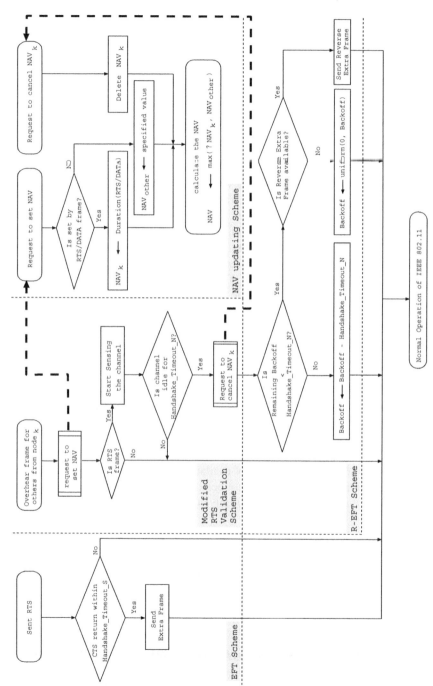

Figure 3.8 Flow chart of combination of schemes.

55

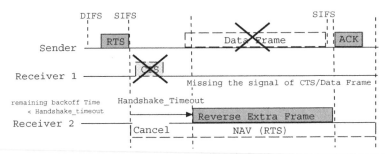

Figure 3.9   Reverse extra frame transmission.

from sender). Then, the node goes back to the normal operation, regardless of the successful transmission of reverse extra frame.

We have earlier seen that even when RTS/CTS handshaking is interrupted, the neighboring nodes will be inhibited from transmitting. RTS validation can release the channel, but the nodes cannot recover from the loss incurred by the interruption. Because when the nodes sensed the channel as busy, their backoff timers were halted and stopped decrementing during RTS/CTS handshaking.

When reverse extra frame is available, nodes can make up the above loss of time. Because without performing RTS/CTS handshaking, node transmits the data frame that is supposed to send in the near future. But due to the restriction imposed to prevent collision, reverse extra frame may not be always available. When there is no reverse extra frame to compensate the loss of time, we allow the nodes to decrement their respective backoff timer.

We allow those nodes to decrement the time equal to the "handshake timeout" from their respective remaining backoff timer. But for those nodes whose remaining time of the backoff timer is less than or equal to the "handshake timeout," to differ their access to avoid collision, it will choose a uniform random backoff time from (0, current backoff time).

So, when reverse extra frame is not available, the nodes will decrease backoff timer for the deferred time as if it had not been interrupted. We can thus reduce the waiting time for the node before transmitting and increase the throughput.

3.4.5   Compatibility with IEEE 802.11

CRTS [11] is releasing the NAV by introducing an extra frame. However, our schemes keep the format unchanged. Thus, our proposed schemes are compatible with the existing IEEE 802.11 standard. Even if there are stations that do not support our enhancement, they will also work as well.

3.5   PERFORMANCE EVALUATION AND DISCUSSIONS

In this section we have discussed on performance evaluation and scenario used for network simulation. We later explained and analyzed the results.

3.5.1 Simulation Scenario

The most widely recognized network simulator, *ns-2* [16], is used to evaluate the effectiveness of our mechanism. We compared the performance of standard IEEE 802.11 [13], RTS validation [5], and our proposed enhancements.

The network model of a multihop wireless topology and routing protocol ad hoc on demand distance vector (AODV) [15]. The link layer is a shared media radio with nominal channel bit rate of 1 Mbps. The antenna is omnidirectional with radio range of 250 m.

Setup parameters are listed here: slot time = 20 μs, SIFS = 10 μs, DIFS = 50 μs, propagation delay = 2 μs, RTS threshold = 0 bytes.

Traffic source and destination pairs are randomly spread over the network. Type of traffic is constant bit rate (CBR) with packet size randomly chosen between 512 and 2048 bytes, to prove that our evaluation process is not affected by the frame size. We have created 30 sets of CBR traffic. The sum of all the senders' transmission rate is represented as offered load. For example, if the offered load is 600 kbps, then the transmission rate of CBR traffic is 20 kbps.

The random waypoint model is used. In this model, a mobile station begins by staying at one location for a certain period of time (we call this pause time). Once this time expires, this station chooses a random location in the simulation field with a random speed between [min speed, max speed], which is randomly selected with uniform probability. Each node moves according to random waypoint model with parameters max speed = 10 (m/s), min speed = 0 (m/s), and pause time = 50(s).

In each experiment, we run the simulation on the 1500×500 m^2 field for 700 s. We start measuring from 100 s and up to 700 s. Each plot in these graphs is the average of at least 50 simulations. As a default parameter of each simulation, we define the number of stations as 50. And each node moves according to random waypoint with parameters max speed=10 (m/s), min speed = 0 (m/s), and pause time = 50 (s).

3.5.2 Results and Analysis

In this section, we compare the performance between our proposed schemes with RTS validation. We ran our simulation in two different scenarios. In one, the offered load varies between 200 and 800 kbps and in another, the number of nodes varies from 10 to 90.

We evaluate the effects of our schemes in terms of MAC layer throughput (Th$_{mac}$) with above two scenarios. Th$_{mac}$ is computed as a summation of data frame size successfully sent by each node per unit time. Suppose, u_i is the data frame size in bit successfully transmitted by node i. If the total transmission time is t, then Th$_{mac}$ in bits per second (bps) is defined as:

$$\text{Th}_{mac} = \frac{\sum_{i=1}^{i=N} u_i}{t} \tag{3.1}$$

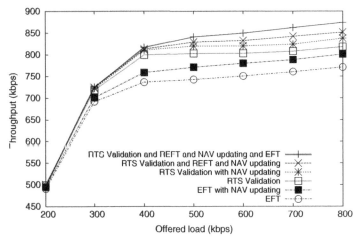

Figure 3.10   MAC layer throughput (Th$_{mac}$) comparison as a function offered load.

Firstly we confirm the comparison of throughput (Th$_{mac}$) as a function of offered load in Mac layer for six different combinations of schemes in Figure 3.10. Figure 3.10 shows that NAV update scheme improves the Th$_{mac}$ of extra frame and RTS validation. RTS validation with NAV updating is more effective than extra frame with NAV updating due to the increase in the occurrence in the channel release. RTS validation including its variants (RTS validation with NAV updating and reverse extra frame) yield high performance on proposed combination schemes.

On the other hand, when the number of nodes is varying, Figure 3.11 shows, Th$_{mac}$ of RTS validation including its variants is rapidly decreasing with the increase

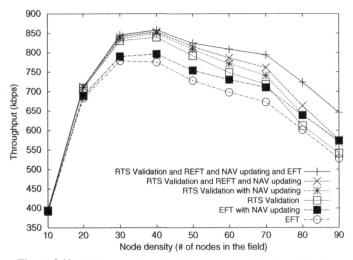

Figure 3.11   MAC layer throughput (Th$_{mac}$) as a function of node density.

in the number of nodes. In worst case, the performance of RTS validation is less than that of extra frame with NAV updating. This is because RTS validation and its variants are based on physical carrier sensing; they are very much sensitive to the number of transmission in the networks.

Even in above severe condition, extra frame can keep improvement level higher due to its two-fold advantage of sending extra frame and releasing the channel. The combination of EFT and R-EFT shows a much steadier nature even with increase in the number of nodes. The improvement is more visible when the number of nodes is 60 or more. Therefore, combination of RTS validation and its variants and extra frame works in a complementary style. This leads to high performance in both scenarios.

In our proposed combination of schemes, we reused the channel as aggressively as possible. So, in order to verify whether there is any degradation in reliability by this enhancement, we measured the delivery rate of each unicast frame.

Packet delivery ratio (PD) is computed as:

$$PD = \frac{\sum_{i=1}^{i=N} R_i}{\sum_{i=1}^{i=N} S_i} \tag{3.2}$$

S_i is the total data size of unicast data frame node i sent and R_i the total data size of ACK frame node i received.

Figure 3.12 and Figure 3.13 show CBR packet delivery rate as a function of offered traffic load and node density for RTS validation and IEEE 802.11.

In both scenarios, when RTS/CTS is not interrupted frequently, all schemes have similar packet delivery ratios. As the interruption of RTS/CTS handshaking becomes more frequent, our combined scheme yields a higher packet delivery ratio than the others in both scenarios. In high nodes density scenario, especially when interrup-

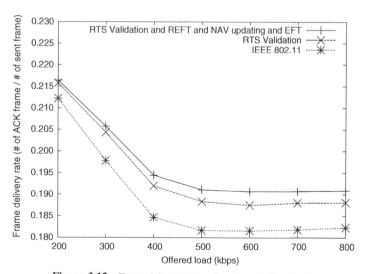

Figure 3.12   Frame delivery rate as function of offered load.

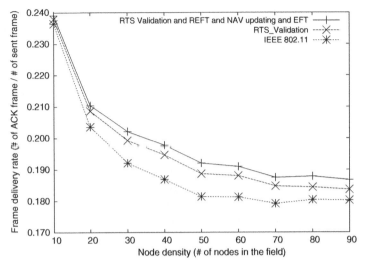

Figure 3.13   Frame delivery rate as a function of node density.

tion of RTS/CTS handshaking occurs more frequently, our scheme's packet delivery ratio is higher than those of other schemes. This is because reusing the wasted channel aggressively enables nodes to deliver more packets in a stable manner, consequently increasing the packet delivery ratio. These results show that regardless of whether channels are released/reused aggressively, our combination of schemes does not adversely effect packet delivery ration.

3.6 SUMMARY

In this chapter we have shown that our proposed schemes can overcome the inherent channel inefficiency of RTS/CTS handshaking in mobile ad hoc networks, especially when node density is high and interruption is frequent. At the same time, our enhancements do not suffer from any compatibility problems and generate no additional overhead, which is vital for smooth deployment. Results obtained from extensive simulations show that our combined method considerably improves the throughput compared to standard IEEE 802.11 and the RTS validation scheme. Thus, we can confirm that such smart approach can improve channel utilization in wireless communication.

REFERENCES

1. P. Karn, MACA: – A new channel access method for packet radio. In *Proceedings the 9th ARRL Computers Networking Conference*, Ontario, Canada, September 1990.
2. V. Bharghavan, A. Demers, S. Shenker, and L. Zhang, MACAW: A media protocol for wireless LANs. In *Proceedings of the ACM SIGCOMM'94*, London, UK, September 1994.

3. S. Xu and T. Saadawi. Does the IEEE 802.11 MAC protocol work well in multihop wireless ad hoc networks? In *Proceeding of IEEE Commun. Mag.*, 39(6):130–137, 2001.

4. Y. Wang and J. J. Garcia-Luna-Aceves. Collision avoidance in multihop ad hoc networks. In *Proceedings of the IEEE/ACM MASCOT'02*, Texas, USA, October 2002.

5. S. Ray, J. Carruthers, and D. Starobinski. Evaluation of the masked node problem in ad-hoc wireless LANs. *IEEE Trans. Mobile Comput.*, 4(5): pp. 430–442, 2005.

6. A. Chayabejara, S.M.S. Zabir, N. Shiratori, An enhancement of the IEEE 802.11 MAC for multihop ad hoc networks, In *Proceedings of IEEE Vehicular Technology, 2003 Fall*, Florida, USA. October 6–9, 2003.

7. L. Kleinrock and F. A. Tobagi, Pakcet switching in radion channels: Part 1. Carrier sense multiple access modes and their throughput-delay characteristics. *IEEE Trans. Commun.*, COM-23(12):1400–1416, 1975.

8. L. Kleinrock and F. A. Tobagi, Pakcet switching in radion channels: Part 2. The hidden node problem in carier sense multiple access modes and the busy tone solution. *IEEE Trans. Commun.*, COM 23(12):1417–1433, 1975.

9. V. Bharghavan, Performance evaluation of algorithms for wireless medium access. In *IEEE Performance and Dependability Symposium '98*, IEEE, Raleign, NC 1998.

10. T. Shigeyasu, T. Hirakawa, H. Matsumo, and N. Morinaga. Two simple modifications for improving IEEE 802.11 DCF throughput performance. In *IEEE Wireless Communication and Networking Conference (WCNC)*, 2004.

11. T. Harada, C. Ohta, M. Morii. Improvement of TCP throughput for IEEE 802.11 DCF in wireless multi-hop networks. *IEICE Trans. Commun.*, J85-B(12):2198–2208, 2002.

12. G. Bianchi, Performance analysis of the IEEE 802.11 distributed coordination function. *IEEE J. Select. Areas Commun.*, 18(3):535–547, 2000.

13. IEEE Standard for Wireless LAN Medium Access Control (MAC) and Physical Layer (PHY) Specication. IEEE Std. 802.11, August 1999. [Online]. Available: http://standards.ieee.org/getieee802/802.11.html

14. H. Wu, Y. Peng, K. Long, S. Cheng, and J. Ma, Performance of reliable transport protocol Over IEEE 802.11 wireless LAN: Analysis and enhancement. In *Proceedings of the IEEE INFOCOM'02*, June 2002.

15. C. E. Perkins, E. M. Royer, and S. Das, Ad hoc on-demand distance vector (AODV) Routing, *RFC 3561*, July 2003. [Online]. Available: ftp://ftp.rfc-editor.org/in-notes/rfc3561.txt

16. The Network Simulator Version 2 (ns–2), [Online]. Available: http://www.isi.edu/nsnam/ns/

Chapter 4

Mobility in Publish/Subscribe Systems

Vinod Muthusamy and Hans-Arno Jacobsen

Department of Electrical and Computer Engineering, University of Toronto, Toronto, Canada

4.1	Introduction	62
4.2	Mobile Applications	63
4.3	Publish/Subscribe	66
4.4	Client Mobility	73
4.5	Example Application	82
4.6	Summary	84
	References	84

4.1 INTRODUCTION

Emerging mobile applications such as location-based services, mobile commerce, games, and entertainment services exhibit communication patterns for which existing transport layer communication primitives are ill-suited. These applications produce massive volumes of data and require sophisticated interaction patterns, such as many-to-many communication, which necessitate powerful and efficient filtering and routing capabilities. Furthermore, mobile devices are typically resource constrained and highly dynamic, which presents a challenge in developing a communication infrastructure that scales with network traffic and size.

The distributed content-based publish/subscribe communication paradigm addresses the requirements of emerging mobile applications. The loose coupling among

Mobile Intelligence. Edited by Laurence T. Yang, Agustinus Borgy Waluyo, Jianhua Ma, Ling Tan, and Bala Srinivasan
Copyright © 2010 John Wiley & Sons, Inc.

participants supports mobility seamlessly at the messaging layer. Also, an expressive, declarative language allows for fine-grained message filtering, complex interactions, and scalable, efficient routing of messages to mobile nodes.

In this chapter, Section 4.2 first provides some context for the discussion by describing a sample of mobile applications of interest and extracts the key characteristics of these applications. These characteristics are used to motivate the requirements of a communication infrastructure for these applications. Section 4.3 describes the publish/subscribe model, which addresses the above requirements, and Section 4.4 presents algorithms to efficiently support mobility in a distributed publish/subscribe system. Then, in Section 4.5, an example scenario is presented to illustrate the filtering and complex interaction patterns possible using a publish/subscribe communication abstraction. Finally, Section 4.6 concludes with a brief summary of the chapter.

4.2 MOBILE APPLICATIONS

Before designing a system architecture for mobile applications, it is important to understand the characteristics of these applications. In this section, we present various mobile applications and outline the properties of these applications and the devices they run on. From this analysis, we argue for certain infrastructure requirements to facilitate mobile applications.

4.2.1 Example Applications

Mobile applications are constantly evolving and it is difficult to outline a definitive set of application requirements. However, there are certain trends and properties we see in emerging applications.

Early applications on mobile devices were often stand-alone programs that did not interact with other applications or with a network. Examples include rudimentary personal information management (PIM) utilities such as calendars and to-do lists; single-player games, such as solitaire card games; and camera applications that let users capture and view photos. In the context of this discussion, the notable property of these applications is that they had no network access and all the data are manually entered. While such applications may provide the ability to synchronize data with a desktop computer, physical access to the computer is required, decreasing the advantage of a mobile device.

Here, we are more concerned with mobile applications that have access to a network while mobile. Most mobile devices today provide some means to access a network, such as the Internet, while the user is mobile. For example, mobile phones allow applications to utilize the SMS capabilities of the phone, or to access the Web. Many PDAs, likewise, support a variety of means to access data outside the device, such as infrared, Bluetooth, GPS, or WiFi interfaces.

Initial network-enabled applications on mobile devices were often traditional network-aware desktop applications that were ported to mobile devices. This includes PIM applications such as e-mail (a popular example of which are the

BlackBerry devices), instant messaging capabilities (such as the short message service), and simple file sharing (using the multimedia messaging service). These applications were typically limited versions of their desktop equivalents but could be used while mobile.

As the mobile device capabilities improved (most notably in terms of their wireless network cost and performance), and as developers gained experience on the various mobile platforms, new classes of mobile applications emerged that had no desktop equivalents. One such class of applications are location-based services. Examples of such services include location-aware alerts (such as information about local traffic advisories or advertisements about nearby merchants), or an enterprise fleet tracking system that can be used to provide real-time locations of the vehicles belonging to an organization. These location-aware applications, naturally, require knowledge about the location of the device. This information can be gathered automatically using GPS receivers or various location-tracking technologies provided by telecom providers, or be input by the user. They may also be deduced in various ways, such as detecting the presence of a signal from a known landmark.

More recently, commerce and financial applications have emerged on mobile devices. Examples of such applications include mobile payment services that allow users to securely purchase goods from their mobile device, or store financial data such as credit card information on their mobile devices. Such applications may be location-aware, network-capable, and above all must be robust to malicious attacks.

All the above examples consist of applications and devices ultimately used by human, which we note does not preclude the applications from performing intelligent processing or automatic actions the user is unaware of. However, another class of applications consist entirely of autonomous agents with little or no human interactions. Perhaps the most representative of such applications are those based on sensor networks, in which a large number of resource-constrained nodes collaborate to perform some common task. Examples of sensor network applications include distributed environment sensing or assembly line automation.

4.2.2 Device and Application Characteristics

On the basis of the sampling of applications above, we summarize some key properties of the mobile devices and applications of interest.

The mobile devices are typically resource constrained. While the computation, memory, bandwidth, and battery life available in such devices are always improving, these devices are still constrained relative to other compute devices such as desktop computers. Also, the user interface of these devices is either nonexistent in the case of sensor network nodes or very limiting in the case of PDAs and mobile phones. We also notice that most mobile devices today support some means to communicate with the outside world while the user is mobile. The technology used to achieve this, however, varies widely and includes WiFi, Bluetooth, and GPRS.

In terms of the applications, the key characteristic we note is that virtually all emerging applications require some sort of wireless network connectivity in order to collaborate with other mobile devices or communicate with other services on the Internet. Also, the applications are, unsurprisingly, becoming more resource intensive as they exploit improvements in the mobile devices. Of particular interest in this discussion is the increasing amount of data the applications must process, much of which are transferred over the limited, unreliable network. For example, a location-aware traffic report service may need to filter and process a large and continuous stream of data in order to find the information relevant to the user.

4.2.3 Infrastructure Requirements

The mobile applications presented above require some sort of infrastructure to operate. We will focus in particular on the communication infrastructure needed by these applications to communicate with other devices or services accessible through the network and outline some of the requirements of such an infrastructure.

- Large scale: The incredible growth in the use of mobile phones and the increasing number of mobile applications available on these devices requires that the infrastructure be able to support a very large number of devices.

- Network constraints: Mobile devices today typically have wireless connections that have limited bandwidth, long latencies, and are unreliable. The infrastructure must be able to tolerate such networks. Most notable here is that unreliable network connections require the disconnected operation of applications, in which applications must continue to operate (in some limited way) while the network is unavailable and "recover" when the network becomes available.

- Dynamic system: A system of mobile nodes is highly dynamic. Not only are the nodes constantly moving, but they also experience ongoing fluctuations in their network connection performance and availability. Applications require a communication abstraction that can simplify dealing with such an environment.

- Data volume: The infrastructure as a whole must be able to support the immense volume of data being transferred among the large number of mobile devices. In addition, the applications on the individual devices need a simple and powerful mechanism to filter and process these data.

- Sophisticated interaction: While much of the communication in traditional Internet applications involves a conversation between two parties (such as Web browsing and e-mail applications), applications are increasingly establishing more complex communication among mobile devices and services. For example, sensor networks collaborate in unpredictable ways, and location-aware mobile applications may communicate with any number of mobile devices within proximity. The interaction patterns here may be arbitrarily complex and include any combination of one-to-one, one-to-many, and many-to-many communication.

- Location support: Location is an important contextual property of mobile devices that applications can exploit. Therefore, the communication abstractions need to expose (or hide) this information as required by the applications. One application may require addressing the services in a location-independent manner, while another may require location-dependent names.

- Flexible naming and addressing: The issue of device naming cuts across several of the points above. Devices or services need to be named in such a way that they can be addressed in a location-dependent or -independent way, or be addressed individually or as a meaningful subset of some group. For example, an application may wish to communicate with John's PDA regardless of its location (or even its network connectivity), with all sensors within some radius around the device or with all mobile phones anywhere that support the SMS service. The communication abstractions must allow such flexible addressing capabilities.

- Responsiveness: Some mobile applications require real-time notifications about external events. For example, a traffic reporting application may need to notify the user as soon as a traffic alert is issued.

While the properties above need to be supported by the physical communication infrastructure, they must also be addressed by the software abstractions made available to the applications and system software architecture. We refer to this software and architecture as the *middleware* and will spend the remainder of this discussion presenting a middleware paradigm that addresses these concerns.

4.3 PUBLISH/SUBSCRIBE

The requirements of a communications infrastructure presented in the previous section are addressed by the *publish/subscribe* model. The publish/subscribe model provides a simple yet powerful abstraction for application interactions. Below, we present the publish/subscribe model, briefly outline its benefits, and describe a distributed implementation of this model.

4.3.1 Publish/Subscribe Model

Publish/subscribe is a data-dissemination model that has many useful properties. There are three main entities in the publish/subscribe model: the *publisher*, *subscriber*, and *broker*. The publisher is the data producer, the subscriber the data consumer, and the broker mediates between the publisher and subscriber. These entities are only logical and do not have to correspond to physical components. For example, it is possible for a computer to be a publisher, subscriber, and broker.

A popular example of the publish/subscribe model is a stock quote information dissemination application. In this example, the publisher would be the stock exchange, such as the New York Stock Exchange, and the consumer could be a stock broker who is interested in tracking the latest prices of certain stocks. A subscriber S expresses his interest in these stocks by sending a *subscription* message s to the broker. The broker stores these subscriptions in an index T as a set of (*subscription*, *subscriber*) tuples.

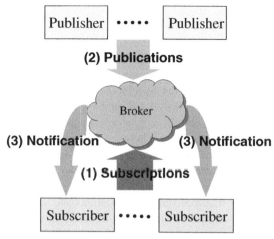

Figure 4.1   Publish/subscribe message sequence.

The publisher *P* communicates the latest stock updates by sending a *publication* message *e* to the broker. Upon receipt of a publication *e*, the broker searches index *T* for the set of subscribers whose subscriptions *match e* and notifies these subscribers of a matching publication (usually by simply forwarding *e* to the subscribers). This sequence of events is illustrated in Figure 4.1.

Notice that messages from publishers (publications) do not contain any address; instead, they are routed through the system based on their content (for a content-based system). This ability to send messages to a set of subscribers without specifying their explicit address decouples the interaction between the publishers and subscribers and is key to achieving many of the infrastructure requirements outlined in Section 4.2.3

The publish/subscribe model is in some ways the dual of the traditional database model. In a database, data are persisted in a database and queries retrieve these data, whereas in the publish/subscribe model, the queries (subscriptions) are stored at the broker, and the data (publications) are pushed to the subscribers. A subscription can be thought of as a long running query, so it may be helpful to think of publish/subscribe model as being similar to database triggers [18]. It is important to note that in most publish/subscribe systems, subscriptions only match future publications. Publications enter the system, are sent to interested subscribers, and are then lost[1]. This is an important difference from the database model. Another point to note is that data only travel from publisher to subscriber. Of course, in a given application, bidirectional communication between two components is possible by having the components act as both publisher and subscriber.

Publish/subscribe systems can be classified based on the expressiveness of their subscription language, such as those based on *topics* (or *subjects*), *types*, or *content*.

---

[1]There are extensions to the publish/subscribe semantics to support matching historic data [15].

In topic-based publish/subscribe, clients can subscribe to several topics and receive notifications about all publications within these topics. These systems usually offer flat or hierarchical addressing. In flat addressing, all topics are disjoint, while in hierarchical addressing, topics are organized in hierarchies; subscriptions can address any node in the hierarchy, implicitly addressing all subtopics of the node. *Type-based* publish/subscribe systems are similar to topic-based, but use publication types instead of topics for matching. *Content-based* publish/subscribe systems increase the expressiveness of subscriptions by allowing more complex queries on the publication content [11] including XML data [16]. There are even systems that support location constraints, such as conditions on the number of mobile users near a fixed landmark, or within the vicinity of themselves [28–30].

Early publish/subscribe systems were centralized system [2, 11]; there was only one broker in the system that received all publications and subscriptions in the system. This is the scenario shown in Figure 4.1. Research in these systems were focused on developing algorithms that can quickly and efficiently match a publication against millions of subscriptions. However, in systems with potentially millions of subscriptions and with subscribers dispersed geographically, a centralized broker can still become a processing, memory, and network bottleneck. Newer systems were developed that employed a distributed set of brokers [7, 10, 12]. The research here was focused on distributed matching and multicasting. Distributed matching refers to storing subscriptions in a subset of the nodes in the system such that matching can be distributed among the nodes. Multicast refers to a technique of disseminating publications to interested subscribers such that total network traffic is minimized.

4.3.2 Publish/Subscribe Benefits

The publish/subscribe model offers many benefits . First, the model has a very simple interface . As shown in Figure 4.1, the only messages involved are publications, and subscriptions, and the only operations are publish, subscribe, and notify.

Another important benefit is the *decoupling* of publisher and subscribers. This decoupling exists along several dimensions as described below.

- *Address decoupling*: Publishers and subscribers do not know the addresses of one other. Instead, publications are routed to subscribers based on the content of the publications and the interests of the subscribers. For this reason, publish/subscribe is also often referred to as *content-based routing*. Another advantage of address decoupling is the anonymity it offers to the publishers and subscribers. Only the brokers know the identity and interest of a subscriber, and the data published by a given publisher.

- *Platform decoupling*: All entities in the publish/subscribe communicate using network messages, and as such can run on heterogeneous platforms. Therefore, a powerful multiprocessor server with a dedicated network connection can seamlessly communicate to a resource-constrained PDA with a wireless link.

- *Space decoupling*: As mentioned above, publishers and subscribers can exist in geographically distant areas of the globe.

- *Time decoupling*: Some publish/subscribe algorithms allow publishers and subscribers to not be connected to the network simultaneously, allowing subscribers that experience network disconnection to retrieve missed publications upon reconnection [4, 20].

- *Representation (semantic) decoupling*: Some publish/subscribe systems can mediate between publications and subscriptions whose predicates have different meaning [23]. For example, in a dating service application, it is possible to match publications that specify a user's age in *years* and subscriptions that request users *born before* a certain date.

Publish/subscribe also provides benefits regarding network bandwidth use. Notably, data are *pushed* to subscribers as it becomes available. This is more efficient than having subscribers periodically poll for new data. A system with millions of subscribers polling a broker can easily overwhelm a broker. Data push also delivers data quicker to subscribers than polling. Furthermore, distributed publish/subscribe systems use efficient multicast techniques to minimize the messages used by subscribers. An example of the benefits of multicast is shown in Figure 4.2 in which the publication only travels a total of three hops to reach both subscribers. *Unicasting* the publication, that is, sending the publication individually from the publisher to each subscriber, would require four hops. In a large system with millions of subscribers, multicast is much more bandwidth efficient than unicast.

In addition to the above inherent benefits of publish/subscribe, there has been much work to augment the basic publish/subscribe model with enterprise-grade features such as security [22, 26, 27], reliability [9, 14], load-balance [8], transactional client movement guarantees [13], and unified access to publications in the future and past [15]. These efforts make publish/subscribe more attractive and useful in mission-critical enterprise applications.

4.3.3 Publish/Subscribe Router

As mentioned in Section 4.3.1, a publish/subscribe broker can be replaced by a broker network for scalability reasons. In a broker network, each broker only has local

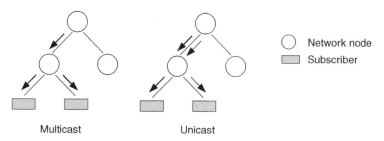

Multicast   Unicast

○ Network node
▨ Subscriber

Figure 4.2 Comparison of multicast and unicast propagation.

knowledge and routes messages to its neighbors. In this section, we describe the main routing operations performed by a broker in a distributed publish/subscribe network.

A publish/subscribe router performs the following three operations: (1) forwarding of advertisements, (2) forwarding of subscriptions, and (3) forwarding of publications.

The details of the three operations are highly dependent on the expressiveness of the subscription and publication representation languages.

To better understand the operation of a publish/subscribe router, we give a more formal definition of a subscription and a publication representation language most commonly used in the research literature. The formal representation of a publication is given by the following expression: $\{(a_1, val_1), (a_2, val_2), \ldots, (a_n, val_n)\}$. Subscriptions are expressed as conjunctions of Boolean predicates. In a formal description, a simple predicate is represented as (*attribute_name relational_operator value*). A predicate (a *rel_op* val) is matched by an attribute–value pair (a, val) if and only if the attribute names are identical ($a = a$) and the (a *rel_op* val) Boolean relation is true. A subscription s is matched by a publication p if and only if all its predicates are matched by some pair in p.

We are now ready to look at the details of the publish/subscribe router operations.

4.3.3.1 *Forwarding of Advertisements*

Advertisements are used by publishers to announce the set of publications they are going to publish. Consequently, advertisements create routing paths for subscriptions from subscribers to publishers, whereas subscriptions build routing paths for publications from publishers to subscribers. Usually, both subscriptions and advertisements have the same formal representation. However, there is an important distinction between the predicates in an advertisement and those in a subscription: the predicate in a subscription is considered to be in a conjunctive construction, while that in an advertisement behaves as in a disjunctive one.

An advertisement a *matches* a publication e if and only if all attribute–value pairs match some predicates in the advertisement. Formally, an advertisement $a = \{p_1, p_2, \ldots, p_n\}$ determines a publication e, if and only if $\forall (attr, val) \in e, \exists p_k \in a$ such that (attr,val) matches p_k.

An advertisement a *intersects* a subscription s if and only if the intersection of the set of the publications determined by the advertisement a and the set of the publications that match s is a nonempty set. Formally, at the predicate level, an advertisement $a = \{a_1, a_2, \ldots, a_n\}$ intersects a subscription $s = \{s_1, s_2, \ldots, s_n\}$ if and only if $\forall s_k \in s, \exists a_j \in a$ and there exists some attribute–value pair (attr, val) such that (attr, val) matches both s_k and a_j. Table 4.1 presents some examples of subscriptions and advertisements and the corresponding intersection relations.

The following are the steps performed by the publish/subscribe router upon receiving an advertisement:

Table 4.1 Examples of Subscriptions, Advertisements, and Intersection Relations

Subscription s	Advertisement a	Intersection relation
(product = "computer", brand = "IBM", price $\leq$ 1600)	(product = "computer", brand = "IBM", price $\leq$ 1500)	a intersects s
(product = "computer", price $\leq$ 1600)	(product = "computer", brand = "IBM", price $\leq$ 1600)	a intersects s
(product = "computer", brand = "IBM", price $\leq$ 1600)	(product = "computer, brand = "Dell", price $\leq$ 1500)	a does not intersect s

1. For the advertisement received, check if there are *covering* advertisements in the advertisement table. If there are, then do not forward the advertisement.

2. If there is no covering advertisement, *insert* incoming advertisement in the advertisement table and forward the advertisement to all neighbors.

3. Check if there are *intersecting* subscriptions in the subscription table. If there are, forward the intersecting subscriptions to the neighbor from which the advertisement was received.

4.3.3.2 Forwarding Subscriptions

Subscription processing is similar to advertisement processing. Given two subscriptions s_1 and s_2, s_1 *covers* s_2 if and only if all the publications that match s_2 also match s_1. In other words, if we denote with E_1 and E_2 the set of publications that match subscription s_1 and s_2, respectively, then $E_2 \subseteq E_1$.

At the predicate level, the covering relation can be expressed as follows: Given two subscriptions $s_1 = \{p_1^1, p_2^1, \ldots, p_n^1\}$ and $s_2 = \{p_1^2, p_2^2, \ldots, p_m^2\}$, s_1 covers s_2 if and only if $\forall p_k^1 \in s_1$, $\exists p_j^2 \in s_2$ such that p_k^1 and p_j^2 refer to the same attribute, and if p_j^2 is matched by some attribute–value pair (a, val), then p_k^1 is also matched by the same (a, val) attribute–value pair. In other words, s_2 has potentially more predicates and they are more restrictive than those in s_1. Table 4.2 presents some examples of subscriptions and covering relations.

Informally, when a broker B receives a subscription s, it will send it to its neighbors if and only if it has not previously sent them another subscription s', that covers s. Broker B will receive all publications that match s, since it receives all publications that match s' and the publications that match s are a subset of the publications that match s'.

Table 4.2 presents some examples of subscriptions and the corresponding covering relations. The goal of subscription covering is to quench subscription propagation, thereby reducing network traffic and trimming the size of subscription (i.e., routing) tables.

Table 4.2   Examples of Subscriptions and Covering Relations

Subscription s_1	Subscription s_2	Covering relation
(product = "computer", brand = "IBM", price ≤1600)	(product = "computer", brand = "IBM", price ≤1500)	s_1 covers s_2
(product = "computer", brand = "IBM", price ≤1600)	(product = "computer", price ≤1600)	s_2 covers s_1
(product = "computer", brand = "IBM", price ≤1600)	(product = "computer", brand = "Dell", price ≤1500)	s_1 does not cover s_2 s_2 does not cover s_1

The processing of the subscriptions at the publish/subscribe router then proceeds as follows:

1. For each incoming subscription, check if there are *covering* subscriptions in the subscription table. If there are, then do not forward the subscription.

2. If there is no matching subscription, *insert* incoming subscription in the subscription table.

3. Check if there are *intersecting* advertisements in the advertisement table. If there are, forward the subscription to those neighbors from which the matching advertisements were received.

4.3.3.3   Forwarding Publications

Finally, publications are processed as follows.

- For each incoming publication, check if there are *matching* subscriptions in the subscription table. If there are, forward the publication to all the neighbors from which each of the matching subscriptions was received.

4.3.4   Publish/Subscribe Broker Network

To summarize, a number of publish/subscribe brokers can be organized into a content-based routing network. In such a network, one of the most important problems is the routing of a publication to interested subscribers based on the content of the publication. There are several routing protocols proposed in the literature, all of which follow the described publish/subscribe broker architecture to some extent [7, 10, 12]. Some of them do not use advertisements and some of them do not perform subscription covering.

Generally, the proposed solutions always involve a network of brokers that collaborate in order to route the information in the network based on its content. As shown in Figure 4.3, advertisements from publishers are flooded throughout the network and stored at each broker's routing table in order to build a distributed advertisement tree. When a subscriber receives an advertisement, it sends intersecting subscriptions along the reverse path of the advertisement tree. These subscriptions

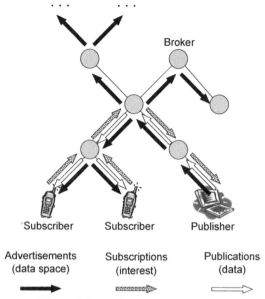

Figure 4.3   Distributed publish/subscribe.

are stored in the routing table of each broker along the subscription path and result in a distributed multicast tree. Finally, publications from a publisher follow the reverse path of all matching subscriptions (i.e., the multicast tree rooted at the publisher) and are delivered to interested subscribers.

In some distributed publish/subscribe systems, the message flow follows Figure 4.3 conceptually but the details may differ. For example, protocols where brokers are implemented in mobile ad hoc networks [3, 24], wireless sensors networks [25], peer-to-peer networks [1, 19], or cyclic overlays [17] tweak their routing protocols to exploit features or overcome limitations in their respective environments. In this chapter, however, we restrict ourselves to a traditional distributed publish/subscribe system in which brokers are assumed to be relatively stable and form an acyclic overlay as shown in Figure 4.3.

4.4 CLIENT MOBILITY

Cugola et al. were among the first to support mobility in distributed publish/subscribe systems [10]. They introduce the "movein" and "moveout" operations that offer clients the ability to disconnect from and reconnect to the system. However, these standard distributed publish/subscribe algorithms may not perform well when the clients in the system experience frequent mobility [5, 20]. Various algorithms to address client mobility are presented below. We distinguish between subscriber and publisher mobility as the solutions to deal with each are different.

4.4.1 Subscriber Mobility

We first describe the subscriber mobility algorithm proposed by Cugola et al., which we refer to as the STANDARD algorithm [10]. Then we present a set of optimizations to this algorithm.

4.4.1.1 Standard Algorithm

Figure 4.4 illustrates a timeline of a subscriber going into a period of disconnection and reconnecting to a different broker. During period t_1, the client is connected to Broker 1 and can receive events. At the end of period t_1, the client disconnects from Broker 1 and reconnects after period t_3 to Broker 2. Period t_2 is used by an optimization described in Section 4.4.1.2. Period t_4 is required to complete the reconnection phase, which involves retrieving and playing back events that the client missed while disconnected. Finally, during period t_5, the client receives newly published events as in period t_1.

The objective of the algorithm is to reconfigure the event multicast tree to account for client mobility. We assume the client always reconnects to its physically closest broker. When a client disconnects from Broker 1, the broker begins to locally store the events that the client would have received if it had been connected. If the client reconnects to Broker 1, then the stored events are simply replayed to the client during period t_4. The more interesting case is when the client reconnects to some other broker, say Broker 2. The following steps take place during period t_4:

1. Upon reconnection, the client notifies Broker 2 that it was previously connected to Broker 1.

2. Broker 2 retrieves the subscriptions associated with the client from Broker 1.

3. Broker 2 subscribes to these subscriptions and then sends a *REQUNSUB* message to Broker 1 requesting it to unsubscribe.

4. Broker 2 stores in a local queue new events it receives for the client.

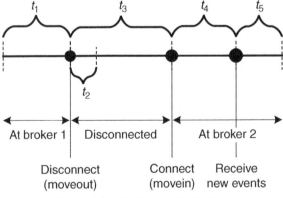

Figure 4.4 Mobile subscriber timeline.

5. Broker 1 forwards stored events to Broker 2.

6. All the state has now been transferred from Broker 1 to Broker 2. Broker 2 now replays both the events received from Broker 1 and the new events stored in its local queue, to the client.

In step 6, duplicates in the set of events transfered from Broker 1 and those in Broker 2's local queue are removed. This assumes duplicate events can be distinguished, with, for example, a publisher-specific event sequence numbers.

In step 3, to guarantee that no events are lost, we require that the common ancestor of Brokers 1 and 2 in the broker hierarchy sees the subscriptions from Broker 2 before the unsubscriptions from Broker 1. To simplify this step, we assume that the overlay broker hierarchy is consistent with the underlying network hierarchy such that the shortest path between any two brokers in the overlay topology is the shortest path in the underlying topology. We also assume that messages between brokers are received in the order they were sent, which is reasonable, as we expect TCP connections between brokers. Since Broker 2 sends the subscriptions followed by *REQUNSUB*, the common ancestor sees them in this order, and since the common ancestor is in the shortest path between Brokers 1 and 2 in the overlay topology, Broker 1 sees *REQUNSUB* after the common ancestor, by our assumption. Thus, our requirement is met, and no events are lost. More elaborate schemes are possible in cases where our assumption is not valid [6]. However, more complex algorithms likely increase the state transfer costs and thus only worsen the overhead costs [20].

All the messages exchanged between the two brokers represent unicast messages. As we will see in the experimental section, the unicast messages contribute significantly to system overhead. Also, it is possible that cost of reconfiguring the multicast tree may be greater than the gains from having events take the shortest path.

We now describe several optimizations to the STANDARD subscriber mobility algorithm.

4.4.1.2 Prefetching Algorithm

The PREFETCHING algorithm exploits knowledge of future mobility patterns. It is similar to the STANDARD algorithm except that steps 2–5 (the transfer of subscriptions and stored events) occur during period t_2. In period t_4, only step 6 (the replay of stored events) takes place. PREFETCHING gains in two ways. First, the latency of the state transfer from Broker 1 to Broker 2 is hidden from the user, since is occurs while the user is disconnected. This results in a shorter t_4 period. Secondly, since state transfer occurs early, there are very few stored events that Broker 1 needs to forward. Therefore, there are fewer messages transferred overall.

The effectiveness of PREFETCHING depends on the successful prediction of the client's destination. To improve the likelihood of success, Broker 1 can predict the *set* of brokers that are likely to be the next destination of the mobile client. Broker 1 can make this prediction based, for example, on statistics of the mobility patterns of its clients. If the client reconnects to one of the predicted brokers, that particular broker replays all stored events and informs the other brokers to discard the stored events and subscriptions for that client. If the client reconnects to a completely different broker, the new broker can contact the closest broker from the set of prefetched brokers, and the state transfer happens as in the STANDARD algorithm. We do not evaluate prefetching to multiple brokers here.

4.4.1.3 Logging Algorithm

The LOGGING algorithm exploits subscription locality (a measure of the similarity of subscriptions) in the system. Here, all brokers maintain a log of recently received events. When the mobile client reconnects, Broker 2 scans its log for events of interest to the mobile client. Any relevant events found in the log do not need to be transferred from Broker 1.

The state transfer is similar to the STANDARD algorithm. After Step 2, however, Broker 1 sends a message with the IDs of its stored events to Broker 2. Broker 2 checks for these events in its log and sends any matched IDs to Broker 1 instructing it to not send these events during Step 5. By only transferring, in Step 5, those events not already at Broker 2, the period t_4 can be shorter than in the STANDARD algorithm. However, if no such events exist, the wasted overhead of sending event IDs can make period t_4 longer. Thus, in terms of both message and latency costs, LOGGING can perform better or worse than STANDARD depending on the movement scenario and subscription locality.

LOGGING requires that events have system-wide unique IDs. This can be easily achieved if we consider that each publisher in the system has a unique ID and that events are ordered within the publisher. Thus, the ID for the event is composed from the ID of the publisher that sent it and a locally incrementing sequence number.

4.4.1.4 Home-Broker Algorithm

In this algorithm each client is assigned a home broker. Upon a `movein` operation, a client reconnects to its physically closest broker. The client, however, reconnects logically to its home broker. Subscriptions remain on the home broker, which keeps receiving events through the regular multicast mechanism. The home broker then forwards these events to the client using unicast messages. The HOME-BROKER algorithm is designed to emulate handling mobility in a way similar to MobileIP [21].

HOME-BROKER gains by not migrating subscriptions and rebuilding the multicast tree, at the expense of not sending events through the shortest path. Note that there is unicast traffic even when the client is connected.

4.4.1.5 Subscriptions-On-Device

Subscription migration is an important component of state transfer. If the mobile device has sufficient resources, it can locally store the subscriptions, and directly send the subscriptions to the new broker upon reconnection. This way, the old broker does not need to transfer the subscriptions to the new broker. The applicability of this method depends on the number of subscriptions that a client has, the resources of the mobile device, and whether the user wishes to use more than one device.

4.4.1.6 Discussion

It should be noted that these optimizations make minimal assumptions about the underlying system. They can be used with any type of distributed publish/subscribe system. Moreover, the optimizations can be combined. For example, subscriptions-on-the-device can be combined with PREFETCHING or LOGGING.

In an experiment with in which 800 subscribers commuting home at the end of the work day were simulated, we saw that the STANDARD algorithm incurred an almost 80% overhead in terms of the overhead of the mobility algorithm [5]. This means that introducing mobility to a network built with enough capacity to service a non-mobile publish/subscribe system will require almost doubling the network capacity. PREFETCHING, by transferring state early, has almost no overhead in this scenario, because the periods of disconnection are large and the traffic required to migrate subscriptions is much smaller than the event traffic during the disconnection period. LOGGING improves on STANDARD by partially replacing unicast state transfer with multicast by leveraging logging and subscription locality. The HOME-BROKER algorithm fared the worst with an average overhead of up nearly 600%. Since the HOME-BROKER algorithm works similar to MobileIP, the latter result indicates that addressing mobility at the network layer is not feasible in such a scenario.

Figure 4.5 plots the total message costs (as opposed to relative overhead) of the above experiment. In the figure, the total messages incurred by each mobility algorithm are plotted in the four curves, and the impulses indicate the number of disconnected subscribers in the corresponding time period. First, we notice that the PREFETCHING algorithm achieves almost optimal performance, with little increase in message load when subscribers are mobile compared to when they are stationary at

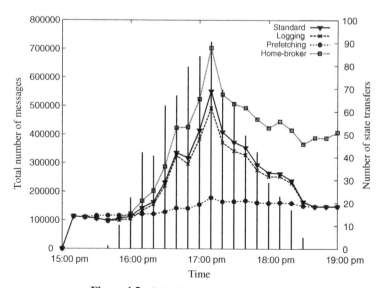

Figure 4.5 Subscriber mobility message cost.

the beginning and end of the experiment. We also note that the message costs of the other algorithms closely track the number of concurrent clients executing state transfers, indicating that these state transfers are the primary cause of the increased message load. This figure helps explain the dismal HOME-BROKER performance. After all clients have reconnected, the other three approaches have identical costs, since they rebuild the multicast tree and events are multicast to clients. With HOME BROKER, however, events are still unicast from the home broker to the client. This large and sustained unicast that persists even after reconnection is why HOME BROKER performs poorly.

The reader is referred to Ref. [3] for more details on the above as well as other experiments. The results indicate that the PREFETCHING algorithm virtually eliminates the mobility overhead, while the LOGGING algorithm can approach the performance of the optimal PREFETCHING algorithm in situations where the subscriptions exhibit a high degree of locality. This makes sense because when subscriptions express interest in similar publications, the publications cached in the LOGGING algorithm for one subscriber are more likely to be needed by other subscribers. The experiments also indicate that the HOME BROKER algorithm is almost always a poor choice.

4.4.2   Publisher Mobility

As described in Section 4.4.1, to support the disconnected operation of subscribers, brokers must store publications for the subscriber and replay them to the subscriber when it reconnects to the network. Unlike disconnected operation with subscribers, there is no information that the publisher has missed while disconnected, and hence it may seem that publisher mobility is not a problem. However, this is not the case. It has been shown that publisher mobility can generate significant overhead as advertisement trees (which are broadcast throughout the network) need to be torn down and rebuilt every time a publisher moves [20].

Below we describe the standard publisher mobility algorithm as well as some optimizations that we propose.

4.4.2.1   Standard Algorithm

The timeline of a publisher going into a period of disconnection and reconnecting to a different broker is identical to Figure 4.4 except that after period t_4 the publisher *publishes* new events as opposed to receiving them. During period t_1, the publisher is connected to Broker 1, the publisher rooted advertisement and multicast trees have been built, and publications are correctly multicast to all interested subscribers. At the end of period t_1, the publisher disconnects from Broker 1 and reconnects after period t_3 to Broker 2. Period t_2 is used by the PREFETCHING and PREFETCH-DELAYED optimizations below. Period t_4 is required to complete the reconnection phase, which involves rebuilding advertisement and multicast trees. Finally, during period t_5 publications from the publisher are again correctly multicast to all interested subscribers.

The objective of the publisher mobility algorithm is to reconfigure the advertisement and multicast trees to account for publisher mobility. We assume the publisher always reconnects to its physically closest broker. When a publisher disconnects from Broker 1, Broker 1 sends an unadvertisement message for any advertisements it has received from the publisher. Note that these unadvertisements may induce unsubscriptions to be propagated in the reverse direction of the unadvertisement propagation (to tear down the multicast tree) just as advertisements induce subscriptions to be sent. This tree teardown occurs during period t_2 after which there is no state associated with the publisher in the system. At the end of period t_3, the publisher connects to Broker 2, and upon reconnection sends its advertisements again. During period t_4, the advertisement and multicast trees are rebuilt, and finally publications sent during period t_5 are delivered to all interested subscribers. It is this mobility-induced tree teardown and reconstruction that makes the traditional assumption that there are advertisements invalid in this context.

Publications sent during period t_4 may not be delivered to interested subscribers, since the multicast tree has not been rebuilt yet. However, in this algorithm, there is no way to know when period t_4 is complete; Broker 2 does not know for certain when the multicast tree has been rebuilt. This is a fundamental problem arising from the decoupling of publishers and subscribers in the publish/subscribe model; publishers do not know the set of subscribers that are interested in their content. In fact, no one node knows the set of participants in a given multicast tree. This problem is exacerbated by the fact that it is difficult to distinguish between new subscriptions that enter the system shortly after the publisher moves in, and old subscriptions that simply arrive slowly at the publisher. The publish/subscribe semantics require the latter subscribers to receive publications from the publisher, while the former is allowed to miss some publications until their subscriptions propagate through the system. Since the length of period t_4 is unknown, we would like to minimize this period so as to minimize the probability that a publication sent soon after reconnection is not delivered to an interested subscriber.

4.4.2.2 Prefetching Algorithm

The PREFETCHING algorithm exploits knowledge of future mobility patterns. It is similar to the STANDARD algorithm except that the advertisement and multicast trees are rebuilt during period t_2 instead of period t_4. Therefore, the length of period t_4 is now zero, and any publications sent immediately after reconnection are delivered to interested subscribers. Note that t_2 is the time it takes to rebuild trees, and is independent of the disconnection time.

This algorithm has the advantage of hiding tree rebuilding time from the publisher, since it occurs while the publisher is disconnected. Also, since the old tree (rooted at Broker 1) is being torn down concurrently with the building of the new tree (rooted at Broker 2), it may occur that the new tree grafts onto the old tree before it is torn down, obviating the need to tear down the old tree completely. The DELAYED algorithm below tries to force this case, which only occurs by chance in the PREFETCHING algorithm.

4.4.2.3 Proxy Algorithm

The PROXY algorithm is an extension of the PREFETCHING algorithm. The assumption here is that publishers tend to move within a restricted area. For example, a taxi driver may service only certain regions of a city. (The taxi may be publishing location updates to a dispatcher or potential customers.) The PROXY algorithm assigns a set of brokers to act as proxies for the publisher. These proxies always maintain a tree for the publisher. This way, there is no teardown or rebuilding of the tree when the publisher disconnects from or connects to one of its proxies. However, tree rebuilding does need to occur if the publisher connects to a non-proxy broker.

4.4.2.4 Delayed Algorithm

The DELAYED algorithm exploits the fact that the old tree (rooted at Broker 1) and the new tree (rooted at Broker 2) have significant overlap. This is especially true if Broker 1 and 2 are nearby, as in the case of an always connected mobile publisher. In the DELAYED algorithm, the teardown of the old tree at Broker 1 is delayed for some time after moveout. This time is to allow the publisher to reconnect to another broker, and graft the new publication tree to the old one. After the delay, the old broker tears down only the extraneous portions of the combined tree.

4.4.2.5 Prefetch-Delayed Algorithm

The PREFETCH-DELAYED algorithm is a combination of the PREFETCHING and DELAYED algorithms. As in the PREFETCHING algorithm, PREFETCH-DELAYED initiates tree rebuilding at Broker 2 when the publisher disconnects from Broker 1, thereby hiding the tree reconstruction time from the publisher. However, since the tree rooted at Broker 1 is torn down concurrently with the construction of the tree rooted at Broker 2, it may occur that the old tree is completely torn down before the new tree is built. In the PREFETCH-DELAYED algorithm, the teardown of the old tree is delayed to allow the new tree to graft onto the old one. In this way, PREFETCH-DELAYED offers the advantages of both the PREFETCHING and DELAYED algorithms.

4.4.2.6 Discussion

These optimizations make minimal assumptions about the underlying system and can be used with any type of distributed publish/subscribe system. Moreover, the optimizations can be combined. For example, PROXY can be combined with DELAYED to potentially achieve even better performance.

It is instructive to notice that the STANDARD algorithm does not distinguish between a moving publisher and a publisher that leaves and enters the system. Therefore, it discards all state (advertisement and multicast trees) associated with a publisher on moveout, and must completely rebuild it on movein. Our optimizations address this issue.

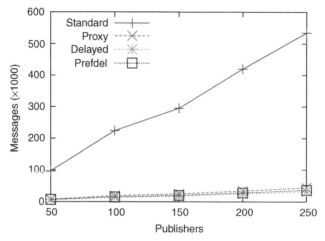

Figure 4.6   Publisher mobility message cost.

Since publisher mobility causes expensive reconstruction of advertisement and multicast trees, it may be tempting to eliminate advertisement flooding and flood subscriptions instead. However, subscribers typically outnumber publishers, so the savings in publisher mobility-induced tree rebuilding cost do not justify subscription flooding. Furthermore, subscriber mobility will now cause multicast tree reconstruction; this reconstruction is much more expensive than when advertisements are used, since the multicast trees now span the whole network rather than being a minimal tree from the subscriber to interesting publishers.

Figure 4.6 shows the tree building message cost for the four mobility algorithms with an increasing number of publishers. The details of this experiment can be found in Ref. [20]. We see that while the cost of tree rebuilding grows approximately linearly for all the algorithms, there is a substantial difference in their costs. The STANDARD algorithm has more than 10 times the message cost of the PROXY, DELAYED, and PREFETCH-DELAYED algorithms. STANDARD performs poorly because every moveout (movein) causes the entire advertisement and multicast trees for the moving publisher to be torn down (rebuilt). While not shown here, the same is true for PREFETCHING but with an additional cost of having the old broker inform the new broker to start rebuilding the tree. Therefore, PREFETCHING is not building the new tree fast enough to graft onto the old tree that is being torn down. Since PREFETCHING does not offer any benefits that are not present in PREFETCH-DELAYED, we do not present results for the PREFETCHING algorithm here in order to simplify the exposition.

Unlike the STANDARD algorithm, the PROXY, DELAYED, and PREFETCH-DELAYED algorithms all graft onto existing trees. The reason DELAYED performs better than PROXY is because in DELAYED the old tree is rooted at a nearby broker (recall that the publishers move along adjacent brokers), so the distance an advertisement must travel to graft onto an existing tree is short. In PROXY, the distance to the old tree depends on the position of the publisher relative to its fixed proxies. PREFETCH-DELAYED has the same message cost as DELAYED. The only difference between the two is that PREFETCH-DELAYED performs tree reconstruction earlier—during the publisher's disconnection period.

More experiments that analyze the various publisher mobility algorithms can be found in Ref. [20]. The results generally indicate that the PREFETCH-DELAYED and PROXY algorithms perform the best, with the caveat that they require precise (in the case of PREFETCH-DELAYED) or approximate (for the PROXY) knowledge of future mobility patterns. The DELAYED algorithm is worse than the above two algorithms, but does not require any mobility knowledge. The STANDARD algorithm, however, performs quite poorly and should not be used to handle publisher mobility.

4.5 EXAMPLE APPLICATION

In this section, we describe how the publish/subscribe model can be used to implement an example scenario. While the previous section illustrated the performance of the publish/subscribe system, here we point out the benefits of the model by showing how the publish/subscribe messaging abstractions provide a simple and powerful mechanism to achieve complex interactions among the various services.

We consider a scenario in which a municipality develops an integrated system where information about its transportation infrastructure, including road conditions, highway traffic patterns, and transit operations. Such systems have been implemented by jurisdictions such as Georgia's Department of Transportation and the Commonwealth of Pennsylvania. The goal of such a system would be to provide an integrated platform where a rich, diverse, and large volumes of data can be filtered and delivered to the interested parties in a flexible and efficient manner.

In our scenario, the information producing entities, or publishers, include sensors that monitor traffic volumes and speeds, citizens who can report on road congestion during their commute, police officers who file electronic reports from the scene of an accident or request for backup at a crime scene, or meteorological organizations that send important weather alerts. Correspondingly, subscribers include emergency response personnel that need to be notified of accidents on roads, taxi drivers that would like to know about congested roads in their vicinity, or transportation departments that wish to monitor traffic patterns in order to manage traffic flow.

Figure 4.7, presents a sample of possible publications and subscriptions that various entities in this system might issue. The entities along the top are the publishers and the subscribers are at the bottom of the figure. In Figure 4.7, a police officer publishes traffic reports, such as publication $P1$, that includes information about the nature and details of the accident. Motorists use their cell phone to notify the system about congestion they experience while driving as shown in publication $P2$, and the local meteorologist sends weather alerts as in publication $P3$. The information published into the system is used by various users such as paramedics who subscribe to $S1$ in order to be notified of any reports within their jurisdiction in which a serious injury has occurred, and to $S2$ in order to be notified of congested routes they should avoid while traveling to an emergency. Automobile dealerships may be interested in

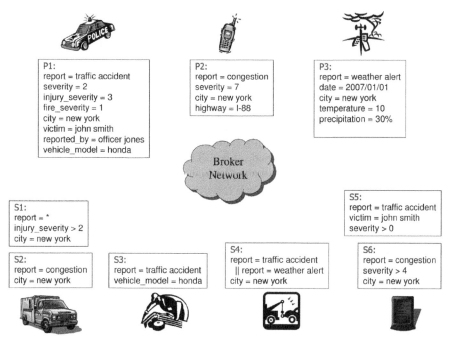

Figure 4.7   Publish/subscribe messages in a traffic reporting scenario.

accidents involving their vehicles, as shown in subscription $S3$. With subscription $S4$, tow truck operators will be notified of potential incidents where a tow truck may be needed and of any weather alerts. Ordinary citizens wishing to be notified of any emergency involving their friends or family would issue subscription $S5$, and send $S6$ to receive severe congestion alerts.

The above example illustrates the complex interactions that are possible with the simple content-based publish/subscribe model. First, we notice that the publications and subscriptions are relatively easy to construct and understand. Also, the decoupling of publishers and subscribers means that the publisher simply publishes information without any regard for who should ultimately receive it, and subscribers indicate their interest without explicitly specifying the publishers from whom they should receive the data. Notice that these simple, descriptive messages give rise to complex messaging patterns. For example, publication $P1$ matches subscriptions issued by all four subscribers and is delivered to all of them, which is essentially a broadcast operation in the scenario depicted in Figure 4.7. Meanwhile, publication $P2$ only matches subscriptions $S2$ and $S6$, and is sent to two subscribers, which is a multicast operation. Finally, publication $P3$ only matches subscription $S4$ and behaves like a unicast transmission. From the subscriber's point of view, we also see that some subscribers receive data from multiple publishers, while other only interact with one publisher. The simple yet powerful publish/subscribe messaging abstraction achieves all these interactions patterns through the interplay among the subscriptions and publications.

We also note that subscribers are able to filter the data they receive, which is important in a system with large volumes of publications. Also notice that subscribers may "address" their publishers in various ways, including based on location. For example, subscription S4 is only interested in accidents within a particular city. Much more complex location-based semantics are supported by extended content-based publish/subscribe languages [28].

All the publishers and subscribers in Figure 4.7 may be mobile, and the mobility algorithms in Section 4.4 can be employed by the broker network to transparently manage disconnected or intermittently connected users. For example, the subscriber on the bottom right may have turned off her PDA, but the publish/subscribe broker network will store any accident reports of interest to her and deliver them to her when she eventually turns on the PDA. Such features provided by the publish/subscribe infrastructure simplify application development.

4.6 SUMMARY

Mobile applications increasingly make use of network connectivity to provide enhanced services such as multiplayer games, mobile payment applications, and location-based services. The constraints of mobile devices, especially in terms of limited or unreliable network connections, the large volumes of data that need to be processed, and the complex interactions exhibited by some mobile applications require a communication infrastructure that addresses the requirements of these applications and platforms.

The publish/subscribe messaging paradigm provides a simple yet powerful abstraction that decouples the communication between providers and consumers of data. This decoupling is especially of benefit in mobile applications by hiding intermitted disconnections by mobile devices from the applications. Also, the powerful filtering capabilities provided by the model allow applications to receive only those messages of interest to them over the often expensive and limited wireless channel available to mobile devices. Furthermore, content-based publish/subscribe routing allows applications to devise sophisticated interactions in which applications can address devices using, among other ways, location-independent or location-dependent addresses in a unified manner. In addition to the benefits of the publish/subscribe model, distributed publish/subscribe systems have been shown to be scalable and algorithms exist to handle the stresses of a highly dynamic mobile environment.

REFERENCES

1. I. Aekaterinidis and P. Triantafillou. Pastrystrings: a comprehensive content-based publish/subscribe DHT network. In *ICDCS '06: Proceedings of the 26th IEEE International Conference on Distributed Computing Systems*, Washington, DC, USA. IEEE Computer Society, 2006, p. 23.
2. M. K. Aguilera, R. E. Strom, D. C. Sturman, M. Astley, and T. D. Chandra. Matching events in a content-based subscription system. In *Symposium on Principles of Distributed Computing*, 1999 pp. 53–61.

3. S. Baehni and R. Guerraoui, and C. S. Chhabra. Frugal event dissemination in a mobile environment. In Gustavo Alfonso, editor, *6th International Middleware Conference (MIDDLEWARE 2005)*, Lecture Notes in Computer Science Grenoble, France, vol. 3790, 2005, pp. 205–224, Springer.

4. S. Bhola, Y Zhao, and J S. Auerbach. Scalably supporting durable subscriptions in a publish/subscribe system. In *Proceedings of the International Conference on Dependable Systems and Networks (DSN 2003)*. IEEE Computer Society, 2003, pp. 57–66.

5. I. Burcea, H.-A. Jacobsen, E. de Lara, V. Muthusamy, and M. Petrovic. Disconnected operation in publish/subscribe middleware. In *International Conference on Mobile Data Management (MDM)*, 2004.

6. M. Caporuscio, A. Carzaniga, and A. L. Wolf. Design and evaluation of a support service for mobile, wireless publish/subscribe applications. *IEEE Trans. Software Eng.*, 29(12):1059–1071, 2003.

7. A. Carzaniga, D. S. Rosenblum, and A. L Wolf. Design and evaluation of a wide-area event notification service. *ACM Trans. Comput. Syst.*, 19(3).332–383, 2001.

8. A. K. Y. Cheung and H.-A. Jacobsen. Dynamic load balancing in distributed content-based publish/subscribe. In *ACM/IFIP/USENIX 7th International Middleware Conference*, Melbourne, Australia, 2006. ACM/IFIP/USENIX, ACM.

9. P. Costa, M. Migliavacca, G. P. Picco, and G. Cugola. Epidemic algorithms for reliable content-based publish-subscribe: an evaluation. In *ICDCS '04: Proceedings of the 24th International Conference on Distributed Computing Systems (ICDCS'04)*, , Washington, DC, USA. IEEE Computer Society, 2004, pp. 552–561.

10. G. Cugola, E. Di Nitto, and A. Fuggetta. The JEDI event-based infrastructure and its application to the development of the OPSS WFMS. *IEEE Trans. Software Eng.*, 27(9):827–850, 2001.

11. F. Fabret, H.-A. Jacobsen, F. Llirbat, J. Pereira, K. Ross, and D. Shasha. Filtering algorithms and implementation for very fast publish/subscribe systems. In *Proceedings of ACM SIGMOD*, 2001.

12. E. Fidler, H.-A. Jacobsen, G. Li, and S. Mankovski. The PADRES distributed publish/subscribe system. In *International Conference on Feature Interactions in Telecommunications and Software Systems (ICFI)*, Leisester, UK, 2005.

13. S. Hu, V. Muthusamy, G. Li, and H.-A. Jacobsen. Transactional mobility in distributed content-based publish/subscribe systems. In *29th IEEE International Conference on Distributed Computing Systems (ICDCS)*, Montreal, Canada, 2009.

14. R. S. Kazemzadeh and H.-A. Jacobsen. Reliable and highly available distributed publish/subscribe service. In *Symposium on Reliable Distributed Systems*, Niagara Falls, New York, 2009.

15. G. Li, A. Cheung, S. Hou, S. Hu, V. Muthusamy, R. Sherafat, A. Wun, H.-A. Jacobsen, and S. Manovski. Historic data access in publish/subscribe. In *Proceedings of the Inaugural Conference on Distributed Event-Based Systems*, , New York, NY, USA, ACM Press, 2007, pp 80–84.

16. G. Li, S. Hou, and H.-A. Jacobsen. Routing of XML and XPath queries in data dissemination networks. In *Proceedings of the 28th International Conference on Distributed Computing Systems (ICDCS'08)*. IEEE Computer Society Press, 2008.

17. G. Li, V. Muthusamy, and H.-A. Jacobsen. Adaptive content-based routing in general overlay topologies. In *ACM Middleware*, Leuven, Belgium, 2008.

18. D. McCarthy and U. Dayal. The architecture of an active database management system. In *Proceedings of the 1989 ACM SIGMOD International Conference on Management of Data*. ACM Press, 1989, pp. 215–224.

19. V. Muthusamy. Infrastructureless Data Dissemination: A Distributed Hash Table Based Publish/Subscribe System. PhD Thesis, University of Toronto, 2005. (Also available as a Technical Report.)

20. V. Muthusamy, M. Petrovic, and H.-A. Jacobsen. Effects of routing computations in content-based routing networks with mobile data sources. In *International Conference on Mobile Computing and Networking (MobiCom)*, Cologne, Germany, 2005.

21. C. E. Perkins and D. B. Johnson. Mobility support in IPV6. In *MobiCom '96: Proceedings of the 2nd annual international conference on Mobile computing and networking*, New York, NY, USA, ACM, 1996, pp. 27–37.

22. L. I. W. Pesonen and D. M. Eyers. Encryption-enforced access control in dynamic multi-domain publish/subscribe networks. In *Proceedings of the Inaugural Conference on Distributed Event-Based Systems*, New York, NY, USA. ACM Press, 2007, pp. 104–115.

23. M. Petrovic, I. Burcea, and H.-A. Jacobsen. S-ToPSS: a semantic publish/subscribe system. In *International Conference on Very Large Databases (VLDB)*, Berlin, Germany 2003.

24. M. Petrovic, V. Muthusamy, and H.-A. Jacobsen. Content-based routing in mobile ad hoc networks. In *MOBIQUITOUS '05: Proceedings of the the Second Annual International Conference on Mobile and Ubiquitous Systems: Networking and Services*, Washington, DC, USA. IEEE Computer Society, 2005, pp. 45–55.

25. M. Petrovic, V. Muthusamy, and H.-A. Jacobsen. Managing automation data flows in sensor/actuator networks. Technical report, Middleware Systems Research Group, University of Toronto, Toronto, Canada, 2007.

26. A. Wun, A. Cheung, and H.-A. Jacobsen. A taxonomy for denial of service attacks in content-based publish/subscribe systems. In *Proceedings of the Inaugural Conference on Distributed Event-Based Systems*, New York, NY, USA. ACM Press, 2006, pp. 116–127.

27. A. Wun and H.-A. Jacobsen. A policy framework for content-based publish/subscribe middleware. In Gustavo Alonso, Eyal de Lara, Indranil Gupta, and Ramsés Morales, editors. *ACM/IFIP/USENIX 8th International Middleware Conference, Lecture Notes in Computer Science (LNCS)*. Springer-Verlag, Vol. 4834, 2007.

28. Z. Xu and H.-A. Jacobsen. Expressive location-based continuous query evaluation with binary decision diagrams. In *IEEE International Conference on Data Engineering (ICDE)*, 2009.

29. Z. Xu and H.-A. Jacobsen. Adaptive location constraint processing. In *Proceedings of the International Conference on Management of Data (SIGMOD 2007)*. ACM, 2007 pp. 581–592.

30. Z. Xu and H.-A. Jacobsen. Evaluating proximity relations under uncertainty. In *Proceedings of 23rd International Conference on Data Engineering (ICDE)*. IEEE Computer Society, 2007.

Chapter 5

Cross-Layer Design Framework for Adaptive Cooperative Caching in Mobile Ad Hoc Networks

Jun Tian and Mieso K. Denko

Department of Computing and Information Science, University of Guelph, Guelph, Ontario, Canada

5.1	Introduction	87
5.2	Cross-Layer Design in Wireless Networks	88
5.3	Cooperative Caching Approaches in MANETs	91
5.4	The Proposed Cluster-based Adaptive Cooperative Caching Scheme	95
5.5	Conclusions and Future Trends	102
	References	103

5.1 INTRODUCTION

A mobile ad hoc network (MANET) consists of a number of mobile devices that form a network on demand without support from any existing network infrastructure or central administration. In a MANET, mobile devices are connected by wireless links, and every device acts as a router, forwarding data packets for other nodes. Its characteristics include dynamic topology, scarce energy, limited bandwidth, and time-varying link rate [1]. In the past, MANETs have mainly been used in battlefields and disaster

Mobile Intelligence. Edited by Laurence T. Yang, Agustinus Borgy Waluyo, Jianhua Ma, Ling Tan, and Bala Srinivasan

areas, where a centralized infrastructure is expensive or even impossible. Nowadays, they are becoming an essential part of the ubiquitous computing environment, but there are several challenges, particularly with respect to improving performance and providing application quality of services (QoS).

Cross-layer design has recently emerged as an important design methodology to cope with the performance issues in the wireless computing environment. To make the concept of cross-layer design clearer, let us review layered network architectures first. One of them is the hybrid reference model, which is used in the Internet [2]. The hybrid reference model is divided into five layers according to the overall networking tasks. Each layer provides certain services to the upper layer while hiding the upper layer from the detailed implementation. Direct communication between nonadjacent layers in the architecture is not allowed and communication between adjacent layers is provided by a set of primitives [3]. This strict layering design has led to the great success of the Internet and has become the default network protocol structure for wireless networks as well. Cross-layer design has been introduced to overcome the problems caused by strict layered architecture. Protocol design through the violation of layered communication architecture gives rise a cross-layer design [3]. One such example of violation is letting nonadjacent layers communicate directly.

In the past few years, most of the research on MANETs focused on the development of routing protocols [4–8] to increase connectivity among mobile hosts in a constantly varying topology. However, information or data access is the ultimate goal for a network. Hence, the design of higher layer protocols is necessary. When a MANET is being integrated into the Internet, or when a data center is placed in a MANET, having mobile hosts effectively access the Internet or the data center becomes a big challenge. Recently, several proposals [9–15] have been introduced to increase data accessibility and reduce query delay in MANETs. These approaches were based on the idea of cooperative caching, in which multiple hosts work cooperatively to share their cached data. Although the above approaches greatly improve the performance of data access in the MANET environment, they do not fully adopt cross-layer design to further improve performance and make the system more adaptive.

The rest of this chapter is organized as follows: Section 5.2 describes the cross-layer based approaches and implementation frameworks in wireless networks. Section 5.3 is a literature review of the current cooperative caching approaches in MANET environment. In Section 5.4, a detailed description of our proposed approach is given and Section 5.5 concludes this chapter and discusses some open challenges and research opportunities.

5.2 CROSS-LAYER DESIGN IN WIRELESS NETWORKS

5.2.1 The Benefits of Cross-Layer Design in Wireless Networks

The main reasons for introducing cross-layer design to wireless networks can be illustrated as follows.

First of all, the assumptions in wireless networks are different from those of the wired networks [16]. For instance, in wired networks, the packet losses are assumed to be a result of network congestion, while in wireless networks, packet losses are often caused by corruption. If TCP still invokes congestion avoidance mechanisms to deal with packet loss in wireless environments, it will make things worse. Explicitly notifying the packet corruption rather than congestion in the signaling from link layer to transport layer will resolve this problem efficiently [17].

Second, due to the resource constraints in MANETs, it is imperative that the various protocols in the protocol stack collaborate to use the limited bandwidth efficiently and reduce power consumption. For instance, poor channel conditions will lead to frame retransmissions and delay at the link layer, which, in turn, leads to TCP retransmissions. To solve this problem, retransmission information can be communicated between the link layer and transport layer; power consumption will be decreased by increasing MAC sublayer retransmission to prevent TCP from retransmissions [18, 19].

Third, with the acceptance of MANETs and wireless mesh networks, WMNs QoS requirements such as acceptable packet loss rate and bounded end-to-end delay have to be taken into account, which requires the application layer and middleware layer to exchange information with the lower layers to improve performance. For example, applications may send the QoS requirements to the TCP layer, and TCP in turn adjusts the receiver windows [18].

Finally, the wireless medium allows modalities of communications that do not exist in wired networks. For example, the physical layer is capable of receiving multiple packets simultaneously. In order to exploit these modalities in wireless networks, cross-layer design is required for protocol design [3].

5.2.2 Cross-Layer Design Proposals and Implementation Frameworks

A number of proposals on cross-layer design in the wireless environment can be found in the literature. According to how the different protocol layers communicate with each other, current cross-layer proposals can be divided into fours types [3].

Creation of new interfaces: New interfaces are created at the specific protocol layers and used for information exchange between different layers. The direction of information flow between different layers can be upward, downward, or back and forth.

Mergence of adjacent layers: If two or more adjacent layers collaborate frequently, these layers can be designed as a new layer, and the layers around this new layer can still use their original interfaces for communication.

Design coupling without new interfaces: Two or more coupling layers exchange information with one another at the runtime without creating any extra interfaces for information exchange. This type requires that the designed layer is familiar with the processing at other layers.

Vertical calibration across layers: In this type, the shared data can be accessed to by the whole protocol stack. This type makes it possible for all layers in the protocol stack to work together to gain maximum performance improvement.

From the literature, there are three methods on how to implement a cross-layer proposal in wireless network architecture.

Method 1: extending or modifying existing protocols. Some proposals extend the functionality of existing internal protocols, such as Internet control message protocol (ICMP), so that they can carry more message types; some proposals modify the packet headers and add extra information into the packet headers. In this way, cross-layer information can flow through protocol layers and be shared by the targeted protocols.

Yang et al. [20] proposed a *rate adaptation* scheme that adapts the data rate at the MAC sublayer according to channel estimation information, such as signal strength, from the physical layer. In order to get physical layer information, they made a minor modification to the current IEEE 802.11 RTS/CTS frame formats. They use two fields: *data rate* (4 bits) and *data packet length* (12 bits) to replace the original *duration* field in the RTS/CTS frame. In this scheme, a source node sends the modified RTS frame including basic data rate and packet length to the destination node before data transmitting. After receiving the frame, the destination node estimates the signal strength and calculates an optimal data rate. Then it puts the new data rate and packet length (from the received RTS packet) into the modified CTS frame and sends it back to the source node. Upon receiving the reply, the source node will use this data rate for the sequent data propagation.

Method 2: creating shortcuts between different layers. Another method to implement cross-layer information propagation is to create shortcuts between adjacent or nonadjacent layers, which need information exchange. Because there is no internal protocol used in this method, shared information does not need to go through the intermediate layers. Cross-layer signalling shortcuts (*CLASS*) [16] is a representative method. By using out-of-bind signaling shortcuts, information can be exchanged between any two adjacent or nonadjacent protocol layers. For instance, if there is cross-layer information between the application layer and network layer, the information communicates directly between them without via transport layer.

Method 3: creating new independent components. In this kind of implementation, shared information extracted from each layer is stored in an independent component. Protocol layers can access the component to fetch the required information [16]. Like Method 2, this method does not need internal protocols for information exchange between different layers. Different layers exchange shared information via the independent component. The implementation proposed in Ref. [21] used an independent component called *system profile* for information exchange between the middleware layer and network layer. In the proposed system framework, the application layer produces and shares video data with other users in the same group in an ad hoc network, the middleware layer provides data accessibility service, and the routing layer searches for feasible routes. The system profile component stores information from middleware layer and routing layer. Through system profile, the middleware layer and routing layer are able to share information with each other and work together to achieve higher data accessibility.

Conti et al. [22] proposed a full cross-layer design scheme, which is similar to the approach proposed in Ref. [21], but is more comprehensive. Unlike most of the proposals on cross-layer design that are specific on information exchange between two or three layers only, this architecture is a full cross-layer design. The architecture creates a core component called *network status*, which works as a database, storing information from all protocol layers. Every protocol can share its own information and access information from other layers through the component network status. It can be considered as a vertical protocol layer and shared by all other protocol layers.

Method 1 extends the functionality of existing protocols or modifies the existing packet headers, and it is not necessary to create a new internal protocol for propagating information between different layers. However, the drawbacks of this kind of method are also obvious. First of all, it is not efficient. If two nonadjacent layers need to exchange information, the intermediate layers have to be involved. Second, it is confined to transmitting information among two or three layers and is difficult to implement a stack-wide cross-layer design. Lastly, only simple messages, such as signaling, can be transmitted.

Method 2 is very efficient and flexible compared to Method 1 because information can be propagated directly between any two layers without going through intermediate layers, so it is a full cross-layer design. However, if layer-specific information needs to be shared by several other layers, the information has to be collected several times, and overhead will be caused by duplicate operations. Furthermore, this architecture may lead to a complex program structure.

Method 3 has a very clear architecture because all shared information is in an independent vertical layer, and it is also a full cross-layer scheme. The main concern of this method is how to implement interactions between every protocol layer and the vertical layer. Because the shared information is kept in the vertical layer, duplicate data collecting can be avoided. On the other hand, it is not as efficient as Method 2 because information exchange between different layers has to be with the aid of network status.

5.3 COOPERATIVE CACHING APPROACHES IN MANETs

Simple and cooperative caching systems have been widely used in the Internet for improving Web service performance [23–30]. A detailed survey on Web caching can be found in Ref. [31]. In general, there are clients, caching proxies, and Web servers in a Web caching system. Caching proxies are placed between servers and clients and work cooperatively to provide caching for clients. When a client requests a Web page, it checks its own cache first. If that fails, it sends the page request to one of the caching proxies. If the requested page cannot be found in this proxy, it sends the request to its cooperative proxy. If the requested page cannot be found among these proxies, the request will be sent to the remote Web server to get the original page. Web caching systems lead to a decrease in bandwidth usage, Web server workload, and client query delay [31].

Web caching systems work well in improving the performance of Web services. However, this scheme cannot be directly used in the MANET environment due to the mobility and resource constraints of MANETs. In a MANET, all nodes are moving, and the network topology changes frequently, so it is impossible to place cache proxies in a MANET. Therefore, a different approach has to be used to implement a cooperative caching system in the MANET environment. Recently, several approaches have been presented for using cooperative caching in MANETs. In these approaches, every mobile node has a certain amount of cache space, which can be accessed by itself or neighboring nodes. Every node gets requested data not only from its own cache space but also from the cache space of its neighbors. In this way, every node owns a much larger cache space than it would by itself.

5.3.1   The Basic Operations in Cooperative Caching Systems

In general, a cooperative caching system includes the following modules.

Information search (or cache resolution): This module deals with where to find the data item requested by the client. The returned data can be either the original data from the data source or the cached copy in a mobile node.

Cache management: It includes three submodules: *cache admission control*, which determines whether a received data item should be cached; *cache replacement*, which determines which cached item in the cache space should be removed when the cache is full and a new data item has to be cached; and *cache consistency*, which keeps the cached data items synchronized with the original data items in data source. Most approaches adopt time-to-live (TTL) based cache consistency strategy, in which every data item is assigned a TTL value, that how long will this data item exists in cache space, and the cached data item is considered as valid before the TTL elapses.

Prefetching: It is responsible for which data item should be prefetched from the data source for future use.

5.3.2   Existing Cooperative Caching Approaches

Current cooperative caching schemes can be grouped by using different criteria such as the underlying routing protocol, content in cache space, and whether it uses broadcasting for information searching. Here, we use the underlying routing protocol to classify them: using general routing protocol and using specific routing protocol.

5.3.2.1   *Approaches Using General Routing Protocols*

[9, 10] Cao and coworkers introduced cooperative cache-based data access in MANETs. It is middleware and stays on top of the routing protocols. There is one or more data centers in the network. They proposed three cache techniques—*CacheData*, *CachePath*, and *HybridCache* to improve data access performance by caching the data

or data path. In the network, every node checks the passing data. If the data are found to be popular, this node will cache this item in its own cache space (CacheData) or cache the path information of this item (CachePath). Next time, when this node has a request, (1) it replies to this request if it has the cached copy. or (2) if it has the path information of the requested item, and the distance between this node and the caching node is less than the distance between this node and the data center, the request will be redirected to the caching node; otherwise, (3) this request packet will be forward to the data center.

CacheData and CachePath use TTL-based cache consistency strategy. For cache admission control, authors believe that data accessibility should have higher priority than the query delay in MANETs and a node should avoid caching data items that have already been cached by their neighbors. Therefore, a received item will not be cached if it is from the neighborhood [9]. When the cache is full, two parameters are used to determine which cached item should be evicted. One is the order of the cached item, which is based on the access interest, and the other is the size of the cached item. The cached item with the largest value of the product of these two factors will be removed.

Lim et al. [11] proposed a caching scheme called aggregate caching for Internet-based MANETs. In their scheme, some hosts have direct link to the Internet and thus serve as access points (APs) or gateways to the Internet. A broadcast-based search algorithm called simple search is used to find the requested date either in a mobile host or in APs. When a mobile host needs a data item, if the data item is unavailable in its own cache and this host cannot directly connect to any AP, it broadcasts a request packet to its neighboring mobile hosts. Upon receiving the request packet, a mobile host either replies the request if it has cached the requested data item or forwards the request to its neighbors. A hop limit is used for the request packet to reduce the traffic caused by flooding. This scheme uses the same cache admission control strategy as that in Ref. [9]. Two factors are considered in their cache replacement policy: distance, the number of hops between the requester and the AP or other nodes caching the requested data item, and access frequency of the requested data item. Based on these two factors, they presented three replacement schemes: distance outweighing access frequency, access frequency outweighing distance, and both having the same weight.

The cooperative caching scheme proposed in [14], focuses on cache resolution and cache management of a caching system. Their cache resolution has three steps. First, profile-base resolution. Every node keeps a record of previously received data requests. If it fails to find the requested data item in local cache, it checks its profile to find a data source and sends the request to that node. Second, limited flooding is used for searching the data item within neighborhood. Third, a data request is sent to a data source after the above two steps fail. Along the route, a forwarding node may reply to the request if it has the copy. The cache management scheme in this approach tries to store more distinct copies of data items within neighborhood, which is similar to the cache admission control used in Cao et al. [9]. LRU is used to for cache replacement, and cache consistency strategy is also based on TTL.

Denko and Tian [15] proposed a scheme that is different from the above schemes in two ways. One difference is that this scheme is cluster-based. Lowest ID clustering

(LIC) algorithm [32] is used for grouping the whole network. In each cluster, there is a cluster head (CH), a data source (DS), caching agents (CAs), and mobile hosts (MHs). The DS generates data items needed by other MHs in the network. There are multiple data sources that store different data items. The hosts that act as the DS are known to the CHs and local CAs. In addition, a secondary cluster head is used to cope with cluster head changes and packet drop [33]. The other difference is that this scheme uses prefetching to further improve the performance of the cooperative caching system. Strategies for proactively prefetching the most frequently accessed data within the cache or prefetching frequently needed data upon the expiry of TTL significantly improve network performance by reducing latency. The prefetch-on-mis scheme is used when the value of TTL expires for a particular data item to reduce communication overhead. TTL is also used for cache consistency purpose, and least recently/frequently used (LRFU) algorithm [34], which combines two metrics of every data item, access frequency and latency, is used for cache replacement.

5.3.2.2 *Approaches Using Specific Routing Protocols*

Sailhan and Issarny's cooperative caching strategy [12] is based on zone routing protocol (ZRP) [8]. For a mobile host, its neighborhood is its zone. When a mobile node requests for a data item that is not found in local cache, this node first broadcasts the request to other hosts within the zone. If this step fails, then the node sends the request packet to a mobile terminal, which is outside the zone, has the copy of requested data, and is nearer to the requester than base stations. Finally, the request packet is sent to the nearest base station. No cache admission control is exploited and every received data item is cached. When a node's local cache is full, the selection of the victim is based on four criteria: popularity, access cost, coherency, and size. The cache consistency strategy is also based on TTL.

A transparent cache-based mechanism is proposed by Wang et al. [13]. First, they introduced a new on-demand routing protocol called dynamic backup routes routing protocol (DBR^{2}P) to support their caching scheme. DBR^{2}P not only discovers a complete route from the source to the destination but also sets up some alternative routes as backup. Then they established their cache mechanism based on DBR^{2}P. In their scheme, the frequently accessed data and data path will be cached in some special mobile nodes. A node can get the requested data item from its neighbors if they have the requested copies, or the path information to the caching node that is nearer to the requester than the data source. They also use LRU for cache replacement.

Flooding is used for information searching within the neighborhood in Refs. [11, 12, 14]. In order to reduce the huge traffic overhead caused by pure flooding, the number of hops of flooding in these approaches is constrained within one or a small number. In Refs. [9, 11, 14, 15], the number of hops is used as a criterion for determining whether or not cache a received data item. If the number of hops between the requester and caching node is less than the threshold, the received item will not be cached in the requester. In this way, nodes within a neighborhood can cache more distinct data items and data accessibility will be increased at the cost of the increase in query delay. However, it is worthwhile because these approaches believe that data accessibility should outweigh query delay in MANET environment. Other schemes just cache every item they receive.

An LRU algorithm is used in approaches Refs. [12–14] for cache replacement and an LRFU algorithm is used in approach Ref. [15]. Some of these schemes [10–12] use their own cache replacement algorithms, which consider more factors. Most of these approaches use TTL-based consistency strategy to keep cached data items coherent with the original items in data source. Approaches [10, 13] cache data or data path information in every node's cache space, while others just cache data item.

5.4 THE PROPOSED CLUSTER-BASED ADAPTIVE COOPERATIVE CACHING SCHEME

In this scheme, we consider a MANET environment in which there is a data center (DC) and a number of MHs. In the following part of this chapter, they are also called nodes). DC, whose address is known to all other MHs, can be regarded as database servers, which store all data items needed by the whole MANET, or can be considered the gateways to the Internet and provide information service for other MHs in the MANET. The data items in DC are regular objects, such as text files and pictures, which have different sizes. All modifications to data items are performed by DC. MHs get data items from DC as they need them. ad-hoc on demand distance vector (AODV) routing protocol [6] is used in the implementation.

5.4.1 Overview of Cross-Layer Based COCA (COoperative CAching)

COCA, illustrated in Figure 5.1, is a cluster-based middleware that stays on top of the network layer and provides caching service for the upper applications. COCA includes *information searching*, *cache management*, and *prefetching* modules. Cross-layer design is exploited to optimize system performance and make the system more adaptive. The *stack profile* module is responsible for cross-layer-related functions. In the MANET, every MH has a certain amount of cache space for caching data items from DC or other MHs, and COCA and stack profile reside in every mobile host.

5.4.2 Descriptions of Proposed Modules

5.4.2.1 Clustering Architecture

Clustering is an effective way to organize MANETs. It reduces traffic overhead, flooding, and collisions in MANETs. Furthermore, it makes the network more scalable. In COCA, we adopt least cluster change (LCC) clustering algorithm [35], which is an improvement over LIC clustering algorithm [32]. LIC works as follows. In a MANET, every node is assigned a unique ID and broadcasts its own ID and IDs it can hear to all its one-hop-away neighbors periodically. A node will become cluster head if all

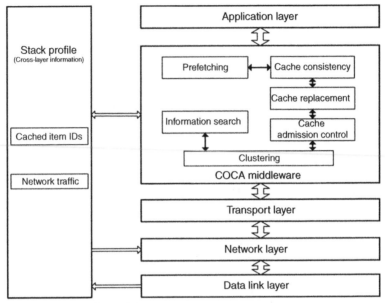

Figure 5.1 Brief system architecture of COCA.

IDs it can hear are larger than its own ID, and other neighboring nodes are its cluster members. If a node can hear two or more cluster heads, it is a gateway as well as a cluster member. In a cluster, any two nodes are at most two hops away. In the whole network, no cluster heads are directly linked. LCC is based on LIC and reduces the cluster head changes, which makes the clustering architecture more stable.

5.4.2.2 Cross-Layer Design

Our implementation of the cross-layer design is similar to Method 3, which is described in Section 5.2.2. The module *stack profile* is independent of the protocol stack and provides a data exchange buffer for layers in the protocol stack. The layers, which have information to be shared, put it in the *stack profile*. Then, other layers, which need the information, fetch it from the *stack profile*. Besides information sharing between different layers, a layer can invoke functions in the other layer via *stack profile* instead of creating interfaces between different layers, which decreases the coupling between layers. The following information needs to be shared between different layers.

Network traffic. It is provided by data link layer. The middleware layer needs it for prefetching process. A nodes initiates prefetching process only when the network traffic is not busy, and DC will not reply the prefetching request if the network traffic is busy.

Cached item IDs. It is provided by the COCA middleware layer. The network layer needs the information. One of the steps of the information search process is sending a request to a DC. When the request packet is passing along the route to DC, it will be checked by forwarding nodes (routing protocol gets ID information from middleware). If a forwarding node has the copy of the requested data item, it drops the request packet and replies the requested data item to the requester.

In our current scheme, we maintain network traffic information and cached items ID. The information is exchanged among middleware, network and data link layers. However, the module *stack profile* can be extended to support information share and function call between any different protocol layers.

5.4.2.3   Information Search

The main duty of information search is to find either the copy of data item or the original data item required by users in MANETs. In a cluster, every cluster member sends the ID list of cached data items to its cluster head periodically. As a result, the cluster head keeps an ID list that records which data item is cached in which node in this cluster. Based on the cluster architecture, information search is a combination of different lookup methods, which are executed at the order from the method having the least communication overhead to the method having the greatest communication overhead.

When a node requests a data item, first, it checks its own local cache. If the data item cannot be found, the node will search the requested data item within the neighborhood. If DC is this node's neighbor, the request is sent to DC. Otherwise, the request is broadcast to its neighbors (within one hop). If a cluster member receives the request packet, it replies to the request if it has cached the requested item; otherwise, it drops the packet. If the cluster head receives the request, it replies to the request if it has cached the requested item; otherwise, it checks its data item ID list. If the ID of the corresponding item is found, the request packet is forwarded to the node that caches the data item. If the cluster head cannot find the data item ID, it forwards the request packet to DC, if DC is its one-hop-away neighbor. After a threshold time, if the requestor does not receive a reply, it sends the request to DC. When the request packet is passing along the route to DC, it will be checked by every forwarding node. If a forwarding node has the requested data item, it drops the request packet and replies to the requested data item to the requester. This method is an example of using a cross-layer design approach in cooperative caching system. Eventually, DC receives the request packet and replies the requested data item to the requester.

5.4.2.4   Cache Management

In our scheme, we adopt the cache admission control strategy proposed in Ref. [9], but with a minor modification. In Ref. [9], when a node receives an item, it will not be cached if there is already a copy of it within neighborhood. This strategy makes nodes cache more distinct data items within the neighborhood and improves data accessibility. Our modification is that a node will cache all received data items until its cache space becomes full. After the cache space becomes full, the received data item will not be cached if the data item has a copy within the cluster, while the node will cache the received data item if this data item is from outside the cluster. TTL-based cache consistency strategy is used in our scheme, and LRU-MIN [36], a variant of LRU, is used for cache replacement. LRU is a widely used replacement algorithm in a variety of areas, which repeatedly removes the least recently referenced data items from the cache space until there is enough space for the arrived item. LRU-MIN biases toward smaller items and tries to minimize the number of removed data items by removing larger items first. It works as follows. When a data item with the size of S needs to be cached and the available cache space is less than S, (1) LRU-MIN selects from cache space data items that are larger than or equal to S, then removes items based on LRU until there is enough space for the new item. (2) If all selected items are removed and there is still not enough space for the arrived item, let $S = S/2$ and repeat step (1).

5.4.2.5 *Prefetching*

The basic idea of prefetching is to anticipate user's requests and to order highly likely data items in advance. Prefetching is an efficient way of reducing access delay, even though it causes network overhead. In COCA, a relatively simple prefetching strategy is used. The idea of this scheme is derived from Ref. [15]. Intuitively, data items in a node's local cache have a high probability of being requested in the future. When some of these data items become invalid because their values of TTL have become zero, they still have the high probability of being requested in the near future. These invalid data items are goals of the prefetching process in COCA. In order to improve the efficiency of prefetching, save local cache space, and save bandwidth, only the invalid data items with high access probability will be prefetched. Although prefetching will help improve the caching performance, it will lead to the increase in network traffic, which has negative effect on wireless networks, especially on MANETs. Hence, the prefetching process should be executed while the network traffic is low. In MANETs, two metrics [20, 37] can be used to indicate whether the network traffic of a node or around it is busy. One is average MAC layer utilization and the other is instantaneous queue length. Here, we choose instantaneous queue length as a traffic indicator. A node sends a prefetching request to the DC only when the network traffic is low. Upon receiving the prefetching request, the DC replies the request if the network traffic is not busy. By using a cross-layer design, we can take advantage of the benefit of prefetching and get rid of the bad effects caused by prefetching, and the system becomes more adaptive.

5.4.3   Experimental Results and Discussions

Our proposed scheme was implemented in an NS2 simulation environment [38]. The moving pattern of every node follows the random way point movement model [39]. We assume that each node generates a sequence of data requests with exponentially distributed time intervals. The request pattern follows Zipf-like distribution [40, 41], which has been used to model various behaviors such as Web page request patterns.

Three performance metrics are used to measure the scheme performance. The first one is *data accessibility ratio*, which is the percentage of successful requests. The second one is *average query delay*, which is the average response time from sending a request till receiving the response. The last one is *average query distance*, which is the average distance (the number of hops) covered by a successful request. The following caching schemes are compared.

Simple caching (SC): The data item request is sent to data source (DC) directly if the requested item cannot be found in local cache.

Cooperative caching (CCNP): The caching scheme we proposed without prefetching.

Cooperative caching with prefetching (CCPF): The cooperative caching scheme combined with prefetching.

The comparison between SC and CCNP is used to demonstrate the performance of cooperative caching and the comparison between CCNP and CCPF is used to demonstrate the effective of prefetching scheme.

In addition, the performance of cooperative caching with cross-layer design (CCCL) and cooperative caching without cross-layer design (CCNCL) is compared to demonstrate the effective of cross-layer design. The simulation was carried out in an area of 3500 m × 500 m with 50–120 nodes roaming in the simulation area. Further simulation parameters are shown in Table 5.1.

Figure 5.2 shows data accessibility ratio as a function of number of nodes. It can be noted from the figure that CCNP outperforms the SC at all levels, clearly showing the usefulness of cooperative caching. Another observation from the graph is that the performance improvement rises with the number of nodes. This implies that as the number of nodes increase, the number neighbors that cache, data also increase.

Figure 5.3 shows the average query delay as a function of the number of nodes. The results show that CCNP outperforms SC. The reason is that CCNP can get requested items from neighboring nodes, which significantly reduces the communication time compared with getting requested items from the remote data source. In MANETs, data packets are multiforwarded from the source to the destination. As a

Table 5.1   Simulation Parameters

Parameter	Values
Transmission range (m)	250
Bandwidth (Mbps)	2
Node speed (m/s)	0–10
Pause time	100
Data item size (kB)	1–10
Cache size (kB)	300
TTL mean (s)	100–3000
Zipf Parameter, θ	0.8
Number of data items	1000
Request interval (s)	10
Simulation time (s)	2000

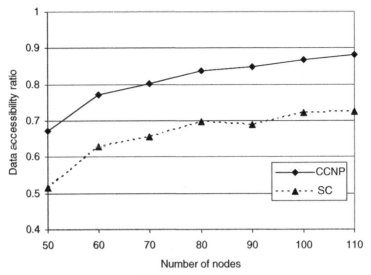

Figure 5.2 Data accessibility ratio.

result, the more hops between the source and the destination, the more communication time it will take.

Figure 5.4 shows the average query delay as a function of mean TTL time. From this figure, we can see that when mean TTL is small, CCPF performs much better than CCNP. However, with the increase in mean TTL, the difference between CCPF and CCNP becomes less and less, although CCPF still outperforms CCNP. The reason is

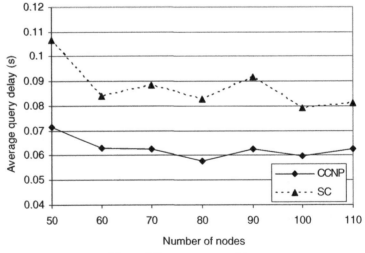

Figure 5.3 Average query delay.

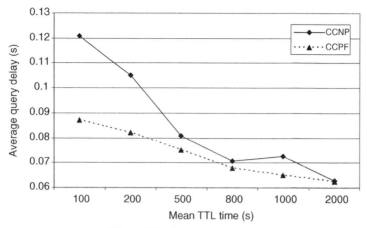

Figure 5.4   Average query delay.

that when mean TTL is small, the expiry of cached items occurs frequently during the simulation period. If prefetching is integrated, these items will be fetched in advance for future use. Therefore, a node can find more requested items in its cache space, which significantly reduces the communication time. However, when mean TTL is large, most of the cached data items stay valid during the simulation time, which leads to the small performance difference between CCPF and CCNP. This figure demonstrates the benefits of prefetching in a MANET environment, especially when the mean TTL is small.

Figures 5.5 and 5.6 show the effectiveness of cross-layer design as a function of the number of nodes. After a node fails to retrieve data from its local cache and its neighboring cache, it sends a request to DC. From Figure 5.5, we can see that CCCL uses less average number of hops than does CCNCL. When the number of

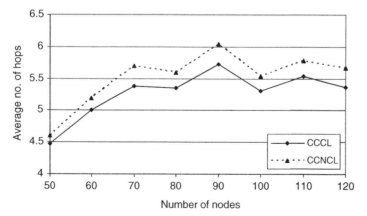

Figure 5.5   Average query distance.

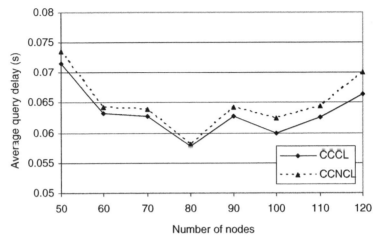

Figure 5.6   Average query delay.

hops is low, the average query delay of CCCL is reduced accordingly. Figure 5.6 shows the average query delay as a function of number of nodes. In this experiment, the cross-layer-based scheme performs relatively better than that without cross-layer.

5.5   CONCLUSIONS AND FUTURE TRENDS

In this chapter, we first gave a review of the current cross-layer design approaches in a wireless environment for performance improvement and then presented a review of recently proposed cooperative caching schemes in a MANET environment for improving data accessibility. After that, we introduced our cluster-based cooperative caching scheme, which fully adopted cross-layer design approach to improve performance and make systems more adaptive. From the above experimental results and other cooperative caching schemes, we can see that cooperative caching is an efficient approach in a MANET environment for reducing data query delay and improving data accessibility, and the employment of cross-layer design approach in a MANET environment makes system more efficient and adaptive. We also presented a prefetching scheme that is integrated into the cooperative caching system to further improving the caching performance. In certain conditions, our prefetching scheme greatly reduces the query delay. However, there are still some challenges.

First of all, although cross-layer design is regarded as an effective way to improve performance in wireless computing environment, Kawadia and Kumar [42] pointed out three problems introduced in using cross-layer design approach. First, due to the great success of strict layered architecture in the Internet, it has become the default network architecture of wireless networks, although cross-layer design is a violation to the layered architecture. Protocol designers should keep in mind the importance of the layered architecture and to what extent it should be violated when designing a cross-layer scheme. Second, since every protocol designer has his own method of implementation, the coexistence of different cross-layer designs and unbridled cross-layer design will probably lead to a complex code. Third, cross-layer design may create unintended interactions between different layers, which, in turn, may lead to undesirable results on overall performance.

Second, more work needs to be done on cache management in the MANET environment. The most important issue in this area is cache replacement algorithm. In MANETs, more factors need to be considered when a cached item has to be evicted. Besides the frequency and latency of an item, we have to consider communication cost, data item size, and energy status to find factors having more weight in determining the eviction of a cached item. In some scenarios, such as battlefields, strong cache consistency strategy is a must, and how to efficiently deploy this strategy is a big challenge.

Third, network traffic has a significant effect on network performance in MANETs. When network traffic is high, packet dropping occurs frequently due to collision, which will worsen network throughput. In our scheme, AODV routing protocol periodically broadcast hello messages for route discovery and maintenance, and clustering protocol also periodically broadcasts these messages for cluster formation and maintenance. We can combine these two kinds of messages together to reduce communication overhead, but we will face the same challenges as we have to face in cross-layer design. In addition, although prefetching improves the performance of caching system, it may cause traffic overhead. We have to find efficient ways to reduce the overhead caused by prefetching. In the future, we plan to deploy multiple DCs in a MANET to reduce the workload of every DC and traffic congestion around every DC. Multiple DCs improve data accessibility as well.

Finally, we need a standardized environment (in terms of simulation tools and parameter settings) for developing and testing cooperative caching systems in MANETs. In the existing cooperative caching approaches, the network area and network density are different. Every approach has its own original source data items and client query model. In the future, there is a need to develop a standard environment for performance evaluation and validation using implementation testbed with real traffic data.

REFERENCES

1. D. P. Agrawal and Q. Zeng. *Introduction to Wireless and Mobile Systems*, 2nd edition, Thomson Brooks/Cole Pacific Grove, 2006.
2. A. S. Tanenbaum. *Computer Networks*, 4th edition, Prentice Hall PTR, 2002.
3. V. Srivastava and M. Motani. Cross-layer design: a survey and the road ahead. *IEEE Commun. Mag.* 43(12): 112–119, 2005.
4. D. B. Johnson and D. A. Maltz. Dynamic source routing in ad hoc wireless networks. In T. Imielinski and H. Korth, eds. *Mobile Computing*. Kluwer Academic Publishers, 1996, pp. 153–181.
5. V. D. Park and M. S. Corson. A highly adaptive distributed routing algorithm for mobile wireless networks. In *Proceedings of the 6th Annual Joint Conference on IEEE Computing and Communication (INFOCOM'97)*. Vol. 3, 1997, pp. 1405–1413.

6. C. E. Perkins and E. M. Royer. Ad-hoc on-demand distance vector routing. In *Proceedings of the 2nd IEEE Workshop on Mobile Computer Systems and Application (WMCSA'99)*. Vol. 2, 1999, pp. 90–100.

7. C. E. Perkins and P. Bhagwat. Highly dynamic destination-sequenced distance-vector routing (DSDV) for mobile computers. In *Proceedings of the Conference on Communications Architectures, Protocols and Applications (SIGCOMM '94)*, 1994, pp. 234–244.

8. Z. J. Haas and M. R. Pearlman. ZRP: a hybrid framework for routing in ad hoc networks. In *Ad Hoc Networking*. Addison-Wesley Longman Publishing Co., Inc., Boston, MA, USA 2001, pp. 221–253.

9. G. Cao, L. Yin, and C. R. Das. Cooperative cache based data access in ad hoc networks. *IEEE Comput. Soc.*, Vol. 37, no. 2, 2004, pp. 32–39.

10. L. Yin and G. Cao. Supporting cooperative caching in ad hoc networks. *IEEE Trans. Mobile Comput.*, 5(1): 77–89.

11. S. Lim, W. C. Lee, G. Cao, and C. R. Das. A novel caching scheme for internet based mobile ad hoc networks. In *Proceedings of the 12th International Conference on Computing and Communication Networks (ICCCN 2003)*, 2003, pp. 38–43.

12. F. Sailhan and V. Issarny. Cooperative caching in ad hoc networks. In *Proceedings of the 4th International Conference on Mobile Data Management*, 2003, pp. 13–28.

13. Y. Wang, J. Chen, C. Chao, and C. Lee. A transparent cache-based mechanism for mobile ad hoc networks. In *Proceedings of the International Conference on Technology and Applications (ICITA'05)*, Vol. 2, 2005, pp. 305–310.

14. Y. Du and S. K. S. Gupta. COOP-A cooperative caching service in MANETs. In *Proceedings of the Joint Conference on Autonomic and Autonomous Systems and International Conference on Networking and Services (ICAS-ICNS 2005)*, 2005, pp. 58–63.

15. M. K. Denko and J. Tian. Cooperative data caching and prefetching in wireless ad hoc networks. In *Proceedings of the 2nd IEEE International Conference on Wireless and Mobile Computing, Networking and Communications (WiMob2006)*. Montreal, Canada, 2006.

16. Q. Wang and M. A. Abu-Rgheff. Cross-layer signalling for next-generation wireless systems. In *Proceedings of the IEEE Wireless Communications and Networking (WCNC2003)*, Vol. 2, 2003, pp. 1084–1089.

17. H. Balakrishnan, *Challenges to reliable data transport over heterogeneous wireless networks*, Ph.D. Thesis, University of California, Berkeley, 1998.

18. V. T. Raisinghani and S. Iyer. Cross-layer design optimizations in wireless protocol stacks. *Comput. Commun.*, 27(8): 720–724.

19. M. Methfessel, K. F. Dombrowski, P. Langendörfer, and H. Frankenfeldt. Vertical optimization of data transmission for mobile wireless terminals. *IEEE Wireless Commun.*, 9(6): 36–43.

20. N. Yang, R. Sankar, and J. Lee. Improving ad hoc network performance using cross-layer information. In *Proceedings of the IEEE International Conference on Communications (ICC2005)*, 2005, pp. 2764–2768.

21. K. Chen, S. H. Shah, and K. Nahrstedt. Cross-layer design for data accessibility in mobile ad hoc networks. *Wireless Personal Commun.*, 21(1): 49–76, 2002.

22. M. Conti, G. Maselli, G. Turi, and S. Giordano. Cross-layering in mobile ad hoc network design. *IEEE Comput. Soc.*, 37(2): 48–51, 2004.

23. C. Aggarwal, J. L. Wolf, and P. S. Yu. Caching on the world wide web. *IEEE Trans. Knowledge Data Eng.*, 11(1): 94–107, 1999.

24. A. Chankhunthod, P. B. Danzig, C. Neerdaels, M. F. Schwartz, and K. J. Worrell. A hierarchical internet object cache. In *Proceedings of the USENIX Annual Technical Conference*. 1996, pp. 153–164.

25. L. Fan, P. Cao, J. Almeida, and A. Broder. Summary cache: a scalable wide-area web cache sharing protocol. *IEEE/ACM Trans. Network.*, 8(3): 281–293, 2000.

26. S. Iyer, A. Rowstron, and P. Druschel. Squirrel: a decentralized, peer-to-peer web cache. In *Proceedings of the 21st Annual Symposium on Principles of Distributed Computing (PODC '02)*, Monterey, California, USA, 2002, pp. 213–222.

27. K. W. Ross. Hash routing for collections of shared web caches. *IEEE Network*, 11(6): 37–44, 1997.

28. A. Rousskov and D. Wessels. Cache digests. *Comput. Networks ISDN Syst.*, 30(22): 2155–2168.

29. D. Wessels and K. Claffy. ICP and the Squid web cache. *IEEE J. Select. Areas Commun.*, 16(3): 345–357.

30. C. M. Bowman, P. B. Danzig, D. R. Hardy, U. Manber, and M. F. Schwartz. The harvest information discovery and access system. *Comput. Networks ISDN Syst.*, 28(1–2): 119–125, 1995.

31. J. Wang. A survey of web caching schemes for the Internet. *ACM SIGCOMM Comput. Commun. Rev.*, 29(5): 36–46, 1999.

32. M. Gerla and T. Tsai. Multiuser, mobile multimedia radio network. *J. Wireless Networks*, 1: 255–265, 1995.

33. H. Lu and M. K. Denko. Replica update strategies in mobile ad hoc networks. In *Proceedings of the 2nd IFIP International Conference on Wireless and Optical Communications Networks (WOCN2005), Montreal, Canada*, 2005, pp. 302–306.

34. D. Lee, J. Choi, J. H. Kim, S. H. Noh, S. L. Min, Y. Cho, and C. S. Kim. LRFU (Least Recently/Frequently Used) replacement policy: a spectrum of block replacement policies. *IEEE Trans. Comput.*, 50(12): 1352–1361, 1996.

35. C. C. Chiang, H. K. Wu, W. Liu, and M. Gerla. Routing in clustered multihop, mobile wireless networks with fading channel. In *Proceedings of the IEEE SICON'97*, 1997, pp. 197–211.

36. M. Abrams, C. Standridge, G. Abdulla, S. Williams, and E. Fox. Caching proxies: limitations and potentials. In *Proceedings of the 4th International World Wide Web Conference, Boston, USA*, 1995, pp. 119–133.

37. Y. C. Hu and D. B. Johnson. Exploiting congestion information in network and higher layer protocols in multihop wireless ad hoc networks. In *Proceedings of the 24th Intternational Conference on Distributed Computer Systems*, 2004, pp. 301–310.

38. Network Simulator 2, version 2.29, http://www.isi.edu/nsnam/ns/.

39. J. Broch, D. A. Maltz, D. B. Johnson, Y. Hu, and J. Jetcheva. A performance comparison of multi-hop wireless ad hoc network routing protocols. In *Proceedings of the 4th Annual ACM/IEEE International Conference on Mobile Computing and Networking (MobiCom '98), Dallas, Texas, USA*, 1998, pp. 85–97.

40. G. K. Zipf. *Human Behavior and the Principle of Least Effort: An Introduction to Human Ecology*, Addison-Wesley, Redwood city, 1949.

41. L. Breslau, P. Cao, L. Fan, G. Phillips, and S. Shenker. Web caching and zipf-like distributions: evidence and implications. In *Proceedings of the 18th Annual Joint Conference of the IEEE Computer and Communications Societies (INFOCOM'99)*, 1999, pp. 126–134.

42. V. Kawadia and P. R. Kumar. A cautionary perspective on cross-layer design. *IEEE Wireless Commun.*, 2(1): 3–11, 2005.

Chapter 6

Recent Advances in Mobile Agent-Oriented Applications

Ali R. Hurson,[1] Evens Jean,[2] Machigar Ongtang,[2] Xing Gao,[2] Yu Jiao,[3] and Thomas E. Potok[3]

[1] *Department of Computer Science, Missouri S&T, Rolla, MO*
[2] *Department of Computer Science and Engineering, The Pennsylvania State University, University Park, PA*
[3] *Applied Software Engineering Research Group, Computational Sciences and Eng. Division, Oak Ridge National Laboratory, Oak Ridge, TN*

6.1	Introduction	106
6.2	Mobile Agents and Distributed Information Retrieval	108
6.3	System Administration	120
6.4	Mobile Agents and Sensor Network	126
6.5	Swarming Intelligence	131
6.6	Agents, Pervasive Computing, and Education: A Case Study	133
6.7	Conclusions and Future Trends	136
	Acknowledgments	137
	References	137

6.1 INTRODUCTION

In recent years, we have witnessed explosive advances in computing technologies leading to the availability of new services to users. At the heart of these advances lies the introduction of networking infrastructures, such as the Internet, to support global computing applications. As a result, a computer is no longer an isolated entity,

Mobile Intelligence. Edited by Laurence T. Yang, Agustinus Borgy Waluyo, Jianhua Ma, Ling Tan, and Bala Srinivasan

but instead interacts with other systems to carry out user tasks. With a new outlook on computing applications comes the need to access data and services, which is heterogeneous and distributed, at anytime and from any location. Under these new requirements, both users and the data of interest may be mobile and subsisting in an environment that is characterized by low bandwidth, limited resources, and frequent disconnection. The development of applications in such an environment requires the adoption of novel techniques to deal with the emerging issues and the technological constraints. Mobile agent technology is one of the new mobile computing application design paradigms that have shown great success in this area.

In general, mobile agent technology refers to a programming model that revolves around the ability for a program to halt its execution and move to a new environment where execution can then be resumed. Roaming the network may require collaboration with other agents even when the agents may not share a common goal. Mobile agents are typically autonomous, and perceivably intelligent, capable of dynamically deciding their next course of action based on their knowledge and current environmental condition. The mobility of agents is not static and can depend either on current computation or on an itinerary specified in advance by the user. Agents often have the ability to clone themselves and hence, execute numerous copies of their code (the agent) in different locations concurrently. Endowed with such abilities, agents have received a great deal of attention from research communities. Their ability to support concurrent execution along with their mobility makes them very suitable for distributed applications. It is worth noting that agents execute independently from the computing device, from which they are originated, until completion of the task at hand, hence, the use of agents in environments with limited resources or intermittent connectivity.

The use of mobile agents covers a wide spectrum of applications ranging from the retrieval of information from multiple sources to the administration of complex distributed systems. Agents are also used in environments where intelligent software can help increase performance and reliability; such scenarios are evident in the use of agents in searches and tracking applications as well as in efforts geared toward improving the educational system. A full appreciation of the technology at hand can only be acquired through an analysis of its numerous applications. Within the scope of this chapter, we will focus on introducing the reader to the broad spectrum of research areas that have benefited from the use of mobile agents. We will begin our discussion by presenting the effect mobile agents have had on distributed information retrieval in Section 6.2. Our discussion will progress to the use of mobile agents to manage complex systems in Section 6.3; Section 6.4 follows with an examination of the recent use of this paradigm in sensor networks. Mobile agents bolster a sizeable influence on swarming intelligence technique, as will be shown in Section 6.5. Section 6.6 will then present readers with a case study of the application of agents in pervasive computing, in which future trends of the technology are highlighted. Finally, we draw conclusions in Section 6.7.

6.2 MOBILE AGENTS AND DISTRIBUTED INFORMATION RETRIEVAL

Distributed information retrieval focuses on ways of locating and retrieving information of interests based on the user's specifications. As mentioned in Section 6.1, mobile agents take the computation to the data. Migration of the code as opposed to the data can highly improve the efficiency of the system in the presence of large sets of data that may not be relevant to the task at hand, a common occurrence in information retrieval systems. This section introduces the most recent work that has been conducted to improve distributed information retrieval through the use of mobile agents; Figure 6.1 depicts an abstract view of the environment. We further divide the section into two distinct parts; while the first one addresses distributed query processing and information retrieval, Section 6.2.2 focuses on the application of mobile agents in transaction management.

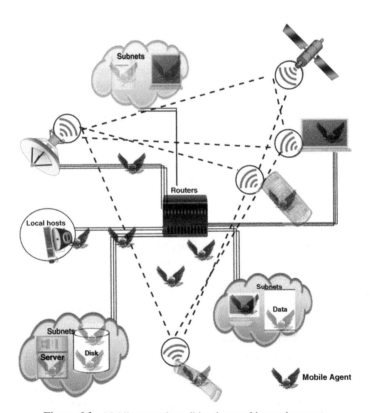

Figure 6.1   Mobile agents in traditional networking environment.

6.2.1 Distributed Query Processing and Information Retrieval

In general, information retrieval systems are concerned with not only the issue of improving the performance of searches but also that of improving the precision and recall of the results. Requests of users may be one-time or periodic. The use of agents to address such issues stems from the fact that agents execute independently from the system that they are originated from and as such can satisfy the requirements of both short- and long-term queries. Agents can also easily abstract themselves from the system environment, whether centralized or totally distributed, as they attempt to locate the information of interest to their tasks. Most importantly, agents may be endowed with intelligence, allowing them to analyze the information collected and dynamically decide the relevancy of the data not only based on the specified request but also on historical data collected or observed. The following discussion will showcase the use of agents in information retrieval starting with the Corporate Memory Management through Agents (CoMMA) project [1] followed by four other agent-based information retrieval systems.

Researchers from the CoMMA aim at supporting corporate memory management through the use of mobile agents. Distribution of information allows organizations to have project-oriented management; however, the information itself now needs to be managed for efficient use by the members of the organization. Managing the information of corporations requires that three distributed and heterogeneous components be taken into consideration. The first such component is the information itself and how it is stored and retrieved. The users that will interact with the information are considered as the constituents of the second component along with their contexts and preferences. The tasks that are to be performed in the system represent the third component and can be thought of as the interface between the first two components. The architecture of CoMMA is modeled based on roles and relationships yielding four communities: ontology and model, interconnection, annotation, and user. The ontology and model society manages the ontology by providing downloads, updates, and querying mechanisms to agents with which they interact. Agents, from the interconnection society, act as matchmakers between other agents in the system. In the annotation society, agents are concerned with the search and retrieval of information requested by users. Within the user's society, agents focus on adapting to users preferences and need for assistance. The internal organization of agents within societies varies from a peer-to-peer configuration to replicated or hierarchical. The interaction between agents is governed by the role of the agents in their respective societies as well as the responsibilities attached to that particular role. The CoMMA prototype was tested and showed that an effective content specialization of the annotation archives results in reduced number of messages involved in solving a query.

Agent-based community-oriented routing network (ACORN) is a matchmaking system designed to realize long-term queries/goals of users [2]. The system's ultimate goal is to achieve a state where agents can be routed to their destination (or relevant information) based on observed behavior of the community consisting of a Web of users. As can be derived from its name, ACORN is a multi-agent system in which a piece of information (piece of a document, music, video, etc.) is represented as an agent containing its globally unique identifier along with the address of a server in the case of thin agents. Fat agents contain the information found in thin agents augmented by all other knowledge of interest to the agent. ACORN consists of a client and a user interface to the system (for the management of incoming and outgoing agents). ACORN contains a main server providing and managing the entry point from a network to a particular site. The system also consists of an agent core containing all information needed for agents to work toward their goals. Furthermore, there exist a directory server and an anonymity server. The former allows agents to be tracked to facilitate real-time communication between users and their agents and the latter makes possible the generation of anonymous agents to preserve users' privacy. As an effort to improve performance of the system, a *k-means* algorithm is used to dynamically cluster agents based on their interests into cafés yielding a minimized "mingling" of agents. ACORN results in reduced network traffic and enhanced privacy of users as per the evaluation of the authors obtained from conducted simulations.

MAMDAS stands for mobile agent-based mobile data access systems [3]. Its design aims to alleviate two major difficulties in large-scale mobile data access systems: heterogeneity and mobility of data sources and/or users. Experimental results have shown that under the same physical configuration, the mobile agent-based computation mode can achieve better performance and robustness than the traditional client/server-based model. In addition, the authors also pointed out that from a software engineering point of view, the use of mobile agents can significantly improve modularity and reusability and simplify the management of large complex systems.

The use of agents to search in a distributed environment is put forth as an attempt to increase the performance of information retrieval and forego a centralized approach [4]. The proposed work deviates from the assumption that there exists a central mediator with knowledge of all resources in the system. The agents in the proposed system are made up of five components. The first one is referred to as a collection, representing the set of information to be shared with other agents. Furthermore, each agent contains:

- a search engine to locally determine whether it can satisfy user queries,
- a control center accepting user queries and conducting searches, and
- an "agent-view" structure maintaining knowledge about the network topology formed by neighboring agents.

The fifth component, collection descriptor, is the signature associated with the collection of the agent. The latter is also used in a distributed manner, thereby allowing agents to acquire knowledge about the information available in the network. The system relies heavily on communication between agents to locate, aggregate, and rank the information requested by users. Through distribution of the search schemes and optimization of the topology based on the search algorithm, the authors reported a significant improvement in the performance of information retrieval.

The implicit project put forth yet another proposal that makes use of mobile agents in information retrieval, specifically in Web searches, as an attempt to improve the relevancy of search results through data mining techniques [5]. Agents in the system exploit observed behaviors from other users conducting similar searches. An agent in the implicit system accomplishes its goals by storing the actions of a user in a database to be analyzed for pattern extractions; the observed patterns can then be used to allow the agent to make suggestions to the user or to other agents. The proposed system claims improved performance, in terms of relevancy of results of information retrieval through preliminary experiments.

As evidenced in the work in information retrieval thus far discussed, mobile agents have helped in addressing the major issues present in the area. Agents, by their mobile, independent, and intelligent nature, have provided a medium to improve information retrieval through their ability to locate information regardless of distribution and/or heterogeneity and to rank results based on various criteria. Agents also play an important role in the system, as it can support the disconnection of users, thereby allowing the execution of long-term queries. We have thus far looked at information retrieval from a global approach. A deeper dive into information retrieval systems reveals the dependency of such systems on query processing. The remainder of this section will thus showcase the impact of the agent paradigm on query processing.

Query processing in distributed information system may require data from multiple data sources. Conventionally, a query processor is used to decompose high-level user query into a set of subqueries, one for each of the involved data sources, and determine an optimal execution plan based on system statistics. As such, query processing faces several challenges. The subqueries executing on remote data sources may generate a substantial volume of intermediate data, leading to excessive network traffic. The heterogeneity of data sources requires the query processor to be aware of the local database schema at different data sources, which leads to more complexity. The query processor determines the query execution plan based on the system statistics; such statistics may be obsolete, leading to a suboptimal execution plan. In a dynamic system, the data or resource at the remote locations may become unavailable, which may lead to the failure of the subquery, even the global query. Moreover, due to local autonomy, local data source may deny the execution of generated subqueries. The potential massive volume of data, its heterogeneity, distribution, and other constraints lead to the need for a reliable, robust, and efficient retrieval technique that enables to access data from distributed heterogeneous sources while maintaining local autonomy. Mobile agents are featured with mobility, autonomy, and ability to process data at remote sites, which makes them a feasible and efficient approach to distributed query processing, as the following discussion will demonstrate.

6.2.1.1 Wired Networks

Agent-based Complex QUerying and Information Retrieval Engine (ACQUIRE) is an agent-based query processing system designed for large, heterogeneous, and distributed data sources in wired network [6]. The central ACQUIRE server has a local database recording site description and domain models on different data sources.

ACQUIRE presents users with the appearance of a single, unified, homogeneous data source despite the heterogeneity of the environment. ACQUIRE directs and controls all mobile agent based generation, plans, and optimizations. A mobile agent is spawned for each subquery generated by the host, and each agent is responsible for retrieving answers from appropriate data sources. The system has three modules: query planning module, query optimization module, and query execution module. ACQUIRE processes a query as follows: Upon receiving a high-level user query, the query-planning module refers to the site and domain database in order to decompose the query into a set of subqueries. The query optimization module creates an optimized plan and orders these subqueries to maximize retrieval efficiency. The query execution module receives a list of subqueries from the query optimization module and generates a series of mobile agents to carry out these subqueries. Each agent is also assigned an itinerary of the database sites to be visited and the data retrieval and processing tasks to be executed at each site. Upon the return of mobile agents, the query execution module filters and merges the returned data into the final query result for the user. Through experimental results, the authors noted that ACQUIRE reduced intermediate data volume and data retrieval time significantly (by 30–70%) compared to standard distributed query processing.

Under local autonomy and distance constraints of data sources, retrieval of statistical information is difficult; yet, it is crucial to the optimization of the query execution plan. Mobile agents have been employed to optimize join operation by implementing a join operator as an agent capable of reacting to estimation errors and new execution status [7]. This approach dynamically adapts to current statistics, resource, and data availability, thereby reducing the response time and leading to effective migration decision. The agent carries the initial execution plan and is first dispatched according to the initial plan. The agent, then, collects the current statistics of the relations involved in the query, checks the estimation errors, and determines whether to follow the original execution plan or adapt to a new one. Their experimental results have shown that using mobile agents leads to significant, up to 60%, reduction in response time compared to implementations without dynamic adjustment to system statistics.

6.2.1.2 Mobile Data Access System

Mobile Data Access System (MDAS) is an extension of distributed information system, where users are mobile with scarce resources in a wireless environment. While providing more flexibility, user's mobility introduces new challenges to query processing beyond what we have discussed to this point. MDAS has several new characteristics; starting with the fact that mobile users connect to the system via intermittent wireless connection which is relatively unreliable and has low bandwidth. Moreover, users may change their locations, which may lead to temporary disconnection or handoff (changing the access points). Finally, portable devices have less computational power, little storage capacity, and limited power supply. MDAS also introduces the concept of location dependent query (LDQ—a query whose result depends on the querying location). These challenges make query processing in mobile data access system more difficult than in distributed system within a fixed network

Mobile agent's autonomy minimizes the user's interaction in query processing, hence users can power off their devices once the query has been submitted. Agent-based approaches efficiently manage the limited resources on portable devices and transfer the computation to the network. The following proposals discussed herein will showcase the contribution of agents to MDAS.

Autonomous ageNT bAsed aRChitecture for cusTomized mobIle Computing Assistance (ANTARCTICA) is a query processing system designed for wireless networks dotted with infrastructure support, such as wireless cellular networks and wireless local area networks (WLANs) [8]. The work demonstrated the efficiency of using mobile agents to track both mobile users and data objects. It assumes that the base station (BSs) store the information for all data objects and mobile users currently in its coverage area. For a simple LDQ, the query processor in ANTARCTICA accesses all BSs in the specified searching area and returns the requested data objects. The LDQ processing procedure is as follows:

1. The mobile user sends a MonitorTracker agent to the current BS covering the querying location. The MonitorTracker agent migrates and follows the user when the user moves to another BS coverage area

2. The MonitorTracker sends a Tracker agent, carrying a subquery to the BS covering each of the referenced objects in the query.

3. Each Tracker agent finds the BSs whose coverage area intersect the searching area and creates one Updater agent for each of those BSs.

4. Each Updater agent executes its subquery against the database of the BS where it resides and sends the collected data to the MonitorTracker, which will combine them into the final answer for the user.

ANTARCTICA can also process continuous queries, where the request needs to be evaluated continuously, using the above steps. Instead of repeatedly issuing the same query, the system monitors the mobile users and referenced data objects. Re-execution of an agent is triggered when the user or the objects of interest move. A Tracker agent is used to update the reference location periodically (at a tracking frequency) and creates Update agents for new BSs. Updater agents execute subqueries with a certain frequency (named refreshment frequency) to update query results when data objects move.

The class of wireless network not supported by any infrastructure is typically referred to as mobile ad hoc network (MANET). It is normally built on a peer-to-peer (P2P) topology, where users act as clients, message routers, as well as service providers. The design of a system to support simple and efficient query processing in such an environment has also been put forth [9]. The referenced system consists of an active database component and a mobile agent structure based on event–condition–action (ECA) rules. Each mobile query agent is defined using a number of ECA rules and a number of datasets. The ECA rules represent the logic of the agent and that of the datasets, which may contain initial parameters as well as agent states along with collected results. The agents migrate through the network, find the requested data from peer databases, and transmit back to the client. In case of a continuous query, more complicated rules can be associated with the mobile query agents to determine when it should stop monitoring the database on one peer and migrate to a new one. The authors demonstrated, through analytical study, that the employment of active database technologies could lead to efficient, simple, and scalable query processing in P2P topology. The work presented does not, however, address the user's handoff or queries with location constraints.

Peer-to-peer Decentralized Information Ecosystem Technologies (P2P-DIET) is an agent-based resource sharing system intended to support continuous queries in super-peer networks [10]. A super-peer network consists of two types of nodes: client nodes and super-peer nodes. All super-peer nodes are equal and have the same responsibilities, thus the super-peer subnetwork is a pure P2P network. Each super-peer node serves as an access point (AP) for a fraction of the clients and keeps indices on the resources of those clients. Clients can run on user computers, and the resources (files or sharing application) are kept at client nodes. P2P-DIET provides support for ad hoc queries by allowing a client to post a query to its access point, the initial AP. The initial AP, then, broadcasts the query to all super-peer nodes and produces the answer using data found in the network. Finally, the initial AP passes the answers to the client. If the client is disconnected, the generated result will be stored in the network to be retrieved by the user upon reconnection. Clients may also subscribe to the AP (super-peer node) with a continuous query expressing their information needs. Super-peer nodes then forward posted queries to other super-peer nodes. Whenever a resource is published P2P-DIET makes sure that all clients, whose continuous queries match this resource's metadata, are notified. P2P-DIET has been implemented and provides support for continuous queries, thereby attaining the system's goal.

Table 6.1 summarizes the benefits of using mobile agent based query processing schemes for fixed network and MDAS. For each environment, the table lists the challenges that are solved by the mobile agents and the representative schemes.

6.2.2 Transaction Management

In the context of database, transaction is the database management system's abstract view of a user program: a sequence of reads and writes [11]. In general, transaction management is concerned with the scheduling of transactions and the maintenance of the ACID properties, which are correctness criteria for transaction execution. ACID properties include the following:

- *atomicity:* The transaction must be entirely completed or aborted.
- *consistency:* Transaction must be coherent and consistent with predefined rules.
- *isolation:* There must not be interference between transactions; intermediate result of one transaction must not be seen by another transaction being executed at the same time.
- *durability:* Transaction execution must be reliable; committed transactions must not be lost.

Table 6.1   Mobile Agent Based Query Processing Systems

Systems	Execution environment	Major challenges	Features of agent solutions
ACQUIRE	Querying heterogeneous and distributed information system in wired network	Data heterogeneity, low bandwidth, intermittent connectivity	Autonomous agents bring computation close to data. Alleviate dependency on bandwidth and connectivity.
MAMDAS	Mobile user access hierarchical wired network, heterogeneous data sources	Data heterogeneity, low bandwidth, intermittent connectivity	Autonomous agents bring computation close to data. Alleviate the dependency on bandwidth and connectivity. Maximized parallelism when searching through agents' cloning capability.
Adaptive optimization	Optimized execution plans in wired network	Low bandwidth, intermittent connectivity, suboptimization of execution plan	Dynamically adjustable execution plan based on current environment.
ANTARCTIC	Mobile users query heterogeneous and distributed system with wired network	Limited resources on mobile devices, intermittent connectivity, users' mobility, location-dependent queries	Support for continuous query with location constraints. Tracking of both mobile users and data objects.
P2P-ECA	Users querying peer-to-peer systems	Limited resources on mobile devices, intermittent connectivity, dynamic network topology	Improved scalability through introduction of active database concept. Support for continuous queries.
P2P-DIET	User querying peer-to-peer systems where certain nodes serve as access points for others	Limited resources on mobile devices, intermittent connectivity, users' mobility, dynamic network topology	Support for continuous queries. Ability to notify reconnected users to retrieve query results from the network.

Since correctness and order of execution of tasks partly depend on how transactions are managed, efficiency of transaction processing greatly impacts system performance. This mission-critical task becomes more difficult in distributed environment. In general, a transaction submitted to the system is split into subtransactions to be executed at different sites; the distribution of tasks brings mobile agents into the picture. This section introduces the concept of transactional agents along with their uses in various database application domains.

6.2.2.1 Transactional Agents

Transactional agents involve multiple autonomous agents collectively performing a transaction, either in cooperation with each other or based on an execution plan generated by a dedicated planning agent [12, 13]. Global transactions are decomposed into subtransactions for each mobile agent under the following constraints [13, 14]:

- Commit or abort should maintain the semantics of transactions across multiple agents.
- Transactions should be executed exactly once regardless of node or network failure.
- Intratransaction parallelism should be supported through synchronization of mobile agents of the same transaction.
- Recovery from failure through preservation of global state of transactions across local sites at which agents execute should be supported.

The literature is abundant with many methodologies to split tasks among agents and to model the way the agents collaborate to accomplish the task while maintaining correctness and reliability of the transaction. The notion of exactly-once execution protocol as well as partial rollback mechanism for mobile agents have been put forth in various proposals [13, 15, 16]. To facilitate interactions between agents to carry out a transaction, usage of commitment rules relying on the concepts of committer, committee, and witness have also been explored [17]. An agent is obligated to keep the commitment it made, otherwise, appropriate cancellation procedures must be performed in the presence of the same participating parties.

Transaction processing is also concerned with the performance, fault-tolerance, and resource utilization of the system. As such, FAult-TOlerant Mobile Agent System (FATOMAS) [18] and nonblocking execution models [19] have been introduced to address these issues. As per their approach, mobile agents travel the network and execute on a sequence of machines; places at which agents execute can be viewed as their logical execution environment. The conventional "place-dependent" execution models are blocking; to preserve atomicity failure of one component prevents the agent from executing on all sites in the system. As a result, locked resources cannot be freed to other transactional agents. Failure of a mobile agent is hard to detect and distinguish from delay due to slow processing. Launching another agent under unreliable failure detection can violate the exactly-once property of transactions, and thus, has motivated the introduction of fault-tolerance protocol that travels with the agent [18]. The proposal exploits agent replication and the notion of a sequence of agreement problems by forwarding replicas of the agent to a set of places in the sequence upon termination of the current stage. If one place fails, another place can take over the execution provided that reliable broadcasting is available. As noted in Ref. [18], the nonblocking agent-based transaction model offers better resource utilization and fault-tolerance at the expense of some overhead due to the communication cost

6.2.2.2 Database Management

One of the most critical issues in database administration is to ensure and enforce integrity as well as consistency of data across multiple databases. Global integrity constraint checking can be achieved efficiently using mobile agents [20]. In the proposed system, global constraints are stored in a global metadatabase, which is constructed from the database description of remote database objects. The constraint checker module accepts insert/update/delete operations from users and checks them against global metadatabase for constraints violation. When a global transaction is submitted at a data source, the constraint checker module at the site sends a mobile agent to global metadatabase to extract all global constraints being affected from the global metadatabase. Next, the mobile agent sends the constraints list back to its origin where subconstraints are generated along with order and plan. Lastly, spawned mobile agents are sent out to all related data sources to check whether any of the subconstraints are violated. After all the checks are conducted, the results are sent back to the original site to check whether the global constraint is violated. Analytical study was done to showcase the ability of the proposed work in offering fast constraint checking [20]. Obviously, the procedure introduced should be applied only on multidatabases in fixed network. In mobile multidatabases, the agents spawned out may not be able to return because of disconnection. As a result, constraint checking in mobile multidatabases is a challenge, which should be explored in the future.

6.2.2.3 Distributed Objects

Mobile agents can be used in multiple object servers to resolve the issue of locking the objects involved while preserving the ACID (atomicity, consistency, isolation, and durability) properties [21]. An object server stores objects consisting of data and methods to manipulate the data. For each transaction, a mobile agent travels from its base computer to locally manipulate objects across multiple object servers. Agents leave a surrogate agent to lock the object in the local object/data base and to locally commit the transaction based on the two-phase commit (2PC) protocol. In case of conflicts between transactions, that is, a mobile agent wants to use the object being locked by a surrogate agent; the former can wait, negotiate, or escape to use a different object server. Compared with the client–server model, the transactional agents take a shorter time to manipulate objects in database servers [21]. When deriving 100 K records, the transactional agent approach yielded 20–50% performance improvement over two-tier client–server approach; and 10–40% better than three-tier client–server architecture. However, the time of loading the agent class is significant; therefore, it was recommended that agent classes be loaded in advance.

Distributed transaction processing may require preknowledge of all resources, as is typically the case for transactions based on object transaction service (OTS), a middleware for building distributed transaction applications. Such requirement renders transaction processing inflexible for many application scenarios, including cases where the transactions are long-lived. X-TRA [14] has been introduced to address such limitations through the use of mobile agents. The proposed transaction model provides support for the following:

- transactions in WAN,
- applications involving rapidly changing resources,
- mobile applications, and
- coordination of various types of resources.

X-TRA also includes components to support control and coordination of agents, migration management, communication, trust service, among others. The model employs two phases of execution, namely preparation and execution. In the preparation phase, agents involved in the transaction inspect the resources in a nontransactional manner, that is bypassing OTS and without concern for the ACID properties because no actual execution occurs. During the execution phase, agents actually perform the transaction using OTS.

The objects participating in the transaction can be both OTS-aware that is, providing prepare-to-commit interface, which enables them to handle rollback. Non-OTS-aware objects are also supported, where the objects do not have the prepare-to-commit interface and hence, require support from agents to compensate/undo partial effect of the transaction. X-TRA supports both models and the interested reader is referred to Ref. [14] for further details.

The proposed framework eliminates the need of prior knowledge of the resources and, thus, allows flexible resource management. The placement of agents at the resource before starting the execution also decreases the probability of failure in long-running multitransactions. Based on the authors' analysis, X-TRA is shown to guarantee exactly-once semantics of transactions.

6.2.2.4 *Web-Based Distributed Database*

To avoid the overhead associated with downloading and initiating Java Database Connectivity (JDBC) driver in Web-based distributed database, the DBMS-Aglet framework [22] has been introduced as an attempt to improve the performance of transactions in the environment considered. The proposed work uses mobile agent between the client and the database server for database connectivity, transaction processing, and communication. Within DBMS-Aglet framework, the mobile agent is deployed and routed through the Web server to initiate JDBC driver at the database server, connect to the database, perform database request, and convey the result back to the client. By moving the transaction to use the execution environment at the server, the performance gain is achieved through elimination of the need to download and setup the execution environment at the client. Moreover, the mobile agent can be scheduled to visit many database servers in a single trip, which can be extended to support multidatabase systems. As a refinement, a service-specific agent can be surrogated to park at the database server during the entire application session, maintain JDBC connection and interface with the database. As such, for every database request within the session, a messenger agent travels between the client and the database server to issue the request through the parked agent and bring the result back to the client. With the refinement, overhead from reloading the JDBC driver and reconnecting to the database on every request is eliminated. Further performance improvement is obtained when the messenger agent is replaced with message communication between Java applet at the client side and the parked agent at the sever side. The message can also include the instruction to direct the parked agent to different database servers. The evaluation shows a decrease in the mean response time of the first query in a session from 8.4 s with client–server scheme to 3.9 s with parked agent and message approach over 10 Mb/s connectivity; and from 249.6 to 13.2 s over 9600 b/s wireless link. This significant improvement for the first query can compensate for small performance drop for subsequent queries, from 0.4 to 0.7 s over 10 Mb/s connectivity and from 3.6 to 4.2 s over 9600 b/s wireless link. The agent-based approach is also proven to be drastically more stable.

6.2.2.5 Multidatabase Systems

Transaction management in multidatabase is more complicated than in traditional centralized or distributed databases as it involves access to multiple heterogeneous autonomous databases. MDBAS [23] is a prototype that exploits the possibility to efficiently perform distributed executions of mobile agents to transparently manage distributed database transactions. Similar to the previous approach [22], it uses JDBC API to access underlying database and takes advantage of being in the same host as the service. On the other hand, MDBAS is designed for multidatabase administration, and thus implements distributed execution of database procedures. The model aims to combine the advantage of easy transfer of data between different local databases from fully interconnected multidatabase architecture with flexible administration. To do so, MDBAS has two types of functional units, central unit and workplace unit. The central unit is the starting point for the global transaction, storing the global schema and code for all agents. Several workplace units are distributed to local database sites or sites that provide fast access to local databases. The workplace units mediate access to local databases using porter agent as a port to the database for mobile worker agents, which submit nonprocedural requests as well as runner agents, which request procedural executions. Under the proposed scheme, heterogeneity of the local databases is hidden behind the uniform workplace units. Furthermore, the system eliminates the need to pre-install the correct integration code to local databases, as

Table 6.2   Mobile Agent Applications in Transaction Management

Application domain	Mobile agent approach	Advantages
Database management	Generation of execution plan controlled by agents. Local constraints are checked to ensure global consistency.	Reduced communication
Distributed objects [14, 21]	Lock objects using surrogate agents. Use of agents to encapsulate non-OTS-aware resources	Decreased probability of failure. Reduced execution time. Removal of resource preknowledge constraint.
Web-based distributed database access [22]	Agents encapsulate interaction with DB server	Reduced overhead. Improved response time. Stability
Multidatabase systems [22, 23]	Transparent execution of requests	Uniform view of heterogeneous data sources. Improved execution time

mobile agents can carry the code to such sites. The continuing paper [24] highlights the profits of mobile procedure execution, including reduced amount of transmitted data and ability to prefetch data at the most beneficial site. The evaluation depicted that when ratio of manipulation statements on remote data becomes higher, the procedure migration with MDBAS gives more advantage over commercial distributed database (Sybase/Oracle). A summary of the applications of mobile agents in transaction management is provided in Table 6.2.

6.3 SYSTEM ADMINISTRATION

As computing systems become more complex, dynamic, and heterogeneous, the task of managing such systems has grown to be nearly impossible for humans to handle. The complexity and dynamicity of the environment requires constant monitoring to detect any fluctuation resulting in an unstable/undesirable state. Moreover, the heterogeneity and distribution of computer systems requires that system administrators be able to independently manage the systems in an intelligent fashion. Mobile agents are endowed with the ability to act independently or cooperatively based on the task at hand; moreover, dotted with intelligence, they can make decisions based on predefined parameters and react to changes. The use of such paradigm to administer complex systems presented itself as the ideal approach not only due to the suitability of the paradigm in any distributed environment but also due to the fact that it addresses the core issues in system administration through its intelligence and independence. The following discussion will highlight the contribution of the paradigm particularly pertaining to system administration.

6.3.1   Resource Management

The use of mobile agents has surfaced as a promising approach to handle modern computer network and telecommunication, since it offers the following advantages:

- Distributed collaborative computing,
- Reduced network traffic and bandwidth requirements by decreasing the number of remote interactions,
- Autonomy and continued operation during disconnection,
- Ease of configuration and upgrading,
- Scalability and dynamicity.

Applications of mobile agents for administrative tasks ranging from database, network and system management, security protection to power system management are outlined in this section to provide the reader with a bird's eye view of the state of research in the area.

6.3.1.1   Network Management and Routing

Mobile agent application allows delegation using the notion of object mobility. An early examination of mobile agent-based solution for network management is performance management part of the MIAMI project [25]. Mobile agents under constrained mobility were used, that is, a static object sends an agent to be executed in a network node guides the mobile agent's movement. The performance management in MIAMI uses static performance negotiation agent (PNA) to receive requests and configuration information from user and negotiate with other parts of the system. PNA creates mobile performance monitor agent (PMA), which migrates to a predefined network element to monitor and summarize performance data received from performance element agent (PEA). PEAs reside in the network node and act as wrapper agent of other network monitoring technologies such as simple network management protocol (SNMP), telecommunication management network with Q3 interface (TMN Q3), and so on. The benefits from mobile agents are object migration and dynamic customization, which can provide a powerful mechanism for intelligence on demand [26].

Although mobile agents can autonomously travel in the network, its itinerary and the agent itself is usually designed to work efficiently in some specific networks and cannot be reused in different network. The aforementioned problem is addressed by introducing a framework separating network-specific itinerary part from task-specific behavioral logic part [27]. The framework builds on the so-called MobileSpaces a hierarchical mobile agent system in which several mobile agents can be contained within one agent during migration. Separation of network-specific and task-specific parts is accomplished with the use of two types of agents: task agents and navigator agents. Task agents perform management task at each network node they visit, although they may not have knowledge about the network. Navigator agents, hosted at agent pools, are familiar with particular subnetwork, as such they carry the task agents throughout the network. They also have routing capability and manage their own routing table The task agents rely on the navigator agents to locate and migrate to nodes of interest using event-based communication. The resulting system provides optimized itineraries; small and simple task agents independent of network types; access control of task agents as permitted by navigator agents; and multiple policies for task deployment defined at navigator agents.

The works thus far discussed in this section have not addressed the issues that are particular to a dynamic environment such as MANET. Mobile agents have been extensively used in MANET to efficiently manage resources. We will bring the reader's attention to some of the most interesting proposals in the area that have made use of mobile agent technology focusing on routing issues in MANET. Within the scope of MANET, ad-hoc on-demand distance vector (AODV) [28] is a routing algorithm that discovers the route between a source and a destination only when the source needs to send the data. As such, the actual data transmission is delayed until a route is discovered, which has led to numerous routing schemes that use mobile agents to maintain the topology of the network. RoyChoudhury et al. [29] have attempted to render nodes topologically aware through the use of mobile agents to periodically update their routing information. As agents move from nodes to node, they increment the node's counter, a recency token before leaving. The recency token is then used by other nodes to determine which routing information about a particular node as maintained by agents is more recent as they update their routing table. The proposed mechanism also attempts to determine the network topology through a predictive algorithm forecasting the lifespan of links, and is able to estimate, as claimed by the authors based on observed simulation results, the deviation of the network topology from that perceived by individual nodes.

Mobile agents typically choose the next node to migrate to randomly as they discover the network topology. The nodes may also have to wait on agents to update their routing table before they can start transmission of data to a destination whose known route may not be recent enough. To address such issues, Marwaha et al. [30] proposed a hybrid routing algorithm that uses "ants" to update the routing table of nodes; but also allows each node to initiate the AODV protocol in cases where the current route to a destination that they wish to communicate with is not fresh enough as opposed to buffering the data while waiting for updated routing information. The proposal has been simulated to prove its efficiency in reducing the latency for topology discovery as well as end-to-end transmission delay.

Determining the best path to route traffic based on the availability of node resources has also been proposed [31]. The proposal assumes the presence of static agents in every node monitoring available connections (local topology), as well as resources of the node. Furthermore, mobile agents traverse the whole network, collecting information from the static agents in order to collectively discover the global topology of the network and subsequently route traffic in an efficient manner taking into account the availability of resources in intermediate nodes. The work is presented as a hypothesis upon which further experiments will be conducted to determine its feasibility and efficiency.

6.3.1.2   Power Systems

Power system is also a type of distributed network environment in which Internet telecommunication, and mobile agent technologies can contribute to its maintenance, control and management. Mobile agents have already been employed in Japan to enhance remote operation and monitoring of system for protection and control of power system [32]. The framework studied the feasibility of automatic collection of information from multiple substations to improve system maintenance. It uses mobile agents moving between relay equipments and the system that collects and processes the required data. A setting agent travels in the network to set protection and control configuration in protection relays, so-called adaptive relays. The agent carries out the tasks according to a script and can clone itself for parallel execution. Alternatively, the adaptive setting may be used where the setting is adjusted based on power system changes. Intelligent analyzing agent collects and analyzes data using inference engine and knowledge base technology. Lastly, patrol agent examines condition of relay equipment for analysis. The proposed work has been implemented and the authors claim an increase in maintenance productivity and effectiveness.

As an attempt to increase system reliability, agents are used for fault-detection at substations through inspection of power/control equipments [33]. The agents travel through the intranet to visit control and protection devices with different characteristics, working environment, and connection pattern to extract data. When abnormal data are detected, the agent develops a plan to acquire more information and to spawn children to other related devices to collaboratively indicate the cause of the fault. The work reported in Ref. [33] also emphasizes the limitation of resources on embedded devices. It uses dynamic loading, that is, classes are downloaded from the server after migration to reduce memory and network resource consumptions. Moreover, the agent migrates to a high-resource server in the power station to perform expensive tasks such as planning. The proposed system was implemented and showed efficient use of network resources as well as reduced response time compared to conventional systems.

In addition to monitoring and maintenance tasks, agents can be used to control cascading failure in electric grids [34] as an attempt to eliminate network constraint violations before they trigger state transition, which may lead to cascading failure. Network of distributed autonomous agents is used to run distributed model predictive control (DMPC). DMPC decomposes the global control problem into subproblems, one for each agent. An agent is deployed to each bus in the network to gather local information and detect network violation. It also collects data from other agents to obtain the state of the entire grid. When violation occurs, the agents collaborate to adjust their local control variables to eliminate the violation. Table 6.3 presents a summary of the work herein discussed to manage resources in fixed network.

6.3.2   System Security

Network and system monitoring can be extended to intrusion detection system (IDS) [35]. In general, traditional distributed intrusion detection systems provide distributed

Table 6.3 Resource Management in Fixed Network

Application domain	Agent-based approach	Advantages
Network management and routing [24, 26, 29, 30]	Migration of agent to collect performance data. Task separation. Migration of agent to discover network topology.	Decreased communication. Dynamic system monitoring. Abstract view of networking environment. Efficient routing of network traffic.
Power system	Use of inference to adjust inspection task. Collection of data and configuration of devices by agents. Agents detect network constraint violations and prevent cascading failure.	Flexible configuration. Improved detection of faults. Reduced communication. Dynamic adjustment of control variables.

data acquisition with central processing and analysis. In contrast, mobile agent-based approach integrates several intrusion detection components into individual agent to be deployed to network nodes and perform autonomous tasks. This section explores several agent-based IDS that have appeared in the literature.

The monitoring of suspicious events such as users' intentional misuse, intrusion, and system inconsistency to defend against potential attacks yields a monitoring framework closely related to IDS [36]. Specifically, the framework attempts to address the need to dynamically introduce new monitoring procedures and correlation functions to monitor events. To adapt to the situation in which there is a suspicious attack, mobile agents can change their monitoring, aggregation, and information-processing policies. In addition, new agents can be easily deployed into the network to add new event types and functionalities. The proposed framework maintains separate prolog-based logic database for particular events and detection procedures. Agents in the system monitor nodes and their coordinates upon detection of specific suspicious events. Such events are stored in a database and may trigger the delegation of other agents, with dynamic itinerary to nodes for correlation and pattern detection. Security administrators are alerted once possible intrusion or abnormalities are detected. Moreover, routine consistency checking as well as system maintenance can be performed by the proposed system. The framework was prototyped to highlight its ability to support the introduction of monitoring procedures and correlation functions dynamically.

The IDS structure proposed in Ref. [37] tackles the issue of detecting complicated attacks while reducing communication load, the existing bottleneck that occurs as a side-effect to central processing. The proposed work introduces a hierarchical architecture that divides monitor agents into three levels:

- node detectors,
- subnet monitors, and
- network monitors

Each agent has its own knowledgebase and analyzer engine. They collaborate both horizontally and vertically to detect collaborative attacks. To ensure robustness, if higher-level monitor finds abnormality in a lower-level monitor, it can ask another detector/monitor at the same level to clone and migrate to the anomalous node. Alternatively, if the detector/monitor detects anomaly at its peer, it can clone and migrate to the node to recover its peer. Lastly, block extensible exchange protocol (BEEP) is used for secure communication between the IDS agents. The framework was prototyped to show the systems' ability to accomplish its design goals.

Distributed intrusion detection system using mobile agents (DIDMA) attempts to detect the origins of attacks [38]. DIDMA has static agents (SAs) to monitor key network nodes. When suspicious activities are detected, the SA sends the event identification and its address to victim host list (VHL) at mobile host dispatcher (MAD). The MAD launches attack-specific mobile agent (MA) to repeatedly visit each node listed in VHL for that type of attack one after another, aggregating and correlating the data from the host with the data it received from previous hosts. The mobile agent analyzes the data and generates alerts to warn agent upon detection of an attack. As the MA is attack-specific, the approach is highly modular and extensible. Having prototyped the system, the authors noted a reduction in network bandwidth and the flexibility of the system in supporting heterogeneous platforms.

Lightweight intelligent agents in which agents carry and perform their task with minimal code have also been developed as an attempt to improve performance of IDS [39]. The proposed hierarchical architecture includes platform-dependent system activity agents, system log routers, and network routers who read system and network activities and feed them to related data-cleaning agents. Each static data-cleaning agent obtains and renders information related to a specific event. Mediator controls the low-level mobile agents. They visit corresponding data-cleaning agents for data, monitor and classify data of the specific event, and return the classified data to the mediator. The mediator routes data to local database and runs data-mining algorithms to connect related events and generate cohesive view. The model also applies distributed knowledge network and data warehouse technologies. It has data fusion agent to combine data from low-level and data mining agents using machine learning to generate predictive rules. Due to the presence of the lightweight agents, the model offers improved performance while providing support for the addition of runtime, communication, and collaboration capabilities.

Host-based monitoring is also an integral part of IDS. The application of mobile agent in port scanning and file integrity checking has been explored in Ref. [40]. Port scanner agent is launched to visit and check availability of different ports on given machines, verify the entries in the service table, and detect illegal server running on particular ports. The agent can then send report back to a control server. The approach can detect opening Trojan ports and close unused ports. File integrity checker agent checks content or some of the key files and system scripts and compare it with a secure copy of system metric. Thus, unauthorized changes to the files and scripts, which might leave a backdoor, can be detected. The prototyped system has shown a minimal effect of the port scanner agent on memory and CPU usage.

6.4 MOBILE AGENTS AND SENSOR NETWORK

Sensor network can be thought of as the resulting network that emerges from the, possibly random, deployment of multiple sensing devices with limited resources, in a particular area, to perform a task through coordination and communication. The sensing devices are typically referred to as sensor nodes. The network may be composed of thousands of nodes with possibly varying computational power, generally communicating over wireless medium. The data collected by the nodes are relayed to a special node in the network referred to as a base station or sink. The base station is usually assumed to not suffer from the resource scarcity present in the other nodes and is not necessarily equipped with any sensing apparatus. Figure 6.2 presents a pictorial abstraction of mobile agents in the sensor network environment. Within the scope of sensor network, applications are generally concerned with the acquisition and aggregation of the data available for processing. Reconfigurability of the sensor nodes for multi tasking that is, to allow a particular network to handle more than one task, is a new emerging trend in this environment. Keeping in line with our discussion thus far, we feel compelled to explore the effect of mobile agent technology on the two main areas of research in sensor network. We will first focus on data aggregation

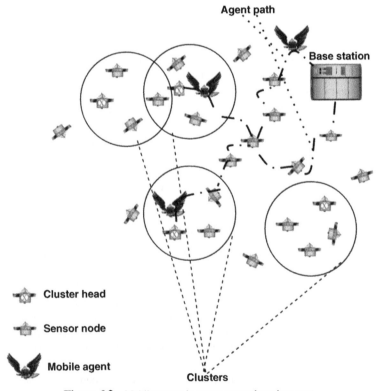

Figure 6.2   Mobile agents in sensor network environment.

while delaying the discussion of research pertaining to reconfigurable sensor network to Section 6.4.2.

6.4.1 Data Aggregation

Within the scope of wireless sensor networks (WSNs), data aggregation typically refers to the task of collecting and fusing the sensed data from the nodes. The manner in which the data from the network nodes are collected typically follows the traditional client–server-based architecture in which individual nodes send the collected data to the base station over the communication channel. Having multiple nodes sending data is inefficient, as the nodes may have sensed the same set of data, thereby introducing redundancy in the information received by the sink. To limit the occurrence of such scenarios, data aggregation focuses on techniques to reduce redundancies in the information traveling over the communication channels. We herein interchangeably refer to data aggregation as data fusion and proceed to discuss what has been thus far put forth to perform data aggregation in wireless sensor networks using mobile agents.

The use of mobile agents has been proposed as an attempt to increase the efficiency of data fusion while reducing the problems that arise due to network connectivity [41, 42]. As per the authors' approach, mobile agents can be sent to individual nodes and collect the information of interest, thereby reducing the amount of data that nodes need to relay to the sink. Moreover, mobile agents can cope with link failures as once they have migrated to a node, should the link fail, the agents can wait for the link to be re-established before submitting the results of the collected data of interest. The proposed work does not make any assumption as to the relative distance between nodes within the network. The model proposes a hierarchical structure in which a group of sensor nodes are clustered under a set of interconnected cluster heads processing elements (PEs) . Furthermore, it is assumed that any node is within one hop from its associated PE. In this model, sensor nodes collect the data and forward it to the designated cluster head where mobile agents aggregate the redundant data. Within the considered environment, the authors claimed a 90% improvement in the transfer time of data and 98% decrease in execution time with the use of mobile agents under idealistic conditions. Such claims were based on simulations and analytical studies; the interested reader is referred to Refs. [41, 42] for further details.

The work presented in Refs. [43, 44] builds on relaxing the assumptions that the network is composed of clusters with associated PEs, as it attempts to study the efficiency of the use of mobile agents in data aggregation. Instead, the environment in the sensor network is composed of sensors that are relatively close to each other, yielding a high amount of redundancy in the sensed data. In this model, sink queries nodes simultaneously and mobile agents are used to collect the results from the nodes of interest in a sequential fashion. The application of mobile agents is evident at three levels in the proposed work to improve the efficiency of data aggregation. At the node level, mobile agents can travel to and from nodes based on the needs of applications;

at the task level, mobile agents are proposed to aggregate data and reduce redundancy of sensed data from neighboring nodes. Lastly, at the combined task level, mobile agents can fuse small data packets destined for the sink into larger ones to avoid the communication overhead associated with multiple small packets, and, therefore, increase the lifetime of the network. The proposed work was shown to decrease execution time while increasing end-to-end latency through simulations.

Computing routes for mobile agents involved in data fusion in wireless sensor network is also an important problem [45]. If mobile agents visit the nodes in a nonoptimal fashion, it can have a significant impact on the lifespan of the network (or individual nodes) due to the communication and computation overheads associated with mobile agents. The accuracy (resolution) of the aggregated data increases as the number of nodes visited by an agent increases. However, not all nodes visited will have relevant data to perform the fusion; yet visiting such nodes will consume network bandwidth and limit node resources. It becomes imperative that mobile agents are able to visit a minimal set of nodes while providing the required resolution for data fusion, this issue is named mobile agent routing problem [45] and is concerned with minimizing the cost of data fusion on energy consumption and path loss. Mobile agent routing is shown to be an NP-complete problem [45], consequently, a genetic algorithm has been proposed to tackle the mobile agent routing problem by using a two-level genetic encoding. The first level represents a numerical encoding of the sensor IDs in the mobile agent's visitation order. The second sequence, in binary format, encodes the visit status of nodes in the same order as the first level. The two sequences are then mapped to determine potential path for the mobile agent; details of the algorithm are, however, beyond the scope of this chapter and interested reader is referred to Ref. [45]. The proposed algorithm has been simulated and shown to have a low overhead in an environment where nodes remain active for 1–2-h sessions.

6.4.2   Reconfigurable Sensor Network

Once deployed, sensor nodes have a static task to accomplish and cannot adapt to changes in the environment that may require new parameters to be taken into account to achieve the task at hand. A new set of sensor nodes have to be redeployed in an environment, even though the environment might already have a deployed set of nodes, should the need arise to conduct two different tasks concurrently or alternately. In order to allow sensors to support multiple tasks, researchers have focused their attention on the issue of reconfigurable sensor network. As the name suggests, reconfigurable sensor network would allow the network (or nodes) to be reprogrammed to react to changes in the environment or to satisfy updated user requirements. A discussion of the systems that have been thus far proposed allowing nodes to be reconfigured follows. Our focus will be on the approaches that have explored the use of mobile agents to accomplish their goals as this represents a very popular trend in attacking the issue.

UC Davis researchers have put forth a mobile agent framework built on top of the Mate virtual machine [46] to study the efficiency of self-directed propagation of agents in sensor network applications [47]. The framework allows agents to execute within an interpreter that attempts to prevent node crashing from corrupted agents. The interpreter implements the basic functionalities of agents, such as agent forwarding, so as to minimize the size of agent code that needs to be transferred from node to node. The interpreter supports agent migration through both uni-casting and broadcasting communication modes, the decision as to which communication mode to be used is left to the agent so as to increase the efficiency based on the needs of an application. Agents can also decide on whether or not to request acknowledgements during migration depending on their fault-tolerant requirements. Agent communication is achieved through "bread crumbs," inspired by ant colony systems (ACS), by writing its state on a node or by reading stored states that are preserved for the duration of the node's lifetime. The proposed agent framework is implemented on Mica2Dot motes having 4 KB of RAM and 128 kB of program flash. The extension to the Mate virtual machine allows agents to discover their neighbors and execute within an agent context, that is not propagated through the network. Agent programmers are given control to emphasize efficiency or reliability within an application. Through simulations, the authors claim agents are suitable to efficiently support sensor network applications in the presence of a large number of nodes [47], especially when only a portion of the network needs to be reconfigured.

Agilla [48] was introduced as an effort to address the lack of flexibility in sensor network due to statically installed software. Agilla allows each node to support multiple agents that may or may not be cooperating to accomplish a task. It allows for agents communication through tuple spaces (an ordered set of fields of types and values). Agilla supports four local atomic tuple space operations and four nonblocking remote tuple space operations. One assumption made by Agilla is that each node knows its geographical location, which is used as its address. Agents can clone themselves or move to another location carrying with them either their code and state or just their code. Agilla agents die at the completion of their task to allow for efficient memory usage. Agilla's architecture consists of three main layers: the mobile agents layer, the Agilla layer, and the TinyOS layer. The Agilla layer is further subdivided into five major components:

- an agent manager,
- a context manager,
- an instruction manager,
- a tuple space manager, and
- an Agilla Engine.

The agent manager handles memory allocation for agents as well as notifying the Agilla engine when an agent is ready to run. The context manager handles context information for a node such as neighbor list and the location of the node; it is responsible for the discovery of neighbors through the use of beacons. The instruction manager is used to handle dynamic memory allocation that is not supported by the TinyOS. Upon arrival at a node, agents specify the amount of instruction memory that they need;

then the memory allocation along with the sequence of instructions to be executed are handled by the instruction manager. The tuple space manager controls memory allocations for tuples. It implements the nonblocking tuple space operations and manages the reaction registry, which keeps track of agents and their tuples of interest. Once an inserted tuple matches the template of interest of one agent, the tuple space manager notifies the agent manager, which then executes the reaction code of the agent. The tuple space manager handles packaging and restoring an agent's reactions as it migrates through the system. The Agilla engine handles the scheduling of agents for execution in a round-robin fashion. It is also responsible for sending and receiving agents between nodes and can be considered to be a virtual kernel controlling the current execution of agents on a node. Agents can react to changes in tuples if they notify the Agilla engine that they are interested in a specific tuple template. Furthermore, the engine handles remote tuple operations through end-to-end communication with at the most two retransmissions. The proposed framework was implemented using MICA2 motes with 128 kb of instruction and 4 kb of data memory running TinyOS, to demonstrate its feasibility in a general-purpose computing grid infrastructure. The authors prototyped a fire tracking application atop of the framework to demonstrate the flexibility offered by Agilla and how the framework simplifies application development in sensor network.

The third and final system that we will present has been named ActorNet [49], which attempts to ease application development through provision of an abstract environment in sensor network. ActorNet provides support for asynchronous communication model, context-switching, multitasking agent coordination, and virtual memory. The agent system can be thought of as two entities: the agent language (actor language) and the platform design. The actor language is based on scheme [49] and provides users with the basic functionalities such as the ability to send and receive messages. Other agent primitives can be implemented using the language. The language allows agents to migrate with their states by considering the state of the agent to be a pair of continuation, which can be obtained by an operator call, innate to scheme, and represent the remaining portion of the program to be executed, as a value to be passed to the continuation. The ActorNet platform is a virtual machine that can support multiple actors (agents) per node. The platform provides a unified environment for actors by hiding the implementation details from the actors. The platform was designed for the Mica2 motes of Berkeley running the TinyOS operating system and has been implemented. The authors note the high level of abstraction for coordinating tasks provided by ActorNet with support for compact and maintainable code. The limitations of the system are also highlighted specifically pertaining to lack of fault-tolerance during communication.

The use of mobile agents is emerging as the dominant approach in addressing the issue of reconfigurable sensor network. Our assertion is evidenced by the use of the technology in the research conducted in the field within the last couple of years. Moreover, sensor networks are inherently distributed, with low connectivity and resources, thereby having special requirements that mobile agents can very suitably satisfy. A comparison of the platforms discussed herein is provided in Table 6.4 based on agent mobility and communication.

Table 6.4   Agent Platforms for Wireless Sensor Network

Agent platform	Mobility	Communication
UC Davis Framework	Weak	Bread crumbs
Agilla	Weak	Tuple space
ActorNet	Strong	Message passing

Weak mobility (migration of an agent code is supported)
Strong mobility (migration of the code and state of an agent is supported)

6.5 SWARMING INTELLIGENCE

Agents provide software designers and developers with a metaphorical way of structuring an application around autonomous, communicative components, and lead to the construction of software tools and infrastructure. In this sense, they offer a new and often more appropriate route to the development of complex computational systems, especially in open and dynamic environment. One interesting example is the natural coupling of software agents and swarm intelligence, which simply put, is the emergent collective intelligence of groups of simple autonomous agents.

The expression "swarm intelligence" was coined by Beni and Wang in 1989, in the context of cellular robotic systems [50]. It is an artificial intelligence technique that studies collective behavior in decentralized, self-organized environments. Swarm intelligence systems are often inspired by social animals in nature: a group of ants forages for food; a colony of termite builds incredibly complex nests; a flock of birds migrates from one location to another; a school of fish swims, forages, and flees together, and so on. In these biological systems, each individual is a simple agent and there is normally no centralized control dictating how each individual agent should behave. Surprisingly, the simple local interactions among agents and the interactions between an agent and its surrounding environment often lead to the emergence of complex and goal-oriented global behavior.

An abstract view of a swarm suggests that the N agents in the swarm are cooperating to achieve some purposeful behavior and achieve some goal. The apparent "collective intelligence" seems to emerge from large groups of relatively simple agents. The agents use simple local rules to govern their actions and via the interactions of the entire group, the swarm achieves its objectives. A type of "self-organization" emerges from the collection of actions of the group. Research in this field has generated many distributed, efficient, heuristic solutions to a variety of difficult problems such as quality of Service (QoS) management optimization in dynamic networks [51], distributed document clustering [52], and the location allocation problem in the geographical information system (GIS) arena [53]. In the rest of this section, we will use the aforementioned research to demonstrate how mobile agents, as a design metaphor, work seamlessly with swarm intelligence

The task of discovering an efficient dynamic resource reservation by optimal rerouting under unpredictable parameters such as traffic and topology changes, network size variations, and disconnection is known to be NP-complete Youssef et al. [51] outlined an effective integration of mobile agents with swarm intelligence in the implementation of an Internet QoS management optimization solution. In this heuristic solution, mobile agents, called explorer agents, are responsible for discovering routes. They use a probabilistic model to guide their movement (next hop) and use the intensity of pheromone to indicate the route quality they discovered. Experimental results have shown that because of its adaptive and distributed nature, the proposed approach converges quickly to an effective route solution with optimal resource allocation for large networks

Cui et al. applied a bio-inspired algorithm as a solution for unsupervised document clustering [52]. It is a distributed flocking algorithm in which each bird (a bird, or a mobile agent) represents a document, and the topic of the document determines the specie of a bird. The goal is to allow birds of the same specie to form a single flock and birds of different species to form different groups of birds. In other words, similar documents should be clustered into a single group and dissimilar documents should be separated. Initially, birds are randomly distributed in the problem space. Each bird follows four simple local rules:

- the alignment rule (to fly to the same direction),
- the separation rule (to avoid collision),
- the cohesion rule (to be close to the neighbors), and
- the similarity rule (to fly with the same specie).

As the birds fly in the virtual space and communicate with others they encounter, flocks start to form when birds of the same specie meet. Experimental results have shown that this algorithm generates high-quality clustering results.

Sharma et al. presented a swarm intelligence based distributed algorithm for the location problem in GIS [53]. The fundamental problem is the selection of a group of locations that best satisfy a set of constraints posed by the selection criteria. In case the problem space is very large, the application of the swarm intelligence based solution allows independent local decisions to be made in parallel. The proposed algorithm in Ref. [53] is inspired by the three steps termites follow to construct a nest:

- identify suitable sites,
- build columns up to a maximum height, and
- build arches connecting the columns.

The algorithm first divides the entire location space into equal sized cells. Mobile agents are then created and randomly distributed on the map. Each mobile agent can sense the eight cells that are surrounding its current cell. The movement of an agent is determined by a ranking function. The parameters of the ranking functions are the location selection criteria, and the mobile agent always tries to move to the cell that has the highest rank. Each agent has a limited lifetime. The algorithm ends when all agents reach the end of their life cycle. The authors concluded that swarm intelligence could be leveraged to provide highly scalable and robust solutions to GIS applications that require intensive computation.

The applications introduced in this section exemplify the use of mobile agents as a design metaphor, rather than a driving technology. This provides us with another angle to view the roles of mobile agents in mobile and distributed system design. As the research in swarm intelligence gains momentum, its indispensable implementation companion, mobile agents, will also win the recognition for being an effective software design paradigm.

6.6 AGENTS, PERVASIVE COMPUTING, AND EDUCATION: A CASE STUDY

Pervasive computing refers to a branch in computer science that focuses on bringing technological advances to the daily activities of users. In another words, pervasive computing explores the task of integrating technology into the user's environment, with the purpose of proactively performing jobs and lightening the user's workload. In pervasive computing, computers work in the background to make pervasively accessed information and services available to users based on some intelligent knowledge. Pervasive computing envisions an environment saturated with computing and communication capability gracefully integrated with human users. Access to information and services such as envisioned by the pervasive computing trend requires the use of intelligent entities as well as cooperation between the devices or systems involved. In a pervasive environment, a user should always be connected to the computing resources, whether it is mobile or stationary. The introduction of small wireless wearable devices, embedded systems, Bluetooth technology, and others, lie at the heart of the recent interests in pervasive computing and represent the driving force behind the advances in the field. The goal of pervasive computing requires any such applications to subsist in an environment that is heterogeneous and distributed. The information and services, which must be made available to users, may reside in any device and location; moreover, the information and services may need to be aggregated in order to be useful. Under such constraints, the use of mobile agents has emerged as an attractive solution to the development of pervasive computing applications. We, herein, present a case study highlighting the benefits offered by mobile agents in pervasive applications specifically pertaining to educational systems.

Researchers in the Computer Science & Engineering at the Pennsylvania State University, Washington State University, and the University of North Texas have joined forces in an effort to break away from the traditional teaching practice. As per their assessment, traditional lecture-based, discrete course structure format, many-to-one teaching mode has been static during the past 800 years. Recent advances in information technology allow one to reform this old tradition by developing a dynamic, continuous, adaptable, and proactive learning environment that supports one-to-one teaching practice.

6.6.1 Static Teaching Environment

Traditional lecture-based, many-to-one student-faculty ratio inherited from a socio-economy platform where higher education was not a necessity but a privilege is no longer effective in a society in which higher education is a necessity. This fact combined with the advances in technology had dramatically increased the student–faculty ratio and expanded the curriculums. Teaching environment is collapsing on itself as class participation suffers from a sharp drop; students typically have Internet access to the course materials, thereby widening the communication gap between faculty and students. The classroom is no longer an active medium to exchange ideas and solutions, courses become more and more voluminous, discrete, and unrelated, and higher education is becoming unaffordable.

Current advances in information technology can partially remedy some of these shortcomings. For example, it is a known fact that different people have different learning styles and the world is full of ear-learners and those who learn by physical practices. Teaching tools and animation techniques and remote access to the information sources can be used to:

- present the same information in many different ways and hence offer all three modes of learning,
- allow spatial and temporal uniformity in presenting the same material,
- allow self-pacing, privacy, and flexibility, and
- ensure efficient utilization of resources and lower cost to students and providers.

Nevertheless, recent practices of the advances in information technology have neither found a solution to the high student–faculty ratio nor have they provided a proactive classroom atmosphere.

6.6.2 Pervasive, Continuous Curriculum

We envision a system in which pervasive computing, mobile agent paradigm, and advances in anytime anywhere access to information are integrated to

- reduce the student faculty ratio,
- develop a proactive, robust, dynamic, and continuous courses and curricula,
- make higher education affordable, available, and accessible,
- reduce inter- and intracurriculum redundancy,
- create reusable curriculum and courses,
- practice proactive and interactive teaching methods, and finally
- develop navigational tools that allow higher accessibility and transparency to curriculum and course contents.

Three sets of entities represent a degree-granting program:

- Set of instructors (I).
- Set of students (S).
- Set of courses (C).

$i \in I$ has expertise in one or several subjects; $s \in S$ is intended to pursue a degree and is required to take courses from C in an orderly fashion to satisfy degree requirements and objectives; and $c \in C$ represents a course in the curriculum. Courses in C are interrelated and the structure of the curriculum determines the interrelationship among the courses. In this view, each set forms a community of software agents that communicate and negotiate with each other according to the defined tasks, that is, advising a student, scheduling courses, individualizing content of a course, and so on. Figure 6.3 portrays the overall system configuration.

The system relies on the pervasive information community organization (PICO) [54] as the medium supporting the interaction between the different actors of the system. PICO allows mobile agents to subsist in a dynamic mission-oriented environment as they accomplish tasks on behalf of users. The use of PICO does not address issues related to searching and aggregation of data that may be of interest to a user.

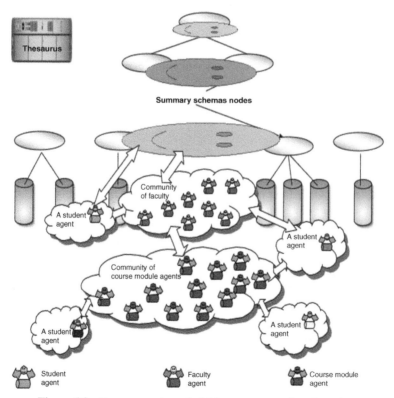

Figure 6.3   Course mentoring and advising system: overall configuration.

The summary schemas model (SSM) [55] is used as the underlying infrastructure to accomplish this task. Each course is subdivided into modules representing the main topics to be covered in the course. Each module is designed to work independently with the ability to cooperate with others; this modular course organization also promotes active rather than passive learning to increase participation of students and absorption of the material, while allowing the individual student to advance at his/her own pace. The division of courses into modules allows the system to view the curriculum at a finer granularity and hence easily minimize redundancy of topics and maximize the adaptability of the course contents as a student progresses toward his/her degree. One or more agents are associated with each module in the course; the modules are linked to other modules of the same course as well as to the ones with related content. When a student registers for a course, the system creates an agent on behalf of the student comprising information about the student's academic background, major as well as interests. The student's agent then negotiates with agents representing the instructors and the courses. Interaction between student agent and the course agents yields a customized set of modules for the student along with the associated assignments (projects, quizzes, etc.) and schedules (meeting times, learning timeline, etc.). The interaction between the student agent, course agents, and the instructor agents designates a "virtual" instructor that monitors and evaluates the student's progress in that course, that is, a virtual one-to-one student–faculty ratio.

The proposed reform to the educational system outlined herein is very promising as it intends to incorporate technology in the classroom at a finer granularity. Upon completion, it is expected that the system will effectively decrease the student–faculty ratio and redundancy in course contents while improving students' participation in lectures. The system will also provide students with a vast amount of information based on their interests and the module being studied through the use of SSM to locate, extract, and process data intelligently.

6.7   CONCLUSIONS AND FUTURE TRENDS

This chapter is intended to introduce the state-of-the-art research that has benefited from the use of mobile agents. As such, we have extensively covered the use of agents to help in distributed information retrieval while highlighting underlying issues in the area pertaining to query processing and transaction management. We have also discussed the contribution of mobile agents to help manage and administer complex systems, when such tasks have become too strenuous for humans to properly conduct on a continuous basis. Wireless sensor network has received a considerable amount of coverage from research communities and has also benefited from the paradigm in aggregating distributed data over the network as well as allowing for WSN nodes to be reconfigured supporting multiple dynamic tasks. As the saying goes, "two heads are better than one"; and such a concept directly applies within the scope of mobile agents. The cooperation and coordination between agent entities is the driving force behind the emergence of swarming intelligence and its wide acceptance as we have shown in this chapter.

At the risk of repeating ourselves, we will reiterate the suitability of the use of mobile agents to foster improvements in any highly distributed field, which may or may not require some form of intelligent processing. Dotted with mobility, independence, and intelligence, mobile agents have to this point revolutionized research in multiple areas and we foresee the continuation of such trends as was demonstrated through the case study. The paradigm is making its way from research onto the daily lives of users and as such will further expand the dependability of applications and research directions alike on its appeal.

ACKNOWLEDGMENTS

National Science Foundation (NSF) under the contract IIS-0324835 in part has supported this work.

Notice: This manuscript has been authored by UT-Battelle, LLC, under contract DE-AC05-00OR22725 with the US Department of Energy. The United States Government retains and the publisher, by accepting the article for publication, acknowledges that the United States Government retains a non-exclusive, paid-up, irrevocable, worldwide license to publish or reproduce the published form of this manuscript, or allow others to do so, for United States Government purposes.

REFERENCES

1. F. Gandon, L. Berthelot, and R. Dieng-Kuntz. A multi-agent platform for a corporate semantic Web. In *AAMAS '02*, 15–19 July 2002, pp. 1025–1032.
2. J. Carter, A. A. Ghorbani, and S. Marsh. Just-in-time information sharing architectures in multiagent systems. In *AAMAS'02*, 15–19 July 2002, Bologna, Italy.
3. Y. Jiao and A.R. Hurson. Application of mobile agents in mobile data access systems: a prototype. *J. Database Manag.*, 15(4), 1–24, 2004.
4. H. Zhang, W. B. Croft, B. Levine, and V. Lesser. A multi-agent approach for peer-to-peer-based information retrieval systems. In *AAMAS '04*, 19–23 July 2004, New York, New York, USA.
5. A. Birukov, E. Blanzieri, and P. Giorgini. Implicit: an agent-based recommendation system for web search. In *AAMAS '05*, 25–29 July 2005, Utrecht, Netherlands.
6. S. Das, K. Shuster, and C. Wu. ACQUIRE: Agent-based Complex QUery and Information Retrieval Engine. In *AAMAS*, 2002.
7. J. Arcangeli, A. Hameurlain, F. Migeon, and F. Morvan. Mobile agent based self-adaptive join for wide-area distributed query processing. *J. Database Manag.*, 15(4), 2004.
8. S. Ilarri, E. Mena, and A. Illarramendi. A system based on mobile agents for tracking objects in a location-dependent query processing environment. In *DEXA Workshop*, 2001.
9. V. Kantere and A. Tsois. Using ECA rules to implement mobile query agents for fast-evolving pure P2P database systems. In *Conference on Mobile Data Management*, 2005.
10. S. Idreos and M. Koubarakis. P2P-DIET: ad-hoc and continuous queries in peer-to-peer networks using mobile agents. *SETN*, 2004.
11. R. Ramakrishnan and J. Gehrke. *Database Management Systems*, 3rd edition. McGrawHill, 2003.
12. K. Nagi. Transactional agents: towards a robust multi-agent system. *Lecture Notes in Computer Science*, Springer-Verlag, No. 2249, 2001.
13. R. Sher, Y. Aridor, and O. Etzion. Mobile transactional agents. In *21st International Conference on Distributed Computing Systems*, 16–19 April 2001, pp. 73–80.

14. H. Vogler and A. Buchmann. Using multiple mobile agents for distributed transactions. In *Proceedings of 3rd IFCIS International Conference on Cooperative Information System*, August 1998, pp. 114–121.

15. K. Rothermel and M. Strasser. A fault-tolerant protocol for providing the exactly-once property of mobile agents. *Proceedings of the 17th IEEE Symposium on Reliable Distributes Systems*, October 1998, pp. 100–108.

16. K. Rothermel and M. Strasser. System mechanisms for partial rollback of mobile agent execution. In *Proceedings 20th International Conference on Distributed Computing Systems (ICDCS'00)*, April 2000, pp. 20–28.

17. H. Little and A. Esterline. Agent-based transaction processing. In *Proceedings of the IEEE Southeast Conference*, April 2000, pp. 64–67.

18. S. Pleisch and A. Schiper. FATOMAS: a fault-tolerant mobile agent system based on the agent-dependent approach. In *Proceedings of the 2001 International Conference on Dependable Systems and Networks*, July 2001, pp. 215–224.

19. S. Pleisch and A Schiper. Non-blocking transactional mobile agent execution, proceedings. In *22nd International Conference on Distributed Computing Systems, (ICDCS' 2002)*, July 2002, pp. 443–444.

20. P. Madiraju and R. Sunderraman. A mobile agent approach for global database constraint checking. In *Proceedings of the 2004 ACM Symposium on Applied Computing, March 2004*.

21. M. Shiraishi, T. Enokido, and M. Takizawa. Agent-based transactions on distributed object server. In *Proceedings of International Conference on Computer Network and Mobile Computing (ICCNMC'03)*, October 2003, pp. 20–23.

22. S. Papastavrou, G. Samaras, and E. Pitoura. Mobile agents for World Wide Web distributed database access. *IEEE Trans. Knowledge Data Eng.*, 12(5), 2000.

23. R. Vlach, J. Lana, J. Marek, and D. Navara. MDBAS: a prototype of a multidatabase management system based on mobile agents. In *SOFSEM 2000, Spring-Verlag, LNCS*, 1963, pp. 440–449.

24. R. Vlach. Mobile database procedures in MDBAS, database and expert systems applications, 2001. In *Proceedings of 12th International Workshop*, September 3–7 2001, pp. 559–563.

25. AC338. *MIAMI: Mobile Intelligent Agents for Managing the Information Infrastructure*. University College London (UCL) http://cordis.europa.eu/infowin/acts/rus/projects/ac338.htm. page: http://www.ee.ucl.ac.uk/~dgriffin/miami/.

26. C. Bohoris, G Pavlou, and H. Cruickshank. Using mobile agents for network performance management. In *Network Operations and Management Symposium (NOMS' 2000)*, April 2000, pp. 637–652.

27. I. Satoh. A framework for building reusable mobile agents for network management. In *Network Operations and Management Symposium (NOMS' 2002)*, 2002, pp. 51–64.

28. C. E. Perkins and E. M. Royer. Ad hoc on-demand distance vector routing. In *Proceedings of the 2nd IEEE Workshop on Mobile Computing Systems and Applications, New Orleans, LA*, February 1999, pp. 90–100.

29. R. RoyChoudhury, S. Bandyopadhyay, and K. Paul. A distributed mechanism for topology discovery in ad hoc wireless networks using mobile agents. In *Proceedings of First Annual Workshop on Mobile Ad Hoc Networking Computing, MobiHOC Mobile Ad Hoc Networking and Computing*, 11 August, 2000.

30. S. Marwaha, C. K. Tham, and D. Spinivasan. Mobile agents based routing protocol for mobile ad hoc networks. In *Symposium on Ad Hoc Wireless Network, National University of Singapore*, 2002.

31. N. Migas, W. J. Buchanan, and K. A. Mc Aartney. Mobile agents for routing, topology discovery, and automatic network reconfiguration in ad-hoc networks. *ECBS*, 2003.

32. T. Shono, K. Sekiguchi, T. Tanaka, and S. Katayama. A remote supervisory system for a power system protection and control unit applying mobile agent technology. In *Transmission and Distribution Conference and Exhibition 2002: Asia Pacific IEEE/PES*, Vol. 1, 2002, pp. 148–153.

33. K. Cho, Y Irie, A. Ohsuga, K. Sekiguchi, and S. Honiden. Application of the uPlangent intelligent mobile agent architecture for embedded systems to the inspection of power systems. In *Systems and Computers in Japan*, Vol. 36. Wiley-Interscience, 2005, pp. 60–70.

34. P. Hines, H. Liao, D Jia, and S. Talukdar. Autonomous agents and cooperation for the control of cascading failures in electric grids. In *Proceedings of the IEEE Networking, Sensing and Control*, March 2005, pp. 273–278.

35. M. Bishop. *Introduction to Computer Security*. Addison-Wesley, 2004.
36. A. Tripathi, T. Ahmed, S. Pathak, M. Carney, and P. Dokas. Paradigms for mobile agent based active monitoring of network systems. In *Network Operations and Management Symposium (NOMS' 2002)*, pp. 1–13.
37. S. Zhicai, J. Zhenzhou, and H. MingZeng. A novel distributed intrusion detection model based on mobile agent. In *Proceedings of 3rd International Conference on Information Security (InfoSecu04)*, November 2004, pp. 155–159.
38. P. Kannadiga and M. Zulkernine. DIDMA: a distributed intrusion detection system using mobile agents. In *Proceedings of the 6th International Conference on Software Engineering, Artificial Intelligence, Networking and Parallel/Distributed Computing and 1st ACIS International Workshop on Self-Assembling Wireless Networks (SNPD/SAWN'05)*, 2005.
39. G. Helmer, J. Wong, V. Honavar, L. Miller, and Y. Wang. Lightweight agents for intrusion detection. *J. Syst. Software*, 67:109–122, 2003.
40. S. Y. Foo and M. Arradondo. Mobile agents for computer intrusion detection. In *Proceedings of the 36th Southeastern Symposium on System Theory*, 2004, pp. 517–521.
41. H. Qi, S. S. Iyengar, and K. Chakrabarty. Multiresolution data integration using mobile agents in distributed sensor networks. *IEEE Trans. Syst. Man Cybernet C Appl. Rev.*, 31(3): 383–391, 2001.
42. H. Qi, Y. Xu, and X. Wang. Mobile-agent-based collaborative signal and information processing in sensor networks. In *Proceedings of the IEEE*, Vol. 91, No. 8, August 2003.
43. M. Chen, T. Kwon, Y. Yuan, and V. C. M. Leung. Mobile agent based wireless sensor networks. *J. Comput.*, 1(1), 2006.
44. M. Chen, T. Kwon, and Y. Choi. Data dissemination based on mobile agent in wireless sensor networks. In *Proceedings of the IEEE Conference on Local Computer Networks 30th Anniversary (LCN'05)*, 2005.
45. Q. Wu, S. V. Rao, J. Barhen, S. S. Iyengar, V. K. Vaishnavi, H. Qi, and K. Chakrabarty. On computing mobile agent routes for data fusion in distributed sensor networks. *IEEE Trans. Knowledge Data Eng.*, 16(6), 2004.
46. P. Levis, D. Gay, and D. Culler. Bridging the gap: programming sensor networks with application specific virtual machines. *UCB//CSD-04-1343*, August 2005.
47. L. Szumel, J. LeBrun, and J. D. Owens. Towards a mobile agent framework for sensor networks. In *Second IEEE Workshop on Embedded Networked Sensors, Sydney, Australia*, 2005, pp. 79–87.
48. C.-L. Fok, G.-C. Roman, and C. Lu. Rapid development and flexible deployment of adaptive wireless sensor network applications. In *Proceedings of the 24th International Conference on Distributed Computing Systems (ICDCS'05), Columbus, Ohio*, 2005 June 6–10, pp. 653–662.
49. Y. Kwon, S Sundresh, K. Mechitov, and G. Agha. ActorNet: an actor plaform for wireless sensor networks. In *Fifth International Joint Conference on Autonomous Agents and Multiagent Systems, AAMAS06*.
50. G. Beni and U. Wang. Swarm intelligence in cellular robotic systems. In *NATO Advanced Workshop on Robots and Biological Systems, Il Ciocco, Tuscany, Italy*, 1989.
51. S. M. Youssef, M. A. Ismail, and S. A. Bassiouny. Integrating mobile agents and swarm optimization for efficient QoS management in dynamic programmable networks. In *IEEE MELECON*, 2002.
52. X. Cui, J. Gao, and T.E. Potok. A flocking based algorithm for document clustering analysis. *J. Syst. Architect.* (in press).
53. A. Sharma, V. Vyas, and D. Deodhare. An algorithm for site selection in GIS based on swarm intelligence. In *2006 IEEE Congress on Evolutionary Computation*, 2006, pp. 1020–1027.
54. M. Kumar, B. A. Shirazi, S. K. Das, B. Y. Sung, D. Levine, and M. Singhal. PICO: a middleware framework for pervasive computing. *IEEE Pervasive Comput. Mobile Ubiquitous Syst.* 2(3): 72–79, 2003.
55. M. W. Bright, A. Hurson, and S. Pakzad. Automated resolution of semantic heterogeneity in multi-databases. *ACM Trans. Database Syst.*, 19(2): 212–253, 1994.

Part II

Location-Based Mobile Information Services

Chapter 7

KCLS: A Cluster-Based Location Service Protocol and Its Applications in Multihop Mobile Networks

Supeng Leng,[1] Yan Zhang,[2] and Liren Zhang[3]

[1] University of Electronic Science and Technology of China, China
[2] Simula Research Laboratory, Norway
[3] University of South Australia, Australia

7.1 Introduction   143
7.2 The KCLS Protocol   145
7.3 Performance Analysis   149
7.4 Location Service and Applications   157
7.5 Conclusion   160
References   161

7.1 INTRODUCTION

The integration and interoperability of mobile communication technologies, along with new broadband wireless innovations and intelligent user-oriented services will lead toward the next-generation mobile systems. The communication infrastructure of the future will be heterogeneous and multihop wireless networks. The examples of such platforms are ad hoc networks and wireless sensor networks. However, many

Mobile Intelligence. Edited by Laurence T. Yang, Agustinus Borgy Waluyo, Jianhua Ma, Ling Tan, and Bala Srinivasan

technical issues such as the design of distributed routing protocols, which are capable of handling the dynamic network environment through multihop mobile communications, are still open for research.

The location information of mobile nodes has recently been applied to improve the performance of routing protocols [1, 2, 7, 8, 12] in multihop wireless networks, in which mobile nodes obtain their own location information either by using the low-power low-cost global positing system (GPS) receivers or by measuring signal strengths and calculating relative coordinates [21]. A location service system may be able to locate objects worldwide, within a metropolitan area, throughout a campus, in a particular building, or within a single room. The main goal of location service in multihop wireless networks is to search the location information for the source node, destination node, and all the possible intermediate nodes. Accordingly, the location service protocol is required for providing the physical location or the logical affiliation of individual nodes.

The efficiency of location service protocols highly depends on the availability of timely and accurate location information. Since frequent topological changes are usually expected in multihop wireless networks, the distribution of up-to-date location information can easily saturate the network. On the other hand, stale arriving location updates caused by long latency can drive network routing into instability. A good technical review on existing location service approaches is presented in Ref. [1].

The location service protocols proposed in recent years can be divided into proactive location services and reactive location services [1]. Proactive location services can be further classified into location database systems and location dissemination systems. A typical location database system uses some specific nodes as location servers to maintain the location information for registered mobile nodes. For example, the distributed virtual backbone mobility management scheme [6], the grid location service (GLS) [10], and doubling circles scheme [12] belong to the location database system approach based on replicating information at multiple nodes, which act as repositories. Another kind of location database system approach is the home region scheme [15, 16], in which each mobile node is associated with a home region. A location database system is able to significantly reduce the communication overheads due to location updates, but its weakness is long latency due to location searching from the location servers. Moreover, the predefined rectangular or cycle region [10, 16] may not be able to match the ad hoc environment well, especially when network mobility is group based, such as military troops moving in battlefield [11].

In contrast, typical location dissemination system approaches include the distance routing effect algorithm for mobility (DREAM) [7], the DREAM location service (DLS) [9], the Simple location service (SLS), and the GPS/ant-like routing algorithm (GPSAL) [9]. In this kind of approaches, nodes periodically exchange their location information, so that each node must maintain a location map of the whole network. The advantage of such system is robust and low cost for location query, since each node is acting as a location server. However, the heavy communication overheads for location updates make such an approach hardly able to support large-scale multihop wireless networks.

On the other hand, using a reactive location service protocol [9, 13], the node location information is obtained either from intermediate nodes or from the desired nodes on request basis. Clearly, this approach has low cost of exchanging location information, but high cost for location searching. However, the network stability is usually weak since there is neither location map maintained in each node nor specific location servers available.

This chapter focuses on a hierarchical cluster-based location service protocol for mobile nodes in multihop wireless networks with good scalability. Different from some typical structures being used in fixed geographic mobile networks such as grid [10] or circle [12], the clustering structure is self-organized and adaptable to build a location service system. Generally, cluster-based location services can be managed by either single level or multilevel. In multilevel clustering, such as the hierarchical state routing (HSR) [20], maintenance of the hierarchical multilevel location map requires heavy communication overheads due to random change of multilevel topology. On the other hand, packets forwarding requires high bandwidth and topology stability on the high-level backbone. Hence, multilevel cluster is unsuitable to ad hoc environment. In contrast, the cluster head of single level clustering is simple, since it only tracks local topology changes due to node mobility. In this case, the size of a cluster can be enlarged by k-hop cluster, which certainly has better scalability than single-hop cluster. Inspired by the distance effect [3] and the virtual backbone quorum scheme [6], a novel k-hop cluster-based location service (KCLS) protocol is proposed in this chapter.

The rest of this chapter is organized as follows. Section 7.2 describes the technical details of the KCLS protocol. The performance evaluation using theoretical analysis and simulations is presented in Section 7.3. Section 7.4 presents some interesting applications based on the KCLS protocol. Finally, Section 7.5 concludes this chapter.

7.2 THE KCLS PROTOCOL

The KCLS protocol deploys a single-level k-hop clustering structure to provide location service. It is supposed that each node has a unique node ID. The ID of the cluster head is defined as the cluster ID. A k-hop cluster C_m is defined as a set of nodes under the same cluster head h_m, and any node in the cluster has a distance of equal to or less than k hops to the cluster head. As shown in Figure 7.1, each cluster consists of one cluster head, ordinary cluster members that are located inside of the cluster, and gateways that are located at the border to connect to neighboring clusters.

A multihop wireless network can be divided into many nonoverlapping clusters by either using a k-hop clustering scheme, such as max–min heuristic [5] and its variations like k-hop compound metric-based clustering (KCMBC) approach [19], or by partitioning the network into several logical groups. If the network is partitioned into several logical groups, each group corresponds to a particular user team with common characteristics and the same mobility pattern, such as tank battalion in the battle field, search team in a rescue operation, moving behavior of the same company, or students within the same class. A logical group can consist of one or several k-hop clusters.

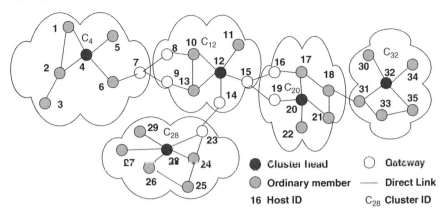

Figure 7.1 An example of k-hop clusters, $k = 2$.

7.2.1 Overview of Location Management

A two-level logical hierarchy including intracluster level and intercluster level is used in the KCLS protocol. The intracluster level is formed by cluster heads, which are acting as a distributed location servers to maintain an intercluster connection table. The cluster ID or the position of the cluster head represents the location information of all the mobile nodes in this cluster. On the other hand, the intracluster level includes the members located in the same cluster, which only hold the local topology information. When a node intends to obtain the location information of a destination, it simply sends a location enquiry packet to its cluster head (or the nearest cluster head). The cluster head, in turn, replies with the location information.

Intracluster location management is based on a link state routing approach [14] within a cluster. After clusters are created, each cluster member maintains a local connection (LC) table and an intracluster routing (IntraR) table. Based on the information provided by the cluster head, cluster members create the LC table, which contains the neighboring nodes of each cluster member, the neighboring clusters connecting with gateways, and the distance between each cluster member and the cluster head. For example, Table 7.1 shows the LC table of cluster members in C_{12} as shown in Figure 7.1. If any intracluster link is broken down or established, local link change

Table 7.1 The LC Table of the Nodes in Cluster C_{12} (Shown in Figure 7.1)

Node	8	9	10	11	12	13	14	15
ID of neighboring nodes	$10, C_4$	$13, C_4$	$8, 12, 13$	12	$10, 11, 13, 14, 15$	$9, 10, 12$	$12, C_{28}$	$12, C_{20}$
Distance to cluster head	2	2	1	1	0	1	1	1

Table 7.2 The IntraR Table of h_{10} (Shown in Figure 7.1)

Destination	8	9	11	12	13	14	15	C_4	C_{20}	C_{28}	C_{32}
ID of the nest hop	8	13	12	12	13	12	12	8	12	12	12

Table 7.3 The LS Table of Cluster Head h_4 (Shown in Figure 7.1)

Cluster ID	Timestamp	Cluster head position	Member ID	Neighboring clusters
C_4	T_{C_4}	Coordinates of h_4	1, 2, 3, 4, 5, 6, 7	C_{12}
C_{12}	$T_{C_{12}}$	Coordinates of h_{12}	8, 9, 10, 11, 12, 13, 14, 15	C_4, C_{20}, C_{28}
C_{20}	$T_{C_{20}}$	Coordinates of h_{20}	16, 17, 18, 19, 20, 21, 22	C_{12}, C_{32}
C_{28}	$T_{C_{28}}$	Coordinates of h_{28}	23, 24, 25, 26, 27, 28, 29	C_{12}
C_{32}	$T_{C_{32}}$	Coordinates of h_{32}	30, 31, 32, 33, 34, 35	C_{20}

(LLC) packet is correspondingly sent to all cluster members for updating their LC table with the link state changes. Based on the existing LC table, each cluster member builds an IntraR table using the shortest path Dijkstra's algorithm to indicate the next hop for each particular cluster member or neighboring cluster. Table 7.2 shows the IntraR table of node h_{10}. Since link state exchange and routing table construction are all performed within the cluster, the intracluster location management is efficient and robust.

In contrast, the management of intercluster location information is controlled by cluster heads. Besides LC table and IntraR table, each cluster head also maintains a location service (LS) table, which lists the membership of each cluster and intercluster connectivity in the network. In the LS table, each row represents the cluster state (CS) of one cluster, including the cluster ID, the sequence number of the location information, the cluster head's planar coordinates (if available), the lists of cluster members, and the neighboring clusters. Table 7.3 shows the LS table in node h_4, which is the cluster head of C_4 in Figure 7.1.

7.2.2 Intercluster Location Update

A cluster head responds to location queries on the basis of its LS table. In order to keep the accuracy of LS table in a dynamic multihop wireless network, the location update rate must be fast enough to reflect topology changes. However, if forwarding of location update packet is based on flooding algorithm, then it produces a large amount of overheads in such a limited bandwidth environment. Thus, the following effective location update mechanism is deployed in the proposed KCLS protocol.

When a new cluster is created, the cluster head broadcasts a cluster state (CS) packet to all the other cluster heads in the network. Based on the CS packets received, a cluster head establishes its LS table or updates the location information of the cluster, since each CS packet carries the location information of the corresponding cluster, including cluster ID, the cluster head's coordinates, cluster member list, and neighboring cluster list. Considering that the multihop wireless network may cover a large area, the sequence numbers of CS packets are used to eliminate duplicate copies and avoid delivery loops.

As the local coordinator, each cluster head maintains and monitors its cluster topology. A cluster head h_m generates and transmits at most one CS packet based on time interval τ_{CSm}, if one of the following events is detected during each τ_{CSm}.

- Any member leaves or joins the cluster.
- A new neighboring cluster is connected or an old neighboring cluster is disconnected.
- The cluster topology is changed due to new cluster creation, cluster head election, or cluster removal.
- Since the moment when the last CS packet was generated by h_m, the accumulated moving distance of h_m exceeds the predefined threshold D_{th}.

In order to reduce the overheads, the CS packet usually contains only the update information rather than the whole image of configurations. Furthermore, based on the cluster mobility pattern such as the average link available time, the time interval τ_{CSm} for cluster head h_m sending CS packet is usually set in the range from 0.1 to 10 s. The cluster head h_m can simply set the transmission interval with:

$$\tau_{CS}m = \sigma T_{Am}, \tag{7.1}$$

where σ is the scaling factor and T_{Am} is the average link available time of cluster C_m. According to the changes of LC table, the cluster head can obtain T_{Am} by calculating the average link available time during a fixed time interval T_{SLOT}.

Let CS_m^u denote a CS packet sent out by the cluster head h_m of C_m with a sequence number u. Then the cluster head h_i of C_i inspects this received packet CS_m^u by the rule as shown in Figure 7.2. Note that the cluster head h_m of C_m sends out one CS packet in

If CS_m^u has been received before
      packet CS_m^u is dropped;
else
      h_i modifies (or creates) the CS item for cluster C_m in its LS table;
      If C_m is not C_i's neighboring cluster or C_m is newly created
            CS_m^u is forwarded to C_i's neighboring clusters;
      else
            If u is an odd sequence number
                  CS_m^u is stored in the memory of h_i;
            else
                  h_i creates a new packet CS_m^u which merges the location information attached in the original CS_m^u and CS_m^{u-1}, and forwards the new CS_m^u to C_i's neighboring clusters.

Figure 7.2   The rule for a cluster head h_m processing a received CS packet CS_m^u.

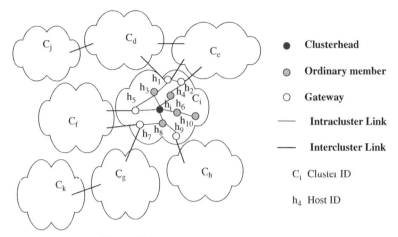

Figure 7.3   Gateway selection for multicast.

each interval τ_{CSm} and its neighboring clusters also receive the CS packet from C_m in each time interval τ_{CSm}, but the other clusters, which are not the neighboring clusters of C_m, receive at most one CS packet of h_m in every two-interval τ_{CSm}. This is because after receiving two consecutive CS packets from h_m, the neighboring clusters of C_m only forward one CS packet to their neighboring clusters for updating the location information of C_m. Because of the distance effect [3], the requirement for accurate location information between clusters decreases as their distance increases.

CS packets can be forwarded by using a multicast mechanism that requires the selected gateways to be able to connect with multiple neighboring clusters simultaneously. This is a well-known set-cover problem [17], which can be done using a greedy algorithm based on the LC table in the cluster head. That is, the node connecting to more neighboring clusters, excluding the neighboring cluster from which the CS packet is received, has the higher priority to be selected as gateway. This gateway selection process continues until all neighboring clusters are covered. Then, the CS packets are multicasted through the selected gateways to the neighboring clusters. For example, as shown in Figure 7.3, upon receiving a CS packet from C_h, cluster head h_i selects gateway h_1 and h_7 to multicast the CS packet to the neighboring clusters C_d, C_e, C_f, and C_g.

In the next section, the performance of KCLS is evaluated in a large-scale multihop wireless network.

7.3   PERFORMANCE ANALYSIS

A discrete-event simulator is conducted to evaluate the performance of the proposed KCLS protocol. The multihop wireless network used in the simulation has N homogeneous mobile nodes, which are randomly distributed in the area with $S = 30 \times 30$ square units. All mobile nodes have the same radio transmission range with a

Table 7.4   Simulation Parameters for Location Service Protocol

Items	Value
Total number of nodes, N	1000–4000
Maximum moving rate, V_m	0.2–1 unit/s
Time interval of node moving epoch, τ_e	1 s
Mean connection arrival rate, λ_{Call}	1/420 s
Scaling factor, σ	0.1–0.5
Time interval for measuring link available time, T_{SLOT}	10 s
Distance threshold, D_{th}	2 units

radius $r = 1$ unit. The connection arrival rate for each node is assumed to be Poisson distributed with a mean value of λ_{Call}. The mobility model deployed in the simulations is the random walk model [32, 33], in which the node moving velocity is changed only at the beginning of each node-moving epoch. The moving rate and direction of a node within an epoch are constant, following uniform distribution among predefined ranges $[0, V_m]$ and $[0, 2\pi]$, respectively, where V_m is the maximum rate of node moving. All clusters are created and maintained by the KCMBC approach [19]. The parameters used in the simulation are illustrated in Table 7.4.

The simulation is independently repeated 10 times. In each simulation run, 50,000 random connection requests are generated excluding a warm-up period of 1000 random connection requests, which is set up to ensure that the results are estimated on the basis of a steady simulation process.

7.3.1   Overheads in the Initial Stage

In the cluster initial stage, each cluster head broadcasts a CS packet to the other cluster heads when clusters are constructed. Let d_C denote the average number of neighboring clusters around one cluster, N_h denote the average number of nodes in a cluster, N_C denote the total number of clusters in the network, and H_C be the average number of hops for an arbitrary intracluster route. Upon receiving a CS packet, the cluster head multicasts the CS packet to other neighboring clusters. Since each CS packet is forwarded by every cluster for one time, then each CS packet is transmitted by nodes totally $d_C H_C N_C$ times. If H_C is approximated as k and $N_C = N/N_h$, the overheads caused by CS packets in the initial stage are given by:

$$O_{CS} = d_C H_C N_C^2 \approx d_C k N^2 / N_h^2$$

Moreover, the overheads generated in the initial stage also include the overheads for cluster formation. According to Ref. [19], the total overhead for cluster formation using the KCMBC approach is $O_{CF} = (2k+3)N$ packets. Consequently, the total overheads created in the initial stage can be obtained as

$$O_{Initial\_KCLS} = O_{CF} + O_{CS} \approx (2k+3)N + d_C k N^2 / N_h^2. \tag{7.2}$$

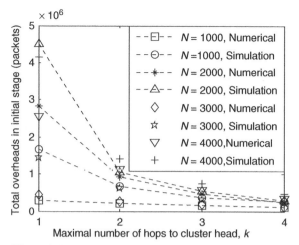

Figure 7.4 The overheads created by KCLS in the initial stage.

Figure 7.4 shows both the simulation results and the numerical results calculated by Eq. (7.2). It can be seen that overheads in the initial stage significantly decrease when the value of k increases, because a large k results in large cluster size and less number of clusters in the network. However, for a fixed value of k, $O_{\text{Initial\_KCLS}}$ increases with the number of nodes N, but a large k can suppress the increasing rate of $O_{\text{Initial\_KCLS}}$ while N increases.

7.3.2 Cost in the Location Maintenance Stage

The overheads in the location maintenance stage include the overheads for intracluster location updates, the overheads for intercluster location updates, and the overheads for periodical beacons. Since beacons have been used in most of routing protocols in multihop wireless networks, from the performance comparison point of view, the overheads of beacons are not necessary to be discussed here. The overheads for intracluster location updates are caused by intracluster link activation and deactivation. Any change in intracluster links triggers a LLC packet to be sent to each cluster member. On the other hand, such link changes may affect cluster membership, cluster structure (such as cluster merger, cluster removal, and cluster re-election), and the intercluster connectivity, so that CS packets are sent out to update location information. In addition, some small amounts of overheads are created by location enquiries. Therefore, the cost of location management, in terms of control packets per second, is a summation of the cost of intercluster location update (C_{inter}), the cost of intracluster location update (C_{intra}), and the cost of location enquiry (C_{enq}). Thus, the total cost of location management is given by

$$C_{\text{KCLS}} = C_{\text{intra}} + C_{\text{inter}} + C_{\text{enq}} \qquad (7.3)$$

In Eq. (7.3), the small cost C_{enq} can be ignored, since every location enquiry packet is only transmitted within one cluster. Table 7.5 illustrates the average values and the 95% confidence intervals of C_{inter} and C_{intra} with different values of k and N. It is obvious that the total cost is dominated by C_{inter} and C_{intra}. From Table 7.5, it can be seen that for a fixed value of N, when k increases, C_{intra} increases, whereas C_{inter} decreases. This is because a large value of k causes large size of cluster and less number of clusters in the network. In this case, the number of transmissions for each LLC packet increases and the number of clusters that forward each CS packet decreases.

It is clear that C_{inter} depends on the transmission interval of CS packets, which is determined by the scaling factor σ and the average link available time of each cluster. Figure 7.5 shows the influence of σ on C_{inter}. It can be seen that for a fixed value of k, C_{inter} increases with the decrease in σ. This is because a small σ results in small latency for transmitting CS packets among clusters. However, frequent intercluster location updates generate a lot of overheads. By contrast, a large σ can reduce the cost of intercluster location update, but a large σ will degrade the accuracy of location service provided by KCLS due to large transmission latency. The effect of σ on the accuracy of location service can be found in section 7.3.3.

Figure 7.6 shows the total cost of location management C_{KCLS} versus k when $V_m = 0.5$ unit/s and $\sigma = 0.2$. It is obvious that the total cost increases with the number of nodes, N. However, a large value of k can suppress the increasing rate of C_{KCLS} when N increases. For example, when $k = 1$, the total cost for $N = 4000$ is about 22.3 times that of for $N = 1000$; but when $k = 3$, the total cost for $N = 4000$ is only 11.6 times of that for $N = 1000$.

Recall Table 7.5 it can be found that for a small value of k, the total cost is mainly determined by C_{inter}. When k increases, C_{inter} decreases, whereas C_{intra} increases and

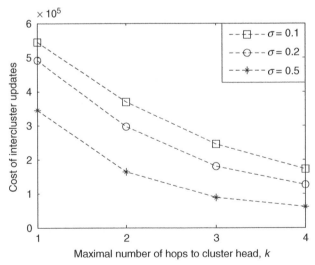

Figure 7.5   The cost of intercluster update C_{inter}, $N = 2000$ and $V_m = 0.5$ unit/s.

Table 7.5 The Cost of Location Management, $V_M = 0.5$ unit/s and $\sigma = 0.2$

k	1000		2000		3000	
	C_{intra}	C_{inter}	C_{intra}	C_{inter}	C_{intra}	C_{inter}
1	$(2.18 \pm 0.26) \times 10^3$	$(8.87 \pm 0.75) \times 10^4$	$(1.09 \pm 0.11) \times 10^4$	$(4.91 \pm 0.53) \times 10^5$	$(3.54 \pm 0.39) \times 10^4$	$(1.15 \pm 0.16) \times 10^6$
2	$(3.15 \pm 0.21) \times 10^3$	$(7.12 \pm 0.54) \times 10^4$	$(2.38 \pm 0.25) \times 10^4$	$(2.97 \pm 0.37) \times 10^5$	$(8.41 \pm 0.73) \times 10^4$	$(4.83 \pm 0.53) \times 10^5$
3	$(3.95 \pm 0.29) \times 10^3$	$(5.15 \pm 0.63) \times 10^4$	$(4.06 \pm 0.29) \times 10^4$	$(1.80 \pm 0.24) \times 10^5$	$(1.44 \pm 0.10) \times 10^5$	$(2.54 \pm 0.39) \times 10^5$
4	$(4.79 \pm 0.33) \times 10^3$	$(3.72 \pm 0.47) \times 10^4$	$(5.61 \pm 0.42) \times 10^4$	$(1.26 \pm 0.21) \times 10^5$	$(2.42 \pm 0.18) \times 10^5$	$(1.27 \pm 0.20) \times 10^5$

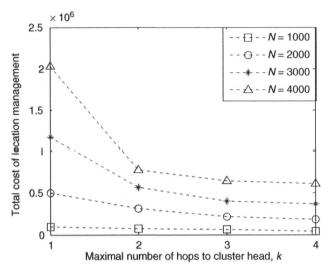

Figure 7.6   The total cost of location management, $V_m = 0.5$ unit/s and $\sigma = 0.2$.

contributes more and more to the total cost. Accordingly, as shown in Figure 7.6, when N has a fixed value, the total cost decreases with the increase in k. It also can be seen that the cluster structure with $k = 1$ is not scalable for a large diameter multihop wireless network. As shown in Figure 7.6, when k increases from 2 to 4, the total cost of location management significantly reduces.

Figure 7.7 shows the effect of node mobility on C_{KCLS}. It is clear that C_{KCLS} increases with V_m, since both link states and cluster structures change frequently

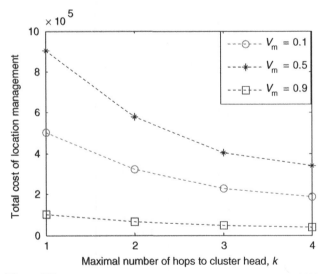

Figure 7.7   The effect of node mobility on C_{KCLS}, $\sigma = 0.2$ and $N = 2000$.

Table 7.6   Cost Comparison Between LSR and KCLS, $V_m = 0.5$ unit/s, $k = 3$, and $\sigma = 0.2$

N	1000	2000	3000	4000
C_{LSR}	8.04×10^5	6.44×10^6	2.17×10^7	5.15×10^7
C_{KCLS}	0.55×10^5	2.21×10^5	3.98×10^5	6.50×10^5

while the rate of node moving becomes large. However, a large value of k is able to suppress the increase in C_{KCLS} while the degree of node mobility increases. From Figure 7.7, we can find that cluster structures with a large k are able to significantly reduce the cost of location management, especially when node moving is drastic.

The KCLS protocol can be considered as a hierarchical link state protocol in which the global intercluster link states are maintained by cluster heads, while the local intracluster link states are stored by each cluster member. Thus, it is interesting to compare the cost of location management under KCLS with the cost of link state updates in the LSR protocol [4, 14], which are denoted as C_{KCLS} and C_{LSR}, respectively. When $V_m = 0.5$ unit/s, $k = 3$ and $\sigma = 0.2$, as shown in Table 7.6, C_{LSR} increases with N following an approximately square rate, which is expected since any link change triggers a location update packet that must be forwarded by every node once. From Table 7.6, it also can be found that C_{KCLS} is much less than C_{LSR} for a fixed N. For example, when $N = 4000$, C_{LSR} is about 80 times of C_{KCLS}. The total cost of KCLS increases with N following a linear rate. The average cost for location management charged by each node is a sublinear function of node density. The simulation results support our claim that the proposed KCLS protocol is scalable to large and dense multihop wireless networks.

7.3.3   Accuracy of Location Service

The accuracy of location service can be evaluated by the average hit probability of the location information obtained by each enquiry. Specially, the hit probability $P_{HC}(\omega)$ is defined as the probability that the response of a location enquiry is able to provide the correct cluster ID of the destination, where ω denotes that ω cluster–hop distance exists between the source–destination pair.

Figure 7.8 illustrates the average hit probability versus k by using the KCLS protocol with $N = 2000$ and $V_m = 0.5$ unit/s. It can be seen that $P_{HC}(\omega)$ increases with an approximately linear rate when k increases, since node sojourn time in a cluster increases while the average cluster size increases. Figure 7.8 also shows that the average hit probability decreases when the cluster–hop distance ω changes from 1 to 5. This is because the neighboring clusters of the destination's cluster C_m can receive at most one CS packet of C_m every τ_{CSm} interval, but other clusters can only get at most one CS packet of C_m every $2\tau_{CSm}$ interval. Thus, stale location information attached in data packets can be revised by the intermediate cluster heads, which are close to the destination.

From Figure 7.8, it also can be found that when both ω and k are fixed, the hit probability decreases with the increase in the scaling factor σ. This is because the

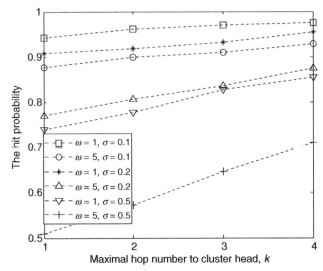

Figure 7.8   The hit probability versus k, $N = 2000$ and $V_m = 0.5$ unit/s.

latency for the transmission of CS packets is positive proportional to σ. It is obvious that there is a tradeoff between the cost of location management and the accuracy of location service. Since the neighboring clusters around the destination maintain the latest location information, the value of σ can be determined by $P_{HC}(\omega = 1)$, the average hit probability of the response for a location enquiry that performs in the destination's neighboring clusters. To ensure $P_{HC}(\omega = 1)$ higher than 90% for any k, σ can be set at 0.2.

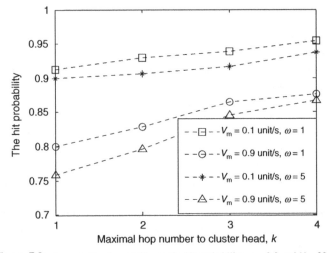

Figure 7.9   Impact of node mobility on the hit probability, $\sigma = 0.2$ and $N = 2000$.

Figure 7.9 shows the hit probability $P_{HC}(\omega)$ when the maximal node moving rate is 0.1 and 0.9 unit/sec, respectively. It can be found that for both of the two moving scenarios, the hit probability increases with k but decreases with the increase in ω. Moreover, when ω has a fixed value, there is a little difference between $P_{HC}(\omega)$ for high moving rate scenario ($V_m = 0.9$ unit/s) and $P_{HC}(\omega)$ for low moving rate scenario ($V_m = 0.1$ unit/s). It is clear that the change in maximal node moving rate has little influence on the hit probability. This is because in the KCLS protocol, the transmission interval for CS packets is proportional to the average link available time of corresponding clusters, which can reflect the degree of node mobility. Thus, KCLS is able to adapt to the frequency of location updates in various dynamic environments.

7.4 LOCATION SERVICE AND APPLICATIONS

On the basis of the clustering architecture, a multihop wireless network using the KCLS protocol is a distributed location service system, where each cluster head takes the responsibility of a location server to provide location information to its cluster members. In this case, the location enquiry from an orphan node is only forwarded to the nearest reachable cluster head. Hence, the KCLS protocol has much shorter latency for location searching than other location service protocols [6, 15], in which it may need to search the destination among several databases.

This distributed location service based on clustering architecture is reliable. Even in a case that some cluster heads are unreachable due to inter-cluster link failures, however, such failure only affects the relevant clusters, and the other clusters in the network are still active. Moreover, the transmission of data packet is still available even in the lack of the destination's accurate location information. This is because the location information in the packet header can be modified by intermediate cluster heads along the path to the destination. For example, when data packet faces inter-cluster link failure on its way to the destination, then this packet is forwarded to the nearest reachable cluster, where the cluster head modifies the location information contained in the packet that includes destination cluster ID, the coordinates of cluster head, and the sequence number. Then, the route of the packet to the destination is modified accordingly.

Location service, when integrated into multihop wireless networks, may provide many potential services. Below we discuss three location-aware applications provided by KCLS: cluster level routing, geocast, and sensor data aggregation.

7.4.1 Cluster-Level Routing

With the aid of the KCLS protocol, cluster-level routing is able to improve the performance of routing protocols by reducing overheads, increasing route stability, and enhancing route recovery capability. Considering proactive routing protocols such as dynamic source routing (DSR) [18], packets must carry a full routing list from source to destination, which generates a large amount of overheads, especially for long routing paths spanning over a large-scale network. In contrast to cluster-level routing, the cluster head can determine a cluster level route to the destination using the LS table based on the shortest path algorithm. Since the response for a location enquiry is sent back to the source node using piggyback along the cluster level path, data packet only needs to carry a cluster-level route, which has much less overheads compared with node-level route.

The cluster-level route is more stable and adaptable for the dynamic multihop topology compared with the node-level route. In this case, the failure of intra-cluster links has very limited effects on the forwarding packet to destination since alternative paths to the next cluster can always be found easily. If the expected cluster on a route is unreachable, instead, the data packet is forwarded to the nearest available cluster for modifying the location information and searching for alternative cluster-level route. Therefore, the KCLS protocol is able to self-recover cluster-level route against link failure. Accordingly, long latency due to rerouting process is avoided.

Figure 7.10 illustrates an example for cluster-level routing and route recovery. When node h_a in cluster C_i intends to communicate with node h_b, h_a sends a location enquiry to its cluster head h_i. According the current LS table, h_i responds the enquiry with h_b's cluster ID (C_h) and cluster head h_h's coordinates. This response can also provide a cluster-level route between h_a and h_b, that is, $C_i \rightarrow C_l \rightarrow C_m \rightarrow C_j \rightarrow C_g \rightarrow C_h$. Containing the cluster-level path and h_b's location information, the forwarding packets are relayed cluster by cluster. It is supposed that the inter-cluster link between C_m and C_j is broken because of node mobility. Upon receiving a forwarding packet, the gateway of C_m finds that this packet cannot be delivered to C_j directly. In this case, the packet will be delivered to cluster head h_m instead. h_m's LS table shows that h_b is still located inside of C_h, then

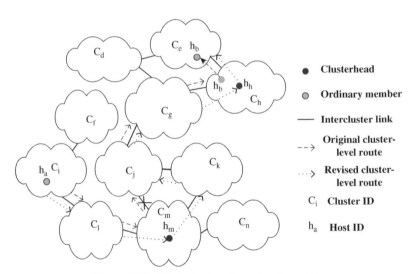

Figure 7.10   Cluster-level routing and route recovery.

h_m designs a new partial route $C_m \rightarrow C_k \rightarrow C_j$ to replace the old one $C_m \rightarrow C_j$, and the packet delivery continues. Furthermore, when the packet is received by C_h's gateway, which detects the destination h_b does not exits in C_h at that moment. Then, the packet is forwarded to cluster head h_h. Since h_h records that h_b has moved into the cluster C_e, the packet is delivered to C_e and is then forwarded to the destination h_b. Lastly, h_b sends the source h_a an acknowledgment packet, which contains the new cluster-level route, that is, $C_i \rightarrow C_l \rightarrow C_m \rightarrow C_k \rightarrow C_j \rightarrow C_g \rightarrow C_h \rightarrow C_e$. The subsequent packet delivery from h_a to h_b continues along the new cluster-level route.

7.4.2 Geocast

Geocast is a kind of location-based multicast. The goal of geocast is to send messages to all mobile nodes within one or a few specified geographical regions [22, 23], that is, Geocast regions. When urgent events occur in a specific area or some information must be advised to mobile nodes in certain areas, geocasting is a convenient way to achieve this goal. To determine the geocast group membership in a multihop wireless network, each node is required to know its own physical location or relative coordinates.

The existing geocast protocols have been categorized into flooding-based, routing-based, and cluster-based protocols. Flooding-based protocols [24, 25] use flooding or broadcasting to forward geocast packets from the source to the geocast regions. A flooding-based protocol compatible node forwards a geocast packet only if it belongs to the forwarding zone and the geocast region. Routing-based protocols [26, 27, 28] create routes from the source to the geocast region. In such a protocol, geocast packets are forwarded by the intermediate nodes along the paths and flooded within the geocast region. Cluster-based geocast protocols [23, 29] geographically partition a network into several disjointed and equally sized cellular/mesh regions and select a cluster head in each region for executing information exchange. The details on these protocols can be found in Ref. [30].

The KCLS protocol is able to support the geocast service in a multihop wireless network. Similar to the routing-based geocast protocols, the cluster head of the source node can easily assign routes to the clusters that cover the geocast regions. In this case, cluster-level routes take the place of node level routes to represent the paths from the source to the geocast regions. On the other hand, the advantage of traditional cluster-based geocast protocols is inherited in ours approach. Instead of having every nodes forward geocast packet, only gateway nodes of a cluster forward data cluster by cluster. Moreover, when the existing cluster-level route for geocasting is broken, the alternative route is immediately designed by the intermediate cluster heads along the path to the destination. Once the geocast reaches the clusters in the geocast region, it is flooded within the clusters.

Combing the characteristics of routing-based geocast protocols and cluster-based geocast protocols, the geocast protocol based on KCLS is able to provide robust service with few packet overheads. The high ability to tolerate link breakage and well scalability makes it competent for large-scale networks.

7.4.3   Sensor Data Aggregation

Recent advancement in wireless communications and electronics has enabled the development of low-cost wireless sensor networks, which can be used for various applications, such as health, military, and home. Wireless sensor networks are composed of nodes with sensing, wireless communication, and computation capabilities. Such networks must carry out not only the task of sensing the environment but also the task of communicating a relevant summary of the data through a sequence of messages passed between nodes and computations at nodes to one or a few designated sink nodes.

Wireless sensor networks improve sensing accuracy by providing distributed processing of vast quantities of sensing information. Networked sensors can aggregate data to provide a rich, multidimensional view of the environment and continue to function accurately in the condition of failure of individual sensors. Hence, data aggregation is an important technique used to solve the implosion and overlap problems in wireless sensor networks [31].

The KCLS protocol can be employed in a wireless sensor network to enhance the performance of data aggregation. During each data-gathering cycle, the sensor nodes send their sensed data to their cluster head that performs data aggregation. Then the cluster head transmits the aggregated data to a sink node via a multihop path assigned by cluster-level routing. In this case, the sensor nodes have simple functionality, since they perform sensing and relatively short-range communication. However, the cluster heads are more complex because they coordinate MAC, perform data fusion, and perform long-range transmissions to the remote sink nodes.

7.5   CONCLUSION

This chapter presents a novel location service protocol based on the clustering architecture, which has the following advantages in support of multihop wireless networks. First, it is able to significantly reduce the overheads by the use of simple cluster-level route combined with self-determined inter-cluster forwarding based on cluster mobility pattern or group characteristics. Next, the clustering architecture based on cluster ID has flexible scalability in support of large-scale multihop wireless networks. Third, it has the capability of cluster-level self-route recovery against interlink failures. Finally, on the basis of the distance effect, it is able to provide more accurate location information within the cluster and nearby neighborhoods, which matches the dynamic nature of multihop wireless networks very well.

The numerical results obtained from simulations indicate that both the overheads in the initial stage and the total cost in the location maintenance stage decrease with the increase in k. A large value of k is not only able to suppress the increasing rate of the total cost when the number of nodes in the network increases, but it is also able to increase the hit probability for location service and reduce the passive effect of node mobility. Moreover, scaling factor σ is also used to balance the cost and the accuracy of location service. With optimal values of k and σ, the total cost for using KCLS is less than 2% of the total cost for using link state protocol.

A wide range of location-based services are currently available for Web, mobile, and handheld devices, including hotel and restaurant information, maps and navigations, travel schedules, weather report, and multimedia entertainment. The scalability, fault tolerance, low cost, and self-discovery characteristics of the KCLS protocol accommodate many exciting location-based applications in multihop mobile networks.

REFERENCES

1. T. Camp. Location information services in mobile ad hoc networks. In *Handbook of Algorithms for Mobile and Wireless Networking and Computing*. 2005, pp. 317–339, Chapter 14.
2. I. Stojmenovic. Position-based routing in ad hoc network. *IEEE Commun. Mag.*, 128–134, 2002.
3. S. Basagni, I. Chlamtac, and V. R. Syrotiuk. Geographic messaging in wireless ad hoc Networks. In *Proceedings of the 49th IEEE VTC*, Vol. 3, pp. 1957–1961, 1999.
4. M. Joa-Ng and I. T. Lu. A peer-to-peer zone-based two-level link state routing for mobile ad hoc networks, In *IEEE JSAC*, Vol. 17, No. 8, pp. 1415–1425, 1999.
5. A. Amis, R. Prakash, T. Vuong, and D. T. Huynh. Max Min D-cluster formation in wireless ad hoc networks. In *Proceedings of the IEEE INFOCOM*, 1999.
6. Z. J. Haas and B. Liang. Virtual backbone generation and maintenance in ad hoc network mobility management. In *Proceedings of the IEEE INFOCOM 2000*, 2000.
7. S. Basagni, I. Chlamtac, and V. R. Syrotiuk, B. A. Woodward. A distance routing effect algorithm for mobility (DREAM). In *Proceedings of the MOBICOM*, 1998, pp. 76–84.
8. Supeng Leng Liren Zhang, L. W. Yu, C. H. Tan. An efficient broadcast relay scheme for MANETs. *Comput. Commun.*, 28(5): 467–476, 2005.
9. T. Camp, J. Boleng, and L. Wilcox. Location information services in mobile ad hoc networks. In *Proceedings of the IEEE International Conference on Communications (ICC)*, 2001, pp. 3318–3324.
10. J. Li, et al., A scalable location service for geographic ad hoc routing. In *Proceedings of the 6th International Conference on Mobile Computing and Networking. ACM*, 2000, pp. 120–130.
11. K. H. Wang, and B. Li. Group mobility and partition prediction on wireless ad-hoc networks. In *Proceedings of IEEE ICC Conference*, 2002, pp. 1017–1021.
12. K. N. Amouris, S. Papavassiliou, and M. Li. A position based multi-zone routing protocol for wide area mobile ad-hoc networks. In *Proceedings of the 49th IEEE VTC*, 1999, pp. 1365–1369.
13. M. Kasemann, et al., A reactive location service for mobile ad hoc networks. Technical Report, Department of Science, University of Mannheim, TR-02-014, 2002.
14. R. Perlman, *Interconnections: Bridges and Routers*. Addison-Wesley, 1992, pp. 149–152, 205–233.
15. L. Blazevic, et al., Self-organization in mobile ad hoc networks: the approach of terminodes. *IEEE Commun. Mag.*, 39(6): 166–175, 2001.
16. S.-C. Woo and S. Singh, Scalable routing protocol for ad hoc networks. *ACM Wireless Networks*, 7(5): 513–529, 2001.
17. T. Cormen, C. Leiserson, and R. Rivest. *Introduction to Algorithms*. MIT Press, Cambridge, MA USA, 1990.
18. D. B. Johnson and D. A. Maltz. In T. Imielinski and H. Korth, editors, *Dynamic source routing in ad hoc wireless networks, mobile computing*, Paper 5, Kluwer Academic, 1996, pp. 153–181.
19. Supeng Leng, Yan Zhang, Hsiao-Hwa Chen, Liren Zhang, and Ke Liu. A novel k-hop compound metric based clustering scheme for ad hoc wireless networks, *IEEE Trans. Wireless Commun.*, 8(1):367–375, 2009.
20. J. Sucec and I. Marsic. Hierarchical routing overhead in mobile ad hoc networks. *IEEE Trans. Mobile Comput.*, 3(1): 46–56, 2004.
21. S. Capkun, M. Hamdi, and J.P. Hubaux, GPS-free positioning in mobile ad-hoc networks. In *Proceedings of the Hawaii International Conference on System Sciences*, 2001.

22. Y.-B. Ko and N. H. Vaidya, Geocasting in mobile ad hoc networks: location-based multicast algorithms. In *IEEE Workshop on Mobile Computing Systems and Applications*, 1999.
23. C.-Y. Chang, C.-T. Chang, and S.-C. Tu, Obstacle-free geocasting protocols for single/multi-destination short message services in ad hoc networks, *Wireless Networks*, 9(2):143–155, 2003.
24. Y. Ko and N. H. Vaidya. Geocasting in mobile ad hoc networks: location-based multicast algorithms. In *Proceedings of WMCSA*, 1999, pp. 101–110.
25. I. Stojmenovic. Voronoi diagram and convex hull based geocasting and routing in wireless networks. Technical Report, University of Ottawa, TR-99-11, 1999.
26. J. Boleng, T. Camp, and V. Tolety. Mesh-based geocast routing protocols in an ad hoc network. In *Proceedings of IPDPS*, 2001, pp. 184–193.
27. T. Camp and Y. Liu. An adaptive mesh-based protocol for geocast routing. *J. Parallel Distribut. Comput.* [Special Issue on Routing in Mobile and Wireless Ad Hoc Networks], 62(2):196–213, 2003.
28. Y. Ko and N. H. Vaidya. GeoTORA: a protocol for geocasting in mobile ad hoc networks. In *Proceedings of ICNP*, 2000, pp. 240–250.
29. W.-H. Liao, Y.-C. Tseng, K.-L. Lo, and J.-P. Sheu. GeoGRID: a geocasting protocol for mobile ad hoc networks based on GRID. *J. Internet Technol.*, 1(2):23–32, 2000.
30. P. Yao, E. Krohne, and T. Camp. Performance comparison of geocast routing protocols for a MANET. Technical report, Department of Mathematics and Computer Sciences, Colorado School of Mines, 2004.
31. Ian F. Akyildiz, et al., A survey on sensor networks. *IEEE Commun. Mag.*, 40(40):102–114, 2002.
32. S. Leng, L. Zhang, H. Fu, J. Yang. Mobility analysis of mobile hosts with random walking in ad hoc networks. In *Computer Networks*. Elsevier Science, Vol. 51, No. 10, pp. 2514–2528, 2007.
33. T. Camp, J. Boleng, and V. Davies, A survey of mobility models for ad hoc network research. *Wireless Commun. Mobile Comput.*, 2(5):483–502, 2002.

Chapter 8

Predictive Location Tracking in Cellular and in Ad Hoc Wireless Networks

Nikos Dimokas,[1] Dimitrios Katsaros,[1,2] Panayiotis Bozanis,[2] and Yannis Manolopoulos[1]

[1] *Department of Informatics, Aristotle University of Thessaloniki, Thessaloniki, Greece*
[2] *Department of Computer & Communication Engineering, University of Thessaly, Thessaly, Greece*

8.1	Introduction	163
8.2	Predictive Location Tracking Techniques	166
8.3	Predictive Location Indexing Techniques	176
8.4	Conclusion	186
	Acknowledgments	187
	References	187

Predicting the future is mostly a matter of managing not to blink as you witness the present.

—William Gibson

8.1 INTRODUCTION

The proliferation of cellular and ad hoc networks and the penetration of Internet services are changing many aspects of ubiquitous mobile computing. Constantly

Mobile Intelligence. Edited by Laurence T. Yang, Agustinus Borgy Waluyo, Jianhua Ma, Ling Tan, and Bala Srinivasan
Copyright © 2010 John Wiley & Sons, Inc.

increasing mobile client populations utilize diverse mobile devices to access the wireless medium and various heterogeneous applications (e.g., streaming video, Web) are being developed to satisfy the eager client requirements. The realization of such a demanding environment requires addressing many technical issues, related to radio management, networking, data management, and so on.

Most of the challenging issues and problems in this area are due to the fact that the underlying environment is extremely resource-starving and inherently uncertain. For instance, the wireless communication channels are bandwidth-limited and error prone. The uncertainty due to node (user) mobility has fundamental impacts, since it induces uncertainly in the network topology and hence causes problems in routing and in data delivery. Additionally, traffic load and resource demands in cellular and in ad hoc wireless networks are also uncertain, depending a lot on the user trajectories.

In this harsh environment, seamless and ubiquitous connectivity is a fundamental goal. This goal calls for smart techniques for determining the current and future location of a mobile. The ability to determine the mobile client's (future) location can significantly improve the wireless network's overall performance. Consider for instance the handover procedure in cellular networks, which is directly related to the design of resource management algorithms in such infrastructured networks; such resources could be bandwidth, MAC frames, packets. Instead of relying on *reactive* approaches, that is, allocating appropriate resources during the handover, we could come up with *proactive* approaches, that is, allocating resources before needed, so as to bypass, instead of correct, the negative effect of handover [28]. Additionally, methods like the Shadow Cluster [31] could benefit from location prediction by refraining from allocating resources to all neighboring cells; with the exploitation of predictions instead, they could allocate resources only to the most probable-to-move cells. Finally, location prediction could be exploited in sequential paging schemes [8] to reduce the combined paging cost and also in techniques for call admission control [60].

Location prediction and tracking is useful not only in cellular networks, but also to other types of wireless networks, such as mobile ad hoc networks (MANETs). A mobile ad hoc network is a wireless network in which a set of mobile nodes with wireless connectivity form a temporary network without the existence and support of any infrastructure, for example, base stations, or centralized administration, for example, switching centers. Communication in an ad hoc network between any two nodes that are out of one another's transmission range is achieved through interme-diate nodes, which relay messages to set up a communication channel between the two nodes. For a MANET node v wishing to communicate with another MANET node u not within its transmission range, knowledge of the future position(s) of node v could help reduce its energy consumption, by postponing its communication to u until it reaches closer to it. Such a technique has been investigated in Ref. [9].

8.1.1 Preliminaries

The present chapter deals with the issue of predictive location tracking in cellular and in ad hoc wireless networks and examines this issue in two different settings. In Section 8.2, we assume a generic symbolic network topology model, like that introduced in Ref. [8], where the existence of "cells" is assumed. The cells need not be hexagonal, but they can be of arbitrary shape. The notion of a wireless cell is well established in cellular networks, and it can be defined in a similar manner for the ad hoc networks, as follows: we overlay a grid of any type [9] over the area, inside which the mobile hosts of the ad hoc network are roaming. In this setting, the positioning of a mobile is performed at the level of a cell. In Section 8.3, we withdraw the assumption of the existence of cells, and the position of each mobile host is determined by its geographic coordinates only.

Connectivity in the presence of mobility is a nontrivial task; the network has to work against the uncertainty created by the mobile's freedom of movement. Thus, the management of mobility is of crucial importance. Depending on whether a mobile terminal is actively communicating or in standby mode, we differentiate between (a) in-session mobility management and (b) out-of-session mobility management. The former is widely known in cellular networks as *handoff* management and deals with mechanisms by which calls and sessions are kept alive while the mobile host is moving from cell to cell, thus changing its network point of attachment. In general, the procedure of handoff management is considered easier than the management of the latter case, which is known as *location management* or *location tracking* and is responsible for keeping track of mobiles in standby mode.

The location tracking problem in generic wireless networks involves two procedures, namely *paging* and *update*. At the one extreme, one can come up with a proposal for this problem with the aid of paging, which is performed by the system; on a call arrival, the network initiates a search for the sought mobile, by (simultaneously) polling every possible site where the mobile can be found. In the case of cellular networks, this is performed by the *mobile switching center*, which broadcasts a page message over a special *forward control channel* via the *base stations*. All the mobiles listen to this paging message, but only the target mobile responses over a *reverse channel*. In the worst case, the system may have to page each cell of the whole service area. Clearly, this approach involves excessive signaling traffic and, thus, is problematic.

At the other extreme, one can come up with a solution that demands from the mobile to report every time it moves from one site (cell) to another. This reporting is called *location registration* and starts with an update message sent by the mobile over a reverse channel, which is then followed by some traffic that takes care of related database maintenance operations at the system's side. Again, this approach may also generate excessive signaling traffic if the mobile changes cells frequently and, thus, is impractical.

In real situations, location tracking is performed as a hybrid between these extreme approaches [47]. Although a lot of (reactive) location management methods have been proposed, the issue of predictive (or proactive) location tracking has lately received significant attention due to its potential to reduce or even eliminate the latency associated with location tracking. Moreover, there are situation where prediction of mobiles' movements, that will eventually lead to network disconnection, may force specific decisions related to routing. Examples of these decisions involve the routing protocols that are suitable for highly mobile ad hoc networks and for delay-tolerant networks [62].

In general, predictive location tracking techniques work by constructing a *mobility model* for each mobile host that models the *mobility history* of the mobile. Clearly, the two notions are different; the former is probabilistic and extends to the future, whereas the latter is deterministic and refers to the past. Location prediction is related to the ability of the underlying network to record, learn, and subsequently *predict* the mobile's movements. The success of the prediction is presupposed and is boost by the fact that *mobile users exhibit some degree of regularity* in their movement [8]. A "smart" network can record the movement history and then construct a mobility model for its clients. The real challenge involved in designing an effective and efficient predictive location tracking method is to quantify the utility of the past in predicting the future.

8.1.2 Chapter Organization

The purpose of this chapter is to provide an overview of techniques suitable for predicting the future locations of mobile hosts in wireless networks. It concentrates on two different scenarios; according to the first scenario, the network coverage area is partitioned in nonoverlapping regions (named cells) and the location tracking is performed at the level of cells; according to the second scenario, there is no such tiling to the coverage area and location tracking is performed at the level of geographical coordinates. The first part of the chapter (Section 8.2) deals with the first scenario and introduces information-theoretic methods suitable for predicting the future locations of mobiles. The second part of the chapter (Section 8.3) deals with the second scenario and presents the issues related to indexing the positions of mobile hosts in order to support predictive queries. For these two broad and significant issues, the chapter surveys, classifies, and compares the state-of-the-art solutions, by discussing the critical issues and challenges of predictive location tracking in wireless networks.

8.2 PREDICTIVE LOCATION TRACKING TECHNIQUES

Owing to the uncertainty inherent in the mobile's movements, we can consider them to be the outcome of an underlying stochastic process, which can be modeled using established information-theoretic concepts and tools [34, 56]. The cornerstone work of Ref. [17] exhibited the possibility of using methods, which have traditionally been used for data compression (thus, characterized as "information-theoretic"), in carrying out prediction. Considering a symbolic network topology model [8], we can model the respective state space as a finite alphabet comprised of discrete symbols. The alphabet consists of all possible sites (cells) where the client has ever visited or might visit (assuming that the number of cells in the coverage area is finite). With this transformation, we can exploit methods that have traditionally been used for data compression (thus, characterized as "information-theoretic") to carry out prediction. In the rest of this section, we elaborate on these methods.

8.2.1 The Discrete Sequence Prediction Problem

In quantifying the utility of the past in predicting the future, a formal definition of the problem is needed, which we provide in the following lines. Let Σ be an alphabet, consisting of a finite number of symbols $s_1, s_2, \ldots, s_{|\Sigma|}$, where $|\cdot|$ stands for the length/cardinality of its argument. A *predictor*, which is an algorithm used to generate prediction models, accumulates sequences of the type $\mathfrak{a}_i = \alpha_i^1, \alpha_i^2, \ldots, \alpha_i^{n_i}$, where $\alpha_i^j \in \Sigma$, $\forall i, j$ and n_i denotes the number of symbols comprising $\mathfrak{a}_i$. Without loss of generality, we can assume that all the knowledge of the predictor consists of a single sequence $\mathfrak{a} = \alpha^1, \alpha^2, \ldots, \alpha^n$. On the basis of $\mathfrak{a}$, the predictor's goal is to construct a model that assigns probabilities for any future outcome given "some" past. Using the characterization of the mobility model as a stochastic process $(\mathbb{X}_t)_{t \in N}$, we can formulate the aforementioned goal as follows.

Definition 8.1 (Discrete Sequence Prediction Problem). At any given time instance t (meaning that t symbols $x_t, x_{t-1}, \ldots, x_1$ have appeared, in reverse order) calculate the conditional probability

$$\widetilde{P}[\mathbb{X}_{t+1} = x_{t+1} | \mathbb{X}_t = x_t, \mathbb{X}_{t-1} = x_{t-1}, \ldots],$$

where $x_i \in \Sigma$, $\forall x_{t+1} \in \Sigma$. This model introduces a *stationary Markov chain*, since the probabilities are not time-dependent. The outcome of the predictor is a ranking of the symbols according to their $\widetilde{P}$. The predictors that use such kind of prediction models are termed *Markov predictors*.

Depending on the application, the predictor may return only the symbol(s), with the highest probability, that is, implementing a "most-probable" prediction policy, or the symbols with the m highest probabilities, that is, implementing a "top-m" prediction policy, where m is an administratively set parameter. In any case, the selection of the policy is a minor issue and will not be considered in this chapter, which is only concerned with methods for inferring the ranking.

The "history" $x_t, x_{t-1}, \ldots$ used in the above definition is called the *context* of the predictor and refers to the portion of the past that influences the next outcome. The history's length (also, called the *length* or *memory* or *order* of the Markov chain/predictor) will be denoted by l. Therefore, a predictor that exploits l past symbols will calculate conditional probabilities of the form

$$\widetilde{P}[\mathbb{X}_{t+1} = x_{t+1} | \mathbb{X}_t = x_t, \mathbb{X}_{t-1} = x_{t-1}, \ldots, \mathbb{X}_{t-l+1} = x_{t-l+1}]. \qquad (8.1)$$

Some Markov predictors fix, in advance of the model creation, the value of l, presetting it in a constant k in order to reduce the size and complexity of the prediction model. These predictors and the respective Markov chains are termed *fixed length Markov chains/predictors* of order k. Therefore, they compute probabilities of the form:

$$\widetilde{P}[\mathbb{X}_{t+1} = x_{t+1} | \mathbb{X}_t = x_t, \mathbb{X}_{t-1} = x_{t-1}, \ldots, \mathbb{X}_{t-k+1} = x_{t-k+1}]. \qquad (8.2)$$

where k is a constant.

Although it is a nice model from a probabilistic point of view, these Markov chains are not very appropriate from the estimation point of view. Their main limitation is related to their structural poverty, since there is no means to set an optimized value for k.

Other Markov predictors deviate from the fixed memory assumption and allow the order of the predictor to be of variable length, that is, to be a function of the values from the past.

$$\widetilde{P}[\mathbb{X}_{t+1} = x_{t+1} | \mathbb{X}_t = x_t, \mathbb{X}_{t-1} = x_{t-1}, \ldots, \mathbb{X}_{t-l+1} = x_{t-l+1}], \qquad (8.3)$$

where $l = l(x_t, x_{t-1}, \ldots)$.

These predictors are termed *variable length Markov chains*; the length l might range from 1 to t. If $l = l(x_t, x_{t-1}, \ldots) \equiv k$ for all $x_t, x_{t-1}, \ldots$, then we obtain the fixed-length Markov chain. The variable length Markov predictors may or may not impose an upper bound on the considered length. The concept of variable memory offers a richness in the prediction model and the ability to adjust itself to the data distribution. If we can choose in a data driven way the function $l = l(\cdot)$, then we can only gain with respect to the ordinary fixed-length Markov chains, but this is not a straightforward problem.

The Markov predictors (fixed or variable length) base their probability calculations $\widetilde{P}$ on counts of the number of appearances of symbols after contexts. They also take special care to deal with the cases of unobserved symbols (i.e., symbols with zero appearance counts after contexts), assigning to them some "minimum probability mass."

8.2.2 The Power of Markov Predictors

The issue of prediction in wireless networks, especially location prediction, has received attention during the past years, and the most important proposed techniques focus on the notions of *learning automata, Kalman filtering,* and *pattern matching*.

Learning automata are finite-state adaptive systems that interact continuously with their environment learning a "behavior." Learning automata have been used in location prediction [28], and, although simple, they are not considered very efficient learners because of the need to devise appropriate penalty/reward policies, which is usually done in an ad hoc way, and their slow convergence to the correct actions.

Kalman filtering is a recursive processing algorithm for producing optimal estimates. Kalman filtering-based methods [33] construct a mobile motion equation relying on specific distributions for its velocity, acceleration, and direction of movement. Therefore, they assume relatively accurate geographic position knowledge via signal strength measurements. Their performance largely depends on the stabilization time of the Kalman filter and knowledge (or estimation) of the system's parameters.

Finally, (approximate) pattern matching techniques have been used for location prediction [33]. They compile (or assume existence of) aggregate or per-user mobility profiles and perform approximate similarity matching between the current and the stored trajectories. The similarity matching is carried out through the computation of the *edit distance* between the current and each stored trajectory in order to derive predictions. Although edit distance computation can be performed quite fast with dynamic programming, it is relatively hard to select the meaningful set of edit operations on the individual symbols (i.e., insert, delete, substitute), to assign weights on them, deal with unequal sequences of symbols, and select as similarity metric the edit distance instead of string alignment.

Consequently, a couple of questions arise regarding (a) why Markov predictors are more appropriate for carrying out location prediction from the technical viewpoint and (b) whether location prediction is amenable to Markovian prediction. Several technical reasons advocate the use of Markov predictors in these problems, but their most profound advantage is their generality; they are domain-independent without any coupling to geographic coordinates or particular assumptions on distributions. A simple mapping from the "entities" of the investigated domain to an alphabet is all that is required. Thus, they are able to support location prediction.

Markovian prediction relies on the *short memory principle*, which, simply stated, says that the (empirical) probability distribution of the next symbol, given the preceding sequence, can be quite accurately approximated by observing no more than the last few symbols in that sequence. This principle fits reasonably and intuitively with how humans are acting when travelling or seeking information. A mobile user usually travels with a specific destination in mind, designing its travel via specific routes (roads or preferred pedestrian paths). This "targeted" traveling is far from a random walk assumption and is confirmed by studies with real mobility traces [45]. Therefore, the power of Markovian prediction stems from its generality and modeling capability and also from its natural accordance with the human behavior.

8.2.3 Families of Markov Predictors

Markov predictors create probabilistic models for their input sequence(s) and use *digital search trees* (*tries*) to keep track of the contexts of interest, along with some counts used in the calculation of the conditional probabilities $\widetilde{P}$. The root node of the trie corresponds to the "null" event/symbol, whereas every other node of the tree corresponds to a sequence of events; the sequence is used to label the node. We will consider a Markov predictor to be equivalent to its respective trie. Each node is accompanied by a counter, which depicts how many times this event has appeared after the sequence of events corresponding to the path from the root to the node's father has been observed.

For our convenience, we present some definitions useful in the sequel of the chapter. We use the sample sequence of events $a = aabacbbabbacbbc$. The *length* of a is the number of symbols it contains, that is, $|a| = 15$. The *appearance count* of subsequence $s = ab$ is $E(s) = E(ab) = 2$ and the *normalized appearance count* of s is equal to $E(s)$ divided by the maximum number of (possibly overlapping) occurrences a subsequence of the same length could have, considering the a's length, that is, $E_n(s) = E(s)/(|a| - |s| + 1)$. The *conditional probability* of observing a symbol after a given subsequence is defined as the number of times that symbol has shown up right after the given subsequence divided by the total number of times that the subsequence has shown up all, followed by any symbol. Therefore, the conditional probability of observing the symbol b after the subsequence a will be denoted as $\widetilde{P}(b|a)$ and at is equal to $\widetilde{P}(b|a) = (E(ab))/(E(a)) = 0.4$. In the rest of this section, we survey the families of Markov predictors.

8.2.3.1 The Prediction by Partial Match Scheme

The prediction by partial match scheme, $\mathcal{PPM}$ for short, is based on the universal compression algorithm reported in Ref. [14]. For the construction of the prediction model, it assumes a pre-determined maximal order, say k, for the generated model. Then, for every possible subsequence of length of 1 up to $k + 1$, it creates or updates the appropriate nodes in the trie. Although, this description implies that the whole input sequence is known in advance, the method works in an online fashion by exploiting a "sliding" window of size $k + 1$ over the sequence as this grows symbol by symbol. The $\mathcal{PPM}$ predictor for the sample sequence *aabacbbabbacbbc* is depicted in Figure 8.1. We can compute the conditional probability of a symbol σ to appear after a context s by detecting the sequence $s\sigma$ as a path in the trie emanating from the root, provided $|s\sigma| \le k$. Prediction is performed in a similar manner. For instance, adopting a "most probable" prediction policy, the predicted symbol for the test context ab is a or b and its conditional probability is 0.50 for either of them (see the gray-shaded nodes in Figure 8.1).

The maximum context that the $\mathcal{PPM}$ predictor can exploit is k; though, all inter-mediate contexts with length from 1 to $k - 1$ can be used. This model is also referred as *all-Kth-Order $\mathcal{PPM}$* model. The interleaving of various length contexts does not mean that this scheme is a variable-length Markov predictor (although sometimes it is referred as such), because the decision on the context length is made beforehand and not in a data-driven way.

Apart from this basic scheme, a number of variations have been proposed, which attempt to reduce the size of the trie by pruning some of its paths on the basis of the statistical information derived from the input data. They set lower bounds for the nor-malized appearance count and for the conditional probabilities of subsequences and then prune any branch that does not exceed these bounds. Characteristic works adopt-ing such an approach are reported in Refs. [10, 16, 37]. Apparently, these schemes are offline, making one or multiple passes over the input sequence in order to gather the required statistical information.

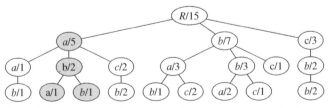

Figure 8.1   A $\mathcal{PPM}$ Markov predictor for the sequence *aabacbbabbacbbc*.

8.2.3.2 The Lempel-Ziv-78 Scheme

The Lempel-Ziv-78 Markov predictor, $\mathcal{LZ}$78 for short, is the second scheme whose virtues in carrying out predictions was investigated very early in the literature [8, 56]. The algorithm $\mathcal{LZ}$78 [64] arose from a need for finding a universal variable to fixed length coding method and construct a prediction model for an input sequence as follows. It makes no assumptions about the maximal order for the generated model. Then, it parses the input sequence into a number of distinct subsequences, say $s_1, s_2, \ldots, s_x$, such that $\forall j$ $(1 \leq j \leq x)$, the maximal prefix of subsequence s_j is equal to some s_i, for some $1 \leq i < j$. The discovered subsequences and the associated statistics are inserted into a trie in a manner identical to that of the $\mathcal{PPM}$ scheme. The computation of conditional probabilities takes place in a manner completely analogous to that of $\mathcal{PPM}$. The $\mathcal{LZ}$78 predictor for the sample sequence *aabacbbabbacbbc* is depicted in the left part of Figure 8.2. However, $\mathcal{LZ}$78 for this example is not able to produce a prediction for the test context *ab* (i.e., there is no subtree under the gray-shaded node).

Apparently, the $\mathcal{LZ}$78 Markov predictor is an online scheme, it lacks administratively tuned parameters, like lower bounds on appearance counts, and is a characteristic paradigm of a variable length Markov predictor. Although, strong results do exist that prove its asymptotic optimality and its superiority over any fixed-length $\mathcal{PPM}$ predictor, in practice, various experimental studies contradict this result because of the finite length of the input sequence. Nevertheless, the $\mathcal{LZ}$78 predictor remains a very popular prediction method. The original $\mathcal{LZ}$78 prediction scheme was enhanced in Ref. [8, 34] in a way such that apart from a considered subsequence that is going to be inserted into the trie, all its suffixes are inserted as well (see right part of Figure 8.2).

8.2.3.3 The Probabilistic Suffix Tree Scheme

The probabilistic suffix tree markov predictor, $\mathcal{PST}$ for short, was introduced in Ref. [41] and presents some similarities to $\mathcal{LZ}$78 and $\mathcal{PPM}$. Although, it specifies a maximum order for the contexts it will consider, it is actually a variable length Markov predictor and constructs its trie for an input sequence as follows. The construction procedure uses five administratively set parameters: k the maximum context length, $P_{\min}$ the minimum normalized appearance count for any subsequence in order to be considered for insertion into the trie, r is a simple measure of the difference between

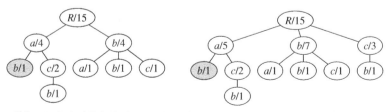

Figure 8.2 (Left) A $\mathcal{LZ}$78 Markov predictor for the sequence *aabacbbabbacbbc*. (Right) A $\mathcal{LZ}$78 predictor enhanced according to Ref. [8].

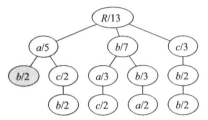

Figure 8.3 A $\mathcal{PST}$ Markov predictor for the sequence *aabacbbabbacbbc*.

the prediction capability of the subsequence at hand and its direct father node, $\gamma_{\min}$ and α together define the significance threshold for a conditional appearance of a symbol. Then, for every subsequence of length of 1 up to k, if it has never been encountered before, a new node is added to the trie labeled by this subsequence, provided a set of three conditions hold. To exhibit the conditions, suppose that the subsequence at hand is *abcd*. Then, this subsequence will be inserted into the trie of the $\mathcal{PST}$ if and only if:

(a) $E_n(abcd) \geq P_{\min}$ and

(b) there exists some symbol, say x, for which the following relations hold:

(b$_1$) $\frac{E(abcdx)}{E(abcd)} \geq (1+a)\gamma_{\min}$ and

(b$_2$) $\frac{\tilde{P}(x|abcd)}{P(x|abc)} \geq r$ or $\leq 1/r \equiv \frac{E(abc)}{E(abcd)} * \frac{E(abcdx)}{E(abcx)} \geq r$ or $\leq 1/r$.

The $\mathcal{PST}$ predictor with the following set of parameters $k = 3$, $P_{\min} = 2/14, r = 1.05$, $\gamma_{\min} = 0.001$, and $\alpha = 0$ for the sample sequence *aabacbbabbacbbc* is depicted in Figure 8.3. Apparently, $\mathcal{PST}$ is a subset of the baseline $\mathcal{PPM}$ scheme, when k is the same. $\mathcal{PST}$ for this example is not able to produce a prediction for the test context *ab* (i.e., there is no subtree under the gray-shaded node).

Apart from this basic scheme, a number of variations have been developed [5], most of them providing improved algorithms, that is, linear, for the procedures of "learning" the input sequence and for making predictions.

8.2.3.4 The Context Tree Weighting Scheme

The context tree weighting Markov predictor [58], $\mathcal{CTW}$ for short, is based on the idea of combining exponentially many Markov chains of bounded order and the original proposition dealt with binary alphabets only. The $\mathcal{CTW}$ assumes a predetermined maximal order, say k, for the generated model and constructs a *complete binary tree* $\mathcal{T}$ of height k. The left and right children of a node s are denoted as $0s$ and $1s$, respectively. Each node s maintains two counters a_s and b_s, which count the number of zeros and ones, respectively, that followed context s in the input sequence so far. Additionally, each context (node) s maintains, apart from the pair (a_s, b_s), two probabilities P_e^s and P_w^s. The former, P_e^s, is the Krichevsky–Trofimov estimator for a sequence to have exactly a_s zeros and b_s ones. The latter probability, P_w^s, is the weighted sum of some values of P_e. The $\mathcal{CTW}$ predictor for the sample binary sequence 010|11010100011 is depicted in the left part of Figure 8.4.

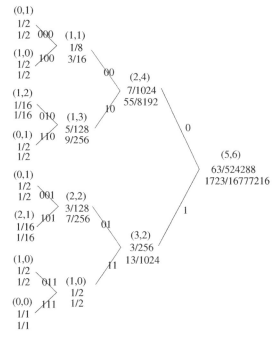

Figure 8.4   A CTW Markov predictor for the binary sequence 010|11010100011.

With P_e^R and P_w^R denoting the Krichevsky–Trofimov estimate and the CTW estimate of the root, respectively, we can predict the next symbol with the aid of a CTW as follows. We make the working hypothesis that the next symbol is a one and we update the T accordingly obtaining a new estimate for the root $P_w'^R$. Then, the ratio $P_w'^R / P_w^R$ is the conditional probability that the next symbol is a one.

For the case of nonbinary alphabets, Volf [57] proposed various extensions; among them, the *decomposed* CTW, $De\,CTW$ for short, is the best compromise between method efficiency and simplicity. First, we assume that the symbols belong to a alphabet Σ with cardinality $|\Sigma|$. We consider a full binary tree with $|\Sigma|$ leaves. Each leaf is uniquely associated with a symbol in Σ. Each internal node v defines the binary problem of predicting whether the next symbol is a leaf on v's left subtree or a leaf on v's right subtree. Then, we "attach" a binary CTW predictor to each internal node. We project the training sequence over the "relevant" symbols (i.e., corresponding to the subtree rooted by v) and translate the symbols on v's left (respectively, right) subtree to 0s (respectively, 1s). A diagram of the $De\,CTW$ is depicted in Figure 8.5.

8.2.4   Comparison of Prediction Schemes

In the preceding section, we described briefly the mechanics of four families of Markov predictors. In this section, we will perform a qualitative comparison of the families; initially, we will comment on some generic features/advantages of the

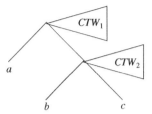

Predictor	Sequence it models
CTW	aa-a---a--a---
CTW	bcbbbbcbbc

Figure 8.5   A sketch of the *De CTW* Markov predictor for the sequence *aabacbbabbacbbc*.

families and then elaborate on the kind of applications that could benefit from the prediction performance of each family.

Implicitly or explicitly, all Markov predictors are based on the *short memory principle*, which says that the probability distribution of the next symbol can be approximated by observing no more than the last k symbols in that sequence. Some methods fix in advance the value of k (e.g., $\mathcal{PPM}, \mathcal{CTW}$). If the selected value for k is too low, then it will not capture all the dependencies between symbols, degrading its prediction efficiency. On the other hand, if the value of k is too large, then the model will overfit the training sequence. Therefore, variable-length Markov predictors (e.g., $\mathcal{LZ}78, \mathcal{PST}$) are in general more appropriate from this point of view. This was the motivation for subsequent enhancements to $\mathcal{PPM}$ and $\mathcal{CTW}$ so as to consider unbounded length contexts, for example, the $\mathcal{PPM}^\star$ algorithm [13].

On the other hand, variable-length predictors face the problem of which sequences and of what length should be considered. $\mathcal{PST}$ attempts to estimate the predictive capability of each subsequence in order to store it in the trie, which results in deploying many tunable parameters. $\mathcal{LZ}78$ employs a prefix-based decorrelation process, which results in some recurrent structures to be excluded from the trie, at least at the first stages. This characteristic is not very important for infinite-length sequences, but may incur performance penalty for short sequences; for instance, the pattern *bba* is missing in both variants of $\mathcal{LZ}78$ of Figure 8.2. Although this example is by no means a kind of proof that $\mathcal{LZ}78$ is inferior to the other algorithms, it is an indication of how an individual algorithm's particularities may affect its prediction performance, especially in short sequences. Despite their superior prediction performance, $\mathcal{PPM}$ schemes are far less commonly applied than algorithms like $\mathcal{LZ}78$, which is favored over $\mathcal{PPM}$ algorithms for its relative efficiency in memory and computational complexity.

In Table 8.1, we summarize the Markov predictor families, their main members, and their main qualitative characteristics.

From Table 8.1, we can gain some insights into which method is more appropriate for which type of application. Although, we emphasize that the choice and performance of a specific model largely depends on the application characteristics, it is the case that some results in the relevant literature show relative gains in the performance of one method with respect the others for specific applications. To the best of our knowledge, we found no study that compares *all* families mentioned in this article for the location prediction issue with both synthetic and real data. In general, the number

Table 8.1 Qualitative Comparison of Discrete Sequence Prediction Models

Prediction method			Overheads			
Family	Variant	Markov class	Train	Parame/tion	Storage	Particularity
$\mathcal{LZ}$78	[8]	Variable	Online	Moderate	Moderate	May miss patterns
	[64]	Variable	Online	Moderate	Moderate	
$\mathcal{PPM}$	[10]	Fixed	Offline	Heavy	Large	Fixed length
	[14]	Fixed	Online	Moderate	Large	High complexity
	[16]	Fixed	Offline	Heavy	Large	
$\mathcal{PST}$	[5]	Variable	Offline	Heavy	Low	Parameterization
	[41]	Variable	Offline	Heavy	Low	
$\mathcal{CTW}$	[57]	Fixed	Online	Moderate	Large	Binary nature
	[58]	Fixed	Online	Moderate	Large	

of such studies is limited and the real data they use (when they use such) come from limited settings (e.g., university campuses) and not from users of commercial wireless systems, for example, cellular networks. Thus, it is not possible to draw safe conclusions from these works. Worthwhile studies containing comprehensive experiments with real data are reported in Refs. [11, 21, 37, 45].

Although alternative approaches could be possible, we prefer to present our suggestions for policy selection along two primary dimensions; the first dimension reflects the type of the problem (i.e., location prediction) and the second dimension reflects the "network part," where the prediction is carried out (i.e., fixed, resource-rich network servers, or resource-starving mobile hosts).

For mobile applications, some very important intuitive results, which have also been confirmed experimentally [25], can be stated: (a) user interests vary significantly with time (not "strong" stationarity), (b) many alternative paths exist that lead to the same target location, thus the regularity patterns are "blurred" by noise. Due to the first observation, the possibility of using $\mathcal{LZ}$78 types of predictors is rather small. Due to the variance in the length of the individual client's trajectories, the rest of the variable-order Markov predictors are more appropriate; $\mathcal{PST}$ would be a perfect choice under the assumptions that the procedure is performed offline and it runs on a resource-rich server, or a relatively powerful laptop.

If energy conservation is the main issue in these applications (small portable devices like PDAs, mobile phones), then the choice of $\mathcal{PPM}$ style predictors seems more appropriate since they are online, but they sacrifice some prediction performance (due to the relatively small and fixed order model employed) for reduced model complexity. The second observation may turn all prediction methods inefficient, since it violates the "consecutiveness" property of appearance of the symbols in the patterns, upon which property all described Markov predictors rely. In cases where "noise symbols" are interleaved in pattern subsequences, the modified Markov predictors described in Refs. [16, 37] can be employed, but these algorithms are offline and require substantial resources (memory, power) to be executed. Therefore, they could only be used by fixed network servers, which collect huge amounts of user trajectories.

Location prediction is considered a relatively manageable problem, because of the few alternatives in possible contexts (i.e., hexagonal architecture of cellular systems, few fixed access points in wireless LANs, and smart home applications) and because of the "strong" stationarity (i.e., few habitual routes in campuses/cities, few travel paths in urban regions—road network).

For location, prediction applications, several families of Markov predictors could be used in some specific scenarios each. For dynamic tracking of mobile hosts (with the tracking application running either in the network server or the mobile host), $\mathcal{PPM}$ and $\mathcal{LZ}78$ methods are appropriate. The small order $\mathcal{PPM}$ model and the enhanced $\mathcal{LZ}78$ [8] are expected to achieve the best performance, because of the undoubted validity of the stationarity assumption. Indeed, the study in Ref. [45] confirmed that intuitive results. These variants are perfect fit for dynamic resource allocation before handovers, as well. For location area design applications, where we are interested in discovering "long-standing" repetitive user routes, the process is offline and therefore methods like $\mathcal{PST}$ or [16, 37] are appropriate and less vulnerable to statistical deviation.

8.3   PREDICTIVE LOCATION INDEXING TECHNIQUES

In the previous section, we presented the dominant approaches for predictive location tracking for the case of a symbolic topology model. In many cases though, we are interested in tracking the location of mobiles at a finer granularity of space and time. Consider, for instance, our interest in tracking the trajectories of birds, airplanes, or satellites, which are considered as points, or our interest in tracking the movement of a tropical storm, of fires. This interests stems from our need to answer queries like, "When two satellites are going to meet?", "Is the fire threatening village Thetidio?", and so on.

This leads to the idea of storing in a database for each moving object not the current position but rather a motion vector, which amounts to describing the position as a function of time. That is, if we record for an object its position at time t_0 together with its speed and direction at that time, we can derive expected positions for all times after t_0. Of course, motion vectors also need to be updated from time to time, but much less frequently than positions. Hence, from the location management perspective, we are interested in maintaining dynamically the locations of a set of currently moving objects and in being able to ask queries about the current positions, the positions in the near future, or any relationships that may develop between the moving entities and static geometries over time.

To support such functionality, the database must build *indexes*; the main task if an index is to ensure fast access to single or several records in the database on the basis of a search key and thus avoid an otherwise necessary naive scan. Numerous indexes able to support continuous movement have appeared in the literature [18].

The indexes capable of accommodating moving objects can be generally categorized into those optimizing queries about past states of movement and those tailored to serving queries about future positions of the moving objects; the first type of queries are called *historical*, while the second one are termed *predictive*. In the following, we will address the predictive location indexing techniques.

8.3.1 Assumptions and Terminology

In the majority of the applications, the size and the shape of the moving objects are insignificant. Consequently, every object is modeled as a geometric point whose position consists of a time function $\mathbf{x}(t)$ of the particular motion parameters. On the basis of $\mathbf{x(t)}$, the applications evaluate the future object locations. Furthermore, it is expected that the involved objects have the capacity and the obligation of periodically reporting any alteration occurring in $\mathbf{x}$ parameters. As an example, in most cases, $\mathbf{x}(t)$ is a linear function of time $\mathbf{x}(t) = \mathbf{x}(t_{\text{ref}}) + \mathbf{v}(t - t_{\text{ref}})$, where the two parameters are the position $\mathbf{x}(t_{\text{ref}})$ at the reference time t_{ref} and the velocity vector $\mathbf{v}$. Generally speaking, the equation parameters specify a dual to the time–location space framework.

As far as the type of queries is concerned, they can be classified as *range queries* and *proximity* or *nearest neighbor* (*NN*) ones. A range query is termed (i) *time-slice* or *snapshot* (r, t) when, given a region r, usually a hyper-rectangle, located at time t, it asks for all moving objects contained in r at that time; (ii) *window* $(r, [t_1, t_2])$ when it asks for all objects crossing a hyper-rectangle r during a time interval $[t_1, t_2]$; (iii) *moving* $(r_1, r_2, [t_1, t_2])$ when it inquires all objects crossing the trapezoid formed by connecting the hyper-rectangle r_1 at time t_1 and the hyper-rectangle r_2 at time t_2; and (iv) *selectivity* or *aggregate* $(r, [t_1, t_2])$ when, given a region r and a time interval $[t_1, t_2]$, it requests an estimation about the number of the objects that will pass through r during $[t_1, t_2]$. Types (i)–(iii) are also known as *range reporting*, whereas type (iv) is well tailored to the case of limited memory and strict real-time processing demand.

On the other hand, a proximity query inquires for the k nearest moving objects to a given location at a time instance t or during a time interval $[t_1, t_2]$, with $k \geq 1$. Sometimes, all objects having a given location as their nearest neighbor are asked for; this kind of searching is termed *reversed nearest neighbor searching* (*RNN*).

In the above definitions, two variations are possible: First, the query range/point also moves which makes the query *continuous*. And second, the answer set may be returned with temporal or spatial validity information, informing thus the user about the future time t the result expires or the validity region r containing the query position within which the answer remains valid.

Concluding this subsection, we briefly discuss the R-tree [19], a general purpose practical indexing structure, since the majority of the solutions to be presented are built upon it. So, an R-tree is a height-balanced tree that can be considered as an extension of the B^+-tree for multidimensional data. The minimum bounding rectangle (MBR) of each geometric object, along with a pointer to the disk address where the object actually resides, are stored into the leaves. Each internal node entry consists of a pair (pointer to a subtree $\mathcal{T}$, MBR of $\mathcal{T}$), with the MBR of a tree $\mathcal{T}$ defined as the MBR

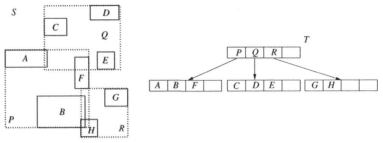

Figure 8.6 An R-tree instance: (a) a set S of rectangles A–H and (b) the corresponding R-tree T.

enclosing all the MBRs stored in T. Like in B^+-trees, each node contains at least m and at most M entries, where $m \geq M/2$. On the other hand, unlike B^+-trees, a search query may activate several search paths from the root to the R-tree leaves, resulting, in the worst case, in a linear to the size of dataset performance just to retrieve a few objects.

Figure 8.6 illustrates an R-tree instance on a set S of rectangles. Since its introduction, several variants of the R-tree have been proposed, each one aiming at improving the performance by tuning some parameters. Among the members of the "R-tree family," the most prominent one is the R^*-tree of Beckmann et al. [6].

8.3.2 Indexing Structures for Range Queries

8.3.2.1 Range Reporting

Tayeb et al. [55] presented one of the first works on indexing mobile objects for snapshot and window queries by reducing the problem to indexing lines in a periodically rebuilded bucket PR Quadtree [44] with high space requirements. Subsequently, Kollios et al. [27] presented solutions for one- and two-dimensional range searching, which index objects either in xt-plane using R-trees [19] or in the two-dimensional dual space deploying dynamic external partition trees [2]; the last proposal is mainly of theoretical interest due to the employed index structures.

In Ref. [43] a time-parameterized version of R^*-trees [6], called *TPR*-tree, was introduced, because none of the members of the R-tree family of indexes is efficient in indexing mobile objects. Let us explain this with the aid of Figure 8.7. The leftmost Figure 8.7 digram shows the positions and velocity vectors of seven point objects at time 0. Assume we create an R-tree at time 0. The second diagram shows one possible assignment of the objects to MBRs assuming a maximum of three objects per node. Previous work has shown that attempting to minimize the quantities known as overlap, dead space, and perimeter leads to an index with good query performance [6], and so the chosen assignment appears to be well chosen. However, although it is good for queries at the present time, the movement of the objects may adversely affect this assignment. The third diagram shows the locations of the objects and the MBRs at time 3, assuming that MBRs grow to stay valid. The grown MBRs adversely affect

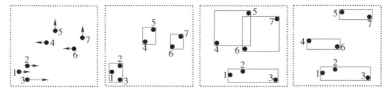

Figure 8.7 Mobile objects and resulting leaf-level MBRs.

query performance; and as time increases, the MBRs will continue to grow, leading to further deterioration. Even though the objects belonging to the same MBR (e.g., objects 4 and 5) were originally close, the different directions of their movement cause their positions to diverge rapidly and hence the MBRs to grow. From the perspective of queries at time 3, it would have been better to assign objects to MBRs as illustrated by the rightmost diagram. Note that at time 0, this assignment will yield worse query performance than the original assignment. Thus, the assignment of objects to MBRs must take into consideration when most queries will arrive.

The *TPR*-tree is capable of accommodating moving objects with constant velocities in one-, two-, and three-dimensional space and is the *de facto* spatial index for future queries. The MBRs in this example illustrate the kind of time-parameterized bounding rectangles-supported by the *TPR*-tree. The novelty of the approach are the TPBRs being employed, associated with a velocity vector (see Figure 8.8): for each coordinate x_i, the lower bound is set to be the minimum observed x_i-coordinate value at time t_{ref}, moving with the minimum observed velocity, and the upper bound is defined to be the maximum x_i-coordinate value at time t_{ref}, moving with the maximum observed velocity. Since the TPBRs never shrink and are conservatively bounded, the index is tuned for H time units; after that, a global rebuilding of the structure is conducted. TPBRs also impose the generalization of the update and rebuilding algorithms of the R^*-trees, so that their respective objective functions are time parameterized. In summary, *TPR*-trees support all types of range queries (time-slice queries, window queries, and moving) while proven to be a very practical solution.

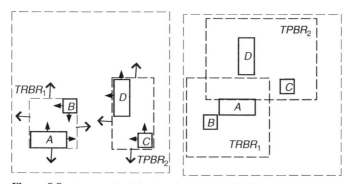

Figure 8.8 An example of forming time-parameterized bounding rectangles.

Agarwal et al. [3] provided solutions of theoretical interest for answering window and moving window queries for one- and two-dimensional moving points with time complexities that depend on the number of events — that is, alteration of the relative with respect to an axis location order between objects or insert/deletion of a moving object—that occurred between the current time and the query time. Their approach is based on a careful segmentation of the plane/space into a logarithmic number of strips (slabs), each one containing a bounded number of events. The arrangements of every slab are then stored in persistent versions of B-trees. As a result, moving window queries have a time complexity of $O(\log_B n + k/B + B^{i-1})$ and $O(\sqrt{n/B^i}(B^{i-1} + \log_B n) + k/B)$, for one- and two-dimensional movements, respectively, B being the page capacity and i the slab number.

R^{EXP}-trees [42] use also TPBRs for bounding moving objects. However, capitalizing on the assumption that objects expire after a time period t_{exp}, within which they have not reported their position, the author derived analytical formulas that produce tighter TPBRs. This fact in conjunction with the lazy removal of expired objects only after an update operation discovers an expired entry make R^{EXP}-trees to exhibit better experimental performance with respect to TPR-trees. STAR-trees [39] were also introduced as an improvement over TPR-trees for the two-dimensional case. The main feature of this proposal is self-adjustment: based upon user specifications concerning space overhead and performance quality, the index self-adjusts without any user interference. Toward this end, the extends of the points are constantly approximated, with the aid of a priority queue, and the children of a node are redistributed whenever they overlap too much. The authors reported the following improvements: a speed-up of 2.3 with respect to TPR-tree was achieved and the deterioration of the scheme over time is quite small.

TPR^*-trees [52] improved TPR-trees by introducing more elaborated decision making processes for insertion path selection, node re-insertion, and children node redistribution that payoff a lot in terms of search performance. Additionally, the authors suggested a cost model for the original TPR-tree that emphasizes the factors that influence its performance and shows the superiority of TPR^*-tree over TPR-trees. This fact was also confirmed by an extensive experimental investigation that demonstrated that with the TPR^*-tree the average query cost is almost five times less, whereas the average update cost is nearly constant.

In Ref. [1] three theoretical indexing schemes for moving points in the plane were introduced. The first one improves upon the approach of Ref. [27] by using a two-level external partition tree and has time complexity of $O(n^{1/2+\epsilon} + k)$, k the output size. The other two support one- and two-dimensional queries referring to the present time or arriving in a strictly chronological order and have logarithmic complexity. The one-dimensional index employs kinetic B-trees while the two-dimensional index employs kinetic range trees.

Patel et al. [38] introduced the STRIPE index, which basically is a multidimensional external PR bucket quadtree supporting all three types of range queries (time slice, window, and moving). Each moving object is represented in dual $2d$-dimensional parametric space. The parametric space is indexed by applying a disjoint regular partitioning mainly imposed by the underlying quadtree. The authors tested the performance of STRIPE against the *TPR**-tree thoroughly and found that their proposal is faster in both update and query time; namely, queries are faster by a factor of four, while the update operations complete by an order of magnitude quicker since TPR*-trees have the disadvantages of poor cache locality and multiple path traversals.

Tao et al. [48] coped with the case of circular static range query during a time interval for moving objects with unknown moving patterns. In order to support arbitrarily movements, Tao et al. suggested a monitor-and-index framework: Each moving object individually and constantly computes the recursive function that best describes its movement using motion matrices introduced by the authors. On the other hand, the server adopts the same coarse polynomial function *m* for every object. The latter fact introduces imprecision that is dealt with a two-step process. During the filtering phase, all objects that either definitely or possibly satisfy the query are determined using STP-tree. STP-tree can be considered as a generalization of both *TPR* and *TPR**-trees to arbitrary polynomial functions *m*. In a second refinement step, the server communicates with the doubtful objects that evaluate the query according to their own accurate moving function and send back the result if necessary along with the corrected parameters of *m*. The applicability of the proposal is thoroughly investigated through a series of experiments that concern both the movement approximation method and the new index.

8.3.2.2 *Range Aggregate.*

In 2002, Choi and Chung [12] presented the first work on rectangular range selectivity for static queries. Their solution for the one-dimensional case is based on the simple observation that a point satisfies the query range *r* if and only if the segment formed by its (tentative) position at the start and the end of the query time interval intersects *r*. The authors proposed a histogram-based counting solution so that the spatial universe is partitioned into a number of time-evolving buckets. This method was also extended to the two-dimensional case by projecting objects and queries on each spatial dimension and evaluating the selectivity as the product of the one-dimensional results.

The previous method suffers from overestimated results since the projection overlooks necessary temporal conditions. Tao et al. [54] achieved better estimation results by dropping the projection step; their solution tackles directly the problem with the deployment of a spatio-temporal histogram, which considers both locations and velocities. The authors report significant improvements in selectivity precision and manipulation of updates. In Ref. [20], one- and two-dimensional movements were also coped with. As a matter of fact, two solutions were introduced: One is also based on multidimensional histograms, which, none the less, are defined on dual space. The accommodation of the moving objects into an index is prerequisite for the second solution. Namely, the summaries of each leaf entry, in the form of number of objects and their bounded rectangular spatial extent, are stored in a hash table that is used for output extrapolation.

In contrast to the previous three solutions, in Ref. [53] random sampling was chosen for selectivity estimation in order to spare space and processing overhead.

Particularly, the introduced method of Venn sampling employs a set S of m pivot queries, which represent the actual distribution and are perfectly estimated. The most interesting part is that, based on S, a weighted sample of m moving objects, non-necessarily belonging to the underlying dataset, is formed and communicated to the moving objects. This sample is queried against any incoming query and the sum of weights of qualifying objects is returned to the user. The system constantly monitors the quality of the sample set in collaboration with the moving objects; in case the estimation error surpluses a threshold, the sample weights but not the respective sample objects are adjusted. The authors present experimental results, rendering the proposal quite appealing.

8.3.3   Indexing Structures for Nearest Neighbor Queries

Kollios et al. [26] introduced practical yet preliminary solutions for locating the nearest moving neighbor of a static query in the plane permitting object movement in fixed line segments either arbitrary or restricted. They consider both the cases of indexing in the xt-plane and in the dual plane. The first case allows indexing with standard spatial indexes, like, for example, R-trees [19], while the second one utilizes B^+-trees [15] and dual plane segmentation in horizontal stripes.

Song and Roussopoulos [46] studied the continuous kNN problem, which inquires for the k-nearest static neighbors of a moving query point. Their approach is based on the observation that as the query point is moving to new locations, some previously reported neighbors are still among the k closest ones. This fact is formally expressed by a series of conditions that allow the employment of standard indexing structures like R-trees.

Zheng and Lee [63] considered enhancing NN queries with validity information. The approach is quite simple: The nearest neighbor of a moving point is calculated using the Voronoi diagram of the static dataset. Since the Voronoi cell c of the nearest neighbor is available, the maximum circle around the query point that does not cross any boundary edge of c consists a safe lower bound for the validity of the query result.

In Ref. [7], two-dimensional nearest neighbor and reverse nearest neighbor queries for a query point q during a time interval t were treated by properly extending *TPR*-tree algorithms. NN queries apply a depth-first search technique that constantly prunes TPBRs with no chance enclosing closest points. The RNN queries are more elaborate: the space around the query point q is divided into six equal sectors s_i by straight lines intersected at q, since there must exist at most six RNN points, at most one in each s_i. Therefore, sectors containing two or more nearest points of q are discarded and the search is restricted among the nearest neighbors of the rest.

Continuous NN queries for static datasets were also treated, albeit from a different perspective, in Ref.[22]. This solution is built upon ellipsoid areas around the moving query point that are generated utilizing current, past, and future trajectory positions and carefully selected metrics. The static dataset, on the other hand, is indexed with a "spatial" structure, like, for example, R-trees. Since the previous answers are not reused as the query point changes position, this method needs involved tuning to be effective.

Tao et al. [51] considered continuous kNN search when the static input dataset is accommodated in an R-tree T and the query point is linearly moving. When $k = 1$, these assumptions guarantee that the output set consists of a sequence of points p_i, which partition the movement line segment into a sequence of disjoint segments s_i. Every point in s_i has p_i as its nearest neighbor. These facts suggest a branch-and-bound investigation of T employing heuristic node pruning. Tao et al. also generalized the pruning rules for the continuous kNN case, providing an extensive experimental evaluation of their proposal.

Aggarwal and Agrawal [1] introduced NN solutions for objects moving in nonlinear trajectories of arbitrary dimensionality whose parametric representation satisfy the so-called convex hull property. The d-dimensional trajectories with constant velocity, d-dimensional parabolic trajectories, elliptic orbits, and trajectories that accept approximate Tailor expansion belong to this category. Due to the convex hull property, the locality in parametric space corresponds to the locality in the positions of objects, and, thus NN search can be accomplished by a branch-and-bound, best-first algorithm conducted in a classical spatial index. The authors demonstrated their method for linear and parabolic trajectories in three- and two-dimensional space, respectively.

TPR-trees algorithms were enhanced in Ref. [40], so that continuous kNN queries on moving points during a time interval $[t_1, t_2]$ can be served. The suggested solution capitalizes on the following geometric fact: the kNN points can be determined by the k levels of the arrangement of the squared distance functions of the moving points with respect to the moving query during $[t_1, t_2]$. Therefore, after collecting at least k points by a depth-first traversal of the underlining TPR-tree according to the minimum squared distance metric between the moving query and the bounding rectangles at t_1, the kNN points of $[t_1, t_2]$ are determined. Then, in a second stage, the TPR-tree is once more traversed, in order to refine and finalize the output set.

Iwerks et al. [23] also presented algorithms for answering continuous kNN queries on a constantly moving point set with respect to either a static or a moving query point. Their approach runs also in two phases. During first phase, input set is filtered with a continuous query that asks about objects within a distance bound d. Then, the qualified points are ranked with a priority queue, which tracks the time instances when points change their distance to the query point or when points change their order with respect to a current kNN point.

Nearest neighbor queries, among others, were also treated in Ref. [1]. Specifically, Agarwal et al. suggested an algorithm that offers approximate results to NN searching by replacing the Euclidean metric with a polyhedral one. The input set is accommodated into a three-level composite index of $O(n^{1/2+\epsilon}/\sqrt{\delta})$ time complexity, $0 < \delta, \epsilon < 1$. The first two levels are external partition trees on dual space. The last level stores the lower envelope of the trajectories in linear lists.

The conceptual partitioning method (CPM) was introduced in Ref. [35] for constantly monitoring of multiple continuous NN queries in highly dynamic environments. In brief, the space is partitioned by a regular grid and indexed in main memory. Every cell of this grid maintains the list of objects residing therein and every posed query along with its current result set is stored in a table. Additionally, CPM imposes a total order on the cells around every query based on proximity criterion. In that way, every update in both the data and the query set is tackled with minimal computational costs without any assumptions about the occurring moving patterns; this is depicted by both qualitative and thorough experimental analysis.

After the previous work, Mouratidis et al. [36] presented a main memory solution for incremental monitoring of continuous kNN queries when the query and the data objects move in road networks. Their approach basically is based on network expansion about the query until k nearest neighbors are collected. The formed shortest path tree is stored along with the query so that any updates are smoothly incorporated. Moreover, the authors proposed a method for computation sharing among queries whose shortest paths are crossed. All suggested solutions are experimentally evaluated.

Finally, Lee et al. [30] treated continuous nearest surrounder (NS) queries that ask for the nearest neighbors at individual distinct angles from the query point. NS queries, thus, monitor the nearest neighbors around the queries by considering both the distance and the angular attributes. The system registers the objects' locations in an R-tree and the queries, along with their current results, in a hash table so that any update of either data or query objects can be incrementally evaluated, capitalizing on the notion of "safe regions" firstly introduced in Ref. [29].

8.3.4 Indexing Structures for Both Window and Nearest Neighbor Queries

In Ref. [50] time-parameterized window and kNN queries (TP) were introduced which, along with the objects that satisfy the spatial conditions, also return the expiration time of the validity of the answer and the change that invalidates the answer at that time. The key concept of this approach is the influence time that is associated with every moving object o and indicates the time o influences the validity of the answer. By definition, the expiration time of the answer equals the minimum influence time of all objects which, in turn, can be evaluated with a NN search where the distance metric is the influence time. This observation is valid for both window and kNN queries. The authors proved analytical formulas for evaluating the influence times of an object. As a result, standard branch-and-bound traversal of the index accommodating the object set can be employed. Their solution also treats continuous spatio-temporal queries by posing a TP query every time the current result expires. Finally, TP queries can also serve earliest event queries that ask for the earliest time in the future a certain event could take place; for example, one may need to figure out the first time a moving query point q meets another moving object. By surrounding q with a time-varying radius cycle, this query reduces to TP by evaluating the earliest time the circle contains a point, which, in turn, equals to determining the smallest radius of such a circle.

Building upon Ref. [50], Zhang et al. [61] dealt also with validity kNN and window queries. In the first case, order-k Voronoi cells comprise the validity regions. These can be found implicitly by, first, generating the nearest neighbors and then issuing time-parameterized kNN queries for locating points that define their border.

In the second case, the maximal rectangle r around the center of the window, within which the result remains unchanged, is first evaluated. Then, r is refined by subtracting the parts that would force the query to mistakenly contain points not in the reported answer. These two steps involve one standard window query, one "holey" window query, and few main memory TP window queries.

Tao et al. [49] proved a number of theoretical bounds on validity range and NN queries. Specifically, when the query's length and movement are chosen from a constant number of combinations and the point set is static, the query cost is logarithmic and the space is linear. When the point set is static, the query length is arbitrarily, and when the movement is axis-parallel, then the time complexity is $O(\log_B^2(n/B)/\log_B \log_B(n/B))$ and the space cost is $O(n/B \log_B(n/B)/\log_B \log_B(n/B))$. On the other hand, in the case of static input point set and queries with arbitrary length and movement, the space complexity is $O(n/B)$ while the query costs are $O((n/B)^{1/2+\epsilon})$. When the data points are dynamic and the query is static, the query has logarithmic complexity and space is bounded by $O(n^2/B \log_B(n/B))$. The case of both dynamic data points and query point is only considered in one-dimensional space, proving a linear space and logarithmic time complexity. As far as the NN search query is concerned, when the input point set is static lying in the plane or comprised of moving points on the line, the solution is of linear space and logarithmic query cost.

The B^x-trees were introduced in Ref. [24], which are capable of serving range, and kNN queries as well as their continuous counterparts. The main ingredient of this method is movement linearization: The time axis is partitioned into intervals of Δ time units, and each interval is further subdivided into n subintervals of equal length. Every object, according to its t_{ref}, is assigned to one subinterval. Within each subinterval, the positions of the objects fallen in are linearized according to a space filling curve and then stored into a B^+-tree. Therefore, the B^x-tree is actually a sequence of B^+-trees evolving as time goes by. The authors conducted extensive experiments, which show the superiority of B^x-trees over TPR-trees for all kinds of queries. Here we must note that BB^x-trees [32] were introduced as a natural extension of B^x ones, able of answering both predictive and historical queries.

After observing that the linearization process of Ref. [24], in Ref. [59] only object's locations were considered, leading thus to excessive false hits, proposed B^{dual}-trees. B^{dual}-trees also deploy B^+-trees indexing, none the less, a space-filling curve that is based on both object locations and velocities. The authors demonstrated the superiority of their method over B^x-trees both analytically (i.e., using derived formulas) and experimentally—the experimental study also gives data for STRIPES and TPR^*-trees. In a nutshell, B^{dual}-trees can be considered as the state-of-the-art solution in its category.

8.3.5 Evaluation of Predictive Indexing

In overall, TPR-trees [43] and its descendant variations, like TPR^*-trees [52], can be considered the most appropriate and practical choice for serving range queries.

In case of limited resources, the Venn sampling technique of Ref. [53] is deemed a natural option. As far as nearest neighbor queries is concerned, the *TPR*-tree variation of Ref. [40] and the CPM of Ref. [35] are both competent candidates. Finally, B^x-trees [24] and B^{dual}-trees [59] proved to be apt solutions when one wants to treat equally well range and nearest neighbor queries.

Regarding future steps toward developing more indexing methods for mobiles, it would be very helpful if the indexes could provide results that capture the uncertainty associated with the location of moving objects due to network delays and the continuous character of motion. It would be also very interesting to efficiently cope with nonlinear trajectories since the scope and the range of indexable moving objects will be significantly extended. Another appealing subject, especially for extending mobile application capabilities, is the design of indexing structures capable of serving mixed queries concerning the past and the future of movement The incremental valuation of validity queries is very intriguing, as well. Finally, from an engineering perspective, it would be very helpful (i) to test all indexes with real datasets, as, until now, every experimental investigation is conducted with semireal ones, where the movement component is actually generated; and (ii) to design efficient updating algorithms for the indexes, different from the usual "deletion and re-insertion" practice, accepting perhaps a tradeoff between either the query time or the accuracy of the result and the update time.

8.4 CONCLUSION

This chapter identified the inherent uncertainty in the movement of mobiles in areas covered by wireless networks and the problems caused to resource allocation due to this uncertainty. Starting from this fundamental observation, it subsequently recognized the benefits of being able to forecast the future locations of the mobile hosts. This ability could be used to act proactively, instead of reactively, to many situations. For instance, having estimates of future positions of mobile hosts, the network could take appropriate decisions regarding the bandwidth that will allocate to the cells containing these locations. In addition, in wireless ad hoc networks, where communication between nodes is performed on a store-and-forward basis for nodes not in close proximity, the communication could be deferred until the nodes come closer to each other, thus saving network resources, like precious bandwidth and storage space in the intermediate nodes, reducing packet collisions, and so on.

However, predictive location tracking can be performed only if the mobiles' movements exhibit some degree of regularity, thus making the construction of mobility models feasible. The generic principle governing location prediction can be summarized in a short sentence: *study the present and project to the future*. Exploiting this principle, the issue of location prediction turned out to be a matter of recording the present mobile trajectories and developing mobility models from these. The storage of the trajectories should be of the kind to allow compact representation and at the same time efficient generation of predictions.

Subsequently, we investigated two different scenarios for predictive location prediction. According to the first scenario, the roaming area of the mobiles can be considered as a union of nonoverlapping cells of arbitrary geometry and according to the second scenario, the location tracking is performed at the granularity of geographical coordinates. For the first scenario, we modeled the problem of predictive location tracking in terms of the discrete sequence prediction problem. For this latter problem, we presented Markov predictors as a practical and high-performance solution, categorized them into four families giving their qualitative characteristics, their strengths, and their weaknesses. For the second scenario, which is more tightly coupled with location databases, we surveyed the state-of-the-art techniques for constructing indexes capable of answering queries that mainly concern various complex future predicates.

Undoubtedly, predictive location tracking is very important for reducing the latency and the resource consumption in any wireless network or for prolonging the lifetime of wireless ad hoc networks. However, the problem is not easily manageable due to the difficulty of constructing models representing the actual mobile trajectories. Although very significant steps have been taken toward achieving this target, work is still needed to characterize the predictability of mobile trajectories, to analyze collections of real mobile trajectories, to develop more effective prediction models, and mainly to develop distributed models of prediction through cooperation, which will be suitable for the emerged area of wireless sensor networks.

ACKNOWLEDGMENTS

Research supported by a ΓΓΕΤ grant in the context of the project "Data Management in Mobile Ad Hoc Networks" funded by ΠΥΘΑΓΟΡΑΣ II national research program.

REFERENCES

1. P. K. Agarwal, L. Arge, and F. Erickson. Indexing moving points. *J. Comput. Syst. Sci.*, 66(1):207–243, 2003.
2. P. K. Agarwal, L. Arge, F. Erickson, P. G. Franciosa, and J. S. Vitter. Efficient searching with linear constraints. *J. Comput. Syst. Sci.*, 61(2):194–216, 2000.
3. P. K. Agarwal, L. Arge, and J. Vahrenhold. Time responsive external data structures for moving points. In *Proceedings of the International Workshop on Distributed Algorithms and Data Structures (WADS)*, Vol. 2125, *Lecture Notes in Computer Science*, pp. 50–61, 2001.
4. C. C. Aggarwal and D. Agrawal. On nearest neighbor indexing of nonlinear trajectories. In *Proceedings of the ACM Symposium on Principles Of Database Systems (PODS)*, pp. 252–259, 2003.
5. A. Apostolico and G. Bejerano. Optimal amnesic probabilistic automata or how to learn and classify proteins in linear time and space. *J. Comput. Bio.*, 7(3–4):381–393, 2000.
6. N. Beckmann, H.-P. Kriegel, R. Schneider, and B. Seeger. The R^*-tree: an efficient and robust access method for points and rectangles. In *Proceedings of the ACM International Conference on Management of Data (SIGMOD)*, pp. 322–331, 1990.
7. R. Benetis, C. S. Jensen, G. Karciauskas, and S. Saltenis. Nearest neighbor and reverse nearest neighbor queries for moving objects. *Very Large Data Bases J.*, 15(3):229–250, 2006.

8. A. Bhattacharya and S. K. Das. LeZi-Update: an information-theoretic framework for personal mobility tracking in PCS networks. *ACM/Kluwer Wireless Networks*, 8(2–3):121–135, 2002.
9. S. Chakraborty, Y. Dong, D. K. Y. Yau, and J. C. S. Lui. On the effectiveness of movement prediction to reduce energy consumption in wireless communication. *IEEE Trans. Mobile Comput.*, 5(2):157–169, 2006.
10. X. Chen and X. Zhang. A popularity-based prediction model for Web prefetching. *IEEE Comput.*, 36(3):63–70, 2003.
11. F. Chinchilla, M. Lindsey, and M. Papadopouli. Analysis of wireless information locality and association patterns in a campus. In *Proceedings of the IEEE International Conference on Computer Communications (INFOCOM)*, Vol. 2, pp. 906–917, 2004.
12. Y.-J. Choi and C.-W. Chung. Selectivity estimation for spatio-temporal queries to moving objects. In *Proceedings of the ACM International Conference on Management of Data (SIGMOD)*, pp. 440–451, 2002.
13. J. G. Cleary and W. J. Teahan. Unbounded length contexts for PPM. *Comput. J.*, 40(2–3):67–75, 1997.
14. J. G. Cleary and I. H. Witten. Data compression using adaptive coding and partial string matching. *IEEE Trans. Commun.*, 32(4):396–402, 1984.
15. D. Comer. The ubiquitous *B*-tree. *ACM Comput. Surv.*, 11(2):121–137, 1979.
16. M. Deshpande and G. Karypis. Selective Markov models for predicting Web page accesses. *ACM Trans. Internet Technol.*, 4(2):163–184, 2004.
17. M. Feder, N. Merhav, and M. Gutman. Universal prediction of individual sequences. *IEEE Trans. Inform. Theory*, 38(4):1258–1270, 1992.
18. R. H. Güting and M. Schneider. *Moving Objects Databases*. Series in Data Management Systems. Morgan-Kaufmann, 2005.
19. A. Guttman. *R*-trees: a dynamic index structure for spatial searching. In *Proceedings of the ACM International Conference on Management of Data (SIGMOD)*, pages 47–57, 1984.
20. M. Hadjieleftheriou, G. Kollios, and V. J. Tsotras. Performance evaluation of spatio-temporal selectivity estimation techniques. In *Proceedings of the IEEE International Conference on Statistical and Scientific Database Management (SSDBM)*, pp. 202–211, 2003.
21. M. Halvey, M. Keane, and B. Smyth. Mobile Web surfing is the same as Web surfing. *Commun. ACM*, 49(3):76–81, 2006.
22. Y. Ishikawa, H. Kitagawa, and T. Kawashima. Continual neighborhood tracking for moving objects using adaptive distances. In *Proceedings of the IEEE International Database Engineering and Applications Symposium (IDEAS)*, pp. 54–63, 2002.
23. G. S. Iwerks, H. Samet, and K. Smith. Continuous *k*-nearest neighbor queries for continuously moving points with updates. In *Proceedings of the International Conference on Very Large Data Bases (VLDB)*, pp. 512–523, 2003.
24. C. S. Jensen, D. Lin, and B. C. Ooi. Query and update efficient B^+-tree based indexing of moving objects. In *Proceedings of the International Conference on Very Large Data Bases (VLDB)*, pp. 768–779, 2004.
25. D. Katsaros and Y. Manolopoulos. Prediction in wireless networks by Markov chains. *IEEE Wireless Commun. Mag.*, in press.
26. G. Kollios, D. Gunopoulos, and V. J. Tsotras. Nearest neighbor queries in a mobile environment. In *Proceedings of the International Workshop on Spatio-Temporal Database Management (STDBM)*, Vol. 1678, *Lecture Notes in Computer Science*, pp. 119–134, 1999.
27. G. Kollios, D. Gunopoulos, and V. J. Tsotras. On indexing mobile objects. In *Proceedings of the ACM Symposium on Principles Of Database Systems (PODS)*, pp. 261–272, 1999.
28. M. Kyriakakos, N. Frangiadakis, L. Merakos, and S. Hadjiefthymiades. Enhanced path prediction for network resource management in wireless LANs. *IEEE Wireless Commun.*, 10(6):62–69, 2003.
29. K. C. K. Lee, W.-C. Lee, and H. V. Leong. Nearest surrounder queries. In *Proceedings of the IEEE International Conference on Data Engineering (ICDE)*, 2006.
30. K. C. K. Lee, J. Schiffman, W.-C. Zheng, B. Lee, and H. V. Leong. Tracking nearest surrounders in moving object environments. In *Proceedings of the IEEE International Conference on Pervasive Services (ICPS)*, pp. 3–12, 2006.

31. D. A. Levine, I. F. Akyildiz, and M. Naghshineh. A resource estimation and call admission algorithm for wireless multimedia networks using the shadow cluster concept. *IEEE/ACM Trans. Network.*, 5(1):1–12, 1997.
32. D. Lin, C. S. Jensen, B. C. Ooi, and S. Saltenis. Efficient indexing of the historical, present, and future positions of moving objects. In *Proceedings of the IEEE International Conference on Mobile Data Management (MDM)*, pp. 59–66, 2005.
33. T. Liu, P. Bahl, and I. Chlamtac. Mobility modeling, location tracking, and trajectory prediction in wireless ATM networks. *IEEE J. Select. Areas Commun.*, 16(6):922–936, 1998.
34. A. Misra, A. Roy, and S. K. Das. An information-theoretic framework for optimal location tracking in multi-system 4G wireless networks. In *Proceedings of the IEEE International Conference on Computer Communications (INFOCOM)*, Vol. 1, pp. 286–297, 2004.
35. K. Mouratidis, M. Hadjieleftheriou, and D. Papadias. Conceptual partitioning: an efficient method for continuous nearest neighbor monitoring. In *Proceedings of the ACM International Conference on Management of Data (SIGMOD)*, pp. 634–645, 2005.
36. K. Mouratidis, M. L. Yiu, D. Papadias, and N. Mamoulis. Continuous nearest neighbor monitoring in road networks. In *Proceedings of the International Conference on Very Large Data Bases (VLDB)*, pp. 43–54, 2006.
37. A. Nanopoulos, D. Katsaros, and Y. Manolopoulos. A data mining algorithm for generalized Web prefetching. *IEEE Trans. Knowledge Data Eng.*, 15(5):1155–1169, 2003.
38. J. M. Patel, Y. Chen, and V. P. Chakka. STRIPES: an efficient index for predicted trajectories. In *Proceedings of the ACM International Conference on Management of Data (SIGMOD)*, pp. 637–646, 2004.
39. C. M. Procopiuc, P. K. Agarwal, and S. Har-Peled. *STAR*-tree: an efficient self-adjusting index for moving objects. In *Proceedings of the Workshop on Algorithm Engineering and Experiments (ALENEX)*, Vol. 2409 *Lecture Notes in Computer Science*, pp. 178–193, 2002.
40. K. Raptopoulou, A. Papadopoulos, and Y. Manolopoulos. Fast nearest-neighbor query processing in moving-objects databases. *Geoinformatica*, 7(2):113–137, 2003.
41. D. Ron, Y. Singer, and N. Tishby. The power of amnesia: learning probabilistic automata with variable memory length. *Mach. Learn.*, 25(2–3):117–149, 1996.
42. S. Saltenis and C. S. Jensen. Indexing of moving objects for location-based services. In *Proceedings of the IEEE International Conference on Data Engineering (ICDE)*, pp. 463–472, 2002.
43. S. Saltenis, C. S. Jensen, S. T. Leutenegger, and M. A. Lopez. Indexing the positions of continuously moving objects. In *Proceedings of the ACM International Conference on Management of Data (SIGMOD)*, pp. 331–342, 2000.
44. H. Samet. *The Design and Analysis of Spatial Data Structures*. Addison-Wesley, 1990.
45. L. Song, D Kotz, R. Jain, and X. He. Evaluating location predictors with extensive Wi-Fi mobility data. In *Proceedings of the IEEE International Conference on Computer Communications (INFOCOM)*, Vol. 2, pp. 1414–1424, 2004.
46. Z. Song and N. Roussopoulos. k-nearest neighbor for moving query point. In *Proceedings of the International Symposium on Advances in Spatial and Temporal Databases (SSTD)*, Vol. 2121 *Lecture Notes in Computer Science*, pp. 79–96, 2001.
47. S. Tabbane. Location management methods for third-generation mobile systems. *IEEE Commun. Mag.*, 35(8):72–78, 83–84, 1997.
48. Y. Tao, C. Faloutsos, D. Papadias, and B. Liu. Prediction and indexing of moving objects with unknown motion patterns. In *Proceedings of the ACM International Conference on Management of Data (SIGMOD)*, pp. 611–622, 2004.
49. Y. Tao, N. Mamoulis, and D Papadias. Validity information retrieval for spatio-temporal queries: Theoretical performance bounds. In *Proceedings of the International Symposium on Advances in Spatial and Temporal Databases (SSTD)*, Vol. 2750 *Lecture Notes in Computer Science*, pp. 159–178, 2003.
50. Y. Tao and D. Papadias. Time-parameterized queries in spatio-temporal databases. In *Proceedings of the ACM International Conference on Management of Data (SIGMOD)*, pp. 334–345, 2002.

51. Y. Tao, D. Papadias, and Q. Shen. Continuous nearest neighbor search. In *Proceedings of the International Conference on Very Large Data Bases (VLDB)*, pp. 287–298, 2002.
52. Y. Tao, D. Papadias, and Q. Sun. The *TPR**-tree: an optimized spatio-temporal access method for predictive queries. In *Proceedings of the International Conference on Very Large Data Bases (VLDB)*, pp. 790–801, 2003.
53. Y. Tao, D. Papadias, J. Zhai, and Q. Li. Venn sampling: a novel prediction technique for moving objects. In *Proceedings of the IEEE International Conference on Data Engineering (ICDE)*, pp. 680–691, 2005.
54. Y. Tao, J. Sun, and D. Papadias. Selectivity estimation for predictive spatio-temporal queries. In *Proceedings of the IEEE International Conference on Data Engineering (ICDE)*, pp. 417–428, 2003.
55. J. Tayeb, O. Ulusoy, and O. Wolfson. A quadtree-based dynamic attribute indexing method. *Comput. J.*, 41(3):185–200, 1998.
56. J. S. Vitter and P. Krishnan. Optimal prefetching via data compression. *J. ACM*, 43(5):771–793, 1996.
57. P. Volf. *Weighting Techniques in Data Compression: Theory and Algorithms*. PhD Thesis, Technische Universiteit Eindhoven, 2002.
58. F. J. Willems, Y. M. Shtarkov, and T. J. Tjalkens. The context-tree weighting method: basic properties. *IEEE Trans. Inform. Theory*, 41(3):653–664, 1995.
59. M. Yiu, Y. Tao, and N. Mamoulis. The B^{dual}-tree: indexing moving objects by space-filling curves in the dual space. *Very Large Data Bases J.*, in press.
60. F. Yu and V. Leung. Mobility-based predictive call admission control and bandwidth reservation in wireless cellular networks. *Comput. Networks*, 38(5):577–589, 2002.
61. J. Zhang, M. Zhu, D. Papadias, Y. Tao, and D. L. Lee. Location-based spatial queries. In *Proceedings of the ACM International Conference on Management of Data (SIGMOD)*, pp. 443–454, 2003.
62. Z. Zhang. Routing in intermittently connected mobile ad hoc networks and delay tolerant networks: Overview and challenges. *IEEE Commun. Surv. Tutorials*, 8(1):24–37, 2006.
63. B. Zheng and D. L. Lee. Semantic caching in location-dependent query processing. In *Proceedings of the International Symposium on Advances in Spatial and Temporal Databases (SSTD)*, Vol. 2121 *Lecture Notes in Computer Science*, pp. 97–113, 2001.
64. J. Ziv and A. Lempel. Compression of individual sequences via variable-rate coding. *IEEE Trans. Inform. Theory*, 24(5):530–536, 1978.

Chapter 9

An Efficient Air Index Scheme for Spatial Data Dissemination in Mobile Computing Environments

SeokJin Im,[1] Chong-Sun Hwang,[1] and MoonBae Song[2]

[1] *Department of Computer Science, Korea University, Seoul, Korea*
[2] *Graduate School of Systems and Information Engineering, University of Tsukuba, Tsukuba, Japan*

9.1	Introduction	191
9.2	Preliminaries	194
9.3	Air Indexes for Nonspatial Data	196
9.4	Air Indexes for Spatial Data	200
9.5	Cell-Based Distributed Air Index for Spatial Data	204
9.6	Summary	210
	References	211

9.1 INTRODUCTION

Mobile computing geared by wireless communication technologies is one of the explosive growth areas. Most wireless communication systems have the characteristics of asymmetry that the bandwidth of the downlink channel is higher than that of the uplink channel. Also, wireless channel has error-prone characteristics caused by

Mobile Intelligence. Edited by Laurence T. Yang, Agustinus Borgy Waluyo, Jianhua Ma, Ling Tan, and Bala Srinivasan

various link error sources such as frequent disconnections [7]. Under mobile computing, mobile clients with limited battery lifetime can get their desired information anytime and anywhere like stock information, and so on. Also, by the advances in high-speed wireless networks and portable devices and location-identification techniques such as GPS, location-dependent information services (LDIS) are enabled [22, 28].

There are two approaches to satisfy the needs of mobile clients for their desired information under mobile computing environments mentioned above. One possible way to support various queries of a huge number of mobile clients is on-demand approach in which an information server takes care of queries from mobile clients [1]. In this approach, each client sends its query that requests its desired data items, then the information server returns results of the query via a one-to-one wireless communication. This may cause scalability problem in which the server is overheated by a huge number of mobile clients. Figure 9.1 shows an example of the on-demand approach. Each mobile client requests the stock price of its interesting company to the information server which maintains DB about 10 companies and then the server returns the stock price requested by the mobile clients.

The other way to support the needs of a huge number of mobile clients is wireless data broadcast approach [5, 12–14]. This is an effective way to deliver information to a huge number of mobile clients because a wireless data broadcast system can accommodate an arbitrary number of clients. In wireless data broadcast, the broadcast server periodically broadcasts a set of data items via the downlink channel and each client filters out its desired data items in the wireless channel after tuning in

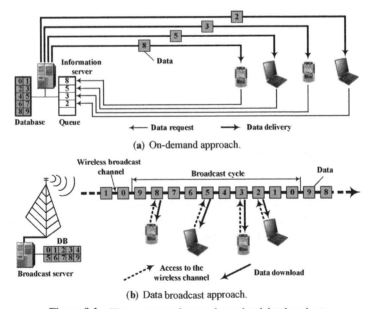

(a) On-demand approach.

(b) Data broadcast approach.

Figure 9.1 The two approaches: on-demand and data broadcast.

the wireless channel. Therefore, the wireless data broadcast approach is free from the scalability problem. Also, data broadcast can alleviate the load on the uplink channel. Figure 9.1(b) shows an example of the wireless data broadcast approach. The broadcast server disseminates stock prices of 10 companies and each mobile client downloads stock prices of its interesting company from the wireless broadcast channel. There are three models that are commonly accepted broadcast, push-based broadcast, on-demand broadcast, and hybrid broadcast. Push-based broadcast model is that the broadcast server disseminates data items without the consideration of the need of mobile clients [1, 29]. In the on-demand broadcast model, the broadcast server disseminates data items, taking into consideration the need of mobile clients [30, 31]. In the hybrid broadcast model, strong points of the push-based model and on-demand model are combined [18, 32].

For the limited battery life time of the clients, they operate in two modes while accessing their desired data items: active mode (the energy-consuming mode) and doze mode (the energy-saving mode). The performance of the system is characterized by access time and tuning time [1]. Access time refers to the elapsed time from the moment when a query is issued to the moment when the query is satisfied and represents access efficiency. Tuning time refers to the amount of time in the active mode during the access time and represents energy consumption of the clients since energy consumption in active mode is larger than that in doze mode.

In order to download the desired data items from the wireless channel, mobile clients have to listen to the channel in active mode until the desired data items appear on the channel. So the mobile clients consume lots of energy during the data retrieval due to the long stay in the active mode. Air indexing in wireless broadcast systems has been proposed for facilitating selectively listening to the desired data items for energy saving [1, 11, 12, 28]. The basic idea of air indexing is that the index information holding the arrival times of all data items is interleaved with data items on the wireless channel. After access to the channel, mobile clients predict the arrival time of the desired data item using index information and switch to doze mode until it arrives on the wireless channel. Then, mobile clients switch to the active mode and listen to the data item. However, air indexing facilitating energy conservation of the mobile clients has the drawback of the lengthened broadcast cycle by the additional index information. This drawback causes the access time of the mobile clients to be come long, while their tuning time is shortened.

Recently, to enable LDISs with the wireless data broadcast system, the systems of spatial data items are proposed that are able to satisfy various spatial queries of the mobile clients. They can get the result of various spatial queries such as window query and k nearest neighbors (kNN) query, where a window query is to find out all data items contained within a given query window, qw, and a kNN query is to find k data items closer to a given query point, q, than other data items among the entire data items [19, 21]. For example, a mobile client could find the nearest restaurant from the current location of the client or all hotels within the radius of 1 km. In the aspect of the filter-and-refinement technique, a kNN query can be processed by finding kNN using the answer to the window query with qw that contains kNN [20]. Therefore, the effective processing of window queries may be effective also to various queries, especially to the *k*NN queries. In this chapter, we explore the air indexing schemes for spatial data in mobile computing environments to efficiently support LDISs. First, we *review simply* existing air indexes for spatial data and nonspatial data. Then, we present distributed air index based on space partitioning with less energy consumption and robustness for link error on the wireless channel than the existing air indexing schemes for spatial data [28].

The rest of this chapter is organized as follows. In Section 9.2, we describe basics and data organization in wireless data broadcast. Various air indexes for nonspatial data are described in Section 9.3. In Section 9.4, air indexes for spatial data are introduced. In Section 9.5, we present cell-based distributed spatial index, called CEDI, for energy conserving and error resilient data search in wireless data broadcast environments. Finally, this chapter is summarized in Section 9.6.

9.2   PRELIMINARIES

9.2.1   Basic Concepts

In air indexing, a broadcast cycle consists of index segments to hold index information and data segments to hold data items. The two segments are organized as a sequence of buckets, where the bucket is the smallest logical unit of information delivery and all buckets are of the same size like disk page [1]. Therefore, mobile clients access data items in the unit of buckets. Index buckets refer to buckets in index segments and data buckets to buckets in data segments. Every bucket contains a bucket header that keeps the following information: *bucket_id* referring to a bucket identifier, *bucket_type* indicating the kind of the bucket, index bucket or data bucket, *bcast_pointer* referring to the offset to the beginning of the next broadcast cycle, and *index_pointer* referring to the pointer to the next index segment. *Bcast* is defined as the length of a broadcast cycle.

Tuning time consists of initial probe, index listening time, and data listening time [1]. Initial probe refers to determining the time when the next index segment appears on the wireless channel and is completed within 1.5 buckets on average after tuning into the wireless broadcast channel. Index listening time is the amount of time the mobile client listens to index segments for predicting the arrival time of the desired data items. Data listening time is the amount of time the mobile client downloads the desired data items from the wireless broadcast channel. Access time is the sum of probe wait and *Bcast* wait. Probe wait is the duration of meeting the next index segment after tuning into the wireless broadcast channel. Bcast wait is the duration from of meeting the first index segment after tuning-in to completing the download of the desired data items.

Air indexes facilitating selective listening to desired data items have to satisfy the following requirements to efficiently support mobile clients to process given queries.

- *Linear structure:* The data structure of air indexes should be linear because the mobile clients access the wireless channel linearly. Therefore, to efficiently support the index search of the mobile clients, linear structure of air indexes is a very important characteristic to be satisfied.

- *Reduction of space cost:* The size of air indexes should be minimized in order to reduce the index listening time in the aspect of tuning time, to minimize *Bcast* increased by additional air indexes on the wireless channel in the aspect of access time, and to strengthen the error resilience against various channel link error sources. Owing to the bigger index size, the larger probability of bucket loss is same as the same link-error probability of the wireless channel.

- *Supporting energy-efficient query processing:* This is the most essential property of air indexes. During the query processing, air indexes should not make the mobile clients listen to redundant data items not contained in the given query for energy conservation.

Especially, the third requirement mentioned above is essential in air indexes for processing the spatial query because the search space determined with air indexes for the given spatial query may contain redundant data items.

9.2.2 Data Organization

The data organization on the wireless broadcast channel is very important as it affects the performances, that is, access time and tuning time. We consider the four methods to interleave index information with data items on the wireless broadcast channel, Latency_Opt, Tune_Opt, $(1, m)$ indexing and distributed indexing as shown in Figure 9.2.

- *Latency_opt:* This technique gives the best access time with the worst tuning time. The best access time is obtained without index information on the wireless channel as shown in Figure 9.2(a). In this technique, *Bcast* is minimal and mobile clients listen to all data items until the desired data items appear on the wireless channel after tuning into the channel.

- *Tune_opt:* This technique gives the best tuning time with the worst access time. The index information is broadcast only once at the beginning of each broadcast cycle as shown in Figure 9.2(b). In this technique, to access the index, mobile clients take the *index_pointer* from the bucket header of a bucket encountered first after tuning into the wireless broadcast channel. Then the mobile clients access the index with *index_pointer* and predict the times the desired data items arrive during the search of the index. Then, mobile clients access the data items at the times determined by the index and download them all.

- $(1, m)$ *Indexing*: In this index allocation method, the index of entire data items is broadcast m times ahead of every $(1/m)$ fraction of data items in a single broadcast cycle as shown Figure 9.2(c) in order to reduce probe wait of access time [1, 16]. The first bucket of each index segment has a tuple of which the

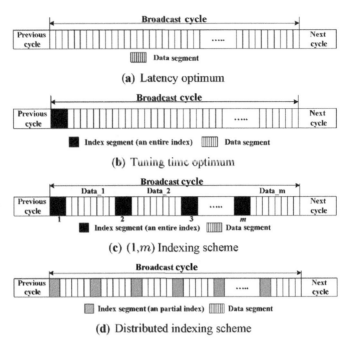

Figure 9.2   The schemes of data organization on the wireless broadcast channel.

first field is the attribute value of the record that was broadcast last and the
second field is the offset to the beginning of the next broadcast cycle. The tuple
guides mobile clients who miss the desired data items in the current broadcast
cycle to tune in to the next broadcast cycle [1].

- *Distributed indexing:* In this index allocation method, indexes of parts of entire
  data items are distributed by being placed ahead of the parts on the wireless
  broadcast channel as shown Figure 9.2(d). Distributed index scheme has shown
  more shortened access time than $(1, m)$ indexing due to reduced probe wait
  and *Bcast* [17]. Also for mobile clients to complete the given query processing
  with this distributed indexing, links between distributed indexes should be
  maintained to efficiently access queried data items.

9.3 AIR INDEXES FOR NONSPATIAL DATA

9.3.1 Tree-Based Index

The tree-based air index is adopted from the tree-structured index in a traditional
disk-based environment [1]. To organize the index of data items to be broadcast
with the form of tree, the unique identifier of each data item is used as a key for
index tree. The index tree keeps the arrival time of each data item on the wireless

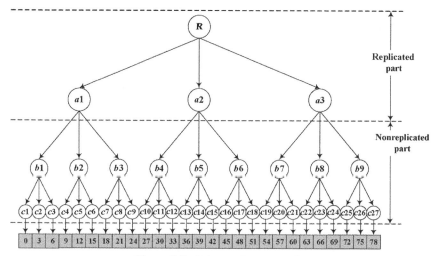

Figure 9.3   Index tree for an example.

broadcast channel, unlike the tree keeping the locations of disk records in the disk-based environments. Figure 9.3 illustrates an example of an index tree for 81 data items to be broadcast, white circles mean the index nodes of the index tree and a gray square means a set of three data items. Every leaf node in the index tree keeps attribute keys of three data items and offset values to them in units of buckets as shown Figure 9.4.

The two indexing schemes based on the index tree, that is, $(1, m)$ indexing and distributed indexing scheme, are depicted in Figure 9.4. For $(1, m)$ indexing, the entire index tree is interleaved with every $(1/m)$ fraction of data items and all nodes in the index tree are placed on the wireless channel according to their level as shown in Figure 9.4(a). For the distributed indexing scheme based on the index tree, a part of the index tree is interleaved with the data items associated with it. For the organization of the distributed indexes, the tree is divided into two parts, replicated part and nonreplicated part as shown in Figure 9.3. The replicated part consists of the upper level nodes of the tree. Each node in the replicated part is repeated d times during a single broadcast cycle, where d is equal to the number of child nodes of the node. For example, in Figure 9.3, d is three and root node, R, and node, $a1$, are replicated three times in a broadcast cycle as shown in Figure 9.4(b). The nonreplicated part is consists of the lower level nodes of the tree. Every node in the nonreplicated part is broadcast only once during a single broadcast cycle. Distributed indexing scheme can greatly reduce the increment of *Bcast* by the addition of index information on the wireless channel. This can improve the access time significantly without sacrificing the tuning time. Also, the replicated part in distributed indexing scheme plays the role of a link between the distributed indexes on the wireless channel and can support multiple access paths to the desired data items.

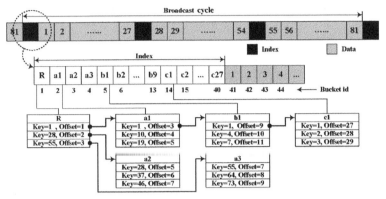

(a) Channel structure by $(1, m)$ indexing scheme based on the tree $(m = 3)$

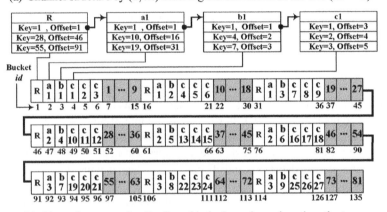

(b) Channel structure by distributed indexing scheme based on the tree

Figure 9.4 Channel structure by various indexing schemes on the index tree.

9.3.2 Hashing-Based Index

For air indexing based on the hashing, data frames consisting of a data part and a control part are used. The data part in every data frame consists of a data item and the control part has the information on a hash function and a shift function for mobile clients to filter out their desired data items. The hash function hashes the key attribute to compute the arrival time of the desired data frame. The shift function is the complementary function of the hash function in such a case of occurrence of collisions due to the imperfectness of the hash function. The shift function keeps the pointer to a bucket that contains keys to occur the collision [2].

9.3.3 Signature-Based Index

In this scheme, a signature of bit string is generated by hashing with an information frame and it is interleaved with the associated information frame as shown in Figure 9.5

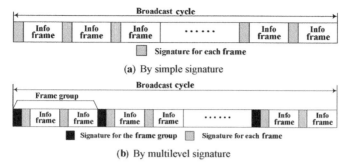

Figure 9.5   Channel structure with signature-based index.

[3]. To filter out the desired information, mobile clients compare a query signature with the signature of a information frame by performing a bitwise *AND* operation. If the query signature matches with the signature of the information frame, mobile clients download the desired data items from the information frame. Otherwise, they move to next signature.

The simple signature-based indexing scheme interleaves signatures ahead of their information frames as shown in Figure 9.5(a), while the multilevel signature-based indexing scheme adds the integrated signature for a group of information frames of simple signature-based indexing scheme. Figure 9.5(b) illustrates the two-level signature-based indexing scheme for groups of two information frames. Every information frame has its own simple signature and the integrated signatures of groups of information frames are placed ahead of the groups.

9.3.4 Hybrid Air Index

Hybrid indexing scheme is the combination of the strong points of the tree-based index good for random data access and the signature-based index good for sequential access such as wireless data broadcast [4]. Hybrid index consists of a sparse index tree and signatures. The sparse index tree, which is the upper *t* levels of the index tree, provides a global view of information frames. Figure 9.6 shows the hybrid index as an example. Mobile clients search the sparse index tree to get the position of the

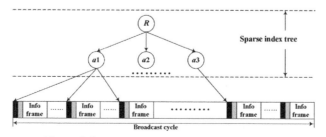

Figure 9.6   Hybrid index and channel structure with it.

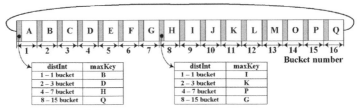

Figure 9.7 Exponential index and channel structure with it.

desire information frame on the wireless broadcast channel. And then, mobile clients compare signatures after accessing the frame to filter out desired data items.

9.3.5 Exponential Index

Exponential index is the index table that allows indexing spaces to be exponentially partitioned at any base value [6, 7]. Exponential indexing scheme allows mobile clients to start to search anywhere on the wireless broadcast by adopting distributed index allocation scheme. Figure 9.7 shows exponential index with 2 as the base value for 16 data items. This indexing scheme supports multiple data access by allowing mobile clients to get multiple pointers through once access to an index table. However, exponential index has a big size due to the number of replications of the identifier and the pointer to a data item as many as the number of entries of a index table. For the large dataset, exponential index makes access time very long due to the rapid increment in the size of exponential index with the number of data items. Also, in the procedure to access intermediate indexes guiding the desired data items, mobile clients cannot avoid listening to redundant data due to the use of chunks, which combine an index table and data items.

9.4 AIR INDEXES FOR SPATIAL DATA

To support various spatial queries in traditional database systems, indexing techniques for spatial data have been proposed such as R-tree, R*-tree, quad-trees, and k–d-trees [8–10, 25]. However, they cannot be easily adopted in wireless data broadcast because of their nonlinear structure not matching with linear access pattern of mobile clients. Recently, some air indexes have been proposed for processing spatial query in wireless broadcast environment [15–17, 22].

9.4.1 Tree-Based Index

In Ref. [24], window query processing is studied on indexes of tree structure. However, the technique proposed in Ref. [24] cannot support kNN query.

9.4.2 Space Partition-Based Indexes

D-tree is a binary search tree based on Voronoi Diagram (VD) to support nearest neighbor (NN) query [26]. It recursively partitions the VD into two subspaces until every space has a region and holds information on the polylines of the subspaces. This makes the size of D-tree big, leading to *Bcast* lengthened and access time deteriorated. Also, D-tree cannot easily extend to kNN query.

Grid partition index is a hybrid index combining VD and grid to efficiently support NN query by reducing the search space [27]. In the index, VD is partitioned into grid cells to map a query point to a grid cell. The scheme has two-leveled indexes: the upper level built on the grid and the lower level built on data items being potential NNs and facilitating the access to them.

9.4.3 Space Filling Curve-Based Indexes

Hilbert curve index (HCI) and DSI have been proposed for processing various spatial query such as window query and kNN query [15–17]. The two indexing schemes use Hilbert curve (HC), a kind of space-filling curve, for the scheduling of spatial data items on the wireless channel and organizing air indexes.

Every data item is given an HC value associated with the location of it. For example, Figure 9.8(a) illustrates a dataset D. The dots in the figure represent the locations of 16 data items in two-dimensional data space DS, $[0, 1]^2$. For generating HC, DS is partitioned into a $n_h \times n_h$ grid until each cell in the grid contains the only one data item, where $n_h = 2^h$ and h means the order of HC. Then, HC is generated on the grid and HC values are given all cell in the grid according to HC. The HC value of a cell in the grid is given to the data item within the cell as an unique identifer of the data item. Figure 9.8 shows the HC with $h = 3$ for D on DS. Figure 9.9(a) illustrates HC values given to data items in D.

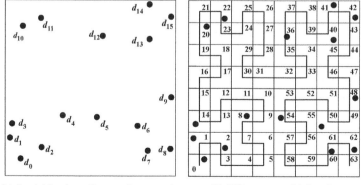

(a) Spatial data items for a running example   **(b)** Hilbert curve with $h = 3$

Figure 9.8   Generation of Hilbert curve.

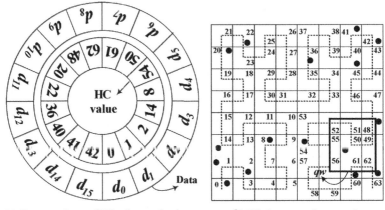

(a) Correspondence of HC values to data items   (b) Window query for example

Figure 9.9   HC values of data items and window query for example.

HCI and DSI are constructed using the HC values of data items. HCI is actually B+ tree that the HC value of a data item plays the role of the key of the tree. Figure 9.10(a) shows B+ tree constructed using the HC values of data items shown in Figure 9.9(a). HCI is broadcast by the allocation scheme of $(1,m)$ indexing for shortening probe wait. Figure 9.10(b) shows the structure of the wireless channel using HCI with $m = 4$. DSI is the index tables in which each table keeps the HC values and pointers to data items in a subset of data set D to be broadcast as shown in Figure 9.10(c) [17]. Each index table of DSI is broadcast by distributed indexing scheme as index allocation method for shorter probe wait than that by $(1,m)$ indexing scheme. HCI and DSI match well with the access pattern of the mobile clients to the channel. However, they have some problems in the view of performances as follows.

First, in the aspect of the access time, reducing the entire size of index becomes a solution of reduction of the access time because the index with reduced size as possible as much make *Bcast* near to *Bcast* of Latency_Opt. Therefore, the entire size of indexes in an broadcast cycle is very important. However, HCI and DSI are inefficient at the size of indexes in a broadcast cycle. In HCI, the entire size of the index is very big due to the m times repetition of the index of entire data items. DSI has more reduced size of indexes in a broadcast cycle than HCI due to distributed indexes of parts of entire data. However, the entire size of indexes in a broadcast cycle increases rapidly with the number of data items to be broadcast because a HC value and a pointer to a data item are repeated as many as the number of entries of an index table in a broadcast cycle.

Second, in the aspect of tuning time, the elimination of listening to redundant data and the reduction of the size of index become solutions to reduce tuning time, that is, energy conservation. HCI and DSI make mobile clients spend much energy during query processing because of filtering out the queried items after accessing too many candidates for the given query determined using HC not their original coordinates.

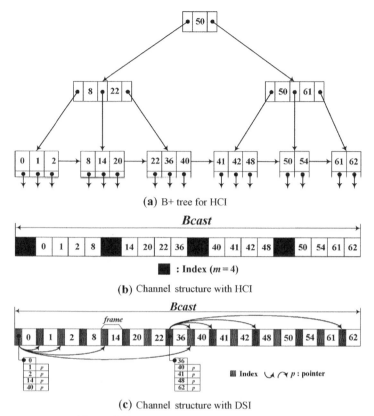

(**a**) B+ tree for HCI

(**b**) Channel structure with HCI

(**c**) Channel structure with DSI

Figure 9.10   Channel structure with HCI and DSI.

For a window query of query window qw, the candidates by HCI are all data items of HC values between the first HC value and the last overlapped with the query window. For example, to process the query window qw in Figure 9.9(b) with HCI, mobile clients have to access the candidates, the data items of HC values from 48 to 62, that is, $d_9(48)$, $d_6(50)$, $d_5(54)$, $d_7(61)$, and $d_8(62)$. Then, they filter out the queried items, d_6 after accessing the five candidates. HCI lets the mobile clients listen to 4 redundant items and it causes them to consume lots of energy.

With DSI the candidates for qw are determined all data items of HC values overlapped with the query window. For example, the candidates for qw in Figure 9.9(b) are determined as $d_9(48)$, $d_6(50)$, $d_7(61)$, and $d_8(62)$. Then, the mobile clients filter out d_6 after accessing the four candidates like by HCI. DSI makes the mobile clients listen to three redundant items. Although DSI decreases more redundant items than does HCI, it still has redundant items. Also with DSI, mobile clients have to listen to redundant data items when they access distributed indexing table to guide them to desired data items since DSI uses chunks that combine the indexing table and data items. This lets the mobile clients consume lots of energy. It becomes a solution of reduction of tuning time that decreasing the number and the size of indexes that the clients have to access for obtaining their desired data items.

Lastly, under the consideration of error-prone wireless channels caused by signal interference, bucket loss, and so on, the larger size of the index, the larger the probability of corruption of index. The size of index directly affects the error robustness. During the recovery from link error, mobile clients have seriously long access time and large energy consumption. In the viewpoint of the error resilience, decreasing the size of the index becomes a solution of error robustness.

9.5   CELL-BASED DISTRIBUTED AIR INDEX FOR SPATIAL DATA

This section presents cell-based distributed air index (CEDI) for energy conservative processing of window query with error resilience against various link error sources on the wireless channel and short data waiting time [28].

9.5.1   Design Goals for CEDI

- *Linear structure:* To support linear access pattern of the mobile clients to the wireless channel, CEDI is designed with table-formed linear structure.
- *Reduction of the size of index:* To reduce the size of index, CEDI is designed to hold only the pointers to data groups by partition of the data space and to eliminate pointers to data items on the wireless channel from it. Instead, CEDI provides the scheme to compute pointers to data items. Also, CEDI is designed with distributed indexing scheme to reduce its size.
- *Energy conservation:* To reduce the energy consumption of mobile clients, CEDI adopts the scheme filtering out queried data items with the original coordinates of data items in the index before accessing data items. Therefore, the filtering scheme removes listening to redundant data items. Also, CEDI is designed to enable the mobile clients to reduce the index listening time by reducing its size and to decrease the number of indexes accessed during query processing.
- *Shrunken access time:* CEDI adopts distributed indexing scheme to shrink access time by reduced probe wait. Also for the same goal, CEDI is designed to support multiple access paths to desired data items by links between distributed indexes.
- *Robust error resilience:* To strengthen resilience against the various channel link errors, CEDI is designed with index size reduced much as possible. For the fast recovery from link errors, CEDI is designed to support multiple access paths.

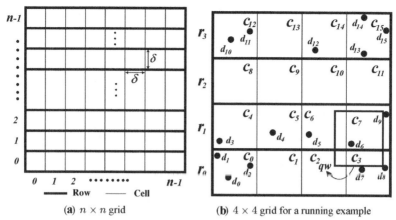

(a) $n \times n$ grid   **(b)** 4×4 grid for a running example

Figure 9.11   Space partition for CEDI.

9.5.2   Index Structure and Broadcast Channel Organization

To reach the goals mentioned above, CEDI is constructed on $n \times n$ grid by partitioning data space DS. For the organization of CEDI, DS of two-dimensional unit square $[0, 1]^2$ with N spatial data items is partitioned into $n \times n$ grid as shown in Figure 9.11(a). Each row of the grid is denoted as r_i ($0 \le i < n$), with r_0 as the bottom row. Each cell of the grid is denoted as c_j ($0 \le j < n^2$) assigned from the left in row major order and has the area of $\delta \times \delta$, where $\delta = (1/n)$. For example, Figure 9.11(b) shows the 4×4 grid of data items in Figure 9.8(a) and r_i and c_j.

CEDI adopts linear tables organized in two levels: the upper level row table denoted as RT_i and the lower level cell table denoted as CT_j. Every row of the grid with data items has its own row table indexing the other rows and the cells in it having data items. Also, the row table carries the global information, that is, the global view of the distribution of data items in the level of grid. Every cell of the grid with data items has its own cell tables carrying local information, that is, the local view of the the original coordinates of data items in a cell and pointers to the neighbor cell and row.

A row table RT_i of r_i is configured as follows:

$$RT_i = \langle i, BR, BC_i, PT_i^r, PT_i^c \rangle, \quad \text{for} \quad 0 \le i < n \qquad (9.1)$$

- Bit string BR: an n bit string indicating whether each row of the grid is empty or not. $BR = (b_0, b_1, \ldots, b_{n-1})$, where, if r_a is with data items, then $b_a = 1$, otherwise 0.

- Bit string BC_i: an n bit string indicating whether each cell in r_i is empty or not. $BC_i = (b_0, b_1, \ldots, b_{n-1})$, where, if c_j in r_i, where $j = i \cdot n + a$ ($0 \le a \le n - 1$), is with data items, then $b_a = 1$, otherwise 0.

- Pointer table PT_i^r: a table that keeps pointers to row tables of rows with data items, that is, times when row tables are broadcast. $PT_i^r = \{t_{Ra} | t_{Ra}$ is the time when RT_a is transmitted for $BR(a) = 1, 0 \le a \le n - 1\}$.

- Pointer table PT_i^c: a table that keeps pointers to cell tables of cells in r_i with data items, that is, times when cell tables are broadcast. $PT_i^c = \{t_{ca} | t_{ca}$ is the time when CT_a is transmitted for $BC_i(a - \lfloor \frac{a}{n} \rfloor n) = 1, n \cdot i \le a < n(i + 1)\}$.

BR and BC_i help the mobile clients to do the following two things: checking out empty rows and cells and deciding the order of the pointers of the rows and cells to be accessed in PT_i^r and PT_i^c. $\sum_{b=0}^{a} BR(b) - 1$ means the order of the pointer to r_a in PT_i^r and $\sum_{b=0}^{a\%n} BC_i(b) - 1$ means the order of the pointer to c_a in r_i in PT_i^r. BR and BC_i contribute to the reduction of the size of CEDI because they do not need to maintain identifiers of rows and cells in PT_i^r and PT_i^c.

The two pointer tables, PT_i^r and PT_i^c, provide multiple access paths for mobile clients to access the desired rows and cells by sharing the links between other row tables and cell tables. Also pointers to the other rows maintained in PT_i^r can decrease the number of the indexing tables accessed during the processing of given query due to the direct access to the desired row with PT_i^r without access to other indexing tables.

A cell table CT_j of c_j is configured as follows:

$$CT_j = \langle j, p_{nc}, p_{nr}, COT_j \rangle, \quad \text{for} \quad 0 \le j < n^2 \tag{9.2}$$

- pointer p_{nr}: the pointer to the row table appearing next to CT_j on the wireless channel.

- pointer p_{nc}: the pointer to the cell table appearing next to CT_j on the wireless channel. If c_j is the last cell in a row, p_{nc} is the same with p_{nr}.

- (COordinate Table COT_j): a table keeping original coordinates of all data items in c_j. $COT_j = \{(d_x, d_y) | (d_x, d_y)$ is the coordinates of data item d, belonging to $c_j\}$.

The two pointers p_{nr} and p_{nc} form a chain of pointers to rows and cells and support for mobile clients to access cell or row sequentially. COT_j enables mobile clients to access only the items within query window by filtering out them before access. This is different from the existing indexing schemes in which the client extract the queried data items after accessing all candidates determined by index. It makes energy consumption of the mobile clients much less than other indexing schemes.

Figure 9.12 illustrates the structure of the wireless broadcast channel with CEDI organized for data items in Figure 9.8(a). Row table RT_i is followed by cell tables of the cell in r_i, while each cell table is followed by data items residing in that cell. PT_0^r of RT_0, denoted by thick solid lines, lets the clients directly access a specific row table without referring to other tables. PT_0^c of RT_0, denoted by thick dotted lines, allows the clients to directly access the cells in r_0. p_{nr} and p_{nc} of CT_6, denoted by thin solid

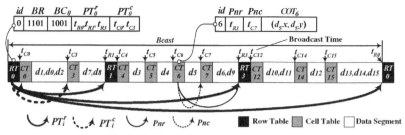

Figure 9.12   Channel structure with CEDI.

line and thin dotted line, respectively, let the clients sequentially access RT_3 and CT_7, respectively.

9.5.3 Implicit Cell Data Filtering

For selective tuning of the mobile clients, CEDI provides *Implicit cell data filtering* (IDF) scheme with which the clients can compute the broadcast time of data items in a cell, instead of providing pointers to data items explicitly in cell tables. IDF scheme reduces the size of the cell table because it does not keep the pointers to data items. To enable IDF scheme, all data items in a cell are broadcast by the order of them in COT_j as shown in Figure 9.13(b).

Lemma 9.1. *For N_c data items with the same size in a cell, the broadcast time of a specific data item is computed by the relative order to be broadcasted and the entire time length for broadcasting the N_c data items.*

Proof. Let t_s and t_e be the starting time and the ending time of broadcasting N_c data items in a cell with the same size, respectively. The length of time for broadcasting N_c data items is $t_e - t_s$. The elapsed time for broadcasting a data item is simply $t_e - t_s/N_c$. Let t_k be the start time of broadcasting the k-th data item. Then t_k is simply computed by $t_k = t_s + (k-1)(t_e - t_s)/N_c$.

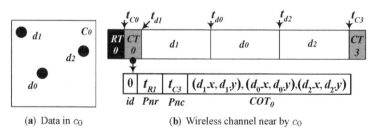

(a) Data in c_0 (b) Wireless channel near by c_0

Figure 9.13   Implicit cell data filtering scheme.

By Lemma 1, the pointers to data items in a cell easily can be computed. The time length to broadcast all data items in a cell is set to the difference between p_{nc} and the time when the broadcast of the cell table is completed. N_c is set to the number of the coordinates in COT_j. The broadcasting order of a specific data item is set to the order of the coordinates of the data item in COT_j. Figure 9.13(a) shows c_0 with three data items, that is, $\{d_0, d_1, d_2\}$. Figure 9.13(b) depicts the structure of the wireless channel near c_0. We consider the case that an mobile client downloads the data item d_0. The client accesses CT_0 on the wireless broadcast channel. Then, with the COT_0, the client confirms that the number of entire data items in c_0 is 3, and the order of the coordinates of the data item d_0 is the second. The client can easily compute t_{d_0}, the broadcast time of d_0. The t_{d_0} is $t_{d_1} + (2 - 1)(t_{CT\_3} - t_{d_1})/3$, where t_{d_1} is the time when the broadcast of CT_0 is completed, that is, the starting time of broadcast of data items in c_0. With the computed t_{d_0}, the client selectively downloads d_0 among three data items in c_0.

The IDF scheme can provide the concept of "indexing data items without index" within a cell. Thus, this scheme can make it possible to significantly decrease the size of entire index. Consequently, *Bcast* will be minimized. Also, it reduces the time to listen to indexing tables and leads to improved energy efficiency.

9.5.4   Query Processing

The client processes a window query for the given query window, qw, as follows. Here, qw is defined by (LL_x, LL_y) and (UR_x, UR_y), the coordinate of lower-left corner and upper-right corner, respectively.

Firstly, the client determines Q, a set of cells overlapped with qw as follows: $Q = \{c_j | j = a \cdot n + b, \lfloor LL_y/\delta \rfloor \leq a \leq \lfloor UR_y/\delta \rfloor$ and $\lfloor LL_x/\delta \rfloor \leq b \leq \lfloor UR_x/\delta \rfloor\}$. While sequentially accessing the cells in Q using the distributed tables, only the data items in qw are downloaded. In sequentially accessing the cells in Q, let c_n denote the cell to be accessed next at ptr, the pointer to guide the client to c_n. The client determines *ptr* using the tables of CEDI as follows:

$$\text{With } RT_i, \; ptr = \begin{cases} PT_i^c[a], & \text{if } \lfloor c_n/n \rfloor == i \text{ , here } a = \sum_{b=0}^{c_n\%n} BC_i(b) - 1 \\ PT_i^r[a], & \text{otherwise} \quad , \text{ here } a = \sum_{b=0}^{\lfloor c_n/n \rfloor} BR(b) - 1 \end{cases}$$

$$\text{With } CT_j, \; ptr = \begin{cases} p_{nc}, & \text{if } (j < c_n) \text{ and } (\lfloor j/n \rfloor) = (\lfloor c_n/n \rfloor) \\ p_{nr}, & \text{otherwise} \end{cases}$$

When the client meets RT_i, it removes the cell $c_j (\in Q)$ satisfying the following condition: $(BR(\lfloor c_j/n \rfloor) = 0)$ or $((BC_i(c_j\%n) = 0)\&\&(\lfloor c_j/n \rfloor = r_i))$ because it is a empty cell. After accessing CT_j, the client removes c_j from Q and extracts the data items in qw using COT_j. Then it computes the broadcast time of the extracted data items using the IDF scheme and accesses them. This operation is repeated until all grid cells in Q are accessed.

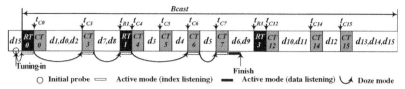

Figure 9.14   Processing of window query with CEDI.

The detailed algorithm of processing of window query is provided in Algorithm 1. The **while**-loop of lines 3–23 is the main loop for the algorithm. In line 10, function *getDataPointer()* returns the queue keeping the pointers to data items in *qw* computed by IDF scheme. The algorithm of the function is provided in Algorithm 2. Then the client selectively downloads the desired data items by invoking *getData()*, which returns data items pointed by *dpQueue*. The main loop is repeated until Q becomes ϕ.

Figure 9.14 illustrates an example of query processing for the given query window *qw* in Figure 9.11(b). In the example, Q are determined into $\{c_2, c_3, c_6, c_7\}$. A client

---

Algorithm 9.1: *WindowQuery*

Input: *qw*, a query window
Output: *Result*, a set of data items belonging to *qw*
1: $Q \leftarrow$ the cells overlapped with *qw*;
2: *Table* $\leftarrow$ the indexing table firstly encountered after tuning-in;
3: **while true do**
4:   **if** (*Table* is CT_j of c_j) **then**
5:     Select c_n from Q, the cell to be accessed next;
6:     **if** $\left((c_j < c_n)\&\&(\lfloor \frac{c_j}{n} \rfloor == \lfloor \frac{c_n}{n} \rfloor)\right)$ **then** $ptr \leftarrow p_{nc}$;
7:     **else** $ptr \leftarrow p_{nr}$;
8:     **if** ($c_j \in Q$) **then**
9:       Remove c_j from Q;
10:       $dpQueue \leftarrow getDataPointer(qw, COT_j)$;
11:       **for** pointer $p_i \in dpQueue$ **do**
12:         $d_i \leftarrow getData(p_i)$; // d_i: data pointed by p_i
13:         $Result \leftarrow Result \cup d_i$;
14:   **else** // when *Table* is RT_i of r_i
15:     **for** all grid cells $c \in Q$ **do**
16:       **if** $\left(BR(\lfloor c/n \rfloor) == 0 \text{ or } \left((BC_i(c\%n) == 0)\&\&(\lfloor c/n \rfloor == r_i)\right)\right)$ **then** Remove c from Q;
17:     Select c_n from Q;
18:     **if**($\lfloor c_n/n \rfloor == r_i$) **then** $ptr \leftarrow PT_i^c \left[\sum_{b=0}^{c_n\%n} BC_i(b) - 1\right]$;
19:     **else** $ptr \leftarrow PT_i^r \left[\sum_{b=0}^{\lfloor c_n/n \rfloor} BR(b) - 1\right]$;
20:   **if**($Q == \phi$) **then break**;
21:   **else** Switch to the doze mode and wake up at ptr;
22:   *Table* $\leftarrow$ the table accessed currently;
23: **endwhile**;
24: **return** *Result* ;

Algorithm 9.2: *getDataPointer(qw, COT_j)*

Input: qw, COT_j
Output: *dpQueue*, a queue keeping pointers to data items contained in qw
1: $N_c \leftarrow$ the number of coordinates in COT_j;
2: **for** $COT_j[a]$, $a = 1$ to N_c **do**
3:    **if** $COT_j[a] \in qw$ **then**
4:       $dpQueue \leftarrow \left(t_s + (a-1)\frac{t_e-t_s}{N_c}\right)$;
       // t_s and t_e are the starting and ending time of broadcasting data items in a cell
5: return *dpQueue*

---

accesses RT_0 using the pointer to the next indexing table kept in the bucket header of the first complete bucket after tuning into the wireless data broadcast channel as shown in Figure 9.14. With RT_0, the client deletes c_2 from Q using BC_0 of RT_0 because it is empty. The client selects c_3 as c_n and determines *ptr* to $PT_0^c[1]$ to access c_3 in Q. With CT_3, the client deletes c_3 from Q and checks whether there are data items in c_3 contained within qw using COT_3. The client does not download any data item because there are no data items within qw. Then, the client selects c_6 as c_n and sets *ptr* to p_{nr} and moves to RT_1. With RT_1, the client sets *ptr* to $PT_0^c[2]$ to access c_6. With CT_6, the client removes c_6 from Q, selects c_7 as c_n and set *ptr* to p_{nc}. The client moves to c_7, not downloading any data items because d_5 of c_6 is not in qw. With CT_7, the client removes c_7 from Q and Q becomes ϕ. The client extracts d_6 residing within qw using COT_7 and computes the broadcast time of it by invoking *getDataPointer()*. Then, the client downloads it and finishes processing qw since Q is ϕ,

9.6 SUMMARY

This chapter has presented air index techniques for energy conservative query processing in wireless data broadcast environments. We briefly reviewed basics of wireless data broadcast and data organization under air indexes. Indexes for nonspatial data, for example, tree-based index, hashing-based index, signature-based index, hybrid index, and exponential index, were described. In the air index for spatial data, HCI and DSI based on Hilbert Curve, a kind of space-filling curves, were described in detail. The indexes have various problems, long broadcast cycle by their big size and listening to redundant data items by filtering queried data items after accessing candidates determined on Hilbert Curve. In this chapter, we presented CEDI. CEDI has much reduced size by keeping pointers of groups of data items by $n \times n$ grid, instead of the pointer of each data item. CEDI has distributed structure and supports multiple access paths by the replication of the pointers of data groups. CEDI eliminates listening to redundant data items during the processing of window query by filtering out the queried data items using original coordinates their before accessing them. CEDI is very efficient in the aspect of energy conservation and access time. Moveover, CEDI has the robustness for link-error in error-prone wireless transmission environments.

REFERENCES

1. T. Imielinski, S. Viswanathan, and B. R. Bardrinath. Data on air: organization and access, *IEEE Trans. Know. Data Eng.*, 9(3):353–372, 1997.
2. T. Imielinski, S. Viswanathan, and B. R. Bardrinath. Power efficiency filtering of data on air. In *Proceedings of the International Conference on Extending Database Technology (EDBT)*, 1994, pp. 245–258.
3. W.-C. Lee and D. L. Lee. Using signature techniques for information filtering in wireless and mobile environments. *Spatial Issue Database Mobile Comput. J. Distribut. Parallel Databases*, 4(3):205–227, 1996.
4. Q. L. Hu, D. L. Lee, and W.-C. Lee. Performance evaluation of a wireless hierarchical data dissemination system. In *Proceedings of the 5th Annual ACM International Conference on Mobile Computing and Networking (MobiCom '99)*, Seatle, WA, USA, Vol. 4, No. 3, August 1999, pp. 163–173.
5. S. Acharya, R. Alonso, M. Franklin, and S. Zdonik. Broadcast disks: data management for asymmetric communications environments. In *Proceedings of ACM SIGMOD Conference on Management of Data*, San Jose, CA, USA, May 1995, pp. 199–210.
6. J. L. Xu, W.-C. Lee, and X. Y. Tang. Exponential index: a parameterized distributed indexing scheme for data on air. In *Proceedings of ACM/USENIX MobiSys*, June 2004, pp. 153–164.
7. J. L. Xu, W.-C. Lee, X. Y. Tang, Q. Gao, and S. P. Li. An error-resilient and tunable distributed indexing scheme for wireless data broadcast. *IEEE Trans. Know. Data Eng.*, 18(3):92–404, 2006.
8. A. Guttman. R-trees: a dynamic index structure for spatial searching. In *Proceedings of the ACM SIGMOD Conference on Management of Data*, 1984, pp. 47–54.
9. T. Sellis, N. Roussopoulos, and C. Faloutsos. R$^+$-trees: a dynamic index for multi-dimensional objects. In *Proceedings of the 13th Internation Conference on VLDB*, 1987, pp. 507–518.
10. N. Beckmann, H. Kriegel, R. Schneider, and B. Seeger. The R*-tree: an efficient and robust access method for points and rectangles. In *Proceedings of the ACM SIGMOD International Conference on Management of Data*, 23–25. May 1990, pp. 322–331.
11. T. Imielinski, S. Viswanathan, and B. R. Badrinath. Energy efficient indexing on air. In *Proceedings of the International Conference on Management of Data*, 1994, pp. 25–36.
12. T. Imielinski, S. Viswanathan, and B. R. Badrinath. Power efficiency filtering of data on air. In *Proceedings of the International Conference on EDBT*, Cambridge, UK, March 1994, pp. 245–258.
13. A. Datta, A. Celik, J. Kim, D. VanderMeer, and V. Kumar. Adaptive broadcast protocols to support power conservation retrieval by mobile users. In *Proceedings of the 13th International Conference on Data Engineering (ICDE97)*, UK, April 1997, pp. 124–133.
14. A. Datta, D. VanderMeer, A. Celik, and V. Kumar. Broadcast protocols to support efficient retrieval from databases by mobile users. *ACM Trans. Database Syst.*, 24(1):1–79, 1999.
15. B. H. Zheng, W.-C. Lee, and D. L. Lee. Spatial index on air. In *Proceedings of the 1st IEEE International Conference on Pervasive Computing and Communications (PerCom'03)*, Dallas-Fort Worth, Texas, 23–26 March 2003, pp. 297–304.
16. B. H. Zheng, W.-C. Lee, and D. L. Lee. Spatial queries in wireless broadcast systems. *Wireless Network*, 10(6):723–736, 2004.
17. W.-C. Lee and B. H. Zheng. DSI: a fully distributed spatial index for location-based wireless broadcast services. In *Proceedings of the 25th IEEE International Conference on Distributed Computing Systems*, 2005.
18. S. Acharya, M. Franklin, and S. Zdonki. Balancing push and pull for data broadcast. In *Proceedings of the ACM SIGMOD Conference on Management of Data (SIGMOD 97)*, 1997, pp. 183–194.
19. N. Roussopoulos, S. Kelley, and F. Vincent. Nearest neighbor queries. In *Proceedings of the ACM SIGMOD International Conference on Management of Data (SIGMOD 95)*, 1995, pp. 71–79.
20. T. Seidl and H. Kriegel. Optimal multi-step k-nearest neighbor search. In *Proceedings of the ACM SIGMOD International Conference on Management of Data (SIGMOD 98)*, 1998, pp. 154–165.
21. B. Zheng, J. H. Xu, W.-C. Lee, and D. L. Lee. Energy-conserving air indexes for nearest neighbor search. In *Proceedings of EDBT04*, Heraklion, Crete, Greece, March 2004.

22. J. L. Xu, B. H. Zheng, W.-C. Lee, and D. L. Lee. Energy efficient index for querying location-dependent data in mobile broadcast environments. In *Proceedings of the 19th International Conference on Data Engineering (ICDE03)*, Banglore, India, March 2003.

23. D. Moore. Hilbert Curve. http://www.caam.rice.edu/ dougm/twiddle/Hilbert.

24. S. Hambrusch, C. L. W. Aref, and S. Prabhakar. Query processing in broadcasted spatial index trees. In *SSTD3*, July 2001.

25. H. Samet. *The Design and Analysis of Spatial Data Structures*. Addison-Wesley, MA, 1990.

26. J. L. Xu, B. H. Zheng, W.-C. Lee, and D. L. Lee. The D-tree: an index structure for planar point queries in location-based wireless services. *IEEE Trans. Know. Data Eng.*, 16(12):1526–1542, 2004.

27. B. H. Zheng, J. L. Xu, W.-C. Lee, and D. L. Lee. Grid-partition index: a hybrid method for nearest neighbor queries in wireless location-based services. *VLDB J.*, 15(1):21–39, 2006.

28. S. J. Im, M. B. Song, J. W. Kim, S.-W. Kang, and C.-S. Hwang. An error-resilient cell-based distributed index for location-based wireless broadcast services. In *Proceedings of the 5th ACM International Workshop on Data Engineering for Wireless and Mobile Access MobiDE06*, Chicago, IL, USA, June 2006 pp. 59–66.

29. S. Hameed and N. H. Vaidya. Efficient alogorithms for scheduling data broadcast. *ACM/Baltzer J. Wireless Networks*, 5(3):183–193, 1999.

30. S. Acharya and S. Muthukrishman. Scheduling on-demand broadcasts: new metrics and algorithms. In *Proceedings of the 4th Annual ACM/IEEE International Conference on Mobile Computing and Networking (MobiCom '98) SIGMOD Conference on Management of Data (SIGMOD 97)*, Dallas, TX, USA, 1998, pp. 43–54.

31. D. Aksoy and M. Franklin. R x W: a scheduling approach for large-scale on demand data broadcast. *IEEE/ACM Trans. Network.*, 7(6):846–860, 1999.

32. T. Imielinski and S. Viswanathan. Adaptive wireless information systems. In *Proceedings of the Special Interest Group in DataBase Systems (SIGDBS) Conference*, Tokyo, Japan, October 1994, pp. 19–41.

Chapter 10

Next Generation Location-based Services: Merging Positioning and Web 2.0

Axel Küpper and Georg Treu

Mobile & Distributed Systems Group, Ludwig Maximilian University, Munich, Germany

10.1 Introduction	213
10.2 LBS: The First Generation	215
10.3 Web 2.0	217
10.4 LBS Classification	220
10.5 A Web 2.0 Supply Chain for LBS	223
10.6 Location Services	227
10.7 Location Privacy	233
10.8 Summary	235
References	235

10.1 INTRODUCTION

Besides time, *location* is one of the major quantities determining our everyday life. People make use of it when navigating through cities or on highways, for making appointments, for ordering goods or services, or simply for informing other people about their whereabouts. Thus, the concept of location is essential for orientation in

Mobile Intelligence. Edited by Laurence T. Yang, Agustinus Borgy Waluyo, Jianhua Ma, Ling Tan, and Bala Srinivasan

the real world. On the other hand, the emergence of the Internet has shown us how to overcome this concept. At least with regard to communication and information exchange between people, it does not play a role where these people are located in the world or by what distances they are separated from each other. The Internet shadows the location of its participants, and that is why it is often also referred to as *global village* or *cyberspace*.

Mobility means change of location (at least in the context considered here). Cellular networks like GSM owe their success to people who want to communicate while being on the move, and provide several mechanisms for supporting mobility, for example, location management for routing incoming telephone calls to the base station the subscriber is connected to. Initially planned for telephony only, cellular networks have been extended with data services like GPRS or UMTS packed-switched in the recent years and are now increasingly used for Internet access. Many services of the global village or modified versions of it, for example, push Email or browsing via WAP, can now be accessed from mobile devices, irregardless of where the user is currently located. However, the added-value of mobile services can be significantly increased if the user's current location is taken into consideration during service execution, and that is what the main idea behind *location-based services* (LBSs) is. By using LBSs, the user's location is not shadowed any longer, but it is used to adapt services to the special situation and needs of a user. Real-world and cyberspace merge by the use of LBSs, and thus they represent a significant core function for achieving *mobile intelligence*.

LBSs can be defined as services that create, compile, select, or filter information based on the current locations of the users or those of other persons or mobile objects. Examples are so-called *finder services*, which show the users a list of nearby points of interest, for example, restaurants or shopping opportunities, *navigation services*, which provide drivers of vehicles with routing instructions, or *child trackers*, which allow parents to follow the movements of their children. An important characteristic of LBSs is that users do not need to enter their location manually, but that they are automatically located and tracked by using a certain positioning technology, for example, the *Cell-Id* technology, where the user's location is derived from the coordinates of the serving base station, or the *global positioning system* (GPS). As these technologies in most cases deliver only geographic coordinates, LBSs must be connected to a *geographic information system* (GIS), which maps the coordinates into a meaningful descriptive location like a street address or which shows the user's position on a map.

Since their introduction around the turn of the millennium, the appearance of LBSs has been dominated by the operators of cellular networks, which have full control about the localization of subscribers and which thus have a unique selling point regarding their location data. Third-party service providers do not get access to location data at all or only against overpriced conditions. As a result, there is less creativity in the LBS market, and the demand for LBSs is not a quarter as good as predicted by many market analysts when the first of these services emerged.

However, the circumstances for LBSs have changed significantly in the recent time, which is basically a result of two developments: first, there is an increasing market penetration of mobile devices with integrated GPS receivers. Thus, the location of a user is not controlled by the serving cellular network operator any longer. Rather, the user is able to locate herself and decide by her own when and in which way to make her location data available to external actors. Second, the *World Wide Web* (WWW) experiences a massive shift toward what is known as *Web 2.0*. This term does not reflect a certain technology, nor is it a standard or service platform. Rather it describes a realignment away from the "plain old Web" with its clearly assigned roles of service providers and consumers toward something where users can produce and publish content by their own (*user-generated content* (UGC)), create new services by combining several existing ones (*mashups*), and organize in communities. Applied with user-centric positioning technologies like GPS, Web 2.0 incorporates a great potential for establishing the next generation of LBSs, which significantly differs from the first one by new innovative services and a large added-value for their users. This development has also been recognized in the Web 2.0 community, which discusses new applications, ideas, and technologies at an annual conference, which, in close analogy to the term "Web 2.0", is called "Where 2.0."

The remainder of this chapter discusses the reasons for the shortcoming of the first-generation LBSs and explains what can be expected from Web 2.0. It provides a classification scheme for identifying the basic mechanisms needed when creating a particular LBS and introduces a supply chain that describes the interactions between the actors of an LBS by applying fundamental principles of Web 2.0. Furthermore, the chapter motivates the need for a common position management and explains its underlying operations. Finally, it addresses the need for privacy protection in the context of LBSs and sketches its basic mechanisms.

10.2 LBS: THE FIRST GENERATION

One of the main origins of LBSs is the E911 mandate in the United States, which was passed by the *Federal Communications Commission* (FCC) in 1996 and which obliges operators of cellular networks to locate the callers of emergency services with a prescribed minimum accuracy and to deliver their geographic position to a nearby *Public Safety Answering Point* (PSAP), the office where emergency calls arrive. According to the emergency number 911 in the United States, this mandate is known as *Enhanced 911* (E911). In 1996, however, the networks were not equipped for meeting the high-accuracy demands imposed by this mandate. The location of an emergency caller could only be determined by mapping the Cell-Id of the serving base station to its geographic coordinates, which, depending on the size of radio cells, results in accuracies not better than 300 m. Also, the use of GPS was not an option, as low-cost GPS receivers for integration into mobile phones were not available at that time. In order to cope with this lack, enormous efforts were launched to extend cellular networks by advanced positioning methods, which are based on lateration between mobile phones and at least three base stations and which achieve accuracies between 50 and 150 m. An overview of these methods can be found in [1, 8]. These advanced methods are called *enhanced observed time difference* (E-OTD) or *advanced forward link trilateration* (A-FLT), and the operators had to spend huge investments for integrating them into their networks. An idea was, therefore, to use the new positioning methods not only for E911 but also to regain these investments by offering commercial LBSs.

As a consequence, cellular network operators in many countries of the world, not only in the United States, then launched a series of LBSs, which in most cases appeared in the form of finder services, which deliver on request a list of nearby points of interest like restaurants, filling stations, or ATMs. The rollout phase of this first-generation LBS was accompanied by several market analyses, which identified LBSs as the new killer application for emerging data services and which predicted revenues in the range of several billions of dollars worldwide. However, it turned out very soon that these predictions could not be fulfilled even approximately, because most subscribers actually were not interested in LBSs, at least not in the form they were offered. As a result, many operators very quickly phased out their LBSs and did stop the development of new ones.

When analyzing the reasons for this failure, a number of reasons can be found, among them the fact that standardization and manufacturers created a network-centric approach that gives operators a unique selling point with regard to location data of their subscribers, see Figure 10.1(a) and [1]. Positioning as well as the exchange

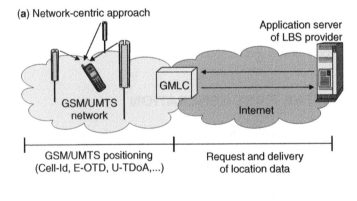

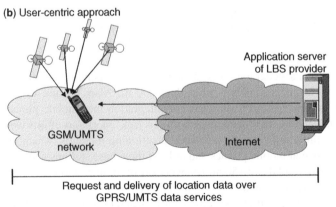

Figure 10.1   User- and network-centric approach for realizing LBSs.

and processing of location data are controlled inside the infrastructure of cellular networks and are insulated from independent third-party service providers. These providers can only get access to location data via a so-called *gateway mobile location center* (GMLC), which, however, is not supported by many operators or only against overpriced conditions. As a result, the LBS market for a long time was a monolithic one, dominated by the operators and with less competition. The application potential of LBSs was reduced to the finder services only, and new applications domains as well as new and sophisticated functions were not developed.

However, the emerging market of mobile devices with integrated GPS receivers developing in the recent time makes it possible to follow another approach for realizing LBSs. As depicted in Figure 10.1(b), the location of a user is determined within her mobile device and can then be passed to external actors, for example, independent LBS providers, by making use of data services like GPRS or UMTS-packet switched. Thus, a user's location is not controlled by the network operator any longer. Rather the user can decide by herself when, how, and to whom to make her current location available. Thus, the network-centric approach as described before is replaced by a user-centric one.

User-centricity is also an important concept of Web 2.0, and that is why both autonomous self-positioning and Web 2.0 may dominate the appearance of the next-generation LBS.

10.3   WEB 2.0

Web 2.0 can be seen as a new paradigm that represents a user-centric approach of how Web services and related content are created and published. While the plain old Web since its emergence in the mid-1990s was dominated by professional service, application, and content providers, Web 2.0 now shifts the focus more toward the user as the main driving force behind the further development of the Web. As a result, Web 2.0 is often also associated with terms like democratization, openness, and social networking.

The Web 2.0 paradigm was essentially formed by a group around the publisher Tim O'Reilly and firstly appeared in the title of a Web conference in 2004. It resulted from a discussion between the people about the reasons of the collapse of the new economy and the conclusion that despite this collapse "the Web was more important than ever, with exciting new applications and sites popping up with surprising regularity" [15]. However, it must be stressed that the hype taking place around Web 2.0 is not without controversy. A lot of people, among them the man who is known as the inventor of the Web, Tim Berners-Lee, argue that Web 2.0 is rather a buzzword than a new innovative approach, because Web 2.0 is based on the same fundamental technology as Web 1.0, and there is basically nothing different between them [2]. However, regardless of which of these views one is willing to accept, when analyzing the history of LBSs and the shortcomings of their first generation, the Web 2.0 approach shapes some interesting concepts, which are worth considering for adoption for the next-generation LBS.

An important design concept of Web 2.0 is to consider "the Web as a platform" [15]. While since the beginning of the PC era, applications needed to be manually installed and maintained on the PC, the idea is now to dynamically load and execute them in the Web browser. The advantage of this approach is that users do not need to pass through cumbersome installation and update procedures, but can request the latest versions of their applications from basically any PC that is equipped with an appropriate Web browser. The enabling technology behind this approach is called *asynchronous Javascript and XML* (Ajax). Client applications written in Ajax are executed within the Web browser. User requests are in most cases locally processed and interactions with the server are hidden from her as far as possible by executing them in the background, that is, *asynchronously*. Thus, in contrast to conventional HTML Web pages, client applications written in Ajax are more interactive and do not suffer from large delays imposed by the client/server communication. However, it is important to note that Web 2.0 is not solely based on the Ajax technology but also incorporates a series of technologies, markup languages, and protocols, for example, Flash players, conventional HTML and HTTP, or the *simple object access protocol* (SOAP).

Another appearance of Web 2.0 are so-called *mashups*, which are defined as combinations of services and content from different sources. The appearance and extent of this combination is not necessarily predetermined by a provider, but by the user herself, who can mix and tailor different services and content sources by her own and according to her special needs. Mashups can be realized by the Ajax technology. For example, a service provider may offer an Ajax script that, when loaded to the user's PC, receives content from different providers and displays it to the user or mixes it in a certain fashion in order to create new functions or information, see Figure 10.2. Google is an example for making extensive use of this approach: the user can combine and

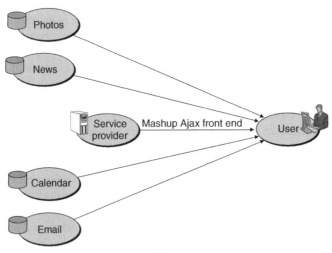

Figure 10.2 Mashup.

interlink searching, calendar, email, and map services. For example, she can display the locations of restaurants, which have been delivered as a search result, on a map, or she can invite participants of a meeting managed by her calendar via email.

Hence, Ajax and mashups are not tools focusing only on professional application developers. Rather, they allow users to assume the role of a service provider by their own and to write applications and offer them to others. These *user-generated services* may compete with professional services or extend them with additional features. However, this approach of user-centricity is not only limited to services but can also be applied for content, which is then known as UGC and represents one of the main concepts Web 2.0 is today known for.

UGC incorporates basically any content form, for example, text, images, music, as well as video, and can be published by different technologies and services. A typical appearance are *blogs*, which are Web sites where users publish news, ideas, or comments to certain subjects and which are listed in chronological order. Another example are so-called *RSS feeds*, where RSS is an acronym with several meanings, for example, *really simple syndication* or *rich site summary*. In contrast to blogs, RSS feeds are accessed by dedicated readers. The user has to subscribe for one or several feeds, and the reader then regularly checks for updates of these feeds and indicates new content to the user. In addition to that, several new services have emerged in the recent years, allowing users to publish and interlink blogs, images, photos, and videos in the Web. The content can be made accessible for all users, or access can be limited to certain users or groups of users. The latter is part of a phenomenon known as *social networking*, which describes the connecting of people that share common interests, follow the same profession, visited the same school in their childhood, or work in the same company. These services may be extended with chat and bulletin board functions and are also referred to as *community services*.

Thus, Web 2.0 incorporates a number of approaches and concepts the next-generation LBS may profit from. Creating LBSs in a Web 2.0 manner means to follow a user-centric approach, where the cellular network operator does not control the access to a user's location, as highlighted in the last section. Following Web 2.0, location data are not simply "disclosed" to other actors. Rather, location becomes a new type of user-generated content, which can be published, refined, and combined with other content forms like photos or videos. An example where this approach is already applied is *Flickr*, a Web site where users can publish their photos together with the geographic coordinates where this photo has been obtained. The photos are usually taken by mobile devices with integrated cameras and GPS receivers and can be uploaded via data services like GPRS or UMTS.

Applying the principle of social networking for LBSs means to share one's own location with other members of a community. For example, the location becomes another attribute in the buddy list of an instant messaging application and is automatically refreshed whenever the location of the respective buddy changes. Also, it is possible to show the spatial distribution of community members on a map or to subscribe for being notified as soon as other members approach. In another example, users can request recommendations for nearby restaurants, which have been generated by other community members that have visited these restaurants in the past.

Finally, users are not dependent on that service providers offer LBSs fulfilling their needs. By applying Ajax and mashup technologies in combination with GPS, they can simply create their own LBSs and offer it to other users, maybe for generating revenues or when having other objectives in mind. This approach then certainly bears the greatest potential for developing new and innovative LBSs in a broad range of different application domains.

The next section presents a classification that is helpful for identifying functions and mechanisms needed when creating such LBSs, followed by an overview of organizational matters, positioning technologies, protocols, and privacy aspects in subsequent sections.

10.4 LBS CLASSIFICATION

The examination of application domains like mobile marketing or mobile gaming is useful for giving people a general idea about the purposes, appearances, and benefits of LBSs, but it does not help in identifying the required functions that constitute a particular LBS. Examples of such functions are positioning methods, strategies for exchanging position data between the actors participating in the supply chain of an LBS, privacy mechanisms, or protocols. They should be arranged in a way that they can be reused for a broad range of different applications and tailored to their special requirements. Figure 10.3 gives an overview of the different criteria for classifying

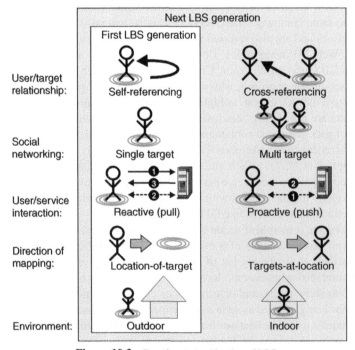

Figure 10.3   Functional classification of LBSs.

LBSs from a functional point of view. The criteria are explained and discussed below.

Self-referencing vs. cross-referencing LBSs.   An important distinction has to be made concerning the roles the participants of an LBS adopt during its execution. Generally, it can be distinguished between the roles of the LBS user and the target. The *LBS user* is the person that requests and consumes the LBS. A *target*, on the other hand, is the individual to be located and tracked. In self-referencing LBSs, user and target are the same individual, that is, the user's location is processed for her own purposes. Typical examples are finder services, which show the user a list of points of interest being in close proximity to her own location, or navigation services, which guide the driver of a vehicle to her destination. In cross-referencing LBSs, the roles of user and target are adopted by different persons. The user requests the location of a target or permanently tracks the location where the target stays. Examples are child tracking services, which show parents the whereabouts of their children, or fleet management services, which show the carrier headquarter the locations of its trucks. Cross-referencing LBSs impose much stronger requirements on mechanisms for privacy protection than self-referencing LBSs. In particular, it must be possible that targets can restrict access to their location data to a limited, well-defined group of users. Optionally, it might be desired that access is controlled by additional constraints, which, for example, limit it to a certain time interval or a geographical region like a company's premises.

Single-target vs. multitarget LBSs.   Another classification concerns the number of targets participating in an LBS service session. In single-target LBSs, the major focus is on tracking the position of a single target. Usually, the target's position is here interrelated with geographic content, for example, in order to show the user the target's position on a map, to show her nearby points of interest, or to create routing data for navigation. Thus, the primary purpose is to convert a geographic position, which is given in terms of latitude, longitude, and altitude, into a meaningful descriptive location, for example, a street address or a city, where the target currently resides. The child-tracking service described before is a typical example for a single-target LBS. In multitarget LBSs, the focus is more on interrelating the positions of several targets among each other. Typical applications determine the distance between targets, show the positions of targets residing in the same city on a map, or detect clusters of several targets. In the context of Web 2.0, multitarget LBSs focus on social networking services. Location data can here be used as a supplement for existing community services or used to establish new social networks.

Reactive vs. proactive LBSs.   LBSs can be classified into reactive and proactive services, see also [4, 8, 16]. An LBS is said to be reactive if location-related information is delivered on request to the user. The interaction between service and user is roughly as follows: the user first invokes the service and establishes a service session, either via a mobile device or a desktop PC. She then requests for certain functions or information, whereupon the service determines the location (either of herself or of another target person), processes it, and returns the location-dependent result to the user. Finder services, as mentioned before, are a typical example for reactive LBSs. Proactive LBSs, on the other hand, are automatically initialized as soon as a predefined *location event* occurs, for example, if the target person enters, approaches, or leaves a certain point of interest or if she approaches, meets, or leaves another target person. As an example, consider an electronic tourist guide that notifies tourists via SMS as soon as they approach a landmark. Thus, a proactive LBS is not explicitly requested by the user. In contrast to reactive LBSs, where the user is only located on demand, proactive LBSs require that the user is permanently tracked and that the obtained position is checked against certain constraints, for example, on proximity to close-by landmarks.

Location-of-target vs. targets-at-location LBSs.   Traditional LBSs derive the current locations of one or several well-known targets, that is, they perform a mapping from the set of targets onto the set of locations. These services are referred to as location-of-target LBSs. However, a mapping between these sets can also be applied in opposite direction, that is, from the set of locations to the (power) set of targets. In this way, it is possible to determine the number and possibly the identity of targets that stay at a particular location. These services can be denoted as targets-at-location services. Usually, positioning methods and control procedures have been designed for supporting location-of-target services only. The target to be located must be known in advance, and positioning is then triggered by signaling between an application or location server and the target's mobile device. For targets-at-location services, on the other hand, appropriate positioning or sensor technology is not available or rarely installed. Alternatively, the targets residing at a particular location can be determined by spatial queries at a central database, which permanently tracks all targets of interest.

Outdoor vs. indoor LBSs.   Outdoor LBSs are available in large or huge geographical areas and make use of satellite or cellular positioning technologies, while indoor LBSs assist the user inside buildings and are based on local-positioning technologies. The distinction between these classes is essential for the underlying positioning technology used. Usually, location-based Web 2.0 services are built around device-based positioning technologies, which allow an autonomous self-positioning in that the device observes transmissions from the surrounding infrastructure (e.g., WiFi access points, cellular base stations, or satellites) and calculates its position from that. Examples are GPS or fingerprinting. The latter is based on observing radio patterns on the spot and comparing them with patterns that have been prerecorded at well-defined positions. Positioning technologies can be classified with regard to outdoor and indoor technologies, which differ in a number of characteristics like the positioning technology used, the coverage area, and the delivered accuracy and format of location data. Outdoor systems usually show accuracies between 10 m (GPS) and several hundreds of meters (e.g., cellular methods) and mostly deliver geographic location data in terms of coordinates based on a spatial reference system. Indoor technologies, on the other hand, achieve accuracies in the range of some meters or even centimeters. In most cases, they deliver symbolic location data, for example, the identifier of an access point or the number of a room inside a building, or location data based on a local coordinate system. An overview of outdoor systems can be found in [8], while indoor technologies are summarized in [7].

The classification criteria are orthogonal to each other, that is, an LBS can be assigned to the classes of different criteria. As can be derived from Figure 10.3, finder services that coined the appearance of the first LBS generation are reactive, self-referencing, and single-target services and were realized by a central application server primarily for outdoor use. Other examples are child-tracker services (reactive, cross-referencing, single target, outdoor, and indoor), or community services (reactive, cross-referencing, multitarget, indoor, and outdoor).

10.5 A WEB 2.0 SUPPLY CHAIN FOR LBS

Generally, there may be many actors involved in the operation of an LBS, and it is therefore useful to describe their interactions in a *supply chain*. An *actor* denotes an individual, organization, department, or enterprise that offers services to other actors, or consumes services from other actors, or does both of that. Actors are classified according to roles, where a *role* represents a certain field of activity of an actor associated with a set of functions for realizing and controlling portions of a service as well as making it accessible to the end user. Figure 10.4 shows a general model for an LBS supply chain, which follows the Web 2.0 approach and which identifies the participating roles and their relationships.

The supply chain follows the approach of user-centricity and mashups as proclaimed by Web 2.0, that is, content from several sources is combined at the user's device to built a new service. Location data of targets are considered to be another kind of content besides images, videos, or text. The supply chain contains the following roles:

- *Target.* As already mentioned in the previous section, a target is a mobile individual or object that is to be located, tracked, or sighted. For this purpose, it is equipped with a mobile device, which besides communication facilities, can perform positioning.

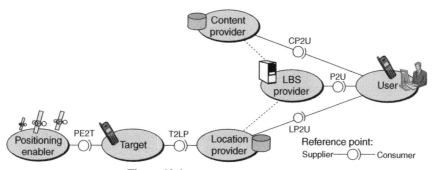

Figure 10.4   Generic LBS supply chain.

- *Positioning enabler.* The positioning enabler maintains a positioning infrastructure for controlling and coordinating the positioning process. In the user-centric approach followed here, this infrastructure enables terminal-based positioning and may be given by the network of GPS satellites or a WiFi network for performing fingerprinting.

- *Content provider.* A content provider offers geographic and non-geographic content. Examples of geographic content are maps, routing data for navigation, or points of interest. Nongeographic content may be available in form of news, blogs, or videos. The latter might be of interest in the context of an LBS if it refers to a certain location.

- *Location provider.* Targets are connected to a location provider for publishing their locations. The location provider is basically another content provider that acts as an intermediary between target and user. It offers so-called *location services* to the users of an LBS for accessing a target's location either in a reactive or proactive fashion and under consideration of the target's privacy rules. Advanced location services may also provide information about the geographic correlation of several targets, for example, about whether or not a pair of targets is located in the same city. Each target is assumed to have a subscription with a location provider that manages and control her location data on behalf of her.

- *LBS provider.* The LBS provider prepares the service logic for realizing an LBS. In the context of Web 2.0, this service logic is given by Ajax scripts, which are passed to the user's Web browser and there interlink content from different providers. For mobile devices, a simplified version might be offered in HTML or as dedicated client application making use of the *Java2 Micro Edition* (J2ME) engine installed in many devices. In order to be independent of a certain technology, scripts, Web pages, or client applications are subsumed under the term *front end* in the following text. The LBS provider maintains subscriptions with LBS users and, if required, also provides auxiliary functions like accounting, session, and identity management.

- *LBS user.* The LBS user is the actor that consumes an LBS. The front end delivered by the LBS provider is executed at her device. Content and location data from different sources are then requested, received, processed, and aggregated according to the instructions given in this front end.

The relationships between roles are referred to as *reference points* in the following and may be of administrative or technical nature. Administrative reference points may be *service level agreements* (SLAs) negotiated between the participating actors, charging conditions, and trust models, while technical relationships cover communication links, interfaces, protocols, and transactions being of relevance during the operation of an LBS. In the following, the reference points are highlighted from their technical point of view:

- *PE2T.* The reference point between positioning enabler and target represents the positioning procedure and depends on the positioning system used. In the case of GPS, this relationship describes the different signals emitted by the GPS satellites. These signals are observed by the target's device for performing range measurements and for calculating its position by lateration. If fingerprinting is used in an indoor WiFi network, this relationship represents the radio beacons emitted by various WiFi access points in the close surrounding as well as control procedures for mapping the observed radio patterns to a geographic (or descriptive) location. If the target is charged for positioning, this relationship also includes related procedures, for example, subscription and key exchange. The European satellite system Galileo, which is expected to go into operation around 2010, is such an example where targets have to pay for high-accuracy positioning.

- *T2LP.* This reference point represents operations of the so-called *position management*, which takes place between target and location provider for offering location services over the LB2U reference point. T2LP incorporate mechanisms for hiding the technical aspects of the used positioning technology from the application, which is referred to as *positioning transparency* in the following. More important, however, the reference point also covers operations for passing location data from the target to the location provider. The location data can be requested by the location provider on demand, or a target can be permanently tracked in that location data are automatically passed to the location provider according to a certain position-update strategy. This is explained in Section 10.6.2.

- *P2U.* This reference point covers the passing of front ends and related data from the LBS provider to the user's device. In the context of an LBS service session, the user has to register and is then provided with a tailored appearance of her service, for example, with regard to her preferred "look-and-feel" and the content sources she wishes to receive information from (which is then requested over the reference point CP2U). Furthermore, the script contains references to targets and associated location services for getting access to their location data (which happens over the reference point LP2U). These data may then be combined with content received from other sources. Changes to a user's service profile are reported back to the LBS provider and are available for future service sessions.

- *LP2U.* This reference point covers the location services offered by a location provider. The user requests location data of one or several targets or advices the location provider to track a target for a certain period of time. In order to meet their privacy concerns, it is required that the user authenticates with the location provider, for example, by using her telephone number, email address, or another pre-negotiated, unambiguous identifier together with a password. Access to location data is then only granted if the target has authorized the requesting user. There may be several options the privacy of a target may be enforced, which is explained in Section 10.7. If the location or tracking request is accepted, the target's location is reported back to the user. This reference point also covers alternative communication links. For example, users can register with a location provider in order to receive a target's location by SMS.

- *CP2U*. Finally, this reference point is used to obtain other content than location, for example, geographic maps by using the API from GoogleMaps. The mechanisms are well known from Web 2.0 and are, therefore, not further explained here.

In addition to these reference points, the LBS provider may be interconnected with content and location providers (represented in Figure 10.4 by dotted lines). These reference points may be administrative or technical ones, the latter, for example, in order to implement an identity management. However, as these reference points are not of special interest for LBSs in particular, they are not covered here.

It is important to note that for concrete realizations of LBSs, this supply chain can be dynamically configured. In particular, an actor may adopt several roles, while other roles can be omitted. Figure 10.5 shows a simple instantiation with five participating actors. In this example, *Bob* wants to find out which vegetarian restaurants are nearby and downloads the corresponding front end from the LBS provider *EatNoMeat*. He is then located by the local GPS receiver of his device and receives a list of restaurants in his close surrounding from the YellowPages content provider. As he also wants to know how other users have rated these restaurants, the front end automatically requests the provider of a recommendation system for ratings. Finally, after he has selected a suitable restaurant under consideration of these ratings, a content provider for maps passes navigation instructions to him for finding the shortest or fastest way to the restaurant. Thus, according to the classification given before, this service falls into the category of a reactive, self-referencing, single-target service. The roles of user and target are adopted by the same person here, and because Bob locates himself, the involvement of a dedicated location provider is not necessary.

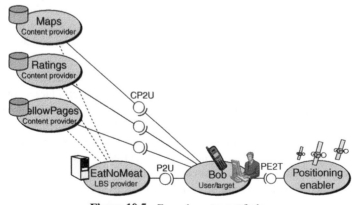

Figure 10.5   Example: restaurant finder.

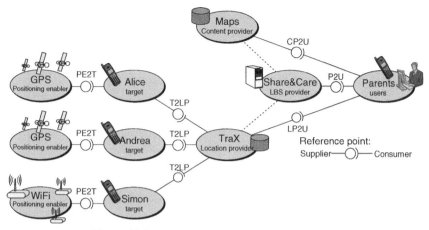

Figure 10.6 Example: child finder and tracking service.

In another scenario, the supply chain is configured for a *child finder and tracking service* offered by the LBS provider Share & Care. Parents can observe the current locations of their children and display them on a map. If desired, they can also continuously track their children or receive an alert via SMS as soon as one of them leaves school. For this purpose, they can configure the service for updating the location whenever a child has covered a preselected distance with regard to the last reported location or leaves a certain geographic zone. The resulting supply chain is depicted in Figure 10.6. *Alice* and *Andrea* are on their way home and are located by GPS, while *Simon* still rests in school and is tracked by the local WiFi network there. The location is passed from the children's devices to their location provider *TraX*. The parents use the front end of Share & Care in order to get connected to TraX and to a content provider for maps. Thus, this service is a cross-referencing LBS with reactive and proactive interaction patterns.

Figure 10.7 shows a sketch of how the front end of such service may look like. It contains elements for requesting location data by entering the name of a child as well as for specifying the desired update distance in case of tracking. The delivered locations are then displayed on a map provided by GoogleMaps.

10.6 LOCATION SERVICES

As stated in previous sections, location services are offered via the LP2U reference point and are realized by the T2LP reference point, which incorporates mechanisms for achieving positioning transparency and the operations of position management. Both functions are based on the architecture depicted in Figure 10.8 and are explained in this section.

The core element of the T2LP reference point is a *positioning daemon* executed at the mobile device of a target. The positioning daemon sits on top of the implemen-

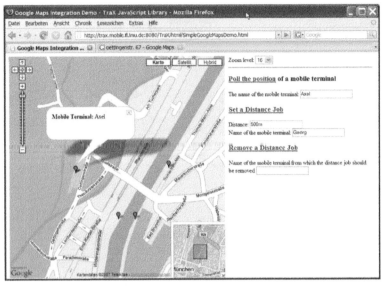

Figure 10.7   Example of an LBS front end.

tations of one or several positioning technologies and publishes location data either to local front ends or to the remote location server maintained by the location provider. The interface to local front ends is necessary for realizing all kinds of self-referencing LBSs, where the user's own location is used to request location-based content. If a dedicated client application is used, it receives location data from the positioning daemon over a set of well-defined APIs. In order to access location data through a local Web or WAP browser, the positioning daemon can also be configured to act as a local proxy and thus appears as a kind of Web service that can be interlinked with Web pages or scripts loaded into the browser. However, in the following, the focus

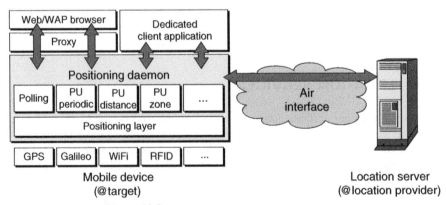

Figure 10.8   Architecture for publishing location data.

is on publishing location data to a remote location server, which is described in the following subsections.

10.6.1 Positioning Transparency

The need for positioning transparency results from the fact that mobile devices may support different positioning technologies, that is, besides GPS also incorporate implementations for the future Galileo satellite system, WiFi fingerprinting or positioning based on radio frequency identification (RFID) tags. Generally, there exist large differences between heterogeneous positioning technologies, for example, in the way they are accessed and activated or in the accuracy and format location data are delivered. Due to these reasons, it is desired to decouple technology-dependent aspects from other functions of the position management, thereby making it possible to reuse its software components at a broad range of mobile devices and independent of the set of positioning technologies applied.

The positioning layer depicted in Figure 10.8 hides the implementations of different positioning technologies from the upper layers. It receives requests for location data, which may either address a certain positioning technology or define requirements on location data in an abstract manner, for example, the desired accuracy and location data format. From the positioning methods being available at the respective mobile device, the positioning layer then selects the most suitable one and activates it. Another task is the mapping between different location formats either by calculation in the case of geographic location data or by the support of local or remote databases when dealing with descriptive locations, see, for example [10]. Optionally, the positioning layer may also support the fusion of several positioning technologies in order to increase the accuracy of location data as well as the execution of *positioning handovers*. The latter refers to the switching between different positioning methods and is needed when a selected method suddenly becomes unavailable, for example, when a target covered by GPS outdoors enters a building and needs to be passed to an indoor WiFi fingerprinting system.

The functions of the positioning layer are accessed by a set of well-defined APIs. A potential candidate for realizing this layer and offering such APIs is the *Location API* for J2ME, see [6].

10.6.2 Position Management

In general, the drawback of device-based positioning is that the mobile device knows where it is, but the location server in the fixed network part does not. The main task of position management is, therefore, to provide a set of operations that allow the mobile devices of targets to pass location data to the remote location server of the location provider. There exist two basic interaction patterns, which are called *polling* and *updating* and which can be further subdivided into several subcategories, see also [13], for example. Furthermore, they can be dynamically applied and parameterized under consideration of the requirements of the respective application.

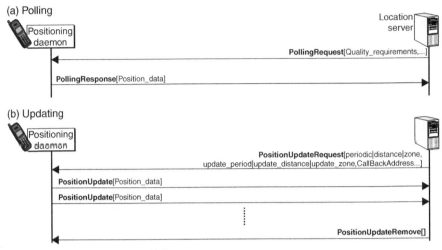

Figure 10.9   Polling and updating of location data.

Polling is based on synchronous communication between the mobile device and the server of the location provider. The server explicitly requests location data from the positioning daemon and is immediately served. The corresponding sequence diagram is depicted in Figure 10.9(a). The request carries requirements concerning the quality of location data, for example, accuracy and latency, and its format. At the positioning daemon, the request is analyzed and passed to the positioning layer, where an appropriate positioning method is selected. After the location has been determined, it is reported to the requesting location provider in a polling response message. Polling may be applied in conjunction with caching strategies at the server site in order to reduce the number of location requests.

As opposed to polling, updating is based on asynchronous communication. It is event-based in that it is initialized by the mobile device when a trigger condition becomes true. In most cases, the server of the location provider has to subscribe first to the mobile device for receiving position updates and specify the trigger conditions. The general procedure of updating is depicted in Figure 10.9(b).

Basically, a trigger condition acts as a filter that is applied on the location data gathered at the mobile device. Each time a new location is delivered from the positioning layer, it is checked against the filter, and only if the condition is true, it is sent to the location server. Updating can be subdivided into the following categories.

- *Immediate updating*. Each time new location data are determined at the mobile device, they are sent to the server. Thus, for this particular category, no trigger condition needs to be specified.

- *Periodic updating*. The mobile device sends a position update message when a predefined time interval, the so-called *update interval*, has elapsed since the last position update.

- *Distance-based updating.* The mobile device sends a position update if the line-of-sight distance between the last reported and the current position exceeds a predefined threshold, which is called *update distance* in the following.

- *Zone-based updating.* A position update is initialized when the target enters or leaves a predefined geographic zone, where an *update zone* can be fixed by a circle with well-defined center and radius or by a polygon.

- *Updating based on dead reckoning.* Another, more advanced strategy is based on *dead reckoning*, where the trigger condition is given by a threshold that refers to the deviation of a location estimate from the actual location. To apply this strategy, the dead reckoning algorithm must be performed on both sites, the mobile device and the server. The mobile device calculates a location estimate based on the location last reported to the server as well as the target's direction and speed of motion. The result is then compared with the current location, and if the difference between them exceeds the threshold, an update with the current location is sent to the server. The server, on the other hand, applies the same algorithm to the location last received from the device whenever it is requested to process it between two updates. A detailed description of this updating scheme can be found in Ref. [12].

As depicted in Figure 10.9(b), the type of updating as well as update period, distance, zone, or the parameters for dead reckoning can be specified in the position update request message passed from the server to the positioning daemon at the mobile device. In addition, the request also contains the callback address of the server to which the update message should be transferred.

The operations described here should be deployed in a manner that the transfer of location data happens as rare as possible and as often as required. There are several reasons why it should happen as rare as possible: first, the excessive transfer of data significantly burdens the battery of mobile devices. Second, the messages of position management are sent over the air interface, which is the most valuable resource in a wireless network and which therefore should be protected from needless or redundant data, which would occur, for example, if the target rests at a certain place and the same location of her is transferred over and over again. In addition, most targets are connected to the location provider via cellular data services like GPRS and UMTS and have to pay for using these services. Finally, the excessive transfer of location data by a large number of targets would heavily burden the location server, which may result in large delays or even failures.

On the other hand, location data should be sent as often as required for meeting the demands of the respective application, for example, with regard to accuracy or up-to-dateness. Polling, for example, is a suitable method to be applied for any kind of reactive LBSs. It has the advantage that location data is as up-to-date as possible, but it suffers from a delay caused by processing the polling request and response messages and obtaining the current location. On average, this delay may be shorter for a caching strategy, which may also reduce network load. However, on the other hand, caching may deliver inaccurate and outdated location data if the target's location has significantly changed since the last polling response. Updating, on the other hand, is useful whenever a target needs to be tracked, which especially holds for proactive services, which are automatically executed or perform any actions if the target triggers a location event. Examples of such events are entering, approaching, or leaving a certain point of interest or approaching or leaving another target. An example for the latter is a function called *proximity detection*, which will be sketched in the following section to demonstrate the usage of the operations of position management.

10.6.3 Example: Proximity Detection

Proximity detection is defined as the capability of a location service to automatically detect when a pair inside a group of mobile targets approaches each other closer than a predefined *proximity distance* d_p. In a simple approach, each target is requested by the location server to perform distance-based updating with a *fixed* update distance r. Thus, each target is surrounded by a circle of static radius r, with the circle's center c being the last reported position, see Figure 10.10(a). If a target leaves its circle, her mobile device performs a position update to the location server, the circle is shifted to the current position, and tracking is continued.

In order to use a reasonable value for the update distance r, the *borderline toler-ance* b is introduced. Let $dist(t_i, t_j)$ be the current distance between a pair of targets t_i and t_j, then the location server has to check for proximity according to the following conditions each time a position update arrives:

- If $dist(t_i, t_j) < d_p$, then proximity *must be* detected.
- If $d_p \leq dist(t_i, t_j) \leq d_p + b$, then proximity *may be* detected.
- If $dist(t_i, t_j) > d_p + b$, then proximity *must not be* detected.

In order to guarantee that proximity events are reliably detected according to these conditions, the update distance must be set to $(b/2)$. For larger values, proximity events would be detected too late in some cases, while smaller values cause unnecessary position updates.

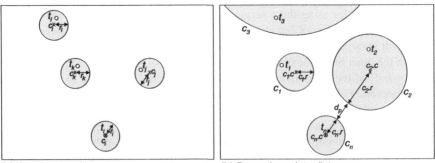

(a) Fixed update distances   (b) Dynamic update distances

Figure 10.10 Proximity detection.

However, a drawback of this strategy is that the update distance of $(b/2)$ is a small value when compared to the expected average distance of targets, and hence it produces a large amount of position updates even if the observed targets are far away from approaching the proximity distance. In a more sophisticated approach, the update distance is therefore *dynamically* determined for each target individually with regard to the distance to her closest neighbor. As a consequence, it is expected that the circles around the targets can be selected to be on average larger than in the static approach, and hence position updates occur less frequently, see Figure 10.10(b). A certain target gets assigned a new update distance from the server either when she has performed a position update or when she firstly registers with the location server for proximity detection. The update distance is determined with regard to the nearest update circle of another target and must be chosen in a way that the border of both circles are separated from each other at least by the update distance d_p.

In Ref. [9], different strategies for proximity detection are explained and compared with each other with regard to the amount of data traffic they cause at the air interface. The operations of position management can also be deployed for detecting other correlations between targets, for example, separation, cluster of targets, or the k nearest neighbors of a target.

10.7 LOCATION PRIVACY

Already the fact that LBSs make it technically possible to determine a target's current location raises manifold ethical concerns that deserve profound public discussing. Doubtless, privacy awareness of an LBS is one key requirement for being accepted by its users and thus for being successful in the marketplace. In short, location privacy refers to the capability of the target person to exercise control about *who* may access her location information in *which situation* and in *which level of detail*. Basically, three different types of access can be distinguished.

First, actors that are not actively involved in the operation of an LBS typically fall into the class of intruders, which can be kept away by applying appropriate security mechanisms for authenticating the LBS actors among each other and for securing communication channels. Fortunately, such mechanisms are widely available and known to be reliable. An exception, however, where access by third parties cannot be circumvented is the so-called *lawful interception* by authorities or intelligence services. In many parts of the world, processing the user's private data in order to prevent or discover criminal activities is in accordance with local law, and service providers are obliged to support the government with that task.

A second class of privacy risks stems from possible malignant behavior of intermediate actors within the LBS supply chain, like the location provider or the LBS provider. For example, the intermediary could illegally sell collected location data to others "behind the target's back." Another possibility is that persons working at the site of the intermediary are motivated to expose the privacy of a specific target person to others or to threaten the target person to do so. For preventing such fraudulent use, classical security mechanisms are not sufficient, because intermediate LBS actors really need to work with the location data and cannot be simply excluded. One possible solution is to *anonymize* the data before handing it out to intermediaries. Meaningful interpretation of the contained location information is then still possible, while the target's true identity is suppressed, for example, by using a pseudonym. The associated problem, however, is that simply stripping user names off the location data does not offer enough protection: by employing background knowledge, for example, about the target person's living or workplace, the data can be de-anonymized. Hence, several methods for verifiably anonymizing location data have been proposed in literature. One of them [5] relies on spatiotemporal cloaking: the spatial and timely accuracy of position fixes is deliberately degraded in order to make the location data of one person indistinguishable from that of at least $k - 1$ other individuals. Unfortunately, this approach, which is based on the formal model of *k-anonymity*, has the disadvantage that it cannot be applied to proactive and multitarget LBSs, which is discussed in Ref. [17].

The user-centric supply chain described before represents a practicable alternative to anonymization techniques, which are by and large quite cumbersome. By composing the LBS in the browser of the user only as described before, the LBS provider does not get in touch with the targets' location information at all. Hence, a target can be tracked by a variety of LBSs without the need for specific protection from the changing LBS providers. Instead, each target is associated with a singular trusted entity, the location provider. Whether users are generally willing to engage in such a trusted relation remains an open question, but it seems so when looking at mobile network subscribers, who trust their operator at a comparable level.

The third type of access a target may want to control is by the LBS user. That is, of course, only the case when the user and the target are in fact different persons and the LBS is cross-referencing. For controlling user access, so-called *authorization policies* can be applied, which can consist of different types of constraints, see also Ref. [14]: *actor constraints* restrict access to certain users only, *time constraints* define at which times location data may be accessed and *location constraints* limit access to certain predefined locations. Furthermore, by specifying *accuracy constraints*, the accuracy of emitted location information can be intentionally degraded. Authorization policies are typically defined by the target and enforced by the location provider on behalf of the target.

An alternative to policies is ad-hoc authorization, which interactively prompts the target for admission when a specific user wants to locate her. Ad-hoc authorization is promoted by work that focuses on social aspects of location disclosure like Ref. [3]. An important factor of social acceptability is the way access refusals are mediated to and perceived by the LBS user. For instance, a user may feel rejected by the target if access is not granted. On the other hand, the target may feel socially pressured to grant the inquirer access in order to avoid possible negative social implications. One general concept to remedy such negative situations is *plausible deniability*, compare also Ref. [11]. Basically, the principle states that the user cannot determine whether a lack of disclosure was intentional or not, which, in turn, reduces social pressure for the target when taking her decision. A simple example, which applies to ad-hoc authorization, is when the target does not respond to a location request until a certain time-out occurs. From the users's point of view, it remains unclear whether the access denial is deliberate or whether the target is currently just too busy to respond.

10.8 SUMMARY

This chapter has described the design and realization of LBSs in the context of Web 2.0. While the first generation of LBSs failed in many countries because of an inherent orientation toward a closed, network-centric solution and the lack of appropriate positioning technologies, the Web 2.0 paradigm incorporates several approaches and concepts, for example, user-centricity, mashups, and social networking, the next generation of LBSs may profit from. Regarding "the web as a platform" and combining it with user-centric positioning methods like GPS is the key for including a large number of users and utilizing their creativity when developing new forms of LBSs. This chapter has given examples for such new forms by proposing a classification scheme for LBSs. It has introduced a generic supply chain for their implementation and explained the need for establishing operations and mechanisms of position management. Finally, this chapter has addressed the need for privacy protection, sketched basic mechanisms for it, and discussed open questions in this context.

REFERENCES

1. 3rd Generation Partnership Project. *TS 23.271 Functional stage 2 description of LCS.* http://www.3gpp.org/.
2. N. Anderson. *Tim Berners-Lee on Web 2.0: nobody even knows what it means,* 2006. http://arstechnica.com/news.ars/post/20060901-7650.html.
3. S. Consolvo, I. Smith, T. Matthews, A. LaMarca, J. Tabert, and P. Powledge. Location disclosure to social relations: why, when, & what people want to share. In *Proceedings of the SIGCHI Conference on Human Factors in Computing Systems,* ACM Press, New York, 2005, pp. 81–90.
4. S. Fischmeister and G. Menkhaus. The dilemma of cell-based proactive location-aware services. *Technical Report TR-C042.* Software Research Lab, University of Constance, 2002.
5. M. Gruteser and D. Grunwald. Anonymous usage of location-based services through spatial and temporal cloaking. In *Proceedings of the First International Conference on Mobile Systems, Applications, and Services,* 2003.
6. Java Community Process, JSR-179 Expert Group. *Location API for Java 2 Micro Edition,* 2003.
7. W. Krzysztof and J. Hjelm. *Local Positioning Systems: LBS Applications and Services,* CRC Press, 2006.
8. A. Küpper. *Location-based Services: Fundamentals and Operation,* Wiley, 2005.
9. A. Küpper and G. Treu. Efficient proximity and separation detection among mobile targets for supporting location-based community services. *Mobile Computing and Communications Review* (MC2R), ACM SIGMOBILE, Vol. 10, No. 3, 2007, pp. 1–12.
10. A. LaMarca, et al. PlaceLab: device positioning using radio beacons in the wild. In *Proceedings of the International Conference on Pervasive Computing,* Munich, Germany, Springer Verlag, 2005, pp. 116–133.
11. S. Lederer, I. Hong, K. Dey, and A. Landay. Personal privacy through understanding and action: five pitfalls for designers. *Personal Ubiquitous Comput.,* 8(6):440–454, 2004.

12. A. Leonhardi. *Architektur eines verteilten skalierbaren Lokationsdienstes*. PhD Thesis, Univerisity of Stuttgart, 2003.
13. A. Leonhardi and K. Rothermel. Protocols for updating highly accurate location information. In A. Behcet, editors. *Geographic Location in the Internet*. Kluwer Academic Publishers, 2002, pp. 111–141.
14. G. Myles, A. Friday, and N. Davies. Preserving privacy in environments with location-based applications. *IEEE Pervasive Comput.*, 2(1):56–64, 2003.
15. T. O'Reilly. *What Is Web 2.0: Design Patterns and Business Models for the Next Generation of Software*, 2005, http://tim.oreilly.com/.
16. G. Popischil, J. Stadler, and I. Miladinovic. A location-based push architecture using SIP. In *Proceedings of the 4th International Symposium on Wireless Personal Multimedia Communications* (WPMC '01), Aalborg, Denmark, 2001.
17. P. Ruppel, G. Treu, A. Küpper, and C. Linnhoff-Popien. Anonymous user tracking for location-based community services. In *Proceedings of the 2nd International Workshop on Location and Context-Awareness* (LoCA), Dublin, Ireland, Springer-Verlag, 2006.

Part III

Mobile Mining

Chapter 11

Data Mining for Moving Object Databases

Yoshiharu Ishikawa

Information Technology Center, Nagoya University,
Furo-cho, Chikusa-ku, Nagoya, Japan

11.1 Introduction	239
11.2 Mobility Prediction Using Movement Histories	240
11.3 Sequential Pattern Mining-Based Approaches	244
11.4 Finding Other Interesting Patterns	247
11.5 Clustering Moving Objects	250
11.6 Dense Regions and Selectivity Estimation	253
11.7 Comparing Moving Object Trajectories	257
11.8 Conclusions	259
Acknowledgments	259
References	260

11.1 INTRODUCTION

Due to the recent developments in mobile devices and GPS systems and the progress of network technology, mobile computing has become a key technology today. Along with the progress in mobile technologies, research on moving object databases is currently being actively investigated in the area of database research [16]. As its name suggests, a *moving object database* is a database that stores and manages information on moving objects such as vehicles, pedestrians with mobile devices, and so on. Various research and development efforts regarding moving object databases have been conducted, including several topics such as data models for representing

Mobile Intelligence. Edited by Laurence T. Yang, Agustinus Borgy Waluyo, Jianhua Ma, Ling Tan, and Bala Srinivasan

movement behaviors appropriately, query processing and indexing methods for answering queries efficiently, and application technologies that utilize the underlying moving object databases effectively.

In this chapter, we focus on *data mining* technology for moving object databases. Since the middle of the 1990s, data mining research has been growing rapidly and it has become one of the main research fields in computer science [25, 56]. Although data mining research has been expanding in a variety of fields in recent years, data mining technology for moving object databases are still in an emerging stage of development. However, it is highly promising because many moving objects can be monitored in real-time using current mobile information technologies. Since data managed in a moving object database are highly dynamic and have spatio-temporal semantics, new data mining technologies should be developed. This chapter presents a brief introduction to current trends in data mining on moving object databases including the author's own efforts in this area. We do not intend a complete survey and omit some topics of moving object databases such as data modeling, query processing, and indexing methods except for the issues related to data mining.

The organization of this article is as follows. Section 11.2 introduces some approaches to mobility prediction that can be considered as special cases of data mining for moving objects. Section 11.3 provides a brief survey of methods applying sequential pattern mining to moving databases. Section 11.4 presents other movement pattern mining techniques. Section 11.5 introduces techniques for clustering moving objects and Section 11.6 provides a short summary of density estimation and query selectivity estimation for moving object databases. Section 11.7 presents some interesting ideas for comparing trajectories. Finally, Section 11.8 concludes the article.

11.2 MOBILITY PREDICTION USING MOVEMENT HISTORIES

Mobility prediction, which is used for predicting the future trajectory of a given moving object, is a widely researched topic in mobile computing. Consider a mobile user in a cell-based mobile phone network who is moving continuously while making a call. The underlying network system has to transfer his calling status between cells [70]. If the next cell to which a mobile user will move can be predicted, then an efficient resource reservation and quick handover between base stations can be achieved. So far, various mobility prediction methods have been proposed. A study [6] provides a good survey of this topic. It roughly classifies the approaches to mobility prediction into two categories:

- *Domain-independent methods*: Locations or cells are treated as *symbols*, and only location names are considered, without taking other semantics into account.

- *Domain-specific methods*: Additional information, such as coordinates, directions, and velocities of moving objects, road networks and map information, and/or facility locations are used.

In this subsection, we introduce some selected mobility prediction models that utilize *movement histories*, since they are related to the concept of data mining. We focus in particular on Markov predictors and their extensions.

11.2.1 Domain-Independent Markov Predictors

We first describe the most basic type of mobility predictors, called Markov predictors, and their variants.

11.2.1.1 Markov Predictors

The underlying idea of *Markov predictors* is simple: the next location is predicted from recent movement history based on the notion of *Markov chains*. In this framework, each location is considered as a *state* and each movement between locations corresponds to a *transition*. In an order-k Markov predictor, the k most recent locations in a movement history are used for prediction.

Suppose $X = (X_1, \ldots, X_n)$ is a sequence of random variables taking values in a finite set of locations $L = \{l_1, \ldots, l_m\}$. L represents the state space. The *Markov properties* are as follows:

$$\Pr(X_{n+1} = l_i | X_1, \ldots, X_n) = \Pr(X_{n+1} = l_i | X_{n-k+1}, \ldots, X_n) \quad (11.1)$$

$$= \Pr(X_{n+1} = l_i | X_{j+1}, \ldots, X_{j+k}) \quad (11.2)$$

Equation (11.1) implies that the probability only depends on the most recent k movements and Eq. (11.2) indicates that the probability is *stationary*, or *time invariant*. The probability is basically estimated according to movement histories. The next location predicted is the location that maximizes the probability.

11.2.1.2 Applying String Compression Techniques

There is an approach to extending Markov predictors using the *string compression* technique to summarize statistics in a compact manner. The underlying idea is that a string compression method has the predictive ability to estimate which characters will follow when an input text is given.

In the following, we briefly illustrate a representative method called the *LZ-based predictor* that is an extended version of the order-k Markov predictor, although k is a variable that changes depending on the input. The method is based on the well-known *Lempel-Zip text compression* method (*LZ78*) [45]. LZ78 reads a text stream sequentially and constructs a *dictionary* with *trie* (or tree) structure to summarize the occurring patterns in an online manner. The dictionary is referenced while the compression is in progress. The idea of LZ78 predictors for mobility prediction is to consider a sequence of location symbols instead of a text stream and to use the constructed dictionary for the purposes of prediction.

For example, suppose the movement history of a moving object is given as "ABCABACBADABCD," where A to D are location symbols. For this input, LZ78

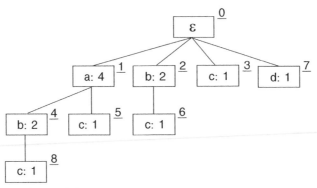

Figure 11.1   Trie-structured dictionary of LZ-based predictors.

constructs a trie as shown in Figure 11.1, where ε represents an empty character. The trie is constructed by partitioning the input into substrings such as "A/B/C/AB/AC/BA/D/ABC" while inserting them. The number shown in the upper-right of each node represents the node identifier and is referenced when the same substring occurs while reading the succeeding text. LZ78 encodes the example text as "0A0B0C1B1C2A0D4C." LZ-based predictors store additional information: each node contains a *counter* to record the number of visits. The count information is used for mobility prediction. For instance, if we want to estimate the probabilities that an object in A next moves to B or C, they are basically given as $Pr(B|A) = 2/4$ and $Pr(C|A) = 1/4$. B is, therefore, predicted as the next location.

The problem of LZ-based predictors is that information for some substrings on the boundaries is lost. For example, a substring "CBA" is contained in the input but does not appear in the trie. To cope with this problem, Bhattacharya and Das [3] proposed the *LeZi-Update* method. When inserting a substring into a trie, LeZi-Update also inserts its suffixes. For example, when we insert "ABC" into the trie for the example text "A/B/C/AB/AC/BA/D/ABC," its proper suffixes "BC" and "C" are also inserted. The method then tries to cope with the cross boundary problem, but note that it does not solve all the boundary problems.

Song et al. [54] collected a large user mobility dataset from an actual mobile wireless network and made an experimental evaluation of domain-independent predictors. Several methods including the Markov-based approach, the LZ-based approach and its variations, and two other related methods were compared. The results are quite interesting: the experiments show that the low-order Markov predictors perform as well or better than the more complex predictors. The best result is obtained by an order-2 Markov predictor with some enhancement. Although the LZ-based approach and other related approaches are effective for text data, the experimental results indicate that they are not necessarily effective for movement prediction. The reason is that the statistical properties of text data and movement histories are quite different.

11.2.2 Markov Chain Model Over Spatial Grid Cells

11.2.2.1 Basic Idea

Markov predictors are basically domain-independent, that is, they treat cells as symbols and do not use other information such as locations. The mobility model proposed by our group [22] extends the basic Markov chain model by incorporating spatial information directly. The fundamental difference is that we consider a spatial grid structure over the target space. Figure 11.2 illustrates this concept. Each dimension of the target space is equally divided into 2^P ranges. The figure shows the case of $P = 2$. We call such partitioning *level-P partitioning*. Based on this partitioning method, there exist $R = 2^{2P}$ grid cells. For each region, a $2P$ bit grid cell number that satisfies the *Z-ordering method* [49] is assigned. The Z-ordering method has the advantage that close cells tend to have similar values. The figure shows that object A located in region 9 at $t = \tau$ moves to region 12 at $t = \tau + 1$ then moves to region 6 at $t = \tau + 2$. We denote the transition by $9^{(2)} \rightarrow 12^{(2)} \rightarrow 6^{(2)}$, where $^{(2)}$ means that the partition level is $P = 2$.

Suppose that another moving object B located in grid cell 9 moves to cell 12 after a unit of time. Supposing that we want to know the probability that object B moves to region 6 next, and that we denote the probability by $\Pr(6|9, 12)$. If we assume the transition between spatial grid cells satisfies the Markov property, we can say that the probability is a second-order Markov transition probability. We can generalize the idea to order-k Markov transition probability $\Pr(c_k|c_0, \ldots, c_{k-1})$, where c_i ($i = 0, \ldots, k$) are cell numbers.

11.2.2.2 Multiple Resolutions

An interesting feature of the model is that it allows multiple resolutions. When analyzing movement data, it is often necessary to view the data at different degrees of coarseness. For example, we may wish to analyze the overall trends at a coarse resolution and then focus on some specific regions and make detailed analyses at a fine resolution. The situation is similar to the case when we use the *drill-down* operation

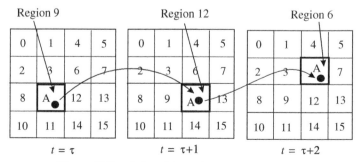

Figure 11.2   Markov chain model over spatial grid cells.

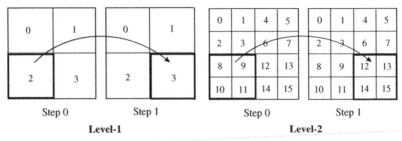

Step 0   Step 1   Step 0   Step 1

Level-1   **Level-2**

Figure 11.3   Roll-up and drill-down.

in *On-Line Analytical Processing (OLAP)*[25, 48]. The opposite operation from fine to coarse resolution corresponds to the *roll-up* operation.

We illustrate how to represent roll-up and drill-down operations in a spatial sense using Figure 11.3, where first-order Markov chains are assumed. The figure on the left-hand side shows the level-1 partitioning and the figure on the right-hand side shows the level-2 partitioning. The level-1 (level-2) partitioning is the "roll-up" ("drill-down") version of the level-2 (level-1) representation. It is easy to map representations at different resolutions because of the property of the Z-ordering method. For example, consider cell number 9 on the level-2 partitioning. Its binary representation is "1001." If we omit the last two digits, we get "10," which is the corresponding cell number 2 in the level-1 partitioning.

On the basis of the above idea, we have demonstrated a mobility histogram construction method that approximates the movement statistics in Ref. [22]. Its details are described in subsection 11.6.3. We have also proposed an efficient algorithm to process queries on mobility statistics based on Markov chains [21]. Its characteristic feature is the effective use of a spatial index to accelerate query processing.

11.3 SEQUENTIAL PATTERN MINING-BASED APPROACHES

Sequential pattern mining [2, 25, 56] is an important topic in data mining and is often applied to basket data analysis and Web usage analysis. There are some interesting applications of sequential data mining to moving object databases. First, we briefly explain the notion of sequential pattern mining.

Suppose that there are four moving objects and their movement histories are given as $h_1 = \langle A, B, C \rangle$, $h_2 = \langle B, C, E \rangle$, $h_3 = \langle A, C \rangle$, and $h_4 = \langle A, C, D, C, E \rangle$, where A to E are location symbols and $\langle \rangle$ represents a sequence. For example, $\langle A, B, C \rangle$ means the moving object visits locations A, B, and C in this order. An example of a *sequential pattern* is $\langle A, C \rangle$ meaning C is visited after A. Note that other places between A and C can be visited. For the above examples, this pattern matches h_1, h_3, and h_4. Since there are two matches for h_4, the total number of matches, called the *support* of the pattern, is four.

A typical objective of sequential pattern mining is to enumerate all the frequent sequential patterns. A *frequent sequential pattern* is defined as a sequential pattern such that its support is greater than or equal to the given threshold, called *min-support*. If we take the domain-independent approach, namely, if we treat locations as symbols, we can directly apply sequential pattern mining to movement histories, but the spatio-temporal nature of moving objects is lost. In the following, we introduce some ideas that incorporate the semantics of moving objects to sequential pattern mining.

11.3.1 TrajPattern: Finding the Top-*k* Trajectory Patterns

TrajPattern was proposed by Yang and Hu [66] and is a framework inspired by sequential pattern mining. The algorithm tries to summarize trajectories into k prominent movement patterns considering imprecision and noise in trajectories.

In their approach, a trajectory has the form $t = \langle (l_1, \sigma_1), (l_2, \sigma_2), \ldots \rangle$, where l_i and σ_i are the expected location and the standard deviation of the mobile object at the ith snapshot. The matching probability $\Pr(p, t)$ between a pattern $p = \langle p_1, \ldots, p_n \rangle$ and a trajectory $t = \langle (l_1, \sigma_1), \ldots, (l_n, \sigma_n) \rangle$ considers the ambiguity of the trajectory (its definition is omitted here). The *match* measure is defined as follows:

$$\mathrm{match}(p, t) = \frac{\log \Pr(p, t)}{|n|}. \tag{11.3}$$

When a trajectory t whose length is longer than pattern p is given, the match measure is extended as follows:

$$\mathrm{match}(p, t) = \max_{\forall t' \subseteq t, |t'| = |p|} \mathrm{match}(p, t'), \tag{11.4}$$

where '$\subseteq$' represents the subsequence relationship. When a trajectory dataset D is given, the match measure is further extended as:

$$\mathrm{match}(p, D) = \sum_{t \in D} \mathrm{match}(p, t). \tag{11.5}$$

The score is regarded as the expected number of occurrences of pattern p in D. Roughly speaking, TrajPattern finds the k patterns with the highest scores.

TrajPattern is based on the sequence mining approach, but clustering is also used. It first identifies short patterns with high scores and then tries to extend the patterns to longer patterns with high scores. To prune nonqualifying candidates, the algorithm utilizes the *min–max property*, which is a similar notion to the *apriori property* [1, 25, 56], but it is a weaker notion due to the definition of the match measure.

11.3.2 Mobility Rules: Consideration of Cell Topologies and Noises

Yavaş et al. [67] proposed a method for mining *mobility rules* from a movement history (a sequence of cell numbers) by extending sequential pattern mining. First, the method extracts *user mobility patterns* from the given sequence. A user mobility pattern means a frequent cell sequence. The algorithm tries to find meaningful user mobility patterns by extending the sequential pattern mining approach. The following two extensions are essential:

- A user mobility pattern is generated such that the pattern consists of *neighboring* cells considering possible movements of mobile objects. That is, the method takes the underlying cell topology into account.

- A movement trajectory often contains noise due to random movements and corruption. They, therefore, provide a *robust* support counting method that utilizes the notion of *string alignment* to enable flexible string matching.

Second, *mobility rules* are generated. On the basis of the user mobility patterns extracted in the previous step, the algorithm generates rules such as $\langle A, C \rangle \rightarrow \langle D, B \rangle$. This rule means that an object that moves from A to C will move from D to B with high support and confidence.

11.3.3 Frequent Mobility Pattern Mining from One Long Trajectory

Cao et al. [4] mine frequent spatio-temporal patterns from *one* long trajectory. An example of such data is a trace of a bus for a single day in a city. The method detects frequent sequential patterns without predefined segmentation of a trajectory. First, the algorithm simplifies the given trajectory data into a list of approximated line segments. Next, similar line segments are grouped to find frequent sequential patterns. Then, each segment contained in each sequential pattern is converted into a sequence of region IDs, where region means the area surrounding a segment. Finally, longer frequent sequential patterns are derived by combining the short sequential patterns. To accelerate the derivation step, they use a tree structure and an apriori-like pruning technique.

11.3.4 Other Work

The algorithm proposed by Peng and Chen [47] identifies movement patterns from movement log data. Given a *support threshold*, the algorithm tries to find long sequential patterns such as "ABDEB," each of which corresponds to a cell in a mobile network. The mined movement patterns are used to allocate data in appropriate mobile sites, so that moving objects can obtain data efficiently while they move in a mobile network.

11.4 FINDING OTHER INTERESTING PATTERNS

The former section focused on the extension of sequential pattern mining. This section introduces other mining approaches to find interesting patterns from moving object databases.

11.4.1 Spatio-temporal Association Rules

Association rule mining is one of the most popular topics in data mining [1, 25, 56]. An example of an *association rule* is {notebook} $\Rightarrow$ {pen}. The rule indicates that if a notebook is purchased, it is likely that a pen is also purchased. So far, various association rule mining methods for large transaction data have been proposed.

Verhein and Chawla [61] extend the notion of association rules to the spatio-temporal context. They call such rules *spatio-temporal association rules*. The most simple type of rule can be written as

$$R = (r_i, \Delta_i) \Rightarrow (r_j, \Delta_j), \tag{11.6}$$

where r_i, r_j are spatial regions and Δ_i, Δ_j are time intervals that satisfy $\Delta_i < \Delta_j$. The above rule says that objects appearing in r_i during time interval Δ_i will appear in region r_j during Δ_j. To find interesting rules, they propose the notion of *spatial support*:

$$\text{spatial\_support}(R) = \frac{\sigma((r_i, \Delta_i) \Rightarrow (r_j, \Delta_j))}{\text{area}(r_i) + \text{area}(r_j)}, \tag{11.7}$$

where $\sigma((r_i, \Delta_i) \Rightarrow (r_j, \Delta_j))$ denotes the conventional *support* of the rule—the number of objects that satisfy the rule and area(r) is the area of region r. The smaller the area covered, the higher the spatial support of the rule.

In addition, several interesting patterns are proposed in Ref. [61]. Some examples are as follows:

- *Dense region*: A region r is called a *dense region* (or *hot spot*) during Δ if density$(r, \Delta) = \sigma(r, \Delta)/\text{area}(r) \geq \delta$, where $\sigma(r, \Delta)$ is the number of objects in r during Δ and δ is a *minimum density threshold*.
- *High traffic region*: A region r is a *high traffic region* if the number of objects entering r (n_r) or leaving r (n_l) during Δ satisfies $\alpha/\text{area}(r) \geq \tau$, where $\alpha = n_e$ or n_l and τ is a *minimum traffic threshold*.
- *Stationary region*: If $\sigma((r, \Delta_i) \Rightarrow (r, \Delta_{i+1}))/\text{area}(r) \geq \tau$, r is called a *stationary region*.

In Ref. [62], the methodology is further refined by one of the authors to represent longer patterns.

11.4.2 Group Patterns

The algorithm proposed by Wang et al. [60] discovers *groupings* of moving objects such that members in the same group are spatially close to one another for a significant amount of time. Such object groupings are called *group patterns*. Given movement histories of objects, the algorithm tries to find appropriate groupings based on the following criteria: (1) the group members should be physically close to one another and (2) the group members should stay together for some meaningful duration.

Two group pattern mining algorithms *apriori-like group pattern mining* (*AGP*) and *valid group-growth* (*VG-Growth*) are proposed. The former is an extension of the well-known *apriori* algorithm [1] and the latter is based on the *FP-growth* algorithm [24]. Some techniques for accelerating the mining process are also introduced.

11.4.3 Periodic Movement Patterns

Moving objects often follow *periodic movements*. For example, a bus run on a regular route shows similar movement patterns every day. Mamoulis et al. [39] try to discover *periodic movement patterns* in spatio-temporal data, including a long movement history of *one* moving object. The approach is an extension of *periodic pattern mining* from event sequences [23].

A *periodic pattern* is defined as a sequence of spatial regions that appears every T timestamp: the pattern should appear at least *min_sup* periodic intervals in the input trajectory. For example, "AB*C*D" is a pattern with length $T = 6$, where "*" is a "don't care" character and matches any region. Namely, the pattern means that the object visits regions from A to D in order and in a cyclic manner. An interesting aspect is that the regions are not predefined; the algorithm *discovers* appropriate regions to form movement patterns. Regions are defined as dense areas and determined using a method inspired by *density-based clustering* [25, 56]. The method is, therefore, prone to small amounts of noise in the trajectory.

Cao et al. [5] further extend the idea and propose variations of periodic patterns and ways to discover them.

11.4.4 Flock, Leadership, Convergence, and Encounter

The *REMO* (RElative MOtion) framework [34, 35] developed by Laube et al. defines various types of behavior pattern for moving object groups. Gudmundsson et al. [14, 15] select some patterns from the framework and then provide formalized definitions:

- *Flock*: At least n objects are within a circle of radius r and they move in the same direction.
- *Leadership*: In addition to the flock pattern condition, the object group should satisfy an additional condition: one of the objects is heading the direction for at least τ time steps.

- *Convergence*: At least *n* objects pass through the same circle of radius *r* without changing direction, but the objects need not arrive at the same time.

- *Encounter*: This is a specialization of the convergence pattern. At least *n* objects are simultaneously inside the same circle of radius *r*.

References [14, 15] provide efficient computation algorithms in terms of computational geometry using approximation techniques.

11.4.5   More Complex Patterns

Some other researchers have proposed using more complex patterns to represent complex behaviors of moving objects. Although their objectives are data representation and query processing, the ideas may be applied to data mining on moving object databases.

Mobility patterns, as proposed by du Mouza and Rigaux [43], is a language to represent movement patterns between location areas. The following are examples of mobility patterns:

- Give all the objects that travel from A to F and from F to C in 10 minutes: `start_at A, follow F, roam 10, follow C`

- Give all the objects that went through F to another area, then went to D or C, and came back to F using the same area: `follow F.@x, follow {D, C}, follow @x.F; @x != F`

Symbols such as A are labels for areas and @x is a variable. In Ref. [42], this idea is generalized to trajectories at multiple resolutions. The target space can be represented at different levels of coarseness and mobility patterns are generalized depending on levels.

Hadjieleftheriou et al. [19] also proposed the notion of complex spatio-temporal pattern queries with their efficient processing methods. Two types of spatio-temporal queries are considered:

- *Spatio-temporal queries with time*: Arbitrary types of spatial predicate may be contained (e.g., range search), and each predicate can be associated with an exact temporal constraint. An example is "find objects that crossed through region A at time T_1, came as close as possible to point B at a later time T_2 and then stopped inside circle C some time during interval (T_3, T_4)."

- *Spatio-temporal queries with order*: The difference with respect to the former is that each predicate is associated with a relative order. In this sense, they are more general than the former. An example is "find objects that first crossed through region A, then passed as close as possible from point B and finally stopped inside circle C."

Efficient query processing methods that use indexing methods have been proposed.

Jin et al. [29] tried to find movement patterns from user movement logs. They considered graph structures in which nodes correspond to locations. Their algorithm considers support counts of node traversals and finds typical movement patterns. An interesting point is that the algorithm finds seldom-visited nodes and random walks from movement logs. A *random walk* consists of multiple nodes that the user moves between frequently in a random manner. Location prediction and location query techniques based on the mined movement behaviors are also proposed.

11.5   CLUSTERING MOVING OBJECTS

Clustering is a technique for grouping a large number of objects and generates *clusters*, which are used to summarize the original dataset. There have been a lot of clustering algorithms proposed in data mining [25, 56]. In the following, we introduce some clustering techniques for moving object databases.

11.5.1   Continual Maintenance of Moving Clusters

Li et al. [36] proposed a real-time and adaptive cluster maintenance method for moving points. The approach is based on the notion of micro-clusters. A *micro-cluster* is a small-sized cluster consisting of nearby objects. After the generation of micro-clusters, some different clustering algorithms can be applied to the micro-clusters by treating each micro-cluster as if it were an individual entity. The idea of micro-clusters was initially proposed in *BIRCH* [69]. The method in Ref. [36] generates *moving micro-clusters* from the target moving objects, and then global clusters are generated using the micro-clusters. Since moving objects change positions and directions, the method maintains clusters adaptively.

The merging and partitioning processes for micro-clusters are performed using *clustering features*, which summarize the clusters. A clustering feature for a (micro-) cluster C_i is defined as

$$cf_i = (sx_i, sy_i, sv_{x_i}, sv_{y_i}, n_i, t_i), \tag{11.8}$$

where t_i is the cluster generation time, n_i the number of elements ($n_i = |C_i|$), and sx_i (sy_i) the sum of the x-axis (y-axis) values of the elements ($sx_i = \sum_{j:o_j \in C_i} x_j$). sv_{x_i} (sv_{y_i}) is the sum of x-axis (y-axis) velocity values for the elements ($sv_{x_i} = \sum_{j:o_j \in C_i} v_{x_j}$).

When two clusters C_i, C_j are merged at time t_k ($t_i, t_j < t_k$), the clustering feature cf_k of the result cluster C_k is defined as

$$cf_k = (sx_k, sy_k, sv_{x_i} + sv_{x_j}, sv_{y_i} + sv_{y_j}, n_i + n_j, t), \tag{11.9}$$

where sx_k is defined as follows:

$$sx_k = sx_i + (t - t_i)sv_{x_i} + sx_j + (t - t_j)sv_{x_j}. \tag{11.10}$$

sy_k is defined in a similar manner. When partitioning a cluster into two clusters, the resulting clustering features can also be easily computed (the calculation method is omitted here).

An interesting point of this approach is its cluster management scheme. It tries to keep the spatial extent of moving micro-clusters small. The compactness of a micro-cluster is measured by its bounding rectangle. If the size of the bounding rectangle exceeds a certain threshold, the microcluster is split.

11.5.2 Detecting Moving Clusters from Object Movement Histories

Kalnis et al. [30] identify moving clusters in long movement histories . Intuitively, a *moving cluster* in their approach is a sequence of spatial clusters that appear in consecutive snapshots of object movements, such as two consecutive spatial clusters that share a large number of common objects.

The basic idea is as follows. The input is the snapshots of moving object positions. If two clusters C_t at time t and and C_{t+1} at time $t + 1$ satisfy the condition

$$\frac{|C_t \cap C_{t+1}|}{|C_t \cup C_{t+1}|} \geq \theta, \tag{11.11}$$

(C_t, C_{t+1}) is called a *moving cluster*. The idea resembles the notion of dense regions which will be described in Section 11.6, but the difference is that clusters move continually while sharing objects. A number of cluster-detection algorithms have been proposed.

Figure 11.4 shows an example illustrating the concept. S_t, S_{t+1}, and S_{t+2} are three consecutive snapshots of movements. Each circle in each snapshot represents a cluster. If a threshold value, say, $\theta = 0.5$, is used, then three clusters are treated as one moving cluster.

11.5.3 Clustering Moving Objects while Considering Positional Uncertainty

The clustering method proposed by Kriegel and Pfeifle [31] considers *uncertainty* in the positions of moving objects. Their focus is the fuzzy nature of the positions of

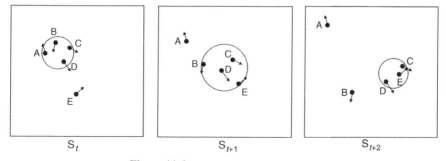

Figure 11.4   Example of a moving cluster.

moving objects. The uncertainty of a moving object is modeled by a *spatial density function* that represents the likelihood that a certain object is located at a certain position. Since the locations of moving objects are uncertain, their algorithm performs several clusterings of points sampled from the probability density functions of the moving objects. A ranking value is assigned to each of the obtained clusterings, which reflects its distance to other sample clusterings. The clustering with the smallest ranking value is called the *medoid clustering* and can be regarded as the average clustering of all the sample clusterings. The method obtains robust clustering results based on this approach.

11.5.4 Other Work

Nanni and Pedreschi [44] employed *density-based clustering* [25, 56] to cluster trajectories. This method groups objects into clusters based on *density*, which is the population within a given region in the space. Typical constraints used in density-based clustering are as follows: for each object in a cluster, its neighborhood, defined by a given radius ε, must contain at least a minimum number of objects, n. Density-based clustering has some advantageous features: it can detect nonspherical clusters of arbitrary shapes and it is robust with respect to noise. Since trajectories may have "snake" shapes and often contain noise, such features are desirable.

Yiu and Mamoulis [68] presented a method to cluster objects on a *spatial network* such as a road network. Each object (not necessarily a moving object) lies on an edge of a large network. The distance between objects is defined by the length of the shortest path between them over the network. Variants of partitioning, density-based, and hierarchical clustering methods have been developed.

The method proposed by Zhang and Lin [71] generates k clusters considering the positions, speeds, and sizes of moving objects. They proposed a special distance function that considers speeds and positions and formalized the clustering as a *k-center optimization* problem. An approximation-based efficient solution method and a cluster refinement method were proposed. The constructed clusters are used to estimate the selectivity of queries on a moving object database.

There exist statistical model-based approaches for trajectory clustering. Gaffney and Smyth [13] applied regression models to cluster similar trajectories. The approach considers two trajectories to be "similar" when they are likely to be generated from a common core trajectory by adding Gaussian noise. A clustering method that is invariant to spatial and temporal shifting of trajectories within clusters was also proposed [10]. The technique was applied to the clustering of human motion trajectories, cyclone trajectories, and so on.

There are some approaches to clustering moving objects from a theoretical perspective. Hershberger [27] proposed a deterministic kinetic data structure for maintaining a covering of moving points in R^d by d-dimensional boxes in an on-line fashion. The number of boxes is always within a factor of 3^d of the best possible static covering. Har-Peled [26] shows how to partition n linearly moving points into k^2 static clusters, so that at any time the diameter of each cluster is at most equal to the maximum cluster diameter in an optimal k-clustering for the current point positions. However, all the movements must be known in advance.

11.6 DENSE REGIONS AND SELECTIVITY ESTIMATION

A region on a space is called a *dense region* if the number of moving objects contained in the region is above some threshold. Detection of dense regions from the underlying moving object database is highly related to density-based clustering (described above). Dense region detection is also related to the estimation problem regarding query selectivity for moving object databases. Some dense region detection methods and query selectivity estimation techniques are briefly reviewed below.

11.6.1 Detecting Dense Regions

Hadjieleftheriou et al. [17] considered processing *density-based queries* on moving object databases. The *density* of region r during time interval Δt is defined as:

$$\text{density}(r, \Delta t) = \frac{\min_{t \in \Delta t} n(r, t)}{\text{area}(r)}, \tag{11.12}$$

where $n(r, t)$ is the number of objects inside r at time t and $\text{area}(r)$ is the area of r. This definition of a dense region is intuitive, but tiny dense regions are also detected. To detect meaningful dense regions, they extend the above basic notion.

For example, a *period density query* is defined as follows. Given movement trajectories, a constant H, and thresholds α_1, α_2, and ξ, find regions $\{r_1, \ldots, r_k\}$ and associated maximal time intervals $\{\Delta t_1, \ldots, \Delta t_k \mid \Delta t_i \subset [t_{\text{now}}, t_{\text{now}} + H]\}$ such that $\alpha_1 \leq \text{area}(r_i) \leq \alpha_2$ and $\text{density}(r_i, \Delta t_i) > \xi$, where t_{now} is the current time. Some algorithms have been provided to find dense areas from a moving object database of linear movements with uniform speeds. To simplify the problem, they partition the data space into disjoint cells instead of arbitrary regions and then find dense regions.

Jensen et al. [28] focused on the identification of dense regions at time $t \in [t_{\text{now}}, t_{\text{now}} + H]$. Objects move continuously and their positions and velocities are updated often. Queries are processed in an online setting, where they assume that objects move linearly until changes are reported. The algorithm computes given queries efficiently using a *density histogram*, which is maintained online.

11.6.2 Selectivity Estimation and Histograms

The *selectivity* of a query is a ratio indicating how many of the objects in the database satisfy the given query [48]. Assume that a query to a moving object database such as "retrieve all the objects that enter a specified region r at time t" is given. To construct an efficient query evaluation plan, the moving object database system needs to estimate the number of objects that qualify for the query. Several methods have been proposed to estimate spatio-temporal query selectivity. A typical approach to estimating query selectivity is to construct a *histogram* [20], which is a compact structure summarizing the underlying database statistics.

11.6.2.1 Histogram for Static Queries on Moving Objects

Choi and Chung [8] extended the traditional histogram technique for spatial databases to cope with linearly moving point objects. The method estimates the selectivity of spatial range queries. Given a rectangular range r and timestamp t that represents some future time, the method estimates the selectivity. In this sense, the method focuses on the case of moving points and static queries. To create a histogram, moving objects are clustered based on their current positions and then organized into buckets. For each bucket, a *spatial bounding rectangle* that covers the objects within the bucket and a *velocity bounding rectangle* that bounds the velocities of the objects within the bucket are constructed. However, the histogram should be rebuilt frequently for accurate estimation. Estimation is performed by assuming the uniformity of velocities. The same authors proposed a further improved method [9]. Nonuniformity of velocities is also considered and a refined histogram is constructed.

Hadjieleftheriou et al. [18] also propose a selectivity estimation method in which they assume linear trajectories. In their approach, the *duality transform* technique is used: the moving points in the primal space–time space are transformed into dual velocity–intercept space. A histogram is then constructed on the dual velocity–intercept space.

11.6.2.2 Histogram for Moving Queries on Moving Objects

In contrast, Tao et al. [57] proposed a further improved method. Their multidimensional *spatio-temporal histogram* supports all types of objects (static/dynamic and points/rectangles) and moving queries. Their method constructs a spatio-temporal histogram, which considers both locations and velocities for partitioning. In addition, an incremental histogram maintenance method was proposed.

11.6.2.3 Other Work

The approach of Sun et al. [55] is to use an adaptive multidimensional histogram to summarize the positions of moving objects for the present time. Past histograms are archived as *historical synopses* and allow users to issue aggregate queries related to the past. In addition, a prediction method is proposed for future movements based on the current movement statistics. Tao et al. [58] present an interesting approach. The method integrates spatio-temporal indexes with sketches in order to aggregate spatio-temporal statistics, including object movements. A *sketch* is a common approach to approximate counting information and is considered as a special kind of histogram. The idea proposed is applicable for finding spatio-temporal association rules such as $(r_i, T, p) \Rightarrow r_j$. This rule means that a user in region r_i at time t will appear in region r_j by time $t + T$ with probability p. Other approaches to spatio-temporal histograms for moving objects can be found in Refs. [12, 46].

11.6.3   Mobility Histograms Based on Markov Chains

Our mobility representation model using Markov chains over spatial grid cells is presented in subsection 11.2.2. In this subsection, we describe its histogram structure for summarizing mobility statistics [22]. The histogram structure has two representation levels: logical and physical.

11.6.3.1   Logical Level: Data Cubes

To represent order-k Markov chain-based movement statistics, a *mobility histogram* is constructed as a $(k + 1)$-dimensional data cube. A *data cube* [25] is a data structure that summarizes the underlying data in a multidimensional array and is often used in OLAP, which provides flexible analysis facilities. Figure 11.5 shows an example of data cube for $k = 2$. The data cube corresponds to the level-1 space partitioning ($P = 1$), mentioned in subsection 11.2.2. Since the two-dimensional target space is partitioned into $R = 2^{2P} = 4$ spatial regions, the data cube contains $R^{k+1} = 64$ cells. For each dimension of the data cube, each step 0, 1, and 2 corresponds to each step of an order-2 Markov chain. For instance, when the sequence $1^{(1)} \rightarrow 1^{(1)} \rightarrow 2^{(1)}$ is given as an input transition sequence, the value of the corresponding cell $(1, 1, 2)$ is incremented.

More intuitive data manipulation is possible using the data cube representation. For example, consider the data cube shown in Figure 11.5. The probability that an object has moved from region 1 to region 2 and then moves to region 4 is calculated as $\Pr(4|1, 2) = \text{val}(1, 2, 4)/\text{val}(1, 2, *)$, where $\text{val}(1, 2, 4)$ is the value of the cube cell $(1, 2, 4)$ and $\text{val}(1, 2, *) = \sum_{i=0}^{2^P-1} \text{val}(1, 2, i)$. Moreover, the probability that an object in region 1 at $t = \tau$ and in region 3 at $t = \tau + 2$ is in region 2 at $t = \tau + 1$ (τ is some arbitrary time) is calculated as $\text{val}(1, 2, 3)/\text{val}(1, *, 3)$.

As described in subsection 11.2.2, our mobility model allows multiple resolutions using different partition level settings. Roll-up and drill-down operations are also supported for data cubes and users can change resolutions when they perform analyses.

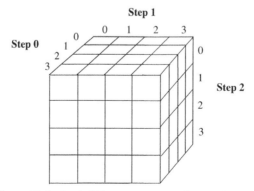

Figure 11.5   Logical histogram representation as a data cube.

Other types of queries can also be supported by the data cube representation. For example, density queries can be processed by a data cube using aggregation and selection.

11.6.3.2 Physical Histogram: Multidimensional Trie

A physical histogram has a structure like a *multidimensional trie* for summarizing mobility statistics with multiple resolutions. It has the following features:

1. Each node of a trie has four branches labeled 00, 01, 10, and 11. Each of the branches corresponds to a quarter region obtained by decomposing the space into 2×2 components. That is, the space decomposition is performed in a similar manner to a *quadtree* [51].

2. The first branch of the trie root corresponds to step 0 of a Markov chain and the next branch to step 1, and so on. Each branch thus corresponds to a Markov transition steps in turn. The idea is inherited from *k-d trees* [51].

3. Each node of a trie has a *counter* for accumulating the number of trajectory patterns visited.

Figure 11.6 shows the structure in the case of the maximal partition level $P = 2$. The dotted lines mean that the edges are not instantiated because the corresponding sequences do not appear. The example shows the situation that a transition sequence $3^{(2)} \rightarrow 6^{(2)} \rightarrow 12^{(2)}$ is inserted. As described in subsection 11.2.2, the model assigns region numbers based on the Z-ordering method. The merit of using this numbering method is that the four-way branching approach corresponds exactly to the Z-ordering numbering. Using this approach, we can accumulate movement patterns adaptively so as not to create unnecessary branches for the trie. Furthermore, in order to reduce the

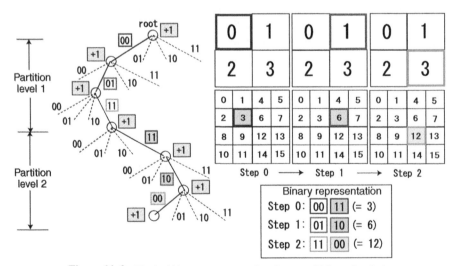

Figure 11.6 Physical histogram structure based on a multidimensional trie.

storage overhead of the histogram, an approximate histogram construction algorithm was proposed [22]. The algorithm receives movement trajectories as a stream and gradually expands the trie considering the statistical properties of the nodes in the trie. Although the histogram constructed does not contain exact count information, it represents the overall statistics approximately with a relatively small storage cost.

11.7 COMPARING MOVING OBJECT TRAJECTORIES

11.7.1 Distance Between Trajectories

To perform clustering or similarity searches on trajectories of moving objects, an appropriate definition of the *distance* (or similarity) between two trajectories is important. Typically, the trajectory of a moving object is represented as a sequence of consecutive locations in a two- or three-dimensional space, in contrast to the one-dimensional case, which is common in timeseries databases [33, 52]. In addition, distances for moving object trajectories should be robust to outliers since measurements in mobile environments are noisy and the movement of objects may contain "gaps." Some distance measures proposed for moving object trajectories are investigated below.

For example, consider two trajectories of moving objects on the (x, y)-plane given as $A = ((a_{x,1}, a_{y,1}), \dots, (a_{x,n}, a_{y,n}))$ and $B = ((b_{x,1}, b_{y,1}), \dots, (b_{x,m}, b_{y,m}))$. Their lengths are n and m, respectively. When $n = m$, we can apply the *Euclidean distance* (L_2 distance), which is simple and also the most popular:

$$L_2(A, B) = \left(\sum_{i=1}^{n} \left[(a_{x,i} - b_{x,i})^2 + (a_{y,i} - b_{y,i})^2 \right] \right)^{\frac{1}{2}}. \tag{11.13}$$

Although the Euclidean distance can be evaluated efficiently, it cannot be applied when two trajectories have different lengths.

11.7.2 Dynamic Time Warping (DTW)

It is often necessary to compute the similarity between two trajectories with different lengths. A well-known approach is to use *dynamic time warping (DTW)* [33, 52], which is defined as follows:

$$DTW(A, B) = |a_n, b_m| + \min\{DTW(head(A), head(B)),$$

$$DTW(head(A), B), DTW(A, head(B))\}, \tag{11.14}$$

where $|a_n, b_m|$ is the distance between the two points $a_n = (a_{x,n}, a_{y,n})$ and $b_n = (b_{x,m}, b_{y,m})$ and is usually measured using the Euclidean distance. $head(A)$ is a sequence A without the last item $(a_{x,n}, a_{y,n})$. An example of DTW matching is shown in Figure 11.7, where A and B are two trajectories. DTW is the accumulated distance between the matching nodes and allows flexible sequence matching, but it is not an

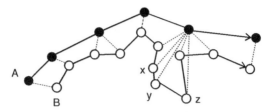

Figure 11.7   DTW matching.

effective distance metric for noisy trajectory data because all the elements in two trajectories must be matched when using DTW. In the figure, three points x, y, and z are outliers, but DTW tries to match these points. Although the two trajectories are similar except for these outliers, DTW returns a large distance value.

11.7.3   More Robust Distance Measures

On the basis of this observation, Vlachos et al. [63] proposed the *least common subsequence (LCSS)* distance for similarity-based retrieval of moving object trajectories. The LCSS score is defined as follows (we have simplified the original definition for simplicity):

$$
\text{LCSS}(A, B) = \begin{cases} 0 & \text{if } A \text{ or } B \text{ is empty} \\ 1 + \text{LCSS}(\text{head}(A), \text{head}(B)) \\ \quad \text{if } |a_n - b_m| < \varepsilon \text{ and } |n - m| \leq \delta \\ \max(\text{LCSS}(\text{head}(A), B), \text{LCSS}(A, \text{head}(B))) \\ \quad \text{otherwise} \end{cases} \tag{11.15}
$$

The parameters ε and δ are given by the user. The condition $|a_n - b_m| < \varepsilon$ means that a_n and b_m can be regarded as if they are the *same symbols*. The LCSS score means the number of matching symbols and takes a large value when two trajectories are similar. Figure 11.8 shows an example of LCSS matching where $\text{LCSS}(A, B) = 6$. Vlachos et al. defined the *LCSS distance* with the following formula:

$$
D(A, B) = 1 - \frac{\text{LCSS}(A, B)}{\min(n, m)}. \tag{11.16}
$$

As shown in the figure, the LCSS matching omits outliers so that it is more robust to noise. They extend the distance considering time stretching and translations.

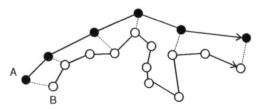

Figure 11.8   LCSS matching.

Chen et al. [7] further extended the idea in this direction. They proposed a new distance called *edit distance on real sequence* (*EDR*) for moving object trajectories. The basic idea of EDR is that it assigns penalties to nonmatched points. EDR is therefore more sensitive than LCSS according to the number of outliers. The distance between two trajectories becomes small when they are similar in the sense of LCSS and have less outliers. This distance function was shown to be more robust than DTW and LCSS over trajectories with noise.

11.7.4 Other Work

Yanagisawa et al. [64] discussed the issues of shape based similarity queries for moving object trajectories. They proposed some similarity measures considering trajectory approximation. Yanagisawa and Satoh [65] define two distance measures for trajectories. They are extensions of the Euclidean distance and DTW respectively and consider the shapes and velocities of moving object trajectories. Lin and Su [37] proposed the "one way distance" function for comparing moving object trajectories. In addition, several techniques for implementation were presented.

11.8 CONCLUSIONS

In this chapter, we have reviewed current trends in data mining technologies on moving object databases. Data mining on moving object databases has different requirements from conventional data mining since moving objects have a dynamic nature and spatio-temporal semantics, and different use is made of mined knowledge. The new requirements give birth to new technologies. As described above, a variety of interesting approaches have appeared in this field of research.

We have also shown some pointers to related areas. The *spatio-temporal database* technology is a closely related topic in moving object databases. It is a generic name for databases that store and manage information regarding objects with temporal and spatial features, and includes the notion of moving objects databases. A spatio-temporal database is, however, not necessarily for moving objects because it is used, for example, for representation of time-varying geographic information. Refs [32, 41] are collections of articles on spatio-temporal data mining. Roddick et al. [50] provides a reference list concerning data mining technology including spatio-temporal databases up to the year 2000. López et al. [38] survey aggregation techniques for spatial, temporal, and spatio-temporal databases. Aggregation is used to accumulate statistics from an underlying database and is useful for data mining and learning from the data. Dunham et al. [11] and Wang et al. [59] review spatio-temporal data mining technologies. There are a number of textbooks on spatial databases [40, 49, 51, 53], and well-known data mining textbooks [25, 56].

ACKNOWLEDGMENTS

This research is partly supported by the Grant-in-Aid for Scientific Research on Priority Areas (19024037, 21013023) from the Ministry of Education, Culture, Sports,

Science and Technology (MEXT, Japan and the Grant-in-Aid for Scientific Research (19300027) from Japan Society for Promotion of Science (JSPS).

REFERENCES

1. R. Agrawal and R. Srikant. Fast algorithms for mining association rules in large databases. In *Proceedings of the International Conference on Very Large Data Bases (VLDB'94)*, 1994, pp. 487–499.

2. R. Agrawal and R. Srikant. Mining sequential patterns. In *Proceedings of the International Conference on Data Engineering (ICDE'95)*, 1995, pp. 3–14.

3. A. Bhattacharya and S. K. Das. LeZi-Update: an information-theoretic framework for personal mobility tracking in PCS networks. *ACM/Kluwer Wireless Networks*, 8(2–3): 121–135, 2002.

4. H. Cao, N. Mamoulis, and D. W. Cheung. Mining frequent spatio-temporal sequential patterns. In *Proceedings of the International Conference on Data Mining (ICDM'05)*, 2005, pp. 82–89.

5. H. Cao, N. Mamoulis, and D. W. Cheung. Discovery of periodic patterns in spatiotemporal sequences. *IEEE Trans. Knowledge Data Eng.*, 19(4):453–467, 2007.

6. C. Cheng, R. Jain, and E. van den Berg. Location prediction algorithms for mobile wireless systems. In B. Furht and M. Ilyas, editors, *Wireless Internet Handbook: Technologies, Standards, and Applications*, CRC Press, pp. 245–263, 2003.

7. L. Chen, M. T. Özsu, and V. Oria. Robust and fast similarity search for moving object trajectories. In *Proceedings of the ACM SIGMOD International Conference on Management of Data (SIGMOD'05)*, 2005, pp. 491–502.

8. Y. -J. Choi and C. -W. Chung. Selectivity estimation for spatio-temporal queries for moving objects. In *Proceedings of the ACM SIGMOD International Conference on Management of Data (SIGMOD'02)*, 2002, pp. 440–451.

9. Y. -J. Choi, H. -H. Park, and C.-W. Chung. Estimating the result size of a query to velocity skewed moving objects. *Inform. Process. Lett.*, 88(6):279–285, 2003.

10. D. Chudova, S. Gaffney, E. Mjolsness, and P. Smyth. Translation-invariant mixture models for curve clustering. In *Proceedings of the ACM SIGKDD International Conference on Knowledge Discovery and Data Mining (KDD'03)*, 2003, pp. 79–88.

11. M. H. Dunham, N. Ayewah, Z. Li, K. Bean, and J. Huang. Spatio-temporal prediction using data mining tools. In *Spatial Databases: Technologies, Techniques and Trends*. Idea Group Publishing, 2004.

12. H. G. Elmongui, M. F. Mokbel, and W. G. Aref. Spatio-temporal histograms. In *Proceedings of the International Symposium on Spatial and Temporal Databases (SSTD'03)*, LNCS 3633, 2005, pp. 19–36.

13. S. Gaffney and P. Smyth. Trajectory clustering with mixtures of regression models. In *Proceedings of the ACM SIGKDD International Conference on Knowledge Discovery and Data Mining (KDD'99)*, 1999, pp. 63–72.

14. J. Gudmundsson, M. van Kreveld, and B. Speckmann. Efficient detection of motion patterns in spatio-temporal data sets. In *Proceedings of the ACM International Workshop on Geographic Information Systems (GIS'04)*, 2004, pp. 250–257.

15. J. Gudmundsson and M. van Kreveld. Computing longest duration flocks in trajectory data. In *Proceedings of the ACM International Workshop on Geographic Information Systems (GIS'06)*, 2006, pp. 35–42.

16. R. H. Güting and M. Schneider. *Moving Objects Databases*. Morgan Kaufmann, 2005.

17. M. Hadjieleftheriou, G. Kollios, D. Gunopulos, and V. J. Tsotras. On-line discovery of dense areas in spatio-temporal databases. In *Proceedings of the International Symposium on Spatial and Temporal Databases (SSTD'03)*, 2003, pp. 306–324.

18. M. Hadjieleftheriou, G. Kollios, and V. J. Tsotras. Performance evaluation of spatio-temporal selectivity estimation techniques. In *Proceedings of the International Conference on Scientific and Statistical Database Management (SSDBM'03)*, 2003, pp. 202–211.

19. M. Hadjieleftheriou, G. Kollios, P. Bakalov, and V. J. Tsotras. Complex spatio-temporal pattern queries. In *Proceedings of the International Conference on Very Large Data Bases (VLDB'05)*, 2005, pp. 877–888.

20. Y. Ioannidis. The history of histograms (abridged). In *Proceedings of the International Conference on Very Large Data Bases (VLDB'03)*, 2003, pp. 19–30.

21. Y. Ishikawa, Y. Tsukamoto, and H. Kitagawa. Extracting mobility statistics from indexed spatio-temporal databases. In *Proceedings of the Workshop on Spatio-temporal Database Management (STDBM'04)*, 1004, pp. 9–16.

22. Y. Ishikawa, Y. Machida, and H. Kitagawa. A dynamic mobility histogram construction method based on Markov chains. In *Proceedings of the International Conference on Scientific and Statistical Database Management (SSDBM'06)*, 2006, pp. 359–368.

23. J. Han, G. Dong, and Y. Yin. Efficient mining of partial periodic patterns in time series database. In *Proceedings of the International Conference on Data Engineering (ICDE'99)*, 1999, pp. 106–115.

24. J. Han, J. Pei, and Y. Yin. Mining frequent patterns without candidate generation. In *Proceedings of the ACM SIGMOD International Conference on Management of Data (SIGMOD'00)*, 2000, pp. 1–12.

25. J. Han and M. Kamber. *Data Mining*. Morgan Kaufmann, 2nd edition, 2005.

26. S. Har-Peled. Clustering motion. *Discrete Comput. Geometry*, 31(4): 545–565, 2004.

27. J. Hershberger. Smooth kinetic maintenance of clusters. *Comput. Geometry*, 31(1–2):3–30, 2005.

28. C. S. Jensen, D. Lin, B. C. Ooi, and R. Zhang. Effective density queries on continuously moving objects. In *Proceedings of the International Conference on Data Engineering (ICDE'06)*, 2006.

29. M. -H. Jin, J. -T. Horng, M. -F. Tsai, and E. H. -K. Wu. Location query based on moving behaviors. *Inform. Syst.*, 32(3):385–401, 2007.

30. P. Kalnis, N. Mamoulis, and S. Bakiras. On discovering moving clusters in spatio-temporal data. In *Proceedings of the International Symposium on Spatial and Temporal Databases (SSTD'05)*, LNCS 3633, 2005, pp. 364–381.

31. H. -P. Kriegel and M. Pfeifle. Clustering moving objects via medoid clusterings. In *Proceedings of the International Conference on Scientific and Statistical Database Management (SSDBM'05)*, 2005, pp. 153–162.

32. R. Ladner, K. Shaw, and M. Abdelguerfi, editors. *Mining Spatio-temporal Information Systems*. Kluwer, 2002.

33. M. Last, A. Kandel, and H. Bunke, editors. *Data Mining in Timeseries Databases*. World Scientific, 2004.

34. P. Laube and S. Imfeld. Analyzing relative motion within groups of trackable moving point objects. In *Proceedings of the International Conference on Geographic Information Science (GIScience'02)*, LNCS 2478, 2002, pp. 132–144.

35. P. Laube, M. van Kreveld, and S. Imfeld. Finding REMO—detecting relative motion patterns in geospatial lifelines. In *Proceedings of the International Symposium on Spatial Data Handling (SDH'04)*, 2004, pp. 201–215.

36. Y. Li, J. Han, and J. Yang. Clustering moving objects. In *Proceedings of the ACM SIGKDD International Conference on Knowledge Discovery and Data Mining (KDD'04)*, 2004, pp. 617–622.

37. B. Lin and J. Su. Shapes based trajectory queries for moving objects. In *Proceedings of the ACM International Workshop on Geographic Information Systems (GIS'05)*, 2005, pp. 21–30.

38. I. F. V. López, R. T. Snodgrass, and B. Moon. Spatiotemporal aggregate computation: a survey. *IEEE Trans. Knowledge Data Eng.*, 17(2):271–286, 2005.

39. N. Mamoulis, H. Cao, G. Kollios, M. Hadjieleftheriou, Y. Tao, and D. W. Cheung. Mining, indexing, and querying historical spatiotemporal data. In *Proceedings of the ACM SIGKDD International Conference on Knowledge Discovery and Data Mining (KDD'04)*, 2004, pp. 236–245.

40. Y. Manolopoulos, A. N. Papadopoulos, and M. G. Vassilakopoulos. *Spatial Databases: Technologies, Techniques and Trends*. Idea Group Publishing, 2004.

41. H. J. Miller and J. Han, editors. *Geographic Data Mining and Knowledge Discovery*. Taylor & Francis, 2001.

42. C. du Mouza and P. Rigaux. Multi-scale classification of moving object trajectories. In *Proceedings of the International Conference on Scientific and Statistical Database Management (SSDBM'04)*, 2004, pp. 307–316.

43. C. du Mouza and P. Rigaux. Mobility patterns. *GeoInformatica*, 9(4):297–319, 2005.

44. M. Nanni and D. Pedreschi. Time-focused clustering of trajectories of moving objects. *J. Intell. Inform. Syst.*, 27(3):267–289, 2006.

45. M. Nelson and J. -L. Gailly. *The Data Compression Book*. M&T Books, 2nd edition, 1995.

46. H. K. Park, J. H. Son, and M. H. Kim. Dynamic histograms for future spatiotemporal range predicates. *Inform. Sci.*, 172(1–2): 195–214, 2005.

47. W. -C. Peng and M. -S. Chen. Developing data allocation schemes by incremental mining of user moving patterns in a mobile computing system. *IEEE Trans. Knowledge Data Eng.*, 15(1): 70–85, 2003.

48. R. Ramakrishnan and J. Gehrke. *Database Management Systems*. McGraw-Hill, 3rd edition, 2002.

49. P. Rigaux, M. Scholl, and A. Voisard. *Spatial Databases: With Application to GIS*. Morgan Kaufmann, 2001.

50. J. F. Roddick, K. Hornsby, and M. Spiliopoulou. An updated bibliography of temporal, spatial, and spatio-temporal data mining research. In *Proceedings of the Workshop on Temporal, Spatial, and Spatio-temporal Data Mining (TSDM'00)*, LNCS 2007, 2000, pp. 147–164.

51. H. Samet. *Foundations of Multidimensional and Metric Data Structures*. Morgan Kaufmann, 2006.

52. D. Shasha and Y. Zhu. *High Performance Discovery in Timeseries*. Springer-Verlag, 2004.

53. S. Shekhar and S. Chawla. *Spatial Databases: A Tour*. Prentice Hall, 2002.

54. L. Song, D. Kotz, and R. Jain. Evaluating next-cell predictors with extensive Wi-Fi mobility data. *IEEE Trans. Mobile Comput.*, 5(12):1633–1649, 2006.

55. J. Sun, D. Papadias, Y. Tao, and B. Liu. Querying about the past, the present, and the future in spatio-temporal databases. In *Proceedings of the International Conference on Data Engineering (ICDE'04)*, 2004, pp. 202–213.

56. P. -N. Tan, M. Steinbach, and V. Kumar. *Introduction to Data Mining*. Addison-Wesley, 2005.

57. Y. Tao, J. Sun, and D. Papadias. Selectivity estimation for predictive spatio-temporal queries. In *Proceedings of the International Conference on Data Engineering (ICDE'03)*, 2003, pp. 417–428.

58. Y. Tao, G. Kollios, J. Considine, F. Li, and D. Papadias. Spatio-temporal aggregation using sketches. In *Proceedings of the International Conference on Data Engineering (ICDE'04)*, 2004, pp. 214–226.

59. J. Wang, W. Hsu, and M. L. Lee. Mining in spatio-temporal databases. In *Spatial Databases: Technologies, Techniques and Trends*. Idea Group Publishing, 2004.

60. Y. Wang, E. -P. Lim, and S. -Y. Hwang. Efficient mining of group patterns from user movement data. *Data Knowledge Eng.*, 57(3):240–282, 2006.

61. F. Verhein and S. Chawla. Mining spatio-temporal association rules, sources, sinks, stationary regions and throughfares in object mobility databases. In *Proceedings of the International Conference on Database Systems for Advanced Applications (DASFAA'06)*, LNCS 3882, 2006, pp. 187–201.

62. F. Verhein. k-STARs: sequences of spatio-temporal association rules. In *Proceedings of the Workshop on Spatial and Spatio-temporal Data Mining (SSTDM'06)*, 2006.

63. M. Vlachos, G. Kollios, and D. Gunopulos. Discovering similar multidimensional trajectories. In *Proceedings of the International Conference on Data Engineering (ICDE'02)*, 2006, pp. 673–684.

64. Y. Yanagisawa, J. Akahani, and T. Satoh. Shape-based similarity query for trajectory of mobile objects. In *Proceedings of the International Conference on Mobile Data Management (MDM'03)*, LNCS 2574, 2003, pp. 63–77.

65. Y. Yanagisawa and T. Satoh. Clustering multidimensional trajectories based on shape and velocity. In *Proceedings of the IEEE International Workshop on Multimedia Databases and Data Management (MDDM'06)*, 2006.

66. J. Yang and M. Hu. TrajPattern: mining sequential patterns from imprecise trajectories of mobile objects. In *Proceedings of the International Conference on Extending Database Technology (EDBT'06)*, LNCS 3896, 2006, pp. 664–681.

67. G. Yavaş, D. Katsaros, Ö. Ulusoy, and Y. Manolopoulos. A data mining approach for location prediction in mobile environments. *Data Knowledge Eng.*, 54(2):121–146, 2005.

68. M. L. Yiu and N. Mamoulis. Clustering objects on a spatial network. In *Proceedings of the ACM SIGMOD International Conference on Management of Data (SIGMOD'04)*, 2004, pp. 443–454.
69. T. Zhang, R. Ramakrishnan, and M. Livny. BIRCH: a new data clustering algorithm and its applications. *Data Mining Knowledge Discov.*, 1(2):141–182, 1997.
70. J. Zhang. Location management in cellular networks. In I. Stojmenovic, editors. *Handbook of Wireless Networks and Mobile Computing*. Wiley, 2002, pp. 27–49.
71. Q. Zhang and X. Lin. Clustering moving objects for spatio-temporal selectivity estimation. In *Proceedings of the Australasian Database Conference (ADC'04)*, 2004, pp. 123–130.

Chapter 12

Mobile Data Mining on Small Devices Through Web Services

Domenico Talia and Paolo Trunfio

DEIS, University of Calabria, Rende (CS), Italy

12.1 Introduction	264
12.2 Mobile Data Mining	265
12.3 Mobile Web Services	266
12.4 System Design and Implementation	270
12.5 Summary	275
References	276

12.1 INTRODUCTION

Analysis of data is a complex process that often involves remote resources (computers, software, databases, files, etc.) and people (analysts, professionals, end users). Recently, distributed data mining techniques are used to analyze dispersed data sets. An advancement in this research area comes from the use of mobile computing technology for supporting new data analysis techniques and new ways to discover knowledge from every place in which people operate.

The availability of client programs on mobile devices that can invoke the remote execution of data mining tasks and show the mining results is a significant added value for nomadic users and organizations that need to perform the analysis of data stored in repositories far away from the site where users are working, allowing them to generate knowledge regardless of their physical location.

This chapter discusses pervasive data mining of databases from mobile devices through the use of Web services. By implementing mobile Web services, we allow

Mobile Intelligence. Edited by Laurence T. Yang, Agustinus Borgy Waluyo, Jianhua Ma, Ling Tan, and Bala Srinivasan

remote users to execute data mining tasks from a mobile phone or a personal digital assistant (PDA) and receive on those devices the results of a data analysis task. A prototype based on a J2ME client will be presented, by describing the data selection task, the server invocation mechanisms, and the result presentation on a mobile device.

The chapter is organized as follows. Section 12.2 introduces mobile data mining. Section 12.3 discusses the use of Web services in mobile environments. Section 12.4 describes the design and implementation of a system we developed for mobile data mining based on the Web service technology. Section 12.5 summarizes and concludes the chapter.

12.2 MOBILE DATA MINING

The goal of mobile data mining is to provide advanced techniques for the analysis and monitoring of critical data from mobile devices.

Mobile data mining has to face with the typical issues of a distributed data mining environment, in addition to technological constraints such as low-bandwidth networks, reduced storage space, limited battery power, slower processors, and small screens to visualize the results [1].

The mobile data mining field may include several application scenarios in which a mobile device can play the role of data producer, data analyzer, client of remote data miners, or a combination of them. More specifically, we can envision three basic scenarios for mobile data mining:

- The mobile device is used as terminal for ubiquitous access to a remote server that provides some data mining services. In this scenario, the server analyzes the data stored in a local or distributed database and sends the results of the data mining task to the mobile device for its visualization. The system we describe in this chapter is based on this approach.

- Data generated in a mobile context are gathered through a mobile device and sent in a stream to a remote server to be stored into a local database. Data can be periodically analyzed by using specific data mining algorithms and the results used for making decisions about a given purpose.

- Mobile devices are used to perform data mining analysis. Due to the limited computing power and storage space of today's mobile devices, currently it is not realistic to perform the whole data mining task on a small device. However, some steps of a data mining task (e.g. data selection and preprocessing) could be run on small devices.

MobiMine [2] is an example of data mining environment designed for intelligent monitoring of stock market from mobile devices. MobiMine is based on a client–server architecture. The clients, running on mobile devices such as PDAs, monitor a stream of financial data coming through a server. The server collects the stock market data from different Web sources in a database and processes then on a regular basis using several data mining techniques.

The clients query the database for the latest information about quotes and other information. A proxy is used for communication among clients and the database. Thus, when a user has to query the database, she/he sends the query to the proxy, which connects to the database, retrieves the results, and sends them to the client. To efficiently communicate data mining models over wireless links with limited bandwidth, MobiMine uses a Fourier-based approach to represent the decision trees, which saves both memory on mobile device and network bandwidth.

Another example of mobile data mining system is proposed in Ref. [3]. Such system considers a single logical database that is split into a number of fragments. Each fragment is stored on one or more computers connected by a communication network, either wiredly or wirelessly. Each site is capable of processing user requests that require access to local or remote data.

Users can access corporate data from their mobile devices. Depending on the particular requirements of mobile applications, in some cases the user of a mobile device may log on to a corporate database server and work with data there. In other cases, the user may download data and work with it on a mobile device or upload data captured at the remote site to the corporate database. The system defines a distributed algorithm for global association rule mining, which does not need to ship all of local data to one site, thereby not causing excessive network communication cost.

Another promising application of mobile data mining is the analysis of streams of data generated from mobile devices. Some possible scenarios are patient health monitoring, environment surveillance, and sensor networks. The VEhicle DAta Stream mining (VEDAS) system [4] is an example of mobile environment for monitoring and mining vehicle data streams in real time. The system is designed to monitor vehicles using on-board PDA-based systems connected through wireless networks. VEDAS continuously analyzes the data generated by the sensors located on most modern vehicles, identifies the emerging patterns, and reports them to a remote control center over a low-bandwidth wireless network connection. The overall objective of VEDAS is to support drivers by characterizing their status and to help the fleet managers by quickly detecting security threats and vehicle problems.

12.3 MOBILE WEB SERVICES

The *service-oriented architecture* (*SOA*) model is widely exploited in modern scientific and business-oriented scenarios to implement distributed systems in which applications and components interact each other independently from platforms and languages.

Currently *Web services* are the most important implementation of the SOA model. Their popularity is mainly due to the adoption of universally accepted Internet technologies such as XML and HTTP. The use of Web services fosters the integration of distributed applications, processes, and data, optimizing the deployment of systems and improving their efficiency. In particular, integration represents an important competitive factor in business-to-business (B2B) scenarios, where information systems can be very heterogeneous and complex.

Recently, a growing interest in the use of Web services in mobile environments has been registered. *Mobile Web services* make it possible to integrate mobile devices with server applications running on different platforms, allowing users to access and compose a variety of distributed services from their personal devices.

The remainder of this section discusses the basic characteristics of the SOA model, introduces the main Web services concepts, and discusses current solutions for the implementation of Web services in mobile environments.

12.3.1 The Service-Oriented Architecture

The *SOA* is a model for building flexible, modular, and interoperable software applications. Concepts behind SOA are derived from component-based software, the object-oriented programming, and some other models. The SOA model enables the composition of distributed applications regardless of their implementation details, deployment location, and initial objective of their development. An important principle of service oriented architectures is, in fact, the reuse of software within different applications and processes.

A service-oriented architecture is essentially based on a collection of services. A *service* is a software building block capable of fulfilling a given task or business function. It does so by adhering to a well-defined interface that specifies required parameters and the nature of the result (a contract between the client of the service and the service itself). A service, along with its interface, must be defined in the most general way to allow utilization in different contexts and for different purposes.

Once defined and deployed, services operate independently of the state of any other service defined within the system. However, service independence does not prohibit to have services cooperating each other to achieve a common goal. The final objective of SOA is to provide for an application architecture where all functions are defined as independent services with well-defined interfaces, which can be called in sequences to form business processes [5].

Service-oriented architectures are based on interactions between three roles: *provider*, which is the entity that provides the service; *consumer*, which requires and uses the service; and *registry*, which publishes information about available services. Three types of interactions are possible among these roles: a provider publishes information about the service to a registry; a consumer queries the registry to find the desired service; and the consumer interacts directly with the service provider.

12.3.2 Web Services

Web services are an Internet-based implementation of the SOA model. Basically, Web services are software services that can be described, discovered, and invoked by using XML formalisms and standard Internet protocols such as HTTP [6]. The use of XML as basic language permits to share data independently from underlying platforms and programming languages. At the same time, the use of standard Internet protocols allows to exploit software and hardware infrastructures that are already available for Internet applications such as the Web.

Web services differ in many respects from classical distributed architectures based on remote components such as RMI, CORBA, and DCOM. While Web services use a platform-independent formalism for message exchange, classical architectures use low-level binary communications, thus data encoding completely depend on specific technologies.

Another important difference is that Web services are thought for coarse-grained services, while classical architectures are mainly designed to support fine-grained components. In other terms, Web services expose their functionalities at a higher level of abstraction, while remote components expose low-level operations that are mainly related to implementation aspects.

Web services exploit a set of XML-based standard technologies for service description (WSDL), communication between clients and services (SOAP), and service discovery (UDDI).

The *Web services description language (WSDL)* [7] is used for describing the interfaces of Web services. Basically, WSDL allows to describe the operations that are provided by a service and the input and output messages to be exchanged with the service for each operation.

The *simple object access protocol (SOAP)* [8] defines a standard formalism for exchanging messages between clients and Web services. SOAP messages can be exchanged over several transport protocols, but typically they are transported by using HTTP.

Finally, *universal description, discovery, and integration* (UDDI) [9] is a registry for publishing and discovering Web services. It is used by service providers to publish their Web services, and by service consumers to locate Web services on the basis of their interface definition.

12.3.3 Web Services in Mobile Environments

The market of mobile devices such as smartphones and PDAs is expanding very fast, with new technologies and functionalities appearing every day. Even if such devices share a common set of functionalities, they run on many different platforms, which makes integration with server applications problematic. As in standard wired scenarios, Web services can be exploited in mobile environments to improve interoperability between clients and server applications independently from the different platforms they execute on.

Basically, there are three architecture models for implementing Web services in mobile environments [10]:

- a *wireless portal network*,
- a *wireless extended Internet*, and
- a *Peer-to-Peer (P2P) network*.

In a wireless portal network, there is a gateway between the mobile client and the Web service provider. The gateway receives the client requests and takes care of issuing corresponding SOAP requests and returning responses in a specific format supported by the mobile device.

In the wireless extended Internet architecture, mobile clients interact directly with the Web service provider. In this case, mobile clients are true Web services clients and can send or receive SOAP messages.

Finally, in a P2P network, mobile devices can act both as Web service clients and providers. This capability of acting both as consumer and provider can be particularly useful in systems such as ad hoc networks. It is not currently implemented in real systems, but it represents the more general model that can offer very interesting opportunities for mobile services in a near future.

In most application scenarios, mobile devices act only as Web service consumers. In these cases, the choice between the wireless portal network and the wireless extended Internet architecture mainly depends on the level of performance required by the application.

The wireless extended Internet configuration requires mobile devices with XML/SOAP processing capabilities. This introduces additional processing load on the device and some traffic overhead for transporting SOAP messages over the wireless network [11]. While the additional processing load could be negligible in most devices, the traffic overhead can affect the response time in the presence of wireless connections with limited bandwidth.

On the other hand, a wireless portal network architecture requires the intermediation of a gateway that acts as proxy between client requests and service providers. This allows to use a set of optimizations (e.g., data compression, binary encodings) for reducing the amount of data transferred over the wireless link, but these methods generally depend on the specific structure of data used by the application [10], so its applicability is limited.

Following either the wireless portal network or the wireless extended Internet architecture, some researchers studied how to improve functionalities and performance of Web services in mobile environments.

Adaçal and Bener [10] proposed an architecture that includes the three standard Web service roles (provider, consumer, and registry) and three new components: a service broker, a workflow engine, and a mobile Web service agent. The mobile Web service agent acts as a gateway to Web services for mobile devices and manages all communication among mobile devices and the service broker or the workflow engine. The agent, which is located inside the mobile network, receives the input parameters required for service execution from the mobile device and returns the executed service. It also selects services according to user preferences and context information such as location, air-link capacity, or access-network type.

Chu et al. [12] proposed an architecture that divides the application components into two groups: local components, which are executed on the mobile device, and remote components, which are executed on the server side. The system is able to dynamically reconfigure application components for local or remote execution to optimize a utility function derived from the user preferences. This approach implements a *smart client* model, which is in contrast with that of *thin client* (which is only capable of rendering a user interface) generally implemented in wired scenarios.

Zahreddine and Mahmoud [13] proposed an approach for Web service composition in which an agent performs the composition on behalf of the mobile user. In the proposed architecture, the client request is sent to a server that creates an agent on behalf of the user. The request is then translated into a workflow to be performed by the agent. The agent looks for services that are published in a UDDI registry, retrieving the locations of multiple services that suit the request requirements. The agent then creates a specific workflow to follow, which entails the agent travelling from one platform to another completing the tasks in the workflow.

Besides these and other research works on architectural aspects, some industries worked on the implementation of a software library, named JSR-172 [14], which provides standard access to Web services from mobile devices. JSR-172 is available as an additional library for the Java 2 Micro Edition (J2ME) platform [15]; thus, it can be used on mobile devices that support the Java technology.

The main goal of JSR-172 is to enable interoperability of J2ME clients with Web services. It does so by providing the following

- APIs for basic manipulation of structured XML data, based on a subset of standard APIs for XML parsing.
- APIs and conventions for enabling XML-based RPC communication from J2ME, including
  - o definition of a strict subset of the standard WSDL-to-Java mapping, suitable for J2ME,
  - o definition of stub APIs based on this mapping for XML-based RPC communication, and
  - o definition of runtime APIs to support stubs generated according to the mapping above.

Our system, described in the next section, implements a wireless extended Internet architecture. We used the JSR-172 library for the implementation of its client applications.

12.4   SYSTEM DESIGN AND IMPLEMENTATION

In this section, we describe the design and implementation of the system. As mentioned before, the goal of the system is to support mobile data mining on small devices, such as cellular phones or PDAs, through the use of Web services. First, we introduce the system architecture and describe the design of system components. Then, we present the functionality of the system and its implementation.

12.4.1   General Architecture

The system is based on the client–server architecture shown in Figure 12.1.

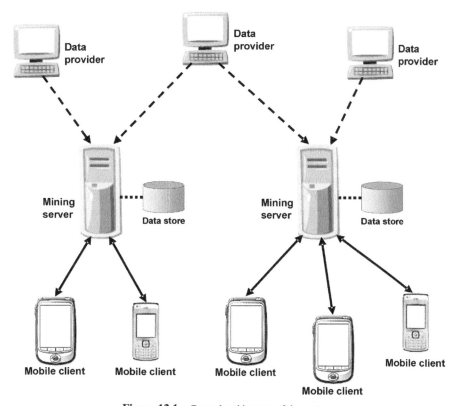

Figure 12.1   General architecture of the system.

The architecture includes three types of components:

- *Data providers*: the applications that generate the data to be mined.
- *Mobile clients*: the applications that require the execution of data mining computations on remote data.
- *Mining servers*: server nodes used for storing the data generated by data providers and for executing the data mining tasks submitted by mobile clients.

As shown in Figure 12.1, data generated by data providers is collected by a set of mining servers that store it in a local data store. Depending on the application requirements, data coming from a given provider could be stored in more than one mining server.

The main role of mining servers is to allow mobile clients to perform the analysis of remote data by using a set of data mining algorithms. Once connected to a given server, the mobile client allows a user to select the remote data to be analyzed and the algorithm to be run. When the data mining task has been completed on the mining server, the results of the computation are visualized on the user device either in textual or visual form.

12.4.2   Software Components

In this section, we describe the software components of mining servers and mobile clients.

12.4.2.1   Mining Server

Each mining server exposes its functionalities through two Web services: the *data collection service* (*DCS*) and the *data mining service* (*DMS*). Figure 12.2 shows the DCS and DMS and the other software components of a mining server.

The DCS is invoked by data providers to store data on the server. The DCS interface defines a set of basic operations for uploading a new dataset, updating an existing dataset with incremental data, or deleting an existing dataset. These operations are cumulatively indicated as *DCS ops* in the figure. Data uploaded through the DCS is stored as plain datasets in the local file system. As shown in the figure, the DCS performs either store or update operations on the local datasets in response to data providers requests.

The DMS is invoked by mobile clients to perform data mining tasks. Its interface defines a set of operations (*DMS ops*) that allow to obtain the list of the available data sets and algorithms, submit a data mining task, get the current status of a computation, and get the result of a given task. Table 12.1 lists the main operations implemented by the DMS.

The data analysis is performed by the DMS using a subset of the algorithms provided by the Weka library [16], which includes a large collection of machine learning algorithms written in Java for data classification, clustering, association rules discovery, and visualization. When a data mining task is submitted to the DMS, the appropriate algorithm of the Weka library is invoked to analyze the local dataset specified by the mobile client.

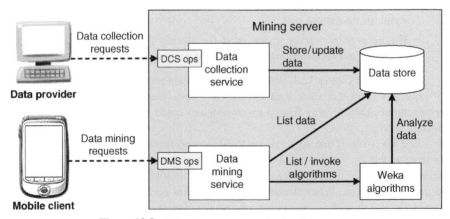

Figure 12.2   The software components of a mining server.

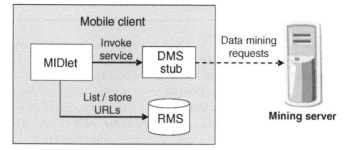

Figure 12.3 The software components of a mobile client.

12.4.2.2 Mobile Client

The mobile client is composed by three components: the *MIDlet*, the *DMS stub*, and the *record management system* (*RMS*) (see Figure 12.3).

The MIDlet is a J2ME application allowing the user to perform data mining operations and visualize their results. The DMS stub is a Web service stub allowing the MIDlet to invoke the operations of a remote DMS. The stub is generated from the DMS interface to conform with the JSR-172 specifications, introduced in the previous section. Even if the DMS stub and the MIDlet are two logically separated components, they are distributed and installed as a single J2ME application.

The RMS is a simple record-oriented database that allows J2ME applications to persistently store data across multiple invocations. In our system, the MIDlet uses the RMS to store the URLs of the remote DMSs that can be invoked by the user. The list of URLs stored in the RMS can be updated by the user using a MIDlet functionality.

12.4.3 Functionality of the System

In the following we describe the typical steps that are executed by the client and server components to perform a data mining task in our system:

1. The user starts the MIDlet on his/her mobile device. After starting, the MIDlet accesses the RMS and gets the list of remote mining servers. The list is shown to the user who selects the mining server to connect with.

2. The MIDlet invokes the `listDatasets` and `listAlgorithms` operations of the remote DMS in order to get the lists of datasets and algorithms that are available on the server. The lists are shown to the user who selects the dataset to be analyzed and the mining algorithm to be used.

3. The MIDlet invokes the `submitTask` operation of the remote DMS, passing the dataset and the algorithm selected by the user with associated parameters. The task is submitted in a batch mode: as soon as the task has been submitted, the DMS returns a unique *id* for it and the connection between client and server is released.

Table 12.1   Data mining service operations

Operation	Description
listDatasets	Returns the list of the local datasets.
listAlgorithms	Returns the list of the available DM algorithms.
submitTask	Submits a DM task for the analysis of a given data set using a specific DM algorithm. Returns a unique *id* for the task.
getStatus	Returns the current status of the task with a given *id*. The status of a task can be *running*, *done*, or *failed*.
getResult	Returns the result of the task with a given *id*, either in textual or visual form.

4. After the task submission, the MIDlet monitors its status by querying the DMS. To this end, the MIDlet periodically invokes the getStatus operation, which receives the *id* of the task and returns its current status (see Table 12.1). The polling interval is an application parameter that can be set by the user.

5. As soon as the getStatus operation returns *done*, the MIDlet invokes the getResult operation to receive the result of the data mining analysis. Depending on the type of data mining task, the MIDlet asks the user to choose how to visualize the result of the computation (e.g., pruned tree, confusion matrix, etc.).

12.4.4 Implementation

All the system components, but the data collection service, have been implemented and tested. The mobile client has been implemented using the Sun Java Wireless Toolkit [17], which is a widely adopted suite for the development of J2ME applications. The data mining service has been implemented using Apache Axis [18], an open source Java platform for creating and deploying Web services applications.

The small size of the screen is one of the main limitations of mobile device applications. In data mining tasks, in particular, a limited screen size can affect the appropriate visualization of complex results representing the discovered model. In our system, we overcome this limitation by splitting the result in different parts and allowing a user to select which part to visualize at one time. Moreover, users can choose to visualize the mining model (e.g., a cluster assignment or a decision tree) either in textual form or as an image. In both cases, if the information does not fit the screen size, the user can scroll it by using the normal navigation facilities of the mobile device.

As an example, Figure 12.4 shows two screenshots of the mobile client taken from a test application. In this example, the MIDlet is executed on the emulator of the Java Wireless Toolkit, while the data mining service is executed on a server machine

Figure 12.4   Two screenshots of the client applications running on the emulator of the Sun Java Wireless Toolkit. On the left, the menu for selecting which part of the result must be visualized; on the right, visualization of the selected classification tree.

using the Web service container provided by Apache Axis. The screenshot on the left shows the menu for selecting which part of the result of a classification task must be visualized, while the screenshot on the right shows the result, in that case the pruned tree resulting from classification.

The system has been tested using some datasets from the the UCI machine learning repository [19], and some data mining algorithms provided by the Weka Library (see section 12.4.2). Our early experiments show that the system performance depends almost entirely on the computing power of the server on which the data mining task is executed. On the contrary, the overhead due to the communication between MIDlet and data mining service does not affect the execution time significantly, since the amount of data exchanged between client and server is very small. In general, when the data mining task is relatively time consuming, the communication overhead is a a negligible percentage of the overall execution time.

12.5   SUMMARY

This chapter discussed pervasive data mining of databases from mobile devices through the use of Web services. By implementing mobile Web services we allow remote users to execute data mining tasks from a mobile phone or a PDA and receive on those devices the results of a data analysis task. A prototype based on a J2ME client has been discussed by describing the data selection task, the server invocation mechanisms, and the result presentation on a mobile device.

Although mobile data mining is not yet in a mature phase, it represents a very promising area for users and professionals who need to analyze data where users, resources and applications are mobile. On the basis of our experience, we can conclude that the combined use of a service-oriented approach with mobile programming technologies makes easier the implementation of mobile knowledge discovery applications that manage heterogeneous and pervasive scenarios where data and processing can migrate across different locations.

REFERENCES

1. S. Pittie, H. Kargupta, and B. Park. Dependency detection in MobiMine: a systems perspective. *Inform. Sci.*, 155(3–4):227–243, 2003.
2. H. Kargupta, B. Park, S. Pitties, L. Liu, D. Kushraj, and K. Sarkar. Mobimine: monitoring the stock marked from a PDA. *ACM SIGKDD Explor.*, 3(2):37–46, 2002.
3. F. Wang, N. Helian, Y. Guo, and H. Jin. A distributed and mobile data mining system. In *Proceedings of the International Conference on Parallel and Distributed Computing, Applications and Technologies*, 2003.
4. H. Kargupta, R. Bhargava, K. Liu, M. Powers, P. Blair. S. Bushra, and J. Dull. VEDAS: a mobile and distributed data stream mining system for real-time vehicle monitoring. In *Proceedings of the SIAM Data Mining Conference*, 2003.
5. K. Channabasavaiah, K. Holley, and E. M. Tuggle. Migrating to a service-oriented architecture, 2007. http://www-106.ibm.com/developerworks/library/ws-migratesoa.
6. Web Services Activity, 2007. http://www.w3.org/2002/ws.
7. Web Services Description Language (WSDL), 2007. http://www.w3.org/TR/wsdl.
8. Simple Object Access Protocol (SOAP), 2007. http://www.w3.org/TR/soap.
9. Universal Description, Discovery, and Integration, 2007. http://www.uddi.org.
10. M. Adaçal, A. B. Bener. Mobile web services: a new agent-based framework. *IEEE Internet Comput.*, 10(3):58–65, 2006.
11. M. Tian, T. Voigt, T. Naumowicz, H. Ritter, and J. Schiller. Performance considerations for mobile web services. *Comput. Commun.*, 27(11):1097–1105, 2004.
12. H. Chu, C. You, and C. Teng. Challenges: wireless Web services. In *Proceedings of the International Conference Parallel and Distributed Systems (ICPADS 04)*, IEEE CS Press, 2004.
13. W. Zahreddine, Q. H. Mahmoud. An agent-based approach to composite mobile Web services. In *Proceedings of the International Conference on Advanced Information Networking and Applications (AINA'05)*, IEEE CS Press, 2005.
14. JSR-172: 2007. *J2ME Web Services Specification*, http://jcp.org/en/jsr/detail?id=172.
15. Java Micro Edition, 2007. http://java.sun.com/javame.
16. H. Witten, E. Frank. *Data Mining: Practical Machine Learning Tools with Java Implementations*. Morgan Kaufmann, 2000.
17. Sun Java Wireless Toolkit, 2007. http://java.sun.com/products/sjwtoolkit.
18. Apache Axis, 2007. http://ws.apache.org/axis.
19. The UCI Machine Learning Repository, 2007. http://www.ics.uci.edu/~mlearn/MLRepository.html.

Part IV

Mobile Context-Aware and Applications

Chapter 13

Context Awareness: A Formal Foundation

Lu Yan[1] and Mats Neovius[2]

[1] *University College London, London, UK*
[2] *Åbo Akademi University, Åbo, Finland*

13.1 Introduction	279
13.2 Background	280
13.3 Related Work	280
13.4 Wireless Sensor Networks	281
13.5 Formalizing Context Awareness and Context Dependency	281
13.6 Case Study: from Specification, via Formalism, to Implementation	290
13.7 Concluding Remarks	292
References	292

13.1 INTRODUCTION

The communication environment surrounding our daily experience is more and more characterized by mobile devices that can exchange multimedia information and provide access to various services of complex nature. The trend is now clear that future consumer computing experience will be based on multiple pervasive communication devices and services, where navigability, context-sensitivity, adaptability, and ubiquity are key characteristics. Several issues have been studied, models and methodologies proposed, and tools and systems implemented. However, we look at the foundation and what we are missing in research, where some of the most relevant issues probably are a formal model of context awareness and context dependency and

Mobile Intelligence. Edited by Laurence T. Yang, Agustinus Borgy Waluyo, Jianhua Ma, Ling Tan, and Bala Srinivasan
Copyright © 2010 John Wiley & Sons, Inc.

a notion of synthesizing reliable complex systems from vast numbers of unreliable components. In this chapter, we discuss a formal foundation and software engineering techniques for mobile context-aware and context-dependent service derivation and application development, emphasizing the relationships between context and system.

13.2 BACKGROUND

With more than 2 billions terminals in commercial operation worldwide, wireless and mobile technologies have facilitated in the first wave of pervasive communication systems and applications. This trend shows several aspects consistent in the evolution of computing including the increasing miniaturization of the computing units and an increasing emphasis of the role of communication between them. Significant research work has been done over recent years on these systems at several levels, from the lowest physical level to the highest information-processing level. However, the latter is less developed than the research at the lower levels. For instance, we think that the most relevant issue for the future perspective of true ubiquitous computing, context awareness and context-dependency has not received justified attention in the research community.

The term context has been extensively studied since the early 1990s; it was mainly associated with the concept of location, but it is much richer than this; some works have categorized context into different aspects, such as computational, user, physical, spatial, and temporal context [1–6]. However, a precise definition of context is still missing. In this chapter, we interpret context as a setting in which an event occurs, and this definition, we believe, is suitable for the system software research.

As the previous work [7], we have described a formal approach to context-aware mobile computing: we offer the context-aware action systems framework, which provides a systematic method for managing and processing context information, defined on a subset of the classical action systems [8]. On the basis of the essential notions and properties of this formalism, we applied this formalism on deriving context-aware services for mobile applications [9] and implemented a smart context-aware kindergarten scenario where kids are supervised unobtrusively with wireless sensor networks [10].

Issues that have been considered are both theoretical and practical: modeling the system requirement rigorously with formal approaches, deriving the software architecture from formal models, and stepwise refinement of the specification, code generation, and verification vs. simulation. While all these research issues have been individually studied in an extensive way, their interaction within the final implementation raises new challenges, which constitutes the focus of this chapter.

13.3 RELATED WORK

Several related works have noticed the importance of seeking a foundation of context-aware computing [22]. Roman *et al.* presented a formal treatment of context awareness via extending the mobile UNITY with context handling part into context UNITY [23]. The context UNITY formalism is similar to our context-aware action systems formalism, but approaching from an agent-like view in modeling context awareness and context dependency.

Henricksen *et al.* showed a conceptual framework and software infrastructure that together address known software engineering challenges in context-aware computing applications [24]. The context model is built on semantic level with the CML language [25], which can be categorized as an extension of the object-role modeling in software engineering process.

UML approach to context models was presented by Hinze *et al.*, where UML diagrams are combined with discrete event systems to facilitate the development of mobile context-aware systems [26]. Due to the limitation of UML, which lacks a rigorous mathematic foundation, this approach can be deemed as a semiformal. The similar UML-like approach can be found elsewhere [27], where a simulation-based paradigm was presented. Besides general aspect of context, fragment aspects of context, such as ontology [28], rational [29], middleware [30], and trust [31], were also considered.

13.4 WIRELESS SENSOR NETWORKS

Wireless sensor networks provide perfect platforms to study context-aware and context-dependent systems. Wireless sensor networks have been an area of active research since the early 1990s [11], accelerated by the advancement of wireless networking and the development of sensors. Only recently, wireless sensor networks have moved from academic research concepts to commercially available products, increasing production quantities.

Although significant research work has been undertaken, most of the research is still very application specific, with security and environmental applications dominating 12. However, it is likely that more generic and comprehensive approach is required, where true system-level problems in wireless sensor networks and their applications can be studied. With such a perspective, we developed a design framework in Figure 13.1 for wireless sensor networks [13].

In this framework, we have distinguished between context-provider and context-utilizer; the former is the reactive part that detects the surroundings and acquires the context and the latter is the proactive part that interprets and responds to the context. The interaction between the context-provider and context-utilizer constitutes a complete context-aware and context-dependent system.

Because the possibly bi-directional communication and the impossibility of restricting context to be a sensor reading, all nodes can potentially act as context-providers as well as context-utilizers. The roles are dependent on whether the data are propagating (an inquiry) or composing (a reply).

13.5 FORMALIZING CONTEXT AWARENESS AND CONTEXT DEPENDENCY

We start by giving a brief overview of the action system formalism and then present how we model context awareness and context dependency within this formalism.

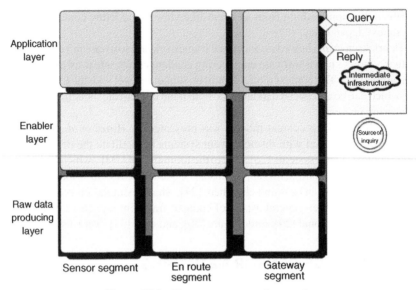

Figure 13.1 The sensornet system framework.

By mapping the formal model back to the software architecture of wireless sensor networks, we show some realistic implementations of this model on system software research.

13.5.1 Action Systems

The action systems formalism is based on Dijkstra's language of guarded commands [14]. This language includes assignment, sequential composition, conditional choice, and iteration.

13.5.1.1 Actions

An *action* is a guarded command, that is, a construct of the form $g \to S$, where g is a predicate, the *guard*, and S is a program statement, the *body*. An action is said to be *enabled* when its guard is evaluated to *true*. If an action does not change the program state, it is called a *stuttering* action.

The body S of an action is defined as follows:

$$S ::= \text{abort}|\text{skip}|x := e|\{x := x'|R\}| \quad \text{if } g \text{ then } S_1 \text{ else } S_2 \text{ fi}|S_1 ; S_2$$

Here x is a list of attributes, e is a corresponding list of expressions, x' is a list of variables standing for unknown values, and R is a relation specified in terms of x and x'. Intuitively, *skip* is an stuttering action, $x := e$ is a multiple assignment, *if g then S_1 else S_2 f_i* is the conditional composition of two statements, and $S_1 ; S_2$ is the sequential composition of two statements. The action *abort* always fails and is used to model disallowed behaviors. Given a relation $R(x, x')$ and a list of attributes x, we denote by $\{x := x'|R\}$ the *nondeterministic assignment* of some value $x' \in R.x$ to x (the effect is the same as *abort*, if $R.x = \phi$).

The semantics of the actions language has been defined in terms of weakest preconditions in a standard way [14]. Thus, for any predicate p, we define:

$$wp(\text{abort}, \ p) = \text{false}$$
$$wp(\text{skip}, \ p) = p$$
$$wp(x := e, p) = p[x := e]$$
$$wp(\{x := x'|R\}, \ p) = \forall x' \in R.x \cdot p[x := x']$$
$$wp(S_1; S_2, \ p) = wp(S_1, wp(S_2, \ p))$$
$$wp(\text{if } g \text{ then } S_1 \text{ else } S_2 \text{ fi}, \ p) = \text{if } g \text{ then } wp(S_1, \ p)$$
$$\text{else } wp(S_2, \ p) \text{ fi}$$

where $p[x := e]$ stands for the result of substituting all the free occurrences of the attributes x in the predicate p.

13.5.1.2 An Action's Building Blocks

An *action system* is a construct of the form:

$$
\begin{aligned}
A = \ |[\ &\text{import} \quad i; \\
&\text{export} \quad e := e_0; \\
&\text{var} \quad\quad v := v_0; \\
&\text{do} \quad\quad A_1[]A_2[]\cdots[]A_n \quad\quad \text{od} \\
&]|
\end{aligned}
$$

The *import* section describes the imported variables i that are not declared, but used in A. The variables i are declared in other action systems and thus they model the communication between action systems. The *export* section describes the exported variables e declared by A. They can be used within A and also within other action systems that import them. Initially, they get the values e_0. If the initialization is missing, arbitrary values from the type sets of e are assigned as initial values. The *var* section describes the local variables of action system A. They can be used only within A. Initially, they are assigned values i_0, or, if the initialization is missing, some arbitrary values from their type set. Technically, all the used variables in import and export sections are global variables, and only variables defined in var section are local. The do $\cdots$ od section describes the computation involved in A. Within the loop, $A_1, \ldots, A_n$ are actions of A.

The behavior of the action system A is as follows: the execution starts by initialization of all variables and then, repeatedly, an enabled action from $A_1, \ldots, A_n$ is nondeterministically selected and executed. If two actions are independent, that is, they do not have any variables in common, they can be executed in parallel [15]. Their parallel execution is then equivalent to executing the actions one after the other, in either order.

13.5.1.3 *Composition of Action Systems*

An action system is not usually regarded in isolation, but as a part of a more complex system. A large action system can be constructed from smaller ones using composition. Consider two action systems A and B below:

$$A = |[\,\text{import}\, i;$$
$$\text{export}\quad e := e_0;$$
$$\text{var}\quad v := v_0;$$
$$\text{do}\quad A_1[]A_2[]\quad []A_n\quad \text{od}$$
$$]|$$

$$B = |[\,\text{import}\, j;$$
$$\text{export}\quad f := f_0;$$
$$\text{var}\quad w := w_0;$$
$$\text{do}\quad B_1[]B_2[]\cdots[]B_m\quad \text{od}$$
$$]|$$

where $v \cap w = \phi$. We define the *parallel composition* of A and B, written $A\|B$, to be the following action system C:

$$A\|B = |[\,\text{import}\, k;$$
$$\text{export}\quad h := h_0;$$
$$\text{var}\quad u := u_0;$$
$$\text{do}\quad A_1[]A_2[]\cdots[]A_n[]B_1[]B_2[]\cdots[]B_m$$
$$\text{od}$$
$$]|$$

where $k = (i \cup j)\setminus h, h = e \cup f$, and $u = v \cup w$. The initial values of the variables and the actions in $A\|B$ consist of the initial variables and actions of the original action systems.

The binary parallel composition operator $\|$ is associative and commutative and thus extends naturally to the parallel composition of a finite set of action systems. The behavior of a parallel composition of action systems is dependent on how the individual action systems interact with each other. The parallel composition operator can also be used in a reverse direction to decompose one action system into a number of those. More on these topics can be found elsewhere [15].

13.5.1.4 *Refinement of Action Systems*

A formal basis for the stepwise development of action systems is the *refinement calculus* [16]. In the refinement calculus, program statements are identified with their weakest precondition predicate transformers. However, the predicate transformer framework is not sufficient to reason about proactive systems. A *trace refinement* extension is described by Back and Wright [17] and *data refinement* extension by Sere and Waldén [18]. Our treatment of the action system refinement is based on the theory presented there.

13.5.2 Context Models

With this formalism, we start modeling the context-aware and context-dependent systems by specifying the context-provider and context-utilizer roles as described in Section 13.3. First we consider a context-dependent system, modeled by the action system CD:

$$CD = |[\text{ import } \ldots$$
$$\text{export} \quad \ldots$$
$$\text{var} \quad \ldots$$
$$\text{do} \quad g \rightarrow S[]\neg g \rightarrow T[]\beta$$
$$\text{od}$$
$$]|$$

Here g is the context guard and S is a statement dependent on the context g: $g \rightarrow S$ models the system behavior with provided context and $\neg g \rightarrow T$ models the system behavior without provided context; β stands for the other actions of CD. The context guard g is a predicate on the local and context variable(s) x. The context variables constitute in a subset of the *import* and *export* variables. The value of g is maintained by some other action system CH, called context-handler. Consequently, the context variable x is an *imported* variable to CD and an *exported* variable in CH.

Hence, we need to introduce the context handler, maintaining g in Figure 13.2. The context handler is a context-provider and can potentially be a context-utilizer, depending on the service. If it were not a context-provider, there would not be anything requiring handling of the context. Thus, the handler is an independent, but necessary part of the system.

The context-handler is modeled by action system CH: where b is a predicate; and $b \rightarrow x := x'|x' \in \{g, \neg g\}$ nondeterministically updates the global context variable x. The nondeterministic update is later refined to a realistic intelligent algorithms. Hence, it models the context provided to CD.

$$CH = |[\text{ import} \ldots$$
$$\text{export} \ldots$$
$$\text{var} \quad \ldots$$
$$\text{do} \quad b \rightarrow x := x'|x' \in \{g, \neg g\}$$
$$\neg b \rightarrow V$$
$$\text{od}$$
$$]|$$

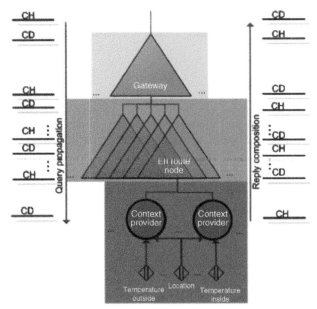

Figure 13.2 Data propagation/composition.

Now, the parallel composition of action systems CD and CH, that is, CD∥CH is a complete context-aware model and it models interactions between the context-provider and context-utilizer.

The implication of this model in the software architecture design can be explained in Figure 13.1, where the gray-shaded areas illustrate the main responsibility for the nodes belonging to them. The dark gray area constitutes the sensing nodes, the gray the en route nodes and the light-gray the gateway node. Moreover, it should be interpreted so each item is considered belonging primarily to *layer* and secondarily to *segment*.

One merit of our model is that we intentionally separate the origin of the context from the whole context-aware system. This separation has one important consequence: the context is the result *after* processing within the context-provider; that is, the action system CH differentiates between *data* and *relevant data* and context is therefore always *refined* raw data.

As the realistic implication, the above idea contributes to a further classification of sensor nodes in wireless sensor networks as Figure 13.2. In this service-oriented view, all sensors do not necessarily provide data needed for replying a query nor do all function as en route nodes. Consequently, if possible, the en route nodes decide based on the context whether their underlying sensing nodes can provide relevant information and, thereby, forward or not. The en route nodes can also, if implemented, compose data for providing relevance and because energy efficiency reasons. In the end, the context information is fused in the gateway node from the en route nodes to provide relevant and accurate answers for the propagated query.

13.5.3 Context Refinement

In this section, we discuss how the refinement principles can be used together with a parallel composition rule in our model. We show how to refine an abstract specification toward a detailed one, as well as the realistic implications of these refinements in system software design.

13.5.3.1 The Context-Utilizer

First, we consider one simple refinement scenario:

$$CD||CH \leq_R CD'||CH$$

where CD' is the refinement result of CD. The realistic implication of this scenario is upgrading the sensor application without touching the sensing part. This kind of refinement could mean the following: suppose we have a supervisory software CD running on top of the wireless sensor network infrastructure, now we update the existing software to a later version with more features CD'.

Since this category of refinement only concerns individual action systems, there should not be any change in the aggregated behavior of the whole system. Thus, we give the refinement rules as follows [17].

Consider two actions systems CD and CD':

$$
\begin{aligned}
CD = |[\; &\text{import}\, i; \\
&\text{export} \quad e := e_0; \\
&\text{var} \quad\;\; a := a_0; \\
&\text{do} \quad\;\;\; A_1[]A_2[]\cdots[]A_n \\
&\text{od} \\
]|&
\end{aligned}
$$

$$
\begin{aligned}
CD' = |[\; &\text{import}\, i; \\
&\text{export}\; e := e_0; \\
&\text{var} \quad\;\; a' := a_0'; \\
&\text{do} \quad\;\;\; A_1'[]A_2'[]\cdots[]A_n'[]X_1[]X_2[]\cdots[]X_m \\
&\text{od} \\
]|&
\end{aligned}
$$

where the local variables a in CD are replaced with new local variables a' in CD'. The actions A_i in CD are replaced with A_i' in CD', and auxiliary actions X_j are added into CD'.

R is mapping a relation between the new local variable a' and the old variable a. Consequently, we can say that the action system CD is refined by the action system CD', if there exists an abstraction relation $R(a, a')$ such that the following conditions hold:

1. Initialization: $R(a_0, a_0')$
2. Main actions: $A_i \leq_R A_i'$, for $i = 1, \ldots, n$
3. Auxiliary actions: skip $\leq_R X_j$, for $j = 1, \ldots, m$
4. Continuation condition: $R \wedge gCD \Rightarrow gCD'$
5. Internal convergence:
6. $R \Rightarrow wp(\text{do } X_1[]X_2[]\cdots[]X_m \text{ od, true})$

Here, the first condition says that the abstraction is established by the initializations. The second condition requires that each action A_i is refined by the corresponding action A_i' using $R(a, a')$. The third condition states that the auxiliary actions X_j behave like *skip* with respect to the global variables $i \cup e$ while preserving $R(a, a')$. The fourth condition requires that an action in CD′ is enabled whenever an action in CD is enabled and $R(a, a')$ holds. The last condition stipulates that the execution of the auxiliary actions taken separately cannot continue forever whenever $R(a, a')$ holds.

13.5.3.2 *Refining the Context Variable*

The other simple refinement scenario considers the context-provider itself:

$$CD||CH \leq_R CD||CH'$$

where CH′ is the refinement result of CH. The realistic implication of this scenario is improving the context-processing unit without touching the upper layer sensor applications. This kind of refinement could be exemplified as follows: suppose we have a supervisory software running on top of the wireless sensor network infrastructure, now we improve the wireless sensor network infrastructure to provide more relevant and precise context information.

This category of refinement also concerns individual action systems and there is no change in the aggregated behavior of the whole system. Therefore, we can use the refinement rules described in Section 13.5.1.1 in this case as well.

Here we consider one common refinement example on refining the context-handling algorithm. In our initial model, the context-handling algorithm is rudimentally expressed as $b \rightarrow x := x'|x' \in \{g, \neg g\}$. There is a need for further refining this algorithm into a realistic intelligent one. Usually this kind of refinement only refines *local actions*, more about this can be found elsewhere [18].

13.5.3.3 *Compositional Refinement*

The last refinement scenario is a complex one, where the context-provider and context-utilizer *co-refines* together:

$$CD||CH \leq_R CD'||CH'$$

where CD′ is the refinement result of CD and CH′ is the refinement result of CH. The realistic implication of this scenario is refining the sensing part and application part simultaneously, interacting with each other. This kind of refinement could be exemplified as follows: suppose we have a supervisory software running on top of the wireless sensor network infrastructure, now we redesign the whole system, touching both the existing upper layer software and lower layer wireless sensor network infrastructure.

Obviously, this category of refinement is complex, because it concerns not only the individual behavior of each action system but also the aggregated behavior of the whole system [19].

We can use the *compositional refinement* extension by Back and Wright [19], together with other refinement rules in Sections 15.5.3.1 and 15.5.3.2, to refine this kind of scenario. In order to make the chapter concise, we do not list down the complete refinement rules (more on these topics can be found [19]) but present an intuitive illustration for understanding this kind of refinement in Figure 13.3, where an arrow represents a refinement step and a line represents an abstraction relation.

Here we show an example on introducing new context to the whole system via compositional refinement: suppose we have the original system modeled as CD∥CH, where CD and CH are defined in Section 15.5.2. In this original setting, we have only g as our context. Now we would like to extend the context part by introducing a new context to the whole system. In reality, this scenario implies the case as utilizing additional data in the system, which usually compels to redesigning of the system.

Using the compositional refinement, we can approach the problem as follows. First we consider the CD′, which is the refinement result of CD. Let this new extra context be d. Assume d is a subset of $\neg g$, that is $d \subseteq \neg g$. Applying the refinement rules in Sections 15.5.3.1 and 15.5.3.2, we can refine the original action $\neg g \rightarrow T$ in Section 15.5.2 into two new actions

$$d \rightarrow R[](\neg g \setminus d) \rightarrow T' \quad \wedge$$

$$\neg b \setminus d \rightarrow V'$$

where R and T' are refined statements satisfying

$$T \leq_R R \quad \text{and} \quad T \leq_R T'$$

Then the new context is evaluated in CH′, which is the refinement result of CH. Now CD′∥CH′ is the refinement result of CD∥CH.

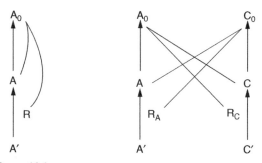

Figure 13.3 Individual refinement vs. compositional refinement.

Actually this is an effective way of stepwise adding new features to the system, when simultaneously touching both the sensing part and the application part is inevitable. If we limit the *context* to *system failure*, this approach is similar to the work done [20] in the fault-tolerant direction to provide means to handle certain faults.

13.6 CASE STUDY: FROM SPECIFICATION, VIA FORMALISM, TO IMPLEMENTATION

We have implemented a smart kindergarten (nursery school) scenario as a case study for the proposed context-role categorization approach. The core concept of this application is illustrated in Figure 13.4, as a smart surveillance system for a kindergarten.

The system consists of stationary base stations, mobile sensor nodes that are attached to the children, and the supervisory application. The children are allowed to move freely in a predefined area (playground) and the supervisor is able to get the location information of all nodes (visually). When a child leaves the predefined area, the alertness level of the system increases and the supervisor is informed. Higher alertness level implies intensified communication. Moreover, intensified location reporting, by the distinct node, is conducted when vibration is detected (the child can be assumed to move).

This scenario is a typical context-aware and context-dependent example consisting of a context-provider and a context-utilizer. The system behavior, the context-utilizer, is critically dependent on different contexts provided by the context-provider, that is, for supervision and localization. Moreover, in this particular example, the base

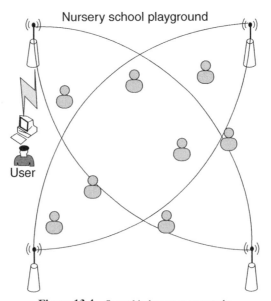

Figure 13.4   Smart kindergarten case study.

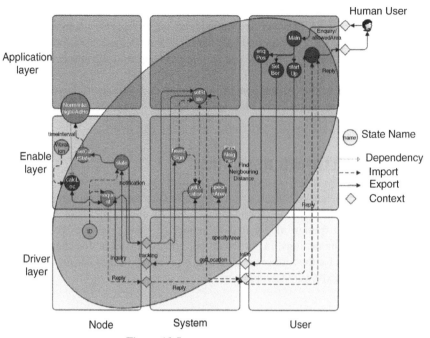

Figure 13.5 Final model of the system.

stations function as context-providers, the beacon, as well as context-utilizers, calculating the position and raising the alertness level.

Using the proposed context model in section 4 and formalism [7], we implemented a variant of ROCRSSI [21] for the localizing service. Here, we show a final
model of the system in Figure 13.5, which is the stepwise developed result of Figure
13.1. This model works as the basis of the kindergarten application. The conclusion
drawn was that the system is hierarchically pushing/pulling context information.

We then take a fraction of the model and show the specification. For instance,
the specification of the tracking part:

getLocationY= Y[|
import   recordHeardSignals, timeInterval, getPos, nowTime;
export   square, checkState, tracking;
var
do
   getPos∧nowTime - recordHeardSignals.time > timeInterval
               -> tracking:= inquiry.calcLocation
   []
   getPos∧square = recordHeardSignals.intersection
       -> checkState:= setState
od
]|

where the context can be viewed as the guard, in this case the variables *getPos*, *checkState* and *tracking*. The imported variables are the services for which this fraction is a context-utilizer for. The exported variables constitute in the context-provider role of the fraction and hence, *checkState* and *tracking* can appear in the guards of systems importing them.

In order to make the chapter concise, we do not elaborate the full system specification here. A description of the kindergarten application and its implementation is available elsewhere [10].

13.7 CONCLUDING REMARKS

By taking a formal view of context-aware computing that integrates different perspectives, we start to understand the foundational relationships that tie them all together, and that provide a framework for understanding the basic principles behind these various forms of interactions. In particular, our context model in this chapter serves as a rigorous basis for the further development of a formal framework for design and evaluation of context-aware technologies.

REFERENCES

1. A. K. Dey and G. D. Abowd. Towards a better understanding of context and context-awareness. In *Proceedings of the CHI 2000 Workshop on the What, Who, Where, When, and How of Context-Awareness, The Hague, The Netherlands*, 2000.
2. M. Raento, A. Oulasvirta, R. Petit, H. Toivonen. ContextPhone: - a prototyping platform for context-aware mobile applications. In *IEEE Pervasive Comput.*, 4(2): 51–59, 2005.
3. G. D. Abowd, M. Ebling, H.-W. Gellersen, G. Hunt, and H. Lei, "Guest Editors' Introduction: Context-Aware Computing," *IEEE Pervasive Comput.*, 1(3): 22–23, July–September, 2002.
4. H. Chen, T. Finin, and A. Joshi. An ontology for contextaware pervasive computing environments. *Special Issue Ontol. Distribut. Syst., Know. Eng. Rev.*, 18(3): 197–207, 2004.
5. A. Schmidt, M. Beigl, and H.-W. Gellersen. There is more to context than location. *Comput. Graphics*, 23(6): 893–901, 1999.
6. G. Chen and D. Kotz. A survey of context-aware mobile computing. Technical Report TR2000-381, Dartmouth College, Department of Computer Science, 2000.
7. L. Yan and K. Sere. A formalism for context-aware mobile computing. In *Proceedings of the Third International Symposium on Parallel and Distributed Computing/Third International Workshop on Algorithms, Models and Tools for Parallel Computing on Heterogeneous Networks*, 2004.
8. R. J. Back and K. Sere. From action systems to modular systems. *Software Concepts Tools*, 17: 26–39, 1996.
9. M. Neovius and C. Beck. From requirements via context-aware formalisation to implementation. In *Proceedings of the 17th Nordic Workshop on Programming Theory, Copenhagen, Denmark*, 2005.
10. C. Beck. An application and evaluation of sensor networks. Master thesis, Åbo Akademi, Finland, 2005.
11. S. S. Iyengar and R. R. Brooks. *Distributed Sensor Networks*. Chapman & Hall/CRC, 2004.
12. E. Yoneki and J. Bacon. A survey of wireless sensor network technologies: research trends and middleware's role. Technical Report UCAM-CL-TR-646, University of Cambridge.
13. M. Neovius and L. Yan. A design framework for wireless sensor networks. In *Proceedings of IFIP 1st International Conference on Ad-Hoc Networking, Santiago De Chile, Chile*, 2006.
14. E. W. Dijkstra. *A Discipline of Programming*. Prentice Hall, 1976.

15. R. J. Back and K. Sere. Stepwise refinement of action systems. In *Struct. Program.*, 12(1): 17–30, 1991.

16. R.-J. Back, J. von Wright. Refinement calculus: a systematic introduction. *Graduate Texts in Computer Science*. Springer-Verlag, 1998.

17. R.-J. Back, J. von Wright. Trace refinement of action systems. In *Proceedings of the 5th International Conference Concurrency Theory*. Lecture Notes in Computer Science, Vol. 836, Springer, 1994.

18. K. Sere, M. A. Waldén. Data refinement and remote procedures. In *Proceedings of the Third International Symposium on Theoretical Aspects of Computer Software*. Lecture Notes in Computer Science, Vol. 1281, Springer, 1997.

19. R. J. Back and J. von Wright. Compositional action system refinement. In *Proceedings of the BCS FACS Refinement Workshop*. Electronic Notes in Theoretical Computer Science, Vol. 70, Elsevier, 2002.

20. K. Sere and E. Troubitsyna. Hazard analysis in formal specification. In *Proceedings of SAFECOMP'99*, Toulouse, France. Lecture Notes in Computer Science, Vol. 1710, Springer Verlag, 1999.

21. C. Liu, K. Wu, and T. He. Sensor localization with ring overlapping based on comparison of received signal strength indicator. In *Proceedings of the IEEE International Conference on Mobile Ad-hoc and Sensor Systems (MASS)*, 2004.

22. P. Dourish. *Where The Action Is: The Foundations of Embodied Interaction*. MIT Press, 2001.

23. G.-C. Roman, C. Julien, and J. Payton. A formal treatment of context-awareness. In *Proceedings of the 7th International Conference Fundamental Approaches to Software Engineering (FASE)*. Lecture Notes in Computer Science, Vol. 2984, Springer, 2004.

24. K. Henricksen and J. Indulska. A software engineering framework for context-aware pervasive computing. In *Proceedings of the 2nd IEEE International Conference on Pervasive Computing and Communications (PerCom)*, 2004.

25. K. Henricksen. A framework for context-aware pervasive computing applications. PhD Thesis, University of Queensland, 2003.

26. A. Hinze, P. Malik, and R. Malik. Interaction design for a mobile context-aware system using discrete event modelling. In *Proceedings of the Twenty-nineth Australian Computer Science Conference (ACSC)*, Hobart, Australia, 2006.

27. P. Guo and R. Heckel. Modeling and simulation of context-aware mobile systems. In *Proceedings of the 19th IEEE International Conference on Automated Software Engineering (ASE)*, 2004.

28. A. Pappas, S. Hailes, and R. Giaffreda. A design model for context-aware services based on primitive contexts. In *Proceedings of the UbiComp*, 2004.

29. Y. Roussos and Y. Stavrakas. Towards a context-aware relational model. Technical Report TR-2005-1, National Technical University of Athens, 2005.

30. E. Katsiri. Middleware support for context-awareness in distributed sensor-driven systems. PhD Thesis, University of Cambridge, 2005.

31. M. Carbone, M. Nielsen, and V. Sassone. A formal model for trust in dynamic networks. BRICS Report RS-03-4, 2003.

Chapter 14

Experiences with a Smart Office Project

F. Bagci, F. Kluge, B. Satzger, A. Pietzowski, W. Trumler, and T. Ungerer

Institute of Computer Science, University of Augsburg, Augsburg, Germany

14.1 Introduction	294
14.2 Smart Office Project	296
14.3 Reflective Mobile Agents Within Smart Offices	300
14.4 Location Tracking and Prediction	304
14.5 Organic Computing Middleware for Ubiquitous Environments	306
14.6 Conclusion	316
References	317

14.1 INTRODUCTION

A long time the technology in offices was limited to paper and ink since the first office environments. Paper is a robust, consistent, and persistent information medium that is easy to use and not fixed to a location. It is still used for communication internally inside the office building and externally with customers or other authorities. Also, the location of offices was a strategic decision that considered traditional circumstances. You can find business areas in nearly each city supporting paper communication and face-to-face meetings. But location is expensive as is generally known. Compared with manufacture plants, where work spaces are used nearly 100% of the day, offices are occupied only 30%. In consulting companies where most of the employees work the most time outside the offices, the usage is even less than 10%. Reducing the amount of offices directly reduces the fixed costs.

Mobile Intelligence. Edited by Laurence T. Yang, Agustinus Borgy Waluyo, Jianhua Ma, Ling Tan, and Bala Srinivasan

Technology progress has fundamentally changed the work situation in offices. Information can be delivered over large distances easily within short time. What was earlier printed on paper became digitized and available online from almost everywhere. New devices were developed changing and supplementing paper work like printer, fax, and photocopier. Today an office without such devices cannot be imagined. Unfortunately, this development did not decrease the number of offices holding the price of placement still very high.

The solution to this unsatisfying situation are flexible offices. In flexible office environments, employees are not bound to a fixed location. The offices can be distributed dynamically among the actual demands. This is not an easy job to do if you want to guarantee a working place for every employee. But visions of ubiquitous and organic systems nurture the hope of automatically accomplishing the distribution and all its implicated management. More than this, ubiquitous systems enable new services that can support office work in an unprecedented manner. Such flexible offices need a sophisticated infrastructure to offer a seamless integration of the employees working environment at different locations. The flexible office concept we envision is described by the following scenarios.

Flexible office scenario: A sales representative needs a meeting room at his company for the next day. He places a reservation online, specifying the time and number of attendees of the meeting. He will also need an office to work there for some hours.

The next day when the sales representative enters the building, he takes his office equipment out of a store. The employee's security badge is used to track his location within the office environment. He is guided to the assigned office by the smart doorplate system. If he passes a smart doorplate, it shows an arrow to the direction of the assigned office. His container can also be localized by the location-tracking system. When he reaches the assigned office, he puts the container near one of the desks and the system reconfigures the environment (e.g., telephone, computer). The doorplate now displays his name to show that he is working in this office for that day.

Some minutes before the meeting, the visitors arrive at the entrance. At the information desk, the visitors receive a security badge and select the name of the person they want to visit. The visitors can also be tracked by the indoor location-tracking system and are guided by the smart doorplates to the meeting room. Meanwhile, the system informs the salesrep that the visitors just arrived and that they are on their way to the meeting room.

A flexible office organization, where office rooms are dynamically assigned to currently present employees, requires a sophisticated software system that is highly dynamic, scalable, and context-aware. The considered computing environment within an office building consists of servers, PCs, laptops, PDAs, and sensor-nodes connected by wired or wireless networks. Because of the heterogeneity of the participating hardware and software systems, we chose a middleware approach, and because of the complexity of the system, we based our middleware on the autonomic [16] and organic [44] computing principles, respectively, trying to overcome the high administration demands of such systems.

On top of the middleware are the services that can be grouped as core services of the middleware and application services. In order to execute highly distributed and mobile applications, we developed a ubiquitous mobile agent system. The paradigm of mobile agents offers a convenient approach to combine personal interests of users and the requirements of a ubiquitous distributed system. Mobile agents basically build the highest level of decentralization. Personal information can be encapsulated by a mobile agent and be used for location-based services on behalf of the user. The idea is that a user is accompanied by a virtual reflection in the form of a mobile agent in the ubiquitous environment.

The next section describes the smart office project in detail. It concentrates on the scenarios and hardware components attached in the offices. Sections 14.3 and 14.4 focus on the application level—the ubiquitous mobile agent system *UbiMAS*, which serves as basis for some application scenario implementions. Section 14.4 introduces location-tracking and location prediction. Section 14.5 introduces OCμ, the underlying organic computing middleware for ubiquitous environments. The middleware realizes organic/autonomic properties like self-configuring, self-healing, self-protecting, and self-optimizing. The chapter closes with a conclusion.

14.2   SMART OFFICE PROJECT

The long-term target of our research is to provide a flexible office environment as a platform for ubiquitous and organic computing applications. We chose our own office floor to build a smart facility. Each office has several devices like PCs, laptops, PDAs, telephone, and sensor-boards. The highlight of the flexible office are the smart doorplates. We replaced the usual doorplates at each office door by electronic displays with touchscreens. Figure 14.1 shows a picture of one smart doorplate and Figure 14.2 shows the floor of our institution with a smart doorplate at each door.

Figure 14.1   Smart doorplate.

Figure 14.2   Floor with smart doorplates.

The smart doorplates can not only display information about the actual office but also provide a convenient possibility to perform user-oriented and location-based services. Using the touchscreen, the person can interact directly with the flexible office system. The following scenarios demonstrate applications of the smart doorplates (see Figures 14.3–14.5) in combination with a visitor/employee tracking system and additional sensors.

14.2.1   Smart Doorplate as Signpost

The smart doorplates can act as signposts within the office building to direct employees to their assigned office and to direct visitors to a sought employee, respectively. We assume that the visitors register at an electronic reception as described in Section 14.1 and that they he select the person they like to meet. The visitors get a security badge, and are directed to the employee's office. The guiding system is implemented by the smart doorplates. As soon as the visitor is in the vicinity of a doorplate, the doorplate points in the direction of the sought office. Assuming that a single visitor passes several smart doorplates on his way to the office, a direction pointer is sufficient and most appropriate. If the employee is not in his office but within the building, the smart doorplates direct the visitor (or colleague) to the current location of the employee.

Figure 14.3 shows the basic function of the smart doorplate presenting two office owners and some additional information. If an employee is in his office, a small icon is shown in front of his name.

14.2.2   Visitor in Front of the Office and Office Owner Present

In case of the presence of the office owner, several possibilities arise. The office owner is on the phone and may not be disturbed. The doorplate displays a phone

Figure 14.3   Smart doorplate in basic function and as signpost.

sign at the employee's icon to prevent disturbances. When the phone call is finished, the smart doorplate notifies the waiting visitor and ushers him in. If two or more colleagues share an office, it may happen that one colleague is busy, whereas the requested employee is ready to meet the visitor. The doorplate allows the visit, if it is for the unoccupied office member using the information provided by the reception.

If the office owner is in a meeting within his room, the smart doorplate displays that the office owner is present, but should not be disturbed. The office owner may receive a notice about the waiting visitor with the name of the visitor on his/her notebook or PDA, and he she may answer by a notice, when the end of the meeting can be expected. All visitor messages are sent and received through the smart doorplate not disturbing the meeting.

If the office is used as a meeting room, the doorplate may display an appropriate message that the room is occupied, which meeting it is, the list of attendance (see Figure 14.4), time when the meeting will end, and so on. Such information may be drawn from an electronic meeting protocol.

14.2.3   Visitor Arrives in Absence of Office Owner

If the office owner has locked his office and a date is shown in his electronic schedule, the doorplate may display his current location (in house, out of house, or more details) and time when he is expected to be back. If the employee is located in a different office, the room number may be displayed, and the surrounding doorplates can be used as direction pointers. Alternatively, the system could predict if the employee is coming back soon and recommend the visitor to wait (see Figure 14.5). The visitor may leave an oral or written message at the doorplate (microphone or touch panel presupposed). On return, the office owner is notified about the message and may read it on the doorplate or within his office. Urgent messages may be forwarded to the

Figure 14.4   Smart doorplate showing currently present people.

current location of the office owner (e.g., by instant messaging over LAN or WLAN, or SMS over GSM).

14.2.4   Smart Doorplate or Foreign PC for E-mail Retrieval

Usually an employee uses his office PC to retrieve e-mails. In this scenario, the employee can arrange to be notified over the nearest smart doorplate if he expects an urgent e-mail. Because of security reasons, he does not want to receive all his incoming mails on a foreign doorplate or a PC in another office. He can use the doorplate or a foreign PC to set up a filter for incoming e-mails. The office system

Figure 14.5   Smart doorplate in case of absence.

Figure 14.6 Keypad for user input.

takes care about security concerns and informs the user when the expected e-mail arrives. The employee can than read the e-mail at the nearest located doorplate or on a foreign PC in an office. For authentication, the user has to enter username and password. We have implemented a keypad (Figure 14.6) that provides a convenient approach to enter these data on the restricted touchscreen.

14.2.5 Smart Doorplate or Foreign PC for File Transfer

In the same way like an e-mail, the employee can choose a file that is placed on his office PC and can transfer it to the actual smart doorplate or a foreign PC in an office. This application is very helpful if the user is, for example, in a meeting that takes place in a distant part of the building. Even if the user has no electronic devices to access his files, he/she can transfer documents from his PC to the meeting room. He can make the document available for the persons in the meeting room, so each one can read or copy it.

14.3 REFLECTIVE MOBILE AGENTS WITHIN SMART OFFICES

As seen in the scenarios above, we propose that the smart environment takes care of storing and sending the personal information. The person is always accompanied by a mobile virtual object in the smart environment. So location-based services adapted to personal profiles can be offered. Certainly the ubiquitous system could be realized as a server-centric approach. But concerning personal movements and data, this would lead to a big-brother-is-watching-you scenario, where entities that gain access to the server would have access to all personal information. Regarding smart environments, a central server could rapidly become a bottleneck because of the amount of clients and services running on the system. Moreover, a failure on the server would endanger the whole system.

The paradigm of mobile agents ideally fits into the decentralized approach. The mobile agent constitutes a virtual reflection of the user. Reflective agents accompany persons in the smart environment and carry user-specific data. Employees have their personal reflective agent, which resides in the environment and contains data about the employee. These data consist of basic user information like name, office room, and so on, and security data like private and public keys, user names, and passwords of the owner used for communication, data security, and access operations. Furthermore, the agent stores context information belonging to the user and updates these data automatically.

The doorplates also serve as interface between user and agent. The users can instruct their reflective agent to perform services on their behalf. The reflective agent communicates with service agents and passes on the instructions. If the user moves to a new location, the reflective agent migrates to the doorplate next to the user. The user location is determined by a combined tracking system that is described in Section 14.4.

This mobile agent model is implemented in Ubiquitous Mobile Agent System (UbiMAS) [2–4]. Few other mobile agent systems exist for ubiquitous environments. Examples are the Hive system [21], which is a distributed agent platform for building applications by networking local system resources, and the Spatial Agents [35], which describes a framework where mobile agents follow their users as they move around and adhere to places as virtual post-its. The UbiMAS framework architecture is described in the following subsection.

14.3.1 UbiMAS Framework

The UbiMAS framework describes a skeleton of a mobile agent system based on a middleware. The agent system is consciously separated from the middleware layer in order to define a generic framework for a broad usage.

The services form the top layer of the middleware. A UbiMAS host is running as a service besides other services as for example a location-tracking service, middleware services, and so on. The UbiMAS host represents a platform for mobile agents. It supports basic functions like starting, performing, and terminating agents and their communication.

We implemented an organic and ubiquitous middleware that uses the peer-to-peer system JXTA [33] as communication infrastructure. The middleware is described in detail in Section 14.5.

UbiMAS hosts consist of two parts: the UbiMAS basic platform and the UbiMAS extensions. The UbiMAS basic platform defines abstract agent hosts and the interface for agent implementations. It realizes the basic communication functions between hosts and agents and several security concepts that cover host and agent security problems.

The UbiMAS extensions implement the application-specific components, that is, the reflective user agents and the service agents. Furthermore, the communication functions are extended with secure agent-to-agent and agent-to-host messaging to fulfill the requirements of the smart doorplate application.

This separation in a basic and an extension part eases the adaption of UbiMAS to a broad range of applications.

14.3.2   The UbiMAS Host Service

To host UbiMAS agents, each peer node has to start at least one UbiMAS agent host as a service on top of the middleware (see Figure 14.7).

UbiMAS implements additional communication protocols for agent and host communication and for agent migration. All messages are acknowledged in UbiMAS, which is not implemented by the base middleware.

If a middleware peer receives a UbiMAS type message, it informs the host by sending an event on which the host listens. UbiMAS hosts implement a message delivery engine that receives the events sent by the middleware and processes the incoming messages. If the host wants to send a message, it forwards it to the message delivery engine, where the appropriate header information is filled in to avoid camouflage attacks.

If the receiver of a message is an agent, the message delivery engine hands it on to the PoBox. The PoBox implements for each agent an interface called PoBox-Adder, which allows the agents to send and receive messages. All communication between agents and hosts is handled over the PoBox using the PoBoxAdder. The PoBoxAdder describes the interface for putting messages into the queues. Besides the

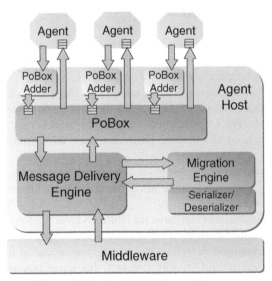

Figure 14.7   The UbiMAS host architecture.

`addMessage` method of the PoBoxAdder, there are no method references between hosts and agents.

The PoBox has a queue for incoming messages for each agent. In the same manner, each agent has a queue for messages sent by the PoBox. The queue lengths are managed dynamically by the possessing entity. This approach offers various security features that are described in more detail in Ref. [2].

The agent hosts can form alliances in form of peer groups using the services of the middleware group manager. If two peers want to build a peer group, they have to arrange new pipes that are built using a pipe advertisement. The advertisement contains a unique pipe ID. Only the peers that know the pipe ID can receive the messages sent over the pipe. Each agent host that wants to join a peer group must send a request. The peer group members can decide if a request will be accepted or rejected. The agent hosts within a peer group communicate over a secure communication protocol.

Different peer groups can build a partnership. Applied to an office building, each floor could build a peer group, and floors of the same institution could form an alliance. This makes it possible for agent hosts to communicate with hosts from other peer groups. Messages to hosts outside the peer group are not secured. Beside UbiMAS, there are other services, for example, location services, that use the same middleware for communication. Mobile agents can register for events of these services and will be informed when new events occur. In this way, agents can be notified by sensor events from the environment or location events of particular persons.

14.3.3   UbiMAS Mobile Agents

Agents in UbiMAS are started as single threads on the actual host. An agent reacts to messages, that is, the agent is in a loop where it waits for incoming messages. A UbiMAS host can manage several agents. There are two types of agents in UbiMAS: user agents and service agents. Both agent types are derived from an abstract agent that defines different basic methods for communication and security functions.

Agents can communicate with each other using messages. An agent must implement functions that are performed when a specific message is received. If an agent does not know the message type, it ignores the message. The abstract agent defines methods for creating and sending messages. The agent hosts serve here as mediators. Messages addressed to a locally available mobile agent are forwarded directly to the recipient by the local PoBox. If the agent is on another host, the message is transferred to this host.

Each agent is identified by a unique ID. UbiMAS supports message encryption. Each host owns a certificate, a public, and a private key. Each agent has an own key pair for message encryption. The abstract agent defines further data structures for security keys of the agent and the public key of the actual host. Furthermore, there are methods for requesting public keys of other agents or hosts.

The security architecture of UbiMAS aims to protect both the agents and hosts against malicious behavior. A secure system is essential for the acceptance of the reflective mobile agent approach because personal data are sensitive. Evaluations showed that in a smart office environment with 10 Mbit/s Ethernet connections, agent transfer with/without security features is adequate [4].

14.4 LOCATION TRACKING AND PREDICTION

14.4.1 Location Tracking

For complete operation, the smart office system needs a service that provides the current positions of the users. This service should at least be able to determine the room a user is currently staying in. Therefore several location-tracking techniques exist. We will give an overview of some alternatives before explaining our own approach and experiences.

Pathbreaking work on positioning systems was done by Want et al. [46] with their *Active Badges*. They employed a room-accurate positioning system based on IR LEDs and sensors. A radio technique, wireless LAN, was employed in the RADAR system by Bahl et al. [5]. Especially this work discusses a wide range of methods for analyzing the measurements. Other radio techniques were analyzed by Feldmann et al. (Bluetooth [10]) or Ni et al. (RFID [22]), among others. For all mentioned techniques, a medium error of about 2–5 m was reported. ARIADNE [18] promises to amend this, and also not to need large overhead for the recording of a reference map through the automatic construction of radio propagation maps. Here, the medium error reported drops to about 1 m.

Definitely better outcomes can be gained by the use of ultrasonic [13, 39]. Here, the error lies at 10 cm resp. 50 cm in 95% resp. 90% of all cases. However, these gains must be seen alongside with the high costs for the technical equipment and its installation.

Due to the high inaccuracy or the high costs for existing systems, we decided to build our own location-tracking system. Therefore, we used the *Embedded Sensor Board* ESB 430 [38], with a TR1001 radio transceiver.

The test bed consisted of three rooms of about 80 m², where we placed three sensor boards as infrastructure. A fourth board was worn by the mobile user. We used a passive infrastructure that acts as a receiver for the signals sent by the mobile board. Thus, the chronological assignment of the signals onto the positions was made easier. With this configuration, we recorded a reference model of over 30,000 data sets. In the rooms, we had 70 measurement points, and at each point, we collected data for four orientations. By recording several datasets at each point, we hoped to outweigh the jitter of the radio transceivers at least a little.

For the calculation of positions, we compared three methods: the *Nearest Neighbour in Signal-strength Space* (NNSS) search that already was used in RADAR [5], a statistical analysis based on the density of a normal distribution (STAT), and a randomized analysis, where a point is chosen randomly for each measurement. The last one was used as a reference method to see how good we can guess the position.

An evaluation using the three methods showed the following [20]: As expected, randomized analysis performed worst. NNSS and STAT nearly performed equal with a median deviation of about 2.5 m, whereas NNSS performed slightly better. In further calculations, we analyzed the influence of the number of neighbouring points on our results. As has shown up, especially when dealing with low numbers of neighbors, improvements can be received with each additional neighbor. Coming to higher numbers, the improvement is negligible. Furthermore, we analyzed the influence of the size of the reference database (number of reference points) and the number of measurements at each point. As has shown up, both only minimally influence the position calculated.

Further analysis of the position calculation showed that only above 60% of all calculations met the correct room. So, if it was possible to estimate first the room with a 100% accuracy, the position calculation could be improved. To confirm this assumption, we used the available IR sender and receiver on the boards to determine the room. Evaluations showed that the median deviation decreased from 2.5 to below 2 m. The accuracy of radio-based systems is very sensitive to changes in the environment, mobile walls, opened doors, and even air moisture. Thus, a combination of two methods by fusion of both sensor systems that are for themselves not that accurate, may increase overall accuracy.

For real use in the smart office project, this method is still too inaccurate. A great problem is the jitter of the received radio signal strength. Here, the use of a better transceiver might bring some improvement.

14.4.2 Location Prediction

Can the movement of people working in an office building be predicted based on room sequences of previous movements? In our opinion, people follow some habits, but interrupt their habits irregularly, and sometimes change their habits. Moreover, moving to another office fundamentally changes habits too. Thus, location-prediction methods need to exhibit some features: high prediction accuracy, a short training time, retention of prediction in case of irregular habitual interrupts, but an appropriate change of prediction in case of habitual changes.

For our smart office application, we used benchmarks with movement data of four persons over several months. The benchmarks are called Augsburg Indoor Location-Tracking Benchmarks. They are publicly available [23] and are applied to evaluate several prediction techniques and to compare the efficiency of these techniques with exactly the same evaluation set up and data.

Our aim is to investigate how far machine-learning techniques can dynamically predict room sequences and time of room entry independent of additional knowledge. Of course, the information could be combined with contextual knowledge as, for example, the office time table or personal schedule of a person, however, we focus on dynamic techniques without contextual knowledge.

Several prediction techniques are proposed in literature—namely Bayesian networks [17], Markov models [6] or Hidden Markov models [34], various neural network approaches [12], and the state predictor methods. The challenge was to transfer these algorithms to work with context information. In our research, we investigated five approaches, a dynamic Bayesian network [29], a multilayer perceptron [45], an Elman net, a Markov predictor, and a state predictor [25]. In the case of the Markov predictor and the state predictor, we additionally used a version that is optimized by confidence estimation [26] and enhanced by various hybrid predictor models [28]. A comparison of these methods can be found in Ref. [24, 27].

Time of arrival at the predicted location depends on the sojourn time at the current location plus the rather constant time to move to the predicted location. The sojourn time can be modeled into a Bayesian network [29]. We tested also a time prediction that calculated the mean and the median of the previous sojourn times within a location. The best results were reached by the median. The time prediction is independent of the location prediction method and can easily be combined with any of the regarded methods.

14.5 ORGANIC COMPUTING MIDDLEWARE FOR UBIQUITOUS ENVIRONMENTS

$OC\mu^1$ [41] is designed with the goal to facilitate the device-independent application of organic computing demands in ubiquitous environments where we expect a heterogeneous collection of devices with diverse capabilities of computing power, memory space, and energy supply.

Beside the design of the middleware, we investigate self-configuration, self-optimization, self-healing, and self-protection within the middleware. The self-x properties that are implemented as services can be used as needed. The overall architecture of $OC\mu$ is shown in Figure 14.8. The architecture of the middleware as well as the self-x properties are described in more detail in the follow sections.

14.5.1 OCμ Middleware Components

The basic architecture of $OC\mu$ is comparable to other state of the art middleware systems. It is comprised of three layers. The lower level (*TransportConnector*) is responsible for the delivery of the messages to other nodes on different communication infrastructures. The middle layer (*EventDispatcher*) is capable of finding the accurate recipient of a message that was sent by another service either locally or from another node. The applications as well as some basic services of the middleware like the configuration service and the discovery service reside on the top level.

There are three additional parts in the middleware that differentiate $OC\mu$ from other middleware systems: first, a sophisticated monitoring on both lower levels of the middleware; second, the organic manager with the self-x services; and third, a typed messaging that adds further freedom in terms of message delivery and service requests.

---

[1]$OC\mu$ is an acronym for Organic Computing Middleware for Ubiquitous Environments.

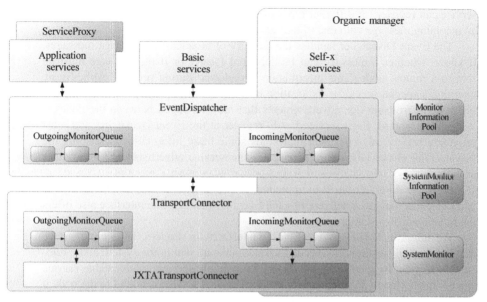

Figure 14.8   OCμ architecture.

To use the offered capabilities of OCμ, the applications should be separated in services that can be distributed on the nodes of the network. Future ubiquitous systems will be comprised of many computing nodes, so applications should be composed of services instead of being a monolithic block of software. The advantage is two-fold. The services of an application can be reused much easier than parts of a code and the services can be distributed over the network to increase the computational performance of the application.

For the smart office scenario, the UbiMAS agent host, location-tracking, and location prediction are some of the implemented application services.

TransportConnector: To decouple the middleware from the underlying commu-nication infrastructure, the TransportConnector employees specific TransportConnec-tors that must be implemented for the different communication infrastructures. In the current implementation, we use JXTA [33] and provide a JXTATransportConnector. The peer-to-peer approach fits best the transport functionality used by OCμ.

The implementation of a TransportConnector can be replaced depending on the given communication infrastructure, which is transparent to the rest of OCμ and the applications built on top of it. It is also possible to use multiple TransportConnector implementations for different communication infrastructures at the same time (i.e., CAN-Bus, Serial Line).

Event Dispatcher: The Event Dispatcher is responsible for the message delivery between the services. The Event Dispatcher offers to services the functionality to send messages and to register themselves as listeners for specified types of messages. A service can register for different types of messages and will be informed in case of an incoming message with one of the registered types.

The Event Dispatcher handles the delivery of broadcast and unicast messages. It knows whether a message can be delivered locally or if the message must be sent to a remote node. This information is collected during runtime explicitly by querying a special service if its location is unknown or implicitly by collecting the information from service advertisements that are exchanged between the nodes if a service is started. Each service has to register at the Event Dispatcher and must provide a service advertisement that holds some basic information describing the service. The Event Dispatcher propagates the service advertisement to the other nodes.

Services: A service needs to implement a special interface to participate in $OC\mu$ and to receive messages delivered by the Event Dispatcher. The interface also offers the functionality to send messages.

We differentiate between two kinds of interfaces for services. The simple service that has the full functionality needed to communicate within the system and a relocatable service. Beyond the functionality of a simple service, a relocatable service can be transferred to another node, whereas a normal service is bound to the node it was started on. The binding on a special node is important for services that need a special hardware or software environment (e.g., fixed sensors or databases).

Service proxy: To forward messages for a service that was recently relocated to another node, the Service Proxy is used. It has a limited lifetime and automatically dies after that time. The lifetime depends on the validity duration of a service advertisement that was created when the service was started on this node. This mechanism prevents the system from sending a surge of messages each time a service was moved to another node. We try to protract this overhead as long as possible. It is not necessary to update the information about a service on all nodes immediately after a service has moved. If another service wants to use the moved service, it sends the request to the Service Proxy, which forwards the message to the new location. The answer is sent back directly to the requesting node, which can now update its entry for this service.

A node, whose information about a moved service exceeds the expiration time of the service advertisement, might be able to reach the Service Proxy that has the same lifetime as the service advertisement. If the Service Proxy cannot be reached, the node must discover the service anyway. So, the overhead of updating all nodes can be shifted to the point where it cannot be avoided anymore, but it is always less then a broadcast to all nodes.

Organic manager: The $OC\mu$ architecture is a observer/controller architecture in which the Organic Manager including the self-x services takes the part of the controller and the monitors are the observers. All relevant information collected by the local monitors about the services and the node itself is stored at the Organic Manager in the Monitor- and SystemMonitor Information Pool, respectively. The Information Pools provide the collected information to other services and the self-x services employing a publisher/subscriber mechanisms. The Information Pool informs the subscriber if new information arises someone has subscribed for.

14.5.2 Monitoring

To get information about the services, resources and the node itself, monitoring, is a vital point in OCμ. To avoid the communication overhead of a centralized monitoring, all monitors collect information locally on every node.

To obtain as much information as possible, it is necessary to use more than just one monitoring point and to use specialized monitors for different tasks. OCμ not only uses a single monitor on every level (System, Transport Interface, and Event Dispatcher) but also uses Monitor Queues to add and remove monitors as needed.

For a fine-grained monitoring, OCμ offers the possibility to add monitors for incoming and outgoing messages at the Transport Interface and the Event Dispatcher. Different Monitor Queues exist for each direction. The advantage of the Monitor Queues is that monitors can be added and removed as needed and that the monitors can be kept simple and fast. Furthermore, the Monitor Queues offer the possibility to add monitors in defined orders to build processing queues for the messages.

Monitoring on the Transport Layer: The monitors at the Transport Layer are responsible for monitoring transport-dependent information. Incoming and outgoing messages, for example, can be used to monitor the latency between peers and the amount of data exchanged between services. This information can be used by metrics to decide if it is better to run a service locally or at a remote node.

Monitoring at the Event Dispatcher: As the Event Dispatcher delivers incoming messages and outgoing messages, it is the perfect instance to monitor service dependent information concerning the message exchange.

If a service requests information from another service, the dispatcher can measure the local and remote response time. It can also collect alive information about the services by monitoring the messages to assure that a service is still available. This prevents the overhead of alive messages by just looking at the ongoing communication.

The System Monitor: The System Monitor is used to gather information about the computing platform a node is running on. The information about memory, processing power, and communication capabilities might be vital for the metrics to reason about the distribution of the services.

Currently, there exists no interface for Java to collect system-dependent information. Various management APIs exist such as SNMP, JMX, and so on, to collect information about a system. To monitor the information about the local node, we use the parts of the JMX and some native code to provide the requested information.

14.5.3 Typed Messaging

The communication of systems like CORBA or JINI is based on method invocation. They use the stub/skeleton principle to guarantee a consistent view on the methods of a remote object. The disadvantage of this approach is that the stub must be known at development (compile) time of a service or application and that a new service must exactly implement that interface.

The communication in OCμ is based on the exchange of messages and not on method calls. An *EventMessage* consists of *MessageElements* where each *MessageElement* has a name and a value it encapsulates.

In OCμ, the parameter types must be defined like in stub/skeleton systems, but the order in which they appear is not relevant. Another advantage is the extensibility of the communication. A service can handle requests with a limited count of parameters as well as a request containing an extended set of parameters. This means that an extended version of a service, which can handle more parameters than the previous version, can be smoothly integrated into the system without notice to the rest.

If an interface is changed, the stub/skeleton systems require a recompilation of all affected components. If an object calls a remote method, stub/skeleton systems identify the desired method through the introspection mechanism, which tries to find a method with the correct name, parameter count as well as the order and types of parameters. This is not the case for OCμ because the services register for the types of the messages they are interested in and they provide the capability to handle messages with different sets of parameters.

OCμ supports two methods for the message delivery. A service can be directly addressed or a message can be typed, resulting in a delivery to all services that have registered for that type of message. If a message is directly addressed to one service, the Event Dispatcher delivers the message only to that service leaving out all other services that might have registered for the type of that message. If a message is sent as a typed message, the Event Dispatcher selects all services that registered for that message type and sends the message to all of them. The message types can be user-defined and hierarchical. Another difference between OCμ and the stubs/skeleton version is the nonblocking nature of requests. All messages are sent asynchronously, thus if a service sends a request to another service, it is not blocked.

14.5.4 Self-Configuration

The self-configuration is used to find a distribution of the services of an application such that the quality of service (QoS) is as high as possible. A configuration description contains all the services and their resource requirements. This configuration description is flooded into the network and every node calculates the QoS for every service locally and orders them in descending order. Those services that cannot be provided due to resource constraints are left out and collected separately.

The configuration description can have so-called constraints. The constraints are requirements for the application that can be described in form of a mathematical forall quantifier, that is, "Every node with a Smart Doorplate should start the smart doorplate service." These requirements are treated prior to the normal service assignments.

After the quality of the services is calculated one node starts to send an assignment message containing the id of the service, the id of the node, and the QoS the node will provide for the service. This message is broadcasted to all other nodes. The receivers of this message mark the service as assigned in their list. Then another node might send an assignment message. This carries on until all services are assigned to the nodes.

During the assignment of the services, one node might provide a service with a higher QoS, thus it sends an additional assignment message that overrides the original assignment in the same manner as a normal service assignment. If a conflict occurs because two nodes want to assign the same service, a conflict-resolution mechanism is employed to solve the conflict without any additional messages. The assignment messages have four extra values that are used in case of a conflict. With this information, every node can decide locally which one of the conflicting nodes should get the service. The conflict-resolution mechanism is a five-stage procedure in which the next stage is used to find a solution if the values of the former stage were equal for both nodes.

After all services are assigned, a verification step is used to assure that all nodes have the same assignments of the services. If a node receives a verification message, it compares the contained information with the local assignments. If the QoS of the local assignments is better, a further verification message is sent to inform the other nodes about the updated information. If the local assignments are worse, the improvements of the received configuration are incorporated. After the verification step, every node can start the services it has assigned.

Evaluation results [19, 42] based on simulations and on running the self-configuration within the $OC\mu$ middleware showed that an average amount of 1.5 messages per service suffices to assign all services to the nodes of the network.

The algorithm is capable of finding all unsatisfiable configurations. All other satisfiable configurations were successfully assigned to the network during the simulations as well as the experiments in a real $OC\mu$ setup. The QoS of the overall assignment is not optimal, but at least good enough to start the application. The effort needed to assign all services depends only linearly on the amount of service given in the configuration.

14.5.5 Self-Optimization

After the initial configuration of the system, the services and the nodes are monitored by the corresponding monitors. The collected information is used to further optimize the system in terms of predefined resource parameters. The parameters used for the self-optimization can differ from those used for the self-configuration. During runtime, parameters like CPU and memory usage, and network bandwidth are of interest. The self-optimization tries to relocate the services in a way that these resources are consumed equally by all nodes of the system.

The idea of the self-optimization is based on the human hormone system. Hormones (information) are produced in the cells (nodes) and released to the blood circuit (communication between the nodes). The hormones bind to cells with a matching receptor (monitor) and may trigger an action inside the cell (relocation of a service).

The digital counterpart of the human hormone system are the digital hormone values, which are piggy-backed on the messages leaving the nodes. The digital hormone values are produced by a metrics that takes the previously mentioned parameters into account. The monitors of the receiving nodes extract the information carried by the messages.

Depending on the received load information of the remote node, the self-optimization of the local node decides whether a service should be relocated to the other node or not.

Simulation results of networks with 1000 nodes showed that the self-optimization scales linearly with the amount of nodes. The self-optimization reaches 98.4% of the theoretical optimum with a minimum of service relocations. Furthermore, the self-configuration is aware of the dynamic behavior of the services and can adopt to changes in the resource consumptions.

Although self-optimization employees a rather simple approach where information is processed only locally and no central component is used to control the self-optimization, it produces excellent results and scales linearly with the size of the network. The last point might be the most important one concerning ubiquitous environments where we expect a huge amount of interconnected devices.

We developed and simulated four different Transfer Strategies [43]. The best one reaches about 98.4% of the theoretical optimum concerning the mean average error. We can show that the amount of transferred services of this is nearly the amount of transfers needed to reach the optimum. So, if all services would be of the same size concerning the resource consumption, we would reach the optimum with a minimal amount of service transfers. The amount of service transfers can be further reduced by an adaptive barrier that suppresses those relocations that do not add a significant gain to the overall optimization.

We also simulated a dynamic service behavior, that is, we considered a change in the resource consumption of the services. First, we simulated a periodic behavior of the services where services change their resource consumption for a given percentage of a period and return to the original value for the rest of the period. Second, we assumed a constant increase in the resource consumption during a predefined time. After that time, the services rest at the new load level for a while and return to the normal level also within the same predefined time. It turned out that the Hybrid Transfer Strategy, which uses the dynamic barrier and an load estimator to calculate the average load of the network, can perfectly adapt to the dynamic behavior of the services and thus suppress unintended service relocations. Experiments in the OCμ middleware [41] showed even better results than the simulation did.

14.5.6   Self-Healing

The task of self-healing is to assure that the smart doorplate system meets some defined conditions as far as possible. In our case, we have to guarantee that all services as determined in the configuration of the system stay available.

Self-healing can be divided into (1) the detection of unwanted conditions and (2) the recovery from these conditions. We identified the following tasks of a self-healing mechanism:

Failure detection: Failure detection provides information on failures of components of distributed systems. A failure detector enables an $OC\mu$ node to detect the crash of a node within a smart doorplate environment.

Grouping: The responsibilities for failure detection and other monitoring tasks have to be arranged intelligently in a way that assures secure monitoring with low overheads. This organization of the failure detection services is called *grouping* and states which $OC\mu$ nodes monitor which other nodes in the system.

Automated planning: Failure detection and grouping refer to the detection of unwanted conditions. Automated planning and scheduling can be used as an intelligent mechanism to recover from these conditions. The planner uses the current situation of the smart doorplate system as input and outputs a solution how to recover to a state of the system that reestablishes the full functionality or abides the main functions of the system as long as possible.

Scheduling: A scheduling component is working hand in hand with the automated planner. It assigns the steps of the generated plan to $OC\mu$ nodes and controls the recovery process.

Distributed datastore: We chose a datastore service to save data in a distributed manner within the $OC\mu$ middleware and refrained from an automatic backup service that might be too resource-consuming for ubiquitous environments. Services can use this *distributed datastore* to save their context persistently, that is, services have to decide themselves when and which information should be stored. Thus, a crashed service has the possibility to recover its last state.

Failure detection and distributed datastore are shortly described. Grouping, automated planning, and scheduling are work in progress.

14.5.6.1 Failure Detection

Failure detectors provide information on failures of components of distributed systems. A failure detector enables an $OC\mu$ node to detect the crash of another node which represents a tough challenge in distributed systems like our smart doorplate environment.

So far, we developed and simulated a new adaptive accrual failure-detection algorithm [36] and also investigated some variations of this algorithm [37].

Failure detectors in distributed systems require a mutual monitoring by heartbeat messages, respectively, by application messages from the monitored node within a certain time interval. Adaptive failure detectors [7, 8] are able to adjust to changing network conditions. The behavior of a network can be significantly different during high traffic times as during low traffic times regarding the probability of message loss, the expected delay for message arrivals, and the variance of this delay. Thus, adaptive failure detectors are highly desirable.

The principle of an accrual failure detector [14] is not to output whether a node is suspected to have crashed or not like common failure detectors. Rather they give a suspicion information on a continuous scale, whereas higher values indicate a higher probability that the monitored node has failed. Thus, accrual failure detectors decouple monitoring and interpretation. That makes them applicable to a wider area of scenarios and more adequate to build generic tools.

The algorithm we developed is based on a contract defining the monitored node sending heartbeat messages to the monitoring node every time step. The monitoring node analyzes the interarrival times of the heartbeat messages using statistical methods. As a result, it outputs the probability that the monitored node has failed based on its previous behavior.

Our algorithm is characterized by very low computational demands on the processing power. This allows an application in a broad variety of hardware settings. It also shows very good results in comparison to other failure detectors, especially in the case of message loss. Comparing our algorithm to other state-of-the-art failure-detection algorithms [7, 8, 14] shows that it significantly outperforms all other algorithms in certain important settings [36, 37]. Furthermore, in contrast to common failure detectors, it outputs a probability that a node has crashed, not only a boolean value. This is an ideal basis to build very flexible failure-detection services in a variety of environments.

14.5.6.2 *Distributed Datastore*

To ensure that a crashed and recovered service can access the last saved state, the $OC\mu$ middleware offers a storage service that itself exposes self-healing features to overcome node failures.

A distributed datastore spreads the data of the services on different nodes to add redundancy to the stored data. In case of a failure, a service asks the distributed datastore about the data and the datastore is responsible to get the latest version of the service's data.

The self-healing of the data storage itself is based on the measurement of some resources like memory consumption and communication bandwidth. The resources are measured during runtime and the data of the services are distributed considering the actual resource consumptions. But not only the resources are of interest. The self-healing also takes the average failure rate and the measured online periods of the nodes into account to raise the dependability of the data storage. The average failure rate is applied to rate a node's trustworthiness. The lower the failure rate of a node, the higher the node is rated in terms of trustworthiness. The online time is especially interesting in environments where nodes are not online 24/7 like, for example, the PC's of employees in an office environment. The online time is measured during the day and the start of the next offline period is predicted. The remaining online time is used as a further parameter to rate the node. The longer the time until the expected offline period, the higher is the node's rate.

We simulated different implementations of the distributed data store [9]. The target of the simulation was to show how many failures are produced within different environmental settings. A failure is either a read operation from the data store, which results in none or outdated data. Simulations with 100 nodes showed that the failure can be reduced to 0 even if we assume that one node of the network crashes every 36. A failure of 0.02% is achieved if the failure rate is raised such that a random node of the network fails every 18s. Assuming an average failure rate of 18s means that every node fails at least after half an hour. The interesting point about the distributed data store is the fact that only two additional nodes must hold a copy of the data to achieve the mentioned results.

14.5.7 Self-Protection

Research on self-protection in $OC\mu$ is so far restricted to the detection of new and, therefore, potentially malicious messages by techniques from computer immunology [11]. Current violation detection systems are normally rule based or use signatures, which rather result in a static detection of intrusions. Computer immunology opens up new ways and methods to recognize new intrusions like our biological immune system does. In the area of self-protection, we are going to design a protection architecture that enables all members in the middleware to identify intrusions without the need of central instances. Once an intrusion is detected by one peer member, the nodes will be able to eliminate threats or exclude malicious services from the middleware on their own by effective communication strategies. As for now, we investigated the usage of artificial antibodies to detect messages from intentionally or unintentionally malicious or yet unknown services. The biologically inspired technique of computer immunology extracts ideas from our human immune system to develop an artificial counterpart [1].

A look at the functionality of the biological immune system shows that the basic requirement of such a system is to distinguish between harmless objects called *selfs* and harmful objects called *nonselfs* [15]. Matching works due to different protein patterns on the surface of the unknown objects and the antibodies. All mammals have a *thymus*, which is aware of all selfs known in the body. Antibodies are continuously created with random protein surfaces and get singled out with *negative selection*. If such a cell is activated in the thymus it will be destroyed [15], otherwise it will be released to the body to protect it against a specific type of intruder. The opposite process called *positive selection* is also used in the body [40] that focuses a T-cell receptor to recognize peptides bound to molecules of the major histocompatibility complex (MHC). It works similar to the negative selection process but in difference it singles out all protein surfaces that *can* bind to selfs in the thymus. Those receptors that do not bind are destroyed. The biological immune system is extremely distributed besides the centralized work of the thymus.

Our middleware services solely communicate over messages and because of the sandbox mechanism in the Java environment, which is the implementation basis of $OC\mu$, those messages are the only way to impact the system right now. Due to that fact, our middleware-based immune system does not aim on proteins but on binary messages. The basic requirement of the artificial immune system is also to distinguish between foreign nonself messages and messages that are tolerated in the system because they belong to self. Instead of storing the huge amount of information about all selfs on every node, we want to filter out messages by comparison to short bit strings. Thus, we first developed antibodies analogous to the r-chunk method [40]. The antibodies are represented by a short bit string of length r, which embodies the receptor. Additionally, they have a specific offset o where they start their comparison to the messages. Before the antibodies get released to the system they are perambulated with the negative selection process. Instead of waiting for all newly created antibodies and probodies until they perambulated the thymus like in the human body, we preferred to generate them in a structured way. This works because the middleware $OC\mu$ is aware of all its self messages and thus it can create all receptor patterns not used by any self message at a specific offset. We also do not compare the whole message because this would lead to an infinite amount of selfs due to the changing data in the message. Therefore, we only consider the header of the messages that consist of sender and message type.

First off, we investigated the relation between the design of the antibodies (e.g., different receptor lengths and offsets) and the resulting detection rate. We designed a simulator to measure the effectiveness and reached detection rates of up to 99.6% of artificial and randomly constructed intrusive messages. Best recognition rates were reached when receptors with a length of $r \simeq \log_2(n)$ are used in a system that contains n self messages [32]. We also recognized that the storage of receptors needed a lot of space and also comparison takes a long time. Therefore, we found optimizations of our technique in both areas. For minimizing the space complexity, we combine similar receptors. This results in receptors with wildcards and in our test runs we were able to eliminate 30% of the space needed for storing the antibodies. This also helps minimizing network usage when exchanging antibodies between nodes. To enhance the comparison itself, we organize the receptors of specific offsets in a tree structure. This results in a constant time complexity for comparison and in simulation runs it gained a speedup of 30 [30].

The opponent of the antibodies is the group of so-called probodies.[2] They are designed the same way as the antibodies but in difference are generated with the use of the positive selection mechanism. We found out that a combined usage of antibodies and probodies reaches better results than only using antibodies but also has some drawbacks. Because of the nature of the selection mechanisms all possible probodies at a specific offset always have to appear completely [31].

14.6 CONCLUSION

Flexible offices can drastically reduce costs by dynamically assigning office rooms to present employees. This organization requires a sophisticated software system

---

[2]We coined the word probody because there is no handy notion in immunology for the item that is required in our artificial immune system.

that is highly dynamic, scalable, context-aware, self-configuring, self-optimizing, self-healing, and self-protecting. In this chapter, we presented our research experiences with a flexible office environment that satisfy these requirements. Due to the organization of our department with fixed offices, we cannot investigate the flexible office paradigm itself. However, we developed and implemented a suitable ubiquitous middleware, a mobile agent system for encapsulation of user context, a location-tracking system, and location-prediction techniques. Office rooms can be dynamically assigned to currently present employees. Applications are realized through mobile agents that offer a possibility to encapsulate personal information of a user and to perform location-based services of the ubiquitous system in the name of the user. Persons and objects are tracked by a location-tracking system that is integrated as an additional ubiquitous service. Moreover, a location-prediction service provides assumptions in which an absent office owner will be next and when he/she will be back. To increase manageability of the complex system, we designed a middleware to fulfill the autonomic or organic computing demands of self-configuration, self-optimization, self-healing, and self-protection. We investigated self-configuration by an social-cooperative behavior, self-optimization by an artificial hormone-based approach, self-healing by dependability approaches, and self-protection by computer immunology. The highlight of our system are the smart doorplates that serve as test and presentation platform. On the smart doorplates we validated the functionalities of an organic computing middleware and implemented several services on basis of mobile agents.

REFERENCES

1. P. S. Andrews and J. Timmis. Inspiration for the next generation of artificial immune systems. In *4th International Conference on Artificial Immune Systems*. LNCS, pp. 126–138. Springer-Verlag, 2005.
2. F. Bagci, H. Schick, J. Petzold, W. Trumler, and T. Ungerer. Communication and security extensions for a ubiquitous mobile agent system (UbiMAS). In *Proceedings of Computing Frontiers (CF 2005)*, Ischia, Italy, 2005.
3. F. Bagci, H. Schick, J. Petzold, W. Trumler, and T. Ungerer. Support of reflective mobile agents in a smart office environment. In *Proceedings of Architecture of Computer Systems (ARCS 2005)*, Hall, Austria, 2005.
4. F. Bagci. Reflektive mobile Agenten in ubiquitären Systemen. PhD thesis, University of Augsburg, December 2005.
5. P. Bahl and V. N. Padmanabhan. RADAR: an in-building RF-based user location and tracking system. In *INFOCOM (2)*, 2000, pp. 775–784.
6. E. Behrends. *Introduction to Marcov Chains*. Vieweg, 1999.
7. M. Bertier, O. Marin, and P. Sens. Implementation and performance evaluation of an adaptable failure detector. In *DSN'02: Proceedings of the 2002 International Conference on Dependable Systems and Networks*, Washington, DC, USA. IEEE Computer Society, 2000, pp. 354–363.
8. W. Chen, S. Toueg, and M. K. Aguilera. On the quality of service of failure detectors. In *Proceedings of the International Conference on Dependable Systems and Networks (DSN 2000)*, New York. IEEE Computer Society Press, 2000.
9. J. Ehrig. Selbstheilung in einem verteilten dienstbasierten Netzwerk. Master's thesis, University of Augsburg, August 2006.
10. S. Feldmann, K. Kyamakya, A. Zapater, and Z. Lue. An indoor bluetooth-based positioning system: concept, implementation and experimental evaluation. Technical report, Universität Hannover, 2003.

11. S. Forrest, S. A. Hofmeyr, and A. Somayaji. Computer immunology. *Commun. ACM*, 40(10):88–96, 1997.

12. K. Gurney. *An Introduction to Neural Networks*. Routledge, 2002.

13. A. Harter, A. Hopper, P. Steggles, A. Ward, and P. Webster. The anatomy of a context-aware application. *Wireless Network*, 8(2/3):187–197, 2002.

14. N. Hayashibara, X. Défago, R. Yared, and T. Katayama. The accrual failure detector. In *SRDS*, 2004, pp. 66–78.

15. S. A. Hofmeyr and S. Forrest. Architecture for an artificial immune system. In *Evolutionary Computation*, Vol. 8, No. 4, Massachusetts Institute of Technology, 2000, pp. 45–68.

16. IBM Corporation. Autonomic computing concepts. http://www.ibm.com/autonomic/, 2001.

17. F. V. Jensen. *An Introduction to Bayesian Networks*. UCL Press, 1996.

18. Y. Ji, S. Biaz, S. Pandey, and P. Agrawal. ARIADNE: a dynamic indoor signal map construction and localization system. In *MobiSys 2006: Proceedings of the 4th international conference on Mobile systems, applications and services*, New York, NY, USA. ACM Press, 2006, pp. 151–164.

19. R. Klaus. Selbstkonfiguration in einem dienstbasierten Peer-to-Peer Netzwerk. Master's thesis, University of Augsburg, 2006.

20. F. Kluge. Untersuchung bestehender Methoden und Entwurf eines Systems zur Ortsbestimmung in Bürogebäuden. Master's thesis, University of Augsburg, October 2005.

21. N. Minar, M. Gray, O. Roup, R. Krikorian, and P. Maes. Hive: distributed agents for networking things. In *Proceedings of Symposium on Agent Systems and Applications/Symposium on Mobile Agents (ASA/MA '99), IEEE Computer Society*, Palm Springs, CA, 1999.

22. L. M. Ni, Y. Liu, Y. C. Lau, and A. P. Patil. LANDMARC: indoor location sensing using active RFID. *Wirel. Netw.*, 10(6):701–710, 2004.

23. J. Petzold. Augsburg Indoor Location-Tracking Benchmarks. Context Database, Institute of Pervasive Computing, University of Linz, Austria. http://www.soft.uni-linz.ac.at/Research/Context_Database/index.php, 2005.

24. J. Petzold. Zustandsprädiktoren zur Kontextvorhersage in ubiquitären Systemen. PhD thesis, University of Augsburg, December 2005.

25. J. Petzold, F. Bagci, W. Trumler, and T. Ungerer. Global and local context prediction. In *Artificial Intelligence in Mobile Systems 2003 (AIMS 2003)*, Seattle, WA, USA, 2003.

26. J. Petzold, F. Bagci, W. Trumler, and T. Ungerer. Confidence estimation of the state predictor method. In *2nd European Symposium on Ambient Intelligence*, Eindhoven, The Netherlands, 2004, pp. 375–386.

27. J. Petzold, F. Bagci, W. Trumler, and T. Ungerer. Comparison of different methods for next location prediction. In *European Conference on Parallel Computing, Euro-Par 2006*, Dresden, Germany, 2006.

28. J. Petzold, F. Bagci, W. Trumler, and T. Ungerer. Hybrid predictors for next location prediction. In *The 3rd International Conference on Ubiquitous Intelligence and Computing (UIC-06)*, Wuhan and Three Gorges, China, 2006.

29. J. Petzold, A. Pietzowski, F. Bagci, W. Trumler, and T. Ungerer. Prediction of indoor movements using bayesian networks. In *Location- and Context-Awareness (LoCA 2005)*, Oberpfaffenhofen, Germany, 2005.

30. A. Pietzowski, B. Satzger, W. Trumler, and T. Ungerer. A bio-inspired approach for self-protecting an organic middleware with artificial antibodies. In *Self-Organising Systems, First International Workshop (IWSOS 2006)*, Passau, Germany, 2006. Springer, Vol. pp. 202–215.

31. A. Pietzowski, B. Satzger, W. Trumler, and T. Ungerer. Using positive and negative selection from immunology for detection of anomalies in a self-protecting middleware. In *Informatik 2006, Informatik für Menschen*, Dresden, Germany, 2006. Gesellschaft für Informatik e.V., LNI, Vol. p-93, pp. 161–168.

32. A. Pietzowski, W. Trumler, and T. Ungerer. An artificial immune system and its integration into an organic middleware for self-protection. In *Genetic and Evolutionary Computation Conference (GECCO 2006)*, Seattle, Washington, USA, 2006. ACM, ACM Press, Vol. 2, pp. 129–130.

33. Project JXTA. http://www.jxta.org, 2002.

34. L. R. Rabiner. A tutorial on hidden Markov models and selected applications in speech recognition. *IEEE*, 77(2), 1989.

35. I. Satoh. Spatialagents: integrating user mobility and program mobility in ubiquitous computing environments. *Wireless Commun. Mobile Comput.*, 3(4), 2003.
36. B. Satzger, A. Pietzowski, W. Trumler, and T. Ungerer. A new adaptive accrual failure detector for dependable distributed systems. In *SAC '07: Proceedings of the 2007 ACM Symposium on Applied Computing*, New York, NY, USA. ACM Press, 2007.
37. B. Satzger, A. Pietzowski, W. Trumler, and T. Ungerer. Variations and evaluations of an adaptive accrual failure detector to enable self-healing properties in distributed systems. In *ARCS '07: Proceedings of the 20th International Conference on Architecture of Computing Systems*, 2007.
38. Website of the Embedded Sensor Board ESB 430 http://www.scatterweb.com.
39. A. Smith, H. Balakrishnan, M. Goraczko, and N. Priyantha. Tracking moving devices with the cricket location system. In *MobiSYS '04: Proceedings of the 2nd international conference on Mobile systems, applications, and services*. ACM Press, 2004, pp. 190–2002.
40. T. Stibor, K. M. Bayarou, and C. Eckert. An investigation of r-chunk detector generation on higher alphabets. In *Genetic and Evolutionary Computation Conference*. Springer-Verlag, 2004, pp. 299–307.
41. W. Trumler. *Organic ubiquitous middleware*. PhD thesis, University of Augsburg, 2006.
42. W. Trumler, R. Klaus, and T. Ungerer. Self-configuration via cooperative social behavior. In *3rd IFIP International Conference on Autonomic and Trusted Computing (ATC-06)*, Wuhan, China. Springer, 2006, pp. 90–99.
43. W. Trumler, T. Thiemann, and T. Ungerer. An artificial hormone system for self-organization of networked nodes. In *IFIP Conference on Biologically Inspired Cooperative Computing*, Santiago de Chile. Springer-Verlag, 2006, pp. 85–94.
44. VDE/ITG/GI. Organic Computing: Computer- und Systemarchitektur im Jahr 2010. http://www.gi-ev.de/download/VDE-ITG-GI-Positionspapier Organic Computing.pdf, 2003.
45. L. Vintan, A. Gellert, J. Petzold, and T. Ungerer. Person movement prediction using neural networks. In *First Workshop on Modeling and Retrieval of Context*, Ulm, Germany, 2004.
46. R. Want, A. Hopper, V. Falcao, and J. Gibbons. The active badge location system. *ACM Trans. Inf. Syst.*, 10(1):91–102, 1992.

Chapter 15

An Agent-Based Architecture for Providing Enhanced Communication Services

Romelia Plesa[1] and Luigi Logrippo[1,2]

[1] School of Information Technology and Engineering, University of Ottawa, Ottawa, Ontario, Canada
[2] Département d'informatique et ingénierie, Université du Québec en Outaouais, Gatineau, Québec, Canada

15.1 Introduction	320
15.2 Motivation	321
15.3 Consolidated Presence Information	322
15.4 Architecture	323
15.5 BDI and AgentSpeak(L)	330
15.6 BDI and Context Aware Communication	332
15.7 Related Work	340
15.8 Conclusions	340
Acknowledgments	341
References	341

15.1 INTRODUCTION

Among the current trends in personal communication, there is a growing need of communication services tailored to the user's specific needs and preferences. With devices that enable mobile communication becoming more popular, there is an

Mobile Intelligence. Edited by Laurence T. Yang, Agustinus Borgy Waluyo, Jianhua Ma, Ling Tan, and Bala Srinivasan
Copyright © 2010 John Wiley & Sons, Inc.

increasing necessity for the users to control and customize their services, making them sensitive to the context in which the communication takes place and influencing the user's availability and reachability for communication.

The literature distinguishes between context-free and context-aware communication systems [1]. *Context-aware* services take advantage of knowledge of real-time context regarding the purpose or the circumstances of an incoming or outgoing call. However, most of today's telephony communication services are *context free*, in the sense that they do not have such knowledge. Context-aware services can be of course much richer and allow the user to manage the use of his communications systems in terms of policies, taking into consideration context information.

Clearly, context-aware systems require much richer architectures than conventional systems. They also require high-level concepts and languages for programming user policies and services. Much work is being done in this area, some of it proposing agent-oriented architectures. This chapter presents the results of a study toward an implementation of a Belief Desire Intention (BDI) agent architecture for context-aware communications systems. AgentSpeak(L), an agent-oriented programming language, is used to implement the architecture. We show how the BDI approach may be used to provide useful, intelligent services to users in realistic settings. In doing this, we discuss different challenges and issues related to the architecture. The elements of this architecture were presented in Ref. [2], with additional details in Ref. [3]. This chapter is an extension of Ref. [3].

Our architecture is applicable to situations where users have the need to define complex policies on how their communication should be handled, based on information about their current context.

Communication requests are handled according to these policies, with the potential of a very high degree of customization.

15.2 MOTIVATION

The advocates of presence technology and contextual services promise a world where people will be connected *when they want, how they want, and with whom they want* and their communication will be tailored on specific desires and preferences, according to circumstances [4].

One hypothetical scenario, described below, illustrates the capabilities of a system that offers converged services, this means services that combine voice, presence, Web, chat, and other elements. Although sounding futuristic, many individual components have been developed and our work proposes an architecture that will enable the realization of such scenarios.

Bill has a conference call at 9:00 am. It is 8:30 and he is stuck in traffic, but he does not worry. He subscribed to a presence management service that knows where he is and how best to reach him. The service forwards the call to his mobile phone so he joins the conference from his car. Once in his office, Bill starts working, but his phone rings continuously, distracting him. A solution is to have a call filtering service that routes all calls, except those from certain clients, to his voice mail. The service recognizes the key callers and routes each one to a personalized message from Bill. Regular callers hear Bill's usual voice message. In the meantime, Bill is able to complete his presentation. He then wants to talk to a customer. He is not sure where the client is (office, lunch, etc.) and which network (home, wireline, or wireless) he should use. The presence management application will locate the client and will use the appropriate device to call. Bill may also want the information under discussion (the context of the conversation) to be provided to himself and to his client. For himself, the information will be presented via the PC. If the client is in his office, he could receive the information on his PC as well. If he is mobile, then the summarized information could be displayed on his Web-enabled cell phone.

The primarily point being made by this example is that the act of communication is always part of a larger context. Communication services will come to exhibit self-awareness: a sense of why they are being used, of what task is being supported, and of the goal to be accomplished.

15.3 CONSOLIDATED PRESENCE INFORMATION

Presence information is information concerning a person's online status, availability, and reachability across different types of communication channels. Presence is defined [5] as the willingness and ability of a user to communicate with others on the network. On most applications, presence has been limited to "on-line" and "off-line" indicators. The notion of presence in our work is much broader. We want it to include the physical location of the user (*at home*, *at the office*), call state (*ready and willing to communicate*, *currently on a call*), the role that the user is currently taking, or the user's willingness to communicate (*available*, *in a meeting*). It can also include indicators that show if a user is logged into a network and whether that user is active, whether a user's mobile phone is switched on, what network it is on, its cellular location and the preferred medium for communication (*voice*, *IM*, *video*, *e-mail*). In this way, presence information becomes relevant to any means of communication, not only instant messaging and buddy lists (applications that are already using it), but also to a large variety of devices and contexts. The notion of context has been discussed in Refs. [6, 7]. According to Ref. [8], the context is defined as information that characterizes the situation of an entity (a person, a place, or a physical or computational object). In most of the work in the area of context-aware computing, the need of awareness of the physical environment surrounding a user and his devices was addressed by implementation of location-awareness, for instance based on global positioning. Therefore, beyond location, there are other features of the environment that contribute to context. Human factor related context includes information about the user (e.g., their habits), the user's social environment (co-location with others, social interaction), and the user's task. Context related to physical environment includes location, but also infrastructure and information about surrounding resources for computation and communication. Information such as what the user has done in the past can also be part of the context. Some context information, such as the role of the user, can be manually keyed in by the user, while other information, such as

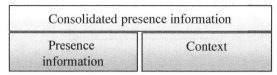

Figure 15.1   Consolidated presence information.

location, time of day, can be easily gathered using sensing devices. Other information, such as the current status of the user, can be gathered from sources such as the calendar of the user or from a meeting attendees list.

A combination of presence information and context information (Figure 15.1) can be used to build an intelligent picture of a user's current situation, status, and accessibility. We call this the *Consolidated Presence Information* or CPI, for the user.

"Consolidated Presence Information" means aggregated information from various devices, applications, and attributes such as networks and locations that results into a unified and aggregated view of an individual's current status. This allows users to dynamically set policies that govern the particulars of how users and applications interact with one another and define their preferred device or application for real-time contact. These policies can be based on attributes such as time-of-day, location, or the people attempting to contact them.

For example, a user's policy may describe the behavior required when the following conditions are met: an individual is at work, located in his office, and is writing an e-mail. With no support for aggregation, a combination of database queries must be used in order to determine when these conditions are met. This is unnecessarily complex and is difficult to modify if changes are required. By aggregation, all the information about a given entity is collected and processed and a single query is sufficient to determine if a combination of complex conditions is met.

The processing of these complex policies (which are supported by the architecture presented in this chapter) needs knowledge about the users and the environment in which their activities take place, organizational activities, as well as complex, intelligent mechanisms for deriving knowledge from the raw information that is available to the system.

Our architecture provides entities responsible for building and maintaining CPI, as well as for storing it. A detailed description of these entities is provided in Section 15.4. The format in which the information is stored can be, for example, presence information description format, PDIF [9].

15.4 ARCHITECTURE

15.4.1 Elements of the Architecture

In order to provide complex services that are tailored to users' specific desires and preferences, the architectural model needs at least the following functional

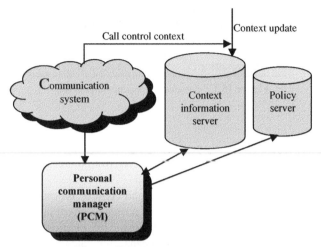

Figure 15.2   The overall architecture.

requirements:

- collection of context information using sensors as well as dissemination of context information, publishing of presence information from users and their devices,
- description of user policies and preferences, and
- preferences-based ubiquitous handling of communication.

Figure 15.2 presents the overall system architecture.

The communication system is responsible for signaling and communication. The communication part of the architecture is a complex system by itself. Our solution is independent of the underlying communication protocol. It can be based on SIP [10], H.323 [11], or other session protocols. The communication system should provide interoperability between communication protocols of different functionalities.

The requirement that we impose on this architecture is that every message that arrives for a user must be intercepted and dealt with such that user's policies and current context will be considered. We will focus here on the processing of such messages.

The Context Information Server (CIS) controls the context updates and stores and distributes the context information. We are not concerned here about how this information arrives. We just assume that there is a mechanism that allows sensors, smart badges, and so on, to collect and forward it as it changes. The dynamic nature of some context information requires a mechanism for keeping up-to-date information in the context information server in order to allow services to adapt to the changing context. Among all possible context information, we define a list of significant events (i.e., a user logging into the system, a resource being added/modified, a device becoming available, etc.). Whenever an event of this type occurs, triggers are fired that determine the rebuilding of the user's CPI.

The *Policy Server* (PS) manages the user's personal policies, as well as the subscription/notification policies. The management of policies includes creating, storing, deleting, retrieving, and fetching policies for end users. Our proposal for a policy language will be presented later.

We propose the introduction of a new architectural entity, a *Personal Communication Manager* (*PCM*) that represents each user and is responsible for deciding the flow of actions for the call, based on personal policies and on information about presence and current context. The PCM has access to up-to-date information about the user's context that is used to influence the call functionality. It is responsible for the proper execution of a call. It is the entity that ultimately receives request messages (such as INVITE or SUBSCRIBE for a SIP-based architecture) and decides the actions that should be taken and how the call should be handled. Moreover, it takes into consideration the current context, which is an important part of the decision. The PCM treats relevant events that occur in the system. Such events can be invitations to a call, but can also be updates in presence information or changes in the current context. Based on the event, PCM will decide the next course of action and how the call will be handled. A further role of the PCM is to detect and resolve (or suggest resolutions) for any conflict that arises among policies. The policies are retrieved by the PCM by querying the PS.

The lifetime of a user's PCM starts with the new user being registered to the presence system and ends with an explicit removal of the user from it. During this time, the PCM's responsibilities will include managing the calls for the user, based on context and personal policies.

The components of the PCM are shown in Figure 15.3.

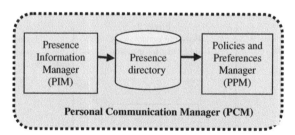

Figure 15.3   Personal communication manager.

The *Presence Information Manager* (*PIM*) is the element responsible for building and maintaining the CPI, see Section 15.3. It does so by aggregating presence and context information from different sources. It manages raw presence data and distils the flow of indicators. The consolidated presence can be maintained by a rule-based process that takes into account presence and context indicators and their ability to reflect the user's state. Several mechanism have been proposed [12, 13]. The actual mechanism to be used for reasoning about context in our architecture is part of our future work. We would like to provide the agents with a choice of reasoning and learning mechanisms that they can use to derive new information from the raw data provided. These mechanisms can be provided in the form of libraries that the agent can use.

The *presence directory* is a repository in which the CPI is deposited and retrieved.

The *Policies and Preferences Manager (PPM)* contains the preference logic and processes that respond to requests to contact an entity. The CPI is interpreted by the PPM to establish the best method for contacting a user at a particular moment, at a given location, based on availability, device capabilities, and personal preferences.

15.4.2 Presence Management Strategy

With this architecture, the process of building the CPI is done in several steps, thus allowing separation of the context acquisition and management from reasoning with context knowledge. We can distinguish three layers (Figure 15.4):

- sources of context and presence information,
- presence management, and
- users of consolidated presence information.

An entity's presence, the CPI, is constructed from a series of presence and context indicators that come from access networks, directly from the user's terminals or from third party information sources. These indicators are combined to form a higher-level view of presence for a user. This combination has to be done in real-time so that the entity's presence can be projected out in advance of any attempt to communicate with the entity.

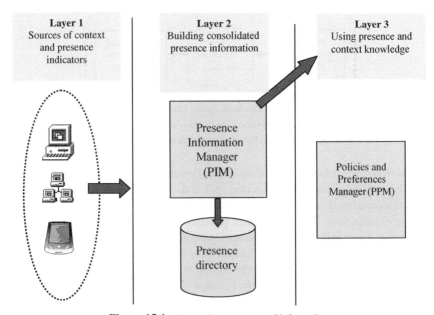

Figure 15.4 Layered management of information.

Presence and context indicators are generated for the PIM from several sources. First, presence information comes as a side-effect of a user utilizing the network. For example, a registration on a GSM network when the user's cell phone is activated constitutes an indicator that the user is currently active in a certain location and ready to use the network. This can generate presence as well as context information. A second source arises directly from presence clients that reside on a user's terminals (PC, PDA, phone, etc.), which will generate presence information. A third source of indicators is from third party services. For example, a hotel registration system can generate an indicator as a side-effect of a guest registering on arrival. This information will be delivered in the CIS and will represent context information about the user. Similarly, an employee logging into the system when he goes to work will constitute an indicator about his location and availability for communication.

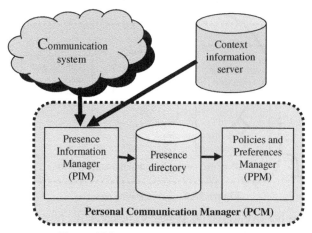

Figure 15.5   Sources of indicators for PCM.

Figure 15.5 shows how the PCM obtains the presence information and the context information. When any change occurs in the presence status of a user, PCM will be informed.

This is possible because, as explained earlier, PCM intercepts the messages exchanged in the communication layer, so it will be aware of any notification of a change in the presence information. In this way, any change in a user's presence will result in the PIM reapplying the rules and rebuilding the user's consolidated presence.

In a SIP-based architecture, for example, the entity that deals with the user's presence is the presence agent [4]. For such architecture, PIM will be informed by the presence agent about the changes that occur in the user's presence. This can be accomplished by allowing the PIM to subscribe to any relevant event that the presence agent receives, that is, any update from presence user agents (PUAs). In this way, any change in a user's presence will result in a notification being sent to PIM, which will have to reapply the rules. In this case, it is necessary to define a new event package that will contain the events that generate presence indicators. This event package is to be defined as an extension of the SIP-specific event framework.

The context information (which resides in the CIS) is obtained by PIM by querying the CIS, which will fetch the relevant information and send it to the PCM. Having the presence information and the context information, PIM will update the CPI in the presence directory.

Figure 15.6 uses use case maps [14], a popular notation for describing causal scenarios, to show the process of building the network presence for a user.

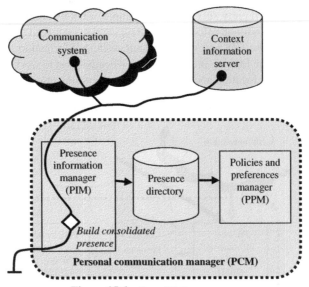

Figure 15.6 Consolidating presence.

The process has two triggering points: a change in the presence information or any event that happens in the CIS, that is, an update in the context information for the user. In case any of these happens, PIM will first obtain presence and context information. After that, it will generate the consolidated presence for the user, which will be deposited into the presence directory. Also, some information about the current context may change after this process, so the CIS must be updated as well.

For simplicity, we used a stub to describe the actions of the PIM. The detailed sequence of responsibilities included in this stub is shown in Figure 15.7.

Figure 15.7 The "build network presence" stub.

15.4.3 Context Aware Call Handling

The feature selection mechanism and the feature execution mechanism will be incorporated into the PCM, more precisely in the PPM.

PPM, which contains the preference logic and rule-based processes to respond to different requests, consults the user's personal policies as well as the information stored in the presence directory and decides about the handling and execution of the call or of any other request that arrives.

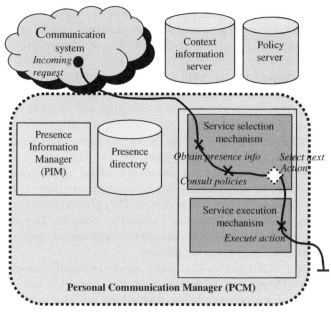

Figure 15.8 Call handling.

In Figure 15.8, we show the actions of the PPM using the use case maps notation. The selection of the next action to be executed is a complex mechanism, which can be implemented in different ways. For this reason, we used a dynamic stub to represent it. The action is then executed by the service execution mechanism of the PPM.

At the communication level, we assume that there are access points for user's communication devices that manage the sessions involving these devices. Conversation devices such as phones or IM clients, as well as messaging devices such as pagers should be supported.

As decisions on how to reach a user are made, the appropriate device is contacted by performing address lookup using the information that is found in the directory service component of the CIS.

There are a number of languages for specifying policies. The call processing language [15] allows users to define how they wish calls to be handled, but is limited in a number of ways that make it unsuitable for expressing complex policies for call control [1]. LESS [16] inherits the basic structure and many elements from CPL and enhance it with elements for end system services, thus allowing end users to program their own communication services. A policy language targeted to policies in the communication domain has been defined in Ref. [17]. We will propose now an approach to define policies based on BDI and AgentSpeak(L).

15.5 BDI AND AGENTSPEAK(L)

One of the most successful theoretical models of rational agents is the BDI model [18, 19]. The concept of BDI agents has also proven useful in practical applications, for example, in the defense industry [20, 21].

The BDI framework was developed in the 1980s by Georgeff and Lansky [22]. It has since been implemented in several software platforms, current examples being Jack Intelligent Agents [21] and Jam Agents [23]. The wide appreciation of the BDI model is witnessed by the development of a BDI logic [24], the definition of BDI-based languages (AgentTalk [25], 3APL [26], AgentSpeak(L) [27]) and the creation of BDI-based development tools such as dMARS [29].

The BDI model provides a convenient terminology and structure for describing intelligent agents. Unlike many other agent systems, the BDI framework has had many practical applications, by which the theory and terminology have been made clearer and more generic than for other system.

BDI agents have been applied and are suited to model highly dynamic and unpredictable situations. By only partially expanding alternative plans of actions, they can remain responsive to changes in system state. They provide ways to recover from failed actions and to customize reasoning for specific situations. They describe also how to handle conflicting actions and goals and to modify already executing actions, all of which are needed in dynamic simulations.

Therefore, the BDI framework provides means to specify complex domain-specific behavior. There are numerous published examples that demonstrate complex domain-specific behavior being captured and modeled in intelligent agent applications.

In the attempt to bridge the gap between theory and practice, the BDI architecture defines a model that shows a one-to-one correspondence between the model theory, proof theory, and the abstract interpreter. The following concepts were developed, as included in the AgentSpeak(L) language.

- An agent's *belief state* is the current state of the agent, which is a model of itself, its environment, and other agents.

- The agent's *desires* are the states that the agent wants to bring about based on its external or internal stimuli. The distinction between desires and goals, while important from a philosophical perspective, is not significant in this context. We shall sometimes use the word *goals* for desires.

- When an agent commits to a particular set of plans to achieve some goal, these partially instantiated plans (i.e., plans where some variables have taken values) are said to be an *intention* associated with that goal. Thus, intentions are active plans that the agent adopts in an attempt to achieve its goals.

The AgentSpeak(L) programming language was introduced in Refs [27, 28], where additional details on it can be found. It is a natural extension of logic programming for the BDI agent architecture and provides an elegant abstract framework for programming BDI agents. The behavior of an agent (i.e., its interaction with the environment) is dictated by a program written in AgentSpeak(L). The beliefs, desires, and intentions of an agent are not explicitly represented. Instead, these notions are programmed in the language AgentSpeak(L).

At run-time an agent can be viewed as consisting of a set of beliefs, a set of intentions, a set of events and a set of selection functions.

A *belief atom* is simply a first-order predicate in the usual notation, and belief atoms or their negations are termed *belief literals*.

A *goal* is a state of the system, which the agent wants to achieve. AgentSpeak(L) distinguishes two types of goals:

- *Achievement goals* are predicates prefixed with the operator "!". They state that the agent wants to achieve a state of the world where the associated predicate is true.

- *Test goals* are predicates prefixed with the operator "?". A *test goal* returns a unification for the associated predicate with one of the agent's beliefs; it fails if no unification is found.

A *triggering* defines which events may initiate the execution of a plan. An *event* can be internal, when a subgoal needs to be achieved, or external, when generated from belief updates as a result of changes in the environment. There are two types of triggering events: those related to the *addition* ("+") and *deletion* ("−") of beliefs or goals.

Plans determine the *basic actions* that an agent will perform on its environment. Such actions are also defined as first-order predicates, but with special predicate symbols (called action symbols) used to distinguish them from other predicates. The syntax for a plan in AgentSpeak(L) is

```
p:: =    te :   ct < -   h
```

A plan is formed by a *triggering event* (te), followed by a conjunction of belief literals representing a *context* (ct). The context must be a logical consequence of that agent's current beliefs for the plan to be *applicable*. The remainder of the plan is a sequence of basic actions or (sub)goals (h) that the agent has to achieve (or test) when the plan, if applicable, is chosen for execution.

Intentions are the plans the agent has chosen for execution (particular courses of actions to which an agent has committed in order to handle certain events). Intentions are executed one step at a time. A step can query or change the beliefs, perform actions on the external world, suspend the execution until a certain condition is met, or submit new goals.

The following example is taken from Ref. [27] and shows some AgentSpeak(L) plans.

```
+concert(A,V): likes(A) < - !book_tickets(A,V).
+!book_tickets(A, V): busy(phone)
<- call(V);....;!choose seats(A,V).
```

The first plan says that when a concert is announced for artist A at venue V (so that, from perception of the environment, a belief concert(A,V) is *added*), then if this agent likes artist A, then it will have the new goal of booking tickets for that concert. The second plan tells that whenever this agent adopts the goal of booking tickets for A's performance at V, if the telephone is not busy, then it can execute a plan consisting of performing the basic action call(V) (assuming that making a phone call is an atomic action that the agent can perform) followed by a certain protocol for booking tickets (indicated by '. . .'), which in this case ends with the execution of a plan for choosing the seats for the performance.

The AgentSpeak(L) interpreter also requires three *selection functions*:

- *SE* selects a single event from the set of events.
- *SO* selects an "option" (i.e., an applicable plan) from a set of applicable plans.
- *SI* selects one particular intention from the set of intentions.

The agents test whether the events generated during belief revision activate any new plan instance. Only one plan can become intended in a single reasoning cycle, and only one intended plan can execute the next part of its body. This can generate new beliefs, goals, or a request for a basic action execution (in the environment) is sent to the interface. New beliefs and goals may serve as triggering events for plans in following reasoning cycles, while actions produce changes in the environment.

Even though there is a body of research on implementing agents on communication devices [29, 30], no attempt has been made, to our knowledge, on using the BDI model to implement a communication system architecture.

15.6   BDI AND CONTEXT AWARE COMMUNICATION

In this section, we show how the architectural elements and functionalities that we have identified for the presence system can be mapped on the BDI model.

15.6.1   BDI Mapping

The selection function of the PPM is its most important function. Its responsibilities are to determine, based on the consolidated presence information available and also based on the user's preferences and policies, the action that should be executed next.

By looking at the BDI model as well as at the PPM component of our PCM and the way we want it to function, we found significant overlap between the requirements of our domain and the concepts discussed in the BDI model of agency.

We consider that the PPM component that we propose as a part of the system's architecture is a BDI agent (Figure 15.9):

- The *CPI*, stored in the presence directory and representing the characteristics of the environment and the user's presence information and which is updated appropriately after each sensing action and each change in a user's status, will represent the *beliefs*.

- The *policies*, which are stored on the PS and seen as objectives to be accomplished, can be considered agent's *desires*.
- The output of the selection function, representing the next course of action, will be the agent's *intentions*.

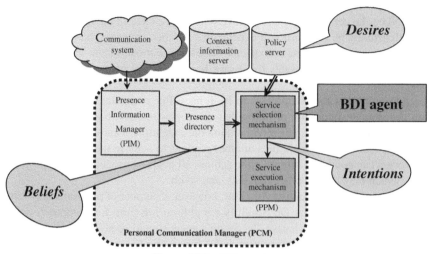

Figure 15.9   BDI mapping.

BDI agents fulfill the requirements needed for our framework, providing a way to interleave deliberation with responsiveness and limit the amount of forward deliberation required to act rationally. BDI agents can partially search and expand planned actions allowing them to select good alternatives, while avoiding constant deliberation and its associated time penalty.

15.6.2 Example

In order to further justify the reasoning behind the adoption of an agent-based approach to support context-aware services and to provide useful, intelligent services to users in realistic settings, we will consider the following situation.

Dr Smith, a doctor at the hospital, needs to achieve the following tasks on a usual day in the hospital:

1. Arrive to the hospital.
2. Log into the hospital's system in order to advertise his presence in the hospital.
3. Attend the patients visits, scheduled every morning, together with other doctors.
4. Get the schedule for his consultations and surgeries that he has to perform that day.
5. Perform his activities, according to his schedule.
6. Time permitting, being able to assist in any emergency situation that can occur.

Within an agent context, the above tasks may be represented as a set of goals that need to be fulfilled in a given sequence. In order to fulfill each goal, a specific sequence of actions must be executed, that is, a plan will be executed.

There may be a number of plans for achieving the same task. For example, in order to achieve the second goal, there can be two possible plans: the doctor can log in from the computer in his office, or, in the case that something occurred in his way to the office, he can use his PDA to advertise his presence in the hospital. The plan that will actually be followed will depend on a number of factors, including doctor's preferences, emergency and unexpected situations that can occur, and so on. As a result, the agent that represents the doctor has to perform *plan selection*, based on some criteria.

Assuming that the doctor logs in from his office computer, the corresponding plan becomes an intention, meaning that the doctor intends to execute the plan in order to achieve his goal. This might require the execution of another set of subtasks, for which, again, there might be a number of plans to achieve them. In this manner, the process can continue, attempting to achieve all goals by executing the appropriate plans, which may trigger other subgoals, and so on.

In the process of executing the plans, however, a number of problems may arise. For example, suppose that Dr. Smith is not able to log from his computer, due to technical problems. In such case, Dr. Smith's agent should be able to find alternative plans for reaching the goal, for example by suggesting to use another computer. In other words, the agent needs to perform *plan failure recovery*. The new alternative plan must take into account the new context.

Another potential problem the doctor might face is *conflict between different goals*. Suppose, for example, that the doctor wishes to play golf with his friends in the morning, which may conflict with his plan to attend a presentation on the latest news in his area of specialty. The agent should be able to *resolve* this conflict by arranging a different time to play golf, or it might have to perform *goal selection* in order to make a selection about which goal is more important.

It can be seen from the scenario we presented above that as the number of tasks and alternative plans involved increases, the complexity of the reasoning that the agent needs to perform increases significantly. This creates the need of providing automated support for the user, in terms of responding to the user's policies as the context changes. However, particular challenges arise due to the dynamic nature of the environment.

Our proposed solution consists of agents, the PCMs, each representing one user, which will be capable of performing tasks as the ones described above. A PCM has access to the consolidated presence information of the user it represents. With a library of plans, representing the user's policies, the PCM is capable of deciding what actions to execute.

With respect to the scenario we have described, the BDI model provides some essential features:

- *Context-awareness.* The choice of plans must take into account the current context in which the user is situated (e.g., the physical location or the latest changes in his schedule), as represented in the consolidated presence information. The agent's beliefs base is updated with all the changes in the environment, so that decision can be made based on the real status of the environment.

- *Plan selection.* The agent is able to make a choice between different plans, in case there are a number of alternative plans for achieving the same goal. The choice may depend on the overall cost, the risk factor, the user preferences, and so on. Appropriate decision procedures must therefore be supplied for supporting plan selection.

- *Plan failure recovery.* If a plan fails, the agent is able to retract properly and select an alternative plan.

- *Conflict resolution.* If a situation occurs where the user has a number of goals that cannot be achieved simultaneously, the agent must be able to make a decision about which goals to try to achieve, based on the importance of the goals or the costs of executing the plans.

15.6.3   Proof of Concept

Given the mapping that we have presented in Section 15.6.2 between the BDI model and the functionalities of our proposed architecture, we have implemented the PCMs as BDI agents. In particular, the agents are programmed in AgentSpeak(L) and we use Jason [27] as the interpreter for AgentSpeak(L).

The BDI mechanism incorporates both the selection mechanism and the execution mechanism that are required by the functionality of the PPM (Figure 15.9). Decisions on call routing and future actions to be executed are based on individual user's policies, expressed as AgentSpeak(L) plans.

In order to illustrate the essential features that need to be provided by the system and show how the BDI model can provide the required functionality, we implemented a pilot demonstration of the framework described here as the first step of implementing a fully functional simulation model based on a real communication system enhanced with context services. The demonstration consists of six agents, each managing the communication for one user. The users are situated in a hospital setting. Initially, the agents share a set of beliefs, which are stored in a relational database. This represents the CIS component of the architecture (Figure 15.1). In addition, each agent can have its own beliefs. Plans, representing user's policies, are defined for each agent in order to cover different scenarios that we envision occurring in the system.

The external events to which an agent must respond are request messages that come from the communication system (Figure 15.1). Such messages can be, for example, INVITE or REGISTER requests on a SIP-based architecture. In our demonstration of the framework, we simulate these requests (the environment) by feeding the agents, periodically, with events that correspond to real requests.

The plans implemented by the agents will affect the current state of the system. We show that agents will respond to contextual information that comes from the system and will be able to behave accordingly, taking into consideration the policies of the users that are defined as plans. We also want the agents to respond to system changes brought about by other agents. Finally, we show that communicating agents can negotiate in order to avoid possible conflicts.

In order to make sure there are no discrepancies between the agents' understanding of the system and of the actions they can perform on it, we need the assumption of a shared data model among all agents. The data model provides a device-independent description of the world. We defined an *entity-relationship data model* that provides the means to hold the context, as well as the semantics to describe the simulation domain. With this, the agent's beliefs will be represented as facts against the data model. The agent's desires, in the form of plans in AgentSpeak(L), are formulated in terms of entities, attributes, and relationships contained in the data model. The advantage of using this data model approach is that each entity class in the data model can be viewed as a finite domain, with the object instances themselves as the elements in that domain. The object's attributes can be used to form the basis for specifying constraints and reasoning in terms of the data model.

Each agent has a set of plans that models the user's policies. We will describe in detail the plans for handling a specific situation.

Dr. Smith is at this moment in his office. A colleague calls him from within the hospital. The presence system will decide, based on the doctor's location, that he is available on several devices: his PDA, his cell phone or his desktop phone. Since the caller is using a phone, and the PDA does not have voice capabilities, the choice is narrowed to two devices. Dr. Smith has a policy that designates his cell phone for calls from his family. Therefore, the presence system decides to transfer the call to Dr. Smith's desktop phone.

Before we start discussing the plans to realize this policy, it is worth discussing the initial belief base that is required in the running version of the program. The beliefs of the agent are based on the data model that we have defined. Thus, the belief base will contain information about the users in the system, the devices available to the users, the locations, and the activities that users are involved in. All these represent the CPI for the user. Relationships between these entities are modeled using databases that contain references to the entities involved in the relationships. (For example, PERSON_PERSON specifies the relationships between two persons identified by their unique ID number. Similarly, PERSON_DEVICE specifies the devices associated with a person.)

All this information is available for the agents in the Presence Directory component of the architecture. From the agent point of view, the information in the database is mapped into predicates. For example,

```
PERSON(john, user00001, TRUE, available, call,
     562-5800, eng, on_the_phone, j, doctor)
DEVICE(fix_phone, dev00001, ip_phone, mitel, 5010,
     eng, call, open, 563-2345)
LOCATION(or1, operating_room, canada, ottawa, 345
     carling, general hospital)
ACTIVITY(a1, meeting, work_related, 11:00:00)
PERSON_PERSON(PERSONID_1,PERSONID_2,RELATIONSHIP)
```

Agents are able to consult, insert, or delete values from this database by simply adding, deleting, or querying the facts as beliefs.

This policy is enforced using a plan for the situation in which incoming call arrives, from X, for Dr Smith. The triggering event for the plan is incoming_call, with a parameter specifying the caller.

```
+incoming_call(X):true < -
    !get_devices(DeviceList);
    !get_relationship(X,Relationship);
    !get_location(Location);
    !get_activity(Activity);
    !process_call(X, DeviceList, Relationship,
        Location, Activity).
```

The first thing to be done when a call request arrives is to obtain the list of devices where the doctor can be reached. This is done by adding the subgoal get_devices, which is achieved using the following plans:

```
+!get_devices(DeviceList): true
< - .findall(X, device(X,Y,Z,_,_,_,"call",_,_),
    DeviceList); ?name(N);
?person(N,ID,_,_,_,_,_,_,_,_);!get_user_devices
    (DeviceList,ID,UserDeviceList).
+!get_user_devices(DeviceList,ID,UserDeviceList) < -
!get_u_devices(DeviceList,ID,[],UserDeviceList).
+!get_u_devices([],ID,L,L).
+!get_u_devices([D|T],ID,L0,L): device(D,
    Did,_,_,_,_,_,_,_) & person_device(ID,Did)
        < -!get_u_devices(T,ID,[D|L0],L).
+!get_u_devices([D|T],ID,L0,L)
        < - !get_u_devices(T,ID,L0,L).
```

First, a list of all devices having the capability "call" is obtained by querying the device table. Recall that this is part of the current context for the user, which is stored in the Presence Directory and is available for the agents to be consulted and queried. After that, the subgoal get_user_devices determines which devices from this list are associated with the user.

The next thing to be done is to determine the relationship that the caller has with the user. This is done by verifying if there is information in the PERSON_PERSON table in the Presence Directory that contains both the ID of the user and of the caller. In case there is no such information, the relationship is said to be unknown.

```
+!get_relationship(X,R):
        name(N)&person(N,ID,_,_,_,_,_,_,_,_) & person
        (X,Xid,_,_,_,_,_,_,_,_) & person_person
        (ID,Xid,R) < - true.
+!get_relationship(X,R) < - R =  "unknown".
```

The current location of the user needs to be determined as well. If there is no information about the user's location, the location is set to unknown. In a similar fashion, the activity that the user is currently engaged in is determined.

```
+!get_location(LName): name(N)&person
     (N,ID,_,_,_,_,_,_,_) & person_location
     (ID,L) & location(L,LName,_,_,_,_) < - true.
+!get_location(L) < - L = "unknown".
+!get_activity(ANAme): name(N)&person
     (N,ID,_,_,_,_,_,_,_)& person_activity(ID,A) &
     activity(A,AName,_,_) < - true.
+!get_activity(A) < - A = "unknown".
```

Having all the information, the actual routing of the call, based on this information, is done by the plan triggered by the addition of the subgoal process_call. Certain conditions are verified and the action to be executed depends on this conditions.

```
+!process_call(X, DeviceList, Relationship,
     Location, Activity): Location == "office" &
     Relationship == "family" < - ring_mobile.
+!process_call(X, DeviceList, Relationship,
     Location, Activity): Location == "office" &
     Relationship == "colleague" < - ring_fixed_
     phone.
```

As stated in the policy, if the doctor is in the office and a member of his family calls, then the call should be routed to his mobile phone. If the call is from a colleague, the fixed phone in his office should ring. The default plan, which is applicable when none of the conditions tested are true, states that his mobile phone should take all other calls.

```
+!process_call(X, DeviceList, Relationship,
     Location, Activity): true < - ring_mobile.
```

In a similar fashion, a set of plans is defined for each agent in order to cover all the policies for the user it represents.

AgentSpeak(L) includes a mechanism that allows agents to communicate, thus sharing plans and consulting each other about the content of their beliefs base. We found that the power of this mechanism can be used to handle conflicting policies.

For illustrative purposes, we use an example with two common features, Originating Call Screening (OCS) and Call Forwarding (CF). These are two classical features in any telephony system. OCS forbids calling numbers on a screening list, while CF forwards incoming calls to another number.

A conflict (feature interaction) occurs if some user *A*, whose OCS screening list includes another user *X*, calls user *B*, who forwards calls to *X* through CF, thus overruling *A*'s policy.

In order to illustrate how this conflict is solved in our simulator, let us assume that Bob has OCS and forbids any calls to Charles and Alice has CF and forwards her call to Charles.

Thus, agent bob will have in its beliefs set a line saying:

```
ocs(charles).
```

Similarly, agent alice will have a belief:

```
call_forward(charles).
```

When Bob tries to call a number, the event that is generated is dial(X), where X is the person that he wants to reach. The event will trigger the execution of a plan. It is checked if the person that is called is on the screening list and only if is not the call is completed.

```
+dial(X): ocs(X)< - .print("You are forbidden to
    call ", X).
+dial(X): true< - .print("Inviting ",X);
    .send(X,tell,incoming_call(bob)).
```

At the other end, when Alice (who forwards all her calls to Charles) receives an incoming call, her agent checks if the call should be forwarded. If this is affirmative, then it will ask the originator of the call if this can be done (it actually asks if the person where the call is about to be forwarded is on the originator's screening list) and upon confirmation, the call will get through.

```
+incoming_call(X): true < - ?call_forward(F);
    .send(X,askIf,ocs(F),Answer); .print("answer
    from a1: ", Answer);
    !ans(Answer,X,F).
+!ans(true,X,Y,F): true < - .print ("not allowed
    to connect ", X, " to ", F).
+!ans(false,X,Y,F): true < - .print("forwarding
    call from ", X, " to", F);
    .send(F,tell,incoming_call(X)).
```

This simple example shows how the mechanism for agent communication can be used in negotiation of conditions and preferences for users and allows for resolution of conflicts. The introduction of context allows for much richer policies that may handle calls depending on presence, availability, role, capability, call type, or call content. Conflicts can occur also between these policies and the built-in mechanism for agent communication in AgentSpeak(L) allows for easy resolution.

This approach has been applied to other features and combinations of features not described here.

15.7 RELATED WORK

In recent years, there has been a lot of interest in service integration, resulting from the increasing interest of users to customize their services. The Universal Inbox project [31] defines an architecture for building personal and service mobility features. The Mobile People Architecture [33] is an architecture based on a Personal Proxy for achieving person-level routing.

Mercury [34] is a system that supports unified communication. It allows a person to initiate a conversation with another party using any available device. The system routes a call with consideration of the device the other party prefers to use in a given context. While Mercury uses SIP as the underlying mechanism for creating, maintaining, and terminating sessions, our goal was to make our architecture protocol independent. The functional entities of our design are independent of the underlying communication mechanisms. Thus, our solutions are compatible with both SIP and H.323.

With respect to Jason and its applications, we can report the work presented in Ref. [34]. The authors present a platform for multiagent-based social simulation. The platform is called MAS-SOC, and the approach to building multiagent-based simulations with it involves the use of Jason.

15.8 CONCLUSIONS

We have proposed an agent-oriented architecture for the provision of enhanced services for users of personal communications systems and ubiquitous computing, as well we have focused on the use of AgentSpeak(L) for the double role of implementation and policy language, following important developments in the area of agent-oriented programming and multiagent systems. Communication and telephony are areas of application where, to the best of our knowledge, the BDI architecture has not been used, yet their complexity requires the sophisticated control that can be readily expressed in AgentSpeak(L). More than other BDI models, AgentSpeak(L) has an exact notation, as well as clear and precise logical semantics, which resulted in the successful implementation of its abstract interpreter [27]. AgentSpeak(L) enables an elegant specification of the BDI agents and allows to model how agents can search and expand planned actions in order to select alternatives. In addition, it allows interagent negotiation for the resolution of interactions.

Another contributions of our work is our concept of "Consolidated Presence Information" [2]. Consolidating all the information that is available for users and their environment provides a unified view of the status of the user at a given time. The architecture allows real-time use of this information, thus simplifying the implementation of a large spectrum of complex services.

ACKNOWLEDGMENTS

This work was funded in part by the Natural Sciences and Engineering Research Council of Canada. We are grateful to Prof. Babak Esfandiari and Prof. Ahmed Karmouch for detailed remarks on earlier versions of this manuscript.

REFERENCES

1. K. Turner, E. Magill, and D. Marples. *Service provision. Technologies for Next Generation Communications. Wiley Series in Communications Networking & Distributed Systems*, 2004.
2. R. Plesa and L. Logrippo. Enhanced communication services through context integration. In T. Magedanz, A. Karmouch, S. Pierre, and I. Venieris, editors. *Mobility Aware Technologies and Applications*, MATA, Montreal, Short papers, Vol. 1–5, 2005.
3. R. Plesa and L. Logrippo. An agent-based Architecture for contex-aware communications. In *Proceedings of the 21st International Conference on Advanced Information Networking and Applications Workshops*, PCAC-07 Niagara Falls, May 2007, IEEE Press, Vol. 2, pp. 133–138.
4. M. Day, J. Rosenberg, and H. Sugano. *A Model for Presence and Instant Messaging*. IETF RFC 2778, February 2000.
5. J. Rosenberg. *SIP Extension for Presence*. IETF Internet Draft, 2001.
6. T. Moran and P. Dourish. Context-aware computing. Introduction to the special issue on context-aware computing. *Hum. Comput. Interact.*, 2001, 16(2–3).
7. N. Ryan, J. Pascoe, and D. Morse. Enhanced reality fieldwork. The context-aware archeological assistant. http://www.cs.ukc.ac.uk/projects/mobilcomp/FieldWork/Papers/CAA97/ERFldwk.html, 1997.
8. A. Schmidt, M. Beigl, and H. Gellersen. There is More to Context than Location, Computers and Graphics, 23/6, 893–902, 1999.
9. H. Sugano and S. Fujimoto. Presence Information Description format, IETF RFC Vol. 3863, 2004
10. J. Rosenberg and H. Schulzrinne. SIP: The Session Initiation Protocol, IETF RFC Vol. 2543, 1999
11. H.323 Information Site. http://www.packetizer.com/voip/h323/.
12. A. Wennlund. Context-Aware wearable device for reconfigurable application networks, Department of Microelectronics and Information Technology (IMIT), Royal Institute of Technology (KTH), Master of Science Thesis at the Royal Institute of Technology (KTH), Stockholm, Sweden, 2003
13. G. Chen and D. Kotz. Solar: a pervasive-computing infrastructure for context-aware mobile applications, Department of Computer Science, Dartmouth College, Hanover, NH, USA, 2002.
14. D. Amyot. Introduction to the user requirements notation: learning by example. *Comput. Networks*, 42 (3), 285–301, 2003.
15. J. Lennox and H. Schulzrinne. Call Processing Language Framework and Requirements, IETF Internet Draft CPL-Framework-02, 2002.
16. X. Wu and H. Schulzrinne. LESS: Language for End System Services in Internet Telephony. IETF Internet Draft, 2005
17. S. Reiff-Marganiec and K. J. Turner. APPEL: The ACCENT project policy environment/language, Technical Report CSM-161, University of Stirling, UK, 2004
18. A. S. Rao and M. P. Georgeff. BDI-agents: from theory to practice. In *Proceedings of the First International Conference on Multiagent Systems*, San Francisco (ICMAS-95), pp. 312–319, 1999.
19. M. J. Wooldridge. Reasoning about Rational Agents. Intelligent Robots and Autonomous Agents Series, The MIT Press, 2000.
20. C. Heinze and S. Goss. Human performance modelling in a BDI system. In *Proceedings of the Australian Computer Human Interaction Conference*, 2000.
21. N. Howden, R. Ronnquist, R. Hodgson, and A. Lucas. JACK intelligent agents: summary of an agent infrastructure. In *Proceedings of the 5th International Conference on Autonomous Agents*, 2001.
22. M. Georgeff and A. Lansky. Procedural knowledge. *Proc. IEEE* 74(10): 1383–1398, 1986.

23. M. Huber. Jam: A BDI-theoretic mobile agent architecture. In *Proceedings of the Third International Conference on Autonomous Agents*, pp. 236–243, 1999.

24. A. S. Rao and M. Georgeff. Decision procedures for BDI logics. *J. Logic Comput.* 8; 293–342, 1998.

25. M. Winikoff. AgentTalk Home Page. http://goanna.cs.rmit.edu.au/~winikoff/agenttalk, 2001.

26. M. d'Inverno, K. V. Hindriks, and M. Luck. A formal architecture for the 3APL agent programming language. In J. P. Bowen, S. Dunne, A. Galloway, and S. King, editors. *Proceedings of the 1st International ZB Conference*, Springer Verlag, pp. 168–187, 2000.

27. A. S. Rao. AgentSpeak(L): BDI agents speak out in a logical computable language. In W. V. de Velde and J. W. Perram, editors. *Agents Breaking Away*. Springer Verlag, LNAI Vol. 1038, pp. 42–55, 1996.

28. Jason - http://jason.sourceforge.net

29. M. d'Inverno, D. Kinny, M. Luck, and M. Wooldridge. A formal specification of dMARS. In M. P. Singh, A. Rao, and M. Wooldridge, eds. *Proceedings of the 4th International ATAL Workshop, Springer Verlag, LNAI*, Vol. 1365, pp. 155–176, 1997.

30. G. Caire, N. Lhuillier, and G. Rimassa. A communication protocol for agents on handheld devices. In *Proceedings of the Workshop on Ubiquitous Agents on Embedded, Wearable and Mobile Devices*, 2002.

31. Z. Maamar, W. Mansoor, and Q. H. Mahmoud. Software agents to support mobile services. In *Proceedings of the First International Joint Conference on Autonomous Agents and Multi-Agent Systems*, pp. 666–667, 2002.

32. B. Raman, R. Katz, and A. Joseph. Universal inbox: providing extensible personal mobility and service mobility, In *Proceedings of the 3th Workshop on Mobile Computing Systems and Applications*, 2000.

33. M. Roussopoulos, et. al. Personal-level routing in the mobile people architecture. In *Proceedings of the USENIX Symposium on Internet Technologies and Systems*, 1999.

34. H. Lei and A. Ranganathan. Context-aware unified communication. In *2004 IEEE International Conference on Mobile Data Management*, pp. 176–186, 2004.

35. R. H. Bordini, A. C. Rocha Costa, J. F. Hubner, A. F. Moreira, F. Y. Okuyama, and R. Vieira. MAS-SOC: a social simulation platform based on agent-oriented programming. *J. Artifi. Soci. Soc. Simul.* 8, (3), 2005.

Part V

Mobile Intelligence Security

Chapter 16

MANET Routing Security

Carlton R. Davis, Claude Crépeau, and
Muthucumaru Maheswaran

School of Computer Science, McGill University, Montréal, Quebec, Canada

16.1 Introduction   345
16.2 MANETs Routing Approaches   346
16.3 Secure MANET Routing Proposals   353
16.4 Robust Source Routing   360
16.5 Summary   376
References   376

16.1 INTRODUCTION

In an ad hoc network, all the nodes may not be within the transmission range of each other; hence, nodes are often required to forward network traffic on behalf of other nodes. Consider for example the scenario in Figure 16.1, if node *S* sends data to node *D*, which is three hops away, the data traffic will get to its destination only of *A* and *B* forward it. The process of forwarding network traffic from source to destination is termed routing.

Secure routing in mobile ad hoc networks (MANETs) has emerged as an important MANET research area. MANETs, by virtue of the fact that they are wireless networks, are more vulnerable to intrusion by malicious agents than wired networks. In wired networks, appropriate physical security measures, such as restriction of physical access to network infrastructures, can be used to attenuate the risk of intrusions. Physical security measures are less effective, however, in limiting access to wireless network media. Consequently, MANETs are much more susceptible to infiltration by malicious agents. Authentication mechanisms can help to prevent unauthorized

Mobile Intelligence. Edited by Laurence T. Yang, Agustinus Borgy Waluyo, Jianhua Ma,
Ling Tan, and Bala Srinivasan
Copyright © 2010 John Wiley & Sons, Inc.

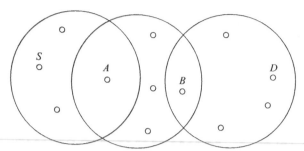

Figure 16.1 Multihop scenario.

access to MANETs. However, considering the high likelihood that nodes with proper authentication credentials can be taken over by malicious entities, there are needs for security protocols that allow MANET nodes to operate in potential adversarial environments.

In this chapter, we provide an in-depth review of routing approaches in MANET; following the review, we present a secure on-demand MANET routing protocol, we named robust source routing (RSR). RSR protocol has been presented in Ref. [11]; this chapter can be distinguished from Ref. [11], in that it provides a more comprehensive review of MANET routing approaches and gives additional analysis of RSR.

16.2 MANETs ROUTING APPROACHES

There are two general categories of MANET routing protocols: topology-based and position-based routing protocols. We present a brief overview of each group. Before proceeding, it is fitting to list some desirable qualitative properties of MANET routing protocols. This list is adopted from an Internet Engineering Task Force (IETF) MANET Working Group memo [10].

- Loop-free: It is desirable that routing protocols prevent packets from circling around in a network for arbitrary time periods.

- Demand-based operation: In order to utilize network energy and bandwidth more efficiently, it is desirable that MANET routing algorithms adapt to the network traffic pattern on a demand or need basis rather than maintaining routing between all nodes at all time.

- Proactive operation: This is the flip-side of demand-based operation. In cases where the additional latency—which demand-based operations incur—may be unacceptable, if there are adequate bandwidth and energy resources, proactive operations may be desirable in these situations.

- "Sleep" period operation: It may be necessary—for reasons such as the need for energy conservation—for nodes to stop transmitting or receiving signals for arbitrary time periods. Routing protocols should be able to accommodate sleep periods without adverse consequences.

- Security: It is desirable that routing protocols provide security mechanisms to prohibit disruption or modification of the protocol operations.

16.2.1 Position-Based Routing Protocols

Position-based routing protocols employ nodes' geographical position to make routing decisions. In order to utilize a position-based routing protocol, a node must be able to ascertain the geographical position of itself and that of all the nodes it wishes to communicate with. This information is typically obtained via global positioning system (GPS) and location services.

The emphasis of this chapter is on topology-based rather than position-based routing; however, we give a brief overview of basic position-based routing algorithms.

16.2.1.1 Greedy

The greedy routing algorithm was developed by Finn [15]. In the greedy forwarding approach, a node selects for the next hop, the node that is closest to the destination of the packet. In Figure 16.2, if S has data traffic to send to D, which is outside of its transmission range, greedy forwarding dictates that S sends the traffic through B since B is the node within S transmission range which is closest to the destination node D.

16.2.1.2 Compass

The compass routing algorithm was developed by Kranakis et al. [27]. In the compass routing scheme, a node S, which has data traffic to send to a destination node D, forwards the traffic to its neighbor N, which has the smallest angle $\angle NSD$, where N is a neighboring node to the forwarding node S and D is the destination. So, for example in Figure 16.3, S forwards the traffic for D to A since the angle $\angle ASD$ is smaller than any other angle $\angle NSD$, where N is a node within S transmission range. Notably, Stojmenovic and Lin [48] showed that the compass algorithm is not loop-free.

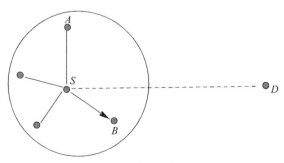

Figure 16.2   Greedy forwarding.

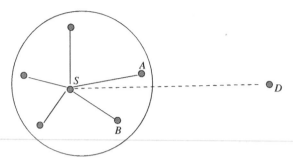

Figure 16.3   Compass forwarding.

16.2.1.3 *Randomized Compass*

The randomized compass routing algorithm [5] is a variation of the compass algorithm that avoids loops with random decisions. Consider a line between a node S and a destination node D. The random compass forwarding approach chooses the next hop for a packet by randomly selecting between the nodes N_i and N_j that has the smallest angle $\angle N_i SD$ and $\angle N_j SD$ between the imaginary line $\overline{NS}$ (between a node N and the forwarding node S) and $\overline{SD}$, above and below the imaginary line $\overline{SD}$, respectively. So, for example in Figure 16.3, node S would randomly select node A or B for forwarding packets to D since $\angle ASD$ is the smallest angle (between a line connecting S and a node that is within S transmission range, and the line $\overline{SD}$) above the line $\overline{SD}$ and $\angle BSD$ is the smallest angle below the line $\overline{SD}$.

16.2.1.4 *Most Forwarded Within Radius*

Takagi and Kleinroc proposed most forwarded within radius (MFR) [49]. Consider an imaginary line $\overline{SD}$ between a node S and a destination node D; in MFR forwarding, S forwards data traffic for D to a node A, which maximizes the progress along the imaginary line $\overline{SD}$. A is, therefore, the node that minimizes the dot product $\overline{DA} \cdot \overline{DS}$. So, in Figure 16.4, S forwards packets for D to A since A is the node within S transmission range that provides the most progress along the line $\overline{SD}$.

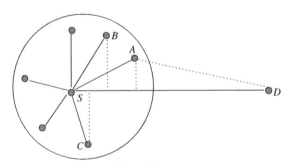

Figure 16.4   MFR forwarding.

16.2.2 Topology-Based Routing Protocols

There are two major categories of topology-based MANET routing protocols: on-demand and proactive protocols. In the section, we briefly describe some of the more prominent existing MANET topology-based routing protocols. We commence with proactive protocols.

16.2.2.1 Proactive Protocols

Proactive protocols are also referred to as periodic protocols. The most prominent proactive MANET routing protocol is dynamic destination-sequenced distance-vector routing (DSDV) [38]. DSDV utilizes the classical distributed Bellman–Ford distance-vector algorithm [3, 23]. In distance-vector algorithms, each node i, for each destination x, maintains a set of distances $\{d_{ij}^x\}$, where j ranges over the neighbors of i. The distances are typically interpreted as the number of hops from i to x via the given neighbor j. Node i designates a neighbor k as the next hop for a packet if d_{ik}^x equals $\min_j \{d_{ij}^x\}$. The succession of the next hop chosen in this manner leads to x along the shortest path. In order to keep the estimated distances up to date, each node monitors the costs of its outgoing links and periodically broadcasts to each of its neighbors, its current estimate of the shortest path to all other nodes in the network. It is well known that the distance-vector routing algorithm outlined above is not loop-free [9]. The main cause of routing loop formation is the fact that nodes choose their next hops in a distributed fashion based on the information that may be stale and therefore incorrect. DSDV avoids the distance-vector looping problem by tagging each routing distance info with a sequence number so nodes can quickly distinguish new routes from stale ones and consequently avoiding the formation of routing loops.

In DSDV routing, each MANET node maintains a routing table that is used for making routing decisions. A DSDV routing table lists all available destinations and the number of hops to each. Each routing table entry is tagged with a sequence number originated from the destination node. DSDV protocol requires each network node to advertise (via broadcasting or multicasting) to each of its current neighbors, its own routing table. Additionally, each node is required to transmit updates immediately when significantly new information is available. The routing information data a node broadcasts, contain a new sequence number and and the following info for each new route:

- the destination's address,
- the number of hops from the source to the destination, and
- the sequence number of the information received regarding the destination, as originally stamped by the destination.

The MANET nodes use the advertised routing tables info and the transmitted updates to update their routing tables, which is utilized by the distance-vector algorithm outlined above to determine the next hop for a packet.

16.2.2.2   On-Demand Protocols

On-demand protocols are also referred to as reactive protocols. Unlike proactive protocols, which seek to maintain routes to all destination in a MANET, on-demand protocols establish routes on a per-need basis. There are a larger collection of existing on-demand protocols compared with proactive protocols. We present brief description of some of the more widely known on-demand protocols below.

Dynamic Source Routing   Dynamic source routing (DSR) was developed by Johnson and Maltz [22]. Its basic operation is as follows: when a node S has a packet to send to a destination D, S checks its routing cache for an entry containing a path to D. If there is no such entry, S broadcasts a routing request (RREQ) packet containing the initiator address, a unique request id , the destination address, and a route record field. The latter is used to accumulate the sequence of hops the RREQ packet takes as it propagates through the network. When a node n_i receives a RREQ packet, if it has previously seen a RREQ packet with the same initiator address and request id, it discards it; otherwise, if n_i is not the destination and its routing cache does not contain a valid path to D, it records the initiator address and request id, appends its address to the route record and forwards the packet. If n_i is the destination, it returns a copy of the route record in a route reply (RREP) packet to the initiator. If n_i is not the destination but it knows of a path to D, it sends a copy of the path in a RREP packet back to the source of the RREQ packet. On receiving the RREP packet, S records the ascertained route to D in its routing cache, writes the route in the source route field of the packet header, and sends the packet to the node that is the next hop in the path to D. The intermediate nodes on the path to D will likewise use the route recorded in the source route field of the packet header to determine the address of the next hop they should forward the packet to, until the packet eventually reaches the intended destination.

Signal Stability-Based Adaptive Routing   Signal stability-based adaptive routing (SSA) was developed by Dube et al. [14]. SSA utilizes signal strength and stability of individual MANET nodes as routing selection criteria. The rational being (in the authors' view) that links that exhibit the strongest signal for the maximum amount of time lead to longer-lived routes and less route maintenance. In SSA routing, a source S sends out a route discovery request when it has data to send to a destination D that is not in its routing table. S broadcasts the route request to all its neighbors. Each neighboring node propagates the route request if it received it over a strong channel and the request has not been propagated previously. A channel is characterized as strong or weak based on the average signal strength at which the packets are exchanged between the nodes at either end of the channel. The route search packet continues to traverse the network until it reaches the destination, and it stores the address of each intermediate node it traversed. The first route search packet that arrives at the destination D is selected and a route reply packet is constructed and returned to S using the selected route. Each intermediate node in the selected route, on receiving the route reply packet, includes the new next-hop, destination pair in its routing table.

Associativity-Based Routing C-H Toh developed the associativity-based routing (ABR) [50]. ABR utilizes the observation that a mobile node's association with its neighbor changes as it migrates and its transiting period can be identified by the associativity "ticks." Associativity ticks are updated by the mobile node's data-link protocol, which periodically transmits beacons identifying itself and updates its associativity ticks in accordance with the mobile nodes in its neighborhood. A mobile node exhibits high associativity ticks (high association stability) with its neighbors when it is in a state of low mobility. Conversely, a state of high mobility is associated with low associativity ticks. In ABR routing, a node S that desires a route to a destination D broadcasts a broadcast query (BQ) message that propagates through the MANET in search of a node that has a route to the given destination. When an intermediate node n_i receives a BQ message it has not previously seen, n_i appends its address, associativity ticks with its neighbors, its relaying load, link propagation delay, and its hop count to the appropriate fields of the BQ and broadcasts the BQ to its neighbors. The next succeeding intermediate node will then erase its upstream node's neighbors' associativity ticks entries and retain only those concerning itself and its upstream nodes. When the destination node D receives the BQ packets, it selects a route based on the following selection criteria: routes consisting of nodes with higher associativity ticks has higher preference even over routes with smaller number of hops. For routes with equal number of associativity ticks, the route with the smaller hop count is selected. If the routes have the equal number of associativity ticks and hop counts, one of the routes is randomly selected. The selected route is used to construct a REPLY packet and returned to the source S via the selected route. The intermediate nodes on the route from D to S will consequently mark their routes to D as valid and subsequently inactivate all other possible routes to D.

Temporally-Ordered Routing Algorithm Temporally-ordered routing algorithm (TORA) was developed by Park and Corson [35]. It is a highly adaptive multipath, loop-free, distributed routing algorithm that was designed for highly dynamic MANET environments. A key design concept of TORA is the localization of routing control messages to a small set of nodes near the topological change. In TORA routing, each node, at any given point in time, has an associated ordered quintuple consisting of the following elements: (1) a logical time of link failure, (2) the unique ID of the node, which defined the new reference level, (3) a single bit that is used to divide each of the unique reference level into two unique sublevels, (4) a propagation-ordering parameter, and (5) the unique ID of the node. Conceptually, the quintuple represents the height of a node defined by a reference level and a delta with respect to the reference level. The reference level is represented by the first three values of the quintuple, while the last two values represent the delta. Each node i (other than the destination) maintains its height H_i, which is initially set to NULL, $H_i = (-, -, -, -, i)$. The height of the destination is always ZERO, $H_{DID} = (0, 0, 0, 0, Did)$, where DID

represents the destination ID. In addition to its own height, each node maintains an height array with an entry $HN_{i,j}$ for each of its neighbor j. Each node i also maintains a link-state array for each of its links. The state of a link is determined by its height H_i and HN_i and is directed from higher node to lower node.

When a node requires a route to a destination D, it sends out a QRY packet. When a node i receives a QRY packet it has not previously seen, it reacts as follows: (a) i rebroadcasts the QRY packet if it has no downstream links; (b) if the receiving node has at least one downstream link and its height is NULL, it sets its height to the minimum height of it non-NULL neighbors and broadcasts a UDP packet (which consists of a destination ID and the height of the node i that is broadcasting the packet); (c) if the receiving node has at least one downstream link and its height is non-NULL, it first compares the time the last UDP packet was broadcast with the time the link over which the QRY packet arrived was active. If the link became active prior to the broadcasting of the UDP packet, i discards the QRY; otherwise, i broadcasts a UDP packet. When a node i receives a UDP packet it has not previously seen from a neighbor j, i updates the entry $HN_{i,j}$ in its height array with the height contained in the UDP packet, then does the following: if its height is NULL, i sets its height to the minimum height of its non-NULL neighbor, updates all the entries in its link-state array and then rebroadcasts the UDP packet that contains its new height. The process (broadcasting of QRY and UDP packets) continues until a directed acyclic graph (DAG) rooted at the destination (i.e., the destination is the only node with no downstream links) is formed. The DAG represents a route from the source S to the destination D.

Ad-Hoc On-Demand Distance Vector Ad-hoc on-demand distance vector (AODV) routing was designed by Perkins and Royer [39]. Its operation can be summarized as follows: Each node using AODV maintains a route table entry for each destination of interest. A route table entry contains the destination D, next hop, number of hops to D, sequence number of the destination, and the expiration time for the route table entry. When a node S has a packet to send to a destination D, S checks its routing table for an entry containing D as the destination with a sequence number equal to or greater than the last known destination sequence number of D. If there is no such entry, S broadcasts a route request (RREQ) packet containing the source address, the source sequence number, broadcast id, destination sequence number, and hop count. The source sequence number and the broadcast id are separate counters that are maintained by each node. A node increments its broadcast id counter each time it constructs a new RREQ packet, whereas the node's sequence number counter is incremented less frequently. The destination sequence number is the last known sequence number of the destination. When a node n_i receives a RREQ packet it has not previously seen, it sets up a reverse path to the source by recording the address of its neighbor from which it received the first copy of the RREQ. If n_i is not the destination and its routing table does not contain an entry for D, it increments the hop count and rebroadcasts the RREQ packet to its neighbors. If n_i, however, is the destination or if its routing table contains an entry with D as its destination with a destination sequence number that is equal to or greater than the destination sequence number in the RREQ packet, it constructs a route reply (RREP) packet and unicasts it to the neighboring node it received the RREQ from. A RREP packet contains the source address, destination address, destination sequence number, hop count, and lifetime. When an intermediate node receives a RREP packet, it updates its routing table with the information the RREP contains and then unicasts the information to the neighbor from which it received the first copy of the associated RREQ packet. The process continues until the RREP packet gets to S. S can now forward its packet to the next hop on the path to D.

16.3 SECURE MANET ROUTING PROPOSALS

The protocols we reviewed in Section 16.2.2 were designed for nonadversarial environments, where the node within a network are nonmalicious, unselfish, and well-behaving. The reality, however, is that in any network, there are likely to be malicious or selfish, miss-behaving nodes that have intentions of disrupting the routing protocol. Security mechanisms are therefore necessary to mitigate against these eventualities. This section reviews some of the routing security schemes that have been proposed to address the security shortcomings of these protocols. For the purpose of the review, we categorized the existing secure MANET routing proposals into the following categories: basic routing security schemes, trust-based routing schemes, incentive-based schemes, and schemes that employ detection and isolation mechanisms. Below, we briefly describe a selection of schemes that fall in these categories.

16.3.1 Basic Routing Security Schemes

The routing schemes that fall in this category provide authentication services that guard against modification and replaying of routing control messages, but they do not attempt to provide solutions for issues such as the dropping of packets by selfish or malicious nodes. We commence the review with one of the earlier proposals.

Binkley and Trost presented an authenticated link-level ad hoc routing protocol [4] that was integrated into the Portland State University implementation of Mobile-IP[1][37]. The protocol uses ICMP router discovery message [13] to discover mobile-IP nodes. It extended the ICMP router discovery packet format to include the media access control (MAC) and IP address of the sender and authentication info that can be used to verify the broadcast beacon. The protocol requires nodes to have shared secret keys for generating message authentication codes that are used to authenticate the routing control messages.

Venkatraman and Agrawal introduced an inter-router authentication scheme [51] for securing AODV [39] routing protocol against external attacks (such as impersonation attacks, replaying of routing control messages, and certain denial of service

---

[1] Mobile-IP is a network-layer protocol that enables a mobile node to retain a fixed IP address even when it changes its point of connectivity to the Internet.

attacks). The scheme is based on the assumption that the nodes in the network mutually trust each other and it employs public key cryptography for providing the security services. The integrity of routing requests are ensured by the originating node hashing the messages and signing the resulted message digest. Recipients of a route request can check its authenticity and integrity by computing the hash of a the message using the agreed upon hash function, comparing the computed hash with that attached to the message, and verifying the signature. "Strong authentication" is provided for adjacent pair of nodes that transmit route replies. The strong authentication procedure is as follows: a node n_i sends a pre-reply plus a random challenge (challenge1) to a neighbor it wishes to send a reply. The neighbor n_j that received the pre-reply generates a random challenge (challenge2), encrypts challenge1 with n_i's public key, and sends the encrypted challenge along with challenge2 to n_i. When n_i receives this message, it encrypts challenge2 with n_j's public key and sends the route reply along with the encrypted value of challenge2 to n_i. This procedure is designed for detecting nodes that attempt to impersonate other nodes.

Papadimitratos and Haas presented secure routing protocol (SRP) [34]. SRP assumes the existence of a security association between a node initiating a route request query and the sought destination. The basic operation is as follows: a source node S initiates a route discovery by constructing and broadcasting a route request packet containing a source and destination address, a query sequence number, a random query identifier, a route record field (for accumulating the traversed intermediate nodes), and the message integrity codes (MICs) of the random query identifier, computed using HMAC [28] and the secret key shared between the S and the destination. Intermediate nodes relay the route request packet so that one or more query packet(s) arrive(s) at the destination. When the route requests reach the destination D, D verifies that (a) the MIC is indeed that of the random query identifier and (b) the sequence number is equal to or greater than the last known sequence number from S. If both (a) and (b) hold, D constructs a corresponding route reply packet containing the source, destination, the accumulated route in the route record field of the request query, the sequence number, the random query identifier, and the computed MIC of the above. D then sends the route reply to S using the reverse path in the route record field. When S receives a route reply packet, it validates the info it contains and verifies the computed MIC. If all is well, it uses the ascertained route to communicate with D.

Hu et al. proposed the secure efficient ad hoc distance vector routing protocol (SEAD) [18]. SEAD is a secure proactive protocol based on the design of DSDV [38]. SEAD uses one-way hash chains [29] for authenticating the hop count values in advertised routes and routing updates. For the authentication of the sender of routing update messages, SEAD allows authentication to be done using broadcast authentication mechanisms such as TESLA [40], HORS [43], or TIK [19], which require the network nodes to have time-synchronized clocks. Alternatively, SEAD allows message authentication codes to be used to authenticate the sender of routing update messages; however, this is based on the assumption that shared secret keys are established among each pair of nodes.

Zapata presented secure AODV (SAODV) [55–57]. SAODV uses two mechanisms to secure AODV: digital signatures to authenticate nonmutable fields of the routing control messages and one-way hash chains (as is the case for SEAD, outlined above) to secure hop count information.

Hu et al. proposed a routing security scheme called Ariadne [17], which is based on the design of Ref. [22]. Ariadne uses message authentication code for authenticating routing control messages and requires time-synchronization hardware for synchronizing the release of the secret keys used for generating the message authentication codes.

Sanzgiri and Dahill presented ARAN [46]. ARAN uses digital certificates to secure the routing control messages. In ARAN route discovery phase, a source node S constructs a route discovery packet (RDP), signs it, attaches its certificate and broadcasts it to its neighbors. When a node A, which is a neighbor of S, receives the RDP message, if it has not previously seen this message, it verifies the signature using the attached certificate, signs the RDP message, attaches its certificate, and broadcasts it to its neighbors. An intermediate node B, which is a neighbor of A, on receiving the RDP message, validates the signature using the attached certificate. B then removes A's certificate and signature, records B as its predecessor, signs the message, and broadcasts it to its neighbors. The process continues in this manner until a RDP message arrives at the destination D. D selects the first RDP message it received, uses it to construct a reply (REP) packet and unicasts it to S using the reverse path. Each node on the reverse path back to S validates its predecessor signature using the attached certificate, removes the signature and the certificate (if the predecessor's certificate does not belong to the destination node D), signs the packet, attaches its certificate, and forwards the packet to the next hop. Eventually, S should receive the REP with the route it seeks.

Hu et al. presented a mechanism called packet leashes for detecting and defending against wormhole attacks [19]. In wormhole attacks, an attacker receives packets at one point in a network, tunnels them to another point in the network, and replays them into the network from that point. The authors proposed two types of packet leashes: geographical leashes and temporal leashes. Geographical leashes require a node to know its own geographical location and all nodes must have loosely synchronized clocks. However, temporal leashes require all nodes to have tightly synchronized clocks. The leash mechanisms add necessary fields to a packet—for example, the time the packet was sent and the sender's geographical location (for geographical leashes)—which allows the receivers to validate whether a node is in its transmission range or not. The authors also proposed a secure broadcast scheme called TIK, which can be used to secure the packet leash mechanisms.

16.3.2 Trust-Based Routing Schemes

The routing security schemes that fall in this category assign quantitative or qualitative trust values to the nodes in the network, based on the observed behavior of the nodes in question. The trust values are then used as additional metrics for the routing protocols. We commence the review with one of the earlier protocols.

Yi et al. proposed security-aware ad hoc routing SAR [54]. SAR classifies nodes based on their trust level. Nodes that have the same classification share a secret group key. In a route discovery process, the source node S can stipulate the minimum security requirement a node must have in order to be an element in the routing path from S to a destination D. S can enforce the stipulation by encrypting the route request packet with the shared key associated with the specified security level. This approach has its virtues; however, key sharing can be problematic, considering the possibility that malicious agents can take over nodes with high-security classifications and gain access to the secret group keys.

Yan et al. proposed a trust model that assigns quantitative trust values to nodes based on the observed behavior of the nodes [53]. The application of this trust evaluation mechanism in routing schemes is similar in principle to SAR [54]. Unlike SAR though, Yan et al.'s proposal does not suggest a means whereby a source node S can prevent a node—which does not meet the trust level requirement—from being on a routing path from S to a given destination.

Pirzada and McDonald presented a model for trust-based communication in ad hoc networks [41]. In this model, each node passively observe other nodes and assigns quantitative values (which range from 0 to +1) to nodes based on observed behavior. The authors proposed an extension of DSR [22] that incorporates the trust model and utilizes trust as an additional routing metric. This scheme is susceptible to malicious accusation attacks in that malicious nodes can selectively drop packets and wrongfully accuse well-behaving nodes of misbehavior.

Nekkanti and Lee presented a trust-based adaptive on demand routing protocol [33]. The authors articulated that the most effective way of preventing certain routing attacks is to totally hide certain routing information from unauthorized nodes. In this regard, the main aim of their proposed scheme is to mask the routing path between a source and a destination from all other node. The scheme is based on AODV [39]. It stipulates that one of three possible encryption levels be applied to a route request packets (RREQ). The encryption levels are high encryption that requires a 128-bit key, low encryption that needs a 32-bit key, and no encryption. The security level of a node and the security level of an application determine which encryption level is utilized. The general idea is that the more trustworthy a node is, the less need there is to hide routing information from this node during a route discovery operation. A summary of the route discovery operation is as follows: A source node S that desires a route to a destination D constructs a RREQ packet. The RREQ has a field where the application can set the security level it requires. The source then utilizes the public key of the destination node D to encrypt (with the appropriate security level) the source ID field of the RREQ packet and broadcasts it to its neighbors. When an intermediate node receives a RREQ packet it has not previously seen, if it is not the destination, it adds its node ID to the packet, signs it then encrypts it using the the public key of D and broadcasts it to its neighbor. Eventually, an RREQ packet should get to D. On receiving an RREQ packet, D verifies the signatures, decrypts the encrypted fields, and verifies that the nodes in the path has the minimum required trust level. If these validation operations succeed, it constructs a route reply (RREP) packet and a flow-id and encrypts the RREP and the flow-id with the public keys of the nodes in the reverse path to S (in the order that the nodes should receive the RREP packet); then D signs the encrypted RREP and broadcasts it to its neighbors. When an intermediate node n_i receives the RREP it will attempt to decrypt it; if the decryption operation fails, n_i discards the packet; otherwise, it updates its routing table, removes its part of the RREP, and broadcasts it to its neighbor. Eventually, the RREP should get to the source S, which will verify the signature and decrypts the RREP to ascertain the route it seeks. This scheme provides a degree of anonymity for nodes in routing paths; but it does not provide protection against misbehaving nodes that selectively drop packets they agreed to forward.

Boukerche et al. proposed secure distributed anonymous routing protocol (SDAR) [6]. The main objective of SDAR is to allow trustworthy intermediate nodes to participate in routing without compromising their anonymity. SDAR utilizes a trust management system that assigns trust values to nodes based on observed behavior of the nodes, along with recommendation from other nodes. SDAR requires each node to construct two symmetric keys and shares one with its neighbors that have high trust values and the other with its neighbors that have medium trust values. When a node S desires to discover a routing path to a destination D, S constructs a routing request packet (RREQ), part of which is un-encrypted and the other part encrypted. The un-encrypted part of the RREQ contains necessary routing information such as the trust level requirement of the message and a one-time public key TPK. The encrypted part of the RREQ packet contains the destination ID, a symmetric key K_s generated by S and the private key TSK for the one-time public key TPK, plus other information. Part of the encrypted portion of the message is encrypted with the public key for the destination D and the other portion is encrypted with the symmetric key K_s. S then encrypts the entire packet with the shared key for the appropriate security level of the message and broadcasts it to its neighbors. When an intermediate node n_i receives the RREQ packet, it discards the message if it is not able to decrypt it. If n_i succeeds in decrypting the message, n_i adds its ID and a session key K_i then signs the portion it added and encrypts it with the one-time public TPK embedded in the un-encrypted portion of the RREQ packet; n_i then encrypts the entire message with the key (of the appropriate security) it shares with it neighbors and broadcasts the message. Eventually, the message should get to D, which decrypts the message with the appropriate keys. After verifying the signatures, D constructs a route reply (RREP) and encrypts it, first using the symmetric key K_s S attached, then encrypts it again using the session keys K_i's in the order that the corresponding intermediate node should receive the RREP packet. D then forwards the RREP to its neighbor. The neighbor that is the intended next hop will decrypt its portion of the packet and forwards it to its neighbors (one of which will be able to partly decrypt it). The process continues until the RREP gets to the source node S, which will be able to decrypt the entire packet and ascertain the route it seeks. The operation of SDAR is similar to that of Marti et al. [31] "Watchdog" operation and is therefore susceptible to the short comings—outline in Section 16.3.4 below—associated with Marti et al. scheme.

Li and Singhal proposed a secure routing scheme [30] that utilizes recommendation and trust evaluation to establish trust relationships between network entities. The scheme uses a distributed authentication model that operates as follows: each network node maintains a trust table that assigns a quantitative trust value to known network entities. If a node S desires to know the trust value of a node n_i and n_i is not in S trust table, S sends out a trust query message—to ascertain n_i's trust value—to all the trustworthy nodes in S trust table. When a node n_j receives the trust query message, if n_i is in its trust table, it sends the indicated trust value to S; otherwise it sends out a trust query message—requesting the trust value of n_i—to all the trustworthy nodes in its trust table. The process continues recursively until eventually a node that has n_i in its trust table forwards the trust value to the node that requested the info, which will in turn forward it to the node that sent it the trust query message, and so on, until eventually the response gets to S. S consequently uses the responses to compute a trust value for the node in question. This distributed authentication model is used to determine the trustworthiness of the network nodes. The end result being that nodes that are considered untrustworthy are excluded from routing paths. The scheme has its merits but malicious agents can thwart the scheme by dropping the trust query messages, and in so doing, renders the scheme ineffective.

16.3.3   Incentive-Based Schemes

In this section, we present a brief description of proposed schemes that attempt to stimulate cooperation among selfish nodes by providing incentives to the network nodes.

Buttyán and Hubaux proposed an incentive-based system for stimulating cooperation in MANETs [7]. The scheme requires each network node to have a tamper-resistant hardware module, called security module. The security module maintains a counter, called nuglet counter, which decreases when a node sends a packet as originator and increases when a node forwards a packet. The operation of the scheme is as follows: when a node S desires to send a packet to a destination D, if the number of intermediate nodes on the path from S to D is n, then S's nuglet counter must be greater than or equal to n in order for S to send the packet. If S has enough nuglets to send the packet, S decreases its nuglet counter by n after sending the packet. On the other hand, S increases its nuglet counter by one each time S forwards a packet on behalf of other nodes. The value of a nuglet counter must be positive; therefore, it is within a node's interest to forward packets on behalf of other nodes and refrain from sending large number of packets to distant destinations. The scheme offers an effective mechanism for discouraging selfishness; however, it may not experience widespread use because of the requirement for a tamper-resistant hardware module.

Zhong et al. presented Sprite: A simple, cheat-proof, credit-based system for MANETs [58]. Sprite provides incentive for MANET nodes to cooperate and report actions honestly. Sprite requires a centralized entity called a credit clearance service (CCS), which determines the charge and credit involved in sending a message. The basic operation of Sprite is as follows: when a node receives a message, the node keeps a receipt of the message. Later when the node has a fast connection to a CCS, it reports to the CCS the message it has received/forwarded by uploading its receipt. The CCS then uses the receipt to determine the charge and credit involved in the transmission of the message. This scheme is based on the assumption that online access to a CCS is available. This assumption may not hold for purely ad hoc networks, which do not guarantee access to online entities.

16.3.4 Schemes that Employ Detection and Isolation Mechanisms

This section contains a brief description of schemes that utilize detection and isolation techniques. We commence the review with an earlier proposal.

Marti et al. [31] proposed a scheme for mitigating against the presence of MANETs nodes that agree to forward packet but fail to do so. The scheme utilizes a "watchdog" for identifying misbehaving nodes and a "pathrater" for avoiding those nodes. Each node has its own watchdog and pathrater modules. Watchdog operation requires the nodes within a MANET to operate in promiscuous mode: meaning that a node n_i that is within the transmission range of a node n_j should be able to over-hear communications to and from n_j even if those communications do not involve n_i. Watchdog is based on the assumption that if a packet was transmitted to node n_i for it to forward the packet to node n_j and a neighboring node to n_i does not hear the transmission going from n_i to n_j, then it is likely that n_i is malicious and should therefore be assigned a lower rating. Pathrater is responsible of assigning ratings. The rating is assigned as follows: when a node n_i becomes known to the pathrater, n_i is assigned a "neutral" rating of 0.5. The ratings of nodes that are on actively used path are consequently incremented by 0.01 every 200 ms; whereas, a node's rating is decremented by 0.05 when a link to the node is surmised to be nonfunctional. "Neutral" ratings are bounded with an upper bound of 0.8 and a lower bound of 0.0; but a node always assigns a rating of 1.0 to itself. Rather than selecting a path to a given destination based on the number of hops in the path, the pathrater selects the path that has the highest average rating. This scheme has several weaknesses. As described in the authors' own words: "watchdog's weakness is that it might not detect a misbehaving node in the presence of (1) ambiguous collisions, (2) receiver collisions, (3) limited transmission power, (4) false misbehavior, (5) collusion, and (6) partial dropping."

Buchegger and Le Boudec proposed a protocol called CONFIDANT [45] that aims to detect and isolate misbehaving nodes in MANETs. CONFIDANT uses a form of reputation systems [42] where the nodes within a MANET rate each other based on observed behaviors. Nodes that are deemed to be misbehaving are placed on black lists and are consequently isolated. The reputation systems, however, do not provide any protection against false accusations. Consequently, the scheme is susceptible to blackmailing.

Awerbuch et al. presented a routing security scheme [2] aimed at providing resilience to byzantine failure caused by individual or colluding MANET nodes. The scheme utilizes digital signature for authentication at each hop, and it requires each node to maintain a weight list consisting of the reliability metric of the nodes within the network. The weight list is used in the route discovery phase to avoid faulty paths. When faults are detected in established paths, an adaptive probing technique is launched in an attempt to detect the faulty links. Faulty links are given decreased rating and are consequently avoided. Probing techniques are useful in identifying faults caused by nonmalicious acts. However, they are ineffective against malicious agents, simply because the probing packets are distinguishable from other packets; therefore, an adversary can choose to behave well when it is being probed, but behave maliciously during intervals when it is not being probed.

Just and Kranakis [24] and Kargl et al. [25] proposed schemes for detecting selfish or malicious nodes in an ad hoc network. The schemes involve probing mechanisms which as is the case with [2], the probing packets are distinguishable from other packets.

Patwardhan and Iorga [36] presented a secure routing protocol called SecAODV. SecAODV is based on AODV but unlike the latter, it requires each node in the MANET to have a static IPv6 address. The scheme allows source and destination nodes to establish secure communication channel based on the concept of statistically unique and cryptographically verifiable (SUCV) identifiers [32], which ensures secure binding between an IPv6 address and a key, without requiring any trusted certificate authority (CA). SecAODV also provides an intrusion detection system (IDS) for monitoring the nodes' activities. The application of this protocol is currently very restrictive because of the requirement that each of the MANET nodes must have a static IPv6 address.

16.4 ROBUST SOURCE ROUTING

In this section, we present a secure on-demand, multipath source routing protocol, called robust source routing (RSR). In addition to providing data origin authentication services and integrity checks, RSR is able to mitigate against intelligent malicious agents that selectively drop or modify packets they agreed to forward. Simulation studies confirm that RSR is capable of maintaining high delivery ratio even when a majority of the MANET nodes are malicious.

RSR has two phases: route discovery and route utilization and maintenance. We give an overview of each phase below.

16.4.1 Route Discovery

In the route discovery phase, a source node S broadcasts a route request indicating that it needs to find a path from S to a destination node D. In the route request, S stipulates that the path it seeks must not contain any node that is listed in its tabu list, or any link that appears in its exclusion links list. We provide a rationale for the tabu list and exclusion links list in Section 16.4.4. Additionally, the path must not contain any node that is found in the tabu list of an element in the path. Each node through which the route request traverses is required to append its identifier and its tabu list to the appropriate field of the route request, and signs the packet. Therefore, the information regarding the identity of the nodes that should be excluded from the path is easily ascertained. When the route request packets arrive at the destination node D, D selects three valid paths, copy each path to a route reply packet, signs the packets, and unicasts them to S using the respective reverse paths. S proceeds with the utilization and maintenance phase when it receives the route reply packets.

16.4.2 Route Utilization and Maintenance

The source node S selects one of the routing path it acquired during the routing discovery phase and sends the data traffic. The destination node D is required to send a signed acknowledgment for each data frame it receives. If S does not get an acknowledgment from D for a data frame after a given number of retries and does not receive a link-layer error message indicating that the destination D is unreachable, it assumes that there are selfish or malicious nodes on the path and proceeds as follows: S constructs and sends a forerunner packet to inform the nodes on the path that they should expect a specified amount of data from the source of the packet within a given time. When the forerunner packet reaches the destination, it sends an acknowledgment to S. If S does not receive an acknowledgment for the forerunner packet, it proceeds as outlined in Section 16.4.4.2, under the heading "no ACK for a FR packet returns from D." Otherwise, S commences the data traffic flow to D. If there are selfish or malicious agents in the path and they choose to drop the data packet or acknowledgment from D, such eventuality is dealt with as outlined in Section 16.4.4.2, under the heading "S commenced data flow to D but the traffic is being dropped."

16.4.3 Problem Definition and Model

In this section we outline the network and security assumptions we utilized in the design of RSR. We also present a more precise description of the problem our protocol addresses.

16.4.3.1 Network Assumptions

RSR utilizes the following assumptions regarding the targeted MANETs:

- Each node has a unique identifier (IP address, MAC address, or certificate serial number).
- Each node has a valid certificate and the public keys of the CAs that issued the certificates of the other network peers.
- The wireless communication links between the nodes are symmetric; that is, if node n_i is in the transmission range of node n_j, then n_j is also in the transmission range of n_i. This is typically the case with most 802.11 [21] compliant network interfaces.

- The link-layer of the MANET nodes provide transmission error detection service. This is a common feature of most 802.11 wireless interfaces.

- Any given intermediate node on a path from a source to a destination may be malicious and therefore cannot be fully trusted. The source node only trusts a destination node, and visa versa, a destination node only trusts a source node.

16.4.3.2 Threat Model

In this work, we do not assume the existence of security association between any pair of nodes. Some previous works, for example [18, 34] rely on the assumption that protocols such as the well-known Diffie–Hellman key exchange protocol [12] can be used to establish secret shared keys on communicating peers. However, in an adversarial environment, malicious entities can easily disrupt these protocols—and prevent nodes from establishing shared keys with other nodes—by simply dropping the key exchange protocol messages, rather than forwarding them. Our threat model does not place any particular limitations on adversarial entities. Adversarial entities can intercept, modify, or fabricate packets, create routing loops, selectively drop packets, artificially delay packets, or attempt denial of service attacks by injecting packets in the network with the goal of consuming network resources. Malicious entities can also collude with other malicious entities in attempts to hide their adversarial behaviors. The goal of our protocol is to detect selfish or adversarial activities and mitigates against them.

One particular type of attacks our protocol cannot prevent is wormhole exploits [19]. In wormhole attacks, an attacker receives packets at one point in a network, tunnels them to another point in the network, and replays them into the network from that point. Colluding adversaries can use this attack, for example, to forward route request packets in an attempt to increase the likelihood of adversarial entities controlling routing paths. If a wormhole exhibits adversarial activities, our protocol mitigates against these exploits by treating the wormhole as a single link and make efforts to avoid utilizing it.

16.4.4 Details of RSR

The protocol requires each node to keep a tabu list containing a list of nodes the owner of the list deems malicious or untrustworthy. The owner of the list will silently drop route request packets originated from any node that is in its tabu list. It is therefore highly likely that the owner of a tabu list will be listed in the tabu lists of the nodes in its tabu list. Hence, it is within a node's best interest to add a node to its tabu list only if it has a high degree of certainty that the given node is malicious or untrustworthy.

As previously indicated, the routing scheme consists of two phases: route discovery and route utilization and maintenance phases. All unicast routing packets transmitted in each phase of the protocol have a common source route header with the following fields:

- *Source address*: The identifier of the node that constructed the packet.
- *Destination address*: The identifier of the destination node.
- *Source route*: The routing path the packet must traverse in transit from the source to the destination.

16.4.4.1 *Route Discovery*

When a node n_i has data to transmit to a destination it does not know of a path to, n_i generates a route request (RREQ) packet containing the following information:

- *Request id*: A unique, random nonce, which, together with the source address, serves as the identifier of a RREQ packet.
- *Exclusion links*: A list of zero or more link(s) that must not be included on a path.
- *Route record*: The list of nodes the RREQ traverses, along with their tabu lists and accompanying signatures.

It should be noted that exclusion links and the tabu lists are separate entities that serve different purposes, namely, when a node n_j is listed in node n_i's tabu list, n_i will silently drop RREQ packets that originated from n_j. If n_i is currently on any of n_j's routing paths, it will continue to forward data traffic along the given path(s); however, n_i will not appear on any new path for n_j since it will not forward any other RREQ packets from n_j. On the other hand, if n_j appears on a link in n_i's exclusion links, n_i will still continue to forward RREQ packets that originated from n_j, since n_i does not know whether it is n_j or n_k (the other node in the problematic link) that is the selfish or adversarial node.

After generating the RREQ, n_i signs the RREQ and broadcasts the packet to its neighbors. When a node n_j receives a set of RREQ packet—it has not previously seen—with the same ⟨source address, request id⟩ identifier, it selects one at random[2] then checks if any of the following holds:

- The source of the RREQ is listed in n_j's tabu list.
- n_j appears in a tabu lists in the route record field.
- There is an exclusion link between n_j and a neighbor that appears in the route record field.

If any of the above holds, n_j discards the packet and records that it has seen a RREQ with the given ⟨source address, request id⟩ identifier. Otherwise, n_j verifies

---

[2]Selecting an RREQ packet at random rather than choosing the first one that arrives provides protection against rushing attack [20].

the initiator's signature[3]; if the verification fails and n_j's link-layer does not report a transmission error, n_j adds the neighbor it received the RREQ packet from to its tabu list and discards the RREQ. The reason being, n_j's neighbor either modified or fabricated the packet, or it did not verify the source's signature before forwarding the RREQ; that is, n_j's neighbor is either malicious or it is not complying with the protocol. If the signature verification succeeds, n_j appends its identifier and its tabu list to the route record field, signs the entire route record field, makes a record indicating that it has seen a RREQ packets with the given ⟨source address, request id⟩ identifier, and broadcasts the packet to its neighbors.

RREQ packets continue to traverse the network in the manner described above until one or more reach the destination node D. On receiving a list of RREQ packets with the same ⟨source address, request id⟩ identifier, node D is expected to select three of the RREQ packets such that the path in their respective route record field has the least number of hops, and no element in the path appears in any of the other path elements' tabu lists and no link is listed in source's exclusion links. Next, D is required to verify the signatures in the route record fields of each of the selected RREQ packets. If the signatures of a selected RREQ packet are all valid, D constructs a route reply (RREP) packet for the given RREQ, signs it and unicasts it—using the reverse path in the RREQ route record field—to the source of the RREQ. If any of the signature verification for a selected RREQ packet fails, the RREQ in question is discarded and another selected using the criteria outline above. The source node S is expected to send a signed acknowledgment for each RREP it receives. If D does not get an acknowledgment from S for a RREP packet after a given number of retries and if there are other RREQ packets remaining, D selects another, processes it as outlined above, and sends the resulted RREP packet to S.

In addition to the common source route header, a RREP packet contains the following information:

- *Request id*: Request id of the corresponding RREQ packet.
- *Path*: The identifiers of the nodes in the routing path, in the order indicated in the route record field of the corresponding RREQ.

When the source of the RREQ receives the RREP packets, it proceeds with the route utilization and maintenance as indicated below.

16.4.4.2 *Route Utilization and Maintenance*

On receiving the RREP packets, the source node S stores the paths, selects the one that has the least number of hops, and proceeds to send the data traffic. The destination node (D) is required to send a signed acknowledgment (ACK) for each data frame it received. If S does not receive a valid ACK for any given data after a certain number of retries nor does it receive a link-layer error message from any of the intermediate

---

[3]Source authentication is utilized to extenuate the effect of denial of service attacks on the network. We discuss the pros and cons of this approach in Section 16.4.5.

nodes, S assumes that there is/are selfish or malicious node(s) on the given path and proceeds with the fault detection and isolation phase below.

Fault Detection and Isolation When there is evidence of misbehaving node(s) on a given path, the protocol utilizes a forerunner (FR) packet to inform the nodes on the path that they should expect a certain data flow rate from S to a specified destination. The intention being that if any of the path elements do not receive the specified data traffic within a configurable time period after receiving a FR packet from S, it will send a negative acknowledgment, informing S that it did not receive the expected data flow. Data flow rate can be obtained from IEEE 802.11 medium access control (MAC) protocol operating in the distributed coordination function (DCF) mode, using mechanisms outlined in [8, 26, 47].

A FR packet has the following fields:

- *FR id*: A unique, random nonce, which, together with the source address (ascertained from the source route header), serves as the identifier of a FR packet.
- *Expected data rate*: Data flow rate that should follow the FR packet.
- *ACK indicator*: This is a 1-bit flag that is set if the intermediate nodes are required to send a signed ACK back to the source of the FR packet.

To avoid unnecessary network traffic, the ACK indicator flag is set to 0 when a FR packet is constructed. The packet is then signed and sent to D using the selected path. When an intermediate node on the path from S to D receives the FR packet, it is expected to verify the signature, if it is valid, it should note the time it received the FR packet then forward the packet to the next hop on the path. When D receives a valid FR packet, it sends a signed ACK back to the source. On receiving the ACK from D, S commences the traffic flow to D.

Selfish or malicious nodes may choose not to forward a FR packet, and they also may not forward data traffic after S commences the traffic flow to D. The protocol deals with these eventualities as indicated below:

No ACK for a FR Packet Returns from D If S does not receive an ACK for a FR packet from D, nor does S receive a link-layer error message from any of the intermediate nodes indicating that the destination D is unreachable, S assumes that a misbehaving node on the given path has dropped the FR packet or the ACK from D and proceeds as follows: if the length of the path from S to D is exactly 3, S adds the link between D and the intermediate node to its exclusion links, discards the path, selects another path to D—if one is available—and repeats the route utilization process indicated above. If there are no more precomputed path to D, S constructs, signs and broadcasts another RREQ packet with the exclusion link field containing all the problematic link(s) it has recorded. If the path length from S to D is greater than 3, S constructs another FR packet, sets the ACK indicator flag to 1, signs the packet and sends it to the first hop on the path to D. When a node n_i receives a FR packet with

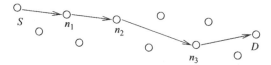

Figure 16.5   A routing path example.

the ACK indicator flag set to 1, n_i is expected to broadcast—via limited flooding—a signed ACK back to S. In the limited flooding broadcast, the time-to-live (TTL) field of the IP header is set to d where d is the number of hops from the node in question to S. If S does not receive a valid ACK from each of the nodes in the path, then the link between the first node n_i—on the given path from which S does not receive a valid ACK—and n_i's upstream path neighbor is added to S exclusion links. For example, in Figure 16.5, if S receives ACKs for the FR packet from n_1 and n_2 but not from n_3, S would add the link between n_2 and n_3 to its exclusion links, select another path to D, or send out a route request as outlined above. A path with a problematic link can be pruned by removing the subpath commencing with the downstream node of the problematic link. For example, in Figure 16.5, n_3 and D would be removed from the path, resulting in a subpath of length 3 from S to n_2. The resulted subpath after the pruning operation is stored if its length is greater than or equal to 3, or discarded otherwise.

S Commenced Data Flow to *D* But the Traffic is Being Dropped

As indicated above, when a node n_i receives a FR packet, it records the time it received the packet. If a configurable time period (which depends on the network latency and available bandwidth) passed and n_i does not receive the expected data flow from S to D, n_i is required to send—via limited flooding—a signed negative ACK to S, indicating that n_i has not received the data flow it expects from S. A negative ACK is similar to an ACK packet, except that it informs the intended recipient S that the source of the negative ACK did not receive the data traffic it expected from S. When S receives a valid negative ACK from a node n_i and it is confirmed by other negative ACKs from downstream nodes on the path to D, S records the link between n_i and n_i's upstream path neighbor as being problematic; S then prunes the given path and repeats the process of selecting or discovering another path, as outlined above.

Rather than dropping data traffic, malicious nodes may choose to tamper with the data. The protocol deals with this eventuality by requiring intermediate nodes to verify the source's signature on packets they received, before forwarding them. If the signature verification fails for node n_i and n_i link-layer does not report a transmission error, n_i is required to add the neighboring node it received the packet from to its tabu list and sends—via limited flooding—a negative ACK to S, informing it that the packet has been modified. On receiving the negative acknowledgment from n_i, S is expected to append the link involving n_i and its upstream neighbor to its exclusion links and prunes the path.

16.4.5   Discussion

In this section, we elaborate on relevant design choices of our protocol. We commence with our choice of using digital signatures for integrity checks and source authentication.

16.4.5.1   Choice of Cryptographic Tools

Most network security schemes utilize message authentication codes, rather than digital signatures, for integrity checks. This is so due to the fact that message authentication codes can be computed much more efficiently than digital signature computations. The drawback for the use of message authentication codes, as is the case for other symmetric-key cryptographic tools, is that it requires shared keys to be established among the communicating peers. As alluded to in Section 16.4.3, our protocol was specifically designed for adversarial MANET environments that contain, or are likely to contain persistent malicious or selfish entities which seek to disrupt the network by perpetrating the adversarial activities outlined in the threat model in Section 16.4.3.2. We argue that it may not be feasible to establish shared keys among communicating peers, using key exchange protocols, since adversarial entities can easily thwart these protocols by dropping the protocol messages, rather than forwarding them. A node S can generate a symmetric key, signs it, encrypts it with the public key of the intended recipient, and sends it via broadcast to the destination D. This will likely allow a shared key to be established between S and D; however, the cost in throughput reduction, due to the extra broadcast messages, may not justify this approach. Alternatively, shared secret keys can be distributed to the network nodes using appropriate out-of-band means; again, this approach is not feasible considering the likelihood of shared keys being compromised if they are not refreshed frequently.

Aside from the problem of establishing shared secret keys among communicating peers in highly adversarial environments, message authentication codes may not be as effective in identifying certain malicious activities. For example, a malicious entity, on a routing path from S to D, which seeks to disrupt the traffic flow on this path, can choose to illicitly modify packets and forward them rather than mere dropping the packets. The end result of these activities is similar to packet dropping, since the destination will discard the packets when it ascertains that they have been illicitly modified. Digital signature can be used to identify malicious entity that modified the packet or identify the colluding malicious entity that forwarded the modified packet; but message authentication code is lacking is this regards, since typically only the source and the destination of a traffic flow know the secret key for computing the message authentication codes. We leveraged the aforementioned feature of digital signature in the design of RSR to help to detect and isolate adversarial entities. RSR source authentication operations serves two main purposes.

1. Consider for example the scenario shown in Figure 16.5. If n_3 (a well-behaving node) receives a packet from n_2 to forward to D, if the signature verification for the packet fails and n_3 link-layer does not report a transmission error, n_3

will add n_2 to its tabu list. The reason being, either n_2 modified the packet or it did not verify the signature on the packet; that is, n_2 is either malicious or it is not complying with the protocol. In addition to adding n_2 to its tabu list, n_3 will discard the packet and send a negative acknowledgment to S informing it that the packet it received has been illegitimately modified. If this info from n_3 is supported by the fact that S does not receive an acknowledgment from D for the given data frame, S will add the link between n_2 and n_3 to its exclusion links and consequently commence the process of isolating n_2.

2. Source authentication can also be used to attenuate certain denial of service exploits. Malicious nodes may attempt to flood the network with fabricated packets in attempts to consume network resources. RSR source authentication operations are partly aimed at reducing the effect of these types of attacks by stipulating that nodes discard unauthenticated packets. It should be noted that adversarial entities can overwhelm individual nodes in their one-hop neighborhood by sending them large number of fabricated packets. However, the fact that the unauthenticated packets will be discarded, the resource consumption exploit will be limited to the one-hop neighborhood of the adversarial entities.

In light of the above possibilities, it is our view that the benefits of using digital signature for source authentication outweighs the associated cost. Digital signature schemes such as RSA [44] allow trade-off between signing and verification operations. If the public exponent of the crypto system is small, verification can be several times faster then signing operations. Example, for a 1024-bit RSA key, if the public exponent (e) is 3, verification operations can be over 700 times faster than signing operations [52]. Verification of signatures can therefore be done fairly efficiently; most of the digital signature operations in RSR are verification activities.

An alternative approach to utilizing cryptographic tools for the operations outlined in item (1) above is to have the nodes' network interfaces operate in promiscuous mode and stipulate that the nodes monitor the traffic that flows in and out of each of their neighbors and report all discrepancies. This operation however, is inefficient and is subjected to the short comings outlined in Section 16.3.4 for Marti et al. scheme [31].

16.4.5.2 Tabu List and Exclusion Links

RSR utilizes tabu lists and exclusion links to record problematic nodes and links, respectively. The consequences of being listed in a node's tabu list is more severe for the following reasons: a node will silently discard route requests from nodes that are listed in its tabu list. Therefore, if a node is listed in the tabu lists of several nodes, it will likely have much difficulties communicating with other nodes that are not in its transmission range. On the other hand, a node's exclusion links list is used solely to exclude problematic links from its routing paths. This design choice is motivated by the fact that a node n_i does not know for sure which element of a problematic link is selfish or adversarial; and n_i wants to avoid the possible of wrongfully isolating well-behaving nodes. Malicious nodes may add well-behaving nodes to their tabu list with the intention of disrupting route discovery processes; however, this eventuality would actually have positive effects on the network, since this reduces the possibility of the given malicious nodes being on routing paths. Similarly, adversarial entities will not achieve any benefit from adding functional links to their exclusion links lists.

16.4.5.3   Forerunner Packets Mechanism

Our FR packet mechanism requires MANET node to be able to determine the flow rate of incoming traffic. As outlined in Section 16.4.4.2, data flow rate can be obtained from IEEE 802.11 MAC protocol quite efficiently using techniques presented in Refs. [8, 26, 47]. The distinguishing feature of our FR packets mechanism compared with other MANET fault detection techniques—such as probing—is the following: FR packets inform the nodes on a path from a source node S to a destination node D that S intends to send a certain amount of data within a given time period; therefore, the nodes should expect the specified data traffic flow rate from S for the time period indicated. If the nodes on the path from S to D do not receive the specified data traffic flow rate within the specified time period, they are required to send negative acknowledgments to S informing S that they did not receive the expected data flow. This mechanism forces selfish or malicious entities on routing paths to cooperate and forward the specified data traffic a FR packet announced would follow or risk being identified as problematic if they choose not to forward the data traffic. The selfish or malicious entities can resume adversarial activities after forwarding the specified data traffic a FR packet announced. However, the end result is that FR packets can force uncooperative entities to forward specified amount of data, or conversely, help to identify links that contains uncooperative nodes. This can be contrasted with schemes such as Refs [2, 24, 25] that utilize probing techniques, in that the probing mechanisms will succeed in enforcing cooperation only if the probing packets are completely indistinguishable from other data packets, which in reality is very difficult to achieve. There are no needs for FR packets to be indistinguishable from other packets, since their purpose is to announce intended traffic flows. Adversarial entities can choose to drop FR packets; however, as outlined Sections 16.4.4.2 and 16.4.6, the protocol operations provide means for identifying these adversarial activities.

16.4.6   Analysis

In this section we give specific examples of malicious behaviors and show how RSR mitigates against these possible exploits.

16.4.6.1   A Single Malicious Node on a Routing Path

Consider the following with respect to the routing path depicted in Figure 16.6:

1. If m drops a data packet sent from S to D, S would not receive an ACK from D for the given packet. Consequently, S sends a FR packet along the path to D. If m drops this packet, no acknowledgment will return from D for the FR packet.

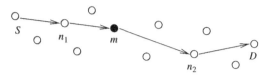

Figure 16.6 One malicious node on a routing path.

S will then send a FR packet with the ACK indicator bit set to 1, along the same path to D. Each node along the path that receives the FR packet (with the ACK indicator bit set to 1) is required to send—via limited flooding—a sign ACK to S. If m drops this packet, S will not receive an ACK from n_2. Therefore, S will classify the link between m and n_2 as problematic and adds it to its exclusion links. The next RREQ packet S sends out will contain information about the faulty link between m and n_2. When n_2 receives this info, if there were at least $N - 1$ (see Section 16.4.5.2 for info related to N) other RREQ packets from different sources that listed this link as problematic, n_2 will add m to its tabu list, thus initiating the process of isolating m.

2. If m acknowledges and forwards the FR packets with the ACK indicator bit set to 1, but succeeded—with the help of other malicious nodes outside the given path—to filter out the ACKs sent by n_2 and D to S, S will not get an ACK from n_2. Therefore, S will add the link between m and n_2 to its exclusion links.

3. If S receives an ACK for the FR packet (with the ACK indicator bit set to 1) from each of the path element, S will start sending the specified data traffic to D. If m drops a data frame, n_2 and D would not receive the data flow the FR packet specified that they should expect. Consequently, they will send a negative ACK—via limited flooding—to S. When S gets the negative ACKs, S adds the link between m and n_2 to its exclusion link.

4. If m, with the help of other malicious nodes outside the given path, succeeds in filtering out the negative ACKs from n_2 and D, S will know that the path has a fault, since it does not receive an ACK from D for the data frame m dropped. Consequently, S will discard the given path. The same holds if m forwards all the data frames from S to D but drops an ACK D sends to S.

16.4.6.2 Colluding Malicious Nodes Adjacent to Each Other

Consider the path shown in Figure 16.7 with the colluding malicious nodes m_1 and m_2. If m_1 or m_2 drops packets it is required to forward, it is trivial to show that the same arguments outlined above hold.

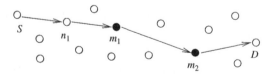

Figure 16.7 Adjacent colluding malicious nodes on a routing path.

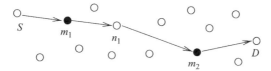

Figure 16.8 Nonadjacent colluding malicious nodes on a routing path.

16.4.6.3 Colluding Malicious Nodes two Hops Away From Each Other

In the path shown in Figure 16.8, if m_1 or m_2 drops packets that were intended to be forwarded to D, it can also be trivially shown that the arguments outlined in (1)–(3) above hold. In this scenario, however, it is unlikely that m_1 will ever succeed—with the help of other malicious nodes outside the given path—in filtering out negative ACKs sent via limited flooding from n_1 to S, unless all of the nodes that are within S transmission range are malicious. It will, therefore, be difficult for m_1 to conceal its malicious behaviors.

16.4.7 Simulation Evaluation

We implemented RSR in NS2 network simulator [1]. For the cryptographic components, we utilized Cryptlib crypto toolkit [16] to generate 1024-bit RSA cryptographic keys for the signing and verification operations. In the simulation implementation, malicious nodes do not comply with the protocol. For example, they do not verify the signatures on the packets they forward, nor do they add nodes to their tabu list or exclusion links or send negative ACKs. In addition, they selectively drop or modify packets they are asked to forward. The exception being that they do not drop or modify RREQ or RREP packets, since their adversarial effects are more pronounced when they are on as many routing paths as possible. Table 16.1 summarises the simulation parameters.

Table 16.1 Simulation parameters values

Parameter	Value
Space	670 m × 670 m
Number of nodes	50
Mobility model	Random waypoint
Speed	20 m/s
Pause time	600 s
Traffic type	CBR
Max number of connections	34
Packet size	512 bytes
Packet generation rate	4 packets/s
Simulation time	170 s

16.4.7.1 Performance Metrics

We used the following metrics to evaluate the performance of our scheme.

1. *Packet delivery ratio*: This is the fraction of data packets generated by constant bit rate (CBR) sources that are delivered to the destinations. This evaluates the ability of RSR to deliver data packets to their destinations in the presence of varying number of malicious agents that selectively drop packets they are required to forward.

2. *Number of data packets delivered*: This metric gives additional insight regarding the effectiveness of the scheme in delivering packets to their destination in the presence of varying number of adversarial entities.

3. *Routing overhead (bytes)*: This is the total number of bytes of routing control messages generated over the length of the simulation.

4. *Routing overhead (packets)*: This is the total number of routing control messages generated over the length of the simulation. We normalized the routing overhead by the number of packets sent and the number of packets received to compensate for the fact that in the simulation implementation adversarial nodes do not sent data packets.

5. *Average end-to-end latency of the data packets*: This is the ratio of the total time it takes all packets to reach their respective destinations and the total number of packets received. This measures the average delays of all packets that were successfully transmitted.

The results of the simulation for RSR is compared with that of DSR [22], which currently is perhaps is the most widely used MANET source routing protocol.

16.4.7.2 Simulation Results

The simulation results confirm that RSR is very effective in delivering data packets to their intended destinations even in the presence of large proportion of malicious entities. As indicated in Figure 16.9, RSR was able to maintain delivery ratio of over 0.8 even when 80% of the nodes are malicious. However, the delivery ratio for DSR was 0.2 when 70% of the nodes are malicious and 0 when 80% of the nodes are malicious.

It should be noted that DSR does not provide any security services, nor does it provide reliable data transfer; whereas RSR provides both. It is, therefore, expected that the overhead associated with RSR will be significantly higher than that associated with DSR. This is the trade-off relating to the overhead of the two protocols. In spite of the higher overhead associated with RSR, Figure 16.10 indicates that over the length of the simulation, RSR on average, delivers more than twice the number of packets DSR delivers when the percentage of malicious nodes in the network is greater than 10. This confirms—as the plot of delivery ratio (Figure 16.9) indicates—that in the presence of active malicious entities, RSR allows much greater throughput than DSR does.

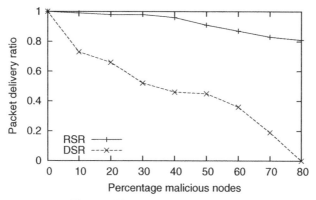

Figure 16.9   Data packet delivery ratio.

RSR employs digital signature to provide data origin authentication and integrity checks. In the RSR simulation implementation, each routing packet is signed and the signature appended to the packet; therefore, RSR packets are much larger than DSR packets. As expected, the simulation results indicate that the routing overhead for RSR, on an average, increases as the percentage of malicious nodes increases. This is due to the following: as malicious activities increase, more FR packets and consequently ACKs for FR packets, and negative ACKs are sent. Figures 16.11–16.14 indicate that the trends are similar whether the overhead in terms of bytes or packets generated is normalized by number of packets sent or number of packets received.

Figure 16.15 shows that there are no clear trends regarding average data packet latency. The fluctuation in data packet latency is likely related to the number of broadcast packets circulating in the network. The higher the number of broadcast packets in the network, the more contention there will be for the wireless access medium, and consequently, the longer it will take for packets to be delivered to their respective destinations. Average data packet latency is also inversely related to

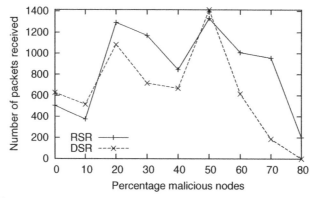

Figure 16.10   Number of packets received over the length of the simulation.

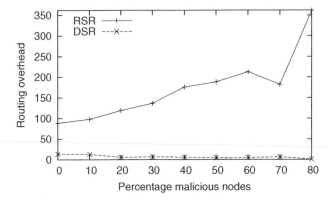

Figure 16.11   Routing overhead (bytes) normalized by number of data packets sent.

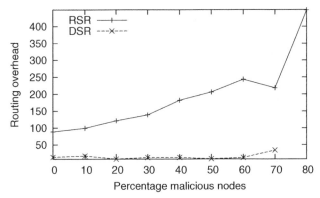

Figure 16.12   Routing overhead (bytes) normalized by number of data packets received.

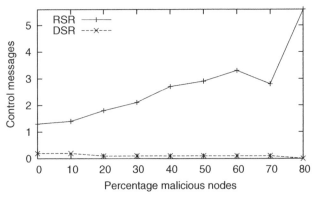

Figure 16.13   Routing overhead (number of packets) normalized by number of data packets sent.

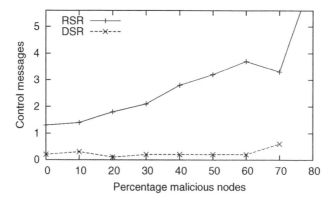

Figure 16.14 Routing overhead (number of packets) normalized by number of data packets received.

data packet size: larger packets, on average, take longer to reach their destinations. Hence, the higher average packet latency for RSR compared with DSR is expected. However, this increase in latency is insignificant compared with the proportionally higher throughput that RSR provides in the presence of increased number of active malicious entities.

One result that is unexpected for RSR is the decrease in the overhead when the percentage of malicious nodes increases from 60 to 70, as indicated in Figures 16.11–16.14. This trend is likely related to the CBR traffic pattern during this time interval in the simulation. One possibility is that the CBR data packets sent—during this time interval in the simulation—traverse fewer malicious nodes; consequently, there may have been a slight decreased in malicious activities during this time interval.

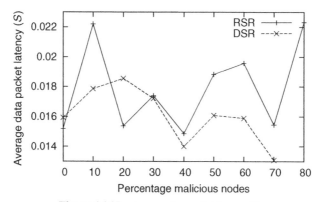

Figure 16.15 Average data packet latency (S).

16.5 SUMMARY

In this chapter, we gave an overview of routing approaches in MANET and analyzed the existing secure MANET routing proposals. We noted that most of these proposals do not mitigate against selfish or malicious entities that selectively drop packets they agreed to foreword. We categorized the proposals that attempt to mitigate against these adversarial activities into three categories: trust-based routing schemes, incentive-based schemes, and schemes that employ detection and isolation mechanisms. We argued that trust-based routing schemes are susceptible to adversarial exploits because they either require group secret keys to enforce trust-level requirements, they do not provide protection against malicious accusation attacks, or they can be thwarted by dropping the trust query messages. Next, we highlighted the point that the incentive-based schemes either require tamper-resistant hardware module or they require online access to a centralized entity. Owing to these requirements, the incentive-based schemes are limited in their applications. Regarding the schemes that employs detection and isolation mechanisms, we asserted that these schemes are inadequate for the various reasons outlined in Section 16.3.4. Finally, we concluded from the review and analysis of the existing MANET secure routing proposals that there are needs for secure routing schemes that adequately mitigate against selfishness and selective packet dropping in MANETs.

As a contribution to this endeavor, we presented a robust, secure MANET on-demand routing protocol that is capable of delivering packets to their destinations even in the presence of large proportions of active malicious or selfish agents that selectively drop packets they agreed to forward. We named this protocol RSR. RSR introduced the concept of FR packets, which inform nodes along a path that they should expect specified data flow within a given time frame. The path elements can therefore be on the look out for the given data flow, and in the event that they do not receive the traffic flow, they can transmit info to the source informing it that the data flow they expected did not arrive. In so doing, links with active malicious agents can be identified and the malicious agents be eventually isolated. Finally, we provided simulation results that attest to the proficiency of RSR being able to deliver packets to their destination even in the presence of large proportions of malicious or selfish entities.

REFERENCES

1. Ns2 network simulator. http://www.isi.edu/nsnam/ns.
2. B. Awerbuch, D. Holmer, C. Nita-Rotaru, and H. Rubens. An on-demand secure routing protocol resilient to byzantine failures. In *Proceedings of the ACM Workshop on Wireless Security (WiSE '02)*, September 2002, pp. 21–30.
3. R. Bellman. On a routing problem. *Quart. Appl. Math.*, 16(1):87–90, 1958.
4. J. Binkley and W. Trost. Authenticated ad hoc routing at the link layer for mobile systems. *Wireless Networks*, 7(2):139–145, 2001.
5. P. Bose and P. Morin. Online routing in triangulations. In *Proceedings of the 10th International Symposium on Algorithms and Computation (ISAAC'99)*, LNCS, Vol. 1741, 1999, pp. 113–122.

6. A. Boukerche, K. El-Khatib, L. Xu, and L. Korba. An efficient secure distributed anonymous routing protocol for mobile and wireless ad hoc networks. *Comput. Commun.*, 28(10):1193–1203, 2005.

7. L. Buttyan and J.-P. Hubaux. Stimulating cooperation in self-organizing mobile ad hoc networks. *ACM/Kluwer Mobile Networks Appli.*, 8(5):579–592, 2003.

8. K. Chen, K. Nahrstedt, and N. Vaidya. The utility of explicit rate-based flow control in mobile ad hoc networks. In *Proceedings of IEEE Wireless Communications and Networking Conference WCNC 2004*, Vol. 3, 2004, pp. 1921–1926.

9. C. Cheng, R. Riley, S. P. R. Kumar, and J. J. Garcia-Luna-Aceves. A loop-free bellman-ford routing protocol without bouncing effect. In *Proceedings of ACM SIGCOMM '89*, September 1989, pp. 229–237.

10. S. Corson and J. Macker. Mobile ad hoc networking (manet): routing protocol performance issues and evaluation considerations. Internet Request for Comments (RFC 2501), January 1999.

11. C. Crépeau, C. R. Davis, and M. Maheswaran. A secure manet routing protocol with resilience against byzantine behaviours of malicious or selfish nodes. In *Proceedings of the 21st International Conference on Advanced Information Networking and Applications Workshops (AINAW)*, Vol. 02, May 2007, pp. 19–26.

12. C. R. Davis. *IPSec: Securing VPNs*. Osborne/McGraw-Hill, New York, 2001.

13. S. Deering. Icmp router discovery messages. Internet Request for Comments (RFC 1256), September 1991.

14. R. Dube, C. D. Rais, K.-Y. Wang, and S. K. Tripathi. Signal stability based adaptive routing (SSA) for ad-hoc mobile networks. *IEEE Personal Commun.*, 4(1):36–45, 1997.

15. G. G. Finn. Routing and addressing problems in large metropolitan-scale internetworks. ISI Research Report ISU/RR-87-180, March 1987.

16. P. Gutmann. Cryptlib encryption toolkit. http://www.cs.auckland.ac.nz/~pgut001/cryptlib.

17. Y. Hu, A. Perrig, and D. Johnson. Ariadne: a secure on-demand routing protocol for ad hoc networks. In *Proceedings of the 8th ACM International Conference on Mobile Computing and Networking (Mobicom 2002)*, September 2002, pp. 12–23.

18. Y. Hu, A. Perrig, and D. Johnson. Sead: secure efficient distance vector routing for mobile wireless ad hoc networks. In *Proceedings of the 4th IEEE Workshop on Mobile Computing Systems and Applications (WMCSA'02)*, June 2002, pp 3–13.

19. Y. Hu, A. Perrig, and D. Johnson. Packet leashes: a defense against wormhole attacks in wireless networks. In *Proceedings of the 22nd Annual Joint Conference of the IEEE Computer and Communications Societies*, April 2003, pp. 1976–1986.

20. Y. Hu, A. Perrig, and D. Johnson. Rushing attacks and defense in wireless ad hoc network routing protocols. In *Proceedings of the 2nd ACM Wireless Security (WiSe'03)*, September 2003, pp. 30–40.

21. IEEE-SA Standards Board. IEEE Std 802.11b-1999, 1999.

22. D. Johnson and D. Maltz. Dynamic source routing in ad-hoc wireless networks routing protocols. In *Mobile Computing*. Kluwer Academic Publishers, 1996, pp. 153–181.

23. L. R. Ford Jr. and D. R Fulkerson. *Flows in Networks*. Princeton University Press, 1962.

24. M. Just, E. Kranakis, and T. Wan. Resisting malicious packet dropping in wireless ad hoc networks. In *Proceeding of ADHOCNOW'03*, October 2003, pp. 151–163.

25. F. Kargl, A. Klenk, S. Schlott, and M. Weber. Advanced detection of selfish or malicious nodes in ad hoc networks. In *Proceedings of the 1st European Workshop on Security in Ad-Hoc and Sensor Networks (ESAS 2004)*, August 2004, pp. 152–165.

26. M. Kazantzidis and M. Gerla. Permissible throughput network feedback for adaptive multimedia in aodv MANETs. In *Proceedings of IEEE International Conference on Communications (ICC 2001)*, Vol.5, June 2001, pp. 1352–1356.

27. E. Kranakis, H. Singh, and J. Urrutia. Compass routing on geometric networks. In *Proceedings of the 11th Canadian Conference on Computational Geometry*, August 1999, pp. 51–54.

28. H. Krawczyk, M. Bellare, and R. Canetti. Hmac: keyed-hashing for message authentication. Internet Request for Comments (RFC 2104), February 1997.

29. L. Lamport. Password authentication with insecure communication. *Commun. ACM*, 24(11):770–772, 1981.

30. H. Li and M. Singhal. A secure routing protocol for wireless ad hoc networks. In *Proceeding of the 39th Hawaii International International Conference on Systems Science (HICSS-39 2006)*, January 2006, pp. 225–234.

31. S. Marti, T. J. Giuli, K. Lai, and M. Baker. Mitigating routing misbehavior in mobile ad hoc networks. In *Mobile Computing and Networking*. August 2000, pp. 255–265.

32. T. S. Messerges, J. Cukier, T. A. M. Kevenaar, L. Puhl, R. Struik, and E. Callaway. A security design for a general purpose, self-organizing, multihop ad hoc wireless network. In *Proceedings of the 1st ACM Workshop on Security of Ad Hoc and Sensor Networks*, October 2003, pp. 1–11.

33. R. K. Nekkanti and C-W. Lee. Trust based adaptive on demand ad hoc routing protocol. In *Proceedings of the 42nd annual Southeast Regional Conference*, April 2004, pp. 88–93.

34. P. Papadimitratos and Z. J. Haas. Secure routing for mobile ad hoc networks. In *Proceedings of the SCS Communication Networks and Distributed Systems Modeling and Simulation Conference (CNDS 2002)*, January 2002.

35. V. D. Park and M. S. Corson. A highly adaptive distributed routing algorithm for mobile wireless networks. In *Proceedings of the 2nd IEEE INFOCOM*, April 1997, pp. 1405–1413.

36. A. Patwardhan, J. Parker, A. Joshi, M. Iorga, and T Karygiannis. Secure routing and intrusion detection in ad hoc networks. In *Proceedings of the 3rd IEEE International Conference on Pervasive Computing and Communications (PerCom 2005)*, March 2005, pp. 191–199.

37. C. Perkins. IP mobility support for IPv4. Internet Request for Comments (RFC 3344), August 2002.

38. C. Perkins and P. Bhagwat. Highly dynamic destination-sequenced distance-vector routing (dsdv) for mobile computers. In *Proceedings of ACM SIGCOMM Conference on Communications Architectures, Protocols and Applications*, October 1994, pp. 234–244.

39. C. Perkins and E. Royer. Ad hoc on-demand distance vector routing. In *Proceedings of the 2nd IEEE Workshop on Mobile Computing Systems and Applications (WMCSA 1999)*, February 1999, pp. 80–100.

40. A. Perrig, R. Canetti, D. Tygar, and D. Song. The tesla broadcast authentication protocol. *Cryptobytes*, 5(2):2–13, 2002.

41. A. A. Pirzada and C. McDonald. Establishing trust in pure ad-hoc networks. In *Proceedings of the 27th Conference on Australasian Computer Science (CRPIT '04)*, January 2004, pp. 47–54.

42. P. Resnick, K. Kuwabara, R. Zeckhauser, and E. Friedman. Reputation systems. *Commun. ACM*, 43(12):45–48, 2000.

43. L. Reyzin and N. Reyzin. Better than biba: short onetime signatures with fast signing and verifying. In *Proceedings of the 7th Australian Conference on Information Security and Privacy*, LNCS Vol. 2384, 2002, pp. 144–153.

44. R. L. Rivest, A. Shamir, and L. M. Adelman. A method for obtaining digital signatures and public-key cryptosystems. *Commun. ACM*, 21(2):120–126, 1978.

45. S. Buchegger and J. Le Boudec. Performance analysis of the CONFIDANT protocol. In *Proceedings of the 3rd ACM International Symposium on Mobile Ad Hoc Networking and Computing (MobiHoc'02)*, June 2002, pp. 226–236.

46. K. Sanzgiri, B. Dahill, B. Levine, and E. Belding-Royer. A secure routing protocol for ad hoc networks. In *Proceedings of 10th IEEE International Conference on Network Protocols (ICNP)*, November 2002, pp. 78–87.

47. S. Shah, K. Chen, and K. Nahrstedt. Dynamic bandwidth management for single-hop ad hoc wireless networks. In *Proceedings of IEEE International Conference on Pervasive Computing and Communications (PerCom 2003)*, March 2003, pp. 195–203.

48. I. Stojmenovic and X. Lin. Loop-free hybrid single-path/ooding routing algorithms with guaranteed delivery for wireless networks. *IEEE Trans. Parallel Distribut. Syst.*, 12(10):1023–1032, 2001.

49. H. Takagi and L. Kleinrock. Optimal transmission ranges for randomly distributed packet radio networks. *IEEE Trans. Commun.*, 32(3):246–257, 1984.

50. C.-K. Toh. Associativity-based routing for ad-hoc mobile networks. *Wireless Personal Commun.*, 4(2):103–139, 1997.

51. L. Venkatraman and D. P. Agrawal. An optimized inter-router authentication scheme for ad hoc networks. In *Proceedings of the Wireless 2001*, July 2001, pp. 129–146.

52. M. J. Wiener. Performance comparison of public-key cryptosystems. *Cryptobytes*, 4(1):1–5, 1998.

53. Z. Yan, P. Zhang, and T. Virtanen. Trust evaluation based security solution in ad hoc networks. In *Proceedings of the 7th Nordic Workshop on Secure IT Systems (NordSec 2003)*, October 2003.

54. S. Yi, P. Naldurg, and R. Kravets. Integrating quality of protection into ad hoc routing protocols. In *Proceedings of the 6th World Multi-Conference on Systemics, Cybernetics and Informatics (SCI 2002)*, August 2002, pp. 286–292.

55. M. Zapata and N. Asokan. Securing ad hoc routing protocols. In *Proceedings of the ACM Workshop on Wireless Security (WiSe'02)*, September 2002, pp. 1–10.

56. M. G. Zapata. Secure ad hoc on-demand distance vector (soadv) routing. INTERNET-DRAFT draft-guerrero-manet-saodv-00.txt, August 2001.

57. M. G. Zapata. Secure ad hoc on-demand distance vector routing. *ACM Mobile Comput. Commun. Rev.*, 6(3):106–107, 2002.

58. S. Zhong, J. Chen, and Y. Yang. Sprite: a simple, cheat-proof, credit-based system for mobile ad hoc networks. In *Proceedings of IEEE INFOCOM*, Vol. 3, March 2003, pp. 1987–1997.

Chapter 17

An Online Scheme for Threat Detection Within Mobile Ad Hoc Networks

A. I. Khan, A. H. Muhamad Amin, and R. A. Raja Mahmood

Clayton School of Information Technology, Monash University, Clayton, Victoria, Australia

17.1 Introduction	380
17.2 Graph Neuron	381
17.3 Hierarchical Graph Neuron	384
17.4 Case Study I: DHGN for Pattern Recognition	389
17.5 Case Study II: GN for Threat Detection in WSN	405
17.6 DHGN Approach for Threat Detection in MANET	407
17.7 Summary	409
Acknowledgments	410
References	410

17.1 INTRODUCTION

In this chapter, we will introduce an online scheme for threat detection within mobile ad hoc networks (MANET) using a single-cycle associative memory algorithm. The scheme implements a pattern-recognition approach, where the states of the network are considered as patterns. These patterns are collected and analyzed in real-time for discovering network intrusions and threat detection.

MANET is a decentralized network that comprises wireless mobile nodes that form a network through self-cooperation [1]. In MANET, there is no node configuration and coordination. The nodes are self-organized to form a mobile network. There

Mobile Intelligence. Edited by Laurence T. Yang, Agustinus Borgy Waluyo, Jianhua Ma, Ling Tan, and Bala Srinivasan

are several issues related to MANET, these include routing in a changing topology, wireless communications, energy constraints, and a general lack of computational resource in the nodes. Being a wireless and mobile network, MANET is prone to security threats such as selfish node, distributed denial-of-service (DDoS), and traffic jamming. Intrusion detection is one of the security measures generally adopted for overcoming these threats. However, implementing intrusion detection within MANET requires a very different strategy to the standard IP-based wired network. There are several intrusion detection schemes (IDS) reported in the literature, which deal with the distributed and decentralized aspects of MANET. Huang and Lee [2] have proposed a cooperative approach that uses a cluster-based detection scheme. Sun et al. [3] have introduced the zone-based intrusion-detection system. Xie and Hui [4] discuss a natural immune system approach for guarding against known and unknown attacks.

One of the core capabilities required for decentralized IDS is the ability to rapidly analyze known and unknown threats within a dynamic and decentralized environment. In this chapter, we will show how an online threat-detection system may be designed using a special form of neural network, known as the graph neuron (GN) [5], which implements single-cycle learning and is highly suited for tiny/resource-constrained devices operating within dynamic networks. The GN implements a single-cycle memorization and recall operation through a novel algorithmic design. Intrinsically, the GN is a lightweight in-network processing algorithm that does not require expensive floating point computations; hence, it is very suitable for real-time applications and tiny devices such as the wireless sensor networks (WSN) [6] and MANET [7]. It is possible to build a virtually unlimited associative memory resource within any network using the GN. In this chapter, we will introduce an associative memory approach as a mean for threat or attack pattern recognition. Furthermore, we will describe in detail a distributed version of GN that would be suitable for use within MANET.

The outline of this chapter is as follows: Section 17.2 briefly describes the GN as a single-cycle learning associative memory approach. Section 17.3 provides an introduction to the distributed hierarchical graph neuron (DHGN)—the proposed online scheme for threat detection within MANET. Section 17.4 comprises a case study relating to the application of GN as the pattern-recognition system for IDS. Section 17.5 deals with the communications aspects through an implementation of GN for WSN, and finally Section 17.6 provides conceptual design with mobility aspects for threat detection within MANET.

17.2 GRAPH NEURON

The GN is a novel approach that uses graph-based representations of patterns to achieve one-shot learning. A highly-scalable associative memory device is thus created, which is capable of handling multiple streams of input that are processed and matched with the historical data stored within the network. The method uses parallel in-network processing to circumvent the pattern-database scalability limitation associated with graph-based techniques [8].

17.2.1 Associative Memory Concept

Associative memory (AM) is derived from the neural network model and has been applied in many different application areas [5]. A widely used unsupervised learning technique is the Hopfield network. This network has been widely used for implementing associative (or content-addressable) memory in pattern analysis and optimization. A study of Hopfield memory model shows that the model is not scalable and is limited by the number of processing/storage nodes in the network [9]. The backpropagation network provides fast recalls, but the training cost becomes excessive for adding newer patterns. Ideally, an associative memory device should include simple one-shot training in the case of discrete time processing and fast retrieval. The GN is an approach that aims to overcome the scalability issues and reduce the training overheads in associative memory devices [10].

17.2.2 Simple Graph Neuron Approach

The GN is a finely distributed *in-network* pattern-recognition algorithm that preserves the data relationships in a graph-like memory structure. The GN structure and its data representations are analogous to a directed graph, the processing nodes of the GN array are mapped as the vertex set V of the graph, and the inter-node connections (i.e., the communication channels) belong to the set of edges, E. The communications are restricted to the adjacent nodes (of the array), hence there is no increase in the communication overheads with corresponding increases in the number of nodes in the network [11]. The information presented to each of the nodes is in the form of a (*value*, *position*) pair. Each of these pairs, in its simplest form, represents a data point in a two-dimensional reference pattern space. Hence, the GN array converts spatial/temporal patterns into a graph-like structural representation, as shown in Figure 17.1, and then compares the edges of the graph with subsequent inputs for memorization or recall.

The pattern-recognition process initially takes place in the following phases.

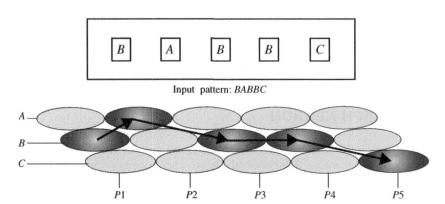

Figure 17.1   An input pattern *BABBC* is stored in a GN array. Each row of the array represents a value and each column represents a position where a value may occur within the pattern.

17.2.2.1 Pattern Input Phase

An input pattern, comprising *p(value, position)* pairs, is sequentially broadcast through the network. Each node based on its pre-defined position and value setting responds to the relevant input pair; disregarding the remainder of the pattern. From Figure 17.2, Node *GN(A,1)* with a predefined *value* = "A" and *position* = 1, will respond to the first letter of pattern *P1*, that is, *"A"BBD*, input as pair *p1(A,1)*. It will ignore the rest of the message. Similarly, node *GN(B,2)* will respond to the second pair *p2(B,2)*, *GN(B,3)* will respond to *p3(B,3)*, and *GN(D,4)* will respond to *p4(D,4)*. All other *GN* nodes will remain inactive during this pattern input phase.

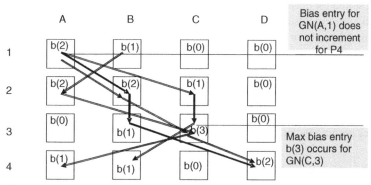

Figure 17.2 This figure shows the storage requirements per GN node when four arbitrarily chosen patterns, that is, *P1: ABBD, P2: ACCB, P3: BACA, P4: ABCD* have been stored in the GN array. The maximum bias size is 3 for storing four patterns, in this case. This suggests that the storage requirement per node would not disproportionably increase with the increase in the stored patterns.

17.2.2.2 Synchronization Phase

A broadcast signal is sent out marking the end of the incoming pattern to all the nodes.

17.2.2.3 Bias Array Update

During this phase, each activated node contacts all the adjacent nodes to find out the ones that responded to the input. It may be seen from Figure 17.2 that for the input pattern *P1 (ABBD)*, Node *GN(A,1)* will update its local bias array with the entry {*GN(B,2)*}. Similarly, Node *GN(B,2)* will update its bias array with the entry {*GN(A,1), GN(B,3)*}, Node *GN(B,3)* will add {*GN(B,2), GN(D,4)*} to its bias array, and *GN(D,4)* will add {*GN(B,3)*}. Thus, each bias array entry records the adjacent nodes being activated within a particular pattern input phase. Thus, a row of the bias array represents a part of the stored pattern. A new pair is defined, by a GN node, as the one that has a different set of adjacent GN nodes to all existing rows of its

Table 17.1   Store and recall responses of a GN array

Input Sequence	Input Pattern	Output
First	*AABA*	#### (store)
Second	*AABA*	*AABA* (recall)

bias array. A new pattern is found when at least one GN, within the list of activated GNs, cannot find a matching entry in its bias array. The new patterns are stored and previously encountered patterns are recalled in this stage. Table 17.1 shows the process where pattern "*AABA*" is stored and then recalled. Note that when the pattern is being stored for the first time, the output from the GN network would be in the form of a null entry; represented by the '*#*' pattern in the table. A null response indicates that no match has been found and that the segments of the pattern have been stored by the GN.

Stages 1 and 2 of the GN learning phase take place in a completely parallel and decentralized manner. It may be seen from Figure 17.2 that the maximum bias array size of three occurs in *GN(B,2)* after the array has stored four patterns. The scalability tests, with up to 16,384 nodes, have shown that the computational complexity only increases nominally with the increases in the size of the network [10].

17.2.3   Cross-talk Phenomenon

A GN network does not possess the information of the entire input pattern. This may lead to the intersection problem, generally known as the cross-talk. The cross-talk in GN may be illustrated as follows. Let us suppose that there is a GN array that can allocate six possible element values, for example, *u, v, w, x, y,* and *z,* for a five-element pattern. We initially introduce pattern *uvwxz*, followed by *zvwxy*. These two patterns would be stored by the GN array. Next, we introduce the pattern *uvwxy*, this will produce a recall. Clearly, the recall is false, as the last pattern does not match the previously stored patterns. The reason for this false recall is that a GN node only knows of its own value and its adjacent GN values. Hence, the input patterns in this case will be stored as segments *uv, uvw, vwx, wxy,* and *xy*. The latest input pattern, although different from the two previous patterns, contain all the segments of the previously stored patterns. Cross-talk problem could be solved by adding higher layers to the GN. This approach is known as hierarchical graph neuron (HGN) [12]. We will briefly outline the HGN approach in the following section.

17.3   HIERARCHICAL GRAPH NEURON

HGN is composed of layers of GN networks arranged in a pyramid-like composition. The base layer corresponds to the size of the pattern to be used in the pattern-recognition application. The size of the pattern is equal to the number of elements in

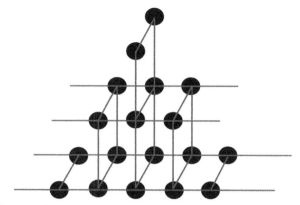

Figure 17.3   HGN Architecture for pattern size 5 and two possible values.

the pattern. Each element may be assigned a fixed number of possible values from the pattern domain. For instance, a black and white bitmap representation would have two possible values, that is, '*1*' or '*0*', for every pixel position. The HGN compositions for pattern size 5 with two possible values and pattern size 7 with three possible values are shown in Figures 17.3 and 17.4, respectively.

HGN algorithm is similar to the simple GN algorithm, with the exception of the hierarchical structure. The benefit of having this structure is that the top GN nodes are able to provide oversight for the base layer, and thus eliminate the cross-talk. Each layer of the HGN acts as a simple GN network with the difference that the higher layers (all layers except the base layer) would store the pair values as $p(index\_left, index\_middle, index\_right)$. Where $index\_left, index\_middle, index\_right$ represent the inputs received from the left, bottom, and right GNs respectively. Each of these indices represents the row number in the bias arrays corresponding to the input pattern.

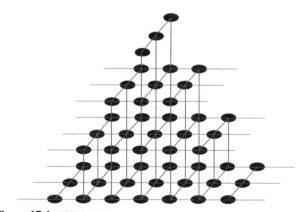

Figure 17.4   HGN Architecture for pattern size 7 and three possible values.

17.3.1 HGN Communications

The communication paths within the HGN layers are similar to that in the simple GN. The HGN communications propagate from the base layer to the top nodes, and consequently, from the top nodes to the base layer. Figure 17.5 shows the communication paths for a base layer where the nodes communicate within the layer and then pass the index values to the higher layer.

The HGN communications occur in the following manner. Each GN node at the base layer receives an input pattern from an outside source, which we refer to as the stimulator and interpreter (SI) module after Nasution and Khan [12]. Figure 17.6 shows the input communications for a five-element pattern, that is *YYXXY* with two possible values of *X* or *Y*, being sent by the SI module to a 10-node HGN base array for mapping.

Note that in Figure 17.6, all the elements with value *X* are processed by the first row of the GNs, while the remaining elements *Y* are processed by the second

To higher level nodes

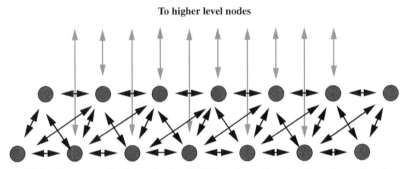

Figure 17.5 The communication paths for a HGN base layer. Note that the nodes on the edges only communicate with their adjacent nodes.

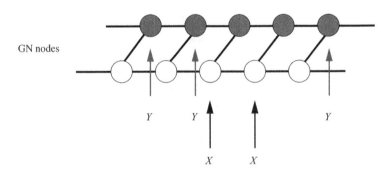

Pattern: *YYXXY*

Figure 17.6 Pattern inputs to a 10-node HGN base layer. In this diagram, the HGN array is capable of accepting five-element patterns with two possible values.

row of the GNs. Each GN node that receives an input is called an active node. An activated node at the base layer would send its *p(column, row)* pair to all the adjacent GNs, acknowledging that it has been activated. The *p(column, row)* pairs make up the GN's bias array entry for the current input pattern. In the end, each node would have received two pairs from its adjacent nodes, with the exception of the nodes on the edges, which will receive a single pair. Each activated GN node must then calculate its bias index. If the incoming pair combination is found in its bias array, then the index of the entry would be noted. Otherwise, a new index would be stored as a reference to the new pattern. Each active GN node would then send its index value to its higher layer node within the same column, except for the nodes on the edges. This process continues until the top most layer has been reached. The top layer GNs decide whether the input is to be treated as a new pattern and stored or it is a previously known pattern which needs to be recalled. A new index value is propagated downward for a stored pattern and an existing index value is propagated downward for a recalled pattern.

17.3.2 Nodes Requirement in the HGN

The main issue with deploying HGN, within a sparse network such as the MANET, is that the number of GN nodes increases substantially for analyzing larger and more complex patterns. Figure 17.7 shows the relationship between the pattern size and the number of GN nodes required for analyzing that pattern. It may be seen from the figure that the number of nodes required would be almost 1000 for a pattern of size 43. The large number of nodes needed by HGN may not be available within a contemporary MANET. This problem is solved by introducing the distributed hierarchical graph neuron (DHGN) approach, where a HGN composition is divided into multiple DHGN compositions. Each of these compositions can be then mapped to a MANET node.

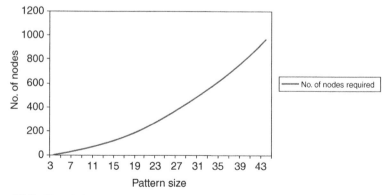

Figure 17.7   The relationship between the pattern size and the number of nodes required when implementing HGN for two possible pattern values.

17.3.3   Distributed Hierarchical Graph Neuron

DHGN is essentially an extension of the HGN algorithm wherein the HGN composition is decomposed into several DHGN compositions. With regards to the communications and the processes involved, DHGN is similar to the HGN with the exception that instead of using the whole pattern as an input, the pattern is segmented into smaller parts and each of the pattern segments is input to the respective DHGN composition. Figure 17.8 shows the decomposition of a HGN structure into several DHGN compositions.

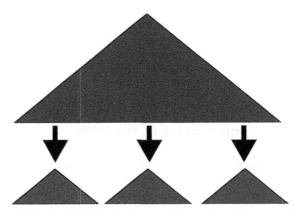

Figure 17.8   HGN decomposition into DHGN arrays. The HGN array is decomposed into three DHGN arrays. Note that the base of HGN array is equal to cumulative bases of DHGN arrays.

Each of the DHGN array is able to handle the pattern segments independently from one another. Hence, DHGN compositions may be independently mapped onto the available nodes in the network without losing the HGN accuracy. Figure 17.9

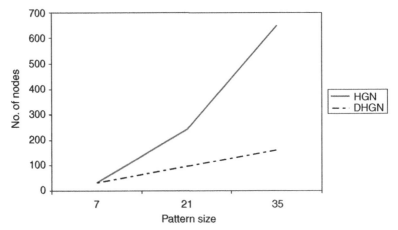

Figure 17.9   The number of nodes comparison between DHGN and HGN using seven-element pattern segments for representing an overall pattern of 35 elements.

shows the comparison between the number of nodes required for HGN and DHGN. The comparison is based on seven-element DHGN pattern segments corresponding to an overall pattern of size 35 for the HGN. It may be seen from the figure that DGHN would require less than 200 nodes for processing a 35-element pattern. The HGN would require about 650 nodes for processing the 35-element pattern.

17.4 CASE STUDY I: DHGN FOR PATTERN RECOGNITION

This case study takes into account two significant factors related to any distributed system, namely the varying capabilities of the participating nodes and the distribution of the computational load. Two different DHGN schemes were simulated to test the effectiveness of our approach as a distributed system. The first test addresses varying processing capabilities within a distributed system through the variable-form DHGN. The second test demonstrates the distributiveness of the approach through the standard-form DHGN.

17.4.1 DHGN Simulator Design

The DHGN simulation program for pattern-recognition application has been developed using C language with message passing interface (MPI) support for internodes communications. Figure 17.10 shows the architecture for the DHGN implementation

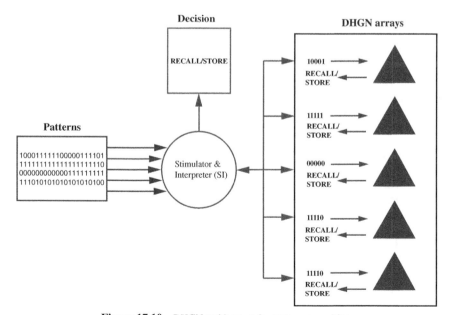

Figure 17.10 DHGN architecture for pattern recognition.

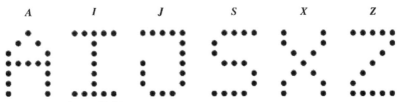

Figure 17.11   Test characters representation in 7 × 5 bit format.

as a pattern-recognition application. The SI acts as the controller for input and output functions for the DHGN architecture.

The test data for this simulation comprise a set of alphabet character patterns that can be visually distinguished. The letters *A*, *I*, *J*, *S*, *X*, and *Z* were selected in this regard. These letters were mapped onto 7 × 5 1-bit image representations, as shown in Figure 17.11.

The letters would then be converted into sequences of 35-bit patterns, as shown in the Table 17.2, using a horizontal scanning approach. These patterns would then be fed into the DHGN simulator. In this regard, a set of commands was implemented. The first command is for the initialization. During initialization, the user inputs the pattern elements into the base arrays. For example, in our bitmap representations, the dot pixel on the image could be represented by 1, while the blank pixel could be represented by 0; Figure 17.12.

The command *I10* would hence initialize the DHGN base-level arrays with pattern element 1 and 0, respectively. The *STORE* command is used to store the original patterns. An example of the command to store a particular pattern would be *S11111110000000011111110000000011111111*, where the pattern after the letter *S* will be stored in the arrays. In response to the *STORE* command, the DHGN arrays would generate a new index in the bias arrays if the pattern store operation was successful. Otherwise, the index under which the pattern was previously stored would be recalled. The *RECALL* command is used to input distorted patterns to the DHGN arrays. The DHGN arrays would respond with the output 0, if a close match is not found by the arrays. The pattern index will be returned if a match was found. The

Table 17.2   Character representations of 35-bit patterns using a horizontal scanning approach

Character	Bit pattern
A	00100010101000111111100011000110001
I	11111000010000110001100011000101110
J	01111100001000001110000010000111110
S	10001100010101000100010101000110001
X	10001100010101000100010101000110001
Z	11111000010001000100010001000011111

A	I	J	S	X	Z
0 0 1 0 0	1 1 1 1 1	1 1 1 1 1	0 1 1 1 1	1 0 0 0 1	1 1 1 1 1
0 1 0 1 0	0 0 1 0 0	0 0 0 0 1	1 0 0 0 0	1 0 0 0 1	0 0 0 0 1
1 0 0 0 1	0 0 1 0 0	0 0 0 0 1	1 0 0 0 0	0 1 0 1 0	0 0 0 1 0
1 1 1 1 1	0 0 1 0 0	1 0 0 0 1	0 1 1 1 0	0 0 1 0 0	0 0 1 0 0
1 0 0 0 1	0 0 1 0 0	1 0 0 0 1	0 0 0 0 1	0 1 0 1 0	0 1 0 0 0
1 0 0 0 1	0 0 1 0 0	1 0 0 0 1	0 0 0 0 1	1 0 0 0 1	1 0 0 0 0
1 0 0 0 1	1 1 1 1 1	0 1 1 1 0	1 1 1 1 0	1 0 0 0 1	1 1 1 1 1

Figure 17.12 Character bitmap image representations.

RECALL command takes the form of *R11111110000000011111110000000111111*, where the letter *R* is used to inform the DHGN arrays that it is a *RECALL* operation. This operation informs the arrays that the pattern is either a new one or it is a distorted version of a previously stored pattern.

17.4.2 Variable-Form DHGN

As mentioned in previous sections, DHGN network takes the form of multiple DHGN compositions. These compositions may then be distributed across the networks. In variable-form DHGN, the compositions may vary in size. For this simulation, we have chosen a 7–21–7 composition, where there are three substructures of DHGN, comprising two seven-element DHGNs and one 21-element DHGN. Figure 17.13 illustrates these compositions.

The variable-form DHGN takes into consideration an environment where some parts of the network would have lower power resources, and hence would only be able to provide limited processing capability as compared to other parts of the network. With this scenario in mind, we would like to analyze the effect of an unbalanced

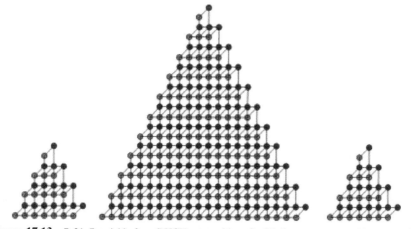

Figure 17.13 7–21–7 variable-form DHGN compositions for 35-element patterns with two possible values. Note that in this diagram, the middle array has been named substantially larger than the other two arrays.

composition on the pattern-recognition accuracy of the DHGN. The results of this simulation have shown that the variable-form DHGN offers almost similar accuracy to the HGN. Furthermore, the DHGN implementation requires less number of nodes. Eq. (17.1) shows the number of nodes N_{HGN} required for a single HGN composition with n possible values and pattern size p.

$$N_{HGN} = n \left(\frac{p+1}{2} \right)^2 \tag{17.1}$$

Eq. (17.2) shows the number of nodes N_{DHGN} required by using DHGN implementation with m compositions for the same sized problem.

$$N_{DHGN} = \sum_{i=1}^{m} n_i \left(\frac{p_i+1}{2} \right)^2 \tag{17.2}$$

It may be readily noted that the squared term in Eq. (17.2) would be substantially smaller that the one in Eq. (17.1) for the same sized problem, resulting in lesser number of GNs being required for the DHGN in comparison with the HGN.

The mapping process within the DHGN simulator begins with the input of the patterns. Each of the patterns, as shown in Table 17.2, is segmented and then loaded into the DHGN arrays, by the SI module. In this regard, Figure 17.14 shows the bitmap of character *I* being analyzed by the DHGN. In this case, the character *I* is stored after the character *A*, which has the index value of 1. Hence, the results show the character *I* as a new pattern with the index value of 2. Note that for this simulation, each segment is sequentially processed. However, in an actual implementation, the processing of these pattern segments would occur in parallel; vastly improving the execution time.

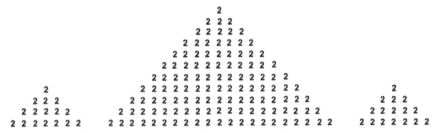

Figure 17.14 The DGHN arrays successfully store the bitmap pattern for character *I* at index value of 2 after having stored the bitmap pattern for character *A* at index value of 1.

17.4.2.1 Pattern Recognition Process

In the DHGN implementation, the overall store or recall decision depends on the collective decision of DHGN compositions. The top-level nodes of each DHGN composition would decide whether the pattern segment would lead to a recall or a store. If the pattern segment has not been identified, then the top nodes would generate index value 0. Otherwise, the recalled index value of the pattern segment would be displayed. Figure 17.15 shows the result of 1-bit distorted character pattern *A* being

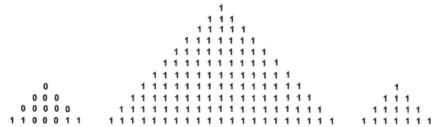

Figure 17.15   Results for introducing 1-bit distortion pattern of character *A*. The first DHGN array shows that a new pattern segment has been found (with assigned index 0), while other compositions correctly recall this as the pattern associated with index 1 (bitmap pattern of *A*).

introduced to the DHGN arrays after the character patterns *A*, *I*, *J*, *S*, *X*, and *Z* have been stored.

The figure shows that only one of the DHGN arrays records the pattern segment as a new pattern. The rest of the DHGN arrays recall the index value of 1, which is the index for the stored character pattern *A*. The decision whether the pattern is a recall or store is made using the cumulative decision among the DHGN arrays, through the determination of recall/store percentages. Eq. (17.3) shows the calculation of recall/store percentage P_{rs} for the DHGN pattern-recognition process for *m* DHGN compositions. Note that n_0 represents the GN nodes with index 0 in a composition, while *n* represents any GN node in a composition.

$$P_{rs} = \frac{\sum_{i=1}^{m} n_0}{\sum_{i=1}^{m} n} \qquad (17.3)$$

The DHGN scheme was tested by applying six levels of distortion to the stored character patterns and then measuring the recall accuracy for these distorted patterns. Figure 17.16 shows the original character patterns and the distorted patterns.

The original character patterns were first stored into the DHGN arrays, and then the distorted patterns were introduced to the arrays. The DHGN simulation was carried out using a SUN grid system where parts of the network were assigned to separate grid nodes for processing.

17.4.2.2 Results and Discussion

The pattern recognition recall rates for DHGN and HGN, with respect to identifying distorted patterns correctly, are shown in Figures 17.17–17.22.

It may be seen from the figures that the recall rates for the DHGN are very similar to the results of the HGN. In fact, some of the values obtained are higher than those of the HGN. For example, the 1-bit distorted patterns data-set shows a significant increase in the recall rates as compared to the HGN. This is owing to the encapsulation effect of DHGN where the effects of the distortion occurring within a particular DHGN array do not affect the other arrays. Figure 17.23 shows the encapsulation effect.

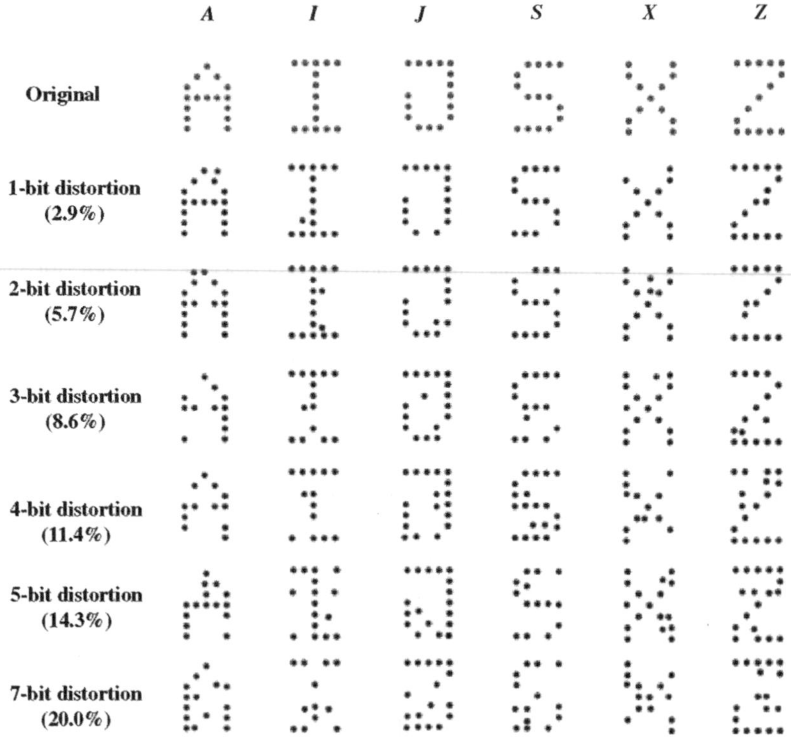

	A	I	J	S	X	Z
Original						
1-bit distortion (2.9%)						
2-bit distortion (5.7%)						
3-bit distortion (8.6%)						
4-bit distortion (11.4%)						
5-bit distortion (14.3%)						
7-bit distortion (20.0%)						

Figure 17.16 Characters set used in DHGN simulation on pattern recognition.

Figure 17.17 also shows the internal state of the arrays from the 1-bit distorted pattern of character A. The effects of the distortion are limited to the first array, where a part of the distorted pattern is analyzed, the remaining DHGN arrays are not affected by the distortion.

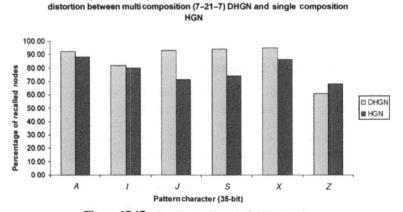

Figure 17.17 Recall rates for 1-bit (2.9%) distortion.

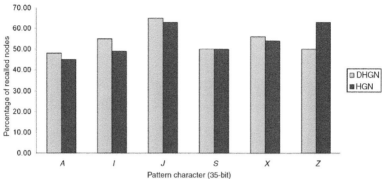

Figure 17.18   Recall rates for 2-bit (5.7%) distortion.

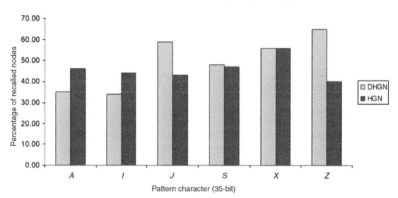

Figure 17.19   Recall rates for 3-bit (8.6%) distortion.

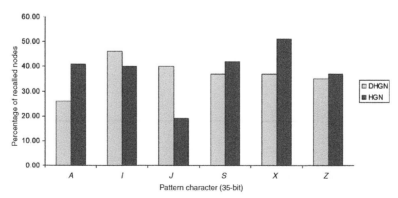

Figure 17.20   Recall rates for 4-bit (11.4%) distortion.

Percentage of recalled pattern for the given pattern sets with 5-bit (14.3%)
distortion between DHGN (7–21–7 composition) and single HGN

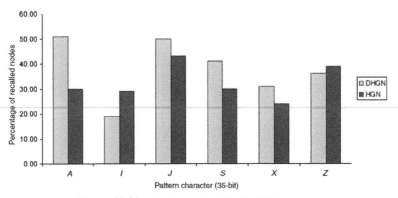

Figure 17.21 Recall rates for 5-bit (14.3%) distortion.

Percentage of recalled pattern for the given pattern sets with 7-bit (20.0%)
distortion between DHGN (7–21–7 composition) and single HGN

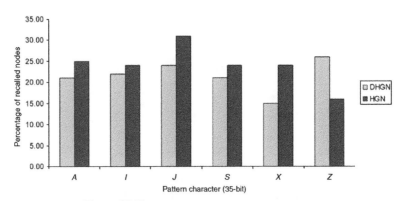

Figure 17.22 Recall rates for 7-bit (20.0%) distortion.

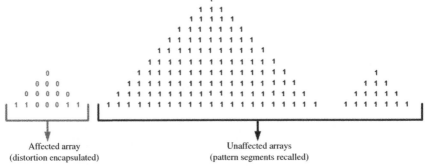

Figure 17.23 A 1-bit distortion occurring within the overall input pattern *A* stays encapsulated within the left composition.

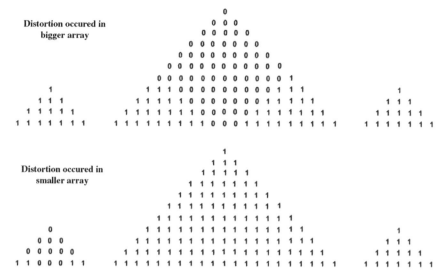

Figure 17.24   The distorted pattern occurring in the larger array (upper part of this figure) leads to the array not being able to recall the input pattern and returns an index of 0. Being the larger array, its decision outweighs the correct decisions reached by the smaller arrays in this case.

The downside of this encapsulation effect is that if the distortion occurs within the larger DHGN array, then the overall recall accuracy may be adversely affected. This phenomenon is called imbalanced recall composition. The upper half of the Figure 17.24 highlights this problem. This problem can be easily resolved if all the DHGN compositions were of similar sizes. The standard-form DHGN, therefore, implements similar sized compositions.

17.4.3   Standard-Form DHGN

The standard-form DHGN is introduced as a measure to delimit the effects of imbalanced recall composition experienced in variable-form DHGN. For the purpose of this pattern-recognition simulation, five of the seven-element pattern DHGN arrays were implemented for analyzing the 35-bit character patterns. Figure 17.25 shows the structure of these 5 × 7 DHGN compositions.

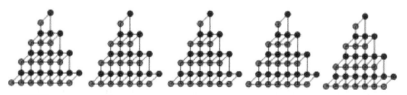

Figure 17.25   A 5 × 7 DHGN compositions for analyzing 35-element patterns with two possible values.

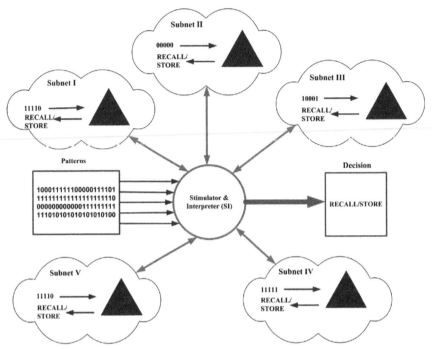

Figure 17.26 Standard-form DHGN implementation for multiple networks or a sparse network such as the MANET.

The standard-form DHGN has been developed to test the distributiveness of the DHGN algorithm for networks comprising small devices and/or limited processing and storage capabilities. With the relatively smaller sized DHGN arrays, each processing node is able to store smaller pattern segments and thus requires lesser processing capability for the pattern-recognition process. Figure 17.26 shows the distributiveness of the standard-form DHGN implementation, where all the compositions are allocated with similar computational loads. Figure 17.27 shows the DHGN index distribution produced by the DHGN arrays after the character *I* has been stored (character *I* is the second character pattern stored, after character *A*).

It may be seen from Figure 17.20 that each composition may be handled by any of the subnets/cluster in the network. The pattern input and output operations can also be decentralized—it is only for the simplicity of representation that we show

```
      2                 2                 2                 2                 2
    2 2 2             2 2 2             2 2 2             2 2 2             2 2 2
  2 2 2 2 2         2 2 2 2 2         2 2 2 2 2         2 2 2 2 2         2 2 2 2 2
2 2 2 2 2 2 2     2 2 2 2 2 2 2     2 2 2 2 2 2 2     2 2 2 2 2 2 2     2 2 2 2 2 2 2
```

Figure 17.27 Outputs from DHGN arrays for character pattern *I* using standard-form DHGN compositions.

the SI module as a separate entity within the network. The standard-form DHGN implementation is similar to the variable-form DHGN implementation. The principal difference is in the larger number of relatively smaller-sized DHGN arrays being deployed in comparison with the variable-form DHGN.

17.4.3.1 *Results and Discussion*

The simulation of standard-form DHGN, using the distorted character patterns from Figure 17.11, shows that it produces greater efficiency in terms of pattern recognition as compared with the variable-form DHGN and the HGN. In this regard, Figures 17.28–17.33 show the recall rates for the standard-form DHGN and the HGN.

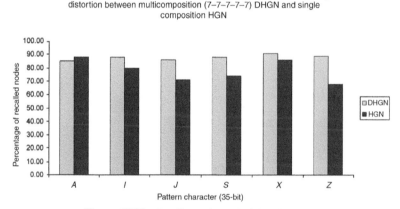

Figure 17.28   Recall rates for 1-bit (2.9%) distortion.

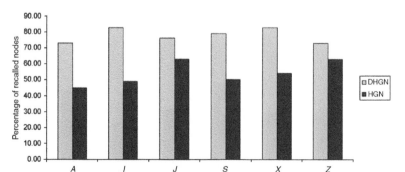

Figure 17.29   Recall rates for 2-bit (5.7%) distortion.

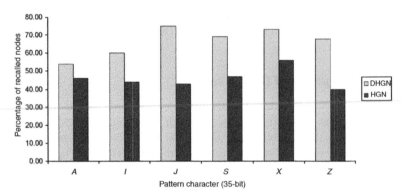

Figure 17.30   Recall rates for 3-bit (8.6%) distortion.

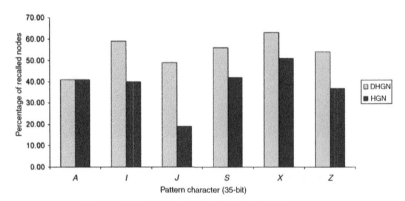

Figure 17.31   Recall rates for 4-bit (11.4%) distortion.

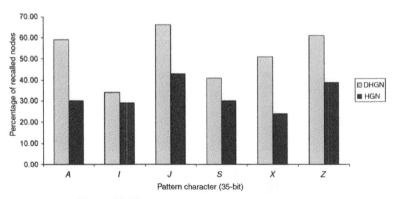

Figure 17.32   Recall rates for 5-bit (14.3%) distortion.

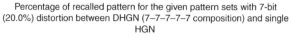

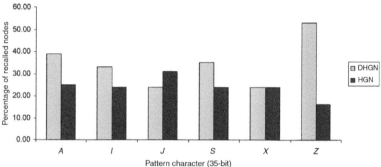

Figure 17.33   Recall rates for 7-bit (20.0%) distortion.

It may be seen from these charts that the standard-form DHGN's recall rates are significantly higher than those of the HGN. The increase in the recall accuracy is due to the encapsulation effect in which the distortions are generally compartmentalized within specific composition(s) and thus do not effect the findings of other compositions. The added benefit of the standard-form DHGN is that all the compositions are of similar sizes, hence the problem of an over-sized composition affecting the accuracy of the results is alleviated. Figure 17.34 shows the encapsulation effect within the standard-form DHGN for the pattern of character *A* with 2-bit distortion.

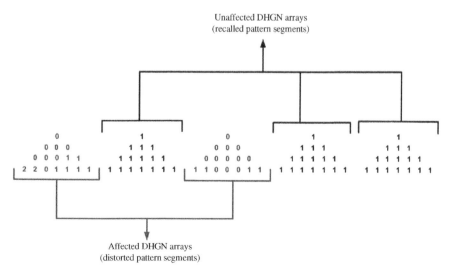

Figure 17.34   Encapsulation effect within the standard-form DHGN when processing a pattern of character *A* with 2-bit distortion. The effects of the distortions are localized within the two compositions on the left and do not influence the findings of the remaining compositions.

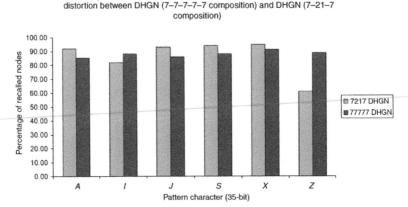

Figure 17.35   Recall rates for 1-bit (2.9%) distortion.

It may be observed from Figure 17.34 that the distorted pattern segments are encapsulated within the first and the third compositions from the left. The rest of the pattern segments are recalled as pattern of character *A* (represented by the bias index entry of 1). A comparison of recall rates for the standard-form DHGN and the variable-form DHGN are shown in Figures 17.35–17.40.

It is apparent from these figures that the standard-form DHGN generally produces better recall rates for the distorted patterns as compared with the variable-form DHGN. Better performance of the standard-form DHGN is due to the uniform encapsulation of the local distortions as compared with the variable-form DHGN. A close-up view of the difference in handling of the distortion by the standard-form DHGN may be observed in Figure 17.41. The figure shows the results for 1-bit distortion using three seven-element DHGN arrays and one 21-element HGN composition for pattern-recognition.

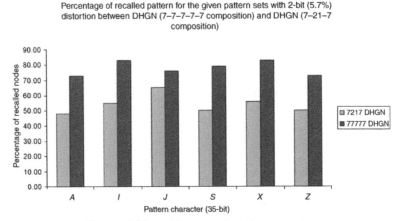

Figure 17.36   Recall rates for 2-bit (5.7%) distortion.

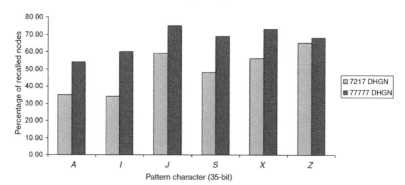

Figure 17.37   Recall rates for 3-bit (8.6%) distortion.

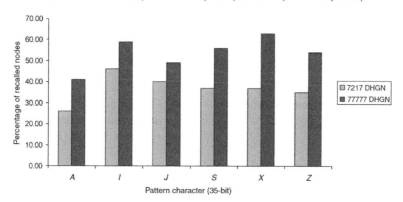

Figure 17.38   Recall rates for 4-bit (11.4%) distortion.

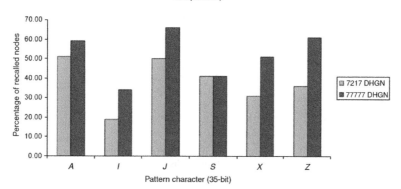

Figure 17.39   Recall rates for 5-bit (14.3%) distortion.

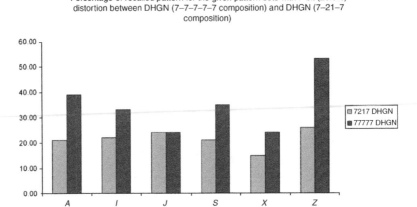

Percentage of recalled pattern for the given pattern sets with 7-bit (20.0%)
distortion between DHGN (7–7–7–7–7 composition) and DHGN (7–21–7
composition)

Figure 17.40   Recall rates for 7-bit (20.0%) distortion.

The distortion effect within the HGN composition cannot be localized and it propagates along the right-hand side of the composition (Figure 17.25), leading to a null recall. It is evident from Figure 17.25 that the smaller and similar sized DHGN compositions have a better chance of discovering the distorted pattern as compared with a single HGN composition.

It may be concluded from the above that DHGN provides a completely decentralized solution for pattern-recognition within distributed and decentralized mobile ad hoc networks. It also retains the single-cycle recognition characteristic of HGN, which makes it highly suitable for IDS within such networks. In the following section,

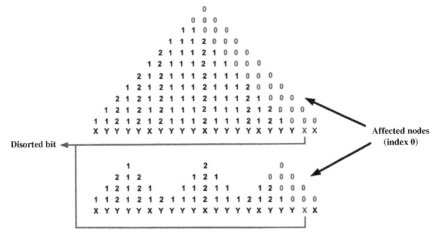

Figure 17.41   The effects of a 1-bit distortion in the pattern get localized within the standard-form DHGN compositions (lower), whereas the effects of the distorted pattern are propagated along the right side of the entire HGN composition, leading to a false conclusion.

we will explore the suitability of DHGN within wireless sensor networks. These networks are characterized by their highly dynamic nature where the nodes may appear or disappear with changes to the communication medium and/or the local conditions such as the energy and the environment.

17.5 CASE STUDY II: GN FOR THREAT DETECTION IN WSN

This case study is based on the work done by Baig et al. [6] for developing an online security scheme for the WSN. In this study, the simple GN algorithm for pattern-recognition has been used to detect the DDoS attack in WSNs. In this scheme, each network node also acts as the GN node within the WSN. The scheme shows that the GN algorithm provides an energy-efficient mechanism for attack pattern detection within the WSN.

17.5.1 The GN Approach for WSN

In this approach, the GN enables the network to process its internal traffic flow patterns in real-time through a process similar to human introspection using very little energy. The GN algorithm is implemented using an energy-efficient scheme, which uses partial updating of patterns within the network for energy conservation. The threat detection scheme is implemented in five stages and is designed for DDoS, which commonly occur within the WSNs:

1. Initialization: At this stage, the GN nodes (i.e., the wireless nodes) would be initialized with the DDoS attack patterns. Therefore, all the GN nodes will store the attack patterns for later detection.

2. Observation: All GN nodes would continuously monitor for attack pattern using the simple GN algorithm described in Section 17.1. The monitoring involves observing the threshold values obtained from the network neighborhood [6].

3. Communication: Each GN node would communicate with its adjacent nodes by exchanging its finding with the others.

4. Verdict: The decision whether the input pattern is an attack or a normal pattern would be simultaneously reached by the nodes within the network.

5. Pattern update: If a normal pattern has been detected, then the pattern would be updated. Otherwise, the attack pattern would be recalled.

The detection for DDoS attack is achieved using a set of threshold values. The threshold values for attack are initially stored in all the GN nodes. If an input pattern is detected with higher threshold value(s) then this triggers an alarm, indicating a possible DDoS attack. Otherwise the input pattern is treated as a normal pattern and no further action is required.

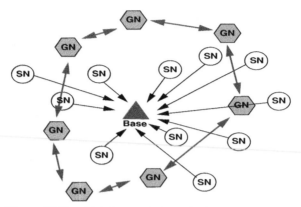

Figure 17.42 GN deployment within WSN flat network topology.

In order to detect DDoS attacks within WSNs, Baig et al. [6] have proposed the use of three different network topologies: (1) flat topology, (2) cluster head (CH) based topology, and (3) data aggregation (DA) based topology. Each topology has its own DDoS attack pattern algorithm. Figures 17.42–17.44 show the network topologies used for the GN deployment within a WSN.

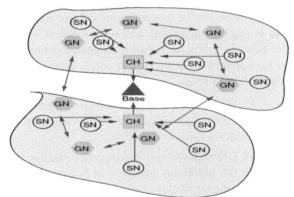

Figure 17.43 GN deployment with WSN CH-based network topology.

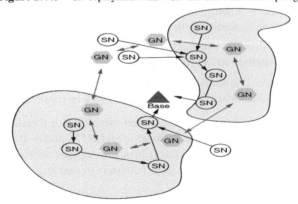

Figure 17.44 GN deployment with WSN DA-based network topology.

17.5.2   DDoS Detection Within Three Main WSN Topologies

In a flat network topology, Figure 17.42, each node in the network directly communicates its readings to the base station using a single-hop mechanism—without intervention from any intermediary nodes. Considering that the loss of any of the sensor nodes has an equal impact on the operations of the network, all sensor nodes are considered to be potential attack targets. The distance between the sensor nodes and the base station is used for generating the thresholds to be stored within the GN nodes.

In a CH-based network topology, Figure 17.43, each CH is responsible for the cluster administration, data gathering, and data forwarding operations within its cluster of operation. Considering the significance of the CHs to the operations of the network, the CHs are considered to be the likely targets for an attack. Individual GN nodes are responsible for observing traffic flow toward the CHs of their respective clusters. All other nodes in the GN array, that is, the ones not belonging to the cluster, are preset with relatively high do not-care threshold values for the CH and thus do not react to the network status queries.

In a DA-based topology, Figure 17.44, the sensor data as it progresses through the network from a source node toward the base station, is aggregated on its way at aggregation points, to minimize the overall traffic in the network. A typical data aggregation topology consists of interconnected trees defining the flow of network traffic from individual source sensor nodes to the base station. It is assumed that all nodes in the aggregation hierarchy are likely targets for a DDoS attack, as the energy resource exhaustion of a single aggregation point on a given source–sink path may cause an entire arm of the network to be incapacitated from participating in further operations. The threshold patterns generated in this case depend on both the proximity of the target nodes to individual GN nodes, as well as the distance between the target nodes and the base station. It may be noted from these figures, that the GN nodes are well-scattered within the WSN. The WSN nodes and the GN processes (executing on these nodes) act collaboratively to detect the DDoS attacks. The WSN nodes continually measure the network traffic metrics and the GN processes compare these measurements with the threshold patterns in real-time. In the event of a positive match, all nodes in the network are simultaneously informed and can start their recovery processes immediately.

The accuracy of this online pattern-recognition scheme depends on the rate of update of individual subpatterns within the GN nodes. By varying the update rate and limiting the updates to regions of interest, higher energy efficiencies may be achieved without substantially affecting the attack-detection rate.

17.6   DHGN APPROACH FOR THREAT DETECTION IN MANET

MANET implements a peer-to-peer multihop concept for supporting node mobility. It has neither fixed communications infrastructure nor any base stations [7].

Figure 17.45 shows the conceptual diagram of a typical MANET. There are quite a number of threats faced within MANETs. These threats are closely related to attacks on the WSN. Being a subset of wireless networks, MANETs are exposed to all the attacks experienced by of the wireless networks. Table 17.3 shows some of the attacks on MANETs according to the protocol stacks (adopted from [13]).

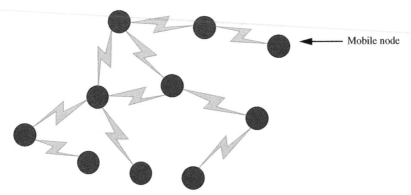

Mobile node

Figure 17.45   A typical MANET representation.

Intrusion detection is one of the countermeasures that may be used in MANETs. One of the important concepts in MANET-based IDS is the cooperation enforcement [13]. In cooperation enforcement, each mobile node is made aware of some of the misbehaving nodes. This gives rise to the self-aware state where the network must be able to recognize any changes in its state. IDS implementations such as Watchdog and Pathrater [14] use self-aware approaches to detect any changes in the behavior of the nodes.

Table 17.3   Security attacks on protocol stacks [13]

Layer	Attacks
Application	Repudiation, data corruption
Transport	Session hijacking, SYN flooding
Network	Wormhole, blackhole, Byzantine, flooding, resource consumption, location disclosure attacks
Data link	Traffic analyzis, monitoring, disruption MAC (802.11), WEP weakness
Physical	Jamming, interceptions, eavesdropping
Multi layer	DoS, impersonation, replay, man-in-the-middle

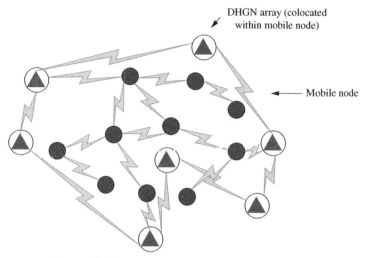

Figure 17.46   A self-aware MANET with a DHGN overlay.

A self-aware network may be easily implemented using a fast and scalable associative memory for storing and recalling its state information. The network memory is able to associate previously stored attack patterns with similar occurrences in future. In this regard, the GN scheme developed for the WSN may be deployed using the DHGN enhancement for IDS within MANETs. This scheme would also be energy-efficient scheme, as individual DHGN arrays would be able to detect the patterns without having to update the pattern information through the entire network. Working within smaller communication domains would provide better energy conservation. DHGN would also provide the much needed mobility support, where a group of nodes may temporarily become unavailable. The pattern distribution over several DHGN compositions in the network allows the matching of the pattern to continue even if part of the network becomes unavailable. Figure 17.46 shows a high-level view of a self-aware MANET being implemented with a DHGN overlay. It may be seen from the figure that each DHGN array is located within a particular mobile node. These mobile nodes act as components of the DHGN. When an attack pattern is detected, an alarm would be raised, providing means to invoke remedial measure throughout the network simultaneously.

17.7 SUMMARY

In this chapter, we have described GN as a new form of neural network that can operate effectively within resource-constrained network and adapts to the changing network conditions automatically. The GN implements a fully distributed single-cycle associative memory within the network. Hence, it is highly suited as an online system for providing rapid responses to the changing network conditions. It is also highly scalable, hence a distributed IDS based on the GN would assimilate vast amount of information on the network traffic patterns without running out of memory resource or becoming sluggish over time. The simple GN approach has been extended, using the HGN model, into DHGN. This distributed form of HGN would be useful for developing self-aware WSN and MANET for threat detection. We have shown through a case study how the simple GN approach could be used for DDoS attack detection in WSNs. The DHGN provides better framework for network deployment and can handle noisy/distorted patterns. Hence, the concept of network thresholds developed for WSNs may be readily applied within MANETs using the DHGN. The mobility consideration and the sparsity of the nodes in a MANET (in contrast with WSNs) are addressed through the standard or the variable forms of the DHGN.

ACKNOWLEDGMENTS

The authors would like to acknowledge the equipment grant from Alphawest, Sun Microsystems, and the support provided by Monash E-Research Centre for performing the network simulations.

REFERENCES

1. S. Kurkowski, T. Camp, and M. Colagrosso, MANET simulation studies: the incredibles. *ACM SIGMOBILE Mobile Comput. Commun. Rev.*, ED-9(4): 50–61, 2005.
2. Y. Huang and W. Lee, A cooperative intrusion detection system for ad hoc networks. In *Proceedings of the 1st ACM workshop on Security of Ad Hoc and Sensor Networks*, Fairfax, Virginia, USA, 2003.
3. B. Sun, K. Wu, and U. W. Pooch, Alert aggregation in mobile ad hoc networks. In *Proceedings of the 2003 ACM Workshop on Wireless Security,* San Diego, CA, USA, 2003.
4. H. Xie and Z. Hui, An intrusion detection architecture for ad hoc network based on artificial immune system. In *Proceedings of the Seventh International Conference on Parallel and Distributed Computing, Applications and Technologies (PDCAT'06)*, Taipei, Taiwan, 2006.
5. A. I. Khan, A peer-to-peer associative memory network for intelligent information systems. In *Proceedings of The Thirteenth Australasian Conference on Information Systems,* Melbourne, Australia, 2002.
6. Z. A. Baig, M. Baqer, and A. I. Khan, A graph neuron-based distributed denial of service (DDoS) attack recognition scheme for wireless sensor networks. In *Proceedings of the International Conference of Pattern Recognition (IPCR'06)*, Hong Kong, 2006.
7. D. P. Agrawal and Q. A. Zeng, Ad hoc and sensor networks. *Introduction to Wireless and Mobile Systems*, Brooks/Cole-Thomson Learning, pp. 297–348, 2003, chapter 13.
8. R. E. Trajan and A. E. Trojanowski, Finding a maximum independent set. *SIAM J. Comput.*, 25(3): 537–546, 1984.
9. E. M. Izhikevich, Weakly pulse-coupled oscillators, FM interactions, synchronization, and oscillatory associative memory. *IEEE Trans. Neural Networks*, 10(3): 508–526, 1999.
10. M. Baqer, A. I. Khan, and Z. A. Baig, Implementing a graph neuron array for pattern recognition within unstructured wireless sensor networks. In *Proceedings of EUC Workshops*, LNCS, Vol. 3823, pp. 208–217, 2005.
11. A. I. Khan, M. Isreb, and R. S. Spindler, A parallel distributed application of the wireless sensor network, In *Proceedings of the Seventh International Conference on Computing and Grid in Asia Pacific Region,* Tokyo, Japan, 2004.

12. B. B. Nasution and A. I. Khan, A hierarchical graph neuron scheme for real-time pattern recognition, *IEEE Trans. Neural Networks,* 19(2): 212–229, 2008.
13. B. Wu, J. Chen, J. Wu, and M. Cardei, A survey on attacks and countermeasures in mobile ad hoc networks. In *Wireless/Mobile Network Security.* Springer, pp. 1–35, 2006.
14. S. Marti, T. J. Giuli, K. Lai, and M. Baker, Mitigating routing misbehaviour in mobile ad hoc networks. In *Proceedings of the Sixth Annual International Conference on Mobile Computing and Networking (MOBICOM),* Boston, USA, 2000.

Chapter 18

SMRTI: Secure Mobile Ad Hoc Network Routing with Trust Intrigue

Venkat Balakrishnan, Vijay Varadharajan, Phillip Lucs, and Uday Tupakula

Information and Networked Systems Security (INSS) Research Group, Department of Computing, Macquarie University, Sydney, Australia

18.1 Introduction	412
18.2 Related Work	414
18.3 Assumptions and Terminologies	415
18.4 SMRTI: Architecture	416
18.5 Simulation Results	426
18.6 Conclusion	435
References	435

18.1 INTRODUCTION

A *mobile ad hoc network* (*MANET*) is a shortlived collection of heterogeneous mobile nodes, in which nodes communicate among themselves by forwarding packets through intermediate nodes. Security is paramount in such networks, as they are not conducive for centralized trusted authorities due to mobility and dynamically changing topologies.

Although several secure routing protocols [1–5] have been proposed to authenticate the intermediary nodes and to verify the integrity of discovered path, they are not

Mobile Intelligence. Edited by Laurence T. Yang, Agustinus Borgy Waluyo, Jianhua Ma, Ling Tan, and Bala Srinivasan
Copyright © 2010 John Wiley & Sons, Inc.

designed to evaluate the trustworthiness of intermediary nodes. This is so required to isolate previously identified malicious nodes and to dynamically choose trustworthy nodes to enhance the security of communications. Consequently, it has resulted in the development of numerous trust and reputation models [6–16], where a node's trust model captures evidence for the trustworthiness of other nodes in order to execute the decisions such as, whether to send a packet to or forward a packet on behalf of other nodes. Note that this approach is collectively known as *detection–reaction* approach. It is also noted that the evidence may be captured through passive monitoring, link layer acknowledgements, and recommendations. However, these models either modify basic routing operations or introduce additional issues to capture evidence for the trustworthiness of other nodes. Especially in the context of recommendations, they either struggle to resolve a recommender's bias or vulnerable to honest-elicitation and free-riding problems. A recommending node is subject to honest-elicitation, if it forwards high recommendation for a malicious node in order to avoid itself from being labeled with low recommendation by the same malicious node. Alternatively, a malicious node may also exhibit honest-elicitation by forwarding low recommendation for a benign node or high recommendation for a colluding malicious node. In the case of free-riding, a node accepts recommendations from other nodes but fails to reciprocate with recommendations when requested by them. Furthermore, exchange of recommendations introduces additional overhead and complexity and ultimately degrades the network performance. In addition, few trust and reputation models [6–8] fail to define efficient policies to isolate malicious nodes.

In this chapter, we detail our reputation-based trust model which is known as secure MANET routing with trust intrigue (SMRTI) and follows detection–reaction approach to evaluate the trustworthiness of other nodes and to isolate malicious nodes. In our model, evidence of trustworthiness is captured in the following ways—*from an interaction with a one-hop node, by observing the interactions between one-hop nodes, and through recommendations*. The captured evidence is then quantified and represented as reputation rating. In the case of interaction with one-hop nodes, the captured evidence enables our model to classify those one-hop nodes as either benign or malicious. Alternatively, the motivation to observe the interactions between one-hop nodes is to shortlist those one-hop nodes that are likely to misbehave in future interactions. Finally, recommendations enable our model to shortlist multihop benign nodes. Unlike related models [6–16], a novel approach is deployed for capturing recommendations, where the deployed approach neither modifies basic routing operations nor disseminates additional packets (or headers) for communicating recommendations. This eventually eliminates free-riding in our model and allows it to operate within the specifications of basic routing protocols. Since nodes never disseminate their reputation ratings as recommendations, our model is also able to defend against both recommender's bias and honest-elicitation problem. It defines effective policies to isolate malicious nodes and to select only benign nodes for communication. The decision for a context such as whether to send a packet to or forward a packet on behalf of another node depends on the policy defined for the context and the trustworthiness held for those nodes involved in the context. In turn, it is important to note that a node's trustworthiness is computed from the reputation ratings held for the node.

Also, our model is independent of centralized and distributed trusted authorities for its operation. A concise version of our model has been published in Ref. [25] and this chapter presents a detailed insight into our model.

The chapter is organized as follows. In Section 18.2, we consider some important related work and discuss their characteristics and limitations. Then we outline the assumptions and context for our model in Section 18.3. We present our trust model and describe its operation in detail in Section 18.4. Section 18.5 discusses the simulation results that demonstrate the efficient performance of our model. Finally, Section 18.6 gives some concluding remarks and future directions.

18.2 RELATED WORK

In this section, we discuss few well-reviewed and recently proposed trust models [6–16]. Liu et al. proposed a trust model [6], in which evidence is collected by monitoring the behavior of one-hop nodes and also through recommendations received from other nodes. They represent trust in discrete values and also explore various approaches for distributing recommendations to other nodes. Pirzada et al. [7] proposed similar approach for establishing trusted routes in *dynamic source routing (DSR)* protocol [17]. However, they represent trust in continuous values. They assign trust weight to each link and then apply shortest path algorithm based on the assigned weights to choose a trusted route. *Ant-based evidence distribution (ABED)* [8] adopted swarm intelligence-based approach, which is modeled from ant colonies for distributing and discovering trust evidence. Mundinger and Boudec proposed a deviation test to handle honest-elicitation problem in Ref. [9], where the trust model at every node accepts recommendation only if it does not differ too much from their reputation ratings. To enforce reputation sharing through recommendations and to address honest-elicitation problem, Liu and Issarny [10] impose a node to rate recommendations based on its experience with the recommended node. They then encourage the node to share recommendations only with nodes that have been sharing honest recommendations. However, their proposal is not foolproof against colluding recommenders and recommended nodes.

Virendra et al. proposed a trust model [11] to establish group keys in MANET using the trust relationship maintained among nodes. They adopt a five-stage procedure to initiate and maintain trust relationship among nodes, for which they accept recommendations only from one-hop trustworthy nodes. *Cooperation Of Nodes: Fairness In Dynamic Ad-hoc NeTworks (CONFIDANT)* [12] collects evidence from direct experiences and recommendations. Later it deploys the collected evidence to make trust-based decisions such as exclusion of malicious nodes and route selection. Although the model excludes incompatible recommendations, it does not penalize the liable recommender in order to encourage an honest recommender to disseminate future recommendations. Rebahi et al. proposed a trust model [13] similar to CONFIDANT, but their model primarily differs from CONFIDANT by enabling every node to broadcast its reputation to all one-hop nodes. Eschenauer et al. [14] recommend peer-to-peer-based strategies for discovering and distributing evidence. Also, they define the requirements for disseminating recommendations in which a recommender has to collect and evaluate evidence for the recommended node in the same manner as the node receiving recommendation collects and evaluates evidence for the recommender. Similar to other trust and reputation models [6-14], Li et al. [15] and Lindsay et al. [16] collect evidence from interactions with one-hop nodes and through recommendations in order to establish trust relationship among nodes. They then utilize the established trust relationships to choose trusted routes and to isolate malicious nodes. However, the former quantifies evidence using subjective logic, while the latter quantifies evidence using entropy.

We unanimously observed from our study that related trust and reputation models [6–16] are often prone to or lack a well-defined approach to completely defend against honest-elicitation problem. Further, the notion of disseminating recommendation in these models is also prone to free-riding, recommender's bias, and overhead. All these motivated us to develop a comprehensive detection-reaction based trust model that can enhance the security of MANET without succumbing to the above-mentioned issues. Interestingly, *Observation-based Cooperation Enforcement in Ad hoc Networks (OCEAN)* [18] collects evidence only from the interactions with neighbors to eliminate the above-mentioned issues. However, our model primarily differs from OCEAN by deploying a novel approach to communicate recommendations, in which the deployed approach derives recommendations without modifying basic routing operations.

18.3 ASSUMPTIONS AND TERMINOLOGIES

SMRTI enhances the security of network layer and hence like related reputation and trust models [14–16], it relies on secure routing protocols [1–5] to authenticate intermediate nodes. Jamming attacks are beyond the scope of this chapter; however, mechanisms such as spread spectrum or routing around jammed area [19] can be deployed to defend against such attacks. Similar to related reputation and trust models [7, 10–12, 15, 16], SMRTI captures evidence for modification attacks such as addition or deletion of nodes in a path, and unauthorized increment of the route request's sequence number through passive monitoring. It relies on cooperation-based fellowship model [20, 21] to defend against denial of service (DoS) attacks such as packet drop and flooding attacks.

We refer to the malicious actions performed by internal attackers as misbehaviors or malicious behaviors. On the other hand, a benign node's actions that confirm to the basic routing operations are known as normal or benign behaviors. We refer to one-hop nodes as neighbors and a node's wireless transmission range as its environment. For ease of explanation, we define the sequence of successful route discovery followed by data flow as *communication flow*. In general, route request (RREQ), route reply (RREP), and route error (RERR) phase of route discovery cycle and data flow are referred as events. We use the terminologies, path and route, interchangeably throughout the paper and they refer to the list of nodes traversed by a packet, including the source and destination. In this chapter, we explore the design and architecture of SMRTI using DSR protocol. Similarly, SMRTI can be overlaid with other reactive protocols such as ad hoc on demand distance vector (AODV) [22], which we leave it for another chapter. An instance of SMRTI is deployed at every node and its operation at each node is distributed and node centric, that is, its operation at each node is decoupled and independent of the operation at other nodes.

18.4 SMRTI: ARCHITECTURE

In SMRTI, reputation is defined as an *"opinion"* held for another node, where the opinion is based on the evidence captured from the node's behavior and it is subjective. Trust is then defined as an "expectation" that a node will behave as predicted, and the factor influencing expectation is then the reputation of node. Hence, trust is subjective and the expectation applies only to a given context and time. Given that the expectation depends on the opinion, trust is expressed as a function of reputation in our model.

18.4.1 Overview

SMRTI asynchronously captures evidence for the trustworthiness of neighbors from passively monitored packets and through recommendations. The evidence is then quantified according to the examined neighbor's behavior and the context of event. Finally, the quantified evidence is represented as reputation rating to express the opinion held for examined neighbor.

In SMRTI, detection component handles the operation of capturing and quantifying evidence into reputation (Figure 18.1). It accomplishes the required operations through its modules—*reputation capture* and *reputation evaluation*, and data structures—*packet buffer*, *reputation table*, and *reject list*. The reputation capture module captures, quantifies, and represents evidence as reputation, while the reputation evaluation module aggregates recent reputation with previously held reputation to revise the opinion of examined neighbor. The evidence for modification attacks is

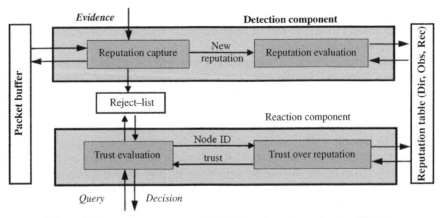

Figure 18.1 Architecture of secure MANET routing with trust intrigue (SMRTI).

captured by comparing promiscuously monitored packets with the packets stored in packet buffer. Alternatively, reputation table lists the reputation rating of other nodes. As detailed in Section 18.4.4.1, reject list contains the identity of nodes that have misbehaved during route discovery stage, so that they can be excluded until the completion of corresponding communication flow regardless of their high reputation ratings.

Recall that detection component collects evidence from passively monitored packets and through recommendations. However, the reputation rating is classified into one of three different categories—*direct, observed*, and *recommended*, depending on the perspective by which evidence is captured for an examined node, that is, whether the evidence is captured through direct interaction, observation, or recommendation.

SMRTI synchronously assists DSR to make decision for the following contexts, whether—*to accept or reject a newly discovered route from route discovery, to record or discard a route from forwarded packet, to send a packet to or forward a packet on behalf of other nodes, and to consider or ignore an evidence.* A decision is made for each of the above context depending on the policy detailed for the context and the trustworthiness held for those nodes that are involved in the context. In turn, a node's trustworthiness is determined from the reputation ratings (direct, observed, and recommended) held for the node.

SMRTI utilizes reaction component to respond back with a decision for the queries received from DSR (Figure 18.1). The reaction component performs required operations through its modules—*trust evaluation* and *trust over reputation*, and data structures— reputation table, and reject list. Depending on the policy detailed for an event's context, the trust evaluation module extracts specific nodes from the route and then makes corresponding decision depending on the trustworthiness held for those nodes. On the other hand, trust over reputation module computes each of the extracted node's trustworthiness using the reputation ratings (direct, observed, and recommended) held for them. Reputation table is utilized by trust over reputation module for its computation, while reject list is utilized by trust evaluation module to exclude nodes that have misbehaved during the route discovery of corresponding communication flow.

Given that discrete values can only provide a small set of possible values, while reputation and trust in MANET evolves continuously, both reputation and trust are represented by continuous real values $[-1, +1]$ as in Ref. [16].

18.4.2 System Operation

Consider the path $S \rightarrow O \rightarrow X \rightarrow N \rightarrow C \rightarrow D$ from Figure 18.2, where S is the source and D is the destination for communication flow. Nodes O, X, N, and C are the intermediaries that form the path from S to D. From here onward, this scenario will be used to explain SMRTI's decision making operation for each of the above-mentioned contexts.

Let us consider the operation of route registry at source, destination, and intermediate nodes. Following DSR protocol, each node discovers a new route either through route discovery or from forwarded packet. SMRTI details the policy to evaluate the

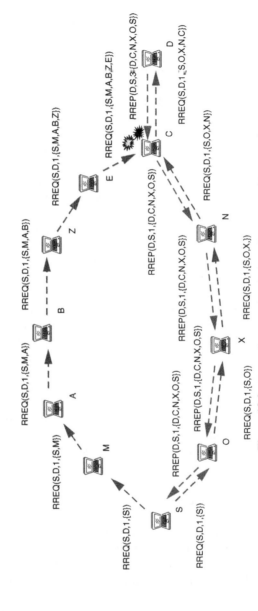

Figure 18.2 Route discovery phase in dynamic source routing protocol.

trustworthiness of all nodes positioned in the route (except for the evaluating node) before registering the route into route cache. SMRTI adopts this strategy to prevent the entry of routes, which may otherwise contain nodes that have been misbehaving except for the current communication flow. This in turn effectively enables the evaluating node to exclude routes that contain selectively misbehaving nodes. However, the evidence for selectively misbehaving node's benign behavior is taken into account as detailed in Section 18.4.4. Note that a route containing selectively misbehaving node will be recorded in future, only if the selectively misbehaving node raises its trustworthiness at evaluating node by exhibiting persistent benign behavior.

Let us now consider the operation of route selection at source S, where S initially checks its route cache for a route to D whenever it wishes to send data packets to D. On finding a route, S then evaluates the route's trustworthiness using SMRTI. Although the route's trustworthiness has been evaluated before registering the route into route cache, SMRTI details the policy to evaluate the route's trustworthiness prior to every deployment. This is due to the fact that trust is nonmonotonic, and hence the trustworthiness held for route may change at anytime between the point of entry and deployment. Given that there is more than one trustworthy route to D, the route with highest trustworthiness is chosen for communication. Alternatively, if only untrustworthy routes exist or routes are unavailable to D, then a new route discovery is initiated to D. In the case of former, SMRTI instigates the route cache to purge untrustworthy routes.

Let us consider in sequence the operation of receiving and forwarding packets at intermediate nodes, source and destination. At intermediate nodes (O, X, N, and C), SMRTI details a set of policies to decide whether to forward packets on behalf of other nodes and to send packets to other nodes. In order to accept a packet, it evaluates the trustworthiness of previous-hop neighbor from whom the upstream packet has been received. Likewise, it also evaluates the trustworthiness of next-hop neighbor to whom the downstream packet would be sent. Remember that the trustworthiness of both previous-hop and next-hop is determined using the evidence captured from their past benign and malicious behaviors. If the previous-hop (or next-hop) has been misbehaving, then SMRTI would have captured sufficient evidence for their past misbehaviors. Hence, this would reduce the trustworthiness held for them and exclude them from the communication flow. Given that these evaluations are performed during route discovery stage and assuming that multiple routes are discovered between S and D, then subsequent data flow is likely to be undeterred and secured with another trustworthy route. Finally, SMRTI evaluates trustworthiness for a packet by evaluating the trustworthiness of both source (S) and destination (D). The reasoning is based on the fact that intermediate nodes (O, X, N, and C) forward packets only for the sake of source (S) and destination (D). Similar to intermediate nodes, SMRTI at S details a policy to evaluate the trustworthiness of both next-hop (O) and destination (D). On the other hand, SMRTI at D details a policy to evaluate the trustworthiness of both previous-hop (C) and source (S). The above set of evaluations vary at source (S) and intermediate nodes (O, X, N, and C) during route request event, as the next-hop is unknown due to the broadcast nature of route request event. In such situation, intermediate nodes (O, X, N, and C) evaluate only the trustworthiness of previous-hop and packet (i.e., both for the source and destination). Similarly, source (S) evaluates only the trustworthiness of destination.

Note that before evaluating a node's trustworthiness for any decision-making operation, SMRTI always confirms that the node is not enlisted in the reject list. In turn, this prevents the inclusion of nodes that have misbehaved in a communication flow, especially during the route discovery stage.

18.4.3   Reaction Component

Reaction component assists DSR to make trust-enhanced decisions based on the policies defined for the contexts, and the trustworthiness held for those nodes that are involved in the contexts. As mentioned earlier, it performs the operation through its modules—trust evaluation and trust over reputation, and data structures—reject list and reputation table.

18.4.3.1   *Trust Evaluation*

To begin with, trust evaluation module extracts the context-oriented policy (such as route registry, route selection, and packet acceptance and forwarding), for which the decision has to be made. It then evaluates the trustworthiness of event-specific context using a predefined threshold value, *threshold-limit* (Δ). The context is declared as trustworthy only if the trustworthiness of context is at least "Δ." Prior to evaluation, trust evaluation module also confirms that nodes identified by the context-oriented policy are not enlisted in the reject list. In the following, we detail the evaluation of trustworthiness of each context, while trust over reputation module is utilized to evaluate the trustworthiness of nodes that are shortlisted by the context-oriented policy. In SMRTI, "Δ" can be varied depending on the context and event. However, for ease of explanation, a uniform "Δ" is assumed for all contexts.

SMRTI's trust evaluation module at node "i" computes trustworthiness for a packet "k," ($T_i\text{Packet}_k(t_{a+1})$), according to (18.1), where $0 < \alpha < 1$ and $t_{a+1} > t_a$. Parameter "a" shifts the priority between the packet's source "Src" and destination 'Dest' depending on the type of event. For instance, source "Src" is given highest priority during route request event, while destination "Dest" takes highest priority during route reply event. Note that trust evaluation module at node "i" computes the trustworthiness of source "Src," ($T_i\text{Node}_{Src}(t_a)$), and destination "Dest", ($T_i\text{Node}_{Dest}(t_a)$), using trust over reputation module.

$$T_i\text{Packet}_k(t_{a+1}) = \{[\alpha T_i\text{Node}_{Src}(t_a)] + [(1-\alpha)T_i\text{Node}_{Dest}(t_a)]\}. \quad (18.1)$$

Likewise in Eq. (18.1), trust evaluation module at node "i" computes the trustworthiness of route "R", ($T_i\text{Route}_R(t_{a+1})$) in Eq. (18.2). Note that trust over reputation module is used to capture the trustworthiness of each node "j," ($T_i\text{Node}_j(t_a)$), in the route.

$$T_i\text{Route}_R(t_{a+1}) = \sum_{j \in R \wedge j \neq i} \{\beta_j T_i\text{Node}_j(t_a)\}, \quad \sum_{j \in R \wedge j \neq i} \{\beta_j\} = 1 \quad (18.2)$$

18.4.3.2 Trust Over Reputation

During the initial stage of deployment, a node initializes its reputation ratings (direct, observed, and recommended) for other nodes at least to "Δ." SMRTI's trust over reputation module at node "i" computes the trustworthiness of node "j," ($T_i \text{Node}_j(t_{a+1})$), by retrieving the reputation ratings (direct, observed, and recommended) of "j" from the reputation table. In Eq. (18.3), "$\omega_{ij}^{RR}(t_a)$" refers to the reputation rating of type "RR" (direct, observed, and recommended) that is held for "j" by "i." Alternatively, the parameter "λ_j^{RR}" signifies the priority given to each type of reputation rating during the computation of trustworthiness of a node. As explained in the next section, reputation ratings are prioritized in ascending order as follows—direct, observed, and recommended. The reasoning is based on the intuition that personal experiences (direct interactions and observations) take higher precedence over the recommendations received from others, and similarly the priority for direct interactions precede the interactions observed between neighbors.

$$T_i \text{Node}_j(t_{a+1}) = \sum \left\{ \lambda_j^{RR} \omega_{ij}^{RR}(t_a) \right\}, \quad \sum \left\{ \lambda_j^{RR} \right\} = 1 \qquad (18.3)$$

18.4.4 Detection Component

The detection component collects, quantifies, and represents evidence as reputation rating for each type of reputation (direct, observed, and recommended). It then aggregates the recent reputation rating with the accrued reputation rating. As mentioned earlier, it performs the operation through its modules—reputation capture and reputation evaluation, and the data structures—packet buffer, reputation table, and reject list.

18.4.4.1 Direct Reputation

A node's direct reputation is defined as the opinion held for other node depending on the summary of evidence captured and quantified from its direct interactions with the other node. The node's reputation capture module collects evidence for direct reputation by validating the passively monitored packet against the duplicate packet stored in packet buffer (Figure 18.3). The collected evidence is then quantified into a positive or negative value depending on whether it accounts for a benign or malicious behavior respectively. The positive (or negative) value, which is the representation of recent direct reputation, is then passed on to the reputation evaluation module for aggregation with the accrued direct reputation.

The evidence is quantified into a positive value, pos(event), only if the packet has been received from upstream previous-hop or the packet has been transmitted by downstream next-hop without any misbehavior. Note that the positive value's magnitude depends on the type of event. Alternatively, a negative value, neg(event, action), is assigned if the evidence confirms to a malicious behavior. Unlike positive value, the negative value not only depends on the type of event but also on the type of malicious action. In the case of misbehavior, the malicious node (previous or next-hop) is appended to reject list so that it can be excluded from the corresponding

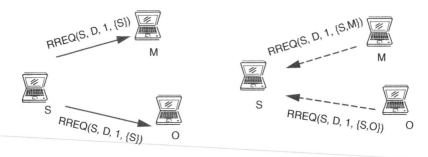

(a) S broadcasts RREQ to O and M   (b) S overhears subsequent broadcasts

Figure 18.3   Evidence collection for direct reputation through passive monitoring.

communication flow. If the malicious node is enlisted into reject list during route discovery stage, then the trust evaluation module would evaluate the route containing malicious node as untrustworthy. This would eventually invalidate the route in route cache, and trigger DSR to choose an alternate route from available valid routes for data flow. However, if the malicious node is enlisted into reject list during data flow, then corresponding invalidation of the route would induce DSR to invoke route maintenance operation.

In Eq. (18.4), the reputation evaluation module at node "i" aggregates the new direct reputation, {pos(event) or neg(event, action)}, with the accrued direct reputation, $\left(\omega_{ij}^{\text{Direct}}(t_a)\right)$, held for node "$j$." The result "$\left(\omega_{ij}^{\text{Direct}}(t_{a+1})\right)$" then becomes the accrued direct reputation for future computations. Note that the quantification of new evidence is not influenced by previously accrued direct reputation. Otherwise, in Eq. (18.4), the quantification of new evidence by "i" is independent of the opinion held for "j." This approach in combination with the limits (-1 and $+1$) applied in Eq. (18.1), prevents continual benign behaviors from concealing a future misbehavior (or continual malicious behaviors from concealing a future benign behavior). For instance, the maximum value that "i" could set to the direct reputation held for consistent benign behaviors of "j" is limited to $+1$, beyond which the continual benign behaviors of "j" fail to contribute. In other words, the state where "i" saturates the direct reputation held for "j" to $+1$, denotes that "i" believes "j" to be absolutely trustworthy with respect to direct interactions. However, an instance of malicious behavior by "j" in near future is adequate enough to disintegrate the absolute trustworthiness.

$$\text{"j" shows benign behavior} \equiv \omega_{ij}^{\text{Direct}}(t_{a+1}) = \min\left\{[+1], \left[\omega_{ij}^{\text{Direct}}(t_a) + \text{pos(event)}\right]\right\}$$

$$\text{"j" shows malicious behavior} \equiv \omega_{ij}^{\text{Direct}}(t_{a+1}) = \max\left\{[-1], \left[\omega_{ij}^{\text{Direct}}(t_a) - \text{pos(event, action)}\right]\right\}$$

$$(18.4)$$

18.4.4.2 Observed Reputation

A node's observed reputation is defined as the opinion held for an observed node depending on the summary of evidence captured and quantified from the observed node's behavior towards a common neighbor. Note that a node that is positioned within the transmission range of both observing and observed node is known as common neighbor. SMRTI derives the notion of observed reputation primarily from the concepts embedded in social psychology. For instance, in a society, an individual's behavior is observed whenever the individual deviates from normal behavior. In turn, it demonstrates the psychology of observers, who are primarily interested in taking note of individuals known for misbehaviors. The objective of observers is to take advantage of their observations, so that they can be cautious if they happen to interact with individuals who are known for misbehaving. It is important to note that the definition of a normal behavior may be subjective from the perspective of an observing individual, even though a generic definition exists in terms of social laws. It is also important to note that the observers fail to consider the individual's normal behavior unless it is of direct benefit to them.

SMRTI observes the interactions between neighbors to capture evidence for modification attacks. Consider Figure 18.4, where node Z collects evidence by observing the interactions between its neighbors D and C. To begin with, Z overhears the packet forwarded by D to C (Figure 18.4(a)) and then the packet is forwarded by C on behalf of D (Figure 18.4(b)). Node Z discards the observed evidence if C forwards the packet without any modification. From Z's perspective, C forwarding D's packet according to DSR's specification is not only an instance of normal behavior in the network but also relatively insignificant. Further, the decision to discard the evidence that is observed for normal behavior assists in counteracting colluding attacks. Otherwise, D and C may exchange dummy packets between them to increase their observed reputation at Z. However, Z assigns a negative value for C if C performs a modification attack. Note that the negative value is proportional to both the type of event and modification attack (such as increased sequence number, addition and deletion of routes). In sequence, node C is appended to reject list for exclusion until the completion of corresponding communication flow. Note that C not only loses direct reputation at previous-hop D for each of its misbehavior but also observed reputation at all observing neighbors including Z. In summary, penalizing a malicious node for multiple times (previous-hop and multiple observing neighbors) due to its misbehavior is believed to discourage the malicious node to deviate from normal behaviors.

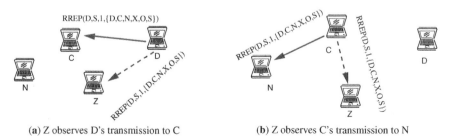

(a) Z observes D's transmission to C   (b) Z observes C's transmission to N

Figure 18.4   Evidence collection for observed reputation through passive monitoring.

If node "*j*" has misbehaved, then the reputation evaluation module at node "*i*" aggregates the new observed reputation, (neg(event, action)), with the accrued observed reputation, $\left(\omega_{ij}^{\text{Observed}}(t_a)\right)$, held for node "*j*," as given in Eq. 18.5. However, if "*j*" exhibits normal behavior, then the accrued observed reputation held for "*j*" remains unaltered. Finally, the result "$\left(\omega_{ij}^{\text{Observed}}(t_{a+1})\right)$" becomes the accrued observed reputation for future computation.

$$\text{"}j\text{" shows benign behavior} \equiv \omega_{ij}^{\text{Observed}}(t_{a+1}) = \left\{\omega_{ij}^{\text{Observed}}(t_a)\right\}$$

$$\text{"}j\text{" shows malicious behavior} \equiv \omega_{ij}^{\text{Observed}}(t_{a+1})$$

$$= \max\left\{[-1], \left[\omega_{ij}^{\text{Observed}}(t_a) - \text{neg}(\text{event, action})\right]\right\}$$

$$(18.5)$$

18.4.4.3 *Recommended Reputation*

A node's recommended reputation is defined as the opinion held for another node depending on the summary of recommendations derived in favor of it. For ease of explanation, the node providing recommendations, that is, the recommending node is referred as recommender. Similarly, the recommended node is referred as recommendee and the node that receives recommendation is known as requesting node. In general, most models [6–16] communicate recommendations by disseminating packets or adding extra headers. However, these models may corrupt their decisions by using recommendations received from other nodes. First, they lack ability to determine the bias of a recommender. Second, they are short of well-developed approaches to investigate the credibility of a recommender, that is, whether the recommender exhibits honest-elicitation and free-riding. In addition, the methodology of disseminating recommendations increases overhead and degrades the network's performance.

Processing Conventional Recommendations   We believe that understanding the steps involved in the methodology of disseminating recommendation is vital to address the associated issues. To begin with, a disseminated recommendation can be defined as an opinion held by a recommender toward a recommendee, which is then forwarded by the recommender to an interested node. Note that the disseminated recommendation reflects recommender's opinion and it may be only based on recommender's direct interactions with the recommended node. Otherwise, it may also include the summary of recommendations received by the recommender in favor of recommended node. Note that the disseminated recommendation presents a snapshot of recommender's relationship with the recommended node. It is, therefore, reasonable to deduce from a disseminated recommendation whether the recommender will forward a packet for the recommended node in future, provided their relationship remain unchanged. This deduction applies only if it is identical with the context to which the disseminated recommendation refers to (e.g., forwarding packets).

Once the disseminated recommendation reaches the requesting node, it is then used by the requesting node to revise the opinion held for recommended node. The reason for evaluation is that the requesting node might not have witnessed the event from which the disseminated recommendation has been crafted. Also, the disseminated recommendation reflects only recommender's evaluation of those events. Hence, the requesting node accepts or rejects the disseminated recommendation based on the trustworthiness held for the recommender. However, if it accepts the disseminated recommendation, then it scales the disseminated recommendation in proportion to the level of trustworthiness held for recommender.

Proposed Approach for Deriving Recommendations   Recall that it is reasonable to deduce from a disseminated recommendation whether the recommender will forward packets on behalf of recommended node in future. In our approach, nodes communicate recommendations by following the inverse of above-mentioned deduction process so that recommendations are free from associated issues.

Consider the scenario from Figure 18.5, where node X unicasts a packet to node N, containing the route $S \rightarrow O \rightarrow X \rightarrow N \rightarrow C \rightarrow D$. In this scenario, N derives an implicit recommendation from its previous-hop X (recommender) for node O (recommended node) depending on whether X has forwarded the packet received from its upstream O. To begin with, N captures evidence for X's recommendation in favor of O from the route contained in upstream packet. N then computes its opinion, that is, recommended reputation for O from the derived recommendation depending on the trustworthiness held for X. Similarly, N derives recommendation from the route until it reaches the end of route, that is, N traverses backward along the upstream route to derive recommendation from O for S.

As recommendations are derived from the route contained in upstream packet, it is necessary to validate the route before deriving recommendations. As explained earlier, nodes that deploy SMRTI will forward a packet, only if the trustworthiness held for—their previous-hop from whom they had received the upstream packet, their next-hop to whom the downstream packet has to be forwarded, and the packet (i.e., the source and destination of packet) are at least "Δ." Hence, it is evident that a node has to confirm that its previous-hop is trustworthy in order to forward an upstream packet. Further, it recursively applies to all upstream nodes that can be traversed backwards along the route until the source of packet.

Revisiting the above scenario, X forwards the upstream packet received from O, only if the trustworthiness held for O (previous-hop), N (next hop), S (source), and

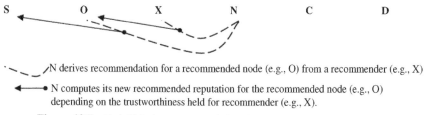

·—— ⁄N derives recommendation for a recommended node (e.g., O) from a recommender (e.g., X)

◀——● N computes its new recommended reputation for the recommended node (e.g., O) depending on the trustworthiness held for recommender (e.g., X).

Figure 18.5   Node N derives recommendations from the route of upstream packet.

D (destination) are at least "Δ." In sequence, N derives X's willingness to forward the packet on behalf of O, as X's recommendation for O. Similarly, N derives O's willingness to forward the packet on behalf of S, as O's recommendation for S. Note that the process of deriving recommendations terminates at S as there is no upstream node (previous-hop) for S.

Let us now consider the evaluation of recommendation by taking into account the recommendation derived from X (recommender) for O (recommended node) by N (requesting node). Initially, N computes the trustworthiness of X according to Eq. (18.3). Depending on whether the trustworthiness held for X is at least "Δ," it then assigns a positive or negative value for the derived recommendation. This indeed demonstrates N's viewpoint on the recommendation derived from X for O. Note that the positive (or negative) value assigned by N for the derived recommendation is also identical with other positive (or negative) values assigned for the rest of recommendations derived from other nodes. In other words, the above operation falls short to present the fact that N's trustworthiness for X need not be the same as the trustworthiness held for other nodes. Hence, the positive (or negative) value representing the recommendation derived from X for O is then scaled by N's trustworthiness for X to give the new recommended reputation for O. The scaling affirms that the derived recommendation is proportional to N's trustworthiness for X. Finally, N revises the accrued recommended reputation of O by aggregating both new and accrued recommended reputations together. Similar operation is then carried out for the recommendation derived from O for S.

Eq. (18.6) summarizes the above operations at time "t_{a+1}", in which node "i" aggregates the accrued recommended reputation, $\left(\omega_{ij}^{\text{Rec}}(t_a)\right)$, with the new recommended reputation, $\{[T_i\text{Node}_h(t_a)\ \text{pos(packet)}]\ \text{or}\ [T_i\text{Node}_h(t_a)\ \text{neg(packet)}]\}$, derived from node "$h$" for node "$j$" to arrive at the result, $\left(\omega_{ij}^{\text{Rec}}(t_{a+1})\right)$. Later, the result becomes the accrued recommended reputation for future computations.

$$T_i\,\text{Node}_h(t_a) \geq \Delta \equiv \omega_{ij}^{\text{Rec}}(t_{a+1}) = \min\left\{[+1], \left[\omega_{ij}^{\text{Rec}}(t_a) + T_i\,\text{Node}_h(t_a)\text{pos(packet)}\right]\right\}$$

$$T_i\text{Node}_h(t_a) < \Delta \equiv \omega_{ij}^{\text{Rec}}(t_{a+1}) = \max\left\{[-1], \left[\omega_{ij}^{\text{Rec}}(t_a) - T_i\,\text{Node}_h(t_a)\text{neg(packet)}\right]\right\}$$

$$(18.6)$$

In summary, the proposed approach prevents a node's recommended reputation from being corrupted by a recommender's recommendation, and this in turn facilitates the node only to believe in its decisions. Hence, the node better resolves the issues concerned with recommender's bias. Given that recommenders do not disseminate their opinions as recommendations, they are eventually prevented from exhibiting both honest-elicitation and free-riding behaviors.

18.5 SIMULATION RESULTS

NS-2 simulator has been widely used for evaluating ad hoc routing protocols. It uses two-ray ground reflection model [23] for radio propagation, which accounts for physical phenomena such as signal strength, propagation delay, capture effect, and interference. The transmission range for a node is set to 250 m. NS-2 deploys random waypoint model for mobility, in which a node starts from a random point, waits for a duration determined by pause time, then chooses another random point, and moves to the new point with a velocity uniformly chosen between 0 and maximum velocity "V_{max}." The medium access control (MAC) protocol is set to IEEE 802.11 distributed coordination function (DCF) and the routing protocol to DSR. In our simulations, the traffic from sender to receiver is a constant bit rate (CBR) flow with data rate of 2 Mbps and packet size of 512 bytes. Instead of maintaining CBR flows between fixed set of senders and receivers for the entire simulation run, CBR flows are dynamically varied between randomly chosen set of senders and receivers. Such a setup increases the total count of discovered routes and in turn exposes benign nodes to more malicious nodes. The simulation run for each run is 300 s, and the maximum and minimum duration for CBR flows are 20 and 40 s, respectively.

18.5.1 SMRTI and Malicious Nodes

SMRTI is implemented as a wrapper to DSR protocol in NS2 and a mobile node with SMRTI enabled is known as *SMRTI node*. During route discovery stage, intermediate SMRTI nodes adapt DSR protocol to accept the new sequence number of route request only when the route is chosen for data flow. The design alternative enables DSR protocol to discard routes with modified sequence number and permits the propagation of subsequent valid route requests. The parameters for SMRTI nodes are given in Table 18.1, which is based on our preliminary simulation study. For simplicity, neg(event, action) is fixed to a single value and adopted as a multiple of pos(event) to make misbehavior unattractive.

Remember that SMRTI relies on cooperation-based fellowship model to defend against flooding and packet drop attacks, respectively. SMRTI nodes are simulated against modification attacks in the absence of fellowship model to study their performance. The modification attacks considered for analysis are addition or deletion of nodes from the route and increase in route request's sequence number. The nodes that modify the routing headers with an aim to embed on the route or to disrupt the data flow are known as *malicious nodes*. We refer to nodes that only follow the specifications laid out by DSR protocol as *DSR nodes*. These nodes neither perform modification attacks nor enable SMRTI to defend against modification attacks.

Table 18.1 SMRTI parameters for simulation

Parameters	Value
Threshold limit (Δ)	0.50
Default reputation value (direct, observed, and recommended)	0.51
Pos(event)	0.02
Neg(event, action)	0.10

18.5.2 Scenarios and Performance Metrics

The total number of nodes in the network is fixed to 100. In our simulations, SMRTI and DSR nodes are independently simulated against malicious nodes under various identical scenarios with same set of parameters for comparison purpose. The set of parameters that are varied for different scenarios is given in Table 18.2.

The simulation scenarios considered for performance analysis and the performance metrics evaluated for each scenario are given below:

Scenario I: The performance of SMRTI and DSR nodes are evaluated against increasing proportion of malicious nodes (i.e., from 0 to 100 in the intervals of 10). Remaining parameters include pause time of 10 s, maximum velocity "V_{max}" of 20 m/s, and simulation area of $1500 \times 1500\,m^2$. This scenario discovers the proportion of malicious nodes beyond which the performance of SMRTI and DSR nodes falls notably.

Scenario II: Following previous scenario, the proportion of malicious nodes to SMRTI (or DSR) nodes is fixed to three uniformly distributed values—25, 50, and 75% of total nodes in the network. For each distribution, the performance of SMRTI (or DSR) nodes is analyzed by varying the maximum velocity "V_{max}" from 0 to 50 m/s in the intervals of 5 m/s. Remaining fixed parameters for the simulation are pause time of 10 s and simulation area of $1500 \times 1500\,m^2$. The purpose of this scenario is to study the impact of mobility in SMRTI's (or DSR) performance against malicious nodes.

Scenario III: In comparison with scenario II, the pause time is varied from 0 to 40 s in the intervals of 4 s and the maximum velocity is fixed to 20 m/s, while other parameters remain unchanged. This scenario is designed to determine the pause times for which SMRTI (or DSR) nodes perform better against malicious nodes.

Scenario IV: In contrast to scenario III, the simulation area is varied from 500×500 to $5000 \times 5000\,m^2$ in the intervals of $500 \times 500\,m^2$ and the maximum velocity is fixed to 20 m/s, while the rest are same as in scenario II. This scenario evaluates the impact of network connectivity and node density in the performance of SMRTI (or DSR) nodes against malicious nodes.

Table 18.2   Variable NS2 parameters for simulation

Parameters	Value
SMRTI nodes	0–100
Malicious nodes	0–100
Maximum velocity "V_{max}"	0–50 m/s
Pause time	0–40 s
Simulation area	500×500 to $5000 \times 5000\,m^2$

- *Performance metric*: Packet delivery ratio (PDR) is the average ratio of total number of CBR data packets received by destination to the total number of CBR packets sent by source.

However, the performance metrics such as packet or byte overhead are not evaluated in our simulations as SMRTI nodes do not generate additional packets or headers in comparison with DSR nodes.

18.5.3 Performance Analysis

From Figures 18.6–18.9, it is evident that SMRTI nodes perform better than DSR nodes. Notably in Figure 18.6, the PDR for DSR nodes falls steeply from 82 to 20% on the introduction of 10% malicious nodes. The steep downfall primarily roots from the modification of route request's sequence number and the other factors responsible for downfall are addition and deletion of nodes in the route. It is evident that a valid route with shortest path can effectively prevent the selection of a maliciously lengthened route for CBR flow. Alternatively, a maliciously shortened route can successfully disrupt corresponding CBR flow. Nonetheless, modification of a route request's sequence number adversely blocks the propagation of corresponding and future route requests, thereby preventing the establishment of valid routes for corresponding and subsequent CBR flows. This continues until the sequence number of a future route request exceeds the sequence number of modified route request. From Figure 18.6, we can observe that further increase in the proportion

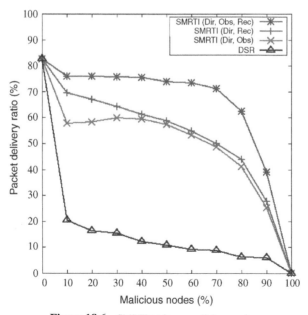

Figure 18.6   SMRTI nodes vs malicious nodes.

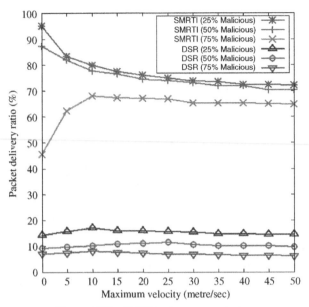

Figure 18.7   SMRTI nodes vs max velocity.

of malicious nodes marginally affects the performance of DSR nodes. Similarly, the performance of DSR nodes fluctuate minimally in the presence of varying maximum velocity "V_{max}" and pause time in Figures 18.7 and 18.8, respectively. Since the rate at which mobility disrupts CBR flows and the probability of locating malicious

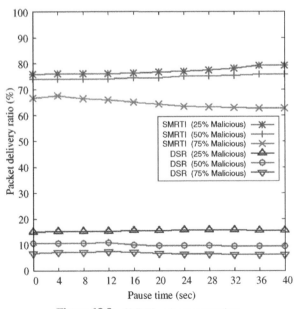

Figure 18.8   SMRTI nodes vs pause time.

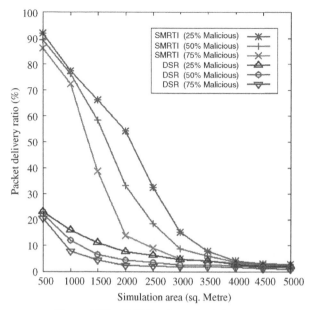

Figure 18.9 SMRTI nodes vs simulation area.

nodes in routes are proportional to the path length, we attribute that the path length of active CBR flows has to be one or two hops. In other words, we assert that the active CBR flows that are devoid of malicious nodes are made up of one or two hops for DSR's performance to reduce marginally. We further substantiate our claim using DSR's remarkable performance at $500 \times 500\,m^2$ in Figure 18.9. This is because CBR flows are predominantly one hop in length as nodes are densely packed at $500 \times 500\,m^2$ due to their 250 m transmission range. In summary, we conclude that DSR nodes fail to establish multihop paths against malicious nodes and deliver PDR only if they are within close proximity such that their CBR flows are one or two hops.

18.5.3.1 SMRTI Nodes Vs. Malicious Nodes

SMRTI nodes are effective against malicious nodes (Figure 18.6) as they *only accept trustworthy routes from route discovery and forwarded packets, record the sequence number of route request only if the route is trustworthy and active, propagate only trustworthy packets, send a packet to or forward a packet on behalf of trustworthy nodes, and exclude nodes that have misbehaved during the route discovery of corresponding communication flow*. Recollect that SMRTI nodes make decisions for each of the above contexts depending on the policy detailed for the context and the trustworthiness held for those nodes that are identified by the context-specific policy. Since a node's trustworthiness is determined by the reputation ratings (direct, observed, and recommended) held for that node, the type of reputation rating (direct, observed, and recommended) considered for evaluating a node's trustworthiness has a binding impact on the performance of SMRTI nodes.

The curve in Figure 18.6, "SMRTI (Dir, Obs)," presents the influence of considering the evidence captured only for direct and observed reputation. In the case of direct reputation, the evidence captured from direct interactions enables SMRTI nodes to classify a neighbor as either malicious or benign. Alternatively, the evidence captured by observing the interactions between neighbors enables SMRTI nodes to shortlist malicious nodes that are likely to misbehave in an interaction. In other words, observed reputation augments the evidence captured for direct reputation and enhances the decisions of reaction component such as exclusion of routes that contain shortlisted malicious nodes, and prevention of packet propagation to shortlisted malicious nodes. PDR reduces notably as the proportion of malicious nodes exceeds half the number of nodes in network. In such situation, the decreasing proportion of SMRTI nodes lowers the availability of trustworthy routes despite the ability of SMRTI nodes to effectively detect malicious nodes using direct and observed reputation.

Similarly another curve in Figure 18.6, "SMRTI (Dir, Rec)," presents the influence of considering the evidence captured for both direct and recommended reputation. Recommendations that are derived by inferring the relationship among nodes enable SMRTI nodes to identify other SMRTI nodes. In other words, recommended reputation not only augments the evidence captured by a SMRTI node for its direct reputation but also associates similar SMRTI nodes into a society within the network. This in turn effectively enhances the decisions of reaction component such as inclusion of trustworthy routes for CBR flow and propagation of packets to only trustworthy nodes. PDR reduces gradually as steadily increasing malicious nodes progressively decrease the proportion of SMRTI nodes in the network.

In summary, both observed and recommended reputations augment direct reputation and mark their influence in enhancing the decisions of reaction component as long as the total count of SMRTI nodes is relatively higher than malicious nodes in the network. In other words, they complement each other, which could be observed in the curve, "SMRTI (Dir, Obs, Rec)," shown in Figure 18.6. In this case, direct reputation in combination with observed and recommended reputation remarkably influences SMRTI nodes to sustain performance up to 75% malicious nodes in the network. Interestingly in Figure 18.6, PDRs for SMRTI and DSR nodes uniformly agree with each other in the absence of malicious nodes. This asserts that SMRTI nodes operate similar to DSR nodes in the absence of malicious nodes and eliminate further overhead such as dissemination of additional packets or headers for recommendations. We also discovered that the performance of SMRTI and DSR nodes never reach 100% even in the absence of malicious nodes due to the following reasons—*few destination nodes are unreachable as the random waypoint mobility disrupts the uniform distribution of nodes, nodes positioned along the shortest paths drop packets as CBR flows opt for congested shortest paths,* and *nodes positioned at the center of network drop packets as most of the shortest paths passing through the network's center experience contention.* Solutions based on load balancing and quality of service (QoS) [24], which are beyond the scope of this chapter, can be employed to improve the performance in such situation.

18.5.3.2 SMRTI Nodes Vs. Maximum Velocity and Pause Time

Depending on the proportion of malicious nodes present in the network, performance of SMRTI nodes differs in the presence of varying maximum velocities "V_{max}", as in Figure 18.7. In situation where the proportion of malicious nodes is at most half the total number of nodes in the network (i.e., 25 and 50% malicious nodes), the performance of SMRTI nodes decreases notably in particular at the lower end of maximum velocities. Alternatively, when three quarter of the network is occupied with malicious nodes, the performance of SMRTI nodes increases remarkably in particular at the lower end of maximum velocities. Although the performance of SMRTI nodes increases with growing maximum velocity when 75% malicious nodes exist in the network, it is comparatively lower than the performance gained against 25 and 50% malicious nodes.

Let us consider the earlier setup with either 25 or 50% malicious nodes in network. From Figure 18.7, it can be seen that SMRTI nodes perform better for such setup only when the maximum velocity is 0 m/s. In this case, immobility anchors SMRTI nodes and their neighbors to fixed positions. In other words, SMRTI nodes capture recurring evidence of trustworthiness for both direct and observed reputations due to the fixed nodal positions and repeating behavior of neighbors. Note that the evidence of trustworthiness captured for recommended reputation may vary during the simulation run due to the dynamic CBR flows. As the simulation time ascends, the recurring evidence influences SMRTI nodes to efficiently deepen the separation between malicious and benign nodes and leads the reaction component to make precise decisions. However, mobility-induced insertion or removal of neighbors in the environment of SMRTI nodes varies the evidence captured during simulation run. Further increase in mobility leads to broken links, which results in loss of data packets and consequently induces new route discovery cycles. All these substantiates for the low performance of SMRTI nodes at the higher end of maximum velocities.

Now let us consider the latter setup with 75% malicious nodes in the network. Unlike the earlier counterpart, it can be seen from Figure 18.7 that SMRTI nodes perform poorer for such setup when the maximum velocity is 0 m/s. Recall that immobility anchors SMRTI nodes and their neighbors to fixed positions, thereby causing SMRTI nodes *to capture recurring evidence of trustworthiness for direct and observed reputations* and *to effectively deepen the separation between malicious and benign nodes*. In comparison with the earlier counterpart, it is important to note that 75% malicious nodes in the network means the environment of SMRTI nodes is populated on average with three fourth of malicious nodes. As a result, the reaction component struggles to find valid routes through high density of malicious nodes regardless of its capacity to effectively distinguish malicious nodes from benign nodes. However, note that the probability for SMRTI nodes to meet similar SMRTI nodes increases considerably once the nodes are mobile. This can be affirmed from the increase in performance for SMRTI nodes at the lower end of maximum velocities as seen in Figure 18.7. Alternatively, higher maximum velocities prevent further increase in the performance of SMRTI nodes as they constrain the duration of data flow between SMRTI nodes and introduce broken links.

The performance of SMRTI nodes against malicious nodes in the presence of varying pause time (Figure 18.8), inversely relates to their performance against malicious nodes in the presence of varying maximum velocities (Figure 18.7). In Figure 18.7, the performance of SMRTI nodes gradually increases with increasing pause time when the proportion of malicious nodes is at most half the total number of nodes in the network (refer to curves, "SMRTI (25% Malicious)," and "SMRTI (50% Malicious)" in Figure 18.8). In other words, the performance of SMRTI nodes at the lower end of pause times (Figure 18.8) is similar to their performance at the higher end of maximum velocities in Figure 18.7. The reason for the similarity rests on the fact that the lifetime of routes is shorter either at higher maximum velocity or lower pause time. Similarly, the performance of SMRTI nodes at the higher end of pause times (Figure 18.8) is analogous to their performance at the lower end of maximum velocities in Figure 18.7. The reason for similarity is derived from the fact that the duration of data flow among SMRTI nodes is longer either at higher pause time or lower maximum velocity.

Alternatively, if three fourth of the network is occupied by malicious nodes, then the performance of SMRTI nodes gradually decreases with growing pause time, which is represented by the curve, "SMRTI (75% Malicious)" in Figure 18.8. Given that the environment of SMRTI nodes is populated by 75% malicious nodes on average, the probability of SMRTI nodes discovering valid routes decreases considerably. In addition, the uniformity between higher pause times (in between 20 and 40 s) and duration for CBR flows (minimum 20 s and maximum 40 s) also increases the likelihood of SMRTI nodes being trapped in the environment of malicious nodes. As expected, the performance of SMRTI nodes against 75% malicious nodes is lower than its performance against 25 and 50% malicious nodes.

In summary, SMRTI nodes perform better in the presence of varying maximum velocities and pause time as long as the proportion of malicious nodes never exceeds their proportion in the network. Alternatively, if the proportion of malicious nodes exceeds the proportion of SMRTI nodes, then SMRTI nodes struggle to establish communication due to their lower density irrespective of their ability to isolate malicious nodes.

18.5.3.3 SMRTI Nodes Vs. Simulation Area

Figure 18.9 represents the scenario in which the performance of SMRTI nodes decreases as the simulation area increases. The high density of nodes at 500×500 and $1000 \times 1000\,m^2$ successfully restricts CBR flows to one hop, while the uniform distribution of nodes at $1500 \times 1500\,m^2$ establishes multihop CBR flows. Further increase in the simulation area not only breaks the network into cluster of nodes but also brings down the performance of SMRTI nodes due to lack of intercluster communication. As the simulation area further distributes nodes away from clusters at $3500 \times 3500\,m^2$, the network settles with sparsely connected SMRTI nodes that could hardly deliver any performance (Figure 18.9). Finally, the expanding simulation area disconnects the network, which can be affirmed from the negligible PDR in Figure 18.9.

In summary, the scenario not only portrays the impact of node density in determining the PDR for SMRTI nodes but also presents the insight that the scope of an attack is relevant as long as the network is operational. In other words, it confirms that the defense mechanism designed against attackers fail to render any significance when the network is inoperative.

18.6 CONCLUSION

We have successfully described our novel detection and reaction model called SMRTI to enhance the security of MANET. SMRTI efficiently captures the evidence of trustworthiness from broader perspectives including direct interactions with one-hop nodes, observing interactions between one-hop nodes and through recommendations received from other nodes. The evidence captured from direct interactions enables SMRTI to classify the one-hop nodes as either benign or malicious, while the evidence captured from the interactions between neighbors enables SMRTI to shortlist malicious nodes that are likely to misbehave in future interactions. Finally, the evidence captured from recommendations enables SMRTI to establish trust relationship with multihop benign nodes. Unlike related models, SMRTI captures recommendations in a novel way in order to eliminate the two well-known problems—honest-elicitation and free-riding. Further, it reduces the complexity in terms of minimal computation and eliminates additional overhead, as it does not disseminate additional packets for communicating recommendations. All these contribute to SMRTI operating within the limitations of MANET. We have also demonstrated the performance and detailed the observed characteristics of SMRTI through extensive NS2 simulations. The simulation results confirm that SMRTI can efficiently deal with modification attacks. In our future work, we foresee to adapt SMRTI for other reactive, proactive, and hybrid protocols, apart from the integration with secure routing and cooperation models.

REFERENCES

1. K. Sanzgiri, B. Dahill, B. N. Levine, C. Shields, and E. M. Belding Royer. A secure routing protocol for ad hoc networks. *Proceedings of the 10th IEEE International Conference on Network Protocols (ICNP'02)*, Paris, France, pp. 78–89, 2002.
2. M. G. Zapata and N. Asokan. Securing ad hoc routing protocols. In *Proceedings of the ACM International Conference on Mobile Computing and Networking, Atlanta*, GA, USA, pp. 1–10, 2002.
3. Y. C. Hu, A. Perrig, and D. B. Johnson. Ariadne: a secure on demand routing protocol for ad hoc networks. *Proceedings of the ACM International Conference on Mobile Computing and Networking*, Atlanta, Georgia, USA, pp. 12–23, 2002.
4. P. Papadimitratos and Z. J. Haas. Secure routing for mobile ad hoc networks. *Proceedings of the SCS Communication Networks and Distributed Systems Modeling and Simulation Conference*, San Antonio, TX, 2002.
5. S. Capkun and J. P. Hubaux. BISS: building secure routing out of an incomplete set of security associations. In *Proceedings of the ACM Workshop on Wireless Security*, San Diego, CA, USA, pp. 21–29, 2003.

6. Y. Liu and Y. R. Yang. Reputation propagation and agreement in mobile ad hoc networks. In *Proceedings of IEEE Wireless Communications and Networking (WCNC'03)*, New Orleans, USA, pp. 1510–1515, 2003.

7. A. A. Pirzada, A. Datta, and C. McDonald. Propagating trust in ad hoc networks for reliable routing. *Proceedings of IEEE International Workshop on Wireless Ad-Hoc Networks*, Oulu, Finland, pp. 58–62, 2004.

8. T. Jiang and J. S. Baras. Ant based adaptive trust evidence distribution in MANET. *Proceedings of 24th International Conference on Distributed Computing Systems Workshops (ICDCSW'04)*, Tokyo, Japan, pp. 588–593, 2004.

9. J. Mundinger and J. Y. Le Boudec. Analysis of a reputation system for mobile ad hoc networks with liars. In *Proceedings of the 3rd International Symposium on Modeling and Optimization in Mobile Ad Hoc and Wireless Networks (WIOPT'05)*, Trentino, Italy, pp. 41–46, 2005.

10. J. Liu and V. Issarny. Enhanced Reputation Mechanism for Mobile Ad Hoc Networks. In *Proceedings of the 2nd International Conference on Trust Management (iTrust'04)*, Oxford, UK, pp. 48–62, 2004.

11. M. Virendra, M. Jadliwala, M. Chandrasekaran, and S. Upadhyaya. Quantifying trust in mobile ad hoc networks. In *Proceedings of International Conference on Integration of Knowledge Intensive Multi-Agent Systems (KIMAS)*, Waltham, Massachusetts, USA, pp. 65–70, 2005.

12. S. Buchegger and J. Y. L. Boudec. Performance analysis of the CONFIDANT protocol. In *Proceedings of 3rd ACM International Symposium on Mobile Ad hoc Networking & Computing*, Lausanne, Switzerland, pp. 226–236, 2002.

13. Y. Rebahi, V. E. Mujicav, and D. Sisalem. A reputation based trust mechanism for ad hoc networks. In *Proceedings of 10th IEEE Symposium on Computers and Communications (ISCC'05)*, Cartagena, Spain, pp. 37–42, 2005.

14. L. Eschenauer, V. D. Gligor, and J. Baras. On trust establishment in mobile ad hoc networks. In *Proceedings of 10th International Security Protocols Workshop*, Cambridge, UK, pp. 47–66, 2004.

15. L. Xiaoqi, M. R. Lyu, and L. Jiangchuan. A trust model based routing protocol for secure ad hoc networks. In *Proceedings of IEEE Aerospace Conference*, Big Sky, Montana, USA, pp. 1286–1295, 2004.

16. S. Yan Lindsay, Y. Wei, H. Zhu, and K. J. R. Liu. Information theoretic framework of trust modeling and evaluation for ad hoc networks. *IEEE J. Selected Areas Commun.*, 24 (2): 305–317, 2006.

17. D. B. Johnson, D. A. Maltz, and J. Broch. DSR: the dynamic source routing protocol for multihop wireless ad hoc networks. In *Ad hoc Networking*, C. E. Perkins, editors, Addison–Wesley Longman Publishing Co., Inc., Boston, MA, USA, 2001.

18. S. Bansal and M. Baker. Observation based cooperation enforcement in ad hoc networks. Technical Report (CoRR cs.NI/0307012), Stanford University, 2003.

19. A. D. Wood and J. A. Stankovic. Denial of service in sensor networks. *IEEE Comput.*, 35(10), pp. 54–62, 2002.

20. V. Balakrishnan and V. Varadharajan. Fellowship in mobile ad hoc networks. In *Proceedings of IEEE Security & Privacy in Emerging Areas (SecureCom2005)*, Athens, Greece, pp. 225–227, 2005.

21. V. Balakrishnan, V. Varadharajan, and U. K. Tupakula. Fellowship: defense against flooding and packet drop attacks in MANET. In *Proceedings of the 10th IEEE/IFIP Network Operations and Management Symposium (NOMS 2006)*, Vancouver, Canada, pp. 1–4, 2006.

22. C. E. Perkins, E. M. Royer, S. R. Das, and M. K. Marina. Performance comparison of two on-demand routing protocols for ad hoc networks. *IEEE Personal Commun.*, 8(1), pp. 16–28, 2001.

23. T. S. Rappaport. *Wireless Communications: Principles and Practice*. Prentice Hall, 1996.

24. E. S. Elmallah, H. S. Hassanein, and H. M. AboElFotoh. Supporting QoS routing in mobile ad hoc networks using probabilistic locality and load balancing. In *IEEE Global Telecommunications Conference (GLOBECOM 2001)*, San Antonio, Texas, pp. 2901–2906, 2001.

25. V. Balakrishnan, V. Varadharajan, P. Lucs, and U. K. Tupakula. Trust enhanced secure mobile ad hoc network Routing. In *2nd IEEE International Symposium on Pervasive Computing and Ad Hoc Communications (PCAC 2007), Proceedings of the 21st IEEE International Conference on Advanced Information Networking and Applications Workshops(AINAW 2007)*, Niagara Falls, Canada, pp. 27–33, 2007.

Chapter 19

Managing Privacy in Location-based Access Control Systems

Claudio Agostino Ardagna, Marco Cremonini, Sabrina De Capitani di Vimercati, and Pierangela Samarati

Dipartimento di Tecnologie dell'Informazione, Università degli Studi di Milano, Milan, Italy

19.1 Introduction	437
19.2 Related Work	438
19.3 Basic Scenario and Concepts	441
19.4 Location-based Access Control	447
19.5 Obfuscation Techniques for User-Privacy	452
19.6 A Privacy-Aware LBAC System	457
19.7 Summary	464
Acknowledgments	465
References	465

19.1 INTRODUCTION

Preserving user data privacy is one of the hottest topics in computer security. Security incidents, faulty data management practices, and unauthorized trading of users personal information have often been reported in recent years, exposing victims to ID theft and unauthorized profiling [46]. These issues are raising the bar of privacy standards, fostering innovative research, and driving new legislations. Some approaches aimed at privacy protection deal with minimizing unnecessary release of personal

Mobile Intelligence. Edited by Laurence T. Yang, Agustinus Borgy Waluyo, Jianhua Ma, Ling Tan, and Bala Srinivasan

information or focus on preventing leakage of personal information while in transit or once it has been released to an authorized party, for example, by delayed enactment of privacy preferences [49]. Our work addresses the latter concern in the framework of location-based services. We consider privacy requirements for location-based access control (LBAC) systems that require, for the provision of an online service, to evaluate conditions depending on users physical locations [6]. In the LBAC area, privacy has been mostly addressed by developing models and techniques that let users access anonymously to online services [10, 12, 23]. Solutions providing different degrees of privacy according to user preferences or business needs are instead less explored. For instance, obfuscation techniques applied to user locations are well suited to degrade the location accuracy for privacy reasons. In this context, however, only solutions based on increasing the granularity of a location measurement have been investigated and implemented in practice [23, 44]. Moreover, the importance of striking a balance between obfuscating locations for privacy reasons and preserving an acceptable accuracy for LBAC policies evaluation is often mentioned but not yet fully supported. In particular, a key aspect for managing such contrasting requirements is the availability of a metric (*relevance*, in our work) measuring, at the same time, the achieved privacy level and the required location accuracy. This metric should be independent from technological details of location measurements and from LBAC systems peculiarities. This way, privacy and accuracy requirements can be evaluated, negotiated, compared, and integrated in a coherent framework.

In this chapter, after a discussion on related work (Section 19.2), we present the scenario and concepts (Section 19.3) that are at the basis of our location-based access control system and authorization language (Section 19.4). We also describe obfuscation techniques that modify location information to provide user privacy protection (Section 19.5). Finally, we illustrate a privacy-aware location-based access control system that integrates the obfuscation techniques (Section 19.6) and conclude the chapter (Section 19.7).

19.2 RELATED WORK

Works related to location privacy techniques can be categorized into three main classes: *anonymity-based*, *obfuscation-based*, and *policy-based*.

Anonymity-based techniques provide solutions for the protection of the identities of the users. This class includes all solutions based on the notion of *anonymity* [10, 12, 19, 23], which is aimed at making an individual (i.e., her identity or personal information) not identifiable. Beresford and Stajano [10, 11] propose a method, called *mix zones*, based on an anonymity service that delays and reorders messages from subscribers within predefined zones. The proposal is based on a trusted middleware that lies between the positioning systems and the third-party applications and is responsible for limiting the information collected by applications. The mix zones model introduces the concepts of *application zones*, which are homogeneous application interested located in a specific geographic area, and *mix zones*, which represent areas where a user cannot be tracked. In particular, within mix zones, a user is anonymous in the sense that the identities of all users coexisting in the same zone are mixed and become indiscernible. The mix zones model is aimed at protecting long-term user movements still allowing the interaction with many location-based services. Other works [19, 23] are based on the concept of location k-anonymity, meaning that a user is indistinguishable from other $k − 1$ users in a given location area or temporal interval. Gruteser and Grunwald [23] define k-anonymity in the context of location obfuscation. They propose a middleware architecture and an adaptive algorithm to adjust location information resolution, in spatial or temporal dimensions, to comply with the specified anonymity requirements. Finally, another strand of research is aimed at protecting the *path privacy* of the users [22, 24, 29]. Path privacy involves the protection of users that are continuously monitored during a time interval. This research area is particularly relevant for location-tracking applications, where data about users moving in a particular area are collected by external services, such as navigation systems, that use them to provide their services effectively. In summary, anonymity-based techniques are suitable for all those contexts that do not need knowledge of the identities of the users and their effectiveness strongly depends on the number of users physically located in the same area.

Obfuscation-based techniques provide solutions for the protection of location privacy. This class includes all the solutions based on the notion of *location obfuscation* [3, 7, 9, 15, 44], which is the process of degrading the accuracy of the location information to provide privacy protection. Differently from anonymity-based techniques, the main goal of obfuscation-based techniques is to perturb the location information still maintaining a binding with the identities of users. Duckham and Kulik [15, 16] define a framework that provides a mechanism for balancing individual needs for high-quality information services and for location privacy. The proposed solution is based on the *imprecision concept*, which indicates the lack of specificity of location information. The authors propose to degrade the quality of location information and provide obfuscation features by adding n points, with same probability, to the real user position. Obfuscation-based solutions also provide mechanisms for specifying privacy preferences in a common and intuitive manner (e.g., as a *minimum distance*), which, however, presents several drawbacks. First, they do not provide a metric for the privacy level, making them difficult to integrate into a full fledged location-based application scenario [6]. Second, they usually implement a single obfuscation technique based on the enlargement of a location area. This way, a possibility that is often neglected by traditional location obfuscation solutions is the definition and composition of different obfuscation techniques to increase their robustness with respect to possible de-obfuscation attempts performed by adversaries. Finally, obfuscation solutions are often meaningful in a specific application context only. Ardagna et al. [3, 5, 7] address the above shortcomings by presenting a novel solution composed by a management process and several techniques aimed at preserving location privacy by artificially perturbing location information measured by sensing technologies. Key aspects of the proposal is, on the one hand, to permit the specification of privacy preferences in a simple and intuitive way, and, on the other hand, to make the enforcement of privacy preferences manageable for location-based services, while preserving the quality of the online service. To this aim, the authors introduce the *rel-*

evance concept as a metric for the accuracy of location information, abstracting from any physical attribute of sensing technology. This permits to quantitatively evaluate the degree of privacy associated with a location information and is adopted by users to define their privacy preferences. On the basis of relevance preferences, different obfuscation-based techniques and their composition are discussed.

Policy-based techniques are based on the notion of *privacy policies* [20, 26, 27, 30, 34]. Privacy policies define restrictions that must be followed when the location of users is used by or released to third parties. Key to policy-based techniques is the definition of policies that can rule location management and disclosure. The definition of complex rule-based policies is, however, difficult to understand and manage for users who often are not familiar with specific policy definition languages. Therefore, although policy-based techniques are powerful and flexible, they can easily result in tools difficult to manage for end users.

Technologies for integrating multiple sources of location information are also investigated [39]. Today, most commercial location platforms include a gateway that mediates between location providers and location-based applications [44]. In these architectures, the location gateway obtains subscriber's location information from multiple sources and delivers them, possibly modified according to privacy requirements, to location-based applications. The increased accuracy and reliability of location technologies have suggested novel ways to exploit location information within location-based services. Some early mobile networking protocols linked the notion of physical position of a terminal device with its capability to access network resources [1]. More recently, an emerging research issue was represented by the inclusion of a negotiation phase of quality of service (QoS) parameters based on service-level agreement (SLA) and privacy preference in LBS [5, 6]. The widespread adoption of wireless networks has also been the subject of some recent studies focused on location information for monitoring users movements [17, 18].

Another strand of research focuses on the underlying description of the architecture and operations of an access control server in a LBS context. For instance, the need for a protocol-independent location technique has been explored by Nord et al. [42], who assume heterogeneous positioning sources like GPS, Bluetooth, and WaveLAN for designing location-aware applications. Given such different sources of location information, a generic positioning protocol for interchanging location information between location sources and client applications is introduced and different techniques for merging location information are presented. Another work [52] studies location-based information and its management in the area of mobile commerce applications and presents an integrated location management architecture to support composite location requirements. However, coordination among multiple wireless networks, location negotiation protocols for mobile commerce, and privacy issues are not considered yet.

Few proposals consider location information as a means for improving security. Sastry et al. [48] exploit location-based access control in sensor networks. Zhang and Parashar [53] propose a location-aware extension to role-based access control (RBAC) suitable for grid-based distributed applications. Ardagna et al. [6] propose a location-based access control model and language together with an evaluation infrastructure.

Other papers take into account time variant information for querying database containing location information [32, 36].

Other works follow a different approach by considering the location information as a resource to be protected against unauthorized access. For instance, Hengartner and Steenkiste [28] present a mechanism to protect user's location information by means of electronic certificates, delegation, and trusted location-based services. The same problem is addressed in [26] by proposing a privacy-aware architecture for a global location service, which should permit users to define rules for accessing their location information.

Finally, several works propose special-purpose *location middlewares* for managing interactions between applications and location providers, while maximizing the QoS [40, 41, 47]. Typically, in these proposals, the location middleware (i) receives requests from LBS components asking for location information, (ii) collects users locations from a pool of location providers, and (iii) produces an answer. Naguib et al. [40] present a middleware framework, called *QoSDREAM*, for managing context-aware multimedia applications. Nahrstedt et al. [41] have designed a QoS middleware for ubiquitous computing environments aimed at maximizing the QoS of distributed applications. Ranganathan et al. [47], instead, consider a middleware that provides a clear separation between business applications and location-detection technologies. Although several middleware components supporting communication and negotiation between location services and applications have been presented, only few proposals try to integrate service quality and privacy protection. For instance, Myles et al. [38] propose an architecture based on a middleware managing the interactions between location-based applications and location providers and on the definition of policies for data release. Hong et al. [30] present an extension of the P3P language for representing user privacy preferences for context-aware applications. Ardagna et al. [3] provide a middleware-based architecture for integrating privacy preferences of the users and location accuracy of LBS in the context of location-based access control systems.

19.3   BASIC SCENARIO AND CONCEPTS

19.3.1   Location-based Access Control Architecture

In a LBAC scenario, there are more parties involved than in conventional access control systems. A LBAC system evaluating a policy does not have direct access to location information; rather, it sends location requests to external services, called *location services* (LSs), and waits for the corresponding answers [6]. The characteristics of these location services will depend on the communication environment where the user transaction takes place. Here, we focus on the mobile network, where location service is provided by mobile phone operators. Typically, a LBAC scenario involves the following three entities (see Figure 19.1).

> *User.* It is the entity whose access request to a service must be authorized by a LBAC system. We make no assumption about users, besides the fact

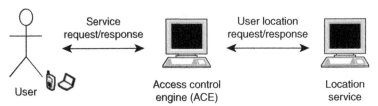

Figure 19.1   Basic location-based access control architecture.

that they carry terminals enabling authentication and some form of location verification.

Access control engine (ACE). It is the entity that implements the LBAC system. It is responsible for evaluating access requests according to some policies containing location-based conditions. The ACE must communicate with a location service for acquiring location information, and it is not restricted to a particular access control model and authorization language.

Location service. It is the entity that provides the location information. The types of location requests that it can satisfy depend on the specific mobile technology, the methods applied for measuring users positions, and environmental conditions.

Note that the functional decomposition between the ACE and the LS is due to the fact that location functionalities are fully encapsulated within remote services that are set up and managed by the mobile operators. Therefore, no assumption can be made on these services besides their interfaces.

The design of privacy-aware systems poses novel architectural and functional issues that were never considered before in the context of traditional access control systems. Among these issues, the problem of protecting location privacy of users stands out and the need of a privacy-aware LBAC system arises. A privacy-aware LBAC architecture must be designed by integrating components logically tied with the applications that need location-based access control enforcement and components providing privacy-aware location services. One typical approach to this problem is to integrate a *location middleware* (LM) that acts as a trusted gateway between an LBAC system and location services. The location middleware should be able to interact with multiple location services and to offer location services to an access control engine. It also should manage low-level communications with location services and should enforce both privacy preferences expressed by users and requirements for location accuracy set by an access control engine. Figure 19.2 shows the reference privacy-aware LBAC architecture.

Communications among logical components are performed via request/response message exchanges. The interaction flow can be logically partitioned in the following six macro-operations.

1. *Initialization,* when user preferences and LBAC policies are defined.
2. *Access request and information negotiation,* when a user submits an access request to the access control engine and a negotiation process resulting in a bidirectional identification between the parties takes place.

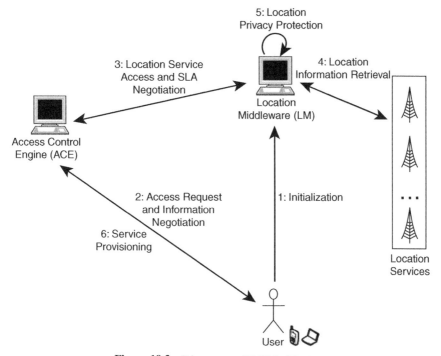

Figure 19.2   Privacy-aware LBAC Architecture.

3. *Location service access and SLA negotiation*, when the access control engine requires location information/service to the location middleware; a service-level agreement could be specified to agree upon QoS attributes.

4. *Location information retrieval*, when the location middleware collects user location information through a communication process with multiple location services.

5. *Location privacy protection*, when obfuscation techniques are used to comply with both user preferences and LBAC accuracy.

6. *Service provision*, when LBAC policies are evaluated and the access request is granted or denied.

19.3.2   Location Measurements

Two characteristics are specific to technologies for location measurements.

- *Interoperability*: location gathering could rely on different sources of location information, depending on availability and cost.
- *Accuracy*: each location measurement exhibits a variable accuracy affected by technological limitations (i.e., measurement errors) and possible environmental effects.

While interoperability largely depends on roaming agreements between mobile phone operators and is more business-oriented, accuracy needs to be carefully considered in the design of LBAC systems.

Today, in the mobile network scenario, no technology is available ensuring perfect user location [31]. The location *accuracy* is always less than 100%, so typically a position is specified as a range, locating the user within a circular area. For a given location request, the location area may depend on the number of nearby antennas and on the surrounding landscape features. Also, a location measurement is often unstable because of changing environmental conditions, such as reflection or interferences that may corrupt the signal. In our model, we take into account these aspects by assuming that the result provided by a location service is always affected by a measurement error. This fact is relevant to the syntax and semantic of the location service interface because the outcome of the evaluation of an access request determined by the access control engine will depend on such an uncertainty, which must then be explicitly represented and processed in terms of accuracy.

It is worth noting that suitability and accuracy of a location service largely depend on the underlying technology. GSM/3G technologies are widespread and recent advancements have sensibly improved location capabilities [2]. 802.11 WiFi and AGPS/GPS [21, 45] could also be exploited, although some limitations reduce their applicability. WiFi has a limited coverage and its usage is restricted to indoor environments or in urban areas covered by hotspots. GPS, on the contrary, does not work indoor or in narrow spaces but has no coverage limitation, a feature which makes it an ideal location technology for open, outdoor environments.

A direct consequence of such a lack of accuracy is that the location position of a user cannot be expressed as a geographical point. We, therefore, introduce a first working assumption that considers the shape of a location measurement returned by a location service.

Assumption 19.1. *A location measurement is represented by a planar and circular area.*

This assumption makes the analysis and the design of LBAC systems more tractable with no loss of generality because (i) it represents a particular case of the general requirement of considering convex areas (areas must be convex to easily compute integrals over them) and (ii) circular areas approximate well the actual shape resulting from many location technologies (e.g., cellular phones). In the following, we use Area (r_i, x_i, y_i) to denote a location measurement centered on coordinates (x_i, y_i) and with radius r_i.

In the same vein of other works in this field [37], we introduce a second assumption as follows.

Assumption 19.2. *Consider a random location within a location measurement Area (r, x, y), where a "random location" is a neighborhood of a random point $(\hat{x}, \hat{y}) \in$ Area (r, x, y). The probability that the real user's position (x_u, y_u) belongs to a neighborhood of a random point $(\hat{x}, \hat{y})$ is uniformly distributed over the whole location measurement.*

Accordingly, the joint probability density function (pdf) of the real user's position can be defined as follows [43].

Definition 19.1 (Joint pdf). Given a location measurement Area (r_i, x_i, y_i), the joint probability density function (joint pdf) $f_r(x, y)$ is

$$f_r(x, y) = \begin{cases} \frac{1}{\pi r^2} & \text{if } x, y \in Area(r_i, x_i, y_i) \\ 0 & otherwise \end{cases}$$

19.3.3 Location Accuracy

The accuracy of a location measurement returned by a sensing technology depends on the radius of the measured circular area, which, in turn, depends on the unavoidable measurement error of the location technology. To evaluate the quality of a given location measurement, its accuracy must then be compared to the best accuracy that location technologies are able to provide.

Several works describe and discuss different location technologies and the best accuracy that can be achieved [25, 50]. In [50], the authors provide a survey of standard positioning solutions for the cellular network such as, *E-OTD* for GSM, *OTDOA* for Wideband Code Division Multiple Access (WCDMA) networks, and *Cell-ID*. Specifically, E-OTD is based on the existing observed time difference (OTD) feature of GSM systems. The accuracy of the E-OTD has been found to range from 50 to 125 m. Observed Time Difference Of Arrival (OTDOA) achieves a location accuracy of 50 m at best. Finally, Cell-ID is a simple positioning method based on cell sector information, where cell size varies from 1 to 3 km in urban areas to 3–20 km in suburban/rural areas.[1]

Therefore, the accuracy of a measured area depends on its radius, which we call r_{meas}. To evaluate the quality of a location measurement, the accuracy of a given measurement must be compared with the best accuracy that the technology can achieve. Let r_{opt} be the radius representing the best accuracy (i.e., the minimum measurement error), ratio r_{opt}^2/r_{meas}^2 is a good estimation of the quality of a location measurement. As an example of a positioning process using the three technologies described above, suppose that a user position is located with radius $r_{meas} = 62.5$ m using the E-OTD method, radius $r_{meas} = 50$ m using the OTDOA and radius $r_{meas} = 1$ km using the Cell-ID. The area corresponding to the best accuracy (minimum radius) has $r_{opt} = 50$ m. The three location measurements result in three different areas. In particular, the area with best accuracy is provided by OTDOA and has a measurement quality of 1, whereas the others have a quality proportionally reduced to 0.8 for the area calculated by means of E-OTD and 0.05 for the one measured through Cell-ID. This way, based on the optimal location accuracy, we can distinguish between different measurement accuracies and reward the best technology available.

---

[1]Other methods [13, 14] are able to further improve the accuracy of standard positioning methods.

19.3.4 Relevance

The notion of relevance is strictly related to the notion of accuracy. The relevance is defined as an adimensional, technology-independent metric of the location position accuracy. A location position could be either a location measured by sensing technology or an obfuscated location. The relevance is a value $\mathcal{R} \in (0, 1]$ that has the following properties.

- It tends to 0 when location information must be considered unreliable. This represents the limit condition of very large values of measurement errors or obfuscations that degrade the information so that no relation with the original measured location is preserved.

- It is equal to 1 when location information has best accuracy. This represents the second limit condition of a measurement error equal to the one introduced by the best sensing technology and no obfuscation applied.

- It falls in $(0,1)$ when the location accuracy is less than optimal either for measurement errors larger than the minimum and/or for degradations artificially introduced by obfuscation techniques. This represents the standard situation where the degradation of the original accuracy provides a certain level of privacy as required by users, while keeping, however, an acceptable degree of accuracy as needed by application providers.

Accordingly, the *location privacy* provided by an obfuscated location is equal to $(1 - \mathcal{R})$. In different terms, the notion of relevance is useful to normalize the accuracy value of a location position by expressing it as an adimensional value, independently from any physical scale or from a reference area if given as a percentage of loss, and to represent the lack of accuracy of obfuscated positions regardless to the specific applied obfuscation techniques. The relevance is the general functional term used to qualify the accuracy (and correspondingly the privacy) of a location position when an LBS interacts with users or application service providers, which, in general, are unaware of the technicalities of both location-sensing technologies and obfuscation techniques.

In our reference scenario, an LBS has to manage locations that, on the one hand, could be perturbed for privacy reasons, while on the other hand could be required to have an accuracy not below a threshold to preserve a certain quality of service. To support such requirements, all location measurements have an associated relevance value and all management decisions, either related to users privacy or to quality of information, are carried out by considering or possibly negotiating relevance values associated with the location measurement.

The following three relevance values characterize our privacy management solution.

- *Initial relevance ($\mathcal{R}_{\text{Init}}$).* It is the relevance of a location measurement as returned by a sensing technology. This is the initial value of the relevance that only depends on the intrinsic measurement error.

- *Final relevance ($\mathcal{R}_{\text{Final}}$)*. It is the relevance of a final obfuscated area produced by satisfying a user's privacy preference. It is derived, starting by the initial relevance, through the application of one or more obfuscation techniques.
- *Required relevance ($\mathcal{R}_{\text{LBAC}}$)*. It is the minimum relevance required by an ACE for a reliable evaluation of a location-based policy. This value represents the threshold for the acceptable accuracy of a location measurement or a location predicate evaluation. Below this threshold, the ACE considers the location information too inaccurate for an access control decision.

The value of $\mathcal{R}_{\text{Init}}$ is calculated by normalizing the best accuracy that could have been achieved with respect to the technical accuracy resulting from the specific measurement. This is represented by the ratio of two measurement errors: the area that would have been returned if the best accuracy was achieved (i.e., having radius r_{opt}) and the actual measured area (i.e., having radius r_{meas}). In other words, $\mathcal{R}_{\text{Init}}$ measures the relative accuracy loss of a given measure—due to, for example, particular environmental conditions—with respect to the best accuracy that the technology would have permitted. This is the only relevance value that is directly calculated from physical values (i.e., measurement errors). $\mathcal{R}_{\text{Final}}$ is derived from $\mathcal{R}_{\text{Init}}$ by considering the accuracy degradation introduced for privacy reason. We use a scalar factor $\lambda \in (0, 1]$ to represent it. Accordingly, the location measurement associated with $\mathcal{R}_{\text{Init}}$ will be perturbed by applying obfuscation techniques so that a resulting area having relevance $\mathcal{R}_{\text{Final}}$ is obtained.

Definition 19.2 ($\mathcal{R}_{\text{Init}}$ and $\mathcal{R}_{\text{Final}}$). Given a location measurement area of radius r_{meas} measured by a sensing technology, a radius r_{opt} representing the best accuracy of sensing technologies, and a degradation $\lambda \in (0, 1]$, initial relevance $\mathcal{R}_{\text{Init}}$ and final relevance $\mathcal{R}_{\text{Final}}$ are calculated as

$$\mathcal{R}_{\text{Init}} = \frac{r_{\text{opt}}^2}{r_{\text{meas}}^2} \tag{19.1}$$

$$\mathcal{R}_{\text{Final}} = \lambda \mathcal{R}_{\text{Init}} \tag{19.2}$$

Differently, the value of $\mathcal{R}_{\text{LBAC}}$ is given, either autonomously defined by the ACE as a requirement for the access control decision, or negotiated as a QoS parameter of the location service.

19.4 LOCATION-BASED ACCESS CONTROL

Conventional access control mechanisms rely on the assumption that requesters' profiles fully determine what they are authorized to do. However, context information and, in particular, physical user locations may also play an important role in determining access rights. We describe the integration of access control policies with location-based conditions, focusing on policy evaluation and enforcement, which represent challenging issues inevitably associated with such an extension to access control policies. LBAC supports access control policies that include conditions based on the physical location of a requester. Difficulties arise from the very nature of location information, which is dynamic, affected by a measurement error and requires a special dedicated infrastructure to be gathered. Rapid advancements in the field of wireless and mobile networking have fostered a new generation of devices suitable for being used as sensors by location technologies able to compute relative position and movement of users. Once a user's location has been gathered, a LBAC policy can be evaluated and the user could be granted access to a particular resource. The location-verification process must be able to tolerate rapid context changes because mobile users can wander freely while initiating transactions by means of terminal devices like cell phones (GSM and 3G) and palmtops with WiFi cards. Regardless to the specific technology, location verification can provide a rich context representation related to both users and resources they access. Location-based information possibly available to access control modules includes the position and mobility of the requester when a certain access request is submitted. In the near future, location-based services are likely to provide a wealth of additional environment-related knowledge (e.g., is the user sitting at her desk or walking toward the door? Is she alone or together with others?). This kind of fine-grained context information potentially supports a new class of location-aware conditions regulating access to and fruition of resources.

19.4.1 Location-based Predicates

The definition of location-based predicates for access control mechanisms requires to specify the conditions that an authorization language can support and today's location technology can verify. Three main classes of conditions could be identified [6]:

- *position-based* conditions on the location of a user, for evaluating, for example, whether a user is within a certain building or city or in the proximity of other entities;
- *movement-based* conditions on the mobility of a user, such as her velocity, acceleration, or direction where she is headed;
- *interaction-based* conditions relating multiple users or entities, for example, the number of users within a given area.

Although we have defined some specific predicates corresponding to specific conditions identified by the classes above, our language is extensible with respect to the predicates that can be added, as the need arises and technology progresses.

Furthermore, the language for location-based predicates assumes the following two elements.

- Users is the set of *user identifiers* (UIDs) that unambiguously identify users known to the location services. This includes both the users of the system (i.e., potential requesters) and any other known physical and/or moving entity that may need to be located (e.g., a vehicle with an on-board GPRS card).

Table 19.1   Examples of location-based predicates

Type	Predicate	Description
Position	inarea(*user, area*)	Evaluate whether *user* is located within *area*.
	disjoint(*user, area*)	Evaluate whether *user* is located outside *area*.
	distance(*user, entity, min_dist, max_dist*)	Evaluate whether the distance between *user* and *entity* is within interval [*min_dist, max_dist*].
Movement	velocity(*user, min_vel, max_vel*)	Evaluate whether *user*'s speed falls within range [*min_vel, max_vel*].
Interaction	density(*area, min_num, max_num*)	Evaluate whether the number of users currently in *area* falls within interval [*min_num, max_num*].
	local_density(*user, area, min_num, max_num*)	Evaluate the density within a "relative" area surrounding *user*.

A typical UID for location-based applications is the SIM number linking the user's identity to a mobile terminal.[2]

- **Areas** is a set of map regions identified either via a geometric model (i.e., a range in a n-dimensional coordinate space) or a symbolic model (i.e., with reference to entities of the real world such as cells, streets, cities, zip code, buildings, etc.) [35].

In the following, we will refer to elements of *users* and of *areas* as *user* and *area* *terms*, respectively. While we assume such elements to be ground in the predicates, a language could be readily extended to support variables for them.

All predicates could be expressed as boolean queries, and therefore have the form *predicate(parameters, value)*. Their evaluation returns a triple [*bool_value, R, timeout*], where the term *bool_value* assumes values *True/False* according to the corresponding access decision, R represents a relevance value that qualifies the accuracy of the predicate evaluation, and *timeout* sets the validity timeframe of the location predicate evaluation. Our core set of location predicates includes the following predicates (see Table 19.1).

- A binary *position* predicate inarea whose first argument is a user term and second argument is an area term. The predicate evaluates whether a user is located within a specific area (e.g., a city, a street, a building).

- A binary *position* predicate disjoint whose first argument is a user term and second argument is an area term. The predicate evaluates whether a user is

---

[2]Individual users may carry multiple SIMs and the same SIMs may be passed over to other users. We shall not elaborate on these issues, since identity management in mobile networks is outside the scope of this chapter.

outside a specific area. Intuitively, disjoint is equivalent to the negation of inarea.

- A four-ary *position* predicate distance whose first argument is a user term, second argument is either a user or area term (identifying an *entity* in the system), while the third and fourth arguments are two numbers specifying the minimum (*min_dist*) and maximum (*max_dist*) distance, respectively. The semantics of this predicate is to request whether the user lies within a given distance from the specified entity. The entity involved in the evaluation can be either stable or moving, physical or symbolic, and can be the resource to which the user is requesting access. Exact distance can be evaluated by setting the same value for *min_dist* and *max_dist*; "closer than" conditions can be evaluated by setting *min_dist* to 0; "farther than" conditions can be evaluated by setting *max_dist* to infinity.

- A ternary *movement* predicate velocity whose first argument is a user term, and the second and third arguments are two numbers specifying a minimum (*min_vel*) and maximum (*max_vel*) velocity, respectively. The semantics of the predicate is to request whether the user speed lies within a given range of velocity. Similarly to what happens for distance, exact velocity can be requested by setting the same value for *min_vel* and *max_vel*, while "smaller than" or "greater than" conditions can be evaluated by setting *min_vel* equal to 0 or *max_vel* equal to infinity, respectively.

- A ternary *interaction* predicate density whose first argument is an area term, while second and third arguments are numbers specifying a minimum (*min_num*) and maximum (*max_num*) number of users. The semantics of the predicate is to request whether the number of users currently in an *area* lies within the interval specified.

- A four-ary *interaction* predicate local_density whose first argument is a user term, the second argument is a "relative" area with respect to the user, and the third and fourth arguments specify a minimum (*min_num*) and maximum (*max_num*) number of users, respectively. The semantics of the predicate is to evaluate the density within an area surrounding the user.

Example 19.1. *Let* Alice *be an element of* Users, *and* Milan *and* Director Office *be two elements of* Areas.

inarea(Alice,Milan) = [True,0.9,2007-08-09_11:10am]
means that the location service assesses as true the fact that Alice *is located in* Milan *with a relevance* $\mathcal{R} = 0.9$; *such an assessment is to be considered valid until* 11:10am *of* August 9, 2007.

velocity(Alice,70,90) = [True,0.7,2007-08-03_03:00pm]
means that the location service assesses as true the fact that Alice *is traveling at a speed included in the range* [70,90] *with a relevance* $\mathcal{R} = 0.7$; *such an assessment is to be considered valid until* 3:00pm *of* August 3, 2007.

density(Director Office,0,1) = [False,0.95,2007-08-21_06:00pm]
means that the location service assesses as false the statement that there is at most one person in the Director Office *with a relevance* $\mathcal{R} = 0.95$; *such an assessment is to be considered valid until* 06:00pm *of* August 21, 2007.

19.4.2 Location-based Access Control Policies

We now discuss how location-based access control policies can be expressed. Note that our goal is not to develop a new language for specifying access control policies. Instead, our proposal can be thought of as a general solution for enriching the expressive power of existing languages (e.g., [8, 33, 51]) by exploiting location information, without increasing the computational complexity of their evaluation. We therefore assume that each user is assigned an identifier or pseudonym. Besides their identifiers/pseudonym, users usually have other properties (e.g., name, address, and date of birth) that can be transmitted through digital certificates and are grouped into a *user profile*. Objects are data/services that users may ask to access to. Properties of an object are grouped into an *object profile*. Each property into user or object profiles are referenced with the traditional dot notation. For instance, let Alice be the identifier of a user and therefore of the corresponding profile. Alice.address denotes the address of Alice. Also, to make it possible to refer to the user and object of the request being evaluated without introducing variables in the language, we rely on the **user** and **object** keywords. For instance, **user**.Affiliation indicates the property Affiliation within the profile of the user whose request is currently processed. A location-based authorization rule is then defined as follows.

Definition 19.3 (Location-based authorization rule). A location-based authorization rule is a triple of the form ⟨*subject_expression, object_expression, actions*⟩, where

- *subject_expression* is a boolean formula of terms that allows referring to a set of subjects depending on whether they satisfy or not certain conditions, where conditions can evaluate the user's profile, location predicates, or the user's membership in groups, active roles, and so on;
- *object_expression* is a boolean formula of terms that allows referring to a set of objects depending on whether or not they satisfy certain conditions, where conditions evaluate membership of the object in categories, values of properties on metadata, and so on; and
- *actions* is the action (or set of actions) to which the policy refers.

Conditions specified in the *subject_expression* field can be classified in two categories: *generic conditions* and *location-based conditions*. Generic conditions evaluate membership of subjects in classes or properties in their profiles and, as for the object expression, they are always of the form predicate_name(*arguments*), where *arguments* is a list, possible empty, of constants or attributes. Location-based conditions corresponds to location predicates.

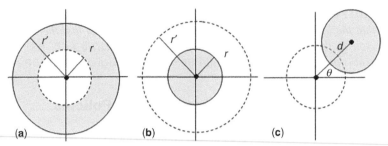

Figure 19.3   Obfuscation by enlarging the radius (a), reducing the radius (b), and shifting the center (c).

Example 19.2. *Consider a healthcare scenario where a hospital provides the mag-netic resonance imaging (MRI) examinations and is responsible for patients data management. Suppose that the MRI machine is the hardware/software that permits to do magnetic relevance tomography. Managing a MRI machine is a critical activity because privileges must be granted to strictly selected medical personnel only and must be performed according to high security standards (see policy 1 in Table 19.2). In addition, access to medical databases must be managed carefully and according to different security standards depending on the level of risk of the data to be accessed. In particular, access to examination data is critical, because they include high sensi-tive information about patients' health condition (see policies 2 and 3 in Table 19.2). Patient-related information needs to be protected, for example, from disclosure to pharmaceutical companies (see policy 4 in Table 19.2). Finally, access to logging and billing data of the patients are usually less critical but still to be handled in a highly secured environment and to be granted only to selected personnel, according to the laws and regulations in force (see policy 5 in Table 19.2).*

19.5   OBFUSCATION TECHNIQUES FOR USER-PRIVACY

To guarantee user location privacy, we introduce three basic obfuscation techniques that modify a user location to reduce the associated relevance (henceforth the accu-racy) until a given level.

19.5.1   Obfuscation by Enlarging the Radius

Obfuscating a location measurement area by increasing its radius (see Figure 19.3(a)) is the technique that most solutions exploit, either explicitly or implicitly by scaling a location to a coarser granularity (e.g., from few meters to hundred of meters, from a city block to the whole town, and so on). The obfuscation is a probabilistic effect due to a decrease in the corresponding joint probability density function (joint pdf), which can be expressed as $\forall r, r', r < r' : f_r(x, y) > f_{r'}(x, y)$. The following proposition allows us to calculate the obfuscated area.

Table 19.2 Examples of access control policies for a healthcare scenario

	Generic conditions	Subject expression Location conditions	Actions	Object_expression
1	equal (**user**.Role, "Doctor") ∧ Valid(**user**.Username, **user**.Password)	inarea(**user**.sim,MRI Control Room) ∧ density(MRI Room,1,1) ∧ velocity(**user**.sim,0,3)	Execute	equal (**object**.name, "MRIMachine")
2	equal (**user**.Role, "Doctor") ∧ Valid(**user**.Username, **user**.Password)	inarea(**user**.sim,Hospital) ∧ local_density(**user**.sim,Close By,1,1) ∧ velocity(**user**.sim,0,3)	Read	equal (**object**.category, "Examination")
3	equal (**user**.Role, "Nurse") ∧ Valid(**user**.Username, **user**.Password)	inarea(**user**.sim,First Aid) ∧ local_density(**user**.sim,Close By,1,1) ∧ velocity(**user**.sim,0,3)	Read	equal (**object**.category, "Examination")
4	equal (**user**.Role, "Doctor") ∧ Valid(**user**.Username, **user**.Password)	local_density(**user**.sim,Close By,1,1) ∧ disjoint(**user**.sim,Pharmaceutical Company)	Read	equal (**object**.category, "Personal Info")
5	equal (**user**.Role, "Secretary") ∧ Valid(**user**.Username, **user**.Password)	local_density(**user**.sim,Close By,1,1) ∧ inarea(**user**.sim,Hospital)	Read	equal (**object**.category, "Log&Bill")

Proposition 19.1. *Given a location area of radius r with relevance $\mathcal{R}_{Init}$ and an obfuscated area of radius r' derived by enlarging the original radius, the relevance $\mathcal{R}_{Final}$ of the obfuscated area is calculated by reducing $\mathcal{R}_{Init}$ of the ratio $f_{r'}(x, y)/f_r(x, y)$ of corresponding joint pdfs.*

From the assumption of uniform distribution over a circular area, the relation between $\mathcal{R}_{Final}$ and $\mathcal{R}_{Init}$ can be written as:

$$\mathcal{R}_{Final} = \frac{f_{r'}(x, y)}{f_r(x, y)} \qquad \mathcal{R}_{Init} = \frac{1/(\pi r'^2)}{1/(\pi r^2)}$$

$$\mathcal{R}_{Init} = \frac{r^2}{r'^2} \times \mathcal{R}_{Init} \begin{cases} = \mathcal{R}_{Init} & r' = r \\ \in (\mathcal{R}_{Init}, 0) & r' > r \\ = 0 & r' \to +\infty \end{cases} \tag{19.3}$$

Therefore, given the two relevances $\mathcal{R}_{Init}$ and $\mathcal{R}_{Final}$, and the radius r of the initial area, an obfuscated area calculated with this technique has a final radius: $r' = r\sqrt{\mathcal{R}_{Init}/\mathcal{R}_{Final}}$.

19.5.2   Obfuscation by Reducing the Radius

Another possible way of obfuscating a user location consists of reducing the radius r of one location to a smaller r', as showed in Figure 19.3(b). The obfuscation effect, in this case, is produced by a corresponding reduction in the probability to find the real user location within the returned area, while the joint pdf is fixed.

To state it formally, consider the unknown real user position coordinates (x_u, y_u). Given a location area of radius r, the probability that the real user position falls in the area is $P((x_u, y_u) \in \text{Area}(r, x, y))$. When we obfuscate by reducing the radius, an area of radius $r' \le r$ is returned, which implies that $P((x_u, y_u) \in \text{Area}(r', x, y)) \le P((x_u, y_u) \in \text{Area}(r, x, y))$, because a circular ring having pdf greater than zero has been excluded.

The obfuscated area, derived by a radius reduction, can be calculated by considering the following proposition.

Proposition 19.2. *Given a location measurement area of radius r with relevance $\mathcal{R}_{Init}$ and an obfuscated area of radius r' derived by reducing the radius, relevance $\mathcal{R}_{Final}$ of the obfuscated area is calculated by reducing $\mathcal{R}_{Init}$ of the ratio $P((x_u, y_u) \in \text{Area}(r', x, y))/P((x_u, y_u) \in \text{Area}(r, x, y))$ of corresponding probabilities.*

In cartesian coordinates, $P((x, y) \in A)$, for all subsets $A \subseteq \mathbb{R}^2$, is calculated as $\iint_A f(x, y) dx dy$, being $f(x, y)$ the corresponding joint pdf. Changing to polar coordinates (s, θ) and solving the double integral requires the transformation $(x, y) \to (s, \theta)$, which gives $dx dy = s ds d\theta$ [43]. The pdf, instead, remains unchanged to the value obtained from the original location measurement, i.e., $f(r, \theta) = 1/\pi r^2$. According to these observations, we have

$$P((x_u, y_u) \in \text{Area}(r', x, y)) = \int_0^{2\pi} \int_0^{r'} f(r, \theta) s\,ds\,d\theta = 2\pi \int_0^{r'} \frac{s}{\pi r^2} ds$$

$$= \frac{2}{r^2} \int_0^{r'} s\,ds = \frac{r'^2}{r^2}$$

Analogously, $P((x_u, y_u) \in \text{Area}(r, x, y))$ can be calculated as

$$P((x_u, y_u) \in \text{Area}(r, x, y)) = \int_0^{2\pi} \int_0^{r'} f(r, \theta) s\,ds\,d\theta + \int_0^{2\pi} \int_{r'}^{r} f(r, \theta) s\,ds\,d\theta - 1$$

resulting in a probability equal to 1 of having the user u inside the location measurement Area(r, x, y). Therefore, the relation stated in the proposition can be written as

$$\mathcal{R}_{\text{Final}} = \frac{Pr((x_u, y_u) \in \text{Area}(r', x_c, y_c))}{Pr((x_u, y_u) \in \text{Area}(r, x_c, y_c))} \times \mathcal{R}_{\text{Init}} = \frac{r'^2}{r^2} \times \mathcal{R}_{\text{Init}} \begin{cases} = \mathcal{R}_{\text{Init}} & r' = r \\ \in (\mathcal{R}_{\text{Init}}, 0) & r' < r \\ = 0 & r' \to 0 \end{cases}$$

$$(19.4)$$

Therefore, given the two relevances $\mathcal{R}_{\text{Init}}$ and $\mathcal{R}_{\text{Final}}$, and the radius r of the initial area, an obfuscated area calculated with this technique has a final radius: $r' = r\sqrt{\mathcal{R}_{\text{Final}}/\mathcal{R}_{\text{Init}}}$.

19.5.3 Obfuscation by Shifting the Center

Location obfuscation can also be achieved by shifting the center of the location measurement area and returning the displaced area, as showed in Figure 19.3(c). Intuitively, the obfuscation effect depends on the intersection of the two areas, that is, the smaller the intersection, the highest the obfuscation. In this case, it should be considered unacceptable to produce obfuscated areas disjoint from the original measured location area. The reason is that all disjoint areas would have probability equal to zero of including the real user location, and then they would be indistinguishable in term of our relevance metric. Such cases are considered as just false location information, which, by design, our system does not produce, assuming that LBS and related applications such as LBAC [6, 7] cannot, in general, deal with false information in the provision of a business service.

We call d the distance between the centers and r the radius of the initial and final (obfuscated) areas. Since the original and the obfuscated areas cannot be disjoint, $d \in [0, 2r]$. In particular, if $d = 0$, there is no privacy gain; if $d = 2r$, there is maximum privacy; and if $0 < d < 2r$, there is an increment of privacy. In addition to distance d, a rotation angle θ must be specified to derive an obfuscated area by shifting the center. For the scope of this chapter, angle θ can be assumed to be generated randomly with no loss of generality. Strategies for selecting a value of angle θ depends on the application context and have been discussed in [3, 7].

Given d and θ, we denote the obfuscated area as $\text{Area}(r, x + d \sin\theta, y + d \cos\theta)$ and the intersection between the original and the obfuscated area as $\text{Area}_{\text{Init}\cap\text{Final}} = \text{Area}(r, x, y) \cap \text{Area}(r, x + d \sin\theta, y + d \cos\theta)$.

To measure the obfuscation effect and define the relation between relevances, two probabilities must be composed. The first is the probability that the real user position falls in the intersection $\text{Area}_{\text{Init}\cap\text{Final}}$, that is, $P((x_u, y_u) \in \text{Area}_{\text{Init}\cap\text{Final}} | (x_u, y_u) \in \text{Area}(r, x, y))$. The second is the probability that one point selected from the whole obfuscated area belongs to the intersection, that is, $P((x', y') \in \text{Area}_{\text{Init}\cap\text{Final}} | (x', y') \in \text{Area}(r, x + d \sin\theta, y + d \cos\theta))$. The product of these two probabilities estimates the reduction of the relevance due to the obfuscation. The obfuscated area, derived by shifting the center, can be calculated by considering the following proposition.

Proposition 19.3. *Given a measured location area of radius $r = r_{meas}$ with initial relevance $\mathcal{R}_{Init}$ and an obfuscated area of same radius derived by shifting the original center of distance d and angle θ, $\mathcal{R}_{Final}$ is calculated by multiplying $\mathcal{R}_{Init}$ by*

$$Pr((x_u, y_u) \in \text{Area}_{\text{Init}\cap\text{Final}} | (x_u, y_u) \in \text{Area}(r, x, y)) Pr((x', y') \in$$
$$\text{Area}_{\text{Init}\cap\text{Final}} | (x', y') \in \text{Area}(r, x + d \sin\theta, y + d \cos\theta)).$$

Since the two probabilities can then be expressed as

$$Pr((x_u, y_u) \in \text{Area}_{\text{Init}\cap\text{Final}} | (x_u, y_u) \in \text{Area}(r, x, y)) = \frac{\text{Area}_{\text{Init}\cap\text{Final}}}{\text{Area}(r, x, y)}$$

$$Pr((x', y') \in \text{Area}_{\text{Init}\cap\text{Final}} | (x', y') \in \text{Area}(r, x + d \sin\theta, y + d \cos\theta))$$
$$= \frac{\text{Area}_{\text{Init}\cap\text{Final}}}{\text{Area}(r, x + d \sin\theta, y + d \cos\theta)}$$

it follows that

$$\mathcal{R}_{\text{Final}} = \frac{\text{Area}_{\text{Init}\cap\text{Final}} \; \text{Area}_{\text{Init}\cap\text{Final}}}{\text{Area}(r, x, y) \; \text{Area}(r, x + d \sin\theta, y + d \cos\theta)}$$

$$\times \mathcal{R}_{\text{Init}} \begin{cases} = \mathcal{R}_{\text{Init}} & d = 0 \\ \in (\mathcal{R}_{\text{Init}}, 0) & 0 < d < 2r \\ = 0 & d = 2r \end{cases} \tag{19.5}$$

Expanding the term $\text{Area}_{\text{Init}\cap\text{Final}}$ as a function of distance d between the centers, distance d can be calculated numerically by solving the following system of equations whose variables are d and σ.

$$\begin{cases} \sigma - \sin\sigma = \sqrt{\delta\pi} & \text{with} \quad \delta = \dfrac{\text{Area}_{\text{Init}\cap\text{Final}} \; \text{Area}_{\text{Init}\cap\text{Final}}}{\text{Area}(r, x, y) \; \text{Area}(r, x + d \sin\theta, y + d \cos\theta)} \\ d = 2r \cos\dfrac{\sigma}{2} \end{cases}$$

$$\tag{19.6}$$

Variable σ represents the central angle of the circular sector identified by the two radii connecting the center of the original area with the intersection points of the original and the obfuscated areas. These two equations represent the solution to the problem of calculating the distance d between the centers of two partially overlapped circumferences, in the special case of same radius.

19.6   A PRIVACY-AWARE LBAC SYSTEM

We now present a privacy-aware LBAC system that integrates the obfuscation techniques with the location-based access control system previously described.

19.6.1   LBAC Predicates Evaluation: $\mathcal{R}_{\text{Eval}}$ Calculation

A major design issue for a privacy-aware LBAC architecture is related to the component in charge of evaluating LBAC predicates. Two choices are possible, which deeply affect how privacy is guaranteed.

- *ACE evaluation*: ACE, the component in charge of evaluating access control policies, asks users locations to the location middleware, without disclosing LBAC predicates. Locations are returned together with a relevance value.

- *LM evaluation*: ACE sends to LM an LBAC predicate for evaluation and receives a boolean answer and a relevance value.

Both choices are viable and well-suited for different set of requirements. On one hand, *ACE evaluation* enforces a clear separation between applications and location services because the location service infrastructure never deals with application-dependent location-based predicates. On the other hand, *LM evaluation* avoids the exchange of user locations, although obfuscated, with applications. This second choice is also more flexible in business terms. For instance, an ACE can subscribe to a location service for a specific set of location predicates and select different QoS according to different needs (e.g., different accuracy levels). The LM could then differentiate prices according to service quality. Since more elaborate, in the following, we focus on this second option.

As previously discussed, we assume that the results returned by LM have the form (*bool_value*,$\mathcal{R}$,*timeout*). However, in the case of ACE evaluation, relevance $\mathcal{R}$ contained into the response is the $\mathcal{R}_{\text{Final}}$ value obtained by obfuscating a measured location, in case of LM evaluation, value $\mathcal{R}$ is the result of an additional elaboration that depends on the type of location predicate. It is important to highlight that the *movement* and *interaction* predicates are intrinsically different from the *position* predicates. Indeed, the *movement* and *interaction* predicates do not release location measurements of the corresponding users and their evaluation requires different location measurements of the same user or of different users, respectively. The *position* predicates release the location measurements of users and involve one location measurement only. For the sake of clarity, we denote $\mathcal{R}_{\text{Eval}}$ the parameter $\mathcal{R}$ contained into a response produced after an LM evaluation.

Position Predicates. Suppose that a location measurement Area(r_{meas}, x_c, y_c) (Area$_{Init}$) with relevance $\mathcal{R}_{Init}$ has been obfuscated producing an area Area$_{Final}$ with relevance $\mathcal{R}_{Final}$. Relevance $\mathcal{R}_{Eval}$ is derived from $\mathcal{R}_{Final}$ by considering both the obfuscated area and the area specified in the LBAC predicate. LM calculates $\mathcal{R}_{Eval}$ of the predicate evaluation as follows:

$$\mathcal{R}_{Eval} = \frac{\text{Area}_{Final \cap LBAC}}{\text{Area}_{Final}} \, \mathcal{R}_{Final}$$

where the scalar factor depends on the intersection, denoted by Area$_{Final \cap LBAC}$, between the obfuscated area and the area specified by the LBAC predicate. For instance, suppose that inarea(*John, Room1*) is the predicate that the ACE component sends to the LM component, which asks whether the user *John* is in room *Room1*. If John's position has an overlap greater than zero with *Room1*, the predicate evaluation returns (true,$\mathcal{R}_{Eval}$,*timeout*), where $\mathcal{R}_{Eval}$ is greater than zero; $\mathcal{R}_{Eval}$ tends to zero, otherwise.

Movement Predicates. Predicate velocity, the only predicate that we currently have defined, is evaluated by first measuring two user positions at different times, and then by calculating her velocity. Relevance $\mathcal{R}_{Eval}$ cannot be generated as for the previous case. Rather, it is generated by considering the mean value of relevances $\mathcal{R}_{Init}$ associated with the positions used to calculate user's velocity.

$$\mathcal{R}_{Eval} = \frac{\mathcal{R}_{Init_1} + \mathcal{R}_{Init_2}}{2}$$

Although estimating a user's velocity, this predicate does not release information about user location. The user can then choose to obfuscate the velocity result. In this case, $\mathcal{R}_{Eval}$ is calculated as the mean of relevances $\mathcal{R}_{Final}$ associated with the obfuscated positions used to calculate the velocity of the user.

$$\mathcal{R}_{Eval} = \frac{\mathcal{R}_{Final_1} + \mathcal{R}_{Final_2}}{2}$$

For Instance, suppose that velocity(*John, 70, 90*) is the predicate that the ACE component sends to the LM component, which asks whether velocity of user *John* is within the range [70,90]. If John's velocity is in the specified interval, the predicate evaluation returns (true,$\mathcal{R}_{Eval}$,*timeout*), where $\mathcal{R}_{Eval}$ is greater than zero; $\mathcal{R}_{Eval}$ tends to zero, otherwise.

Interaction Predicates. Relevance $\mathcal{R}_{Eval}$ is calculated by using location measurements of all users locations intersecting a reference area called Area$_{LBAC}$. Two predicates have been defined. The density predicate, which requires that Area$_{LBAC}$ is geographically identified (e.g., a city), and predicate local_density, which, instead, considers as Area$_{LBAC}$ a given area around a user. $\mathcal{R}_{Eval}$ is calculated from Eq. (19.7) as follows:

$$\mathcal{R}_{Eval} = \frac{\sum_{i=1}^{n} \frac{\text{Area}_{i,Init \cap LBAC}}{\text{Area}_{i,Init}} \mathcal{R}_{Init_i}}{n} \qquad \forall \text{Area}_{i,Init} : \text{Area}_{i,Init \cap LBAC} \neq 0$$

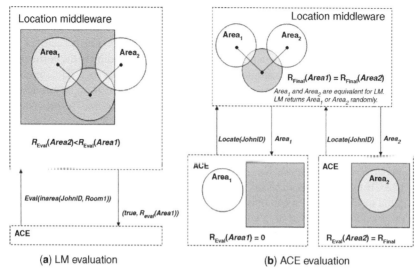

(a) LM evaluation   **(b)** ACE evaluation

Figure 19.4   An example of LM evaluation (a) and ACE evaluation (b).

where $\text{Area}_{i,\text{Init}\cap\text{LBAC}}$ represents the intersection between the ith location measurement $\text{Area}_{i,\text{Init}}$ and the area identified by the LBAC predicate, $\mathcal{R}_{\text{Init}i}$ is the initial relevance of the ith location measurement, and n the number of users involved in the predicate evaluation. Whether obfuscated areas are considered, that is, all the areas generated by location measurements of users that have an intersection with $\text{Area}_{\text{LBAC}}$, $\mathcal{R}_{\text{Eval}}$ is calculated starting from Eq. (19.7) as follows:

$$\mathcal{R}_{\text{Eval}} = \frac{\sum_{i=1}^{n} \frac{\text{Area}_{i,\text{Final}\cap\text{LBAC}}}{\text{Area}_{i,\text{Final}}} \cdot \mathcal{R}_{\text{Final}i}}{n}$$

Suppose that $\texttt{density}(Room1, 0, 3)$ is the predicate that the ACE component sends to the LM component, which asks whether the number of users in *Room1* is between 0 and 3. If the number of users is in the interval, the predicate evaluation returns $(\texttt{true}, \mathcal{R}_{\text{Eval}}, timeout)$. Otherwise, the evaluation returns $(\texttt{true}, \mathcal{R}_{\text{Eval}} \to 0, timeout)$.

19.6.1.1 LM vs. ACE Evaluation

To further analyze the differences between the adoption of ACE or LM evaluation, we focus on a scenario where an obfuscation by shifting the center is applied and a position predicate is evaluated (the same discussion holds for movement and interaction predicates). In this case, the ACE vs. LM choice has a significant impact. Consider the examples in Figure 19.4(a) and 19.4(b) that show the evaluation of predicate $\texttt{inarea}(John, Room1)$ in case of LM evaluation and ACE evaluation, respectively. When obfuscation by shifting the center is applied, there are infinite values of angle θ that could be chosen, all equivalent with respect to the relevance value $\mathcal{R}_{\text{Final}}$. Here, for the sake of simplicity, we consider just two possible obfuscated areas: *Area1* and *Area2*.

If LM evaluation is performed, LM computes $\mathcal{R}_{\text{Eval}}$, as previously seen, for each area and establishes an ordering among obfuscated areas according to the corresponding values of $\mathcal{R}_{\text{Eval}}$. In our example, it is easy to see that relevance $\mathcal{R}_{\text{Eval}}$ resulting from *Area1*, denoted as $\mathcal{R}_{\text{Eval}}(Area1)$, is greater than relevance $\mathcal{R}_{\text{Eval}}$ resulting from *Area2*, denoted as $\mathcal{R}_{\text{Eval}}(Area2)$. This information is important for the provision of the location service, because when returned to ACE, the value $\mathcal{R}_{\text{Eval}}$ is matched with $\mathcal{R}_{\text{LBAC}}$, the minimum relevance required by ACE for LBAC evaluation. The best strategy for LM is therefore to select the angle θ that produces the obfuscated area that, given $\mathcal{R}_{\text{Final}}$, maximizes $\mathcal{R}_{\text{Eval}}$.[3]

If ACE evaluation is in place, LM does not calculate any $\mathcal{R}_{\text{Eval}}$ (i.e., $\mathcal{R}_{\text{Eval}}$ is just equal to $\mathcal{R}_{\text{Final}}$), and it can only select randomly one value for θ among all those that produce an obfuscated area with same $\mathcal{R}_{\text{Final}}$. In this way, random selection of the obfuscated area (in our example, *Area1* or *Area2*) may cause an unpredictable result during ACE evaluation, ranging from relevance equal to zero (e.g., when *Area1* in Figure 19.4(b) is returned) to relevance equal to $\mathcal{R}_{\text{Final}}$ (e.g., when *Area2* in Figure 19.4(b) is returned). As a consequence, also the matching with the condition over $\mathcal{R}_{\text{LBAC}}$ results in random rejection or acceptance of the predicate evaluation. Therefore, obfuscation by shifting the center is incompatible with the ACE evaluation. This result supports architectures including location middleware capable of autonomously evaluating LBAC predicates.

Finally, there is a subtlety to consider when the obfuscation by shifting the center is applied. When a LBAC predicate is evaluated, the choice of θ is relevant because, according to the position of the obfuscated area, the value of $\mathcal{R}_{\text{Eval}}$ may change. Therefore, LM could try to select the θ angle that maximizes $\mathcal{R}_{\text{Eval}}$. Figure 19.5 shows an example with three obfuscated areas, namely *Area1*, *Area2*, and *Area3*, which provide the same $\mathcal{R}_{\text{Final}}$ value and different $\mathcal{R}_{\text{Eval}}$ values, denoted $\mathcal{R}_{\text{Eval}}(Area1)$, $\mathcal{R}_{\text{Eval}}(Area2)$, and $\mathcal{R}_{\text{Eval}}(Area3)$, respectively. It is easy to see that $\mathcal{R}_{\text{Eval}}(Area1)$ is greater than $\mathcal{R}_{\text{Eval}}(Area2)$ (i.e., the overlap between *Area1* and *Milan* is larger than the overlap between *Area2* and *Milan*) and, correspondingly, the value of angle θ that LM should take into consideration is the one that produces *Area1*.

A problem could arises with *Area3*, which has clearly the greatest overlap with *Milan*. *Area3* could provide a $\mathcal{R}_{\text{Eval}}$ greater than the one that would have provided the original area $Area_{\text{Init}}$. This would lead to an inconsistent LBAC predicate evaluation. The reason is that LM would have an incentive to configure obfuscation as a way to artificially increase the odds of satisfying the $\mathcal{R}_{\text{LBAC}}$ threshold. To avoid such a side-effect, we introduce the following additional constraint: *relevance $\mathcal{R}_{\text{Eval}}$ derived from the obfuscated area with relevance $\mathcal{R}_{\text{Final}}$ must be less than or equal to the one provided by the original area with relevance $\mathcal{R}_{\text{Init}}$, which is* $\mathcal{R}_{\text{Eval}}(Area_{\text{Final}}) \leq \mathcal{R}_{\text{Eval}}(Area_{\text{Init}})$. In other terms, areas must not be manipulated with

---

[3]In addition to the strategy that selects the θ angle that maximizes $\mathcal{R}_{\text{Eval}}$, other strategies can be exploited for selecting θ (e.g., a random choice).

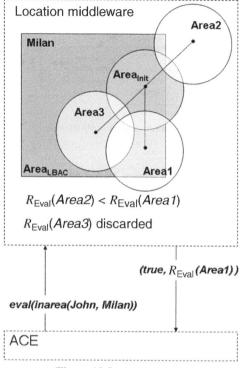

Figure 19.5   Area selection.

obfuscation techniques just to increase the odds of satisfying LBAC quality require-
ments. Our constraint ensures that, given an infinite set Θ of angles, a set $\Theta_f \subseteq \Theta$ is
generated, containing all the valid angles $\theta_1, \ldots, \theta_n$ that produce a relevance $\mathcal{R}_{Eval}$
at most equal to the relevance produced by considering the original area.

In the example of **inarea** evaluation in Figure 19.5, the following restriction is
introduced:

$$\mathcal{R}_{Eval} \leq \frac{\text{Area}_{\text{Init} \cap \text{LBAC}}}{\text{Area}_{\text{Init}}} \, \mathcal{R}_{\text{Init}} \qquad (19.7)$$

and *Area3*, which does not satisfy this constraint, is discarded in favor of *Area1*.

19.6.2   The Privacy-Aware Middleware

Currently available middleware components are mostly in charge of managing interactions between applications and location providers, and communication and negotiation protocols aimed at maximizing the QoS [30, 38, 40, 41, 47]. In a privacy-aware
LBAC, a middleware component is also responsible for balancing users privacy and
location-based services accuracy. To this end, our LM provides functionalities for
both the obfuscation of the location measurements of users and the location-based

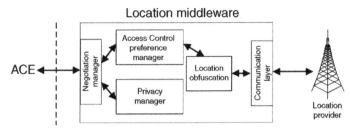

Figure 19.6   Location Middleware.

predicates evaluation. As shown in Figure 19.6, LM is functionally divided into the following five logical components.

- *Communication layer.* It manages the communication process with LPs. It hides low-level communication details intended meaning.
- *Negotiation manager.* It acts as an interface with ACE for negotiating QoS attributes (some negotiation strategies are described in [4]).
- *Location obfuscation.* It applies obfuscation techniques for users privacy.
- *Access control preference manager.* It manages location service attributes by interacting with the location obfuscation component.
- *Privacy manager.* It manages privacy preferences and location-based predicate evaluation.

It is important to highlight that the architecture of our location middleware can be extended to include the important case of users setting *multiple privacy preferences* according to different contexts. For instance, there could be users wishing to set (i) no privacy preferences for location services dedicated to the social network of their relatives and close friends; (ii) a certain level of privacy for business location services aimed at finding point of interests (e.g., shops, or monuments), and for location services whose goal is to find their position while at work; and (iii) strong privacy requirements in high sensitive contexts.

To conclude, we provide two examples of how the `inarea` and `distance` predicates are evaluated.

Example 19.3. *Suppose that ACE requires users to be located in Milan with a relevance $\mathcal{R}_{LBAC} = 0.5$ to access a service. Also, suppose that the privacy preference of user John requires a relevance $\mathcal{R}_{Final} = 0.8$. To enforce John's access request, the ACE asks the LM to evaluate the predicate* `inarea(John, Milan)`*, where John represents the located user. Let the location measurement of John be $Area_{Init}$ with $\mathcal{R}_{Init} = 1$. Figure 19.7 shows graphically an example of $\mathcal{R}_{Eval}$ computation when the obfuscation by enlarging the radius is applied. The scalar factor $Area_{Final \cap LBAC} / Area_{Final}$ is equal to 0.75. We can then produce the final relevance $\mathcal{R}_{Eval}$ associated with the predicate evaluation: $\mathcal{R}_{Eval} = 0.75 \cdot \mathcal{R}_{Final} = 0.6$. The predicate evaluation process is concluded and the result (True, 0.6, timeout) is returned to the ACE.*

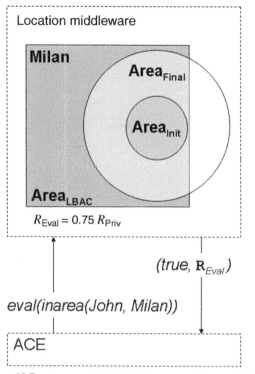

Figure 19.7 An example of the LM `inarea` predicate evaluation.

Finally, the ACE compares $\mathcal{R}_{Eval}$ with $\mathcal{R}_{LBAC}$. Since $\mathcal{R}_{LBAC} < \mathcal{R}_{Eval}$, the quality of the evaluation satisfies the ACE requirements, and John gains the access.

Suppose that the ACE requires users to stay at least 1000 m away from the Dangerous area of Figure 19.8 used for stocking dangerous material, with a relevance $\mathcal{R}_{LBAC} = 0.8$ to access a service. Suppose that John's privacy preference requires a relevance $\mathcal{R}_{Final} = 0.2$. Whenever John submits an access request, the ACE asks the LM to evaluate the predicate distance*(John,Dangerous,d_{min},d_{max}), where John represents the located user, $d_{min} = 1000$ m and $d_{max} = +\infty$. The predicate distance identifies an area $Area_{LBAC}$ (see grey area in Figure 19.8), around the Dangerous area, which contains all the points outside the Dangerous area that have a distance between d_{min} and d_{max}. Let the location measurement of John be $Area_{Init}$ with $\mathcal{R}_{Init} = 0.9$. Figure 19.8 shows graphically an example of $\mathcal{R}_{Eval}$ computation when the obfuscation by shifting the center is applied. Since the intersection between the obfuscated area $Area_{Final}$ and $Area_{LBAC}$ is equal to half of the $Area_{Final}$, the scalar factor $Area_{Final \cap LBAC} / Area_{Final}$ is equal to 0.5. We calculate the final relevance $\mathcal{R}_{Eval}$ associated with the predicate evaluation: $\mathcal{R}_{Eval} = 0.5\,\mathcal{R}_{Final} = 0.1$. The predicate evaluation process is concluded and the result (True, 0.1, timeout) is returned to the ACE,*

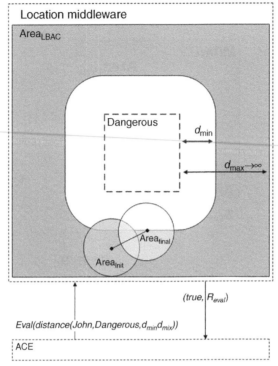

Figure 19.8   An example of the LM `distance` predicate evaluation.

meaning that John is far from the Dangerous area of at least d_{min} with a relevance of 0.1. Finally, since $\mathcal{R}_{LBAC} > \mathcal{R}_{Eval}$, the ACE denies John's request.

19.7   SUMMARY

This chapter has discussed requirements for the design of location-based access control systems and their main differences with respect to traditional access control solutions. We have shown how an access control language can be extended to support the definition and evaluation of location-based conditions. Privacy requirements for protecting location information have also been described. In particular, the trade-off between the need of information accuracy required by LBAC systems and the obfuscation of the same information for privacy reasons has been considered. Some basic obfuscation techniques have been defined together with a general metric, called relevance, that can be used for both measuring the degree of location privacy and the degree of accuracy required. Examples and case studies enriched the presentation of issues and concepts. Many research issues need to be further investigated, such as the analysis of secondary effects of location predicate evaluation, de-obfuscation attacks, and strategies for the negotiation of QoS attributes.

ACKNOWLEDGMENTS

The research leading to these results has received funding from the European Union within the 6FP project PRIME under contract no. IST-2002-507591, the European Community's Seventh Framework Programme (FP7/2007-2013) under grant agreement no. 216483, and the Italian MIUR within PRIN 2006 under project no. 2006099978.

REFERENCES

1. I. F. Akyildiz and J. S. M. Ho. Dynamic mobile user location update for wireless pcs networks. *Wireless Networks*, 1(2):187–196, 1995.
2. M. Anisetti, C. A. Ardagna, V. Bellandi, E. Damiani, and S. Reale. Method, system, network and computer program product for positioning in a mobile communications network. In *European Patent No. EP1765031*, 2007.
3. C. A. Ardagna, M. Cremonini, E. Damiani, S. De Capitani di Vimercati, and P. Samarati. A middleware architecture for integrating privacy preferences and location accuracy. In *Proceedings of the 22nd IFIP TC-11 International Information Security Conference (SEC 2007)*, Sandton, South Africa, May 2007.
4. C. A. Ardagna, M. Cremonini, E. Damiani, S. De Capitani di Vimercati, and P. Samarati. Location-based metadata and negotiation protocols for LBAC in a one-to-many scenario. In *Proceedings of the Workshop on Security and Privacy in Mobile and Wireless Networking (SecPri_MobiWi 2006)*, Coimbra, Portugal, May 2006.
5. C. A. Ardagna, M. Cremonini, E. Damiani, S. De Capitani di Vimercati, and P. Samarati. Location privacy protection through obfuscation-based techniques. In *Proceedings of the 21st Annual IFIP WG 11.3 Working Conference on Data and Applications Security*, Redondo Beach, CA, USA, July 2007.
6. C. A. Ardagna, M. Cremonini, E. Damiani, S. De Capitani di Vimercati, and P. Samarati. Supporting location-based conditions in access control policies. In *Proceedings of the ACM Symposium on Information, Computer and Communications Security (ASIACCS'06)*, Taipei, Taiwan, March 2006.
7. C. A. Ardagna, M. Cremonini, E. Damiani, S. De Capitani di Vimercati, and P. Samarati. Managing privacy in LBAC systems. In *Proceedings of the IEEE 21st International Conference on Advanced Information Networking and Applications Workshops (AINAW 2007)*, Niagara Falls, Canada, May 2007.
8. C. A. Ardagna, E. Damiani, S. De Capitani di Vimercati, and P. Samarati. Towards privacy-enhanced authorization policies and languages. In *Proceedings of the 19th IFIP WG11.3 Working Conference on Data and Application Security*, Nathan Hale Inn, University of Connecticut, Storrs, USA, August 2005.
9. P. Bellavista, A. Corradi, and C. Giannelli. Efficiently managing location information with privacy requirements in Wi-Fi networks: a middleware approach. In *Proceedings of the International Symposium on Wireless Communication Systems (ISWCS'05)*, Siena, Italy, September 2005.
10. A. R. Beresford and F. Stajano. Location privacy in pervasive computing. *IEEE Pervasive Comput.*, 2(1):46–55, 2003.
11. A. R. Beresford and F. Stajano. Mix zones: user privacy in location-aware services. In *Proceedings of the 2nd IEEE Annual Conference on Pervasive Computing and Communications Workshops (PER-COMW04)*, Orlando, FL, March 2004.
12. C. Bettini, X. S. Wang, and S. Jajodia. Protecting privacy against location-based personal identification. In *Proceedings of the 2nd VLDB Workshop on Secure Data Management (SDM 2005)*, Trondheim, Norway, September 2005.
13. J. Borkowski, J. Niemelä, and J. Lempiäinen. Performance of Cell ID+RTT hybrid positioning method for UMTS radio networks. In *Proceedings of the 5th European Wireless Conference*, Barcelona, Spain, February 2004.

14. L. Cong and W. Zhuang. Hybrid TDOA/AOA mobile user location for wideband CDMA cellular systems. *IEEE Trans. Wireless Commun.*, 1(5):439–447, 2002.
15. M. Duckham and L. Kulik. A formal model of obfuscation and negotiation for location privacy. In *Proceedings of the 3rd International Conference on Pervasive Computing (PERVASIVE 2005)*, Munich, Germany, May 2005.
16. M. Duckham and L. Kulik. Simulation of obfuscation and negotiation for location privacy. In *Proceedings of Conference on Spatial Information Theory (COSIT 2005)*, Ellicottville, NY, September 2005.
17. D. Faria and D. Cheriton. No long-term secrets: location-based security in over-provisioned wireless LANS. In *Proceedings of the 3rd ACM Workshop on Hot Topics in Networks (HotNets-III)*, San Diego, CA, November 2004.
18. S. Garg, M. Kappes, and M. Mani. Wireless access server for quality of service and location based access control in 802.11 networks. In *Proceedings of the 7th IEEE Symposium on Computers and Communications (ISCC 2002)*, Taormina/Giardini Naxos, Italy, July 2002.
19. B. Gedik and L. Liu. Location privacy in mobile systems: a personalized anonymization model. In *Proceedings of the 25th International Conference on Distributed Computing Systems (ICDCS 2005)*, Columbus, OH, June 2005.
20. Geographic Location/Privacy (geopriv). September 2006. http://www.ietf.org/html.charters/geopriv-charter.html.
21. I. Getting. The global positioning system. *IEEE Spectrum*, 30(12):36–47, 1993.
22. M. Gruteser, J. Bredin, and D. Grunwald. Path privacy in location-aware computing. In *Proceedings of the Second International Conference on Mobile Systems, Application and Services (MobiSys2004)*, Boston, MA, June 2004.
23. M. Gruteser and D. Grunwald. Anonymous usage of location-based services through spatial and temporal cloaking. In *Proceedings of the 1st International Conference on Mobile Systems, Applications, and Services (MobiSys2003)*, San Francisco, CA, May 2003.
24. M. Gruteser and Xuan Liu. Protecting privacy in continuous location-tracking applications. *IEEE Security Privacy Mag.*, 2(2):28–34, 2004.
25. F. Gustafsson and F. Gunnarsson. Mobile positioning using wireless networks: possibilities and fundamental limitations based on available wireless network measurements. *IEEE Signal Process. Mag.*, 22(4):41–53, 2005.
26. C. Hauser and M. Kabatnik. Towards privacy support in a global location service. In *Proceedings of the IFIP Workshop on IP and ATM Traffic Management (WATM/EUNICE 2001)*, Paris, France, September 2001.
27. U. Hengartner and P. Steenkiste. Protecting access to people location information. In *Proceedings of the First International Conference on Security in Pervasive Computing Security in Pervasive Computing*, Boppard, Germany, March 2003.
28. U. Hengartner and P. Steenkiste. Implementing access control to people location information. In *Proceedings of the 9th ACM Symposium on Access Control Models and Technologies 2004 (SACMAT 2004)*, Yorktown Heights, NY, June 2004.
29. B. Ho and M. Gruteser. Protecting location privacy through path confusion. In *Proceedings of IEEE/CreateNet International Conference on Security and Privacy for Emerging Areas in Communication Networks (SecureComm)*, Athens, Greece, September 2005.
30. D. Hong, M. Yuan, and V. Y. Shen. Dynamic privacy management: a plug-in service for the middleware in pervasive computing. In *Proceedings of the 7th International Conference on Human Computer Interaction with Mobile Devices & Services (MobileHCI'05)*, Salzburg, Austria, September 2005.
31. S. Horsmanheimo, H. Jormakka, and J. Lahteenmaki. Location-aided planning in mobile network – trial results. *Wireless Personal Commun. Int. J.*, 30(2–4):207–216, 2004.
32. H. Hu and D. L. Lee. Energy-efficient monitoring of spatial predicates over moving objects. *Bull. IEEE Comput. Soc. Tech. Committee Data Eng.*, 28(3):19–26, 2005.
33. S. Jajodia, P. Samarati, M. L. Sapino, and V. S. Subrahmanian. Flexible support for multiple access control policies. *ACM Trans. Database Syst.*, 26(2):214–260, 2001.

34. M. Langheinrich. A privacy awareness system for ubiquitous computing environments. In *Proceedings of the 4th International Conference on Ubiquitous Computing (Ubicomp 2002)*, Göteborg, Sweden, September 2002.

35. N. Marsit, A. Hameurlain, Z. Mammeri, and F. Morvan. Query processing in mobile environments: a survey and open problems. In *Proceedings of the 1st International Conference on Distributed Framework for Multimedia Applications (DFMA'05)*, Besancon, France, February 2005.

36. M. F. Mokbel and W. G. Aref. GPAC: generic and progressive processing of mobile queries over mobile data. In *Proceedings of the 6th International Conference on Mobile Data Management (MDM 2005)*, Ayia Napa, Cyprus, May 2005.

37. M. Mokbel, C.-Y. Chow, and W. Aref. The new Casper: query processing for location services without compromising privacy. In *Proceedings of the 32nd International Conference on Very Large Data Bases (VLDB 2006)*, Seoul, Korea, September 2006.

38. G. Myles, A. Friday, and N. Davies. Preserving privacy in environments with location-based applications. *IEEE Pervasive Comput.*, 2(1):56–64, 2003.

39. J. Myllymaki and S. Edlund. Location aggregation from multiple sources. In *Proceedings of the 3rd IEEE International Conference on Mobile Data Management (MDM 2002)*, Singapore, January 2002.

40. H. Naguib, G. Coulouris, and S. Mitchell. Middleware support for context-aware multimedia applications. In *Proceedings of the IFIP TC6 / WG6.1 3rd International Working Conference on New Developments in Distributed Applications and Interoperable Systems*, Deventer, The Netherlands, September 2001.

41. K. Nahrstedt, D. Xu, D. Wichadakul, and B. Li. QoS-aware middleware for ubiquitous and heterogeneous environments. *IEEE Commun. Mag.*, 39(11):140–148, 2001.

42. J. Nord, K. Synnes, and P. Parnes. An architecture for location aware applications. In *Proceedings of the 35th Hawaii International Conference on System Sciences*, Big Island, HA, USA, January 2002.

43. P. Olofsson. *Probability, Statistics and Stochastic Processes*. Wiley, 2005.

44. Openwave. *Openwave Location Manager*, 2006. http://www.openwave.com/.

45. B. Parkinson, J. Spilker, P. Axelrad, and P. Enge, editors. *Global Positioning System: Theory and Application*. Vol. II, Progress in Austronautics and Aerounautics Series, V-164. American Institute of Astronautics and Aeronautics (AIAA), Reston, Virginia, USA, 1996.

46. Privacy Rights Clearinghouse/UCAN. *A Chronology of Data Breaches*, 2006. http://www.privacyrights.org/ar/ChronDataBreaches.htm.

47. A. Ranganathan, J. Al-Muhtadi, S. Chetan, R. H. Campbell, and M. D. Mickunas. Middlewhere: a Middleware for location awareness in ubiquitous computing applications. In *Proceedings of the ACM/IFIP/USENIX 5th International Middleware Conference (Middleware 2004)*, Toronto, Ontario, Canada, October 2004.

48. N. Sastry, U. Shankar, and S. Wagner. Secure verification of location claims. In *Proceedings of the ACM Workshop on Wireless Security (WiSe 2003)*, San Diego, CA, USA, September 2003.

49. K. E. Seamons, M. Winslett, T. Yu, L. Yu, and R. Jarvis. Protecting privacy during on-line trust negotiation. In *Proceedings of the 2nd Workshop on Privacy Enhancing Technologies*, San Francisco, CA, April 2002.

50. G. Sun, J. Chen, W. Guo, and K. R. Liu. Signal processing techniques in network-aided positioning: a survey of state-of-the-art positioning designs. *IEEE Signal Process. Mag.*, 22(4):12–23, 2005.

51. T. W. van der Horst, T. Sundelin, K. E. Seamons, and C. D. Knutson. Mobile trust negotiation: authentication and authorization in dynamic mobile networks. In *Proceedings of the 8th IFIP Conference on Communications and Multimedia Security*, Lake Windermere, England, September 2004.

52. U. Varshney. Location management for mobile commerce applications in wireless internet environment. *ACM Trans. Internet Technol.*, 3(3):236–255, 2003.

53. G. Zhang and M. Parashar. Dynamic context-aware access control for grid applications. In *Proceedings of the 4th International Workshop on Grid Computing (Grid 2003)*, Phoenix, AZ, November 2003.

Part VI

Mobile Multimedia

Chapter 20

VoiceXML-Enabled Intelligent Mobile Services

Stan Kurkovsky and Adam Sharp

Central Connecticut State University, New Britain, CT

20.1 Introduction	471
20.2 Related Work	472
20.3 A Scenario of Using the VOICE System	476
20.4 Architecture of the VOICE System	477
20.5 User Session States	480
20.6 Application Variables	481
20.7 Available User Session States	482
20.8 Usability Experiments	484
20.9 Conclusion and Future Work	485
References	486

20.1 INTRODUCTION

In the last two decades, two technological innovations have proved to provide a very significant impact on our everyday lives and habits. The first innovation, the Internet, provides new means and ways to access the vast amounts of information and an exploding number of online services available today. The Internet has revolutionized the way people communicate with each other, how they receive news, shop, and conduct other day-to-day activities. The second innovation, a mobile phone, provides users with a simple anytime and anywhere communication tool. Originally designed for interpersonal communication, today mobile phones are capable of connecting their users to a wide variety of Internet-enabled services and applications, which can vary from a

Mobile Intelligence. Edited by Laurence T. Yang, Agustinus Borgy Waluyo, Jianhua Ma, Ling Tan, and Bala Srinivasan

simplified Web browser to a GPS-enabled navigation system. Researchers and prac-
titioners agree that a combination of these two innovations (online Internet-enabled
services that can be accessed via a mobile phone) should have a revolutionary impact
on mobile commerce [1, 3, 5, 9, 20]. So far, current research has been focusing mostly
on mobile applications designed for smart phones, in which the application logic is
usually placed within the mobile device. Speech, however, remains the most basic
and the most efficient form of communication, and providing a form of speech-based
communication remains the prevalent function of any mobile phone. Given these
two facts, mobile applications with a voice-based interface for accessing Interned-
enabled services have a very strong potential for increased growth in their popularity
and availability [14].

In this chapter, we present the voice operated interchangeable control environment (VOICE). The objective of VOICE is to create a highly customizable platform
for developing mobile context-aware applications with a voice interface. VOICE-
based applications can be accessed through any telephone as no portion of their logic
resides within the mobile device. The context-aware features of VOICE are achieved
by analyzing the patterns of requests made by the same user and by matching the
similarities among these patterns across a group of users. To test the viability of
this environment, we designed and implemented a voice-operated Restaurant Search
Guide, a simple recommender system that allows its users to call in, specify any number of search parameters, and receive suggestions from the system that are based on
the search criteria specified by the user, as well as on the previously gathered user
context and location context information.

The chapter is organized as follows. We briefly introduce VoiceXML, describe
the related work in the area of recommender systems, and highlight some existing
research and industry projects. The chapter then describes a hypothetical scenario of
using the Restaurant Search Guide and explains in details the architecture of VOICE.
The chapter is concluded with the analysis of our experiments on the usability of the
system and directions of the future work.

20.2 RELATED WORK

This section briefly describes some of the research topics, technologies, existing
projects, and applications closely related to VOICE.

20.2.1 VoiceXML

VoiceXML was introduced by the World Wide Web Consortium [24]; it is a domain-
specific application of XML that describes a voice-based interface between the human
users and computer systems. VoiceXML-enabled applications typically interact with
the user by playing pre-recorded audio fragments and converting text to speech;
these applications can accept user input by recognizing or recording their voice
input and accepting touch-tone input produced by a telephone. VoiceXML appeals
to the developers of applications with a voice-based interface because a VoiceXML

service isolates the application from many complexities, which include provisioning of resources, concurrent threads of control, and platform-specific APIs [5].

A VoiceXML-based application includes one or more VoiceXML documents that specify the structure of the dialog between the application and its user and a set of actions that need to be taken in response to different inputs of the user. VoiceXML documents along with all objects they use (audio files, input grammars, and custom scripts) are Web-based and are specified by URLs.

Today, there are a very large number of commercially deployed VoiceXML applications, which include customer relationship management, refilling of medical prescriptions, providing driving directions, flight tracking, and many others. VoiceXML services play a central role in enabling the speech-based interface and many other functionalities of VOICE.

20.2.2 Recommender Systems

Recommender systems are used to suggest products or services to consumers, and therefore they play a very important role in enabling many e-commerce and social networking services available on the Internet. Typically, items can be recommended based on the overall popularity of an item, based on the demographics or other context information about the user, or as a prediction of the future behavior of the user based on his/her past actions.

Recommender systems used by e-commerce Web sites are intended to increase revenues by making connections between customers and products or services that they may potentially be interested in. Generally, recommender systems can benefit e-commerce systems in the following ways [18]: recommender systems can help convert casual browsers of an e-commerce Web site to buyers; they may aid in cross-selling by making additional recommendations based on the products that the customer has placed in the shopping cart; recommender systems may also increase the customer loyalty by providing a personalized interface and creating a value-added relationship between the customer and the e-commerce Web site as well as among groups of customers who may have something in common.

Techniques employed by recommender systems can be broadly divided into two categories: information filtering and collaborative filtering. Information-filtering methods usually require having a user profile indicating the user's preferences or needs; these profiles can be either constructed by the user or created by the system itself upon observing the user's actions. Such profiles or user context information can be effective when the system needs to predict the level of the user's interest to a previously unseen or unrated content item. Information filtering mechanisms are relatively easy to implement and they are not dependent on the number of users in the systems; on the other hand, the quality of recommendations greatly depends on the quality of information about the content items.

Collaborative filtering techniques, another mechanism used by recommender systems, require a database of user opinions about each of the content items. The system can gather these opinions by directly asking the users to rate the content items or by observing the user's actions, such as viewing a content item description, placing a product in the shopping cart or in the wish list, or purchasing a product. Collaborative filtering mechanisms make recommendations by identifying groups of users with similar opinions on a particular content item and then making a prediction by combining opinions within that group of users. Collaborative filtering does not require a special high-quality source of information about each of the items (it is replaced by the database of the user opinions); on the other hand, the quality of recommendations made by collaborative filtering techniques dramatically increases with the number of user opinions about each of the content items. Collaborative filtering systems naturally suffer from the scarcity and cold-start problems stemming from a small number of users or a low number of opinions about a specific content item, which may result in the inability to make recommendations [15].

Recommendation systems may seem to be a natural fit within the domain of mobile devices due to their convenience and ubiquity. Making a recommender system accessible on a small device constantly carried by a customer who roams around and often moves into new environments could be a significant factor for an increased deployment of such systems by retailers and other commercial outlets. Mobile recommendation systems [11, 25] differ from their full-scale counterparts primarily due to the restrictions imposed on them by the mobile devices on which they run. Smart phones and PDAs have screens that are often too small to display the information needed to convey all details of a specific recommendation. These features of mobile devices necessitate a compromise between the accuracy of recommendation and simplicity of the interface of a mobile recommender system. Furthermore, current infrastructure of the mobile communications industry makes it very difficult for anyone other than the mobile network carrier (such as Verizon, T-Mobile, or Sprint) or a large information service provider (such as AOL, Google, or MSN) to supply and distribute the application software needed to run and use a particular mobile software system on a large scale. On the other hand, mobile systems with speech-based interface have no special distribution requirements and can be accessed by a large range of users of different classes of mobile devices.

VOICE is well suited for developing mobile recommender systems with speech interfaces. The Restaurant Search Guide developed in the VOICE framework includes a simple recommender component. One of the objectives of the VOICE framework is to provide the necessary architectural infrastructure to implement mobile recommendation systems, which is demonstrated in this paper.

20.2.3 Speech-Enabled and Context-Aware Applications

Dey [6] defines context as "any information that can be used to characterize the situation of an entity." Here, an entity can represent a user, location or any object that is relevant to the system/user interaction. In practice, context-aware applications use context information to deliver services to their users whose relevancy depends on the current task of the user. Application context may refer to current location, time, user identity, as well as other users and objects [7, 8, 17].

Current analytical research confirms an intuitive suggestion that a speech-based interface improves the end-user acceptance of mobile applications, both in the area of mobile commerce and in the area of mobile information access [3, 9]. In recent years, as the voice-recognition technology matures, a large number of applications have been developed to take advantage of VoiceXML-enabled speech interfaces that are readily available through any telephone connection. Below is a brief summary of an exemplary selection of some of these applications.

Careflow's Dictation application system [4] provides physicians with a speech-based interface to a service for transcription of patient records. The system's middle-ware generates VoiceXML dynamically through an interface with the enterprise-level CORBA services. Together with other medical transcription and healthcare documentation management services, this system has been deployed to a number of healthcare providers.

History Calls [16] is a VoiceXML-enabled system deployed at an art museum in Utah that is used as an automated audio tour guide. History Calls uses a combination of VoiceXML-generated speech and pre-recorded audio, which are mostly excerpts from interviews related to museum exhibits.

Vocera communication system [22] is deployed at hospitals and provides a single communications solution for mobile workers, such as nurses. The system operates via a specially designed WiFi-capable 2 oz. device that has a hands-free speech interface and can be worn as a badge by each user. System servers provide proprietary voice recognition and enterprise functionality, such as access to directories, paging, and voice mail.

Voice Law Enforcement Tactical System (VoiceLETS) [10] allows police officers in Alabama to make routine queries and returns search results without a human dispatcher. Queries handled by VoiceLETS deal primarily with driver and vehicle information. VoiceLETS is driven by VoiceXML; all of its data come from the databases run by the motor vehicle and corrections departments in Alabama. The viability of such a system is undeniable, especially taking into account the fact that police officers are free to use their hands and do not have to look at a communication device while dealing with a potentially dangerous situation.

Several multimodal Web browsers [12, 23], along with a number of systems for transcoding HTML content into audio via VoiceXML [13, 19] have been developed in recent year. Such systems hold a great promise for people with visual impairments; however, while such systems are technologically possible, their real-world usability is currently very limited.

At the time of this writing, the authors of this chapter could not identify any existing projects aimed at developing VoiceXML-enabled context-aware systems, which would be similar to VOICE in their functionality. However, many existing systems [4, 10, 22] employ architectural solutions that overall are similar to VOICE: they integrate back-end speech recognition services (VoiceXML or proprietary) with enterprise databases and provide anytime access over a tether-less infrastructure (mobile phone or WiFi).

20.3 A SCENARIO OF USING THE VOICE SYSTEM

John is a newcomer to the town of Enfield, Connecticut, and therefore knows very little about the area, especially what restaurants are available for him to eat. It is coming up on dinner time, and John has gotten rather hungry—fortunately John knows of a restaurant guide available through his mobile phone, so he decides to give it a call to decide where to go. As John connects to the Restaurant Search Guide driven by VOICE, the system automatically checks the history of prior calls and determines that John has not previously used the system. Because this is the first time that John calls the Restaurant Search Guide, the VOICE system has no user context information for John and it cannot offer any personalized features; therefore, VOICE automatically extends all of the spoken prompts to make John's first call an easy experience.

After checking its database of previous sessions that contains the user context information, and determining that John is a new user, VOICE responds to John with a greeting as shown in the first system prompt in Figure 20.1.

John says "Zip Code" and is then prompted to "Please state or type your zip code," to which John responds, "06082." After informing the VOICE system that he

---

System: *Welcome to the Restaurant Search Guide, where you can search for restaurants in any area. Since it seems as though this is your first time calling us, we will give you extended information on each of the available options. Please select from the following options: Say 'Zip Code' to select the zip code in which to search. Say 'Type' to select the restaurant type. Say 'Cuisine' to select the cuisine that you desire. If you need further assistance at any time, you may say 'Help,' and if you ever want to start over, say 'Start Over'.*

User: *Zip Code*

System: *Please state or type your zip code.*

User: *06082*

System: *You have entered 06082. Please select what to complete next: Say 'Type' to select the restaurant type, say 'Cuisine' to select the cuisine that you desire, or say 'Start Over' to start from the beginning.*

User: *Type*

System: *Would you like an eat-in or take-out restaurant?*

User: *Take out*

System: *You have selected take out. Please select what to complete next: Say 'Cuisine' to select the cuisine that you desire, or say 'Start Over' to start from the beginning.*

User: *Cuisine*

System: *What type of cuisine would you like?*

User: *Chinese*

System: *You have selected Chinese. Please select from the following restaurants in or around 06082:*
   1) China Garden
   2) Lee's Chao
   3) Cheng's Palace
   ...

User: *Three*

System: *Cheng's Palace. 123 Main Street, Enfield, Connecticut, 06082. Phone number: 860-555-1234. Please say 'repeat' to have this information repeated to you. Would you like to get directions to this restaurant?*

User: *No*

System: *Thank you for using the Restaurant Search Guide.*

Figure 20.1   A typical dialog of a user with the VOICE-based Restaurant Search Guide.

is looking for a "Take-Out" restaurant serving "Chinese" cuisine, John is given a list of Chinese take-out restaurants in the zip code of 06082. John selects the third option, "Cheng's Palace," and is given more information about the restaurant including its address: 123 Main Street, Enfield, CT, and its phone number: 860-555-1234. As John will be ordering his dinner to be delivered to his home address, he responds "No" when asked whether he would like directions to this restaurant and hangs up the phone. A transcript of a possible complete dialog of John's interaction with the system is shown in Figure 20.1.

A few days later, John decides that he wants to go with his friend Tom to dinner at some Italian restaurant in town. After John calls the Restaurant Search Guide once more, VOICE determines that he has called in before, retrieves his context information, and automatically sets his zip code to 06082, responding to John with "Welcome back to the Restaurant Search Guide. Are you calling for restaurants in 06082?" to which John responds in the affirmative. After selecting an eat-in Italian restaurant in the area, John opts to have the system give him the directions from his current location to the restaurant (and possibly repeat a few times, just so he can make sure that he copies down the directions exactly).

Over the coming weeks, John makes multiple calls to the Restaurant Search Guide, where the VOICE system keeps track of the user context information, such as his preferences for zip code, cuisine, and restaurant type (noting that John really does eat a lot of Chinese take-out). After some time, if John stops asking for Italian restaurants in the area, the system will "forget" that John likes this type of food and will not automatically suggest Italian restaurants in his area.

20.4 ARCHITECTURE OF THE VOICE SYSTEM

The VOICE system has a multitiered architecture that is typically distributed across multiple servers corresponding to different tiers of the system's architecture, as shown in Figure 20.2. The VoiceXML service provider is the first tier of the system, which contains a VoiceXML processor and any applicable audio files (in our example, we used the service provided by BeVocal [2]). The VOICE System tier and the VOICE Application tier contain the bulk of the VOICE system and application logic, including the VoiceXML Converter, Flow Control Module, User Session State/Context Manager, XML Application Schema, Query Manager(s), Query Designer(s), and Query Parser(s). The VOICE System tier also hosts the Context Information database. The Data Provider tier contains one or more servers providing the Data Source(s) that the Data Source Adapter(s) will query. The Restaurant Search Guide described in the previous section is an application developed within the VOICE framework and will be used throughout this section to illustrate many of the concepts presented herein.

20.4.1 User Session State/Context Manager

The User Session State/Context Manager (USSCM) controls the states of the current session given its XML Application Schema, determining what states are available

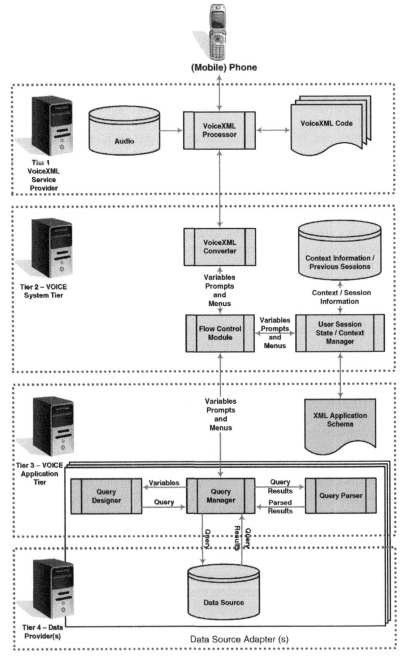

Figure 20.2 Architecture of the VOICE framework.

to transition to next, as well as maintaining the values of each variable in the current state. Given the current values for each of the variables, the USSCM determines which states are available to transition to and what information is required before this state can be exited and a new state can be entered into. The USSCM also queries the Context Information database to retrieve the user context information; it determines if the current caller has used the system before, automatically filling out global variables that indicate whether they have called and what their most likely inputs will be.

The USSCM also creates and passes the list of variables and the internal representation of the menus and prompts to the Flow Control Module and receives updates to these variables from the Flow Control Module.

During the course of the user interaction with the system, the USSCM may also execute queries on the Context Information database in order to keep track of the user context including the user's preferences and usage habits, which will be used to determine information in further sessions with the VOICE system. In the Restaurant Search Guide example, the user context information includes John's zip code and restaurant preferences that are stored in the Previous Sessions database each time John selects a restaurant or changes his zip code, even if he decides to change his preferences later.

20.4.2 Flow Control Module

The Flow Control Module is the central part of the VOICE system, which parses all of the user input and interacts with the USSCM to determine the state of the current session. The Flow Control Module also interacts with the Data Source Adapters, which have been defined in the XML Application Schema for the current state (shown in Figure 20.3), as well as sending all internal menu and prompt representations to the VoiceXML Converter. When a user enters information that needs to be stored in a variable, the Flow Control Module checks to make sure that the information is valid (i.e., fits the given type of the variable) and, if necessary, will automatically issue error messages to the VoiceXML converter to be spoken to the user.

The Flow Control Module issues all queries for information to the appropriate Data Source Adapter and then can return the results of those queries to the VoiceXML Converter (if used in a prompt) or fills in some internal variables with the results from the query.

```
1  <state name="findrestaurant">
2    ...
3    <adapter src="search.cgi"
            variables="zipcode rest-type" />
4  </state>
```

Figure 20.3 Sample Adapter as used in the Restaurant Search Guide example.

20.4.3 VoiceXML Converter

The VoiceXML Converter takes all user input interpreted by the VoiceXML processor and converts it into usable internal representations (to be checked for validity in the Flow Control Module). The VoiceXML Converter also transforms the internal menus and prompts into valid VoiceXML, which, using available audio files and a text-to-speech converter, creates the audio that will be spoken aloud through the user's phone. After the VoiceXML Converter translates the user input into usable representations, it passes the data to the Flow Control Module for further validation and updating, to ensure that the text entered is, indeed, valid.

20.4.4 Context Information Database

The Context Information/Previous Sessions Database holds all of context information derived from the analysis of data gathered during the previous VOICE sessions, including the calling number, what options were selected, and what information was provided to the VOICE system.

20.4.5 XML Application Schema

The VOICE XML Application Schema is integral to the functionality of VOICE, as it describes each of the prompts from which the audio spoken to the user is created, each of the variables and variable types that can (or need to) be filled out, as well as describing each of the internal states along with the transitions to and from each state, including what the user must do in order to have the VOICE system traverse the different states. Subsequent sections describe the integral components of the XML Application Schema and the roles they play in more detail. A fragment of the application schema used in the Restaurant Search Guide is shown in Figure 20.4.

20.5 USER SESSION STATES

At any given moment during the user's interaction with the system, VOICE is said to be in one of the user session states, as shown in Figure 20.5. A user session state describes the different prompts and available variables for that state along with a list of other states to which the system can transition from this state. In each state, the audio prompt should be described, either in a form usable by the VoiceXML text-to-speech translator or as an audio file that will be played to the user. Each state should also contain a list of preconditions that must be met before this state can be entered into. The user session state description should also have any applicable Data Source Adapters defined as shown in Figure 20.5, which variables to pass through to those adapters, and what to do with the results of the queries that are passed back from the adapters. Figure 20.4 shows an example of a user session state "firsttime", which is activated when the user has never called the system before and has yet to fill

```
1 <state name="firsttime">
2   <required>
3     <!-- Require that the previous number of
4         calls is 0, and that no information
5         has been filled out. -->
6     <var name="previouscalls" value="no" />
7     <var name="zipcode" value="" />
8     ...
9   </required>
10  <prompt>
11    Welcome to the Restaurant Search Guide,
12    where you can search for restaurants in any
13    area. Since it seems as though this is your
14    first time calling us, we will give you
15    extended information on each of the
16    available options. Please select from the
17    following options: Say 'Zip Code' to select
18    the zip code in which to search. Say 'Type'
19    to select the restaurant type. Say
20    'Cuisine' to select the cuisine that you
21    desire. If you need further assistance at
22    any time, you may say 'Help,' and if you
23    ever want to start over, say 'Start Over'.
24  </prompt>
25  <variables>
26    <var name="option">
27      <grammar type="application/x-nuance-gsl">
28        [ zip-code type cuisine ... ]
29      </grammar>
30    </var>
31    ...
32  </variables>
33  <next>
34    <moveto state="enterzip">
35      <if cond="option=='zip-code'" />
36    </moveto>
37    ...
38  </next>
39 </state>
```

Figure 20.4   Sample State section in the XML Application Schema, as used in the Restaurant Search Guide example.

out any information. When John calls the VOICE system for the first time, this is the first user session state that is activated, which is then used to introduce John to the system, telling him what his options are.

20.6 APPLICATION VARIABLES

The type of an application variable is determined by the common VoiceXML variable types and can also be given specific "grammars", which allow VOICE to accept a wide range of specialized input. These variables can be either state-specific (those that are not retained and passed through the different VOICE modules) or global (every module and state in the system has access to their values). In our Restaurant Search

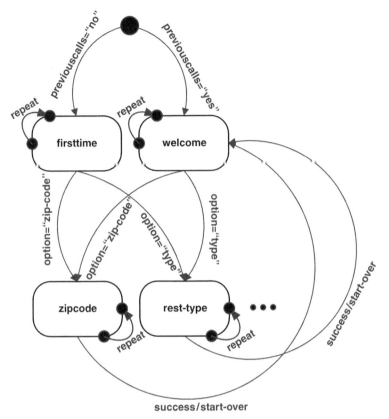

Figure 20.5   Typical user session states used by the Restaurant Search Guide.

Guide example, the "`previouscalls`" and "`zipcode`" variables are global variables, while the "option" variable is state-specific. Once a type is set for a given variable, the VOICE system automatically ensures that each user input matches the type of the given application variable or will prompt the user to enter information that correctly fills the application variable in question. Lines 26–30 in Figure 20.4 outline the "`option`" variable used in the example given above; "option" uses a grammar that defines precise words that can fill this variable ("`zip-code`", "`type`", "`cuisine`", etc.). When John has heard the first prompt from the system and has decided what information to input into the system, only one of these words will be accepted by VOICE; in our example, this variable determines the next prompt that John will hear, as well as which variables will be filled out.

20.7   AVAILABLE USER SESSION STATES

A list of one or more user session states that are available to move to once this state has been passed are also included in the description of each user session states and are referenced by state name. The available states may have preconditions attached to them (see Figure 20.4, lines 2–9), which are described in VoiceXML notation. Lines 33–38 in Figure 20.4 describe the user session states that are available to transition to once the "`option`" variable has been filled in. In our example, specifically, it shows that if John says "zip-code" in response to the first prompt, then VOICE will transition to the "`enterzip`" state.

20.7.1 Data Source Adapter

The Data Source adapter consists of four parts: the Query Manager, the Query Designer, the Query Parser, and the Data Source. The data source adapter allows VOICE to connect and query a specific source of information, which could be a local database, an XML Web Service, or an application API. In our Restaurant Search Guide example, there are two Data Source Adapters: one to receive a listing of all available restaurants in a given zip code (using the Google API), and the other to get directions from a caller's location to the restaurant. An XML description of one of Data Source Adapters used in the Restaurant Search Guide is shown in Figure 20.5.

20.7.2 Query Manager

The Query Manager is essentially a middleman; it merely passes information between the Flow Control Module and the Query Designer/Parser. The Query Manager receives the list of available variables from the Flow Control Module and then sends the list to the Query Designer. Once the Query Designer has created a query for use with the Data Source, the Query Manager executes this query and returns the results to the Query Parser. Once these results have been analyzed by the Query Parser, it sends updated variables and information back to the Flow Control Module for further use.

20.7.3 Query Designer

The Query Designer uses the list of available variables given to it by the Query Manager to create a query that will be executed against the Data Source in the appropriate format. For example, the Query Designer could form an SQL query to be used with a local database or form a URL to be passed to a Web-based application. Once the Query Designer has created the query, it is passed back to the Query Manager.

20.7.4 Query Parser

The Query Parser takes the results received from the Data Source in its native format and given to it by the Query Manager, analyzes the results, and parses the results into usable data. In some cases, these data are in the form of internal variables, while other data can be made into user prompts. In the case of our Restaurant Search Guide example, the results of one query are parsed into a prompt from which the user selects a restaurant about which to get more information as well as a new variable with a maximum and minimum to allow the user to select restaurants by number rather than name, while another query (on a separate Data Source Adapter) returns the driving directions from John's current location to the restaurant.

20.8 USABILITY EXPERIMENTS

We conducted a brief experiment to study the usability of the VOICE system and the Restaurant Search Guide. We selected 12 people to participate, none of whom had any prior experience with the VOICE system. We gave each participant a set of instructions outlining their goals, as well as the tasks required to complete each goal. Each of the participants used VOICE to enter specific queries and retrieve the results. The participants' goals were

1. to find a take-out Chinese restaurant in the 06082 ZIP code area and get the phone number for this restaurant and
2. to find an eat-in Italian restaurant in the 06082 ZIP code area and get directions to this restaurant.

Each participant worked by themselves using only a mobile phone, and all of their attempts were logged electronically. After the experiment, we asked each of the participants to fill out a questionnaire, which included some biographical questions as well as questions to evaluate their experience with VOICE:

1. To what extent did the VOICE system understand what you said?
2. To what extent was the VOICE system's speech easy to understand?
3. To what extent did you understand what to say at each of the prompts?
4. Were you able to complete the tasks and get the information that you required?
5. After using VOICE, how likely are you to use a similar system in the future?

The questions were answered on a five-point Likert scale, where 1 indicated never, unlikely, or very difficult, and 5 indicated always, likely, or very easy.

We used two metrics to measure the effectiveness of the VOICE system: the time required to complete a goal and the number of attempts per goal. On average, our study participants required 62.3 s to complete the first goal and 85.0 s to complete the second goal. All study participants were able to meet each of their goals in a single attempt. Additional performance indicators were used to assess the VOICE system included the average number of attempts per goal (1.1 attempts) and the percentage of error-free performance by the system (92%). As shown in Figure 20.6, the VOICE system received consistently high scores from the participants of the usability study in response to the post-test survey.

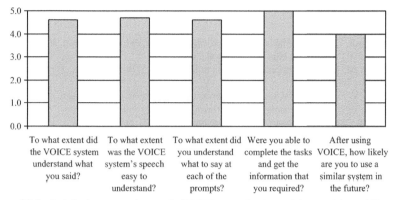

Figure 20.6   Satisfaction scores given to the VOICE system by the participants of the usability study.

Additionally, the participants were encouraged to provide feedback about possible enhancements and extensions to the system. The users suggested the incorporation of an ability to pause the Restaurant Search Guide while it reads the directions to the restaurant or to repeat the directions from a specific step number; they also suggested having more options for food choices incorporated to the system.

20.9 CONCLUSION AND FUTURE WORK

In this chapter, we presented VOICE, a framework for developing mobile context-aware systems with speech interface accessible over a telephone connection. To prove the feasibility of the architecture of VOICE, we implemented a Restaurant Search Guide, a simple context-aware recommender system that helps its users with finding a restaurant according to multiple combinations of criteria; the system connects to multiple data source providers to obtain restaurant data and to generate driving directions to the chosen restaurant. We plan to further validate this solution by conducting a usability study based on the analysis of the system usage patterns.

There are several directions for adding functionality to VOICE architecture. Mixed-initiative dialogs could be used to simplify the number of prompts presented to the user in order to obtain a minimally necessary amount of data needed for a successful execution of queries. To streamline the process of customization of VOICE during the development of specific applications, special templates could be developed for different types of Data Source Adapters corresponding to a local database, an XML Web Service, or an application-specific or platform-specific API. Furthermore, it could be possible to mix some portions of the users' voice-based interface with SMS-based interface; for example, the users might want to query the system using the voice interface but receive the results as a text message on their mobile phones.

REFERENCES

1. N. Anerousis, E. Panagos. Making voice knowledge pervasive. *Pervasive Comput.*, 1(2):42–48, 2002.
2. BeVocal, 2008 Available at http://www.bevocal.com.
3. S. Chang, M. Heng. An empirical study on voice-enabled Web applications. *Pervasive Comput.*, 5(3):76–81, 2006.
4. J. Chugh, V. Jagannathan. Voice-enabling enterprise applications. In *Proceedings of the 11th IEEE Workshop on Enabling Technologies: Infrastructure for Collaborative Enterprises*, Washington, DC, 10–12 June 2002 IEEE Computer Society, Sebastool, CA, 2002, pp. 188–189.
5. P. Danielsen. The promise of a voice-enabled Web. *IEEE Comput.*, 33(8):104–106, 2000.
6. A. Dey. Understanding and using context. *Personal Ubiquitous Comput.*, 5(1):4–7, 2001.
7. A. Dix, T. Rodden, N. Davies, J. Trevor, J. Friday, and K. Palfreyman. Exploiting space and location as a design framework for interactive mobile systems. *ACM Trans. Comput. Hum. Interact.*, 7(3):185–321, 2000.
8. P. Dourish. What we talk about when we talk about context. *Personal Ubiquitous Comput.*, 8(1):19–30, 2004.
9. Y. Fan, A. Saliba, E. A. Kendall, and J. Newmarch. Speech interface: an enhancer to the acceptance of M-commerce applications. In *Proceedings 2005 International Conference on Mobile Business, Sydney, Australia*, 11–13 July 2005. IEEE Computer Society Sebastool, CA, 2005, pp. 445–451.
10. J. E. Gilbert, R. Chapman, and S. Garhyan. VoiceLETS Backs Up First Responders. *Pervasive Comput.* 4(3):92–96, 2005.
11. H. van der Heijden, G. Kotsis, and R. Kronsteiner. Mobile recommendation systems for decision making 'on the go'. In *Proceedings of the 2005 International Conference on Mobile Business*, Sydney, Australia, 11–13 July 2005. IEEE Computer Society Sebastool, CA, 2005, pp. 137–143.
12. J. Kleindienst, L. Seredi, P. Kapanen, and J. Bergman. CATCH-2004 multi-modal browser: overview description with usability analysis. In *Proceedings of the 4th IEEE International Conference on Multimodal Interfaces*, Pittsburg, PA, 14–16 October 2002. IEEE Computer Society, Sebastool, CA, 2002, pp. 442–447.
13. J. Kong. Browsing Web through audio. In *Proceedings of the 2004 IEEE Symposium on Visual Languages and Human Centric Computing*, Rome, Italy, 26–29 September 2004. IEEE Computer Society, Sebastool, CA, 2004, pp. 279–280.
14. M. Lucente. Conversational interfaces for e-commerce applications. *Commun. ACM*, 43(9):59–61, 2000.
15. S. Middleton, N. Shadbolt, D. De Roure. Ontological user profiling in recommender systems. *ACM Trans. Inform. Syst.*, 22(1):54–88, 2004.
16. M. Nickerson. History calls: delivering automated audio tours to visitors' cell phones. In *Proceedings of the 2005 Information Technology: Coding and Computing Conference*, Las Vegas, NV, 4–6 April 2004. IEEE Computer Society, Sebastool, CA, Vol. 2, pp. 30–34.
17. M. Satyanarayanan. Challenges in implementing a context-aware system. *Pervasive Comput.*, 1(3):2, 2002.
18. J. B. Schafer, J. Konstan, and J. Riedi. Recommender systems in e-commerce. In *Proceedings of the 1999 ACM Conference on Electronic Commerce*, Denver, CO, 3–5 November 1999. ACM, New York, pp. 158–166.
19. Z. Shao, R. G. Capra III, and M. A. Pérez-Quiñones. Transcoding HTML to VoiceXML using annotation. In *Proceedings of the 15th IEEE International Conference on Tools with Artificial Intelligence*, Sacramento, CA, 3–5 November 2005. IEEE Computer Society, Sebastool, CA, pp. 249–258.
20. A. Srinivasan, E. Brown. Is speech recognition becoming mainstream? *Computer* 35(4):38–41, 2002.
21. W. Srisa-an, C. T. D. Lo, and J. M. Chang. Putting voice into wireless communications. *IT Professional*, 4(1):62–64, 2002.
22. V. Stanford. Beam me up, doctor McCoy. *Pervasive Comput.* 2(3):13–18, 2003.

23. A. Tiwari, R. A. Hosn, and S. H. Maes. Conversational multi-modal browser: an integrated multi-modal browser and dialog manager. In *Proceedings of the 2003 Symposium on Applications and the Internet*, Orlando, FL, 27–31 January 2003. IEEE Computer Society, Sebastool, CA, pp. 348–351.

24. Voice Extensible Markup Language (VoiceXML) Version 2.0. World Wide Web Consortium, 2008. http://www.w3.org/TR/voicexml20.

25. B. Zhou, S. C. Hui, and K. Chang. Enhancing mobile Web access using intelligent recommendations. *Intell. Syst.*, 21(1):28–34, 2006.

Chapter 21

User Adaptive Video Retrieval on Mobile Devices

Na Zhao,[1] Min Chen,[2] Shu-Ching Chen,[1] and Mei-Ling Shyu[3]

[1] *School of Computing and Information Sciences, Florida International University, Miami, FL*
[2] *Department of Computer Science, University of Montana, Missoula, MT*
[3] *Department of Electrical and Computer Engineering, University of Miami, Coral Gables, FL*

21.1 Introduction   488
21.2 Related Work   490
21.3 System Architecture   492
21.4 Video Database Modeling   494
21.5 MoVR: Mobile-based Video Retrieval   496
21.6 Implementation and Experiments   503
21.7 Summary   507
Acknowledgments   508
References   508

21.1 INTRODUCTION

Nowadays, handheld mobile devices including cell phones and personal digital assistants (PDAs) have become increasingly popular and capable, which creates new possibilities for accessing pervasive multimedia information. The new generation of mobile devices are no longer used only for voice communication, they are also used frequently to capture, manipulate, and display different audiovisual media contents.

With the rapid emergence of wireless network technologies, such as GSM, Satellite, Wireless Local Area Network (WLAN), and 3G, it is much easier now to

Mobile Intelligence. Edited by Laurence T. Yang, Agustinus Borgy Waluyo, Jianhua Ma, Ling Tan, and Bala Srinivasan

transfer large size multimedia items to mobile clients. However, multimedia mobile services still suffer not only the constraints of small display sizes but also the limitation in terms of power supply, storage space, processing speed, among others. The navigation of multimedia contents on handheld devices is always restricted in the limited time periods with minimized amounts of interactions. Meanwhile, the large size multimedia data such as video cannot be stored permanently on the mobile devices due to the limited memory.

Take the following typical scenario in the mobile-based multimedia application. The sport fans wish to watch sports video through their cell phones. However, it is both unaffordable and sometime unnecessary for them to watch the whole game, which will take a long time, occupy huge memory, and exhaust the power supply. Thus, a better solution is to offer them the capability of browsing and retrieving only short video clips containing an interesting event shot or a temporal event sequence. Given a huge collection of sports videos, there exist lots of challenges to accomplish this task.

- It is quite difficult to segment the videos properly and annotate the semantic events automatically. Although advanced techniques offer great capabilities in extracting multimodal visual and audio features from all kinds of videos, the "semantic gap" remains a crucial problem when bridging these low-level or mid-level features with the high-level rich semantics. Even with the best event annotation algorithms, there is hardly sufficient guarantee in terms of the correctness and completeness of the semantic interpretation results. This motivates the modeling of high-level semantic abstractions by utilizing the existing annotation results, their features, and user feedbacks.

- It is critical to address the database modeling issue, especially when considering the temporal and/or spatial relationships between the multimedia objects. It should be able to support not only the basic retrieval methods but also the complicated temporal event pattern queries.

- There is an emerging need to support individual user preferences in multimedia applications. It is well-known that people have diverse interests and perceptions toward media data. Thus, it is desirable to incorporate user feedbacks with the purpose of training the retrieval system.

- In the meanwhile, we may want to reduce the number of user interactions to alleviate the burden on the users and to accommodate the restrictions of the mobile devices. It is thus desirable to keep track of user actions and accumulate the knowledge about user preferences.

- The system architecture should be designed to reduce the size of data to be transferred and to minimize the requirement of data storage for the mobile devices.

- The mobile-based retrieval interface should be user friendly, easy to operate, and capable to offer sufficient information and choices for the users.

Therefore, an efficient and effective multimedia content management and retrieval framework will be essential for the evolution of mobile-based multimedia services.

This chapter addresses the issues of designing and implementing a user adaptive video retrieval system, called mobile-based video retrieval (MoVR), in a mobile wireless environment. Innovative solutions are developed for personal video retrieval and browsing through mobile devices with the support of content analysis, semantic extraction, as well as user interactions. First, a stochastic database modeling mechanism called hierarchical Markov model mediator (HMMM) is deployed to model and organize the videos, along with their associated video shots and clusters, in a multimedia database to offer the supports for both event and complicated temporal pattern queries. Second, HMMM-based profiles are designed to capture and store individual user's access histories and preferences such that the system can provide the "personalized recommendation." Third, the fuzzy association concept is employed to empower the framework so that the users can make their choices of retrieving content based solely on their personal interests, general users' preferences, or anywhere in between. Consequently, the users gain the control in determining the desirable level of tradeoff between retrieval accuracy and processing speed. In addition, to improve the processing performance and enhance the portability of client-side applications, the storage consumption information and computationally intensive operations are supported in the server side, while mobile clients mainly target to manage the retrieved media and user feedbacks for the current query. In order to provide more efficient accessing and information caching for the mobile devices, we also designed the virtual clients at the server-side computers to keep some relevant information that mobile users require. To demonstrate the performances of the proposed MoVR framework, a mobile-based soccer video navigation and retrieval system is implemented and tested.

The remainder of this chapter is organized as follows. Section 21.2 reviews the related research approaches of content-based multimedia retrieval systems, especially the mobile-based systems. Section 21.3 is concerned with the system architecture. In Section 21.4, the multimedia database modeling mechanism, HMMM, and the associated construction methods are discussed. Section 21.5 details the overall framework of the proposed MoVR system. Particularly, the user adaptive solutions are proposed by adopting user profiling, feature weight learning, and fuzzy association techniques. The system implementation and experimental tests are presented in Section 21.6. Finally, concluding remarks and future works are summarized in Section 21.7.

21.2 RELATED WORK

Video browsing and retrieval in mobile devices is an emerging research area. Due to the constraints of mobile devices in terms of their power consumption, processing speed, and display capability, more challenges have been encountered than in traditional multimedia applications and many research studies have been conducted to address various issues.

To reduce the viewing time and to minimize the amount of interaction and navigation processes, a variety of studies in academia and industry have been carried out on the summarization of video contents. For instance, in Ref. [11], singular value decomposition (SVD) of attribute matrix was proposed to reduce the redundancy of video segments and thus generate video summaries. Clustering techniques were also used to optimize key frame selection based on visual or motion features to enhance video summarization [2]. In industry, Virage has implemented preliminary video summarization systems for NHL hockey videos using multimodal features [29]. However, it remains a big issue in terms of the semantic gap between computable video features and the meaning of the content as perceived by the users. For this purpose, metadata about the content was utilized and played an active role in video retrieval [25]. For instance, ontologies have been proposed in Ref. [17] to perform intelligent queries and video summarization from metadata. In Ref. [28], a video semantic summarization system in the wireless/mobile environments was presented, which includes an MPEG-7-compliant annotation interface, a semantic summarization middleware, a real-time MPEG-1/2 video transcoder for Palm-OS devices, and an application interface on color/black-and-white Palm-OS PDA. Metadata selection component was also developed in Refs [19, 20] to facilitate annotation. However, automatic media analysis and annotation is still far from mature and purely manual media annotation is extremely time-consuming and error prone [7]. Alternatively, semantic event-detection frameworks have been proposed to facilitate video summarizations. Hu et al. used similar video clips cross different sources of news stations to identify interesting news events [12]. In our earlier studies [4, 5], an effective approach was proposed for video event detection facilitating multimedia data mining technique and multimodal feature analysis.

In terms of video retrieval in mobile devices, "query by example" (QBE) is a well-known query scheme used for content-based audio/visual retrieval and many systems have been developed accordingly [1, 25]. However, in most existing approaches, the similarity estimation in the query process is generally based on the computation of the (dis-)similarity distance between a query and each object in the database and followed by a rank operation [1]. Therefore, especially for large databases, it may turn out to be a costly operation and the retrieval time becomes unreasonably long for a mobile device. In addition, temporal pattern query, where a sequence of temporal events is of interest, is not well supported in these studies.

In essence, the aforementioned approaches contribute to address some restrictions of video browsing and retrieval in mobile devices. However, they fail to accommodate individual user preferences with their diverse interests and perceptions toward video data. In the literature, relevance feedback [24] has been widely adopted in content-based retrieval research society to address user preference issue. Ref. [6] presents a mechanism for learning the requested feature weights based on user feedbacks to improve the recommendation ranking. In addition, Ref. [8] studied the application of fuzzy logic in the representation of flexible queries and in expressing user's preference learned from feedback in a gradual and qualitative way. In Ref. [9], fuzzy classification and relevance feedback techniques were applied for processing the video content to capture user preference. A similar idea was also proposed in Ref. [18], where a fuzzy ranking model was developed based on user preference contained in feedbacks. However, a common weakness of most relevance feedback methods is that the feedback process has no "memory" [21], so the user feedbacks conducted in the past fail to help the future queries. Therefore, the retrieval accuracy does not improve over time in a long run.

In addition, mobile devices have limited luxury to support frequent interactions and real-time feedback learning. Alternatively, user profiling has been extensively used in information filtering and recommendation. In Ref. [3], a personal agent called WebMate is devised, which learns the user profile incrementally and facilitates browsing and searching in the Web. Ref. [22] also presented a study of the role of user profiles using fuzzy logic in Web retrieval processes. John et al. [16] developed a prototype information retrieval system using a combination of user modeling and fuzzy logic. Ref. [10] presented the idea of extracting relevant video clips based on a user profile of interests and creating personalized information delivery systems to reduce the storage, bandwidth, and processing power requirements and to simplify user interactions.

In our framework, a common profile that represents the general knowledge on the semantics of video data is constructed. Such a common profile serves as the semantic indexes of video data to speed up the retrieval process. Meanwhile, a user profile is set up for each user to characterize the personalized interest. In addition, multilevel video modeling and temporal pattern queries are well supported in our approach.

21.3 SYSTEM ARCHITECTURE

In this proposed research, the traditional client–server system architecture is adopted but enhanced to accommodate the requirements for the mobile-based multimedia services. In order to provide the maximum support and optimized solution, the following criteria are strictly followed in the system design. First, storage consumption information and computationally intensive operations are handled in the server side. Second, mobile clients are solely required to maintain the minimized data to enable the retrieval process. Third, the system should reduce the load for the wireless network, and at the same time increase the data transfer speed for the multimedia data.

In the server-side database, the huge amount of multimedia data are stored and managed by employing the HMMM mechanism. The video database contains not only the archived videos, video shots, and clusters, but also the numerical values that represent their affinity relationships, features, and access histories, and so on. As shown in Figure 21.1, the database for general user feedbacks is developed, which consists of the positive access events or patterns from the whole group of different users. The individual user feedbacks can also be extracted from this database to architect the HMMM-based individual user profiles, which will be explained in the later sections. These access histories are utilized by the system to learn both the general user perceptions and individual user interests. Based on the fuzzy weight provided by the mobile user, the fuzzy associated retrieval algorithm is able to make the compromise between these two models of perceptions intelligently, retrieve the video clips, and make the ranking recommendations accordingly. The request handler is designed to interpret the request packages and to respond to the mobile devices by sending back the retrieved and ranked results.

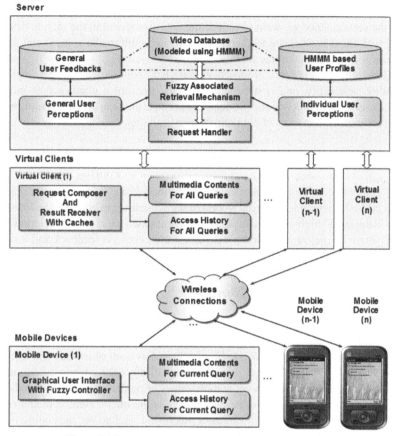

Figure 21.1   Mobile-based video retrieval system architecture.

The client-side applications on mobile devices do not need to keep storage of all the accessed media data. Alternatively, they mainly target to manage the retrieved media and user feedbacks for the current query, which includes the key frames of video shots shown on the current screen, and the video clips requested for the current operation. The mobile-based graphical interface is designed for the video retrieval system, which allows the user to easily compose and issue the event or temporal pattern based queries, to navigate through and watch the collection of retrieved results, and to provide the feedbacks.

In order to promote the mobility and manage ability of this system, a new layer with "virtual clients" is designed and incorporated in the server-side applications to extend the dynamic computing and storage capability to the mobile clients. The virtual client is designed to represent the mobile user state in the mobile-based video retrieval system. Each virtual client is customized to a distinct mobile user who accesses the video retrieval system. It contains a communication component that consists of the requests by checking and collecting the messages and commands sent from the mobile devices. The communication component can also receive the multimedia data results from the server. Since we want to reduce the data size stored in the mobile devices, the virtual client is designed to cache all the related multimedia content and access histories for the corresponding mobile user.

Generally speaking, the proposed virtual client solution can deliver improved flexibility, scalability, and cost benefits over the traditional client–server models. Mobile users gain efficiency and productivity because they can access the multimedia resources without worrying about their storage limitation.

21.4 VIDEO DATABASE MODELING

A key component in this proposed study is the video database modeling mechanism called HMMM, which is deployed to bridge the semantic gap between the low-level video features and high-level concepts, to represent video temporal characteristics for event pattern queries, and to incorporate user preferences via feedback and learning strategies.

21.4.1 Markov Model Mediator

In essence, HMMM is developed based on Markov model mediator (MMM), a well-established mathematical model proposed in our earlier work [27]. MMM is represented by a five-tuple $\lambda = (S, F, A, B, \pi)$, where S is a set of media objects called states, F denotes the feature set, A indicates the affinity relationships among media objects, B represents the low-level feature values of media objects, and π indicates the likelihood of a media object being selected as the query. Here, a media object may refer to an image, a salient object, a video shot, and so on, depending on the modeling perspective and the data source. In addition, A and π are used to model user preference and to bridge the semantic gap, which are trained via the affinity-based data mining process based on the query logs. The basic idea of the affinity-based data mining process is that the more two media objects m and n are accessed together, the higher relative affinity relationship they have, that is, the probability that a traversal choice to state (media object) n given the current state (media object) is in m (or vice versa) is higher. The details about the training and construction processes of the MMM parameters can be found in Ref.[27].

21.4.2 Hierarchical Markov Model Mediator

Although MMM has been successfully applied in content-based image retrieval [27] and Web document clustering [26] in our recent studies, it was constructed to model a single level of media objects (i.e., image or Web document) without temporal constraints. Therefore, for video database modeling, MMM is extended to a multilevel modeling scheme, HMMM [30], which models various levels of multimedia objects, their temporal relationships, the detected semantic concepts, and the high-level user perceptions.

3rd Level MMM Model
for Clusters
$S_3^g(i)$

2nd Level MMM
Models for Videos
$S_2^g(i)$

1st Level
MMM Models
for Video shots
$S_1^g(i)$

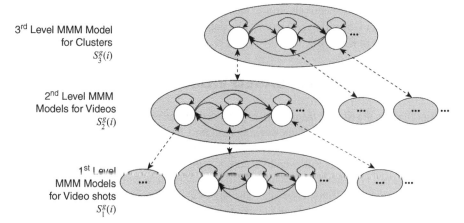

Figure 21.2   Three-level HMMM model.

As indicated by the name, HMMM consists of multiple analysis levels, which in turn might contain one or more MMMs. Given a video database as an example, we can construct a three-level HMMM model where the MMMs in the top, middle, and bottom levels model video clusters, videos, and video shots, respectively. Here, a video shot is considered as the elementary unit in the video database to describe the continuous action between the start and the end of a camera operation. As shown in Figure 21.2, each state of the higher-level MMM corresponds to and is linked to a distinct lower-level MMM.

Definition 21.1. An HMMM is defined as an eight-tuple $\Lambda = (d, \mathbf{S}, \mathbf{F}, \mathbf{A}, \mathbf{B}, \Pi, \mathbf{O}, \mathbf{L})$. It can also be represented as $\left(d, \left\{\lambda_i^j\right\}, \mathbf{O}, \mathbf{L}\right)$, where

- d is the number of levels in HMMM, for example, $d = 3$ in Figure 21.2,
- $\lambda_i^j = \left(\mathcal{S}_i^j, \mathcal{F}_i^j, \mathcal{A}_i^j, \mathcal{B}_i^j, \pi_i^j\right)$ the jth MMM in the ith level of HMMM, where $i = 1, \ldots, d$ and $j = 1, \ldots, |\lambda_i|$, $|\lambda_i|$ is the number of MMMs in the ith level. Consequently, we have $\mathbf{S} = \left\{\mathcal{S}_i^j\right\}, \mathbf{F} = \left\{\mathcal{F}_i^j\right\}, \mathbf{A} = \left\{\mathcal{A}_i^j\right\}, \mathbf{B} = \left\{\mathcal{B}_i^j\right\}$, and $\Pi = \left\{\pi_i^j\right\}$,
- $\mathbf{O} = \left\{O_{i,i+1}\right\}$ the weights of the importance for the ith level features in describing $i + 1$th feature concepts, where $i = 1, \ldots, d - 1$, and
- $\mathbf{L} = \{L_{i,i+1}\}$ the link conditions between the higher level states and the lower level states, where $i = 1, \ldots, d - 1$.

It is worth mentioning that though we use λ_i^j as a general notation to represent the MMMs in the constructed HMMM, the meanings of its parameters vary slightly among different levels to reflect the various natures of distinct media objects. For instance, as shown in Figure 21.2, $\mathcal{S}_1^j$, the states in the first level of HMMM, represent the video shots, and $\mathcal{F}_1^j$ consists of low-level or mid-level visual/audio features in our study. In contrast, the states in the second and third levels denote the sets of videos and video clusters, respectively, in the video database, and $\mathcal{F}_2^j$ contains the semantic events detected in the video collection. The details about the construction of HMMM were presented in Ref. [30].

21.5   MoVR: MOBILE-BASED VIDEO RETRIEVAL

In this chapter, the MoVR framework is proposed for the mobile-based video retrieval system development. This framework can support not only basic event queries but also the complicated queries toward some temporal event pattern, which consists of a set of important events followed by a certain temporal sequence. More importantly, this framework is capable of providing both personalized recommendations and generalized recommendations. Users can also specify a fuzzy weight parameter if their query interests have not yet been clearly formed. The system will make the adjustment and generate different retrieval results based on the fuzzy-associated queries. In essence, MoVR is designed to provide not only powerful retrieval capabilities but also a portable and flexible solution to the mobile users.

21.5.1   Overall Framework of MoVR

As shown in Figure 21.3, the overall framework of MoVR includes three main processing phases on the server side.

- Phase 1 is for video data preprocessing. It consists of the following steps. The first step is to process the source video data to detect the video shot boundaries, segment the video, and extract the shot-based features. Data cleaning and event annotation algorithms are then applied to detect the anticipated semantic events by employing the extracted shot-level features and multimodal data mining scheme. The components in Phase 1 are processed offline, which is not the focus of this chapter.

- Phase 2 is to model the video databases. As shown in the top-center box, the HMMM mechanism is deployed to model the multilevel video entities, all kinds of features, along with their associated temporal and affinity relationships. These process steps are also performed offline. Such HMMM database model will be updated periodically during the learning process by utilizing user feedbacks.

- Phase 3 includes the system retrieval and learning processes that are mainly performed online in real time to interact frequently with the virtual clients and the mobile clients.

Once a user issues a query requesting a certain semantic event or a temporal event pattern, such information will be sent to the virtual client, where it is packed and passed to the server for processing. After this point, the process will be slightly different for a first time user or a re-visiting user on the server side. For the former case, the HMMM model with initial settings will be adopted and the system performs general similarity

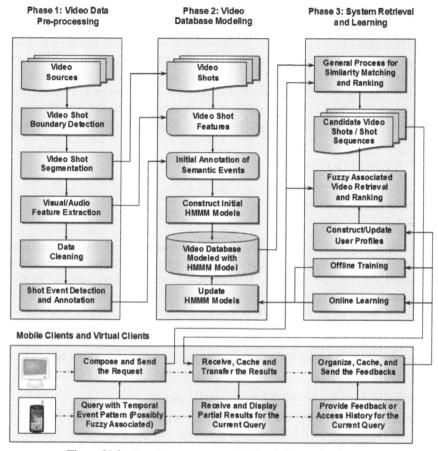

Figure 21.3 Overall framework of mobile-based video retrieval system.

matching and ranking processes. In contrast, for a re-visiting user, his/her user profile stored in the server side will be retrieved and utilized for more advanced retrieval functionalities. Accordingly, an enhanced algorithm is developed on the server side to handle these fuzzy associated video retrieval and ranking tasks.

The retrieved video clips are ranked and sent back to the virtual clients. Although all the results are cached for fast retrieval, only a portion of them are actually delivered to the mobile devices in default. Users may issue feedbacks toward the query results through their mobile devices, which will be sent to the virtual client so that these feedbacks can be organized and temporarily stored. After that, the feedbacks will be delivered to the server for the construction or update of the user profiles. Moreover, real-time online learning is also supported so that the system will yield refined results based solely on the feedbacks for the current query.

Essentially, two innovative techniques, user profiling and fuzzy association, are adopted and integrated with HMMM intelligently for server side applications, which are introduced in the following sections.

21.5.2 User Feedbacks and Profiles

One of the major challenges in multimedia retrieval is to identify and learn person-alized user interests. The underlying reason of this challenge is that the user's query interests can hardly be expressed precisely using query examples or keywords. In addition, different users tend to have diverse opinions or perceptions toward even the same query and intend to seek for different results and rankings. For example, given a query for soccer goal shots, different users may be interested in different retrieval requirements:

- followed by a corner kick,
- in the female soccer videos,
- with exciting screams,
- and so on.

In this research study, the constructed HMMM model can serve as a "general user profile" that represents common knowledge on the multimedia data and the related semantics. On the other hand, the HMMM-based "individual user profile" is also constructed for each mobile user, which is mainly constructed based on learning the individual user's query history and access patterns. The definition is described as follows.

Definition 21.2. An HMMM-based user profile is defined as a four-tuple: $\Phi = \{\tau, \widehat{A}, \widehat{B}, \widehat{O}\}$, where

- τ represents the identification of a mobile user.
- $\widehat{A}$ is affinity profile, which incorporates a set of affinity matrices $\widehat{A} = \{\widehat{A}_n^g\}$ that describes the relationships between the user accessed media objects and all the media objects. Here, $1 \leq n \leq d$ and $1 \leq g \leq |\lambda_n|$.
- $\widehat{B}$ is the feature profile, which represents the feature measurements based on the positive feedbacks of certain events and/or event patterns.
- $\widehat{O}$ is the feature weight profile, which consists of the feature weights obtained by mining and evaluating the users access history.

21.5.2.1 Affinity Profile

The affinity profiles $\widehat{A}$ is designed to model the affinity relationships among the multimedia objects that are related to user's historic query/feedback logs. The proposed solution tries to minimize the memory size that a user profile would occupy. As illustrated in Figure 21.4, the system will check the query logs and access histories for the purpose of constructing the affinity profile. Taking affinity matrix A_1^j as an example, it describes the temporal based affinity relationships among the video shots of the jth video. The system will find the "positive" video shots that the user accessed before

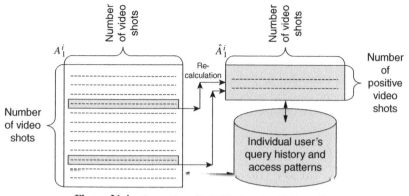

Figure 21.4 Generation of individual user's affinity profile.

and the corresponding rows are extracted from the original matrix (A_1^i). These values are then updated and utilized to architect a new matrix $\widehat{A}_1^j$ in this user's affinity profile. Similarly, in the second-level user affinity profiles, the rows represent the accessed videos that include at least one positive video shot, and the column includes all the videos in the cluster.

For a mobile user, his/her query log and access history include the set of issued queries as well as the associated positive feedbacks. We define matrices UF_n to capture the individual user's access frequencies for the nth level objects in the multilevel HMMM database. For example, let $UF_1(i, j)$ represents the number of positive feedback patterns that contain the temporal sequence as $\{..., S_1^g(i), S_1^g(j), ...\}$. $UF_2(i, j)$ denotes the number of positive patterns where both video v_i and v_j are accessed together and $UF_3(i, j)$ denotes the number of positive patterns that contain the video shots across both video cluster CC_i and CC_j. For the affinity matrix $\widehat{A}_n^g$, the corresponding affinity profile are computed and updated as below.

$$\widehat{A}_n^g(i, j) = \frac{A_n^g(i, j) \times (1 + UF_n(i, j))}{\sum_x A_n^g(i, x) \times (1 + UF_n(i, x))}, \tag{21.1}$$

where $1 \leq n \leq d, d = 3$, and $S_n^g(x)$ represent all the possible states in the same MMM model with $S_n^g(i)$ and $S_n^g(j)$. In addition, when $n = 1$, states $S_1^g(i)$ and $S_1^g(j)$ also need to follow the certain temporal sequence where $T_{S_1^g(i)} \leq T_{S_1^g(j)}$.

21.5.2.2 Feature Profile

Feature profiles are constructed to describe the distinct searching interests for each user by modifying the target feature values. As discussed in our previous paper [30], an event feature matrix B_1' was computed based on the annotated events. However, the annotated results may not be fully correct or complete. Further, the users may have their particular interests when looking for a certain event. To address these issues, a feature profile $\widehat{B}_1$ is proposed. Specifically, in the profile matrix $\widehat{B}_1$, each row represents an event, and each column represents a feature. Let f_j represent the jth feature, where $1 \leq j \leq K$, and K is the total number of features. Given $\widetilde{z}_m$ as a subset of all the positive shots with event type e_m and let $B_1(\widetilde{z}_m(i), f_j)$ denote the feature values for video shot $\widetilde{z}_m(i)$, Eq. (21.2) defines $\widehat{B}_1$. In case if there is no positive shot with event type e_m ($|\widetilde{z}_m| = 0$), the corresponding row is copied from the event feature matrix B_1' in the constructed HMMM model.

$$
\widehat{B}_1(e_m, f_j) = \begin{cases} \dfrac{\sum_{i=1}^{|\widetilde{z}_m|} B_1(\widetilde{z}_m(i), f_j)}{|\widetilde{z}_m|}, & \text{where } |\widetilde{z}_m| \geq 1, 1 \leq i \leq |\widetilde{z}_m|, 1 \leq j \leq K \\ B_1'(e_m, f_j), & \text{where } |\widetilde{z}_m| = 0, 1 \leq j \leq K \end{cases} \quad (21.2)
$$

21.5.2.3   Feature Weight Profile

In the literature, many approaches used Euclidean distance, relational coefficients, and so on, to determine the similarity measure between two data items in terms of their feature values. However, the effectiveness of different features might vary greatly from each other in expressing the media content so it is essential to apply feature weights in measuring the similarity between multimedia data objects. In HMMM, a matrix $O_{1,2}$ is utilized to describe the importance of lower level visual/audio features $\mathcal{F}_1$ when describing the event concepts $\mathcal{F}_2$. The initial values of all its entries are set to be equal, which indicates that all the features are considered to be equally important before any user feedbacks are collected and any learning process is performed. Once we get the annotated event set, the feature weights will then be updated as introduced in our previous work [30].

This research mainly focuses on the feature weight profile, which is constructed based on the mobile user's individual access and feedback histories. As the users could provide positive feedbacks on their favorite video shots, the basic idea is to increase the weight of similar features among the positive video shots, while decrease the weight of dissimilar features among them. For this purpose, we use the standard deviation $Std(e_m, f_j)$ to measure the distribution condition of feature f_j ($1 \leq j \leq K$) on the video shots containing event e_m ($1 \leq m \leq C$), where C represents the number of distinct event concepts. A large standard deviation indicates greater scatter of the data points. Accordingly, when there is more than one positive shot for event e_m, ($\widetilde{z_m(i)} > 1$), the value of $1/Std(e_m, f_j)$ can be employed to measure the similarity of the features, which in turn indicates the importance of this feature in terms of evaluating event e_m. However, this solution does not apply when there is no positive shot or only one positive shot ($\widetilde{z_m}(i) \leq 1$), so we would borrow the corresponding feature weights for event e_m from matrix $O_{1,2}$. The feature weight profile is thus defined as follows.

$$
Std(e_m, f_j) = \sqrt{\frac{\sum_{i=1}^{|\widetilde{z}_m|} (B_1(\widetilde{z}_m(i), f_j) - \widehat{B}_1(e_m, f_j))^2}{|\widetilde{z}_m| - 1}} \quad (21.3)
$$

where $\widetilde{z_m} > 1$, $1 \le m \le C$, and $1 \le j \le K$.

$$\hat{O}_{1,2}(e_m, f_j) = \begin{cases} \dfrac{1/\text{Std}(e_m, f_j)}{\sum_{k=1}^{K}(1/\text{Std}(e_m, f_k))}, & \text{where } |\tilde{z}_m| > 1, 1 \le m \le C, \text{and } 1 \le j \le K \\ O_{1,2}(e_m, f_j), & \text{where } |\tilde{z}_m| \le 1, 1 \le m \le C, \text{and } 1 \le j \le K \end{cases}$$

(21.4)

21.5.3 Fuzzy Associated Retrieval

Fuzzy logic has been remarked for its ability to describe and model the vagueness and uncertainty, which is inherent in multimedia information retrieval. In this proposed framework, fuzzy logic is adopted to model the uncertainty of users' retrieval interests. Specifically, the users are allowed to make their choices of retrieving content based solely on general user perceptions, the personalized interests, or anywhere in between, which leads to a different level of tradeoff between retrieval accuracy and processing speed. We utilize a fuzzy weight parameter $\rho \in [0, 1]$ to measure the uncertainty that users may pose when issuing the video queries. As shown in Figure 21.5, we utilize an interactive gauge on the mobile device interface for the users to adjust ρ.

By choosing the personalized interest (as shown in Figure 21.5(b)), $\rho = 0$, the system will evaluate the user's profile and retrieve the video clips based on the learned knowledge with respect to the user's previous access patterns. On the other hand, if the generalized recommendation mode is selected (Figure 21.5(a)), that is, $\rho = 1$, the system will comply with the common knowledge learned from the complete query log collected across different users. Therefore, the most popular video clips will be retrieved with higher ranks. Assuming that we have already performed video clustering [31] through the database, the generalized recommendation mode is normally more efficient as satisfactory results can generally be retrieved by checking the related clusters.

Let $Q = \{e_1, e_2, \ldots, e_C\}$ $(T_{e_1} \le T_{e_2} \le \ldots \le T_{e_C})$ be a query pattern and $\overline{s_t} \in S$ be a candidate video shot for the event e_t $(1 \le t \le C)$, the system can adjust the retrieving algorithm and provide three kinds of recommendations, namely, generalized recommendation, personalized recommendation, and fuzzy weighted recommendation according to the fuzzy weight issued by the user. The details are addressed below.

21.5.3.1 Generalized Recommendation

In the case when the generalized recommendation (Figure 21.5(a)) is selected, the matrices in the constructed HMMM model will be used as the common user profile

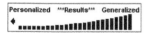 

(a) Generalized recommendation　(b) Personalized recommendation　(c) Fuzzy associated recommendation

Figure 21.5　Fuzzy weight adjustment tool.

to perform the stochastic retrieval process. First, as shown in Eq. (21.5), the weighted Euclidean distance $\mathrm{dis}(\overline{s_t}, e_t)$ is calculated by adopting the general feature weight $O_{1,2}(e_t, f_y)$, which is then used to derive the similarity measurements (see Eq. (21.6)). Here, $B_1'(e_t, f_y)$ denotes the extracted mean value of feature f_y with respect to event e_t based on the learned general users' common knowledge.

$$\mathrm{dis}(\overline{s_t}, e_t) = \sqrt{\sum_{y=1}^{K}(O_{1,2}(e_t, f_y) \times (B_1(\overline{s_t}, f_y) - B_1'(e_t, f_y))^2)} \qquad (21.5)$$

where $\overline{s_t} \in S_1, 1 \leq y \leq K, 1 \leq t \leq C$.

$$\mathrm{sim}(\overline{s_t}, e_t) = \frac{1}{1 + \mathrm{dis}(\overline{s_t}, e_t)}, \qquad (21.6)$$

where $\overline{s_t} \in S_1, 1 \leq t \leq C$.

Next, the edge weights are calculated based on Eqs (21.7) and (21.8). When $t = 0$, the initial edge weight is calculated using the initial state probability and similarity measure between state $\overline{s_1}$ and event e_1. It is worth mentioning that the system tries to evaluate the optimized path to access the next possible video shot state that is similar to the next anticipated events. Therefore, the edge weight from state $\overline{s_t}$ to $\overline{s_{t+1}}$ ($1 \leq t < C$) is calculated by adopting the affinity relationship as well as the similarity between the candidate shot $\overline{s_{t+1}}$ with the event concept e_{t+1}.

$$w_1(\overline{s_1}, e_1) = \Pi_1(\overline{s_1}) \times \mathrm{sim}(\overline{s_1}, e_1) \qquad (21.7)$$

$$w_{t+1}(\overline{s_{t+1}}, e_{t+1}) = w_t(\overline{s_t}, e_t) \times A_1(\overline{s_t}, \overline{s_{t+1}})$$
$$\times \mathrm{sim}(\overline{s_{t+1}}, e_{t+1}), \quad 1 \leq t < C \qquad (21.8)$$

After one round of traversal, the system retrieves a sequence of video shots R_k that match the desired event pattern Q. The next step would be the calculation of the similarity score. Here, $SS(Q, R_k)$ is computed by summing up all the edge weights, where a greater similarity score indicates a closer match.

$$SS(Q, R_k) = \sum_{t=1}^{C} w_t(\overline{s_t}, e_t). \qquad (21.9)$$

21.5.3.2 Personalized Recommendation

In case the user prefers personalized recommendation (Figure 21.5(a)), the overall process steps are similar except that the matrices used are mainly from HMMM-based user profiles.

$$\widehat{\mathrm{dis}}(\overline{s_t}, e_t) = \sqrt{\sum_{y=1}^{K}(\widehat{O_{1,2}}(e_t, f_y) \times (B_1(\overline{s_t}, f_y) - \widehat{B_1}(e_t, f_y))^2)}, \qquad (21.10)$$

$$\widehat{\mathrm{sim}}(\overline{s_t}, e_t) = \frac{1}{1 + \widehat{\mathrm{dis}}(\overline{s_t}, e_t)}, \qquad (21.11)$$

where $\overline{s_t} \in S_1, 1 \leq t \leq C$.

When calculating the edge weight $\widehat{w_{t+1}}(\overline{s_{t+1}}, e_{t+1})$, there could be two conditions. If the candidate video shot has been accessed by this user and was marked as "positive" ($\overline{s_{t+1}} \in R_k$), the user's affinity profiles should include this video shot and therefore, the formula takes the affinity value from the user's personal affinity profile ($\widehat{A_1}$). Otherwise, there is no record for the video shot in the user profile ($\overline{s_{t+1}} \notin R_k$), and the system will pick the value from the affinity matrices (A_1) of the constructed HMMM.

$$\widehat{w_1}(\overline{s_t}, e_t) = \Pi_1(\overline{s_1}) \times \widehat{sim}(\overline{s_1}, e_1) \tag{21.12}$$

$$\widehat{w_{t+1}}(\overline{s_{t+1}}, e_{t+1}) = \begin{cases} \widehat{w_t}(\overline{s_t}, e_t) \times \widehat{A_1}(\overline{s_t}, \overline{s_{t+1}}) \times \widehat{sim}(\overline{s_{t+1}}, e_{t+1}), \\ \qquad \text{where } 1 \leq t < C, \overline{s_{t+1}} \in R_k. \\ \widehat{w_t}(\overline{s_t}, e_t) \times A_1(\overline{s_t}, \overline{s_{t+1}}) \times \widehat{sim}(\overline{s_{t+1}}, e_{t+1}), \\ \qquad \text{where } 1 \leq t < C, \overline{s_{t+1}} \notin R_k. \end{cases} \tag{21.13}$$

$$\widehat{SS}(Q, R_k) = \sum_{t=1}^{C} \widehat{w_t}(\overline{s_t}, e_t). \tag{21.14}$$

21.5.3.3 Fuzzy-Associated Recommendation

Alternatively, if the user is uncertain and thus chooses a fuzzy weight parameter $\rho \in (0, 1)$ to describe his/her interest (as illustrated in Figure 21.5(c)), the system will make the adjustment on the edge weights and accordingly, the optimized path and the similarity score would possibly be changed as defined in the following equations.

$$\widetilde{w_t}(\overline{s_t}, e_t) = \rho \times w_t(\overline{s_t}, e_t) + (1 - \rho) \times \widehat{w_t}(\overline{s_t}, e_t) \tag{21.15}$$

where $1 \leq t \leq C$.

$$\widetilde{SS}(Q, R_k) = \sum_{t=1}^{C} \widetilde{w_t}(\overline{s_t}, e_t). \tag{21.16}$$

After getting the candidate video shot sequences, they will be ranked based on their similarity scores and sent back to the client.

21.6 IMPLEMENTATION AND EXPERIMENTS

A mobile-based soccer video retrieval system is developed based on the proposed MoVR framework, which consists of the following components:

- A soccer video database is constructed and maintained in the server side by using PostgreSQL [23]. Totally 45 soccer videos along with 8977 segmented video shots and corresponding key frames are stored and managed in the database.
- Server-side engine is implemented using C++. This module contains not only the searching and ranking algorithms but also a set of other computationally

intensive techniques, including video shot segmentation, HMMM database modeling, user profile generation and updating, and so on.

- The virtual client application is implemented using Java J2SE [14]. It works as a middleware between server engine and mobile clients, where data communication is mainly fulfilled by using UDP and TCP.

- The user interface on the mobile device is developed by using Sun Java J2ME [13] Wireless Toolkit [15]. We try to make it portable, flexible, and user friendly with simple but effective functions. The user can easily issue event/pattern queries, navigate key frames, play interested video clips, and provide feedbacks.

Figures 21.6 and 21.7 show the user query interfaces of the MoVR soccer video retrieval system.

- In Figure 21.6(a), the initial choices are displayed, which include "Soccer Video Browsing," "Soccer Video Retrieval by Event," and "Soccer Video Retrieval by Event Pattern," and so on. The user can use the upper-center button to move up/down to select the target menu and then push the left-upper button to launch the selected application.

- Figure 21.6(b) shows the query interface for the single-event queries. It allows the user to choose one or more events with no temporal constraints. For instance, in this figure, the user chooses "Goal," "Free Kick," and "Corner Kick," which means the video clips with either one of these three events are of the

  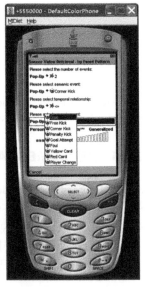

(a) initial choices   (b) retrieval by event   (c) retrieval by pattern

Figure 21.6   Mobile-based soccer video retrieval interfaces.

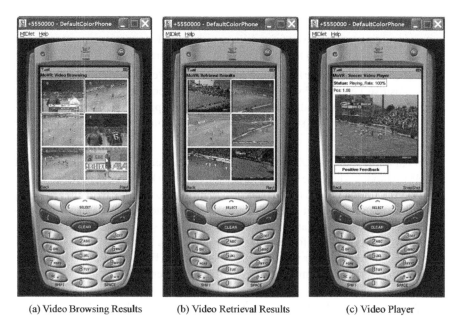

(a) Video Browsing Results   (b) Video Retrieval Results   (c) Video Player

Figure 21.7   Mobile-based soccer video retrieval results.

interest. Under the event list, there is a gauge control that allows the user to change the fuzzy weight parameter between two extremes: personalized recommendations and generalized recommendations. The upper-left button can be used to exit this component and go back to the main menu, while the upper-right button can be used to issue the query.

- Figure 21.6(c) illustrates the interface for the temporal event pattern retrieval. The user can use the popup lists to choose the event number to define the size of the query pattern. Then it is allowed to choose the events one by one, along with the temporal relationship between two adjacent events. Taking this figure as an example, the user first sets the event number as 2 and then chooses the pattern as "Corner Kick <= Goal," which means that the user wants to search for the video clips with a "Corner Kick" followed by a "Goal." These two events could also occur in the same video shot (when temporal relationship is set to "="), which can be called a "Corner Goal."

- The returned key frames are displayed as shown in Figure 21.7(a) and (b). Due to the limitation of screen display size, there are only six key frames shown in the first screen, where each key frame represents the first frame for each of the returned video clips. The user can choose their interested key frame and then trigger the "Play!" button to display the corresponding video clip, which may include one (for event query) or more video shots (for event pattern query). Note that Figure 21.7(a) shows the video browsing results with the key frames displayed for consecutive video shots in one video. Figure 21.7(b) illustrates the results for an event query targeting the video shots with corner kicks. These video shots are retrieved from different soccer videos and are ranked from left to right and from top to down based on their similarity scores.

- The video player interface is shown in Figure 21.7(c). It contains a button called "Positive Feedback," which can be selected to send back a positive feedback if the user is satisfied with the current video clip. A "Snapshot" functionality is also provided such that the user can capture a video frame from the video.

In our experiments, totally 300 historical queries were used for the construction of HMMM model, where the system learned the common knowledge from the general users. As shown in Figure 21.8, we performed the test for 10 sets of distinct queries, including three single-event queries, five two-event pattern queries, and two three-event pattern queries. For example, Query 9 "Corner Kick <= Goal < Free Kick" means a pattern with a corner kick, followed by a goal, and then a free kick, where corner kick and goal may possibly occur within the same video shot. For each of these 10 queries, three sets of tests are performed and each of them represents distinct user interests. In each of these tests, the user profile is constructed based on 30 historical queries and all three possible recommendation methods are testified. Twelve top-ranked video clips shown in first two screens (with six results each) are checked, which is called "scope." Thus, "accuracy" here is defined as the percentage of the number of user satisfied video clips within the scope. Finally, the average accuracy is computed based on these tests. Figure 21.9 illustrates the comparison of the average accuracy values across these three kinds of recommendations. As we can see from this figure, the results based on "Generalized" recommendation have the lowest average accuracy; "Personalized" recommendation offers the best results; while "Fuzzy Weighted" recommendation has the performance in between. In general, the personalized recommendation does offer better results by learning individual user's preferences using the HMMM-based profile. Meanwhile, although the generalized recommendation may not fully satisfy the individual users, it represents the common knowledge learnt from the general users and requires shorter processing time. By adopting fuzzy-weighted recommendations, the users are offered more flexibilities in video retrieval.

ID	Query	Generalized recommendations	Fuzzy weighted recommendations	Personalized recommendations
1	Goal	30.6%	38.9%	61.1%
2	Free Kick	19.4%	50.0%	58.3%
3	Corner Kick	27.8%	58.3%	72.2%
4	Goal < Goal	36.1%	55.6%	86.1%
5	Free Kick <= Goal	36.1%	50.0%	66.7%
6	Corner Kick <= Goal	33.3%	47.2%	72.2%
7	Free Kick < Corner Kick	25.0%	36.1%	52.8%
8	Corner Kick < Free Kick	36.1%	44.4%	52.8%
9	Corner Kick <= Goal < Free Kick	16.7%	38.9%	63.9%
10	Free Kick <= Goal < Goal	25.0%	36.1%	55.6%

Figure 21.8   Average accuracy for the different recommendations.

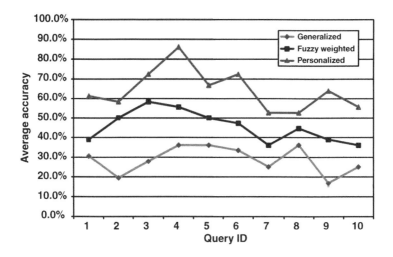

Figure 21.9   Experimental comparison of different recommendations.

21.7 SUMMARY

With the proliferation of mobile devices and multimedia data sources, it is in a great need of effective mobile multimedia services. However, with their unique constraints in display size, power supply, storage space as well as processing speed, the multimedia applications in mobile devices encounter great challenges. In this chapter, we present MoVR—a user adaptive video retrieval framework in the mobile wireless environment. While accommodating various constraints of the mobile devices, a set of advanced techniques are developed and deployed to address essential issues, such as the semantic gap between low-level video features and high-level concepts, the temporal characteristics of video events, individual user preference, and so on.

Specifically, HMMM scheme is proposed to model various levels of media objects, their temporal relationships, the semantic concepts, and high-level user perceptions. HMMM-based user profile is defined, which is also integrated seamlessly with a novel learning mechanism to enable the "personalized recommendation" for individual user by evaluating his/her personal histories and feedbacks. In addition, the fuzzy association concept is employed in the retrieval process such that the users gain the control of the preference selections to achieve reasonable tradeoff between the retrieval performance and processing speed. Furthermore, to improve the processing performance and enhance the portability of client-side applications, storage consumption information, and computationally intensive operations are supported in the server side, whereas the mobile clients are solely required to maintain the minimized size of data to enable the retrieval process. The virtual clients are designed to perform as a middleware between server applications and mobile clients. This design helps to reduce the storage load of mobile devices and to provide greater accessibility with their cached media files. Finally, a mobile-based soccer video retrieval system is developed and testified to demonstrate the effectiveness of the proposed framework.

ACKNOWLEDGMENTS

For Shu-Ching Chen, this research was supported in part by NSF EIA-0220562, HRD-0317692 and Florida Hurricane Alliance Research Program sponsored by the National Oceanic and Atmospheric Administration. For Mei-Ling Shyu, this research was supported in part by NSF ITR (Medium) IIS-0325260. For Min Chen and Na Zhao, this research was supported in part by Florida International University Dissertation Year Fellowship.

REFERENCES

1. I. Ahmad, S. Kiranyaz, F. A. Cheikh, and M. Gabouj, Audio-based queries for video retrieval over Java enabled mobile devices. In *Proceedings of SPIE (Multimedia on Mobile Devices II), Electronic Imaging Symposium 2006*, San Jose, California, USA, 2006, pp. 83–93.
2. N. Babaguchi, Y. Kawai, and Y. Kitahashi, Generation of personalized abstract of sports video, In *Proceedings of IEEE International Conference on Multimedia and Expo*, Tokyo, Japan, 2001, pp. 800–803.
3. L. Chen and K. Sycara, WebMate: a personal agent for browsing and searching. In *Proceedings of the 2nd International Conference on Autonomous Agents and Multi-Agent Systems*, 1998, pp. 132–139.
4. S.-C. Chen, M.-L. Shyu, C. Zhang, L. Luo, and M. Chen, Detection of soccer goal shots using joint multimedia features and classification rules, In *Proceedings of the Fourth International Workshop on Multimedia Data Mining (MDM/KDD), in Conjunction with the ACM International Conference on Knowledge Discovery & Data Mining (SIGKDD)*, Washington, DC, USA, 2003, pp. 36–44.
5. S.-C. Chen, M.-L. Shyu, M. Chen, and C. Zhang, A decision tree-based multimodal data mining framework for soccer goal detection. In *Proceedings of the IEEE International Conference on Multimedia and Expo*, Taipei, Taiwan, R.O.C., 2004, pp. 265–268.
6. L. Coyle and P. Cunningham, Improving recommendation ranking by learning personal feature weights. In *Proceedings of the 7th European Conference on Case-Based Reasoning*, Madrid, Spain, 2004, pp. 560–572.
7. M. Davis and R. Sarvas, Mobile media metadata for mobile imaging. In *Proceedings of the IEEE International Conference on Multimedia and Expo*, Taipei, Taiwan, R.O.C., 2004, pp. 1707–1710.
8. D. Dubois, H. Prade, and F. Sedes, Fuzzy logic techniques in multimedia database querying: a preliminary investigation of the potentials. *IEEE Trans. Knowledge Data Eng.*, 13(3):383–392, 2001.
9. A. D. Doulamis, Y. S. Avrithis, N. D. Doulamis, and S. D. Kollias, Interactive content-based retrieval in video databases using fuzzy classification and relevance feedback. In *Proceedings IEEE Multimedia Computing and Systems (ICMCS)*, Florence, Italy, 1999, pp. 954–958.
10. D. Gibbon, L. Begeja, Z. Liu, B. Renger, and B. Shahraray, Multimedia processing for enhanced information delivery on mobile devices. In *Proceedings of the Workshop on Emerging Applications for Wireless and Mobile Access*, New York, USA, 2004.
11. Y. Gong and X. Liu, Summarizing video by minimizing visual content redundancies. In *Proceedings IEEE International Conference on Multimedia and Expo*, Tokyo, Japan, 2001, pp. 788–791.
12. J. Hu, J. Zhong, and A. Bagga, Combined-media video tracking for summarization. In *Proceedings of ACM Multimedia*, Ottawa, Canada, 2001, pp. 502–505.
13. Java 2 Platform, Micro Edition (J2ME). http://java.sun.com/javame/.
14. Java 2 Platform, Standard Edition (J2SE). http://java.sun.com/javase/.

15. Sun Java Wireless Toolkit. http://java.sun.com/products/sjwtoolkit/.
16. R. I. John and G. J. Mooney, Fuzzy user modeling for information retrieval on the World Wide Web. *Knowledge Inform. Syst.*, 3(1):81–95, 2001.
17. S. Jokela, M. Turpeinen, and R. Sulonen, Ontology development for flexible content. In *Proceedings of the 33rd Hawaii International Conference on System Sciences*, 2000, pp. 160–169.
18. B.-Y. Kang, D.-W. Kim, and Q. Li, Fuzzy ranking model based on user preference. *IEICE Trans. Inform. Syst.*, E89–D(6):1971–1974, 2006.
19. J. Lahti, M. Palola, J. Korva, U. Westermann, K. Pentikousis, and P. Pietarila, A mobile phone-based context-aware video management application. In *Proceedings of SPIE-IS&T Electronic Imaging (Multimedia on Mobile Devices II)*, San Jose, California, USA, Vol. 6074, 2006, pp. 83–194.
20. J. Lahti, K. Pentikousis, and M. Palola, MobiCon: mobile video recording with integrated annotations and DRM. In *Proceedings of IEEE Consumer Communications and Networking Conference (IEEE CCNC)*, Las Vegas, Nevada, USA, 2006, pp. 233–237.
21. Q. Li, J. Yang, and Y. T. Zhuang, Web-based multimedia retrieval: balancing out between common knowledge and personalized views. In *Proceedings of 2 International Conference on Web Information System and Engineering*, 2001, pp. 100–109.
22. M. J. Martin-Bautista, D. H. Kraft, M. A. Vila, J. Chen, and J. Cruz, User profiles and fuzzy logic for Web retrieval issues. *Special Issue J. Soft Comput.*, 6:365–372, 2002.
23. PostgreSQL, an open source object-relational database. http://www.postgresql.org/.
24. Y. Rui, T. S., Huang, M. Ortega, and S. Mehrotra, Relevance feedback: a power tool for interactive content-based image retrieval. *IEEE Trans. Circuits Syst. Video Technol. Special Issue Segment. Descript. Retriev. Video Content*, 8:644–655, 1998.
25. A. Sachinopoulou, S.-M. Mäkelä, S. Järvinen, U. Westermann1, J. Peltola, and P. Pietarila, Personal video retrieval and browsing for mobile users. In *Proceedings of SPIE—Multimedia on Mobile Devices*, San Jose, California, USA, 2005, pp. 219–230.
26. M.-L. Shyu, S.-C. Chen, M. Chen, and S. H. Rubin, Affinity-based similarity measure for Web document clustering. In *Proceedings of the 2004 IEEE International Conference on Information Reuse and Integration (IRI)*, Las Vegas, Nevada, USA, 2004, pp. 247–252.
27. M.-L. Shyu, S.-C. Chen, M. Chen, and C. Zhang, A unified framework for image database clustering and content-based retrieval. In *Proceedings of the Second ACM International Workshop on Multimedia Databases (ACM MMDB)*, Arlington, VA, USA, 2004, pp. 19–27.
28. B. L. Tseng, C. Lin, and J. Smith, Video summarization and personalization for pervasive mobile devices. In *Proceedings of the IS&T/SPIE Symposium on Electronic Imaging: Science and Technology—Storage & Retrieval for Image and Video Databases*, SPIE Vol. 4676, 2002, pp. 359–370.
29. Virage. http://www.virage.com.
30. N. Zhao, S.-C. Chen, and M.-L. Shyu, Video database modeling and temporal pattern retrieval using hierarchical Markov model mediator. In *Proceedings of the First IEEE International Workshop on Multimedia Databases and Data Management (IEEE-MDDM), in Conjunction with the 22nd IEEE International Conference on Data Engineering (ICDE)*, Atlanta, GA, USA, 2006.
31. N. Zhao, S.-C. Chen, and M.-L. Shyu, An integrated and interactive video retrieval framework with hierarchical learning models and semantic clustering strategy. In *Proceedings of IEEE International Conference on Information Reuse and Integration (IEEE IRI)*, Hawaii, USA, 2006, pp. 438–443.

Chapter 22

A Ubiquitous Fashionable Computer with an i-Throw Device on a Location-based Service Environment

Jupyung Lee, Jong-Woon Yoo, Seung-Ho Lim,
Ki-Woong Park, and Kyu Ho Park

Korea Advanced Institute of Science and Technology, Daejeon, Korea

22.1 Introduction	510
22.2 Ubiquitous Fashionable Computer	512
22.3 Location-Based Service Environment	515
22.4 Application Example: User-Friendly Interaction	516
with Ubiquitous Environment Using i-Throw	
22.5 Related Work	526
22.6 Conclusion	528
References	528

22.1 INTRODUCTION

In recent years, wearable computing and ubiquitous computing system environment have realized due to the rapid progress of computing and communication technology. Wearable computing system can be broadly defined as mobile electronic devices that can be unobtrusively embedded as part of garment or accessory [2]. Unlike

Mobile Intelligence. Edited by Laurence T. Yang, Agustinus Borgy Waluyo, Jianhua Ma, Ling Tan, and Bala Srinivasan
Copyright © 2010 John Wiley & Sons, Inc.

conventional mobile devices, it is always active and running without user's attention and gives services to a user with the support of ubiquitous environment. In the ubiquitous environment, using the wearable computer, we get necessary information anytime and anywhere with a minimal constraint.

In a wearable computer, the design of new paradigm has concentrated on the problems that arises when user has to put up with considerable inconvenience carrying around devices [3–6]. The success of wearable computer will heavily rely on good wearability, usability, aesthetic appearance, and social acceptance. In addition, not merely considering about these uncomfortable carrying issues, the reflection about the exploit of ubiquitous computing environment should be done. The emerging wearable ubiquitous computer should have various communication interfaces, such as WLAN, Bluetooth, and ZigBee communication to support various ubiquitous network environment.

When these various communication interfaces are integrated and used in one converged device, these heterogeneous communication interfaces may interfere with each other because these are usually operated in 2.4 GHz ISM band. Therefore, the wearable computer might not operate well for these interference problems. To solve these problem, coexistence and interoperability algorithm should be designed with appropriate strategies.

In addition to that, a novel user interface should be considered to minimize user inconvenience while users access and utilize the system resource. The uncomfortable interface means that users should get training its function with burdensome input or output devices using wearable keyboard and mouse. In wearable computing, it is desired to develop comfortable and user friendly input devices for wearable computing. A novel user interface is required to be simple, easy and intuitive by recognizing human friendly gestures, activities, or senses. Intuitive interface can be described as the mechanism of controlling devices in a ubiquitous environment by human friendly gestures that everyone can easily accept and recognize. Human voice, eyes, and hand gesture are good means to realize input devices as user interfaces in wearable computer.

In this chapter, we present the design approach and the philosophy of our wearable computing system that is wearable, aesthetic, and intuitive. The main features are as follows. It supports the coexistence and interoperability of various communication interfaces between WLAN, Bluetooth, and ZigBee devices that can be operated freely in any ubiquitous system environment. This can be done using dynamic channel allocation mechanism between communication devices that operate in the same ISM band. In addition, in our wearable computer system, novel user interfaces are developed and integrated to help the intuitive use of it. Among possible user interfaces, in this chapter, we focus on what we call *i*-Throw, which means intuitive input devices. This input device can recognize a user's hand gestures and direction, and allows the user to control ubiquitous devices with the gestures. Our wearable computer is called ubiquitous fashionable computer (UFC), which is named based on our special emphasis on its wearability, aesthetic design, and close interaction with ubiquitous environment. Our UFC system is realized and tested with actual testbed environment that has various network interfaces, sensor nodes, and ubiquitous components, and also various applications are implemented and presented.

This chapter is an extended version of a previously published paper [1], additionally including an explanation about the adaptive angle assignment and a detailed description about a sequence of interactive operations.

22.2 UBIQUITOUS FASHIONABLE COMPUTER

A new physical form of a portable computer has emerged in the form of a wearable computing system [10]. The wearable computing system can be broadly defined as mobile electronic devices that can be unobtrusively embedded as part of garment or accessory [2]. Unlike conventional mobile devices, it is always active and running without user's attention and gives services to a user with the support of ubiquitous environment. In the ubiquitous environment, using the wearable computer, we get necessary information in anytime and anywhere with a minimal constraint.

22.2.1 UFC Design Philosophy

The new paradigm of wearable computer design has concentrated on the problems that arises when user has to put up with considerable inconvenience carrying around devices. The wearable computer platform should be lightweight, easy to carry, easy to use, and should have aesthetic appearance and social acceptance. We chose to design and implement a wearable platform from the scratch and make use of it as a main user device. Deliberating on the aforementioned requirement, we present the design philosophy of our wearable computing system as follows.

- *Wearability and usability:* Our wearable platform, in contrast to either laptop PC or handheld device, allows a user to carry the computing device in a comfortable and natural manner, because clothes have been already a essential component of our daily lives. Large surface area of clothes can be utilized for various purposes and thus, I/O interface does not have to be located in only a small-sized computing device. Moreover, the wearable platform makes it easier to measure and gather biosignal data such as temperature and heart rate by integrating body-attached sensors and computing devices upon a same clothes interface.

- *Aesthetic appearance and social acceptance:* We tried to find the solution to fulfill the requirements by repeating the prototyping bodystorming progresses. We defined the target users as young university students and drew design concepts by analyzing their activities in everyday life and fashion trend. In addition, we have made effort for each part of the UFC platform to look like familiar fashionable components: for example, an attachable/detachable module is comparable to a button of clothing and an i-Throw device is comparable to a ring. Also, PANDA can be worn as a form of a necklace.

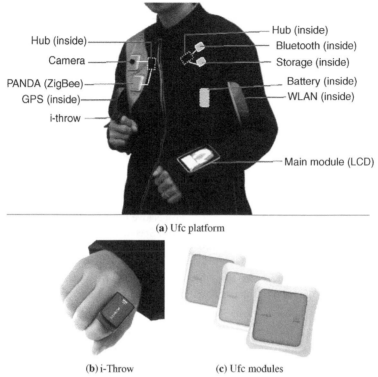

Hub (inside)
Camera
PANDA (ZigBee)
GPS (inside)
i-throw

Hub (inside)
Bluetooth (inside)
Storage (inside)
Battery (inside)
WLAN (inside)

Main module (LCD)

(a) Ufc platform

(b) i-Throw   (c) Ufc modules

Figure 22.1   UFC platform design and implementation.

22.2.2 UFC Platform

The implemented UFC platform is shown in Figure 22.1. Our UFC consists of several module parts: main module including CPU and memory, communication modules including various communication interfaces, and user interface modules with I/O interfaces. UFC modules are distributed on a garment, considering the distribution of weight and aesthetic design. Moreover, each UFC module can be attached and detached easily on a garment, allowing users to construct one's own UFC platform. Since we utilized a standard USB protocol to communicate between the main module and various UFC modules, due to the hotswap capability of USB devices, each UFC module can be attached and detached while the system is running. Main specification of UFC platform is described in Table 22.1. In the main module, the core of the UFC system is an ARM-based Intel XScale processor: the PXA270. Main features of this processor include clock scaling and dynamic voltage scaling up to 624 MHz. With this, power management of wearable computer can prolong the lifetime of the platform. Main memory of UFC is 256 MB, which is relatively large capacity for a mobile devices. However, with this large capacity, we can support a wide range of

Table 22.1 UFC Platform Specifications

Modules		Specifications
Main module	CPU	XScale PXA270 624 MHz
	Memory	Mobile SDRAM 256 MB
	I/O Interfaces	RS232, USB 1.1/2.0, Mini-PCI and PCMCIA
	WLAN	IEEE 802.11 a/b/g
Communicatnon module	Bluetooth	IEEE 802.15.1
	ZigBee	IEEE 802.15.4
	i-Throw	Intuitive Input Device using hand motion
User interface module	Audio In/Out	Microphone for Voice Command / Earphone
	Video In/Out	Camera / VGA with HMD, 2.5" LCD
Software part	Operating System	Embedded Linux 2.6.9
	VM	Java Native Interface
	Middleware	OSGI Specification

services such as audio and video transmission, and Java virtual machine based middleware services. We present the main features of our wearable computing system as follows.

- *Novel user interface:* In our wearable computer system, novel user interfaces are developed and integrated to help the intuitive use of it. Among possible user interfaces, in this chapter, we focus on what we call *i*-Throw, which means intuitive input devices. This input device can recognize a user's hand gestures and direction and allows the user to control ubiquitous devices with the gestures. Our wearable computer is called UFC, which is named based on our special emphasis on its wearability, aesthetic design and close interaction with ubiquitous environment. Our UFC system is realized and tested with actual testbed environment that has various network interfaces, sensor nodes, and ubiquitous components, and also various applications are implemented and presented.

- *Multimodal communications:* It supports the coexistence and interoperability of various communication interfaces between WLAN, Bluetooth, and ZigBee devices that can be operated freely in any ubiquitous system environment. This can be done using dynamic channel allocation mechanism between communication devices that operate in the same ISM band.

- *System software:* Operating system running on UFC is GNU/embedded Linux with kernel 2.6. Linux 2.6 with ARM processor shows more deemed feasible performance in real-time embedded system than lower versions. Efficient middleware platform is implemented with UFC to provide various helpful services with low overhead and power. The middleware interface can be implemented with standard Java execution environment called Java Native Interface (JNI) [16]. Useful functions that are covered by middleware include context management, service discovery, and local file sharing.

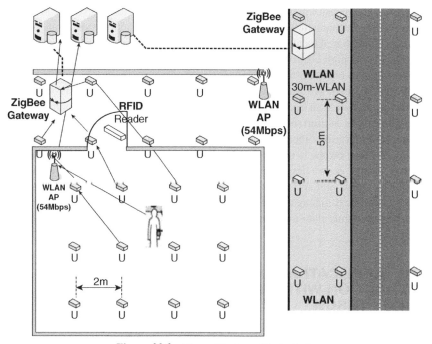

Figure 22.2   The architecture of the testbed.

22.3   LOCATION-BASED SERVICE ENVIRONMENT

Since our location-based service environment aims for a campus-wide environment, it is essential to build an intelligent testbed that various services can be operated. Figure 22.2 illustrates the architecture of our testbed. Three important components of the target testbed are a communication infrastructure, a location-tracking infrastructure, and a middleware.

Communication infrastructure lays the groundwork for ubiquitous computing. Two well-known standards are used to support anytime and anywhere wireless communication services; IEEE 802.11 (WLAN) [24] and IEEE 802.15.4 (ZigBee) [25] We installed an enough number of ZigBee sensor nodes and WLAN access points in a wide mesh manner. In this environment, communication could be done via multihop sensor nodes for low-speed data transmission and WLAN for high-speed data transmission.

Location-tracking infrastructure is necessary for location-based services. The ZigBee sensor nodes are also used for location tracking. Every nodes periodically broadcast beacon signals, which uses 2.4 GHz band as a physical channel. A moving user receives beacon signals from the ZigBee communication interface. When the user receives multiple beacons from the multiple sensor nodes, the users who have received the beacons can identify his location by calculating each Received Signal Strength Indicator (RSSI) value from each sensor node [15].

During the measurement, however, we found that the resolution of location sensing using this mechanism was not sufficient for our target application. Thus, we also utilized a UWB-based location tracking device [26] whose typical accuracy is 6 in. (15 cm). Due to the high cost of this solution, the UWB-based location-tracking device have installed in only two rooms inside the testbed.

In this testbed, we assume a situation where thousands of users move here and there, interact with each other or a location-based service environment, share information with authorized other users, access to diverse devices for diverse purposes, and run various location-based applications. In this situation, an extensible middleware framework is necessary to keep up with highly variable dynamic environment. Therefore, we developed a middleware, called μ-ware [22]. μ-ware is composed of a lightweight service discovery protocol, a distributed information sharing function, a context manager, and an instance service loader, all of which are useful to manage dynamic data and develop new application utilizing various ubiquitous resources.

22.4 APPLICATION EXAMPLE: USER-FRIENDLY INTERACTION WITH UBIQUITOUS ENVIRONMENT USING I-THROW

To explain the practical use of our UFC platform and the user-friendly interaction with ubiquitous environment, we have implemented a ubiquitous testbed where multiple UFC users interact with various ubiquitous devices or other UFC users. Figure 22.3 illustrates the concept of the ubiquitous testbed room. In addition, we

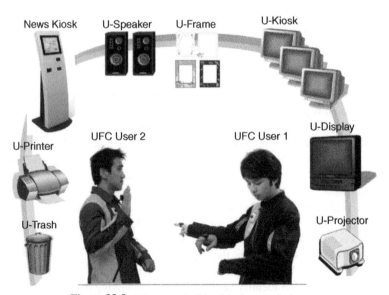

Figure 22.3   The concept of the ubiquitous testbed room.

have implemented a practical application that runs upon the UFC platform and the ubiquitous devices, which makes it possible to exchange the various objects and control ubiquitous devices very easily.

22.4.1 Motivation

Due to its small form factor, most portable devices, including our UFC platform, have only small-sized display and limited input devices. The UFC platform has 2.5 in. LCD display and 12 input buttons, which are definitely insufficient to monitor the status of a UFC main module and various peripheral modules, control the modules, and send a user's intention to the UFC platform.

This problem is exacerbated when a UFC user tries to control various ubiquitous devices using one's UFC platform: as the number of controllable ubiquitous devices increases, it becomes more inconvenient to find one among them and exchange information with it, due to the small-sized display and limited input devices of the UFC platform. Efficient utilization of a small-sized display and intelligent mapping of various commands on the input buttons can partially solve this problem. However, such an approach usually makes it difficult to learn how to use the device, which degrades the usability of the UFC platform. One recent workshop underscored that usability is one of the primary challenges in a next-generation "smart" room that is full of various ubiquitous devices [19].

We attempt to resolve this problem by making full use of spatial resources inside the testbed room: given that various ubiquitous devices are spatially distributed inside the testbed room, a UFC user can express one's intention easily by using one's spatial movement and gesture. For example, let us assume that one UFC user takes a picture and intends to put it on a public display so that other people can see the picture he takes. From the perspective of the user, the most natural way of reflecting one's intention on the environment is pointing his finger at the public display and throwing one's picture at the public display. If this kind of a user-friendly spatial gesture interface is supported, the limitation of the I/O resources of the UFC platform can be overcome by fully utilizing abundant spatial resources.

In order to explain the components necessary to realize the user-friendly interaction in the ubiquitous testbed room, we will take the following scenario: there are two UFC users, "user 1" and "user 2." "User 1" takes a picture and "user 2" wants to see the picture by downloading it from "user 1." A sequence of interactive operations necessary to realize the scenario is shown in Figure 22.4. In the figure, each operation is denoted by the number and each operation corresponding to each number is detailed below.

- Each UFC user's location is recognized by the *location tracking device*. We utilized UWB-based location tracking device [26] whose typical accuracy is 6 in. (15 cm). Each UFC user has a UWB tag so that the UWB-based location tracking device can obtain the signal that the tag transmits and estimate the location of the tag.

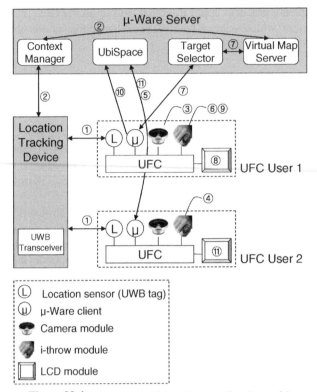

Figure 22.4 A sequence of interactive operations (example).

- The location information of UFC users estimated by the location-tracking device is sent to the *context manager* of the μ-ware server. On the basis of the location information the context manager obtains, the *virtual map server* of the μ-ware updates the virtual map of the ubiquitous testbed room.

- "User 1" takes a picture using the camera module attached to the UFC platform. The picture appears on the LCD screen.

- "User 2" wants to receive the picture that "user 1" takes from "user 1." To do so, "user 2" makes a 'ready-to-receive' gesture to express one's intention to receive other users' objects.

- On recognizing the "ready-to-receive" gesture, the μ-ware of "user 2" sends the "subscription" message to the *UbiSpace* of the μ-ware server to setup a data channel through which the user can receive other objects later.

- "User 1" wants to send the picture to "user 2." To express one's intention, "user 1" points a finger at "user 2."

- On recognizing the "pointing" gesture, the μ-ware of "user 1" sends a query to the *target selector* of the μ-ware server in order to identify the device the user is pointing at. The target selector, together with the virtual map server, selects an appropriate target device based on the target selection mechanism. In this scenario, "user 2" will be selected as a target device. The information about the selected target device will be finally sent to "user 1."

- On receiving the information, it appears on the LCD screen of "user 1" in the form of either a textual or graphical feedback, so the user can identify the currently selected target device.

- After identifying that the selected device is "user 2," "user 1" makes a throwing gesture at the "user 2," expressing one's intention to send the most recently generated object, that is, the picture one has just taken.

- On recognizing the "throwing" gesture, the μ-ware of "user 1" publishes the picture one has just taken to the UbiSpace of the μ-ware server, so the subscriber, "user 2," can download the picture.

- Finally, "user 2," a subscriber, will download the picture via the UbiSpace of the μ-ware server. The picture will automatically appear on the LCD screen.

In summary, for the UFC platform to support such a gesture interface, the following components are necessary:

- *Gesture recognition device* recognizes the target device that a UFC user is pointing at and the gesture such as "throwing" and "receiving."

- *Location-tracking device* keeps track of a UFC user's location. This is necessary because finding the target device that a UFC user is pointing at is dependent on the absolute location of the UFC user. We utilized UWB-based location tracking device [26] whose typical accuracy is 6 in. (15 cm).

- *Location server* gathers and manages the location information of both UFC users and ubiquitous devices. When one UFC user points at a specific device, the recognized gesture information is sent to the location server and it finally decides what the target device is.

- *Service discovery platform*: For a UFC platform to exchange information with one ubiquitous device, the UFC platform should be able to obtain the interface through which the communication is made possible. The interface includes IP address, port number and simple properties of the device. We have been working with middleware team and they developed ubiquitous service discovery (USD) protocol as part of KAIST Ubiquitous Service Platform (KUSP) [22]. USD protocol was originally based on UPnP [23], which is widespread as a service discovery, and this protocol is simplified to avoid XML parsing overhead. In this study, the USD protocol and KUSP was used as a service discovery platform.

- *Application that runs upon a UFC platform* infers a UFC user's intention based on a gesture, a target device, and previous operations and conducts a corresponding operation.

Among these necessary components, in this chapter, we focus on the gesture-recognition device and the application that runs upon an UFC platform.

22.4.2   i-Throw

In order to add a gesture interface to the UFC platform, we developed a wireless gesture recognition device, called *i-Throw*. This device is small enough to be worn on one's finger like a ring. It has a three-axis accelerometer and a three-axis magneto-resistive sensor for recognizing a gesture and the direction of the finger. It also has a ZigBee transceiver for transmitting the recognized gesture information to the UFC platform. Figure 22.1(b) shows its appearance.

22.4.3   Recognition of Gesture

The gesture recognition performed by the i-Throw device consists of two stages: *feature extraction* stage and *testing* stage. The feature extraction stage is a prepro-cessing stage to find reference features of each gesture. A feature f is represented by a four-dimensional vector as follows:

$$f = (A_{THx}, A_{THy}, A_{THz}, T_H) \tag{22.1}$$

where A_{THx}, A_{THy}, and A_{THz} are the acceleration thresholds of each axis and T_H is time duration threshold.

In the feature extraction stage, we should find appropriate thresholds for each possible input gesture. Due to the limitation of space, we omit the detailed explanation of the feature extraction stage. Table 22.2 shows the extracted features of various input gestures.

In the testing stage, i-Throw device compares the output of the accelerometer with each reference feature for over T_H seconds. If one of the features is matched, then i-Throw transmits the recognized gesture to the UFC platform via ZigBee interface.

The gesture-recognition algorithm is designed to be simple enough to run on a microcontroller inside the i-Throw by extracting the minimum set of required features and using threshold-based simple features.

We summarize and illustrate the gesture sets that the i-Throw recognizes in Figure 22.5. Other possible gestures, scrolling up/down and canceling, are inten-tionally omitted here.

Table 22.2   Features of Various Gestures

Gestures	Features
Throwing	(2 g+, X, 2 g+, 150 ms)
Increasing	(X, 0.5 g+, X, 500 ms)
Decreasing	(X, 0.5 g−, X, 500 ms)
Scrolling up	(0.5 g−, X, X, 500 ms)
Scrolling down	(0.5 g+, X, X, 500 ms)
Selecting	(X, X, 0.5 g−, 70 ms) and (X, X, 1.5 g+, 70 ms)
Scanning	(1.7 g+, X, X, 100 ms) and (X, 1.4 g+, X, 100 ms)

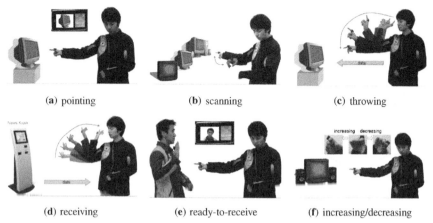

(a) pointing	**(b)** scanning	**(c)** throwing
(d) receiving	**(e)** ready-to-receive	**(f)** increasing/decreasing

Figure 22.5   Gesture sets of i-throw.

Every time a UFC user points to a device, the UFC platform displays the selected target device upon its screen. This feedback information helps the UFC user find the correct target device. Similarly, a scanning gesture allows the user to investigate controllable devices inside the room. This scanning operation is similar to the operation of moving a mouse pointer across several icons in a typical PC desktop environment.

"Ready-to-receive" gesture is necessary for a UFC user to express one's intention to receive other UFC users' objects. When one user makes a pointing or scanning gesture, only limited users who makes the "ready-to-receive" gesture can be selected.

22.4.4 Detection of Target Device

Our target-detection algorithm is based on Cone selection, which is used in virtual computing environments [20]. A cone is cast from i-Throw and a set of devices that intersect with it are chosen. Additionally, we have modified that typical cone selection algorithm to vary the area of the cone adaptively to improve the overall target detection accuracy. To do this, the orientation of i-Throw and the position of the UFC user and devices should be known. The location information is gathered and managed by the location server, as mentioned in Section 22.1. And the orientation of i-Throw can be obtained by combining the accelerometer and magnetic sensor outputs. The accelerometer is used for tilt compensation. By using the orientation and position information, target detection can be performed properly.

We define the object sets that can be sent and received inside our ubiquitous testbed room as follows:

- *music*: mp3 format, music objects can be played either by a UFC platform or public speaker, called *u-speaker*;
- *news*: html format, news objects can be obtained from *news kiosk*; and
- *photo*: jpg format, photo objects are generated when UFC users take picture.

Table 22.3 Target Device Sets

Target device	Supported objects	Flow of objects
u-display, u-projector	news, photo	input, output
news kiosk	news	output
u-printer	news, photo	input
u-trash	news, photo, music	input
u-speaker	music	input, output
UFC	news, photo, music	input

Next, we further define the target device sets that consist of possible target devices in the ubiquitous testbed and the characteristics of each one, which are summarized in Table 22.3.

Among these devices, the *news kiosk* automatically gathers recent news from Internet Web sites and displays each one, that is refreshed every 10 s. When one UFC user sees the interesting news upon the news kiosk, he or she can obtain the news by making "receiving" gesture toward the news kiosk.

The *u-trash* functions as a symbol of "deleting a file." Similarly to the natural way of using an actual trash, throwing something that is useless anymore at the trash, if one UFC user makes a throwing gesture toward the u-trash, the current object will be deleted automatically. The u-trash device is different from other devices in that it is not an electric device; it acts only as a marker standing for a particular operation and thus actual operation is not conducted inside it. This example gives us insight as to how to fully utilize spatial resources inside the room. If various markers whose symbolic meaning can be easily interpreted are added inside the room and each corresponding operation is efficiently conducted, the spatial gesture interface allows UFC users to conduct various operations in a user-friendly manner.

Table 22.3 points out that some target devices do not support all kinds of objects: music objects can be supported by the device that is capable of playing the music such as u-speaker or UFCs and the news kiosk only supports the news object. Table 22.3 also shows that some devices only allow either an input channel or an output channel: a u-trash only opens an input channel, while a news kiosk only opens an output channel.

22.4.5 Adaptive Angle Assignment for Resolving the Difficulty in the Target Selection Procedure

As mentioned in the previous subsection, the graphical feedback allows a UFC user to find a correct target device efficiently and intuitively. However, what if an intended target device is too small or too adjacent to other devices? This situation leads to a difficulty in the target selection procedure.

The difficulty of pointing or selecting a target device is closely related to its physically assigned angular width, which depends on the device size, its relational location, and the user's location. Fitts' law describes well the relationship of the assigned angle and the difficulty with selection [8, 9]. According to the law, the index of difficulty is expressed as a log function of angular movement and angular target width and it is proportional to the time for selection.

For this, we propose an algorithm, called adaptive angle assignment, which makes all of the assigned angular width of devices from the user in a given space bigger than the specific threshold, A_{TH}. This algorithm solves the problem of the lagged selection time by reassigning the angular width. When he starts the target selection at the specific location, the location server calculates the physically assigned angular widths of the devices and then, if necessary, it reassigns adaptively the angular widths. The process is following.

- Grouping of target devices: After the calculation of physically assigned angular width, the location server creates an angle table as shown in Figure 22.6. The contiguous angles are regarded as a group angle and the location server reassigns each angle within a fixed group angle.

- Adaptive angle assignment: The server performs the reassignment for each group according to our algorithm which is represented as a pseudo code in Figure 22.8.

G_k is the kth group angle and A_i is the ith angle in each G_k. A_{lack} is the sum of required angles for expanding angles, which are smaller than A_{TH}, to A_{TH}. A group with zero of A_{lack} is not necessary to be reassigned. A_{res} is the sum of excessive angles to A_{TH} within a group. The group cannot be reassigned when the A_{res} is zero. In addition, A_{don} is a sum of the angles that are donated from the excessive angles. If A_{res} is smaller than A_{lack}, A_{don} becomes A_{lack} and otherwise, it becomes A_{res}. After deciding the value of A_{don}, all angles are expanded or shrunken with a proportion to the gap angle with A_{TH}. At this time, the group angle is consistent

```
For all A, in Gₖ {
    If |A| < A_TH    then A_lack += (A_TH − |A|) ;
    Else             then A_res += |A| − A_TH ;
}

If A_lack > A_res then A_don = A_res ;
Else then A_don = A_lack ;

For all A, in Gₖ {
    If A < A_TH then
        |A| += A_don × (A_TH − |A|) / A_lack ;
    Else then
        |A| −= A_don × (|A| − A_TH) / A_res ;
}
```

Figure 22.6 Adaptive angle assignment.

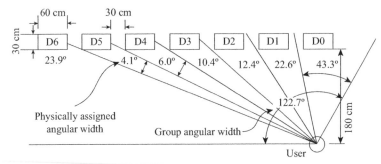

Figure 22.7 Virtual space of the experiment environment.

because the expanding angle and the shrunken angle are same. Figure 22.9 shows the reassignment result when A_{TH} is set to $10°$.

Even though a little gap between the original angular region, which is a region between the start degree and the end degree, and the reassigned angular region exists, it does not affect the performance of selection much because of general operating pattern of users; when a user wants to select a device with i-Throw, she stares and point to it. The right selection is finalized by the user's correction through the graphical feedback on the LCD. It means that the feedback lessens the user's confusion caused by the gap.

In order to adapt our adaptive angle assignment algorithm and verify it as well as to decide the threshold value, A_{TH}, we set up an experiment environment whose virtual space is represented in Figure 22.7. In the corresponding real space, seven same-sized LCD monitors were deployed at equal intervals. Thirteen males ranged in ages from 23 to 31 years participated in this experiment and they were requested to make 70 correct selections, 10 times per each device, in a randomly generated order. The location of user is fixed at the designated point which was 180 cm away from D_0.

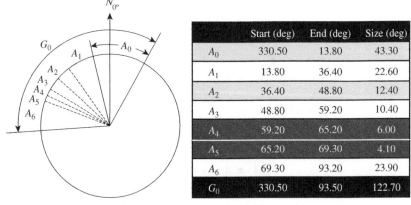

	Start (deg)	End (deg)	Size (deg)
A_0	330.50	13.80	43.30
A_1	13.80	36.40	22.60
A_2	36.40	48.80	12.40
A_3	48.80	59.20	10.40
A_4	59.20	65.20	6.00
A_5	65.20	69.30	4.10
A_6	69.30	93.20	23.90
G_0	330.50	93.50	122.70

Figure 22.8 Angle table for the situation shown in Figure 22.7.

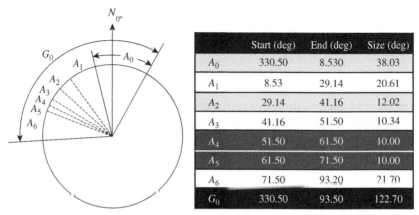

	Start (deg)	End (deg)	Size (deg)
A_0	330.50	8.530	38.03
A_1	8.53	29.14	20.61
A_2	29.14	41.16	12.02
A_3	41.16	51.50	10.34
A_4	51.50	61.50	10.00
A_5	61.50	71.50	10.00
A_6	71.50	93.20	21.70
G_0	330.50	93.50	122.70

Figure 22.9   After the angle reassignment.

We measured average time to select each device in both cases of using the ray-based technique and our algorithm.

Figures 22.7 and 22.8 show how the angular widths were physically assigned at this situation and Figure 22.9 shows the reassigned angular width according to our algorithm. Due to the effect of marginal angle, A_{mar}, which was set to 20° in this experiment, both end devices (D_0 and D_6) were assigned larger angles than others, while the assigned angles of D_4 and D_5 were relatively small. The result of the experiment is shown in Figure 22.10. Figure 22.10 shows the average time spend to select each device correctly. In case of using the ray-based technique, selection time significantly increased when the angular width is less than 10°. From the result, we convinced that the selection action with i-Throw follows Fitts' law and we decided a reasonable value of A_{TH} as 10°. This value is a parameter determined by the user's experiences and can vary with the user characteristics.

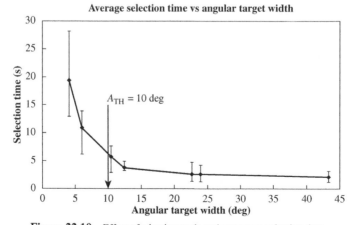

Figure 22.10   Effect of adaptive angle assignment on selection time.

The proposed algorithm prevented the rapid increase in selection time when a physically assigned angular width became lower than A_{TH}. Adaptive angle assignment reduced average time for selecting D_5 whose physically assigned angle was $4.1°$ about 62.6%. However, in case of D_2 and D_3, even though their angular widths were nearly unchanged, selections took slightly longer time than before. The reason is that a little gap between the original angular region and the reassigned angular region exists; the start and end degree of each angle may be changed. Unlike classic ray-based selection which does have any gap between original and reassigned angular regions, the proposed algorithm requires some additional movements to compensate the gap. However, it does not affect the selection performance much because of general operating pattern of users.

22.4.6 UFC Operation Sets

Until now, we have summarized the gesture sets, the object sets, and the target device sets that our ubiquitous testbed supports. When a UFC user makes one gesture among the given gesture sets, the UFC platform should be able to decide which object and which operation needs to be processed. That decision depends on the type of a target device and the type of a recent operation. For example, if one UFC user takes a picture ("taking a picture" operation) and made a "throwing" gesture toward the u-printer, it is highly likely that the user intends to print the photo he has just taken. On the other hand, if the user reads a news using his UFC terminal ("reading a news" operation) and makes a "throwing" gesture toward the u-printer, the user may intend to print the news he has just read rather than photos or other objects. These examples show that the UFC platform has to keep track of the recent operations, more specifically, recently selected objects. To do this, we define the followings: *mrso* indicates the most recently selected objects among various music, news, and photo objects. *mrso_music*, *mrso_news*, and *mrso_photo* indicate the most recently selected music, news, and photo objects, respectively.

Finally, Table 22.4 summarizes the UFC operation sets that were actually utilized in our demonstration. In this table, "(1)" indicates that the corresponding object depends on the status of the selected target device. The demonstration video clip is available in Ref. [27].

22.5 RELATED WORK

Wearable computing systems have already become one of the main research area. Some of the currently available wearable computers include PDA-based systems relying on industrial standard architecture concepts. MITthril [7] is one of these examples that integrate computation, sensing, networking into cloths, and all devices are connected using body bus, which is single-cable hardwired connection. These PC-or PDA-based solutions focus on transforming commercially available devices into a wearable device, thus they do not reach a high degree of wearability and usability.

Table 22.4   UFC Operation Sets

Operation	Selected object	Condition target_device	gesture	etc.
Increase volume	None	u-speaker	Increasing	None
Decrease volume	None	u-speaker	Decreasing	None
Delete a file	mrso	u-trash	Throwing	None
Send a file	mrso	u-display	Throwing	mrso_type = photo or news
		u-printer		mrso_type = photo or news
		UFC		None
	mrso_music	u-speaker		None
Read news	(1)	News kiosk	Receiving	None
	other.mrso	None	Ready-to-receive	other.mrso_type==news and other.target_device==this and other.gesture==throwing
Listen to music	(1)	u-speaker	Receiving	None
	other.mrso	None	Ready-to-receive	other.mrso_type==music and other.target_device==this and other.gesture==throwing
View a photo	(1)	u-display	Receiving	None
	other.mrso	None	Ready-to-receive	other.mrso_type==photo and other.target_device==this and other.gesture==throwing

(1) indicates that the corresponding object depends on the status of the selected target device.

Wearable systems based on custom design include low-end special-purpose appliances such as watchs [13] and garments [14]. IBM's Linux Watch [13] packed a fair amount of hardware into wristwatch size and showed the model of a wearable computer form. With this small watch form and communication interface with Bluetooth and Irda, it supports short-range communication applications. However, it lacks other communication interfaces and user interfaces. Therefore, it should be considered what kind of service can be supported with its limited interfaces. These low-end custom designs are bound to their specific task due to lack of system specification and interfaces.

The custom-designed multiple-purpose systems have been developed that provide more functionality and flexibility. WearARM [11], QBIC [2], and Xybernaut wearable computer system [12] are designed for multiple purposes. These are integrated into a shoulder, a belt, and a waist, respectively. These systems establish a compromise between high and low end in providing processing performance and a wearable system design. QBIC shows a case that addresses ergonomic aspects as well as provides sufficient connectivity and computational performance in hardware level. However, these systems should consider ubiquitous environment to realize real necessity of wearable platform in any time and any where. It means that wearable system should be realized for supporting various ubiquitous systems environment without any interference and disturbance. Also, intuitive user interface should be applied to these system to support the comfortable use of it.

22.6 CONCLUSION

The ubiquitous fashionable computer, introduced in this chapter, is a wearable computer that exploits ubiquitous computing environment. The success of wearable computer will heavily rely on good wearability, usability, aesthetic appearance, and social acceptance. Therefore, its external design should be aesthetic and comfortable for it to be popular and let more people use for their real lives. In addition, not merely considering about these uncomfortable carrying issues, the reflection the exploit of ubiquitous computing environment should be done. In the ubiquitous environment, wearable computing systems integrate various types of devices like communication interfaces to get services in any time and any where, and user interfaces to facilitate the computing system without any burdensomeness.

In this chapter, we present the design approach and the philosophy of our wearable computing system that is wearable, aesthetic, and intuitive. The main features are as follows. It supports the coexistence and interoperability of various communication interfaces between WLAN, Bluetooth, and ZigBee devices that can be operated freely in any ubiquitous system environment. This can be done using dynamic channel allocation mechanism between communication devices that operates same ISM band. In addition, novel user interfaces are developed and integrated with the UFC to help the intuitive use of it in our wearable computer system. Among possible user interfaces, in this chapter, we focus on what we call *i*-Throw, which means intuitive input devices. This input device can recognize human's hand gestures and direction, and control ubiquitous devices with the gestures. Our UFC system is realized and tested with real testbed environment that has various network interfaces, sensor nodes, and ubiquitous components, and also many applications are implemented and presented.

REFERENCES

1. J. Lee, S. Lim, J. Yoo, K. Park, H. Choi, and K. Park, A ubiquitous fashionable computer with an i-Throw device on a location-based service environment. In *Proceedings of the 2nd IEEE International Symposium on Pervasive Computing and Ad-Hoc Communications*, May 2007.

2. O. Amft, M. Lauffer, and S. Ossevoort, Degisn of the QBIC wearable computing platform. In *Proceedings of the 15th IEEE International Conference on Application-Specific Systems, Architectures and Processors*, September 2004, pp. 398–410.

3. S. Mann, Wearable computing as means for personal empowerment. In *Proceedings 3rd International Conference on Wearable Computing*, May 1998.

4. A. Pentland, Wearable intelligence. *Sci. Am.*, 276(1es1), 1998.

5. T. Starner, The challenges of wearable computing: Part 1. *IEEE Micro*, July 2001.

6. M. Weiser, The computer for the 21st century. *Sci. Am.*, 265(3):66–75, 1991.

7. R. Devaul, M. Sung, J. Gips, and A. pentland, MITthril 2003: applications and architecture. In *Proceedings of 7th International Symposium on Wearable Computers*, 2003.

8. P. M. Fitts, The information capacity of the human motor system in controlling the amplitude of movement. *J. Exp. Psychol.*, 47:381–391, 1954.

9. G. V. Kondraske, An angular motion fitts' law for human performance modeling and prediction. *IEEE Eng. Med. Bio. Soc.*, 207–308, 1994.

10. S. Mann, Smart clothing: the shift to wearable coputing. *Proc. Commun. ACM*, 39:23–24, 1996.

11. U. Anliker, et al., The WearARM: modular, high performance, low power computing platform designed for integration into everyday clothing. In *Proceedings of 5th International Symposium on Wearable Computers*, 2001.

12. Xybernaut Corp. Xybernaut wearable systems. Mobile Assistant wearable computer, www.xybernaut. com.

13. C. Narayanaswami, et al., IBM's Linux Watch: the challenge of miniaturization. *IEEE Comp.*, 35(1):33–41, 2002.

14. J. Rantanen, et al., Smart Clothing prototype for the arctic environment. *Personal Ubiquitous Comput.*, 6(1):3–16, 2002.

15. H. Cho, M. Kang, J. Park, B. Park, H. Kim, Performance analysis of location estimation algorithm in ZigBee networks using received signal strength. In *Proceedings of the 21st International Conference on Advanced Information Networking and Applications Workshops (AINAW'07)*, pp. 302–306, 2007.

16. S. Liang, *The Java Native Interface: Programmer's Guide and Specification.* Addison Wesley, 1999.

17. IEEE Standard 802, part 15.4. Wireless Medium Access Control (MAC) and Physical Layer (PHY) Specifications for Low Rate Wireless Personal Area Networks (WPANs), 2003.

18. IEEE802.15.2 Specification. Coexistence of Wireless Personal Area Networks with Wireless Devices Operating in Unlicensed Frequency Bands, 2003.

19. M. Back, S. Lahlow, R. Ballagas, S. Letsithichai, M. Inagaki, K. Horikira, and J. Huang, Usable ubiquitous computing in next-generation conference rooms: design, evaluation, and architecture. *UbiComp 2006 Workshop*, 2006.

20. J. Liang and M. Green, JDCAD: a highly interactive 3D modeling system. *Comput. Graphics,* 18(4): 499–506, 1994.

21. M. Kang, J. Chong, H. Hyun, S. Kim, B. Jung, and D. Sung, Adaptive interference-aware multi-channel clustering algorithm in a ZigBee network in the presence of WLAN interference. *IEEE International Symposium on Wireless Pervasive Computing*, Puerto Rice, February 2007.

22. Y. Song, S. Moon, G. Shim, and D. Park, Mu-ware: a middleware framework for wearable computer and ubiquitous computing environment. *A Middleware Support for Pervasive Computing Workshop at the 5th Conference on Pervasive Computing & Communications (PerCom 2007)*, New York, March 2007.

23. UPnP Forum. *UPnP Device Architecture 1.0*, Version 1.0.1, 2003.

24. IEEE 802.11 Specification. Wireless LAN medium access control (MAC) and physical layer (PHY) specifications. *IEEE Specifications*, June 1997.

25. IEEE 802.15.4 Specification. Wireless LAN medium access control (MAC) and physical layer (PHY) specifications for low rate wireless personal area networks (LR-WPANs). *IEEE Specifications*, October 2003.

26. Ubisense. http://www.ubisense.net.

27. KAIST UFC Project. http://core.kaist.ac.kr/UFC.

Chapter 23

Energy Efficiency for Mobile Multimedia Replay

Chu-Hsing Lin and Jung-Chun Liu

Department of Computer Science and Information Engineering, Tunghai University, Taichung, Taiwan

23.1 Introduction	530
23.2 Video Codec	531
23.3 Experiment Setting	532
23.4 Experiment Results	535
23.5 Summary	543
Acknowledgments	544
References	544

23.1 INTRODUCTION

Mobile handheld devices are the key players for applications in a ubiquitous computing environment. One characteristic of the ubiquitous computing environment is the limited resources. Ubiquitous engineering needs to deal with limitations of the mobile handheld devices, such as memory space, processing time, and battery capacity.

Furthermore, along with the progress in information technology, consumers are attracted to innovative and convenient electronic products. Mobile handheld devices have become a trend and all kinds of applications on them follow suit, including audio-visual, event recording, Web surfing, phone calls, and so on. Among these applications, audio-visual is most indispensable for entertainments. Because of the high complexity of operations, video applications consume tremendous amount of energy that is very challenging for battery-powered handheld devices. The capacity

Mobile Intelligence. Edited by Laurence T. Yang, Agustinus Borgy Waluyo, Jianhua Ma, Ling Tan, and Bala Srinivasan

of battery grows up very slowly, only about 5–10% every year, which is insufficient for the electricity demanded by the handheld devices [1].

In recent years, lots of new compression codecs have been issued for the two major video codec standards: MPEG and ITU-T. What is the best codec by which energy-efficient video codes can be encoded? How to encode video by any arbitrary codec to have good enough picture quality, and at the same time, to be economical in energy consumption for replaying on the handheld devices? These are typical and bewildering questions perplexing the users.

In this chapter, we try to answer the above-mentioned questions. Experiments have been conducted over various genres of video compressed by various codecs. Energy consumption analysis is performed to give users suggestions about choices of suitable codecs to encode video to be replayed on mobile handheld devices.

The remainder of this chapter is structured as follows. Section 23.2 introduces the popular video codecs in use today. We describe the experimental design and setup in Section 23.3. Experimental results will be given and energy analyses will be performed in Section 23.4. Finally, we conclude this chapter in Section 24.5.

An earlier version of this chapter was presented at MUE 2007 [2]. This version is substantially expanded, in terms of both the presentation of experimental results and depth of analysis. New plots of energy consumption in all experimental scenarios give insight into effects on energy efficiency of codecs and their parameters.

23.2 VIDEO CODEC

For fast transmission and easy storage, video is encoded or compressed by codecs. Here we briefly introduce the popular codecs that are used to encode video to be replayed on the handheld device in the experiment.

23.2.1 MPEG-4

MPEG-4 was issued in 1998. It consists of a series of audio and video code standards and a set of relevant technology. Its formulation organization is ISO/IEC Moving Pictures Experts Group, namely MPEG. MPEG-4 includes most functions of MPEG-1 and MPEG-2 and strong features of other formats. It adds and expands supports for the Virtual Reality Model Language, the object-oriented synthetic files (including audio, video, and VRML Object), and the Digital Rights Management and other interactive functions. Most functions of MPEG-4 are decided or adopted by the developer. This means that the whole function is not concluded by certain procedure. So, this form uses the so-called "profiles" and "levels" to define some specific sets of functions for certain applications of MPEG-4 [3, 4].

23.2.2 DivX

DivX is the name of products by Div, Inc. Basically Microsoft's encoders do not allow users to save MPEG-4 streams into the AVI format and force users to use ASF instead.

They also have some other limitations that are overridden in DivX. Many newer "DivX Certified" DVD players are able to play video coded by DivX. The Quarter pixel (QPEL) and Global Motion Compensation (GMC) features are often omitted to reduce processing requirements, so for compatibility reasons, they are excluded from the base DivX encoding profiles. We adopt DivX3 and DivX5 in our experiment [5].

23.2.3 XviD

Xvid (formerly "XviD") is an open-source MPEG-4 video codec. Xvid was created by a group of volunteer programmers in 2001. The Xvid video codec implements MPEG-4 Simple Profile and Advanced Simple Profile standards, such as b-frames, lumi masking, trellis quantization, QPEL, GMC, and H.263, MPEG and custom quantization matrices. It allows compressing and decompressing digital video to reduce the required bandwidth of video data for transmission over computer networks and efficient storage on CDs or DVDs [6].

23.2.4 H.263

H.263 is a video compression algorithm and protocol that is standardized by ITU. It was published in 1995. The video source coding algorithm of H.263 is based on Recommendation H.261. It is a hybrid of interpicture prediction, exploiting temporal redundancy, and transform coding of the remaining signal to reduce spatial redundancy, with some changes to improve performance and error recovery. It has been further enhanced in projects known as H.263v2 (also known as H.263+ or H.263 1998) and H.263v3 (also known as H.263++ or H.263 2000). We adopt H.263v2 in this experiment [7].

23.2.5 WMV

WMV is generally packed into an Advanced Systems Format (ASF) container format with .wmv file extention. It can also be put into AVI containers. The resulting files may be named .avi if it is an AVI-contained file. WMV is a popular codec for distributing video on the Internet and is also used to distribute high-definition video on standard DVDs in WMV HD format. This WMV HD content can be replayed on computers or on compatible DVD players [8].

23.3 EXPERIMENT SETTING

23.3.1 Experiment Environment

We used Xvid, DivX5, DivX3 (open source edition), MS MPEG-4 v2 (Microsoft MPEG-4 version2), H.263+, WMV2 (Windows Media Video V8), and WMV3 (Windows Media Video V9) as compression encoders on SUPER©(Simplified Universal

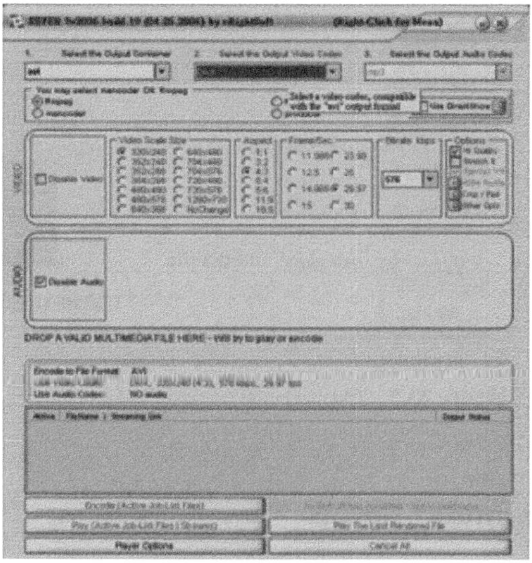

Figure 23.1 SUPER9©.

Player Encoder & Renderer) (Figure 23.1). SUPER© is free software developed by eRightSoft and it supports a lot of media formats [9].

We used TCPMP (The Core Pocket Media Player) shown in Figure 23.2 as the media player 10. The process of encoding and decoding is shown in Figure 23.3.

The video file is encoded from the source of the film (DVD) by SUPER©, loaded on the handheld device and then replayed or decoded by TCPMP. We measured the electricity consumed while playing the video stream.

Figure 23.2 TCPMP.

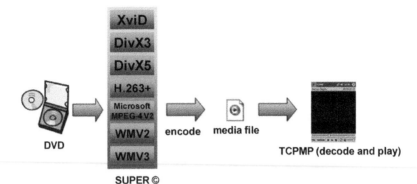

Figure 23.3   The process of encoding and decoding.

23.3.2   Experiment Setup

The experiment setup is shown in Figure 23.4. In the experiments, an Acer n300 PDA was used, which has a 400 MHz Samsung S3C2440 processor (with 64 MB ROM and 64 MB SDRAM) and has Microsoft® Windows Mobile™ Version 5.0 as its operating system 11. In order to measure the power consumption, we removed the battery from the Acer n300 PDA and powered the PDA by a DC 5 V power supply in series with a 1 Ω resistor [12–14]. We used a National Instruments PCI DAQ data acquisition board to sample the voltage drop across the resistor (to calculate the current) at 1000 samples/s. The energy measurement was done by using LabVIEW 8, a GUI-based data acquisition, measurement analysis, and presentation software [15]. We calculated the instantaneous power consumption corresponding to each sample by Ohm's Law in Eq. (23.1), and sum up the energy consumption per sample to obtain the total energy

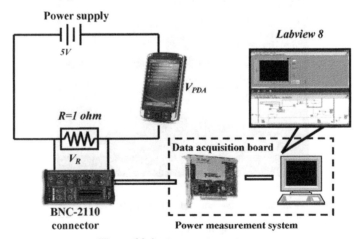

Figure 23.4   The experimental setup.

Table 23.1 Films Tested in the Experiment

Animation films	The Simpsons	Kiki's Delivery Service	Porco Rosso	Laputa: Castle in the Sky	Nausicaa of the Valley of the Wind
Action thriller	Courage Under Fire	X-Men 3	Spider Man 1	Spider Man 2	Superman Returns
Romance films	Broken-back Mountain	Goodbye Koru	The Lake house	Blue Gate Crossing	Now, I want to see you

consumption by Eq. (23.2):

$$P_{\text{Inst}} = \frac{V_R}{R} \times V_{\text{PDA}} \qquad (23.1)$$

$$E = \sum P_{\text{Inst}} \times T \qquad (23.2)$$

where $T = 1/1000$ s.

In the experiment, we measured energy consumption of three genres of films, with five different films for each genre listed in Table 23.1. In experiment 1 and 2, all films have resolution of 352×288, the frame rate of 29 frames/s, the bit rate of 512 kbps and the playing time of 5 min. In experiment 3, we changed the resolution, the frame rate, and the bit rate one by one to investigate the effects of these parameters.

23.4 EXPERIMENT RESULTS

We investigate energy efficiency of video decoding on the handheld devices by various video encoding scenarios. We measured energy consumption of video encoded (a) by various types of codecs in experiment 1, (b) in various video file formats, and (c) by changing coding parameters: the bit rate, the frame rate, and the resolution one by one.

23.4.1 Experiment 1

We used XviD, DivX3 (open source version), DivX5, MS MPEG-4 V2, H.263+, WMV2, and WMV3 as codecs to encode the films, five video clippings for each genres. The output formats were in AVI. We replayed the video clippings on the PDA and measured the energy consumption. Plots of the average energy consumptions in Joules for the five video clippings versus various codecs are shown in Figure 23.5 for the animation films, in Figure 23.6 for the action thrillers, and in Figure 23.7 for the romance films.

From these three energy consumption plots, we observe that energy consumptions of films encoded by XviD, DivX3, DivX5, and H.263+ are similar, and those encoded by MS MPEG-4 V2, WMV2, and WMV3 are similar. We also observe that the films encoded by XviD, DivX3, DivX5, and H.263+ consume less electricity than those encoded by MS MPEG-4, WMV2, and WMV3.

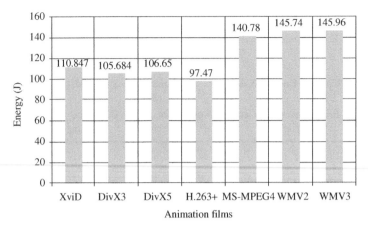

Figure 23.5 Average energy consumption (unit in Joules) of animation films encoded with various codecs: XviD, DivX3, DivX5, H.263+, MS MPEG-4, WMV2, and WMV3.

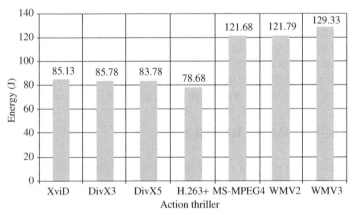

Figure 23.6 Average energy consumption (unit in Joules) of action thriller encoded with various codecs: XviD, DivX3, DivX5, H.263+, MS MPEG-4, WMV2, and WMV3.

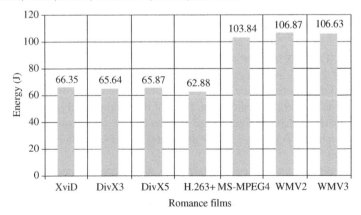

Figure 23.7 E Average energy consumption (unit in Joules) of romance films encoded with various codecs: XviD, DivX3, DivX5, H.263+, MS MPEG-4, WMV2, and WMV3.

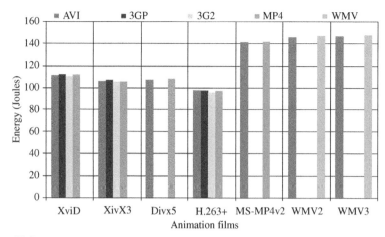

Figure 23.8 Average energy consumption of animation films in various file formats: XviD, DivX3, DivX5, H.263+, MS MPEG-4v2, WMV2, WMV3. Missed color bars indicate nonapplicable file formats for those codecs.

23.4.2 Experiment 2

In experiment 2, we encoded films in different video formats: 3GP, 3G2, AVI, MP4, and WMV by the following codecs: XviD, DivX3, DivX5, MS MPEG-4v2, H.263+, WMV2, and WMV3. We replayed the video clippings on the PDA and measured energy consumptions. Plots of the average energy consumptions of films in various file formats versus codecs are shown in Figure 23.8 for the animation films, in Figure 23.9 for the action thrillers, in Figure 23.10 for the romance films. Note that when the file format is not applicable for some codec, the energy consumption of it is not

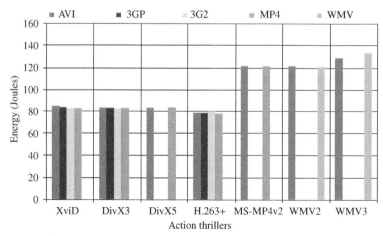

Figure 23.9 Average energy consumption of action thriller in various file formats: XviD, DivX3, DivX5, H.263+, MS MPEG-4v2, WMV2, and WMV3.

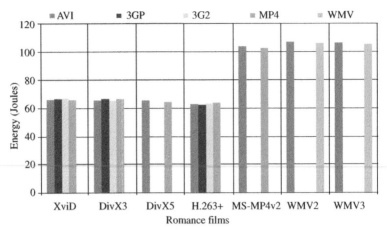

Figure 23.10 Average energy consumption of romance films in various file formats: XviD, DivX3, DivX5, H.263+, MS MPEG-4v2, WMV2, and WMV3.

measured. Missed color bar in the plots means that the file format is not applicable for the particular codecs.

According to plots in Figures 23.8–23.10, we observe that films encoded by the same codec consume similar amount of energy and that the file format does not have significant effect on energy consumption.

23.4.3 Experiment 3

In experiment 3, we encoded films with various setting of parameters of codecs. Three kinds of codec parameters were tuned: the bit rate, the frame rate, and the resolution. We adjusted these three parameters to observe their influence on energy consumption in order to decide which parameter has the greatest influence.

The film used in this experiment was "The Simpsons," and the playing time was 5 min. We replayed the video clippings on the PDA and measured energy consumptions.

Effect on Energy Consumption of Bit Rates Figure 23.11 shows plots of the energy consumption of the film with various bit rates.

Figure 23.12 shows the growth rate of energy consumption when the bit rate is doubled. The growth rate of electric energy consumption is calculated by.

$$\frac{\text{latter's energy} - \text{former's}}{\text{former's}} \times 100\%$$

For example, if the film is encoded by DivX3, the energy consumption is 62.04 J when the bit rate is 144 kbps and 71.18 J when the bit rate is 288 kbps. We compute the growth rate of energy consumption for this example to be 14.73% with the bit rate changed from 144 to 288 kbps. In the same way, we calculate the other growth

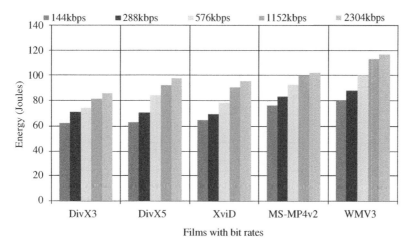

Figure 23.11 Energy consumption of videos with various bit rates: 144, 288, 576, 1152, and 2304 kpbs.

rates of the energy consumption and average them to find the average growth rate of energy consumption when the bit rate is doubled to be 11.85%.

Effect on Energy Consumption of Frame Rates Figure 23.13 shows the energy consumptions of films encoded with a resolution of 320×240, and 25 frames/s, whereas Figure 23.14 shows energy consumptions of films encoded with the same resolution but the frame rate being changed to 30 frames/s.

Figure 23.15 shows the growth rate of energy consumption by changing frame rates. We find the average growth rate of the energy consumption is 15.76% when frame rate is changed from 25 to 30 frames/s.

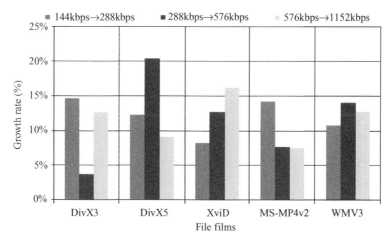

Figure 23.12 Growth rate of energy consumption when video's bit rate is doubled: 144–288, 288–576, and 576–1152.

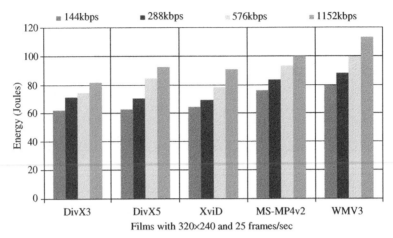

Figure 23.13   Energy consumption of videos with 25 frames/s.

Effect on Energy Consumption of Film Resolution   Figure 23.16 shows the energy consumptions of films with the resolution of 320×240, and 25 frames/s, while Figure 23.17 shows those with the resolution of 384×288, and 25 frames/s.

Figure 23.18 shows the growth rate of energy consumption of experiments with different resolutions. We find that the average growth rate of the energy consumption is 22.44% when the resolution of the film is changed from 320×240 to 384×288.

On the basis of the values of average growth rates for changing parameters of the bit rate (11.85%), the frame rate (15.76%), and the resolution (22.44%) of the film, we observe that the resolution has most significant influence of the energy consumption.

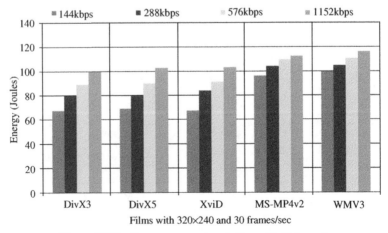

Figure 23.14   Energy consumption of videos with 30 frames/s.

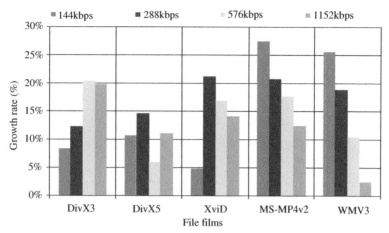

Figure 23.15 Growth rate of energy consumption of videos with the frame rate changed from 25 to 30 frames/s.

Discussion We observe that when the video's bit rate is doubled, the average growth rate of energy consumption is 11.85%, which is about half of the average growth rate of energy consumption by adjusting video's resolution from 320×240 to 384×288. The average growth rate of energy consumption from increasing the video's frame rate lies between these two.

In regard to the picture quality of the film, we can clearly see the degradation of it when the bit rate is decreased. Figure 23.19 shows three pictures of a film encoded by DivX5 with resolution of 320×240, the frame rate of 25 frames/s, but with different bit rates. We observe that the clock (in the enlarged windows) in the second picture (288 kbps) has better picture quality than the one in the first picture (144 kbps). And

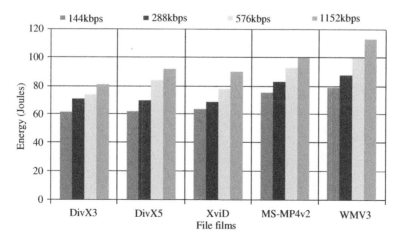

Figure 23.16 Energy consumption of videos with resolution of 320×240.

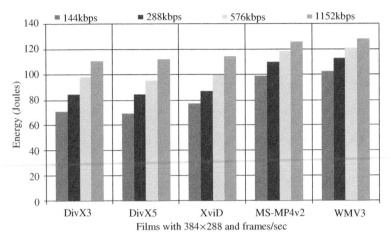

Figure 23.17 Energy consumption of videos with resolution of 384×288.

the third one (576 kbps) has better picture quality than the second one. The degrading effect of films with low bit rate is even more obvious when watching them playing continuously on the screen.

Figure 23.20 illustrates the effect on picture quality when the resolution of the film is changed. As one can observe that the picture quality of the film does not improve too much when the resolution is increased from 320×240 to 384×288.

We conclude that we can encode films with higher bit rates instead of higher resolution to have better picture quality, and at the same time, have energy-efficient video code suitable for replaying on handheld devices.

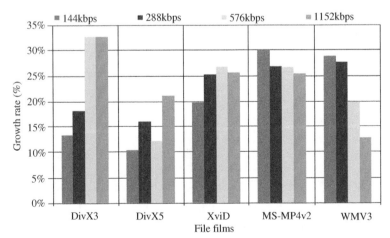

Figure 23.18 Growth rate of energy consumption of videos with resolution changed from 320×240 to 384×288.

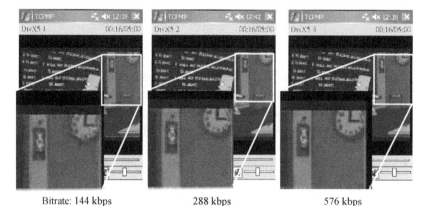

Bitrate: 144 kbps   288 kbps   576 kbps

Figure 23.19   Picture quality of videos encoded with various bit rates: 144, 288, and 576 kbps.

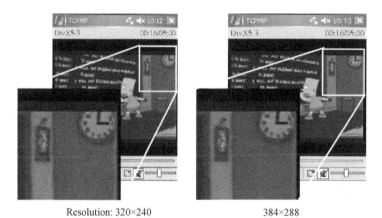

Resolution: 320×240   384×288

Figure 23.20   Picture quality of videos encoded with different resolutions: 320×240, and 384×288.

23.5 SUMMARY

The ubiquitous computing environment has strict constraints on the resources, such as processing time and battery life. How to minimize energy consumption while ensuring a desirable level of video quality becomes a challenge. In this chapter, we encode films by several popular codecs nowadays and measure the electric energy consumption of them to find out which video codec is most suitable for mobile handheld devices.

Constraints of resources of the ubiquitous computing environment call for wise strategy for energy consumption management. Codecs and their encoding parameters have effects on energy consumptions when replaying or decoding video code encoded by them. We summarize their influences as follows.

- *Type of codecs:* From experimental results, we find that when playing films encoded by one group of codecs, including XviD, DivX3, DivX5, and H.263+, less electricity is consumed than playing films encoded by the other group of codecs, including MS MPEG-4, WMV2, and WMV3. When watching films encoded by all kinds of popular codecs on small screens of handheld devices, the picture quality are found to be nearly the same. The mobile handheld device users are recommended to encode films by the group of codecs consuming less electricity.

- *Video file format:* In view of the file formats, experiments show that as long as the same codec is used, it makes not much difference as what file format is chosen.

- *Encoding parameters:* To have better picture quality, we can encode video with higher bit rates, frame rates, or resolution. Experimental results show that video encoded with higher resolution results in sharp increase in energy consumption when replaying. However, increasing the bit rate to encode video will not induce too much energy consumption when replaying it. The effect of increasing the frame rate on energy consumption is in between of the other two parameters. So, mobile handheld device users are encouraged to use higher bit rates to encode films if better picture quality is called for.

ACKNOWLEDGMENTS

This work was supported in part by Taiwan Information Security Center (TWISC), National Science Council, under the grants NSC 95-2218-E-001-001, NSC95-2218-E-011-015 and NSC95-2221-E-029-020-MY3.

REFERENCES

1. K. Lahiri, A. Raghunathan, S. Dey, and D. Panigrahi. Battery driven system design: a new frontier in low power design. In: *Proceedings of ASP-DAC/VLSI Design 2002, Bangalore, India, January 2002,* pp. 261–267.
2. C.-H. Lin, J.-C. Liu, C.-W. Liao, Energy analysis of multimedia video decoding on mobile handheld devices. In: *Proceedings of the International Conference on Multimedia and Ubiquitous Engineering, Korea, 26 April, 2007,* pp. 120–125.
3. Applications and Requirements for Scalable Video Coding. MPEG-document ISO/IEC JTC1/SC29/ WG11 N5540, 2003.
4. Overview of the MPEG-4 Standard: http://www.chiariglione.org/mpeg/standards/mpeg-4/mpeg-4.htm.
5. DivXNetworks, Inc. [online]. http://www.divx.com/divx/.
6. XviD Software Package [online]. http://www.xvid.org/.
7. G. Cote, B. Erol, M. Gallant, and F. Kossentini, H.263+: video coding at low bit rates. *IEEE Trans. Circuits Syst. Video Technol.,* 8(7): 849–866, 1998.
8. W. Ashmawi, R. Guerin, S. Wolf, and M. Pinson, On the impact of policing and rate guarantees in DiffServ networks: a video streaming application perspective. In: *Proceedings of the 2001 Conference on Applications, Technologies, Architectures, and Protocols for Computer Communications,* California, United States, 2001, pp. 83–95.
9. SUPER©. http://www.erightsoft.com/SUPER.html.
10. TCPMP. http://tcpmp.corecodec.org/.
11. Acer n300. http://www.acer.com.tw/PRODUCTS/pda/n300.htm.

12. N. R. Potlapally, S. Ravi, A. Raghunathan, and N. K. Jha. Analyzing the energy consumption of security protocols, In: *Proceedings of the 2003 International Symposium on Low Power Electronics and Design*, Seoul, Korea, 2003, pp. 30–35.

13. A. Kejariwal, S. Gupta, A. Nicolau, N. Dutt, and R. Gupta, Energy efficient watermarking on mobile devices using proxy-based partitioning. *IEEE Trans. Very Large Scale Integrat. Syst.*, 14(6): 625–636, 2006.

14. T. K. Tan, A. Raghunathan, and N. K. Jha. A simulation framework for energy-consumption analysis of OS-driven embedded applications. *IEEE Trans. Comput. Aided Design Integrat. Circuits Syst.*, 22(9): 1284–1294, 2003.

15. National Instruments Corp. http://www.ni.com.

Part VII

Intelligent Network

Chapter 24

Efficient Data-Centric Storage Mechanisms in Wireless Sensor Networks

Chih-Yung Chang,[1] Gwo-Jong Yu,[2] Kuei-Ping Shih,[1] and Sheng-Shih Wang[3]

[1] *Department of Computer Science and Information Engineering, Tamkang University*
[2] *Department of Computer Information Science, Aletheia University*
[3] *Department of Information Network Technology, Chihlee Institute of Technology*

24.1 Introduction	549
24.2 Basic Data-Centric Mechanisms	551
24.3 Data-Replica Mechanisms	552
24.4 Load-Balanced Mechanisms	560
24.5 Hierarchical Management Mechanisms	563
24.6 Future Directions and Open Issues	568
24.7 Summary	570
References	571

24.1 INTRODUCTION

Wireless sensor networks (WSNs) have a wide range of potential applications, including environment monitoring, military, smart house, and remote medical system. A WSN comprises a sink node and an extremely large of sensor node that communicates with each other in order to perform a broader sensing task. Sensor node is a tiny device with capabilities of sensing, data processing and storing, and communication

Mobile Intelligence. Edited by Laurence T. Yang, Agustinus Borgy Waluyo, Jianhua Ma, Ling Tan, and Bala Srinivasan
Copyright © 2010 John Wiley & Sons, Inc.

but is constraint in energy. The sink node is a control center that typically initiates a request demand for collecting interested information. Linked by a wireless medium, the sensor nodes perform distributed sensing tasks and store particular sensing information for query. The WSN can be virtually treated as a database system that provides sensing and data-storing functions for all active sensor nodes and provides query function for the sink nodes. One critical problem in sensor networks is how to make the effective use of the vast amount of data, providing sink and sensor nodes with efficient data retrieving and storing, respectively. Previous solutions to this problem can be classified into three categories: *local storage* (LS), *external storage* (ES), and *data-centric storage* (DCS).

In *local storage* mechanism, the data are stored in sensor nodes local memory when the event is detected. Since the sink node does not know which sensor nodes store the interested data, it typically executes the blind flooding over the whole WSN for sending a query packet that defines the criteria of its interests. The *external storage* proposes an alternative mechanism. Once a sensor node detects events, the data are stored at the external sink. Although there is no cost for sink query, it may waste much energy in transmitting to the sink node the data that the sink node is not interested. In the *data-centric storage* mechanism, there are some data-centric nodes selected from the WSN that are responsible for handling the data storing and retrieving. When an event is detected by a sensor node, the data are stored by name at the corresponding data-centric node. Because all sensor nodes and sink nodes are aware of the information of data-centric nodes, they do not use the blind flooding for sending data or queries to the data-centric nodes. Therefore, data-centric architecture can certainly conserve the energy for data storing and retrieving in the WSN.

In recent year, a number of data-centric mechanisms have been proposed from different aspects. According to their goals and designing concepts, these mechanisms can be categorized into four classes, namely the *basic*, *data-replica*, *load-balanced*, and *hierarchical management* strategies. Mechanisms that fall in the *basic data-centric* category usually apply a hashing function that maps an event type and the time of event occurrence to a physical location of the monitoring region. Then the GPSR[1] routing protocol is applied for each sensor to store its readings at the node that is closest to the mapped location. When the sink intends to query a specific event type, it applies the same hashing function to derive a location and then applies GPSR to deliver the query packet to the node closest to the location.

Some other mechanisms that fall in the *data-replica* category intend to store the same event information at different data-centric nodes. Since the sensor node is battery powered, it may be failure due to energy exhaustion. The fault-tolerant capability for storing the event information increased with the number of data-centric nodes that store the same event. Another purpose of maintaining same event information at several data-centric nodes is to speed up the sink's query and reduce its cost. With the *data-replica* mechanisms, the sink can construct an efficient route to the closest data-centric node and, therefore, reduce the cost for a query.

In addition to the aspects of fault-tolerance and efficient query, a number of mechanisms that fall in the *load-balanced* category aim to distribute the workload of data storage at different data-centric nodes. Different from the mechanisms belonging to the *data-replica* category, mechanisms fallen in the *load-balanced* category store the same event information at exactly one data-centric node. These mechanisms treat the storages of all sensor nodes as a huge database storage and aim to develop strategies to map the possible event types and the time of event occurrence onto the locations of all sensor nodes. Therefore, the workloads for storing the event information and handling the sink's query can be balanced. Since the lifetime of a WSN highly depends on the lifetime of the first failure node, the *load-balanced* aspect is even important in the WSN.

Another category, namely *hierarchical management*, aims to construct a hierarchical management structure among the data-centric nodes so that the sink's query can be more efficient. The data-centric nodes can be partitioned into several zones according to the geographic information or they can be virtually partitioned into several clusters. In a zone or a cluster, a node will be elected as the manager to provide an efficient query from the sink node to the data-centric node.

This chapter classifies the existing data-centric mechanisms into the abovementioned four categories, reviews the basic concepts and designing principles of the existing mechanisms, and points out some possible future directions and open issues related to the data-centric storage problem.

24.2 BASIC DATA-CENTRIC MECHANISMS

The *basic data-centric* mechanism simply applies hashing function that maps the event type to a location of the monitoring region. The geographic hash table (GHT)[2] mechanism is the most famous approach that stores the detected events at the sensor closest to the mapped location. The GHT operates on the network environment that consists of a large number of connected sensors and a small number of static sinks that connect to the outside world. To achieve this, the GPSR routing protocol is applied to find a route from the sensor that detects the event to the target location for storing or retrieving data. By using a uniform geographical hashing function, the GHT spreads storage and communication loads over the entire network.

Two operations, Put() and Get(), are supported by GHT. The Put(*key, value*) operation stores value at the sensor closest to the location that is generated by hashing on the key. On the other hand, the Get(key) operation retrieves all values associated with the key. In general, a key corresponds to a special event. For example, as shown in Figure 24.1, when the sensor node detects a lion, it stores the event using Put("lion," data) operation. Then the sensor node hashes the key, "lion," to the corresponding locations, uses GPSR to reach the storage location, and puts the sensing data to the target sensor node. Since it is possible that no sensor node exists in the storage location, a node that is geographically closest to the storage location can be chosen to store the data. In executing the GPSR routing protocol, if there exists no neighbor of the sensor node closer to the destination coordinates of the packet, the packet will switch to the perimeter mode. In the perimeter mode, the packet traverses the entire perimeter that encloses the destination location until it arrives at the sensor node that initiates the perimeter mode. The data packet can be stored in the data-centric node

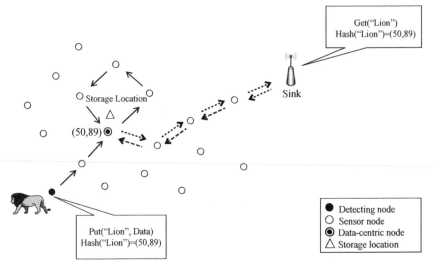

Figure 24.1   An example that illustrates the Put() and Get() operations of the GHT mechanism.

that is geographically closest to the storage location. When the sink node intends to retrieve the data related to the event "lion," it uses Get("lion") operation for finding the storage location and obtaining the data. On the basis of the same hashing function, the key "lion" will be hashed to the same location as Put("lion", data). When the query packet is forwarded to the data-centric node by applying GPSR, the data-centric node returns the data to the sink.

Although GHT develops a simple and efficient data-centric storage architecture, however, using static hash table to determine the location of data-centric node might raise communication overhead that highly relies on the locations of sink nodes and the frequencies of data delivery, especially in a multisink environment. Moreover, if the data of specific event are stored in a fixed data-centric node for a long time, sensors nodes that are nearby the data-centric node likely exhaust their energy due to the frequently data forwarding, resulting in unbalanced power consumption among WSN.

24.3   DATA-REPLICA MECHANISMS

The *basic data-centric* mechanism aims to develop data-centric storage architecture and reduce the query overheads. However, the data stored in exactly one data-centric node would cause the failure access when the data-centric node is failure. Another consideration is the query efficiency. In the *data-replica* mechanisms, some backup nodes surrounding the data-centric node store the same data. In this way, the sink node may access one of the backup nodes and thus has more opportunities to improve the query efficiency. To achieve this, the *data-replica* mechanisms provide some architectures for storing and accessing the backup data, and hence increases the fault-tolerance capability. Most of the existing *data-replica* mechanisms backup the same data based on a particular topology, making ease of the data accessing. According to their topologies, these mechanisms can be further classified into ring-based, tree-based, and zone-based classes.

24.3.1 Ring-Based Mechanisms

The basic idea of the *ring-based* approach is to backup the same data on those sensor nodes that are located on the ring. Zhang et al. [3] proposed a *ring-based* mechanism to implement the *data-replica* concept. The application scenario is that the sensor network is used for detecting targets within a sensing field. Each sensor detects nearby target and periodically generates reports. When the operator intends to know the status of target, he sends a query packet to the network for the purpose of target tracking. The entire network is initially partitioned into a number of grids. The operations within each grid are coordinated by a leader node. Through leader nodes, the sink node can collect information of target object. To avoid unnecessary message flooding during query operations, a ring-based index mechanism is designed in this chapter.

When a sensor detects a target, it calculates a relative coordinate of target using geographical hashing function and constructs a ring containing forwarding and index nodes surrounding the relative coordinate of the index center. After that, the detecting node forwards the target information to those nodes on the ring. All index nodes on the ring will record the target information while forwarding nodes on the ring just forward the received target information. When the sink node intends to query the target location, it uses the same hashing function to calculate the relative coordinate of index center and then forwards a query packet to the index center. The routing path from the sink node to the index center will intersect with the ring surrounding the index center. The ring member that receives query packet will forward the packet to closest index node on the ring. The index node then replies the target information to the sink node. Figure 24.2 illustrates a ring structure of the proposed approach. Through geographical hashing function, the sink node and detecting node can derive the same relative coordinate of index center and the corresponding storage ring. When the event or query packets reach this ring, they will be forwarded along this ring and the corresponding information will be stored or retrieved, respectively.

Figure 24.2 shows an example for illustrating the mechanism proposed in Ref. [3]. When the sensor node *a* detects an event, it calculates the index center for the target "lion" using a predefined hash function that maps the target to a location of index center. The node *a* further constructs a ring containing forwarding and index nodes around the index center. After that, node *a* sends the target information to those nodes on the constructed ring. Since nodes *b*, *c*, *d*, and *e* play the role of index nodes, they record the location of the detecting node *a*. When the sink intends to query the target information, it sends the query request to the forwarding nodes or the index nodes on the ring. Upon receiving the query request message, the forwarding node *f* further forwards query request message to the index node *d*. Afterward, the index node *d* asks the detecting node *a* to report the target information to the sink node.

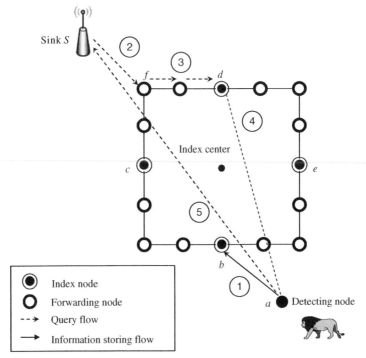

Figure 24.2 The *ring-based* mechanism proposed in [3] stores the same event information at several index nodes (or data-centric nodes) that are located on the ring to improve the query efficiency.

The main advantages of the *ring-based* storage structure are load balance and fault tolerance.

Storing the copies of the sensing information on several nodes can speed up the query operation. Xing et al. Li proposed a *data-replica* mechanism, called LCS[4], which stores the same sensing information at several data-centric nodes. The number of data-centric nodes that store the specific information increases with the importance of this information. The LCS assumes that each sensor node is aware of its own location and there exists a reliable broadcasting mechanism that can successfully deliver the packet to all nodes in the WSN. Consider a source sensor s with location (x_s, y_s). When the sensor s detects an event, it assigns the event with a value of σ, which represents the importance of that event. Then the sensor stores and broadcasts the event information with the value of σ over the WSN. Upon receiving the information, each receiver determines whether or not it should store the information according to the distance between the source sensor and itself and the value of σ. Let the receiver be r and its location be (x_r, y_r). In case the receiver r satisfies both of following equations, it stores and forwards the packet.

$$x_r = (x_s + 2^1 - 1, x_s + 2^2 - 1, \ldots, x_s + 2^\sigma - 1) \text{ and}$$
$$y_r = (y_s + 2^1 - 1, y_s + 2^2 - 1, \ldots, y_s + 2^\sigma - 1)$$

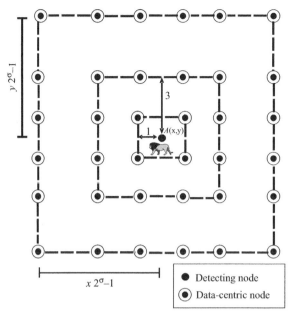

Figure 24.3   The LCS data-replica mechanism uses ring topology to store the same event information on multiple rings that surround the event detected sensor.

There is a TTL field in the broadcasted packet. After storing the event information, the receiver r decreases the TTL value by 1 and then broadcasts the packet to its neighbors. On the other hand, if the receiver only satisfies one of the two equations, it simply stores the information but does not broadcast the packet to its neighbors. In case any of the above two equations is not satisfied, the receiver simply discards the packet and does nothing.

Figure 24.3 shows the data-replica by applying the proposed LCS mechanism. When the source node A detects an event, it stores and broadcasts the packet to its neighbors. The event information will be stored in the solid nodes that are located on the surrounding rings of the node A. For any two nodes whose locations are totally different in x and y coordinates but they detect the same event, it is proved that LCS stores at most 16 copies of this event information at different data-centric nodes. Even though the locations of the two nodes have the same value in x or y coordinates, LCS at most stores $4(2\sigma+1)$ copies at different data-centric nodes. As a result, LCS prevents a vast number of data-centric nodes from storing the same copy of an event information.

24.3.2   Tree-Based Mechanism

The *data-replica* mechanism not only increases capability of fault tolerance but also reduces the query overhead initiated from the sink node. Fang et al. [5] proposed a data-centric mechanism to speed up the sink's query and reduce the control overhead for finding the data-centric node, based on the GLIDER[6] framework. Before reviewing their data-centric mechanism, we firstly introduce how GLIDER works. In GLIDER, the wireless sensor network consists of a sink node, a few landmarks that are aware of their own locations, and a large amount of sensor nodes. The network region is geographically partitioned into many vonoroi cells that are constructed by the landmarks. Each sensor does not necessary know its own location but knows the hop counts from itself to both the closest landmark and the landmark's voronoi neighboring landmarks. This can be easily achieved by a controlled flooding from each landmark over the WSN. Let $T(u)$ denote the vonoroi cell of the landmark u. Let N_u be the set of landmarks that are voronoi neighbors of u. Assume that a source sensor $s \in T(u)$ intends to send a data to the destination $d \in T(v)$. The source s will select a neighboring landmark from N_u, say w, such that w is closest to v. Then source s delivers its data packet to the boundaries of $T(u)$ and $T(w)$, according to the maintained hop count information. After that, the sensor in $T(w)$ further forwards the received packet to the boundaries in a similar way. When the packet reaches the boundary of $T(v)$, it is routed to d greedily [6].

After reviewing the GILDER routing mechanism, the following presents the data-centric mechanism proposed in Ref. [5]. We illustrate how a producer stores its data reading on several data-centric nodes. Let $p \in T(u)$ be the producer that generates a sensing data. Then producer p will hash a location $h \in T(v)$ for storing the sensing data. Here we assume that the sensing data sent from producer p will pass through $T_1, T_2, T_3, ..., T_k$, where $T_1 = T(u)$ and $T_k = T(v)$. For each cell T_i, $1 \leq i \leq k$, the sensor in T_i that receives the data packet will randomly select a sensor $a \in T_i$ as a separate point that is responsible for constructing a *finger tree* in this cell, as shown in Figure 24.4. The construction of finger tree is described in below. Firstly, the sensor a replicates the data along the paths from itself toward three vonoroi cells, including T_{i+1} and two common vonoroi neighbors, say n and m, of T_i and T_{i+1}, respectively. The data will be replicated on the three paths until they encounter the boundary of T_i. As a result, the finger tree is constructed as shown in Figure 24.4. Figure 24.5 depicts an

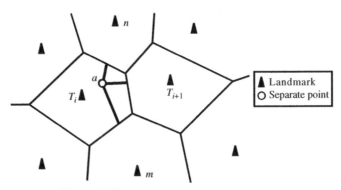

Figure 24.4   The constructed *finger tree* in cell T_i.

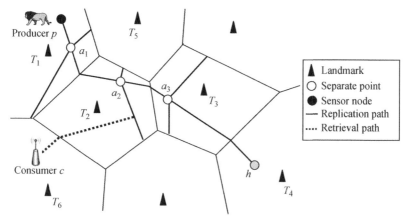

Figure 24.5 An example of the replication and retrieval paths constructed by the *Tree-based* mechanism [5].

example of the constructed finger tree. In this example, the producer p intends to send its sensing data to the sensor closest to location h. The data packet passes through voronoi cells T_1, T_2, T_3, T_4. Initially, the producer p arbitrary selects the sensor node a_1 as the separate point. Then sensor a_1 constructs a finger tree that consists of three paths, from a_1 to the boundaries of T_1 and T_5, T_1 and T_2, as well as T_1 and T_6. Then the packet will be forwarded to cell T_2 and node a_2 will be selected to play the role of separate point. The node a_2 will execute the procedure similar to a_1 and thus construct a finger tree in T_2. Finally, the finger tree is completely constructed and the producer's data are replicated at those sensors that are located on the finger tree.

When a consumer c intends to query the data generated by some producer, it calculates a hashed location h and then sends the query packet to h by applying GILDER routing mechanism. Since the same data are stored on the finger tree, the consumer can retrieve the data from the intersection of replication path and retrieval path, instead of h, as shown in Figure 24.5. The advantage of constructing a finger tree for data replication is that the producer and consumer are distance-sensitive. More specifically, the finger tree is constructed in those cells that are passed by the path from the producer to the hashed destination location. Therefore, a consumer closer to the producer can earlier obtain the data by traveling the shorter path from itself to the intersection of the replication path and retrieval path.

24.3.3 Zone-Based Mechanisms

Some other *data-replica* approach stores the same data at several data-centric nodes based on the zone-based topology. Seada and Helmy proposed the concept of rendezvous regions (RRs) [7], which extends the storage location from a single point coordinates in GHT to a region. This extension resolves the data loss problem caused by node mobility or node failure. Initially, the entire network is partitioned into several zones of equal size. The size of each zone is set as a multiple of communication range. Each sensor node is aware of the zone to which it belongs, but the detail coordinate information is not necessary. A mapping function is used to associate a key to a zone. The sensing data are stored at multiple sensors within the same zone so that the probability of data loss due to node mobility or node failure can be reduced. This fault-tolerant mechanism also avoids unnecessary power consumption caused by large amount of data transfer, which is the main disadvantage of the single storage mechanism.

Each node periodically monitors its location and decides the zone it belongs to. If a sensor node recognizes the change of its zone, it forwards its data to the other nodes in the old zone. Assume an event is detected and data are generated by a sensor node s. The sensor s uses the key corresponding to the event and the mapping function, and a corresponding zone RR_i can be obtained. Applying the existing geographical routing protocol, such as GPSR, the sensing data can be forwarded from the source sensor s to the zone RR_i. Each sensor node receiving the event data needs to check whether it is in RR_i or not. If it is not the case, it directly forwards the packet to the destination RR_i. When the first sensor node within the zone RR_i receives event packet, it acts as a flooder of the zone. Then the flooder broadcasts the event packet to all sensor nodes within the zone. After server nodes received the event packet, it stores the data in its internal memory and reports its status to the flooder. The flooder needs to check the number of returned reports. If the number of backup servers is not enough, it will broadcast a *server addition request* that includes a probability p to all sensor nodes within the zone. Any sensor node that receives the *server addition request* will generate a random number and compared the number with p. In this way, a node can decide whether or not it should become a server. If a node decides to play the server role, it will store the data and report its status to the flooder. The flooder checks the number of server reports, changes probability p, and broadcasts the *server addition request* until sufficient number of server reports. The flooder needs to maintain the locations of the servers. When the sink node intends to query the data, it simply uses the key and mapping function to obtain the corresponding zone RR_j. By applying the geographical routing protocol, a query packet can be delivered to the zone RR_j. The node that receives the query packet and knows the location of server will forward the query packet to the closest server. Otherwise, it simply floods the query packet to all nodes within the zone. Any server receiving the query packet will return data to the sink node.

Figure 24.6 shows an example for illustrating the data storing and query operations proposed in Ref. [7]. In Figure 24.6, sensor node A in zone RR_{13} detects an event and calculates that the storage zone is RR_4 by using a predefined hash function. By applying the geographic routing protocol, node A sends the event information to RR_4. In the zone RR_4, since node B is the first node that receives this event information, it switches its role to the flooder and broadcasts the received event information to all nodes in its zone. Herein, we assume that nodes C and D are data-centric node in zone RR_4. Upon receiving the event information, nodes C and D store the event information. When the sink node intends to query the event information, it uses hash

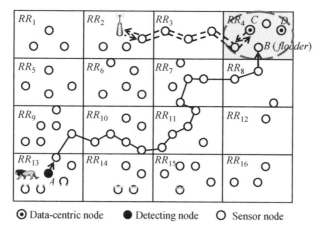

⊙ Data-centric node   ● Detecting node   ○ Sensor node

Figure 24.6 An example of data storing and query operations in Ref. [7].

function to derive the event storage zone RR_4 and sends the query request packet to this zone by applying the existing geographic routing protocol. In the zone RR_4, any node that receives the query request packet should further forward the packet to the closest data-centric node D. Finally, node D reports the event information to the sink node along the reversed request path.

Some other *zone-based data-replica* approach was proposed for the large-scale sensor networks. Sadagopan et al. [8] considered the traditional flooding-based query as a pull-based strategy and proposed a zone-based mechanism that combines the push-based strategy with the pull-based strategy to increase the query efficiency. The following describes the basic concept of the proposed comb-needle model nodes that detect the event will push (or send) the information vertically in the networks and the information will be stored at the traversed nodes. An entry point that is interested in such events will disseminate the query request horizontally, instead of broadcasting the query packet over the network. That is, the query node dynamically constructs a route that resembles a comb and the sensor nodes that detect the event data push the data duplication structure like a needle. As shown in Figure 24.7, node A detects the event and broadcasts the event information in a vertical direction. The mobile sink that intends to query the event information will send the query request in a horizontal direction. Simulation results reveal that the comb-needle strategy is more efficient than both pure push and pure pull strategies in most cases.

The existing *data-replica* mechanisms apply backup nodes to improve the fault-tolerance capability and query efficiency. However, in case that some events are occurred more frequent than the others, the loads for storing the event information and handling the query requests in data-centric nodes are not balanced. The next subsection reviews some *load-balanced* mechanisms that are proposed for balancing the loads among data-centric nodes.

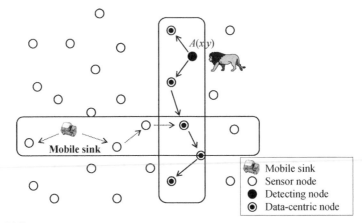

Figure 24.7   An example of combining push and pull strategies in Ref. [8]. The node A detects an event and stores the same event at several data-centric nodes along the vertical direction. The sink node sends the query request along the horizontal direction for obtaining the event information.

24.4 LOAD-BALANCED MECHANISMS

The network lifetime is one of the most important issues in a WSN. However, the lifetime of a WSN highly depends on the degree of load-balance among all sensor nodes. A load-balanced WSN also implies that the WSN is energy-balanced so that the lifetime of each sensor is about the same. The main source of workloads in a WSN is the information sensing, storing, retrieving, and routing. A data-centric node that is responsible for storing a sensing information is also responsible for handling the query request for this information and, therefore, consumes more energy than the other nodes. Other than the above-mentioned *basic* and *data-replica* mechanisms, the *load-balanced* mechanisms aim to distribute the workloads of data storing and retrieving to all sensor nodes in a load-balanced manner. The following illustrates some important *data-centric* mechanisms that consider the load-balanced issue in their designs.

In a WSN, an event can be generally described as a tuple of k attribute values, $A_1, A_2, \ldots, A_k$, where each attribute A_i represents a sensor reading, or some value corresponding to a detection. Li et al. developed a *distributed index for multidimensional data* (DIM) [9], which takes the load-balance issue into account. The main idea behind the DIM is to partition the WSN region according to a specific coding method and assign each region with a code. According to the values of the multi-attribute event, the event information can be mapped to a code that determines the zone that stores this event. A query node may also use the multi-attribute values to derive the coded zone, reducing the query overhead, and improving the query efficiency.

The following presents the details of DIM. The monitoring region is divided into 2^i zones. The monitoring region is assumed to be a rectangle shape. In the zone creation process, the bounding rectangle R is initially divided into two zones at level 0 by a vertical line that splits R into two equal pieces. Each of the two subzones can be split into two smaller zones at level 1 by a horizontal line. The dividing operation can be repeatedly performed to partition the whole region into several small zones. If the dividing level i is an odd number, the split line is parallel to the y-axis. The codes 0 and 1 are assigned to the right-side and left-side zones, respectively. On the contrary, if the dividing level is an even number, the split line is parallel to the x-axis and the codes 1 and 0 are assigned to the up-side and down-side zones, respectively. As a result, each partitioned zone will be assigned with an unique code.

A geographic locality-preserving hash is provided for mapping a multi-attribute event to a geographic zone. When a sensor node detects an event, it maps the values of the event's attributes to the code and sends the event information to the zone with this code for storing by applying GPSR routing mechanism. Similarly, the sink node intending to query the event information also hashes the multi-attribute values to a code and applies GPSR routing mechanism to forward the query request to the zone that stores the event. Consequently, the event information can be stored in a load-balanced manner and the event information can be queried efficiently.

Figure 24.8 shows an example of *DIM*. In this example, each event consists of two attributes: temperature and light. Initially, the zone creation process is applied. During the process, whenever a zone is partitioned into two smaller zones, the interval of the corresponding attribute will be divided into two subintervals. The degree of load-balancing is decreased with the zone size. That is, the zone creation process can be executed repeatedly until the number of zones matches the requirement of load-balancing. The right-down zone is assigned with the code of 10 because it lies

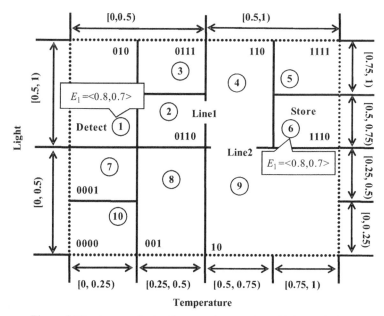

Figure 24.8   An example that illustrates the operations of DIM mechanism.

on the right side of the Line 1 (first bit=1) and the down side of the Line 2 (second bit=2). Moreover, the zones labeled by 5 and 6 have been partitioned by two vertical and two horizontal lines. Therefore, zones labeled by 5 and 6 have been assigned with the codes 1111 and 1110, respectively.

In each zone, there is exactly one sensor node responsible for storing the event information with specific attributes values. According to the interval in which each attribute value falls, the event information is stored in some zone according to the multidimensional range values of the event's attributes. For example, let A_1 and A_2 denote the values of temperature and light of an event E, respectively. In the zone labeled with 1, a sensor node detects an event E with values $(A_1, A_2)=(0.8, 0.7)$. Since $A_1=0.8>0.5$, the first bit of the event's code is 1. Similarly, the value $A_2=0.7$ >0.5, the second bit of this event's code is 1. Continuously, the value $A_1=0.8$ falls in interval (0.75, 1), hence the third bit of the event's code is 1. Finally, the value $A_2=0.7$ falls in interval (0.5, 0.75), hence the fourth bit of the event's code is 0. As a result, the event $E=(0.8, 0.7)$ will be stored at the sensor in zone 6, which is assigned with the same code 1110 with this event.

In literature, Liu et al. [10] also aim to improve the network lifetime of GHT[2]. The GHT using static hashing function may lead to the problem that a large number of sensors access the same geographical location. In this circumstance, network resources closer to the geographical location are depleted earlier. This article proposed a dynamic GHT mechanism that makes the following improvements. First, a temporal-based GHT is adopted. As time changes, the storage location also changes accordingly for the same event type. Second, a location selection scheme that searches possible storage locations based on access frequencies is proposed to migrate the load of storage node and to avoid unbalance power consumption. Figure 24.9 illustrates the basic concept of the proposed approach. In Figure 24.9, the event circle denotes

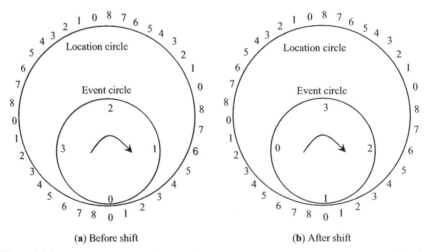

(a) Before shift   (b) After shift

Figure 24.9   The temporal-based GHT use the event and location circles to distribute the workload of data-centric nodes.

various event types and the location circle denotes different storage locations. In each slot, the rotation of event circle depends on the rotation of location circle. In other words, event 0 maps to cell 0 (location 0) in the first slot; event 1 maps to cell 1 (location 1) in the second slot, and so on. This mapping continues until the event circle returns to the original location. At this moment, the event circle shifts one location as shown in Figure 24.9(b). In this circumstance, event 0 maps to cell 8 (location 8) and event 1 maps to cell 0 (location 0). After several slots, an event will be mapped to the same location as the first slot again. In this way, data for the same event will be distributed to each cell evenly, achieving the objective of load balance.

The rule of selecting a sensor node for returning data to the sink node is quite straightforward. When a sensor node detects an event, it calculates the corresponding event node based on the event type and the current slot number. The remaining storage space and remaining energy are then used to select a sensor node within the same cell. The sensor node that has the most resources is responsible to store and return data to the sink node. The sink node also uses the event type and the time slot number to find the location of the sensor that stores the interested event information.

24.5 HIERARCHICAL MANAGEMENT MECHANISMS

To reduce the network traffic and provide an efficient query, some other researchers proposed *hierarchical management* mechanisms that support multilayer storage architecture to efficiently store the sensing information and response the queries. This subsection reviews the zone-based and cluster-based architectures.

24.5.1 Zone-Based Mechanism

Several zone-based mechanisms have been proposed for hierarchical management of data storing and retrieving. The main idea of the zone-based mechanisms is to partition the whole WSN region into several zones and elect a leader in each zone for efficient management of data storage and query requests. Le et al. [11] proposed a zone-based storage architecture where each sensor node needs not to be aware of its own location information. On the basis of the zone-based management, the proposed mechanism provides a hierarchical management of data queries.

The design of the architecture relies on a novel concept of node contribution potential and a distributed clustering algorithm operating on this concept. They assume that the sensors nodes are uniformly deployed over the monitoring region. All sensors have equal transmission range, though they may have different initial energy or storage spaces. The following describes the operations of data storage and data query. A distributed zone division algorithm is proposed to partition the WSN region into several zones and select a *leader* and a *storage sensor* in each zone. During the zone-partitioning process, the leader will construct a tree in that zone and notify all sensors in that zone about the routing information to the storage sensor. Furthermore, routing paths among leaders have been constructed. All sensors in a zone will transmit their sensing information to the storage sensor in that zone. The leader will maintain the members and storage sensor information in its zone. The query packet will be initially sent to all leaders. The leader then forwards the query packet to the storage sensor in that zone. The storage sensor will return the required information if the data stored in its storage match the query's requirement.

The details of the proposed zone division algorithm are described in below. Let *contribution potential* $P(s)$ of a sensor s denote its potential contributing to the network. The potential value can be evaluated according to the degree, the remaining energy, the storage space, or the serving-slot of that node. Each node s will exchange its contribution potential $P(s)$ with its neighbors. A node s_i will select a neighbor s_j as an uplink node if $P(s_i)$ is smaller than $P(s_j)$. Nodes without an uplink node will become zone leaders and broadcast a zone formation message containing the zone *ID* (its *ID*) with *TTL* and the maximum zone radius (R) to its downlink nodes. These nodes decreases *TTL* by 1 and recursively repeat the process until *TTL* becomes 0. Finally, nodes without zone will reinitiate the zone finding process until all nodes belong to a certain zone. As a result, each zone will contain a zone leader nearby the zone's central point. Figure 24.10 depicts that nodes a and b are the zone leaders of sets A and B, since $P(a)$ and $P(b)$ are the largest values in sets A and B, respectively. If the value of $P(a)$ is smaller than $P(b)$, sets A and B will be combined into a larger set, say C, and node b will be the zone leader of set C. The sets will be repeatedly combined into a larger one until the radius of the set is larger than the predefined value R. Finally, the node that has largest potential value in that zone will become the zone leader of that zone. Figure 24.11 shows the constructed zones and their corresponding zone leaders.

24.5.2   Cluster-Based Mechanisms

The cluster architectue has been widely used in the management of ad hoc network. Using the cluster architecture may reduce the flooding overhead, since only the cluster headers need to participate the packet forwarding in the ad hoc network. Some other data-centric mechanisms utilize the *cluster-based* architecture for efficient data storing and query. Most of the existing data-centric mechanisms apply geographic routing

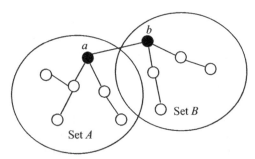

Figure 24.10   An example that illustrates the zone leader election process in Ref. [11]. Nodes a and b are the leaders in sets A and B, respectively, as they have the largest potential values in their sets.

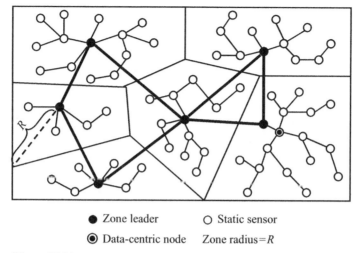

● Zone leader   ○ Static sensor

◉ Data-centric node   Zone radius$=R$

Figure 24.11 The constructed zones and their zone leaders and storage nodes.

to efficiently forward the sensing information and query request to the data-centric nodes. However, the accurate location information of each sensor node is difficult to be known. Newsome and Song [12] proposed a cluster-based data-centric mechanism, called GEM, which efficiently copes with the information store and query problems without location information. GEM initially constructs a virtual tree topology and then maps the tree onto the physical wireless sensor network. In the tree construction phase, the sink node floods a packet over the WSN to identify the levels from each sensor node to the sink node. A field, named *level*, of the packet is used to count the level from the sink to the receiver and is increased by 1 whenever the packet is forwarded to the next node. A receiver will treat the sender that has minimal level as its parent and thus the virtual tree can be constructed when the packet is flooded over the WSN. After that, each node reports to its parent the number of nodes in its subtrees from leaves to the root.

The next step in GEM is to assign each node of the constructed virtual tree with a virtual angle range. Figure 24.12 depicts the constructed virtual tree and the virtual angle for each node. Initially, the root is assigned with an angle range. In this example, the root's angle range is [0,100]. Then the root partitions its angle range into several intervals, so that the number of intervals equals to the number of its children. Then the root assigns each interval to each of its children. The partitioning of the angle range highly depends on the subtree size of its children. For example, the partitioning of root A's virtual angle range depends on the subtree size of nodes B, C, and D. Since subtree sizes of nodes B, C, and D are 3, 4, and 3, respectively, node A's angle range [0, 100] can be partitioned into three subsets [0, 30], [31, 70], and [71, 100] as well. Similarly, nodes B, C, and D will partition their angle range into several intervals and assign each interval to each of their children, until the leaf nodes are assigned with angle ranges.

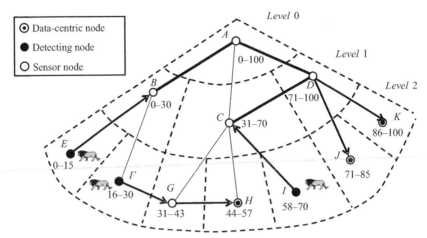

Figure 24.12   The virtual tree construction and angle assignment in GEM.

When an event information is generated, a hashing function will be applied to hash the event value to an angle range. The data-centric node that is responsible for storing this event is the one whose angle range covers the hashed angle range. Since there is no location information, a routing mechanism based on the angle range is further developed for sending the event information to data-centric node for data storing purpose or for relaying the query request to the data-centric node for information query purpose. For instance, in Figure 24.12, an event information is generated at node F. We assume that this event information is mapped to the angle range [60, 70]. As a result, the event information will be automatically sent to node I according to the shortest route (F, G, H, I). Since the angle range labeled on node I is [58, 70], which covers the subrange [60, 70], the event information will be stored at node I.

Xu et al. also proposed a *hierarchical management* mechanism, called EASE, based on the cluster-based topology. They proposed an energy-efficient data storage policy that efficiently reports target location, update target location with low cost, and increases network lifetime. The proposed mechanism first chooses local storage node to store the detected target information. When the local storage node receives new target location, it can update data-centric node according to the requirement of precision range. Accurate location information is stored in the local storage node, while rough location information is stored in the data-centric node. When sink node intends to query the target location, it sends a query request to the data-centric node. If a high-precision query is required, the data-centric node further retrieves the detail information from the local storage nodes.

As shown in Figure 24.13, when node A detects a mobile target, it switches its roles to the detected node and the local storage node. Node A also sends an update message to the data-centric node for updating the corresponding information. If the target moves closer to another node B and is detected by node B, node B will switch its role to the detected node and send the detected event information to the local storage

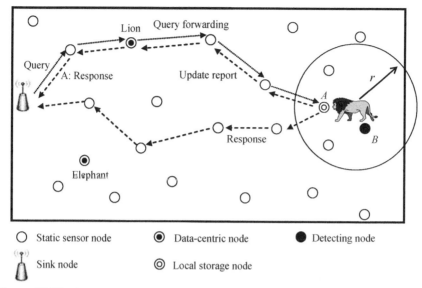

Figure 24.13   An example that illustrates the hierarchical management strategy in the EASE mechanism.

node A. Upon receiving the new message from node B, node A checks whether or not the moving range is within the predefined precision range. If the moving range is larger than the tolerable precision range, node A will send an update message to the data-centric node to make sure that the precision of moving target is within the precision range. When the sink node intends to query the information about this target, it sends a query message to the data-centric node. If the precision in data-centric node is not sufficient, the data-centric node further sends a *location update message* to the local storage node for obtaining the detail information of the target. The *location update message* includes the location of sink node so that the local storage node can update the location information at both the sink node and the data-centric node. If only rough information is queried, the data-centric node directly provides the required information to the sink node. On the contrary, if the detail information is queried, the data-centric node updates its information from the local storage node on demand. In this way, the number of unnecessary update message can be improved, significantly reducing the power and bandwidth consumptions.

In literature, Desnoyers et al. [14] proposed a cluster-based management that aims to exploit the multitier nature of emerging sensor networks. They proposed a Two-tier Sensor storage ARchitecture, called TSAR. In the proposed architecture, the WSN comprises several *inxproxy nodes* and a large amount of sensor nodes. Compared with the sensor nodes, the proxy nodes typically have better computing and communication powers, larger storage space, and more remaining energy. Each proxy node manages tens to hundreds of sensor nodes. At the sensor tier, each sensor stores its sensing information independently and reports to the proxy node a summary

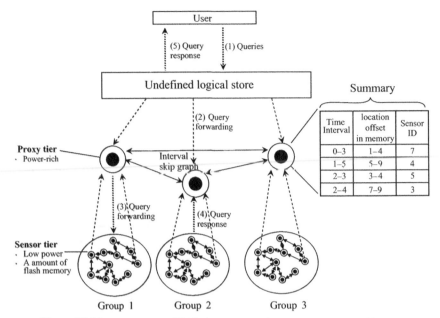

Figure 24.14   The two-layer TSAR data storage and query management architecture.

about its information. At the proxy tier, TSAR employs an index structure for efficient query.

Figure 24.14 shows the architecture of TSAR. In Figure 24.14, the sensor nodes that belong to the same group are managed by a proxy node. Each sensor node stores its sensing information in its own storage and reports to the proxy node the summary of this information, including its ID, the time interval, and the location offset in its memory. Upon receiving the summary information, the proxy node records the summary information as a summary table. All proxy nodes periodically exchange their information to create an interval skip graph. When a user intends to query an event information, it firstly sends the query request to the proxy nodes. According to the query request, the proxy node checks its summary table and further forwards the query request to the managed sensor if the abstract information recorded in the summary table matches the request. Upon receiving the query request, the sensor reports the detailed information about the queried event to the user. The TSAR proposes a hierarchical management for data storing and search and significantly reduces the search time.

24.6   FUTURE DIRECTIONS AND OPEN ISSUES

In the previous subsections, several data-centric storage articles have been introduced. Although the above-mentioned articles devote themselves to develop data-centric storage architecture in different environments, most of them did not consider the multiple sinks environment and the factor of query frequencies. Using static hash table to determine the location of data-centric node might raise communication overhead, which highly relies on the locations of the sink nodes and the frequencies of data delivery, especially in a multisink environment. Moreover, if the information of specific event is stored in a fixed data-centric node for a long time, sensor nodes that are nearby the data-centric node likely exhaust their energy due to frequent data forwarding, resulting in unbalanced power consumption.

One of the open issues is to develop path sharing for those sink nodes that have the similar or even same queries. The developing of path-sharing mechanism should consider the arriving of a new query, initiated by a new sink node, which has the same query to the existing queries except the query frequency. In addition, the optimal sharing paths from the same data-centric node to multiple sink nodes also should be changed if there are some sink nodes completing their queries. The changes of queries would raise the problems that the data-centric nodes might be changed and hence the optimal sharing paths for the existing queries should be changed accordingly. The challenge of creating optimal sharing paths for multiple sinks is how to dynamically determine the location of data-centric node according to the locations and requested frequencies of the multiple sink nodes so that the redundant packet transmission and the number of forwarding nodes can be reduced.

The selection of data-centric node without considering the locations and requested data collection frequencies of sinks nodes might raise the problem of inefficient communication. Figure 24.15 is an example that depicts this situation. There are three different sinks, X, Y, and Z, intending to query the information of event A with required reply frequencies, 1/50s, 1/10s, and 1/20s, respectively. Figure 24.15(a) depicts the communication path using the basic data-centric algorithm. The sink X requires the largest reply frequency, but its location is farthest away from the data-centric node of event A. Therefore, large communication overhead will be spent for transmitting interested information from the data-centric node to sink X. According to the locations and requested reply frequencies of sink nodes X, Y, and Z, a better mechanism might dynamically change the location of the data-centric node as shown in Figure 24.15(b). In Figure 24.15(b), the location of new data-centric node is closer to the sink X, since the request of X has a high frequency. In comparison, Figure 24.15(a) and 24.15(b), respectively, requires 6 and 3 forwarding packets to reply each data from the data-centric node to the sink X. Although sinks Y and Z increase the forwarding overhead for replying data in Figure 24.15(b), however, the total cost of data reply decreases because that their requests have low frequencies. The change of data-centric node will reduce the total forwarding overhead, especially for application of long time data collection, conserving much energy for periodic data collecting.

Additionally, applying the GPSR to construct separate routes from the data-centric node to each sink node will result a large amount of forwarding nodes participating the routes. Figure 24.15 depicts the paths individually constructed from data-centric node to each sink node. A new path sharing routing algorithm is required to construct a shared path from the data-centric node to multiple sink nodes. As shown in Figure 24.15(b), a shared path is constructed for sinks Y and Z according to their

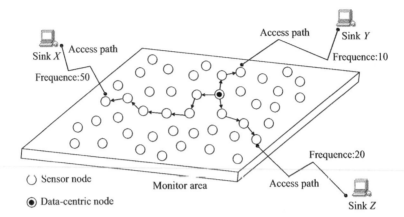

(a) The static location of data-centric node and the constructed routes by applying *basic data-centric* mechanism

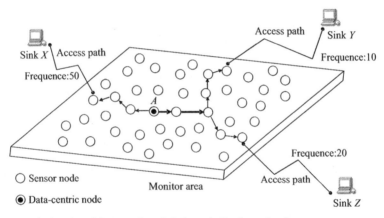

(b) The location of data-centric node is dynamically changed and a shared path is constructed from new data-centric node to multiple sink nodes

Figure 24.15   An example to illustrate the obtained benefits by changing the location of data-centric, node and establishing a shared route.

locations. Compared with Figure 24.15(b), Figure 24.15(a) depicts that several individual paths result in duplicate transmission and consumes energy of forwarding nodes and the network bandwidth.

24.7 SUMMARY

This chapter classifies the existing data-centric mechanisms into *basic, data-replica, load-balanced,* and *hierarchical management* categories, according to their goals and designing concepts. Then the chapter reviews the basic concept and the strategy of each mechanism. Finally, we point out some possible future directions and open issues related to the data-centric storage problem. In the *basic data-centric* category, the GHT mechanism that simply applies hashing function to derive the location of data-centric node is reviewed. To improve the fault-tolerance capability and the query efficiency, some mechanisms that fall in the *data-replica* category are reviewed. These mechanisms are further classified into ring-based, tree-based, and zone-based classes, according to the applied topologies. Since the network lifetime is one of the most important concerns in designing any protocol for WSN, some other data-centric mechanisms that fall in the *load-balanced* category are reviewed. Finally, the chapter reviews four data-centric mechanisms that aim to reduce the network traffic and provide efficient queries. These mechanisms belonging to the *hierarchical management* category are further classified into *zone-based* and *cluster-based* classes.

Although there are a number of existing data-centric mechanisms proposed in the literature, however, none of them considers the network environment with multiple sinks, which might initiate the same or similar query requests to the same data-centric node. This chapter finally points out the open issues and the possible challenges to the data-centric storage problem in the multisink environment. In the future, more efforts are required to develop dynamic data-centric storage mechanisms and path-sharing algorithms for improving the query efficiency in a complicated wireless sensor network environment.

REFERENCES

1. B. Karp and H. T. Kung. GPSR: greedy perimeter stateless routing for wireless networks. In: *International Conference on Mobile Computing and Networking (MobiCom)*, 2000, pp. 243–254.
2. S. Ratnasamy, B. Karp, S. Shenker, D. Estrin, R. Gvoindan, L. Yin, and F. Yu. Data-centric storage in sensornets with GHT, a geographic hash table. *Mobile Networks Appli.*, 8(4):427–442, 2003.
3. W. S. Zhang, G. H. Cao, and T. L. Porta. Data dissemination with ring-based index for wireless sensor networks. In: *IEEE International Conference on Network Protocols (ICNP)*, 2003, pp. 305–314.
4. K. Xing, X. Z. Cheng, and J. Li. Location-centric storage for sensor networks. In: *IEEE International Conference on Mobile Adhoc and Sensor Systems Conference*, 2005.
5. Q. Fang, J. Gao, and L. J. Guibas. Landmark-based information storage and retrieval in sensor networks. In: *Proceeding of the 25th Conference of the IEEE Communication Society (INFOCOM)*, April 2006.
6. Q. Fang, J. Gao, L. Guibas, V. de Silva, and L. Zhang. GLIDER: gradient landmark-based distributed routing for sensor networks. In: *Proceeding of the 24th Conference of the IEEE Communication Society (INFOCOM)*, March 2005.
7. K. Seada and A. Helmy. Rendezvous regions: a scalable architecture for service location and data-centric storage in large-scale wireless networks. In: *IEEE/ACM IPDPS International Workshop on Algorithms for Wireless, Mobile, Ad Hoc and Sensor Networks (WMAN)*, April 2004.
8. N. Sadagopan, B. Krishnamachari, and A. Helmy. Active query forwarding in sensor networks. *Elsevier J. Ad Hoc Networks*, 3:91–113, 2003.
9. X. Li, Y. J. Kim, R. Govindan, and W. Hong. Multi-dimensional range queries in sensor networks. In: *Proceeding of the 1st International Conference on Embedded Networked Sensor Systems*. 2003, pp. 509–517.

10. T. N. Le, W. Yu, X. Bai, and D. Xuan. A dynamic geographic hash table for data-centric storage in sensor networks. In: *Wireless Communications and Networking Conference (WCNC)*, 2006, pp. 2168–2174.

11. T. N. Le, D. Xuan, and W. Yu. An adaptive zone-based storage architecture for wireless sensor networks. In: *IEEE Global Telecommunications Conference (GLOBECOM)*, 2005, pp. 2782–2786.

12. J. Newsome and D. Song. Gem: graph embedding for routing and data-centric storage in sensor networks without geographic information, *ACM SenSys*, 2003, pp. 76–88.

13. J. L. Xu, X. Tang, and W. C. Lee. EASE: an energy-efficient in-network storage scheme for object tracking in sensor networks. In: *IEEE Communications Society Conference on Sensor and Ad Hoc Communications and Networks(SECON)*, 2005, pp. 396–405.

14. P. Desnoyers, D. Ganesan, and P. Shenoy. Tsar: a two tier sensor storage architecture using interval skip graphs. *ACM SenSys*, 2005, pp. 39–50.

Chapter 25

Tracking in Wireless Sensor Networks

Sheng-Shih Wang,[1] Kuei-Ping Shih,[2] Chih-Yung Chang,[2]
and Gwo-Jong Yu[3]

[1] *Department of Information Management, Minghsin University of Science
and Technology*
[2] *Department of Computer Science and Information Engineering, Tamkang University*
[3] *Department of Computer and Information Science, Aletheia University*

25.1 Introduction   573
25.2 Tracking Scenarios in WSNs   574
25.3 Tracking Techniques   575
25.4 Tracking in WHSNs   582
25.5 Summary   592
References   593

25.1 INTRODUCTION

The wireless sensor network (WSN) is an application-specific networking framework, which is widely used in battlefields, disaster areas, and nuclear power plants for environmental monitoring, target detection, event tracking, security surveillance, and so on. Among these applications, tracking is one of the most significant tasks especially for the critical scenario, in which a user intends to realize the event in the region of interest. In general, a sensor in WSNs needs to continuously sense the attributes of the event. Once detecting the event, the sensor conveys the sensor data either toward the sink or to the data-centric sensors for external storage. In WSNs, tracking is to achieve

Mobile Intelligence. Edited by Laurence T. Yang, Agustinus Borgy Waluyo, Jianhua Ma,
Ling Tan, and Bala Srinivasan
Copyright © 2010 John Wiley & Sons, Inc.

two goals: target tracking and event boundary determination. The former focuses on identifying one or multiple static or mobile targets, whereas the latter concentrates on recognizing the edge of the event of interest by using the sensors near the event boundary.

The majority of previous work in terms of tracking not only considers a wireless homogeneous sensor network, in which only one type of sensors equipped with the same sensing units are deployed but also focuses on the event formed by only one attribute. If an event is composed of multiple attributes, the existing approaches are unsuitable for the event tracking because any one of the attributes cannot be perceived by only one type of sensor. Thus, different types of sensors with various sensing units are necessary for such application. Here, a network formed by various types of sensors is called the wireless heterogeneous sensor network, termed WHSN.

In the chapter, we will start by giving an introduction to tracking task in WSNs, followed by numerous approaches to target tracking and boundary determination. The progress of the advance technology enables a sensor to equip with multiple sensing units; thus, the sensor is able to be widely used in WHSN to fulfill the tracking task. Then, we will look at the challenges to be tackled in designing the tracking protocol for WHSNs, followed by a novel collaborative event-tracking scheme to WHSNs. Finally, we will present the open issues in terms of tracking in WSNs.

25.2 TRACKING SCENARIOS IN WSNs

In general, two categories, target tracking and event boundary determination are mainly concerned for tracking tasks in WSNs. Target tracking aims at identifying the location of the target at any moment. For example, in Figure 25.1, a lion existing

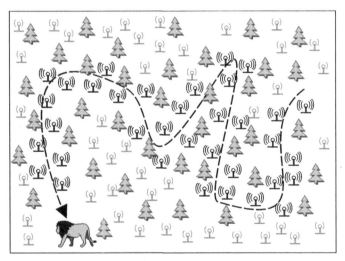

$\longrightarrow$ lion's route $((\underset{\varrho}{\varrho}))$ active sensor $(\underset{\varrho}{\varrho})$ inactive sensor

Figure 25.1 Illustration of target (lion) tracking.

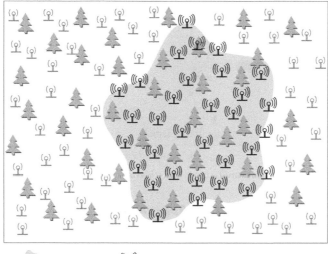

event region ((ρ)) active sensor (ρ) inactive sensor

Figure 25.2   Illustration of event (fire) determination.

in the national park is viewed as a mobile target. The sensors in the vicinity of the route of the lion are able to detect the lion, so that these sensors are regarded as the active sensors to be responsible for lion tracking.

Alternatively, the main goal of event determination is to enable the sensors within the event region to track the event. Figure 25.2 illustrates a national park, in which the scene of a fire is represented by the shaded region. Here, the fire can be viewed as an event. Obviously, once detecting the fire, the sensors within the event region should become active sensors for fire tracking.

25.3 TRACKING TECHNIQUES

Recently, numerous techniques to tracking in WSNs have been paying much attention to two goals: target tracking [1, 2, 7–9, 12, 16, 18, 19, 21–23] and event boundary determination [3, 4, 10, 11, 13, 15]. The approaches to target tracking focus on identifying the location of the static or mobile target, such as a car in the highway, a tank in the battlefield, or a wild animal in the national park. Alternatively, the approaches to event boundary determination concentrate on determining the edge of the event of interest by using the sensors around the exact boundary of such event.

25.3.1 Target Tracking

Overall, the majority of the existing approaches to target tracking in WSNs are classified into the *tree construction*, *clustering*, and *stochastic* techniques. Here, we present the representative protocols corresponding to the individual category.

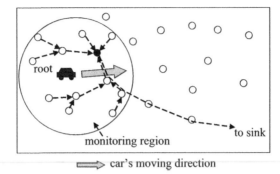

car's moving direction

Figure 25.3 Illustration of a convoy tree, whose root is a car (adapted from [21]).

25.3.1.1 Tree Construction Approaches

Zhang and Cao [21] propose a tree-based collaboration (DCTC) framework to target tracking in WSNs. Basically, DCTC constructs a tree, called the *convoy tree* to form an efficient topology for moving target tracking. Two significant properties are involved in such convoy tree. The first property is that a convoy tree is composed of the sensors around the target and the second is that a convoy tree is rooted at the most closest sensors to the target. Figure 25.3 illustrates an example of a convoy tree. According to the aforementioned properties, the convoy tree is able to efficiently and timely reflect the movement of the target. That is, when the target moves, a convoy tree will be dynamically configured by adding or pruning some sensors. For example, once the car moves, the convoy tree is changed and a new convoy tree structure shown in Figure 25.4 will form. In DCTC, communication overhead and energy consumption are apparently reduced because only the sensors near the target take charge of aggregating the sensor data and invoking the data transmission.

On the basis of the DCTC framework, Zhang and Cao [22] further mention the optimization of a tree reconfiguration. The problem of optimizing tree reconfiguration for tracking the moving target is formulated as finding a min-cost convoy tree sequence. Basically, two steps are involved when a convoy tree is reconfigured: (1) root replacement and (2) reconfiguration of the remaining part of the tree.

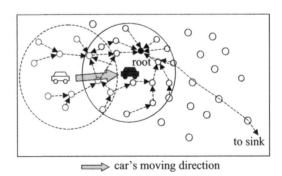

car's moving direction

Figure 25.4 A new convoy tree forms when the car moves (adapted from [21]).

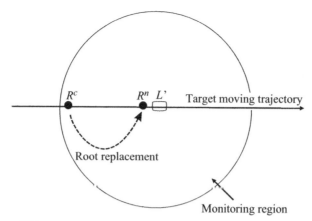

Figure 25.5 Illustration of root replacement of a convoy tree (adapted from [22]).

- Root replacement: The main idea of root replacement is to use the movement-prediction techniques such as in Ref. [20] to predict the new location of the target. Figure 25.5 shows an example of the root replacement rule. Suppose the current root of the dynamic convoy tree is denoted as R^c. Let L' be the location of the target at the next time predicted by R^c. If the distance between R^c and L' is larger than a predetermined threshold δ, R^c is replaced by a sensor closest to L', and the sensor closest to L' will be selected as the new root, termed R^n.

- Tree reconfiguration: Two optimized schemes, the optimized complete recon-figuration (OCR), and the optimized interception reconfiguration (OIR), are proposed in reconfiguration of an optimal convoy tree. In the OCR scheme, all sensors in the tree are involved in the course of reconfiguration. However, in the OIR scheme, only a small part of sensors in the tree needs to be reconfigured. Obviously, the OIR scheme results in less overhead than the OCR one.

Kung and Vlah [8] introduce a scalable architecture, called scalable tracking using networked sensors (STUN), whose main goal is to enable a sensor to communicate the location of the detected target to a querying point (e.g., sink). An efficient technique to construct such hierarchical structure is called drain-and-balance (DAB), which is based on the expected properties of the target movement patterns. In the DAB method, when the target moves, only the sensors in the subtree rooted at the common ancestor of such sensors, which detect the departure and arrival of the target have to relay the updating message. Thus, the updating traffic of DAB is significantly reduced. However, the physical structure of the network is unlikely to be reflected by using the DAB tree, so the tree constructed may be unsuitable for some scenarios.

Motivated by the DAB tree structure, Lin et al. [12] propose a novel tree struc-ture for in-network target tracking in WSNs. In Ref. [12], the network topology is regarded as an undirected weighted Voronoi graph $G = (V_G, E_G)$, in which V_G and E_G represent the sensors and the links between the neighboring sensors, respectively. The weight of each link $(u, v) \in E_G$ indicates the sum of the event rates, including the arrival and departure rates from u to v and v to u. The Delaunay triangulation corresponding to the generated weighted Voronoi graph is formed as well.

Like the DAB approach, in Ref. [12], given a graph G, a logical weighted tree termed T is constructed. Each sensor $u \in T$ not only maintains the target currently within its sensing range but also keeps track of the targets currently within the sensing ranges of the sensors in the subtree rooted at the corresponding children of sensor u. Additionally, once the target moves from the sensing range of sensor u to that of sensor v, the arrival and departure event messages will be forwarded to the lowest common ancestor of both sensors u and v.

A centralized algorithm, called deviation-avoidance tree (DAT) is proposed, in which the update cost of a tree is obtained by counting the average number of both the arrival and departure event messages transmitted in the network per unit time. To reduce the update and query cost, in DAT, the following properties are taken into account.

Property 1: No sensor should not deviate from its shortest path to the sink.
Property 2: Each sensor's parent should be one of its neighbors.
Property 3: The link with a higher weight should be near the sink.

Basically, the DAT algorithm prefers to first select the link with the higher weight to be included into the final logical weighted tree (i.e., T) because of less updating cost. In addition, the DAT algorithm also avoids a sensor to deviate from its shortest path to the sink in the physical structure of the network due to less query cost. Overall, under the consideration of the aforementioned properties, the DAT method significantly constructs an efficient tree structure. The complete algorithm please refer to Ref. [12].

25.3.1.2 Clustering Approaches

Yang and Sikdar [19], taking the scalability and robustness into account, exploit a cluster-based architecture to develop a distributed predictive tracking (DPT) algorithm to moving target. The main idea of DPT is to determine the optimal set of sensors, which are selected by the cluster heads (CHs) approaching the target to achieve target tracking. Obviously, by using the clustering technique, energy conservation is guaranteed because the number of sensors required to track the target is efficiently reduced.

In DPT, to obtain the accurate information, at least three sensors are assumed to sense the target jointly at any time. A CH is assumed to be aware of the location information of the sensors in its cluster. Additionally, each sensor has two sensing radii, *normal beam* and *high beam*, corresponding to the normal and hibernation operation modes, respectively. The operation of DPT can be viewed as a sequence of tasks in the order of *sense–predict–communicate–sense* performed by the sensors located along the track of the moving target. DPT has two major components: *target descriptor formulation algorithm* and *sensor selection algorithm*.

In the target descriptor formulation algorithm, a target descriptor (TD) is exploited for a CH to maintain the information of the target. A CH's TD involves the identity of the CH, the current location of the CH, the next predicted location of the CH, and the time stamp. In addition, with the aid of the linear predictor that is able to predict the nth location depending on the previous $n - 1$ actual locations, the next location of the moving target can be efficiently determined according to the locations at which the target has traveled.

In the sensor selection algorithm, once the cluster head, CH_i, predicts the location of the target, the downstream cluster head, CH_{i+1}, will intend to select three sensors to take charge of sensing the target. In principle, to accurately identify the target, the distance of any one of these sensors to the predicted location of the target should be less than the normal beam of the sensor. If the downstream CH is unable to find enough sensors for sensing task, it intends to use the high beam to search for suitable sensors. In case not enough sensors are found, the downstream CH will consequently ask its neighboring CHs for help.

In general, the theme of energy efficiency should be seriously considered in WSNs, so the DPT algorithm uses the sensor-hibernation mechanism to not only conserve energy but also prolong the network lifetime. Namely, only the selected sensors need to wake up to sense the target, whereas all other sensors can hibernate. Additionally, a sensor also uses the different beams to sense the target according to the location of the target so as to efficiently reduce the energy consumption.

25.3.1.3 Stochastic Approaches

Aslam *et al.* [1] develop a binary sensor model in which the sensor data are converted into 1 bit of information and propose an algorithm based on the particle filtering to moving target tracking. Due to the utilization of a single bit to represent the tracking information, such mechanism not only achieves energy efficiency but also provides high accuracy in case of low sensor density. In Ref. [23], an information-driven approach based on sensor collaboration and sensor data aggregation is introduced for target tracking. A leader assumed to be active at any moment takes charge of determining and routing the tracking information to the next leader. Basically, the tracking task in Ref. [23] is regarded as a sequential Bayesian estimation problem. The proposed technique enables a sensor to determine the state of a target by using a Bayesian filter, which primarily combines its current measurement with the previous estimates. Instead of relying on the measurement of only one sensor, the approach considers multiple measurements related to the spatial and temporal properties. Such sensor collaboration efficiently improves the accuracy of event tracking.

25.3.2 Event Boundary Determination

In addition to target tracking, event boundary determination is another practical manner for tracking applications in WSNs. Specially, the manner is widely used in the scenario in which the phenomenon (a.k.a. event) of interest occupies a large area. Much research to event boundary determination primarily focuses on selecting a set of sensors, termed edge sensors or border sensors to stand for the exact boundary of the phenomenon.

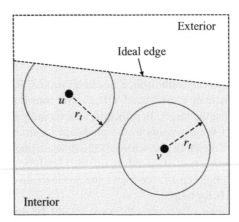

Figure 25.6 Illustration of the edge and nonedge sensors. Sensors u and v are an edge and an nonedge sensors, respectively (adapted from [3]).

In Ref. [3], three approaches, categorized as the statistical, image processing, and classifier based approaches are introduced to determine the edge of the event. Overall, such approaches concentrate on the determination of edge sensors. The edge sensor is defined as a sensor, which is not only in the interior of the event region but also less than the predetermined distance (namely, radius of tolerance) to the ideal edge. As shown in Figure 25.6, suppose the shaded region is the event region. Let r_t be the radius of tolerance and the distance between sensor u and the ideal edge be d_u. Obviously, sensor u is an edge sensor because it is in the event region and $d_u < r_t$. However, since $d_v > r_t$, sensor v is not an edge sensor, although sensor v is within the event region. Basically, the main idea of the proposed approaches is that each sensor gathers the information from its neighborhood and then determines whether it should become an edge sensor in a distributed manner.

In the statistical approach, a sensor collects information from its neighbors and uses a boolean decision function to either become an edge sensor or remain a nonedge sensor. The proposed image-processing mechanism, based on the filtering techniques [6], largely considers the high frequencies (e.g., abrupt change) present in the image and removes all the uniformities for localized edge detection. The high frequency is regarded as the edge of the phenomenon. Motivated by the concept of pattern recognition, the classifier-based approach uses a simple linear classifier for a sensor to partition the gathered data into either similar data or dissimilar data. By means of the partitioning line provided by the linear classifier, a sensor is able to self-determine whether it is in the interior or the exterior of the phenomenon according to the spatial relationship between its location and its corresponding partitioning line.

Ji et al. [7] investigate the detection and tracking of the continuous object, such as chemical liquid, wild fire, or poison gas. Such continuous object here can be viewed as an event. Compared to one or multiple individual objects (e.g., vehicles or animals), the continuous object usually occupies a large area. Additionally, such continuous object is most likely to tend to increase or decrease in size, change in shape, or

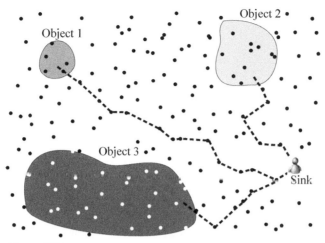

Figure 25.7 Network with three continuous objects (adapted from [7]).

split into multiple smaller objects with time. For example, illustrated in Figure 25.7, there are three continuous objects in the network. Sensors around the object are responsible for not only detecting and tracking the boundary but also sending the boundary information to the sink if the object is detected.

In Ref. [7], a dynamic cluster-based structure is introduced to track the object boundary. The proposed clustering scheme mainly involves three components, *boundary sensor selection*, *boundary sensor clustering*, and *boundary tracking*.

The main objective of boundary sensor selection is to determine the boundary sensor to efficiently identify the object boundary. Here, if a sensor detects the object within its sensing range but its one or more one-hop neighbors do not detect the object, the sensor will be regarded as a boundary sensor. For example, in Figure 25.8, suppose the shaded region is the event region. Sensors E, F, and G are self-determined as the boundary sensors if detecting no object in the former time slot but detecting the existence of object in the current time slot. However, sensors A, B, C, and D are not the boundary sensors because they do not detect the object in the former and the current time slots. Additionally, sensor H is also a nonboundary sensor because it always detects the object in both the former and the current time slots.

In boundary sensor clustering, part of the determined boundary sensors are selected as the cluster heads. An effortless method that the boundary sensor closest to the center of the area covered by the cluster structure will become a cluster head is used to cluster the boundary sensors. Once the cluster head is determined, all boundary sensors are organized into a cluster structure depending on their locations. Then, the cluster head broadcasts a connection request to enable the boundary sensors in the cluster to form a connected route so as to efficiently track the continuous object.

Recall that the continuous object is likely to change in size or in shape with time elapsed. Thus, in the course of boundary-tracking procedure, the boundary sensor selection procedure mentioned above will be invoked to determine new boundary

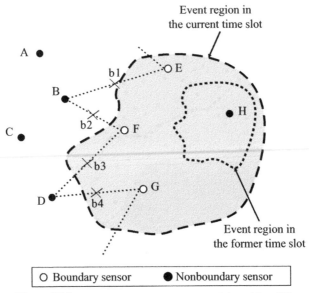

Event region in
the current time slot

Event region in
the former time slot

| O Boundary sensor | ● Nonboundary sensor |

Figure 25.8   Illustration of boundary sensors (adapted from [7]).

sensors once the boundary changes. Meanwhile, the cluster structure needs to be updated as well.

Nowak and Mitra [13] investigate a strategy to boundary detection and estimation in WSNs. The study considers a broad class of boundaries, such as linear and other curves. The main objective of the study is to devise a boundary-estimation mechanism based on multiscale partitioning methods. The network considered is partitioned into multiple square grids of equal size by means of a recursive dynamic partition (RDP) process. Each grid is viewed as a cluster, in which sensors communicate their sensing measurements to a clusterhead. To efficiently distinguish between the boundary and the nonboundary region, the proposed approach further produces an RDP with nonuniform resolution. That is, the partition is of high resolution along the boundary, whereas the it is of low resolution within the nonboundary region.

25.4   TRACKING IN WHSNs

25.4.1   Problem Statement

Previous work mainly aims at event detection and tracking in wireless homogeneous sensor networks, in which a considerable number of one type of sensors equipped with the same sensing units are deployed. In addition, the event formed by only one attribute is mainly concerned. Due to the limitation of sensing capability, the existing approaches are obviously unsuitable for the event, which is a compound of multiple attributes and any one of the attributes cannot be perceived by only one type of sensor.

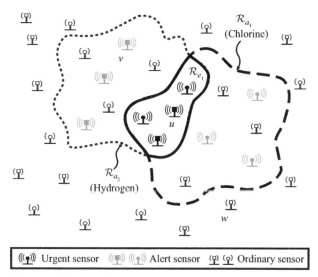

Figure 25.9   Illustration of a WHSN, in which two attribute regions, $\mathcal{R}_{a_1}$ and $\mathcal{R}_{a_2}$, exist and form event region, $\mathcal{R}_{e_1}$. The circle and the square sensors are able to, respectively, perceive attributes a_1 and a_2.

Without loss of generality, a network composed of various types of sensors is called the WHSN.

Figure 25.9 illustrates an example of a WHSN, in which two types of sensors, circle and square sensors, are assumed to be capable of sensing the concentrations of chlorine and hydrogen, respectively. Consider a tracking application in which the hazardous gas (event) is only composed of either chlorine or hydrogen. The corresponding type of sensor deployed in the network is able to determine such hazardous gas. However, if we are concerned with the event, required to be determined by the existence of both chlorine and hydrogen, any type of sensor is obviously unable to realize the event for lack of sensing capability.

Note that the formidable challenge of event detection and tracking in the WHSN is the constraint on sensor's sensing capability, especially in a distributed manner. With the characteristics of low power, low cost, and short communication range, a sensor in WHSNs has the potentiality to collaborate with other sensors to fulfill various tasks. Additionally, the variation of the attribute region is most likely to enlarge or narrow the event region with time, so the capability of adaptation also has to be taken into account in designing tracking approaches.

A Collaborative Event deteCtion and Tracking protocol (CollECT) based on sensor collaboration is proposed. Specially, CollECT is well suited to a scenario in which a user (e.g., rescuer or fireman) requires to be aware of the knowledge of event boundary if the user enters the communication range of the border sensor. In the section, the network model is first introduced, followed by details of CollECT.

25.4.2 Network Model

Let a WHSN be represented by an undirected simple graph $G = (V(G), E(G))$, in which $V(G)$ is the set of all sensors (vertices) and $E(G)=\{(u, v) : u$ and v can communicate with each other$\}$ is the set of links (edges). Let N_u^I be the set of sensors within the communication range of sensor u, that is, $N_u^I = \{v \in V(G) : (u, v) \in E(G)\}$. For each sensor $u \in V(G)$, the number of sensors in N_u^I is denoted as $|N_u^I|$. All sensors are stationary, time-synchronized, and randomly scattered. Each sensor is aware of its own location information via the GPS receiver [17] or existing localization schemes [5, 14].

The network we are concerned with is a connected network (i.e., for each sensor $u \in V(G)$, $|N_u^I| \geq 1$). Let N_T be the number of types of sensors and N_i be the number of sensors with type i, where $1 \leq i \leq N_T$. Let a_i and e_j denote attribute i and event j, respectively. Additionally, let A_u be the set of attributes that sensor u is able to detect. Sensors u and v are the same type if $A_u = A_v$. The attributes related to each event are given and are aware of by each sensor. Attribute spread is assumed to be slower than packet dissemination. In CollECT, an *actual attribute region*, $\mathcal{R}_{a_i}$, is viewed as a continuous area in which attribute a_i exists. An *actual event region*, $\mathcal{R}_{e_i}$, is regarded as an overlapping area among multiple attribute regions and all of these attributes comprise event e_i. For example, in Figure 25.9, suppose a hazardous gas, termed e_1, is a compound of two attributes, a_1 and a_2. Thus, the overlapping area between $\mathcal{R}_{a_1}$ and $\mathcal{R}_{a_2}$ is viewed as an actual region $\mathcal{R}_{e_1}$.

The main goal of CollECT is to first estimate the attribute regions and consequently determine the event region. Here, $\mathfrak{R}_{a_i}$ and $\mathfrak{R}_{e_i}$ are used to indicate the *estimated attribute region* of a_i and the *estimated event region* of e_i, respectively. In CollECT, the estimated attribute region of each attribute is represented by a two-connected plane graph, called *triangulation* in which all faces are bounded by a triangle whose vertices are the sensors of the same types. Formally, an estimated attribute region, $\mathfrak{R}_{a_i}$, is defined as an area composed of multiple nonoverlapping triangles, each triangle has at least one vertex that is able to detect attribute a_i.

Because any two vertices of the triangle in an estimated attribute region may not be within the communication range of each other, such triangle is called a *logical triangle*. For a logical triangle, any two vertices are viewed as the *logical neighbors* of each other. Formally, sensor $v \in V(G)$ is defined as the logical neighbor of sensor $u \in V(G)$ if edge (u, v) belongs to any logical triangle of $\mathfrak{R}_{a_i}$, where $a_i \in A_u$ and $a_i \in A_v$.

Let LT_{uvw} be the logical triangle whose vertices are u, v, and w. Suppose sensor x is different from sensors u, v, or w in type. Sensor x within LT_{uvw} is denoted as $x \in C_{LT_{uvw}}$, where $C_{LT_{uvw}}$ is the face bounded by edges (u, v), (v, w), and (w, u). The *logical neighborhood*, termed N_u^L, is the set of logical neighbors of sensor u, and maintained by sensor u. For a sensor u, there may exist no any sensor $v \in N_u^I$, where $A_u = A_v$; thus, any one of sensor u's logical neighbors is likely to be several hops from sensor u.

In CollECT, a sensor has to be self-recognized a suitable role for the sake of event tracking. The sensor either in the event region, in the attribute region but outside the event region, or outside the attribute region is, respectively, regarded as an ordinary, alert, or urgent sensor. Let C_u be the sensing coverage of sensor u, and A_{e_i} be the set of attributes comprising event e_i. Thus, the ordinary, alert, and urgent sensors are elaborately discussed below.

- *Ordinary sensor*: A sensor $u \in V(G)$ is regarded as an ordinary sensor if u does not detect attribute a_i, where $a_i \in A_u$.
- *Alert sensor*: A sensor $u \in V(G)$ is regarded as an alert sensor if u detects attribute a_i, where $a_i \in A_u$.
- *Urgent sensor*: A sensor $u \in V(G)$ is regarded as an urgent sensor of event e_j if u is within the overlap of all $\Re_{a_i}$, where $a_i \in A_{e_j}$.

As shown in Figure 25.9, sensors u, v, and w are the urgent, the alert, and the ordinary sensors, respectively, in accordance with the above descriptions.

25.4.3 Collaborative Event deteCtion and Tracking Protocol (CollECT)

CollECT generally involves three major procedures, including the *vicinity triangulation*, *event determination*, and *border sensor selection* procedures. The vicinity triangulation procedure enables the same type of sensors to dynamically construct the respective attribute region, which is viewed as a triangulation comprised of multiple logical triangles. During the event determination procedure, a sensor locally determines the existence of the event according to the received messages from different types of sensors within its logical triangles. Like most existing protocols [3, 4, 11], in the course of the border sensor selection procedure, CollECT aims to identify the event boundary represented by the border sensors. Overall, the above procedures perform repeatedly to timely detect and track the event, since the event spreads out from a small region with time elapsed.

25.4.3.1 Basic Idea

Recall that, in CollECT, the estimated attribute region $\Re_{a_i}$ can be viewed as a convex polygon, whose vertices are the sensors with the same type. Figure 25.10 illustrates an estimated attribute region $\Re_{a_i}$ whose vertices are alert sensors u, v, w, x, y, and z. Suppose alert sensor s in $\Re_{a_i}$ is another type of sensor and is able to collaborate with vertices of $\Re_{a_i}$ to determine the event. Thus, we reason that the event is likely to exist in $\Re_{a_i}$. However, such rough inference most probably incurs erroneous event determination. Note that sensor s is actually near u, y, and z rather than other sensors. According to the above inference, the event actually existing in the vicinity of sensors u, y, and z may be considered near other sensors.

To improve the accuracy of event determination, CollECT intends to minimize the region used for a sensor to make a decision of the existence of the event. As shown in Figure 25.10(b), if $\Re_{a_i}$ is divided into multiple nonoverlapping quadrilaterals. Sensor s will be viewed as in the quadrilateral (Q_1) formed by u, v, y, and z instead of in

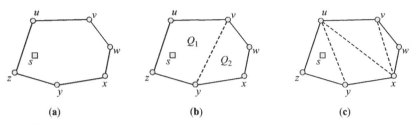

Figure 25.10   Basic idea of triangulation. (a) Sensor s is in attribute region $\mathfrak{R}_{a_i}$. (b) Sensor s is in the quadrilateral formed by u, v, y, and z. (c) Sensor s is in the logical triangle formed by u, y, and z.

the one (Q_2) formed by v, w, x, and y. Apparently, the event is likely to exist in Q_1; thus, the inaccurate level of event determination is improved compared to that in Figure 25.10(a). In general, triangle is the elementary face among all faces. Therefore, any convex polygon can be separated into multiple nonoverlapping triangles. As Figure 25.10(c) illustrates, for example, $\mathfrak{R}_{a_i}$ is divided into four logical triangles, LT_{uyz}, LT_{uxy}, LT_{uvx}, and LT_{vwx}. Significantly, the event is likely to exist in logical triangle LT_{uyz}, so the inaccurate level of event determination is further improved in comparison with that in Figure 25.10(b). On the basis of the above discussion, in CollECT, we adopt triangulation to not only construct the individual attribute region discussion but also determine the event.

25.4.3.2 Vicinity Triangulation

The main goal of vicinity triangulation is to construct the estimated attribute corresponding to each attribute. Recall that an estimated attribute region, $\mathfrak{R}_{a_i}$, is viewed as a convex polygon composed of multiple logical triangles. Thus, a sensor requires determining its logical neighbors so as to cooperate with the logical neighbors to construct the logical triangles.

Initially, all sensors are ordinary ones after deployment, and $N_u^N = \varnothing$ for each sensor u. Once detecting attribute $a_i \in A_u$, an ordinary sensor, u, becomes an alert sensor and then issues a VTREQ packet to all of the sensors in N_u^I. For ease of explanation, the sensor that initiates a VTREQ packet is called the *originator* of the VTREQ packet.

Upon the receipt of a VTREQ packet from originator u, sensor v will perform *ReceiveVTREQ* shown in Algorithm 25.1. In principle, sensor v makes a decision of maintaining the information of the new logical neighbor depending on the VTREQ packets already received. If required, such information is maintained in a table, each of which is a five-tuple with the format $(id(u), loc(u), attr(u), t(u), role(u))$, where $id(u)$ is sensor u's identifier, $loc(u)$ is sensor u's location, $attr(u)$ is the attribute that sensor u detected, $t(u)$ is the timestamp when sensor u detects $attr(u)$, and $role(u)$ is sensor u's role. Then, the sensor also determines whether it should forward the received VTREQ packet depending on its sensing capability. Alternatively, if $attr(u) \notin A_v$, sensor v only forwards the VTREQ packet because it is unable to cooperate with sensor u to construct the logical triangle.

Algorithm 25.1: *ReceiveVTREQ*

{**Input**: u: originator of the VTREQ packet}
if $attr(u) \in A_v$ **then**
    if $u \notin N_v^L$ **then**
        if I am an ordinary or alert sensor **then**
            Create a new entry $(u,w,loc(u),attr(u),t(u),role(u))$ for the received VTREQ
            packet;
            $N_v^L \leftarrow N_v^L \cup \{u\}$;
            Forward the received VTREQ packet;
        else
            Discard the received VTREQ packet;
        end if
    else
        Discard the received VTREQ packet;
    end if
    if I maintain at least one logical triangle **then**
        Perform *ReOrganizeLT*;   /*Algorithm 2 */
    end if
else
    Forward the received VTREQ packet;
end if

Significantly, for a sensor, the operation of forwarding the received VTREQ packet is mainly to guarantee that the VTREQ packet can be received by a sensor w, where $A_u = A_w$, to construct the logical triangle. In CollECT, each sensor, say v, needs to maintain all of its logical triangles in a set termed $S_{LT}(v)$. The number of logical triangles maintained by sensor v is denoted as $|S_{LT}(v)|$. For example, $S_{LT}(v) = \{LT_{uvw}, LT_{vwx}\}$ if LT_{uvw} and LT_{vwx} are sensor v's logical triangles. Actually, each logical triangle in $S_{LT}(v)$ is denoted by all of its vertices. Namely, $S_{LT}(v) = \{(u, v, w), (v, w, x)\}$.

In the course of attribute dissemination, numerous sensors able to perceive a certain attribute will detect such attribute and then issue VTREQ packets; thus, a sensor may receive multiple VTREQ packets. CollECT uses Algorithm 25.2 for a sensor to reorganize its logical triangle when it receives multiple VTREQ packets.

Overall, during the vicinity triangulation procedure, CollECT intends to minimize the logical triangle in size for the reduction of computation overhead and improvement of the accuracy of event determination. For the example shown in Figure 25.11, without using Algorithm 25.2, sensor w will regard LT_{uvw}, LT_{uwx}, and LT_{vwx} as its logical triangles due to the receipt of VTREQ packets from sensors u, v, and x. Sensors u and v also regard such three logical triangles as their logical triangles (namely, $S_{LT}(u)$ and $S_{LT}(v)$ both include LT_{uvw}, LT_{uwx}, and LT_{vwx}). In this case, LT_{uvw} is maintained by all vertices of itself (namely, sensors u, v, and w), so that the computation overhead significantly increases in event determination. Obviously, if more sensors responsible for event determination exist, more messages to announce

---

Algorithm 25.2: *ReOrganizeLT*

{**Input**: i, j, and x: originators of VTREQ packets received by w}
{**Condition**: w receives VTREQ packets from i and j prior to x}
$S_{LT}(w) \leftarrow S_{LT}(w) \cup \{LT_{iwx}\}$;
$S_{LT}(w) \leftarrow S_{LT}(w) \cup \{LT_{jwx}\}$;
if $x \in C_{LT_{uvw}}$ **then**
   $S_{LT}(w) \leftarrow S_{LT}(w) - \{LT_{uvw}\}$;
else
   if w is one of the end point of the shorter diagonal of the quadrilateral with vertices
   u, v, w, and x **then**
      if $\overline{uw}$ is the shorter diagonal **then**
         $S_{LT}(w) \leftarrow S_{LT}(w) - \{LT_{vwx}\}$;
      else
         if $\overline{vw}$ is the shorter diagonal **then**
            $S_{LT}(w) \leftarrow S_{LT}(w) - \{LT_{uwx}\}$;
         else
            if $\overline{xw}$ is the shorter diagonal **then**
               $S_{LT}(w) \leftarrow S_{LT}(w) - \{LT_{uvw}\}$;
            end if
         end if
      end if
   else
      $N_w^L \leftarrow N_w^L - \{x\}$;
      Remove entry corresponding to originator x from my table;
   end if
end if

---

the existence of the event generate and probably cause severe collisions. As a result, CollECT aims at enabling only one vertex of a logical triangle to determine the existence of the event. Consequently, sensor w in Figure 25.11 only maintains LT_{uwx} and LT_{vwx}.

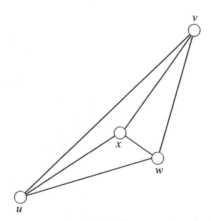

Figure 25.11   Logical triangle reorganization. Sensor w only maintains LT_{uwx} and LT_{vwx}.

25.4.3.3 Event Determination

Motivated by the collaboration of various types of sensors, CollECT adopts the following alert-in-triangulation (AIT) test for an alert sensor to determine whether the event exists within its logical triangle.

AIT test: Once receiving VTREQ packets, alert sensor u regards event e_i occurs within its logical triangle if all of the originators of the received VTREQ packets are within such logical triangle, and the attributes that these originators as well as sensor u detect are able to form event e_i.

Once passing the AIT test, an alert sensor becomes an urgent sensor, and then transmits an *Event Discovery* (EVT) packet to inform other sensors of the existence of the event. An EVT packet sent from urgent sensor u involves the identifier of the event, the identifier of sensor u, and the set of logical triangles of sensor u (namely, $S_{LT}(u)$).

Note that a considerable number of EVT packets will flood in the network if all of the vertices of a logical triangle issue EVT packets once the AIT test passes. Thus, to reduce the computation overhead and avoid packet collisions, CollECT intends to select one alert sensor, called the *leading sensor* in a logical triangle to take charge of event determination. Since the event spreads out from a small region and event spread is assumed to be slower than packet dissemination, the alert sensor that issues the VTREQ packet with the largest value of timestamp may be near the event boundary. Such alert sensor is reasonably viewed as the leading alert sensor of its logical triangle so as to timely inform the sensors in the exterior of the event region of the existence of the event.

In CollECT, a leading sensor actively performs the AIT test to make a decision to become an urgent sensor, whereas a nonleading sensor passively becomes an urgent sensor if it receives an EVT packet and locates in the logical triangle whose leading sensor is the originator of the received EVT packet.

An example of the AIT test is illustrated in Figure 25.12. Suppose that the circle and the square sensors are able to detect attributes a_1 and a_2, respectively. Event e_1 is assumed to be composed of attributes a_1 and a_2. Suppose LT_{abc}, LT_{ace}, LT_{cde}, and LT_{xyz} be logical triangles. Let $t(u)$ denote the timestamp when sensor u detects the attribute. Here, we assume that sensors b, c, and y issue VTREQ packets posterior the other vertices of their logical triangles, LT_{abc}, LT_{ace}, and LT_{xyz}, respectively. That is, sensors b, c, and y are all leading sensors. Once receiving a VTREQ packet issued from y, c will regard event e_1 as have occurred in LT_{ace} because the AIT test is passed. Sensor c actively becomes an urgent sensor and then sends out an EVT packet. Similarly, sensor y will actively become an urgent sensor and then transmit an EVT packet as well when receiving the VTREQ packet from sensor a.

Algorithm 25.3 shows how sensor v performs when receiving an EVT packet. As shown in Figure 25.12, in LT_{ace}, the nonleading sensors a and e will receive the EVT packets from c and then passively become urgent sensors. Likewise, in LT_{xyz}, sensors x and z also become urgent sensors in a passive manner. Figure 25.13 illustrates the result of event determination.

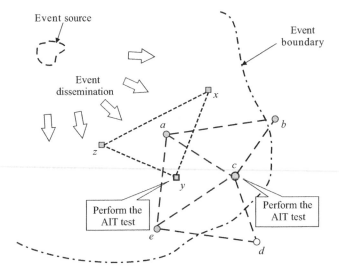

Figure 25.12   Alert-in-triangulation test. Suppose event spreads out from a small region (i.e., event source) at the upper left of the figure. Event e_1 is composed of attributes a_1 and a_2. Gray sensors are alert sensors. Sensors c (blue border) and y (red border) are, respectively, the leading sensors of LT_{cde} and LT_{xyz} and perform the AIT test when receiving the VTREQ packets.

25.4.3.4 Border Sensor Selection

In addition to urgent sensors, CollECT also concentrates on border sensor selection to identify the actual event boundary. Recall that vicinity triangulation is most likely to minimize the logical triangle in size if sensors are densely deployed in the network. As a result, based on the concept, a logical triangle only comprising one

---

Algorithm 25.3: *ReceiveEVT*

{**Input**: u: originator of the EVT packet}
if v is in the logical triangles in $S_{LT}(u)$ and the leading sensor of such logical triangle is u
  then
    if I am a non-urgent sensor **then**
      Set my role as an urgent sensor;
    end if
    Forward the received EVT packet;
  else
    if $v \in N_u^L$ **then**
      Set my role as an urgent sensor;
      Discard the received EVT packet;
    else
      Forward the received EVT packet;
    end if
  end if

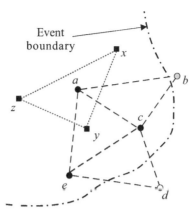

Figure 25.13   Illustration of event determination. Sensors c and y actively become urgent sensors. Additionally, sensors a, e, x, and z passively become urgent sensors.

or two urgent vertices is probably near the event boundary. To efficiently represent the logical triangle around the event boundary, in CollECT, we define the logical triangle as a boundary logical triangle if it has one or two urgent vertices. A boundary logical triangle whose vertices are sensors u, v, and w is denoted as LT^B_{uvw}. Under the consideration of attribute distribution and dissemination, only the ordinary or the alert vertex of a boundary logical triangle has a chance to become a border sensor.

Consider a boundary logical triangle with one urgent and two alert vertices. Obviously, an urgent sensor is within the event region, while the alert ones are only within the attribute region. Namely, in such triangle, some sensors are inside the event region, whereas the others are in the exterior of the event region. Thus, an alert vertex of a boundary logical triangle should determine whether it should become a border sensor or not. Alternatively, consider a boundary logical triangle that has at least one ordinary and at least one urgent vertices. Basically, some sensors around the urgent vertex are inside the event region. However, some other sensors around the ordinary vertex are only in the attribute region but not in the event region. Apparently, such ordinary vertex most probably become a border sensor. Thus, an ordinary vertex of a boundary logical triangle should make a decision of becoming a border sensor. consequently, the following principles are respectively used for an alert and an ordinary sensor to determine whether it should become an urgent sensor or not.

> *Principle 1*: An alert sensor $u \in LT^B_{uvw}$ will become a border sensor if either sensor v or w is an urgent sensor.
> *Principle 2*: An ordinary sensor $u \in LT^B_{uvw}$ will become a border sensor if both sensors v and w are urgent sensors.

In CollECT, a sensor invokes the border sensor selection procedure after receiving an EVT packet. In Figure 25.14, suppose that u, v, and w are urgent sensors, x and y are alert sensors, and sensor z is an ordinary sensor. Assume LT^B_{uxy}, LT^B_{uvy}, and LT^B_{vwz} are boundary logical triangles. According to Principle 1, alert sensors x and y will become border sensors because u is an urgent sensor. Additionally, ordinary

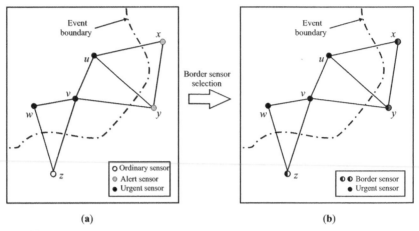

(a)   **(b)**

Figure 25.14 Illustration of border sensor selection. (a) Before border sensor selection. (b) After border sensor selection. Alert sensors x and y become border sensors according to Principle 1. Ordinary sensor z becomes a border sensor according to Principle 2.

sensor z, according to Principle 2, will become a border senor since both v and w are urgent ones. The result of border sensor selection is illustrated in Figure 25.14(b).

25.4.3.5 Event Tracking

For realistic applications, the actual attribute region is most likely to vary with time elapsed so that the actual event region may be changed. As a result, to track the event, sensors require changing their roles timely if the event changes in size or in sharp. For example, at a certain time, an ordinary sensor is most probably about to detect the attribute, so it will become an oncoming alert sensor. Additionally, a border sensor probably becomes a nonborder one if it is not near the event boundary. Obviously, an ordinary or alert border sensor has to change its role so as to adapt efficiently to the variance in the event. Consequently, the aforementioned procedures used in CollECT should be invoked repeatedly to dynamically detect and track the event.

25.5 SUMMARY

In this chapter, the classification of tracking scenarios and numerous solutions proposed in the literature for tracking in wireless sensor networks were introduced. These approaches are categorized as target tracking and event boundary determination. The focus on target tracking is the identification of the target, while the objective of boundary determination is the selection of sensors near the event boundary to stand for the exact boundary. Then, a novel distributed tracking method, called CollECT, for WHSNs was mentioned. CollECT involves vicinity triangulation, event determination, and border sensor selection procedures to not only construct the vicinity triangulation for the determination of the event but also select several reasonable border sensors for the identification of the exact event boundary. Overall, motivated by the collaboration of both the same and the different types of sensors, CollECT is able to achieve event detection and tracking promptly. Note that because the wireless heterogeneous sensor network differs from the wireless homogeneous sensor network in numerous characteristics, various issues discussed in the wireless homogeneous sensor network should be worthy to be revisited. Thus, future studies can also explore the solutions to deployment, routing, coverage, and active/asleep scheduling in WHSNs.

REFERENCES

1. J. Aslam, Z. Butler, F. Constantin, V. Crespi, G. Cybenko, and D. Rus. Tracking a moving object with a binary sensor network. In *ACM International Conference on Embedded Networked Sensor Systems (SenSys)*, 2003, pp. 150–161.
2. W.-P. Chen, C.-J. C. Hou, and L. Sha. Dynamic clustering for acoustic target tracking in wireless sensor networks. In *IEEE International Conference on Network Protocols (ICNP)*, 2003, pp. 284–294.
3. K. Chintalapudi and R. Govindan. Localized edge detection in sensor fields. *IEEE International Workshop on Sensor Network Protocols and Applications (SNPA)*, 2003, pp. 59–70.
4. M. Ding, D. Chen, K. Xing, and X. Cheng. Localized fault-tolerant event boundary detection in sensor networks. In *IEEE INFOCOM, The Annual Joint Conference of the IEEE Computer and Communications Societies*, 2005, pp. 902–913.
5. T. He, C. Huang, B. M. Blum, J. A. Stankovic, and T. Abdelzaher. Range-free localization schemes for large scale sensor networks. In *ACM International Conference on Mobile Computing and Networking (MOBICOM)*, 2003, pp. 81–95.
6. B. Jähne. *Digital Image Processing*. Springer, 1997.
7. X. Ji, H. Zha, J. J. Metzner, and G. Kesidis. Dynamic cluster structure for object detection and tracking in wireless ad-hoc sensor networks. In *IEEE International Conference on Communications (ICC)*, 2004, pp. 3807–3811.
8. H. T. Kung and D. Vlah. Efficient location tracking using sensor networks. In *IEEE Wireless Communications and Networking Conference (WCNC)*, 2003, pp. 1954–1961.
9. S.-M. Lee, H. Cha, and R. Ha. Energy-aware location error handling for object tracking applications in wireless sensor networks. *Comput. Commun.*, 30(7):1443–1450, 2007.
10. J. Lian, L. Chen, K. Naik, Y. Liu, and G. B. Agnew. Gradient boundary detection for time series snapshot construction in sensor networks. *IEEE Trans. Parallel Distribut. Syst.*, 18(10):1462–1475, 2007.
11. P.-K. Liao, M.-K. Chang, and C.-C. Jay Kuo. Distributed edge detection with composite hypothesis test in wireless sensor networks. In *IEEE Global Telecommunications Conference (GLOBECOM)*, 2004, pp. 129–133.
12. C.-Y. Lin, W.-C. Peng, and Y.-C. Tseng. Efficient in-network moving object tracking in wireless sensor networks. *IEEE Trans. Mobile Comput.*, 5(8):1044–1056, 2006.
13. R. Nowak and U. Mitra. Boundary estimation in sensor networks: theory and methods. *Lect. Notes Comput. Sci.*, 2634:80–95, 2004.
14. A. Savvides, C.-C. Han, and M. B. Strivastava. Dynamic fine-grained localization in ad-hoc networks of sensors. In *ACM International Symposium on Mobile Ad Hoc Networking and Computing (MOBIHOC)*, 2001, pp. 8–14.
15. S. Susca, F. Bullo, and S. Martínez. Monitoring environmental boundaries with a robotic sensor network. *IEEE Trans. Control Syst. Technol.*, 16(2):288–296, 2008.
16. V. S. Tseng and K. W. Lin. Energy efficient strategies for object tracking in sensor networks: A data mining approach. *J. Syst. Software*, 80(10):1678–1698, 2007.
17. B. H. Wellenhof, H. Lichtenegger, and J. Collins. *Global Positioning System: Theory and Practice*. Springer-Verlag, 1997.

18. J. Xu, X. Tang, and W.-C. Lee. EASE: an energy-efficient in-network storage scheme for object tracking in sensor networks. In *IEEE International Conference on Sensor and Ad Hoc Communications and Networks (SECON)*, 2005, pp. 396–405.

19. H. Yang and B. Sikdar. A protocol for tracking mobile targets using sensor networks. In *IEEE International Workshop on Sensor Network Protocols and Applications (SNPA)*, 2003, pp. 71–81.

20. Z. Yang and X. Wang. Joint mobility tracking and hard handoff in cellular networks via sequential monte carlo filtering. In *IEEE INFOCOM, the Annual Joint Conference of the IEEE Computer and Communications Societies*, 2002, pp. 968–975.

21. W. Zhang and G. Cao. DCTC: dynamic convoy tree-based collaboration for target tracking in sensor networks. *IEEE Trans. Wireless Commun.*, 3(5):1689–1701, 2004.

22. W. Zhang and G. Cao. Optimizing tree reconfiguration for mobile target tracking in sensor networks. In *IEEE INFOCOM, the Annual Joint Conference of the IEEE Computer and Communications Societies*, 2004, pp. 54–61.

23. F. Zhao, J. Shin, and J. Reich. Information-driven dynamic sensor collaboration for target tracking. *IEEE Signal Process. Mag.*, 19(2):61–72, 2002.

Chapter 26

DDoS Attack Modeling and Detection in Wireless Sensor Networks

Zubair A. Baig[1] and Asad I. Khan[2]

[1] Department of Computer Engineering, King Fahd University of Petroleum & Minerals, Dhahran, Saudi Arabia
[2] Faculty of Information Technology, Monash University, Clayton, Victoria, Australia

26.1 Introduction   595
26.2 Attack Models in Wireless Sensor Network   597
26.3 Requirements for DDoS Attack Detection in   603
     Wireless Sensor Networks
26.4 Adversary Model   604
26.5 Network Model   607
26.6 Threshold Pattern Modeling   609
26.7 Traffic Flow Observation Table   613
26.8 Centralized Attack Detection   614
26.9 Conclusions   625
References   626

26.1 INTRODUCTION

In this chapter, we model distributed denial of service attacks in wireless sensor networks and define a centralized neural network-based scheme for detection of such attacks. The scheme involves the training and subsequent clustering of patterns of network traffic flow for attack classification purposes.

Wireless sensor networks (WSNs) consist of a collection of hundreds to thousands of tiny devices called sensors or sensor nodes. The limited on-board memory resources of such tiny devices restricts the size of applications, program codes, and actual data that can be stored in their memory. Therefore, most applications and programs designed for high-performance computing devices cannot be accommodated unaltered into the small memory space of sensor nodes.

The on-chip processing capability of a typical sensor node is several orders of magnitude less than that of a standard desktop processor. These networks generally have a rooted topology with a computing device called the *base station* at the root of the network. The base station has several orders of magnitude more power and a longer lifetime as compared to a standard sensor node [18]. Moreover, the base station has larger storage capacity and higher communication bandwidth links. Operations of the base station include information dissemination, network initialization, node activation and revocation tasks, and interfacing with other WSNs.

Sensor nodes are generally deployed in harsh and inaccessible environments to monitor and report real-world events. Common applications of these networks include bush fire monitoring, building structure monitoring, battlefield monitoring, and surveillance. Each sensor node is prone to a plethora of possible attacks that may be launched by the adversary class from either within or outside the network. Deployment of sensor nodes over a larger geographical area makes them even more vulnerable to any of these attacks [3]. The mission-critical nature of applications of such networks demands that they are protected from malicious attacks launched by the adversary class.

Denial of service (DoS) attacks are defined as attacks that are launched by a set of malicious entities toward a victim with the aim of incapacitating it from providing further service to legitimate clients. The objectives of the attack are achieved by exploiting either system/protocol-level vulnerabilities or by forcing the victim to undertake computationally intensive tasks such as exponentiating large integers for applications such as Diffie–Hellman key exchanges [1].

On the contrary, distributed denial of service (DDoS) attacks are defined as flooding attacks that do not rely on any particular network or system-level weaknesses. Rather, they tend to exploit the asymmetry that exists between the network line rate and the victim's processing capabilities. DDoS attacks are based on the principal that power of many is greater than "the power of few" [14]. Such attacks are launched subsequent to subversion and/or compromise of client machines of the network. The malicious nodes launch such an attack by amassing a large clan of hosts to simultaneously send useless packets toward the victim, leading to a flood of requests at the victim's end (Figure 26.1). The intensity of the traffic is high enough to incapacitate the victim or its network.

The wireless nature of sensor networks, accompanied with the resource-constrained nature of the sensor nodes, makes the success of such attacks easily accomplishable by the adversary class. We model distributed denial of service attacks in wireless sensor networks and derive such attacks from other known attacks in these networks. The primary step toward protection against such attacks is the

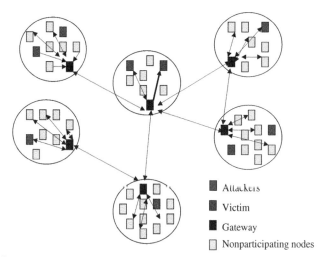

Figure 26.1   Distributed denial of service attack launched by adversarial nodes towarda victim node.

attack-detection process. We define a centralized neural network-based approach toward attack detection.

26.2 ATTACK MODELS IN WIRELESS SENSOR NETWORK

Sensor nodes deployed in a battlefield may have intelligent adversaries operating in their surroundings, intending to subvert, damage, or hijack messages exchanged in the network. The compromise of a sensor node can lead to greater damage to the network. All security solutions proposed for sensor networks need to operate with minimal energy usage, without affecting the security of the network. We classify sensor network attacks into three main categories:

- identity attacks,
- physical-layer attacks,
- denial of service, and
- distributed denial of service.

26.2.1 Identity Attacks

Identity attacks intend to steal the identities of legitimate nodes operating in the sensor network. The purpose of these attacks is to facilitate rogue node participation to either deny the base station access to sensor readings or to tamper with node readings.

26.2.1.1 Sybil Attack

A *Sybil attack* is defined as an identity attack wherein malicious devices illegitimately take on multiple identities in the network [15]. The malicious device's additional identities resulting from such an attack are termed as Sybil nodes. Messages received by a Sybil node are in actuality received by the malicious device and all messages transmitted by the Sybil nodes are actually sent by the malicious device. Another version of such an attack is when the Sybil nodes are inaccessible for direct communication by legitimate operating nodes of the network. In such scenarios, the malicious device will act as intermediary node, receive the messages, and pretend to forward them to the Sybil nodes.

Once the Sybil nodes have been successfully created by the malicious device, the actual attack is launched in one of several ways. For peer-to-peer networks involving replication and storage of distributed data across the network, such an attack will entail toward the storage of data on Sybil nodes. A Sybil attack if launched against the routing topology of a network can have catastrophic consequences [15]. For instance, a multipath routing channel may in fact be passing through multiple Sybil nodes representing a single malicious entity. A geographical approach to such an attack is when the attacker places multiple Sybil nodes at various locations of the network.

A defence mechanism against such an attack is *validation*, defined as a process of verifying that the identity given by a node is true, and is the only identity presented by its corresponding physical sensor node. The damage that is incurred on the network as a consequence of a Sybil attack can be appeased by validating all message transfers, using cryptographic techniques. Such techniques incur a high cost associated with large-scale key generation, distribution, and subsequent use for neighbor verification.

26.2.1.2 Node Replication Attacks

A *node replication attack* is defined as an attack wherein an adversary injects one or more nodes into the network with the same identity as an existing node. Unlike a Sybil attack, where a set of fictitious nodes are created by the adversary, node replication attacks involve physical insertion of rogue nodes into the network. This attack assumes that the adversary nodes have the capabilities for changing and subverting existing topological information in the network, such as route and trust in the network [16]. The centralized approach toward detecting such an attack is to have every node generate and transmit a list of its neighbors and their claimed identities to the base station. The base station does the verification and subsequent revocation, if need be, of the replicated nodes.

A randomized multicast mechanism [16] for detecting such attacks performs node replication detection by having each neighbor node of a location-declaring node to multicast a copy of the node location, confidentially, to a set of randomly selected witness nodes. Based on the birthday paradox [5], for a network with n nodes, if each location produces $\sqrt{n}$ witnesses, at least one collision will occur with high probability. In other words, the probability of atleast one of the witnesses receiving conflicting location claims (replicate) is high. However, because of the constraints on the maximum program size that can be stored in a sensor node's memory [19], the protocol becomes inefficient and less feasible for deployment.

26.2.2 Physical Layer Attacks

A jamming attack [23] against a sensor node is defined as a physical layer attack, wherein the radio frequencies of the victim node are disrupted. A node can observe the constant energy of its neighbors to conclude on a jamming attack as opposed to node failure. The standard defense against jamming involves various forms of spread-spectrum communication techniques. If the adversary can permanently jam the entire network, effective and complete denial of service is achieved. An alternate but costly strategy toward protection against such an attack is to use any available alternate modes of communication, such as infrared or optical, if the attacker has not jammed them as well.

26.2.3 Attacks Against the Base Station

The base station is central to all activity of the sensor network. Therefore, it is imperative to protect it from attacks that are intended to isolate and/or incapacitate the base station from participation in the activities of the network. Traffic analysis attacks launched by the adversary class against the sensor base station is done in one of three ways: (a) flooding of hoax requests to the base station, (b) remote spoofing of the base station for traffic misdirection (a.k.a. sinkhole attacks), and (c) message eavesdropping to locate and subsequently jam or destroy the base station. An inaccessible base station denies service to sensor nodes, and therefore traffic analysis attacks against a base station may be classified as a denial of service attack in a sensor network.

Several approaches have been proposed in the literature to thwart such attacks. A multibase station, redundant path setup mechanism is proposed in Ref. [6], so as to facilitate tolerance to failure of single base stations. The scheme assumes that messages are routed on several paths from the source node to different base stations, and therefore multiple copies of messages are stored in multiple base stations at any given time. The vulnerability of the multibase station setup phase to spoofing attacks is countered by having a one-way hash function applied to all base station-generated messages. One-way hashes are initially defined upto the nth place by the base station $(h_n, h_{n-1}, \ldots, h_0)$, and are then revealed in the reverse order, that is h_0 is revealed first. Any hash value in the sequence is verifiable by the previously revealed hash values. For instance, the second hash value in the chain h_1 is equal to $f(h_0)$. A sensor node upon receiving a multihop setup message from the base station, verifies the message-origin authenticity by comparing the hash value with the outcome of the one-way hash verification process. Such an approach helps protect the scheme against spoofing attacks.

26.2.4 Denial of Service: Wireless Sensor Networks

The capabilities needed by an adversary to initiate a DoS attack in a wireless sensor network are minimal. Denial of service attacks in WSNs aim at diminishing and/or exhausting the limited battery power of the sensor nodes. If the adversary class includes laptop-class adversaries with higher processing and communication capabilities than standard sensor nodes, the outcome of such an attack can be disastrous for the entire sensor network. Very little research has been done in the area of denial of service attack detection and defence in wireless sensor networks. A detailed classification of possible DoS attacks at various layers of operation has been elaborated upon in Ref. [23].

Spam attacks are one such form of denial of service attacks[22], wherein the attacks are launched by a set of nodes called antinodes injected into the sensor network by an adversary. The total number of antinodes, a, is much smaller than the actual network size n. The antinodes initiate a spam attack by generating frequent unsolicited dummy messages to their legitimate neighbor nodes of the sensor network. Considering the rooted topology of a sensor network, the amount of traffic accumulating at the nodes closer to the sink that is, the base station, is much larger than that accumulating at the leaf nodes. Consequently, nodes up the tree hierarchy will exhaust sooner than other nodes.

26.2.5 Distributed Denial of Service: Wireless Sensor Networks

Distributed denial of service attacks do not exploit any particular vulnerability in the system, but rather the asymmetry that exists between the network line-rate and the server processing rate [4, 8, 9]. As part of a DDoS attack, the adversary amasses a large clan of hosts, called zombies, to simultaneously send useless packets toward the victim, leading to a flood of requests at the victim's end. The intensity of the traffic is high enough to incapacitate either the victim or its network from further operations. The DDoS attack process consists of two stages, namely, zombie initiation and attack launch. During the zombie initiation process, the adversary compromises vulnerable nodes in a network and installs on them attacker source code, possibly in the form of a script. The code is written such that it awaits a "trigger" call from the adversary to participate in the actual attack process, wherein all zombies generate a large set of useless packets toward a set of victim nodes in the network [7, 17]. The zombie nodes may either exist in the same network as the victim or be part of another network. The attacker script may instruct the zombies to generate packets with randomly selected source addresses. The intent being to hide the identities of the zombie nodes.

In high-performance networks, DDoS attacks can be classified into two categories:

(i) Direct attacks: In a direct attack, the attacker arranges to send a large number of attack packets directly to the victim. SYN flooding is the most common attack case, in which TCP SYN packets are sent to the victim's server port. The victim will respond by sending back a SYN-ACK response to the source address of the packet. Since the source address of the packet was spoofed, the victim will not receive the third message of the three-way handshake required for connection establishment in TCP. Thus, the number of half-open connections at the victim's end consume all the available memory, forcing the victim to deny service to subsequent clients (including legitimate clients).

(ii) Reflector attacks: In a reflector attack, intermediate nodes (reflectors) are used as innocent attack launchers. The attacker sends packets with source addresses set to the victim's address. Without realizing that the packets had spoofed source addresses, the reflectors send the response to the requests to the victim. As a result, the victim's link is flooded with responses to reflected packets [4].

On certain occasions, it may happen by coincidence that a large number of legitimate packets are generated in a small time span, for transfer toward a certain set of destination nodes. Such a large influx of legitimate packets is referred to as a flash crowd [10]. The process of distinctly identifying DDoS attack traffic and differentiating it from flash crowds is nontrivial.

The sensor nodes or the base station of a wireless sensor network are analogous to the server of an IP-based network, being a victim of a flooding-based attack. Due to the resource-constrained nature of sensor nodes, any proposed solution that may demand heavy usage of memory, power, and communication resources of the sensor nodes is impractical for deployment on sensor nodes.

The various attack models described above can culminate into DoS or DDoS attacks and vice versa. In Figure 26.2, a relationship between the various attacks in a sensor network is illustrated. The probabilities of the described attacks to culminate into a denial of service attacks, along with their need for having colluding adversaries or rogue nodes in the network, are given in Table 26.1. A typical DoS/DDoS attack can be launched by a malicious entity by instigating a set of Sybil nodes to simultaneously generate malicious traffic packets toward a set of victim nodes on multiple routing paths. A successful Sybil attack is easily detectable by traffic packet validation, as defined earlier. However, when the attacker injects nodes into the network as part of a

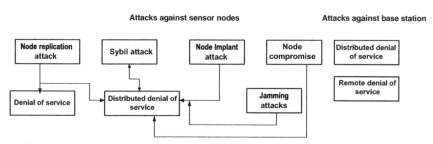

Figure 26.2 Derivation of distributed denial of service attacks in wireless sensor networks.

Table 26.1 Probabilities for Various Wireless Sensor Network Attacks to Culminate into Distributed Denial of Service Attacks

Attack	Consequence probability	Colluding adversaries	Detection/ defence options
Sybil	High	No	Probabilistic
Node replication	High	No	Probabilistic
Wormhole	Med	Yes	Antijamming techniques
Network intrusion	High	Yes	Tamper-resistant nodes
Node implant	High	No	Cryptosecrets
Node compromise	High	No	Cryptosecrets/Validation

node injection or replication attack and launches a DDoS attack against target nodes in the network, the resources of the target nodes will exhaust soon and consequently the attacker can steal the identities of these nodes and reallot them to the injected rogue nodes initially operating as fictitious Sybil nodes. A network wherein the uniqueness of node identities is verified at regular intervals, the probability of detecting Sybil attacks is diminished in the event of DDoS attacks. This is because legitimate neighbor nodes of a Sybil node will be flooded with large traffic inflow, incapacitating the total number of monitoring nodes of the network. The damage caused by a DDoS attack in such scenarios is irreversible and potentially catastrophic to all network operations.

Sensor network routes connecting the various sensor nodes and the base station in the form of a tree can be affected by a large influx of traffic owing to DDoS attacks. The limited-bandwidth wireless channels will eventually drop legitimate packets traversing the network, owing to the large number of useless data packets generated and transmitted by the rogue nodes.

Compromised nodes can generate enormous amount of traffic in a short span of time toward a set of target nodes in the network. The net traffic influx associated with the compromise of i nodes in a network with n nodes, where $i \ll n$, is aggravated in the event where all nodes further participate in a collusion-based flooding attack against critical sensory resources. Sensor networks operating without a mechanism in place to detect node replication attacks will succumb to the large inflow of traffic flow toward the critical node set. The previously described techniques to detect node replication attacks have a reasonable degree of uncertainty in the detection process, as they rely on probabilistic assumptions for conducting the detection process and are therefore not ideal for detecting flooding attacks. Moreover, the overhead incurred by the proposed schemes in Ref. [16] make such schemes less practical for detection of collusion-based attacks, which will necessitate their extensions to collaboration and extensive communications. It may be observed here that the success of the node replication attack is increased manifold if it results in a flooding attack. The resulting victim nodes can have their identities compromised by their rogue node replicas to ascertain a greater degree of damage to the entire network.

26.3   REQUIREMENTS FOR DDoS ATTACK DETECTION IN WIRELESS SENSOR NETWORKS

In wireless sensor networks, the wireless nature of the communication media, accompanied by the limited energy resources of sensor nodes, differentiates distributed denial of service attack modeling and detection in them. The adversary class monitors the flow of traffic in the network and labels the more active nodes, in terms of transmitting and receiving data packets, as critical nodes, which need to be targeted as part of the distributed denial of service attack. We refer to all such critical nodes as target or victim nodes. The distributed denial of service attack is launched by the adversarial nodes toward these critical sensor nodes from multiple ends of the network. The purpose of such attacks is to deplete the limited energy resources of the victim nodes. Furthermore, injected malicious nodes in the network steal the identities of the energy-depleted victim nodes and participate, with malicious intent, in the network operations. The lack of a single entry point to the network makes the task of detecting these attacks more cumbersome.

The topology of the wireless sensor network defines the network data delivery model. The topological designation of individual sensor nodes of the network, together with their placement, imply different expected traffic flow observations by each of the detector nodes. Each traffic threshold value (subpattern) defines the maximum numbers of packets a victim node may receive during a given period of time under normal circumstances. We refer to these threshold values interchangeably as subpatterns or threshold subpatterns. A systematic concatenation of these subpattern values will generate an entire pattern of threshold values, defining the maximum numbers of traffic flow packets that may be destined for a given victim node from various regions of the network in a given period of time. The attack detection process thus depends on the coordinated effort of a set of attack detector nodes to reconstruct a complete pattern of observed traffic flow values for attack classification purposes.

A single centralized entity can be designated the task of detecting anomalous traffic flow in the network. However, such an approach suffers from several drawbacks:

- the traffic flow may be outside the observation range of the detector node;
- A large set of threshold patterns will have to be stored and processed by the detector node for each victim node of the network; and
- the lack of multiple interfaces on the detector node implies that the attack traffic flow may overwhelm the detector node itself and thus disrupt the entire detection process.

The solutions proposed for distributed denial of service attack detection in high-performance networks are not directly applicable to wireless sensor networks because

- the lack of a single entry point to the wireless network demands the presence of multiple attack detector points to cover the entire network,
- the adversary class consists of adversarial nodes of varying capabilities, which need to be modeled individually,

- the limited energy resources of sensor nodes cannot sustain any resource-demanding attack detection techniques on them, and

- the distinct topologies of wireless sensor networks demand the need for definition of distinct patterns of normal network traffic, based on specific network topologies, to facilitate attack pattern detection.

26.4 ADVERSARY MODEL

The adversary class is defined as a set of malicious entities intending to inflict loss either directly or through other entities on the network. It is responsible for defining, and if need be, introducing malicious nodes into the network, with the purpose of launching a distributed denial of service attack. The set of malicious nodes intending to launch a distribute denial of service attack in a wireless sensor network can be classified into the following categories: (a) injected sensor nodes, (b) compromised sensor nodes, and/or (c) laptop-class nodes. Injected sensor nodes again may consist of either sensor nodes with normal sensor capabilities or more powerful sensor nodes with the capabilities of say the base station. Laptop-class nodes are defined as nodes with more communication resources in terms of transmitting and receiving capabilities that is, stronger antennas as compared to standard sensor nodes. In addition, laptop-class nodes have a battery supply sustaining the node for a longer lifetime as compared to normal sensor nodes. Compromised nodes are defined as legitimate sensor nodes whose operations are taken over by the adversary class for purposes of disrupting normal network operations. In Figure 26.3, we illustrate a distributed denial of service attack model in the presence of various types of nodes in the network. The legitimate nodes of the network include intermediary data aggregation (DA) nodes, cluster heads, noncluster heads, and the base station. The malicious nodes of the network fall into three classes, namely, compromised nodes, malicious (injected) nodes, and laptop-class adversarial nodes. The cluster heads are labeled as target nodes in this particular example scenario. All other legitimate nodes are also vulnerable to a distributed denial of service attack launched by the adversarial nodes of the network.

We address the problem of detecting attacks launched by the adversary-injected nodes in the network and also scenarios wherein a set of sensor nodes in the network may be compromised to launch flooding attacks against other legitimate nodes of the network.

The adversary class launches the distributed flooding attack by instigating the malicious nodes in the network to generate a large set of attack packets from multiple ends of the network towards the victim nodes. The success of the attack is achieved by the collusion feature of such an attack, where participation of multiple malicious nodes takes place. As a result, the per-node overhead incurred due to participation in the attack, that is, generation and transmission of large volumes of hoax packets by the adversarial nodes is lowered significantly. The distributed flooding attack being launched from multiple ends of the network by multiple adversarial nodes incurs energy usage on the adversarial node, with total updated energy content of an adversary node a_k at time t_1 given by $E_{a_k}(t_1) = E_{a_k}(t_0) - E_{\text{trans}}(p/k)$, for a k-adversary

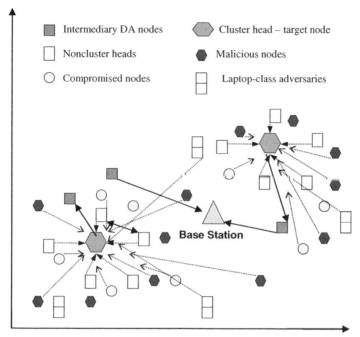

Figure 26.3 Distributed denial of service attack model: wireless sensor networks.

node network, with the summation of attack packets generated by all adversarial nodes $= p$. $E_{\text{trans}}(p)$ is defined as the energy usage for transmission of p packets by an adversarial node. In case a single adversarial node is launching a flooding attack, the total amount of energy needed for transmission of p packets by this single adversarial node is given by: $E_{\text{trans}}(p)$. The added saving of the total energy contents of the malicious nodes of the network when more than one adversarial node is present facilitates the subsequent use of the adversarial nodes by the adversary class, for participation in further disruption activities in the network. These activities may include continuous transmission of flooding attack packets toward other nodes of the network, routing path disruptions, as well as message injection and tampering attacks.

The set of malicious nodes injected into the network by the adversary class need to communicate with each other for synchronous launch of a flooding attack. All such communication for adversary control operations takes place outside the communication band of standard sensor node communication channel to avoid monitoring of adversarial activity as anomalies in communication channel usage by the sensor network. The adversary class monitors the activity of the sensor network to handpick the most active nodes of the network. Therefore, sensor nodes participating in frequent reception and transmission of messages in the network are tagged as "critical" nodes by the adversary class. We refer to these critical nodes as target (victim) nodes, and denote them as $T = \{T_1, T_2, \ldots, T_r\}$. For instance, nodes closer to the base station, responsible for data forwarding to the base station from other nodes, will be more active in the reception and delivery of aggregated messages and are, therefore, more likely to be labeled as target nodes by the adversary class. The purpose of our proposed attack detection scheme is to detect distributed flooding traffic toward this set of r nodes of the network.

The tasks assigned to a sensor node, along with its topological placement in the network, define the level of its criticality to the operations of the network. If the sensor node has a high criticality index, implying that its availability is essential for ensuring uninterrupted operations of the network, its identification and labeling as "critical" by the adversary class can have catastrophic consequences for smooth network operation. The attackers upon identification of critical nodes will launch a distributed flooding attack against them. The adversary class intends to exhaust the energy resources of the r identified target nodes belonging to the set T, of the network, by simultaneous launch of flooding attack traffic toward the victim nodes. All other nodes $N \notin T$ are less significant to the operations of the network and, therefore, can be safely neglected for purposes of attack detection.

A distributed flooding attack can be considered successful if the energy resources of the victim node(s) are exhausted due to the processing demands incurred on them for all operations related to the processing of this large-scale influx of attack traffic packets. Upon complete exhaustion of energy in the victim nodes, the adversarial nodes may steal the identities of legitimate nodes in the network to generate redundant or incorrect sensory data for delivery to the base station, degrade network performance by increasing packet drop rates, and add to the delays associated with packet delivery to the base station. In addition, the adversary nodes may also launch further flooding attacks against other unaffected legitimate nodes of the network.

A second attack scenario involves the adversarial nodes launching the flooding attacks by masquerading as legitimate sensor nodes and flooding the immediate network, that is, neighbor nodes, with large number of hoax requests. An ideal attack situation is when a set of colluding adversaries simultaneously send a large number of hoax requests to the target node from multiple ends of the network. The attackers can thus act stealthily, where otherwise heavy traffic flow intended for a particular target node launched from a single end point of the network can be easily detected as a localized traffic anomaly by a single attack detector node operational in the specified region of the network. We denote the set of adversarial nodes in the network as: $A = \{A_0, A_1, \ldots, A_{k-1}\}$.

The reachability matrix of size $a \times r$ defines the distances that need to be traversed by adversary-initiated messages for traversal toward the r victim nodes. The energy used for message traversal, $E(a, r)$, is proportional to distance(a, r), $\forall \{a \in A$ and $r \in T\}$. The average energy utilisation rate, E_a, of an adversary node $a \in A$ is given by

$$E_a = \frac{1}{|A|} \sum_{i=1}^{|A|} \left(\frac{1}{v(i)} \sum_{j=1}^{|v(i)|} (E_{\text{util}}(i, v(i))) \right) \tag{26.1}$$

where $v(i)$ is the set of victim nodes targeted by node a and $E_{\text{util}}(i, v(i))$ is the energy consumption for transmission of a malicious packet by a malicious node i toward a victim node $v(i)$. The larger the victim node set for a given adversarial node a, the higher the energy consumption rate will be. However, in the presence of a large number of adversarial nodes in the network, the per-node energy consumption rates associated with launching such an attack will be reduced. As a result, the lifetimes of the adversarial nodes will also be extended.

The average energy usage by a sensor node n intending to transmit a message to a destination node d in the network is given by

$$E_n = \frac{1}{|N|} \sum_{i=1}^{|N|} E_{util}(n_i, d) \tag{26.2}$$

where both n_i and d are both sensor nodes or either one of them is the base station.

The lifetime of the n sensor nodes of the network is given by $G(n)$, where $G(n) \propto 1/E_n$ and the average lifetime of the adversary node in the network, $G'(a)$ is given by $G'(a) \propto 1/E_a$. Considering the frequent nature of packet generation and transmission by the adversarial nodes participating in the attack, for scenarios with fewer number of adversarial nodes, the individual lifetimes of the adversarial nodes, $G'(a)$, can be considered to be $\ll G_n$. On the other hand, if a large set of adversarial nodes participate in the attack, we can expect $G'(a) \gg G(n)$. It may thus be observed that the adversary class participating in a distributed flooding attack will survive for longer postattack success as compared with scenarios where centralized attacks are launched by the adversary class through a single attacker node. As a result, the adversarial nodes can successfully masquerade as legitimate but victimized nodes and operate unaffected in disrupting the operations of the network.

The distributed denial of service attack is thus more successful, if launched, in distributed fashion by multiple adversarial nodes. It is, therefore, imperative to have multiple attack detector nodes in place to detect traffic flow for attack classification purposes.

26.5 NETWORK MODEL

The wireless sensor network model consists of a finite set of sensor nodes given by $N = \{N_1, \ldots, N_n\}$, where $|N| = n$. The network also consists of a centralized base station in addition to the sensor nodes. The n sensor nodes of the network consist of sensors with added capabilities and/or administrative and control tasks of the network (cluster heads and data aggregation points), as will be explained in the next paragraph. Victim nodes are defined as a set of nodes $T = \{T_1, \ldots, T_r\}$, where $T \subset N$, such that, each target node r of set T is a critical node of the network, and $|T| = r \ll n$. The adversary class is defined as the set of malicious nodes in the network and are denoted as $A = \{A_0, A_1, \ldots, A_{k-1}\}$, where $|A| = k \leq n$.

Sensor nodes of a typical wireless sensor network operate with the purpose of monitoring and detecting events in their environments for subsequent delivery of their respective observations and readings to a centralized base station. The data can either be delivered to the base station directly by the sensor nodes or through a chain of defined intermediary nodes. We refer to the frequency of communication of messages by the nodes to the base station as the network taxonomy. We classify the data delivery model, that is, network topologies, for transfer of data from sensors to the base station into three most common sensor network classes, namely, flat, cluster-based, and data aggregation. These network topologies are defined based on the data delivery model of the network. In the source–sink model of communication, the traffic packets originating from a source node can either be forwarded by the sensor nodes directly to the base station or through a set of intermediary nodes. These intermediary nodes are referred to as aggregation nodes. We further classify the latter case by dividing it into two subcategories of network topologies, namely, cluster-based network topology and data aggregation-based network topology. The cluster-based topology relies on a two-hop approach for data delivery, whereas the data aggregation-based topology uses multiple hops for packet delivery from the sensor nodes to the base station.

These network topologies, together with their traffic flow models, are defined as follows:

1. **Flat topology:** In a flat topology, each sensor node in the network directly communicates its sensor readings to the base station using a single-hop mechanism, without intermediate message transfer nodes to aid in the communication process. Every sensor node has equal priority designated for such networks. The traffic flow from the sensor nodes to the base station here can be expressed as $f = \{f_1\}$, depicting a single hop transmission to the base station. It is assumed that for the flat topology to operate successfully, all sensor nodes must have sufficient transmission ranges to facilitate their communication with the centralized base station.

2. **Cluster-based topology:** In a cluster-based network topology, a set of sensor nodes with added capabilities are defined as cluster heads. These cluster-head nodes act as control and administrative centres for a set of predefined clusters of sensor nodes in the network. Cluster heads are responsible for the administration of their respective clusters, data aggregation from sensor nodes of their clusters, and data forwarding to the base station. In addition, cluster heads are also responsible for monitoring the status of sensor nodes in their clusters and reporting of faults and losses to the base station. Cluster-based networks generally follow a two-hop traffic flow path to reach the base station. This flow can be expressed as $f = \{f_{f,\text{ch}(f)}, f_{\text{ch}(f),\text{bs}}\}$, where $f_{f,\text{ch}(f)}$ is the flow from node f to its cluster head $\text{ch}(f)$, and $f_{\text{ch}(f),\text{bs}}$ is the flow from $\text{ch}(f)$ to the base station. Certain special-case cluster-based topologies rely on multiple hops for data transfer between the cluster heads before being forwarded to the base station. We classify such topologies as data aggregation network topologies.

3. **Data aggregation topology:** In a data aggregation topology, sensory readings from individual sensor nodes progress through the network from the source node toward the base station through a well-defined tree of interconnected intermediary nodes. The data along the path is aggregated at specific nodes in the network called aggregation points, defined as nodes with the numbers

of incoming edges to the nodes exceeding their total outgoing edges (usually equal to unity). The purpose of aggregating intermediary data is to reduce the total traffic flow in the network and to miminize the energy consumption associated with frequent and large-scale data transfer operations to the base station by individual sensor nodes of the network. A typical data aggregation topology consists of interconnected trees defining the flow of network traffic from individual source sensor nodes to the base station. The traffic flow from the sensor nodes of a data aggregation tree through aggregation nodes can be expressed as $f = \{f_1, f_2, \ldots, f_{L(f)}\}$, where $L(f)$ is the length of path from node f to the base station.

The network model defined in this section is crucial for the generation of traffic threshold patterns for attack detection purposes. The threshold pattern generation process relies on the underlying topology of the network for generation of subpattern values for storage and subsequent comparison by the attack detector nodes.

26.6 THRESHOLD PATTERN MODELING

In the previous section, we have classified wireless sensor networks into three most network topologies based on the data delivery models. In this section, we propose the generation of threshold subpattern values for storage and comparison within the attack detector nodes based on these defined network topologies.

The analytical model of a sensor network undergoing a DDoS attack consists of two types of network traffic, namely, normal and attack. The flow of traffic in a typical sensor network is directed from the sensors to the base station. During normal operations mode, a sensor node may receive traffic from several sources, such as from nodes within its immediate vicinity. The volume of traffic and, in essence, the numbers of traffic flows are higher if the sensor node is a cluster-head or a data aggregation node. The traffic constituting a distributed denial of service attack can also be categorised as a flow, albeit with a different label. We assume that each adversarial node generates a single flow of traffic toward a victim node r. In the presence of attack traffic, the total traffic received by a target node r in a given time epoch, and that needs to be monitored by the attack detection scheme, is given by

$$\lambda_r = \sum_{i=1}^{f} \lambda_{r,i}^i + \sum_{j=1}^{k} \lambda_{r,j}^j \tag{26.3}$$

where $\lambda_{r,i}^i$ is the normal traffic rate belonging to traffic flow from node i and $\lambda_{r,j}^j$ is the attack traffic rate originating from an attacker node j belonging to the attacker set A. Each node in the network is considered to bear a single queue, with average time for packet processing and transmission at node i being s_i. The intensity of the arriving traffic at node r is thus given by

$$\rho_r = s_i \left(\sum_{i=1}^{f} I_{r,i}^i + \sum_{j=1}^{k} I_{r,j}^j \right) \tag{26.4}$$

$I_{r,i}^i$ is defined as normal traffic intensity, whereas $I_{r,j}^j$ is defined as the attack traffic intensity for all attack nodes $k \in A$. We consider the case of attack detection by means of studying the overall traffic intensity toward a set of target nodes in the network. The traffic arrival intensity at node r thus is a function of the individual arrival intensities of both the normal sensory traffic and the attack traffic.

A distributed denial of service attack flow is launched by several attacker nodes from multiple ends of the network. The attack packets may arrive at the target node(s) from different regions of the network and, therefore, a collaborative effort is needed to detect distributed anomalous traffic flow toward the target node set. We define a set of sensor nodes called attack detector nodes as nodes that observe traffic flow of the network toward the target node set T. These nodes are notated as $G = \{g_0, g_1, \ldots, g_{d-1}\}$, where $|G| = d$. The broadcast nature of traffic in sensor networks facilitates the promiscuous monitoring of traffic flows in the network toward the target nodes. The threshold values are defined as the maximum numbers of packets a node r is willing to accept from a particular network region during a constant time interval Δ from the region of operation of the observer (detector) node. One of the factors for generating these threshold values is the topological designation of a target node in the network. We define distinct attack patterns based on this topological placement of the sensor nodes. Considering that different threshold (subpattern) values will be stored for each target node, the complete threshold pattern vector is a unique pattern defining a set of bounds on the receivable traffic by a node r during a given time interval, from all regions of the network. For a constant network taxonomy, the total traffic that a particular node r in the network can expect in a given time interval is denoted by P_r. This expected traffic inflow value depends on the network taxonomy, node r's initial energy content, its expected lifetime, and the average energy resource usage by node r for processing of each received packet. These values are generated beforehand at network initialization time.

The pattern generation criteria varies for the three network topologies defined earlier. Each network topology has a different set of selected potential target nodes T in the network. Each target node has a different set of traffic flow patterns toward them, which need to be observed. In a flat topology, all sensor nodes in the network are at the same level of criticality. The loss of any of the n nodes in this topology is likely to have an equal impact on the operations of the network. Subpattern (threshold) values for a flat topology are generated based on Eq. (26.5). The threshold subpattern values for target node r, associated with the detector node d, is denoted as th_d^r.

Sensor nodes are deployed at network initialization time. The base station has a record of the total number of nodes in the network, as well as an estimate on the distances to each of these nodes. The above parameters facilitate computation of the density of node deployment. For instance, a network spanning a large geographic area with fewer numbers of nodes will have a low node deployment density and a network covering a smaller geographic area, with large numbers of deployed nodes, will have a higher node deployment density. The density of deployment of nodes in a flat topology defines the extent of loss that may be incurred on the network due to the loss of a single target node. For denser networks, the loss of a few nodes will be less significant as compared to a network with low node deployment densities. Therefore, the threshold value, th_d^r, is high for denser networks, implying that a larger set of target nodes can be lost before an alarm can be raised.

$$th_d^r = \left[P_r + \text{nw(density)} + \frac{1.0}{d_{G(d)}(r)} \right] \tag{26.5}$$

where nw(density) = normalized node deployment density of the network, $d_{G(d)}(r)$ = normalized Euclidean distance from detector node d to target node r, and P_r = normalized number of expected packets by node r in a fixed time interval, Δ.

The Euclidean distance of a target node from a particular detector node is another factor used for computation of the threshold pattern value. Target nodes outside the observation range of a detector node, d, need to be monitored by other closely located detector nodes in the network. A lower threshold value implies fewer numbers of traffic packets are expected from this particular region toward the target node.

In a cluster-based network topology, the cluster heads play a crucial role in the operations of the network and, therefore, need constant monitoring of traffic flow toward them. We, therefore, consider the cluster heads to be critical nodes in this topology. The threshold subpatterns for this network topology are generated from Eq. (26.6).

$$th_d^r = \left[P_r + \text{num}_{ch} + \frac{1.0}{d_{G(d)(r)}} \right] \tag{26.6}$$

where num_{ch} = normalized number of clusters in the network, and $d_{G(d)(r)}$ = normalized Euclidean distance from detector node d to the node r.

The value of $d_{G(d)(r)}$ is the normalized distance between the cluster heads and the detector nodes. This value defines the expected traffic flow intensity toward the cluster heads from different regions of the network. Cluster heads distantly located from the base station are generally at the end of a tree routing hierarchy and thus accumulate fewer numbers of traffic packets from leaf-end sensor nodes in a given time interval. The values of P_r are lower for such nodes. On the contrary, cluster heads closer to the base station are responsible for aggregation of packets, in addition to their cluster head operations and therefore expect higher traffic inflows due to the influx of large cumulative traffic payload. The values set for P_r are higher for such nodes. The normalized Euclidean distances between detector nodes closer to cluster nodes yield higher threshold values, depicting more numbers of expected requests toward these cluster nodes, whereas detector nodes farther away from a cluster node are considered to be outside their respective regions of monitoring, and therefore lead to reduced threshold values. The node deployment density in cluster-based networks defines the total numbers of operating clusters in the network. Therefore, for higher node deployment densities, higher values of th_d^r are generated, indicating lesser significance given to each cluster node.

In a data aggregation topology, the data aggregation nodes in the network are significant in the aggregation and forwarding of sensory data up the tree hierarchy. The loss of these nodes may lead to the inactivity of a complete arm of operation (sensor region) of the network. Data aggregation nodes are considered critical target nodes in this topology. The pattern generation equation for a data aggregation topology is

$$\text{th}_d^r = \left[P_r + \frac{1.0}{d_{G(d)(r)}} \right] \tag{26.7}$$

where $d_{G(d)}(r) = $ Euclidean distance from detector node d to the target node r.

The density of node deployment plays a less significant role in data aggregation networks, as the tree paths for routing of sensory data are fixed at network initialization time, and remain unaltered. The number of hops separating a data aggregation node from the base station define its level of significance. Aggregation nodes closer to the base station will expect more inflows of network traffic toward them, and therefore will have higher associated P_r values, thus leading to higher threshold values. On the other hand, aggregation nodes closer to the leaf-end sensor nodes will expect lesser traffic inflow from sensor nodes lower in the hierarchies, and therefore, will have smaller associated P_r values, thus leading to smaller threshold subpattern values. Detector nodes in proximity to the data aggregator nodes are expected to observe higher traffic flow toward them, whereas detector nodes farther away from the aggregation nodes set lower th_d^r values, indicating fewer traffic flow rates toward the target node.

All communication packets in the network are assumed to have a node identification tag appended to them for identifying both the source as well as the intended destination of the traffic packets. Node identification can be generated using unique knowledge possessed by a sensor node. Such knowledge can be the relative geographic location of the sensor node, which can be preset into the sensor memory at network initialization time. The ID for node n at location $\langle l_x(n), l_y(n) \rangle$ is given by $\Lambda : N \rightarrow N(l_x, l_y)$, where the function Λ uses the geographic coordinates of a node n to derive its unique location coordinate identifiers.

In Table 26.2, we have illustrated the subpattern (threshold) values associated with each of the d detector nodes of the network, along with the location coordinates of the target nodes $l_x(r), l_y(r)$, to facilitate feature comparison associated with real-time traffic flow, with stored threshold subpattern values. The complete pattern vector, if to be analyzed by a centralized entity, for a given target node t_1, in the presence of d detector nodes, is given by $\langle l_x(t_1), l_y(t_1), \text{th}_0^1, \text{th}_1^1, \ldots, \text{th}_d^1 \rangle$.

Although the individual observation of a single detector node will not depict an entire flooding attack scenario, the coordinated reconstruction of the complete pattern of observed traffic readings, by all detector nodes, facilitates achieving the same.

Table 26.2 Threshold Subpatterns for a Set of Two Example Target Nodes, to be Stored One Each Within the d Detector Nodes

(Detector node, node ID)	t_1	t_2
$(1, \text{ID}(1))$	$\text{th}_1^1, l_x(t_1), l_y(t_1)$	$\text{th}_1^2, l_x(t_2), l_y(t_2)$
$(2, \text{ID}(2))$	$\text{th}_2^1, l_x(t_1), l_y(t_1)$	$\text{th}_2^2, l_x(t_2), l_y(t_2)$
.	.	.
.	.	.
$(d, \text{ID}(d))$	$\text{th}_d^1, l_x(t_1), l_y(t_1)$	$\text{th}_d^2, l_x(t_2), l_y(t_2)$

26.7 TRAFFIC FLOW OBSERVATION TABLE

After the expected initial threshold values for a set of target nodes are generated, the attack detection scheme is trained with these values to facilitate learning of normal network traffic flow patterns. Subsequently, comparisons of statistical features extracted from observed traffic flow in the network help the base station to make a decision on whether an attack is in progress or not. These features define the intensity of traffic flow in the network toward a set of r target nodes for classification of flooding attacks by the attack detection scheme. The features to be extracted from the traffic constitute the pattern vectors that need to be compared during the pattern-matching process of the detection scheme. These traffic features are given by:

- percentage of packets with destination address $= d$, where $d \in T$,
- percentage of packets with source address $= \{s \mid \forall_r, \text{Euclidean}(s, r) > \text{thr}_{euc}\}$, where thr_{euc} is the threshold on maximum permissible distance between the detector and the target nodes, and
- percentage of packets with source address $= \{s \mid s \notin \text{cluster}_d, \text{where } d \in T, s \in N\}$.

Definition 26.1. $\forall$ patterns p_r, length$(p_r) = 2r$.

For a centralized approach toward attack detection, that is, without the presence of localized decision making in the network, the total number of pattern vectors expected by a base station for classification purposes at the end of a time epoch Δ is equal to n. The length of each pattern vector, as can be seen from Figure 26.4, is equal to $2r$. For each of the r target nodes, a pattern vector, p_r, will be reconstituted at the base station based on the receipt of individual subpatterns from each of the n detector nodes. The pattern vector for target node r is given by $p_r = \{p_1^r, pe_1^r, p_2^r, pe_2^r, \ldots, p_n^r, pe_n^r\}$, where p_n^r is the percentage of packets destined for target node r observed by detector node n and pe_n^r is the percentage of packets observed by node n as to possessing a source address outside the Euclidean threshold defined by thr_{euc}, as satisfying the second rule for feature extraction defined above.

For cluster-based wireless sensor networks, the pattern vector for a target node r is given by $p_r = \{p_1^r, pc_1^r, p_2^r, pc_2^r, \ldots, p_n^r, pc_n^r\}$, where p_n^r is the same as the previous scenario. However, for cluster-based networks, we define the subpattern pc_n^r as a value indicating the percentage of packets observed by detector node n as being directed to a target r from outside its cluster of operation, cluster_r.

We have illustrated above the techniques for generation of pattern vectors from observed real-time network traffic flow. These pattern vectors are compared with the threshold subpattern values, generated and stored in each of the attack detector nodes. The attack subpatterns vary based on several parameters including the proximity of the target nodes to the detector nodes. Therefore, the threshold values generated are different for each detector node region and cannot be modeled simplistically as a cumulative sum of all subpatterns, that is, traffic flow values, toward a target node during a given time epoch Δ.

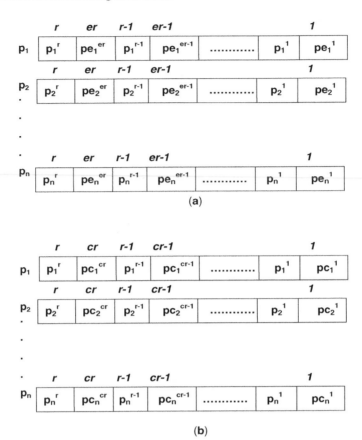

(a)

(b)

Figure 26.4   Reconstituted pattern vectors depicting bounds on network traffic flow toward the victim node set.

26.8 CENTRALIZED ATTACK DETECTION

Neural networks are known to be a very powerful tool in detecting anomalous network traffic in high-performance networks. One such class of neural networks that has been used extensively for intrusion detection and attack detection is the self-organising map (SOM). A SOM is a nonlinear, ordered, smooth mapping of high-dimensional input data manifolds onto the elements of a regular, low-dimensional array [20]. From an intrusion detection perspective, the resulting geometric map of neurons depicts patterns of actual network traffic flow. By constructing a lattice mapping of higher dimensional data, the SOM facilitates visualization and subsequent analysis of data required for detecting anomalies in network traffic. The SOM algorithm is topology preserving in nature, that is, input pattern vectors close to each other in terms of similarity are mapped on neurons of the map, which are in close proximity to each other [12]. This characteristic of the SOM neural network makes it more practical for accurate differentiation between normal and anomalous network behavior. Several

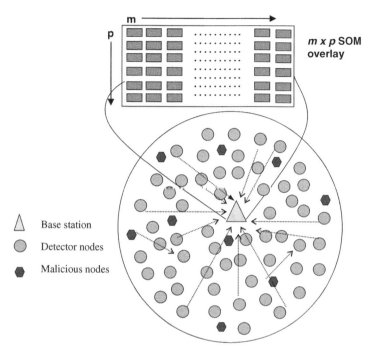

Figure 26.5   SOM overlay on base station.

SOM-based intrusion detection schemes have been proposed in Refs [11, 12]. In wireless sensor networks, self-organizing maps have been introduced for generation of optimal data-aggregation trees [13] and context classification [2].

We propose an attack detection scheme that uses the clustering capabilities of the SOM algorithm to cluster DDoS attack patterns into neurons on the map and extend it to incorporate a decision-making layer, operating on the base station of the wireless sensor network, to generate traffic classification outcomes. The SOM Overlay for the proposed attack detection scheme is illustrated in Figure 26.5.

The proposed scheme operates at two layers, namely SOM layer and the decision layer. The SOM application operating as part of the SOM layer is responsible for clustering of network traffic packets into one of l neurons of the lattice map, based on the proximity of the neuron weight to the weight of the input vector, in terms of Euclidean or Hamiltonian distance between them. The decision layer does the actual classification of network traffic into attack or normal based on the inputs received from the SOM layer.

The SOM-based attack-detection scheme has two phases of operation, namely learning and classification.

26.8.1   Learning Phase

During this phase, a.k.a training phase, the SOM application is introduced with a set of learning patterns to train the l map neurons to map data points of the input vector onto the array of neurons. The mapping process is competitive. It is performed by introducing data points of the input vector to each of the l neurons of the map, one vector at a time. For each input vector, the neuron with the closest weight in distance (Euclidean or Hamiltonian), to the input data point, is declared the winner. Subsequently, the weight of the winner neuron is adjusted to ensure that its values are inclined more toward data points similar in characteristics to the current input data point. In addition, for each of the input vectors, neighbors of the winning neuron have their weights updated as well. A neighborhood function needs to be defined to calculate the neighbors of a given winner neuron. Typically, the neighborhood function is taken as either Gaussian or Bubble. The k-dimensional values within the neighboring neurons of the winner are adjusted accordingly.

26.8.1.1 Neighborhood Function

The most common neighborhood functions proposed for SOMs are Gaussian and Bubble. We refer to neighborhood function for winner n at a given time t, as $h_{n_i}(t)$. The time factor t is a monotonous discrete time variable, incremented with each iteration of the SOM training process. The update factor α of the neighborhood function is defined as the rate at which the weights of the neighbor neurons n_i of a winner n are updated. The radius of the neighborhood of the winner in the map is defined as σ. The Bubble function specifies a constant learning update factor α for each of the neighbors n_i. The Bubble neighborhood update function is given by

$$h_{n_i}(t) = \begin{cases} \alpha(t) & \|d_n, d_i\| < \sigma(t) \\ 0 & \text{otherwise} \end{cases}$$

where d_n and d_i are the positions of the winner n and its respective neighbors n_i, with $\|d_n, d_i\|$ defined as the Euclidean or Hamiltonian distance between the node pair (n, i).

On the other hand, the Gaussian neighborhood function follows the standard bell-shaped Gaussian model for updating the neighbor weights. The update factor α here is inversely proportional to the distance of the neighbor nodes n_i from the winner n. Therefore, for distant neighbors, the α has lower values and for neighbors within close proximity of n, the update rate is significantly higher. The Gaussian neighborhood update function is given by

$$h_{n_i}(t) = \begin{cases} \alpha(t)e^{-\frac{\|d_n, d_i\|^2}{2\sigma^2(t)}} & \|d_n, d_i\| < \sigma(t) \\ 0 & \text{otherwise} \end{cases}$$

26.8.1.2 Update Function

The update function of the SOM algorithm performs the update of the neuron weights of the neighbors of the winning neuron, subsequent to the successful determination of the winning neuron n, and the neighbor set n_i. The update function is given by

$$m_i(t+1) = m_i(t) + h_{n_i}(t)[\mathrm{sd}(t) - m_i(t)] \tag{26.8}$$

where $m_i(t)$ is the set of values of the k-dimensional vector of neuron i at time t and $\mathrm{sd}(t)$ is the k-dimensional vector of the sample (training) data.

The SOM is trained for a period of τ training cycles, with a single training vector of length k introduced to the SOM during each iteration. The value of σ (neighborhood radius) is initially chosen to be large and is progressively reduced in value with a certain factor (r_0), with each training cycle (iteration).

At the end of training phase, each neuron of the map is labeled as an attack or normal class on the basis of a majority count of the classes of input patterns for which the neuron was declared the winner. The labeling function is defined as

$$\mathrm{label}_l = \begin{cases} \mathrm{attack}, & t^l_{\mathrm{attack}} > \mathrm{thresh}_{\mathrm{attack}} \\ \mathrm{normal}, & t^l_{\mathrm{normal}} > \mathrm{thresh}_{\mathrm{normal}} \end{cases}$$

where t^l_{attack} is the total number of attack packets for which neuron l was declared the winner and t^l_{normal} is the total number of normal packets for which neuron l was declared the winner. The $\mathrm{thresh}_{\mathrm{attack}}$ is the threshold of attack packets, if observed by a neuron i, will lead to it being labeled an attack class. Similarly, $\mathrm{thresh}_{\mathrm{normal}}$ is the threshold of normal packets, if observed by a neuron i, will lead to it being labeled a normal class, where $\mathrm{thresh}_{\mathrm{normal}} = 1 - \mathrm{thresh}_{\mathrm{attack}}$.

26.8.2 Data Classification

During the classification phase of the scheme, the k-dimensional weight arrays associated with the input vectors are compared with the weight vectors of the l neurons of the map. The neuron with the closest match is declared a winner, and the corresponding input vector is classified accordingly. The decision-making layer of the scheme generates the final verdict on the classification of the observed input pattern vector into attack or normal traffic flow.

26.8.3 Traffic Features

Distributed denial of service attack patterns are defined as increases in traffic flow toward a set of r target nodes during an epoch of time of t. The rate of packet flow in the network tends to change in the event of a flooding attack. The statistical features defining the intensity of traffic flow in the network toward a set of r target nodes are extracted and used for neural network training for classification of flooding attacks. During the learning phase of the SOM application, the k neurons are introduced with extracted statistical features of traffic in a fixed interval of time of length τ. This traffic is defined as the *training* traffic for the SOM application and consists of both attack as well as normal traffic vectors. The training traffic may contain random repetitions of pattern vectors of both attack and normal traffic. The total number of pattern vectors introduced to the SOM application during the training phase also defines the total number of epochs of time for which the application was trained. The features to be extracted from the traffic constitute the input vector and are used both during the training phase of the detection scheme and during the traffic classification phase. These traffic features are given by

- percentage of packets with destination address $= d$, where $d \in r$,
- percentage of packets with source address $= \{s \mid \forall_r, \text{Euclidean}(s, r) > \text{thr}_{\text{euc}}\}$, and
- percentage of packets with source address $= \{s \mid s \notin \text{cluster}_d, \text{where } d \in r, s \in n\}$.

Definition 26.2. $\forall$ patterns p_r, length(p_r) $= 2r$.

The total number of pattern vectors expected by a base station at the end of a time epoch τ is equal to n. The length of the pattern vector, as can be seen from Figure 26.4, is equal to $2r$. For each of the r target nodes, a pattern vector, p_r, is reconstituted at the base station based on the receipt of individual subpatterns from each of the n detector nodes. The pattern vector for target node r is given by $p_r = \{p_1^r, pe_1^r, p_2^r, pe_2^r, \ldots, p_n^r, pe_n^r\}$, where p_n^r is the percentage of packets destined for target node r observed by detector node n and pe_n^r is the percentage of packets observed by node n as to possessing a source address outside the Euclidean threshold defined by thr_{euc}, as satisfying the second rule for feature extraction defined above.

For cluster-based wireless sensor networks, the pattern vector for a target node r is given by $p_r = \{p_1^r, pc_1^r, p_2^r, pc_2^r, \ldots, p_n^r, pc_n^r\}$, where p_n^r is the same as the previous scenario. However, for cluster-based networks, we define the subpattern pc_n^r as a value indicating the percentage of packets observed by detector node n as being directed to a target r from outside its cluster of operation, cluster_r.

26.8.4 Detection Algorithm

The training phase of the SOM algorithm is performed offline on the base station, using the patterns generated as part of the sample data. Prior to executing the training phase, the SOM application is initialized with the selected SOM training parameter values. The patterns generated as part of the training dataset depend on the length of the convergence time window τ of the application. The classification phase of the application takes place at the end of each time interval τ as follows:

1. Observation phase: During this phase, the sensor nodes constituting the set of attack detectors observe traffic flow within their respective neighborhoods towards any of the r target nodes of the network.

2. Convergence phase: The observer sensor nodes construct r vectors based on the observed traffic flow toward the r target nodes.

3. Communication phase: At the end of the current time epoch t_i, the observer nodes communicate the constructed observation vectors in the form of patterns to the base station.

4. Verdict: The base station decides on declaring the traffic as attack or normalcy toward any of the r target nodes for a given time epoch t.

26.8.4.1 Map Initialization

The initial values selected for the map are crucial in defining a good-quality map layout at the end of the training phase. Initial neuron weights can be defined either linearly or randomly. The weights must be within the range of values of the r-dimensional pattern vectors in the sample dataset. We generate values for the initial map dimensions, based on the sample data consisting of both attack and normal network traffic, generated based on the definitions given in the previous section. The process of map generation is performed by computing the eigenvalues of the square matrix given by AA^T, where A is the matrix of training patterns of size $m \times p$ and A^T is the transpose of the matrix. The two largest eigenvalues are then selected and the ratio between the largest and smallest is calculated. The map dimensions are selected such that the ratio of the map dimensions is proportional to the square root of the calculated ratio.

26.8.4.2 Optimal Time Epoch

The value of τ, the optimal time epoch length, has a significant effect on the performance outcomes of the scheme. Smaller values of τ will incur more communication overhead due to more frequent exchange of messages on the the detector node–base station channel. In addition, the accuracy in attack detection will be lower, since fewer observed packets will have a significant chance of generating greater numbers of false alarms. On the contrary, very large values of τ will reduce the overall communication overhead of the scheme and increase the accuracy in the detection rates albeit at the cost of delayed response, and false negative rates are higher where the attack packets will cause damage before the actual scheme convergence.

The optimal value of τ for the SOM-based attack detection scheme is given by

$$\tau = \sqrt{\frac{c1}{c2} \frac{n_{\text{total}}}{TI}} \qquad (26.9)$$

where c_1 is the cost incurred due to a single communication on a wireless channel $\approx$ 1 ms, c_2 the cost incurred due to loss of target node in the network $\propto 1/\text{NodeDensity}$ $\approx \{0.5–1.0\}$, n_{total} is the total detector nodes, and TI the Attack traffic intensity (packets/s).

26.8.5 Evaluation

We performed simulations to generate results for the attack detection rates, false positive rates, and the false negative rates for varying values of N and varying network traffic intensities. The simulator was written in a Java Integrated Development Environment. The traffic flow follows the Poisson distribution.

26.8.5.1 Detection Rates

In Figure 26.6, the attack detection rate during initial lifetime of the network (postinitialization), is plotted for varying node deployment densities. The detection rate is nearly 92% for high node deployment densities and low traffic intensities, whereas for $N = 128$, the detection rate is only 65%, even for low traffic intensities. The lower node deployment density networks have lesser detection rates as compared with the higher density networks. The presence of fewer detector nodes in the low-density network scenarios lead to fewer successes in the attack-detection process due to the incompleteness in the pattern vectors generated for transfer to the base station for subsequent clustering by the SOM application.

Higher traffic intensities imply more numbers of packets penetrating the network, unnoticed, during the convergence of the detection scheme, and therefore larger values of TI lead to lower detection rates for all values of TI.

In Figure 26.7, we plot the average detection rate of the SOM-scheme vs. the rate of decline of energy content within the target nodes, for TI = 500. The rate of decline is defined as the percentage reduction in the energy content, as compared with the initial energy values of the target nodes. It may be noted that the rate of decline of energy content in the target nodes depends on the traffic intensity as well as the node's topological commitments. Timely detection of distributed denial of service attacks will facilitate resource reallocation by the base station, and hence avoid disruptions in the operations of the network.

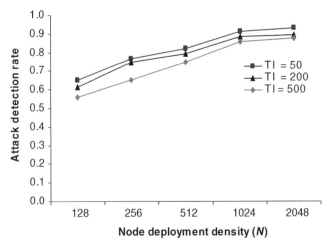

Figure 26.6 Initial attack detection rate vs. network types for varying traffic intensities. a peak value of 92% is achieved for $N = 2048$ and TI = 50. The lowest detection rate is for $N = 128$ and TI = 500.

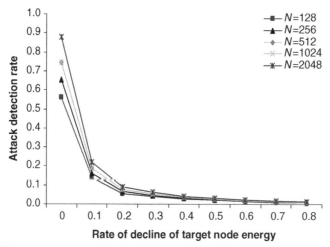

Figure 26.7   Average attack-detection rate vs. rate of decline of energy content in the target nodes.

As can be observed, the average detection rate drops significantly with the corresponding decay in the energy content of target nodes. This occurs due to the inability of the SOM application to update pattern values depicting total number of receivable requests by the target nodes, based on the nodes' energy decline rate. The continuous decay of energy content within the target nodes demands a corresponding update of pattern values, which need to be observed. Considering the inability of the SOM-based approach to update pattern vector values in real time, the detection rates fall considerably with the passage of time. At the 0.9T mark, roughly 90% of the detector nodes die, leaving incomplete pattern vectors for analysis until the 1.0T mark, where nearly all detector nodes die.

26.8.5.2 False Positive Rates

The false positives for the detection scheme are the total number of normal packets clustered by the detection scheme as attack packets. Considering the lack of detector nodes in certain regions of the network, for smaller values of N, incomplete pattern vectors are generated for clustering at the base station. Therefore, the false positive rates are higher for such networks. Figure 26.8 illustrates the false positive rate for the attack detection scheme.

The false positive rate increases with increasing traffic intensities. This is because the pattern vectors generated for traffic analysis are more accurate when fewer numbers of packets penetrate the network at time of scheme convergence. As a consequence, the accuracy of the classification performed at the base station is higher for low TI values. Therefore, fewer false positives are observed for lower intensities of traffic flow.

From Figure 26.9, we can observe that the average false positive rate increases with the increasing rates of decline of target node energy content. As with the false

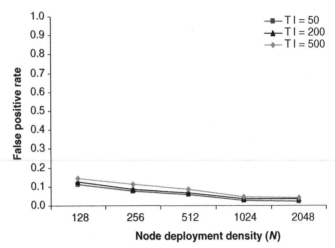

Figure 26.8   Initial false positive rate vs. network types for varying traffic intensities. A high false positive rate of nearly 14% is observed for $N = 128$ and TI $= 500$, whereas a very low false positive rate of approximately 2% is observed for $N = 2048$ and TI $= 50$.

negative rates, the reducing energy contents of the target nodes require corresponding updates in the pattern values, unaccomplished by the self-organising map application. Therefore, the false positive rates show a significant and steady increase with the decline in node resources over time.

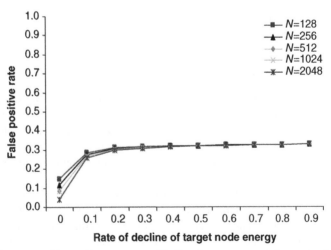

Figure 26.9   Average false positive rate vs. rate of decline of energy content in the target nodes. TI $= 500$. A peak false positive rate of 30% is observable for all N values when 10% of the target node's energy content is depleted.

26.8.5.3 False Negative Rates

The false negatives for the scheme are defined as the total number of attack packets classified by the detection scheme as legitimate traffic packets. The false negatives for the detection scheme are a summation of the total number of attack packets clustered by the SOM application into a normal cluster and the total number of attack packets that remain undetected at the time of application convergence (i.e., node-to-base station communication). The comparison of the initial false negative rates for varying network types is illustrated in Figure 26.10. The false negative rates are lower ($<10\%$) for larger values of N as compared with corresponding false negative rates from smaller values of N. The absence of detector nodes leads to the generation of incomplete pattern vectors for subsequent analysis by the SOM application, and therefore causes an increase in the false negative rates of the scheme.

The false negative rate increases with increasing values of TI, as higher traffic intensities also lead to higher numbers of attack packets entering the network within the same convergence time period, and therefore the total number of attack packets that remain unobserved during the same time epoch length is higher. Therefore, higher traffic intensities will lead to higher rates of false negatives.

In Figure 26.11, the false negative rate for progressing lifetimes of the detector nodes is given

As can be observed, the average false negative rate increases with corresponding decrease in the lifetime of the target nodes. A peak false negative rate of 62% is observable when nearly 12% of a target node's lifetime is depleted. It may be observed that after a certain lifetime of a target node is reached, 12% in this case, the role of

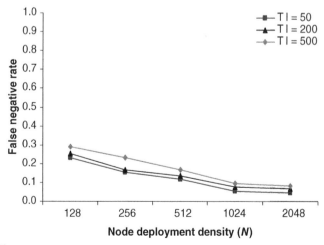

Figure 26.10   Initial false negative rate vs. node deployment density (N) for varying traffic intensities. The highest false negative rate value observed is 30% for $N = 128$ and TI = 500 and the lowest value observed is 5% for $N = 2048$ and TI = 50.

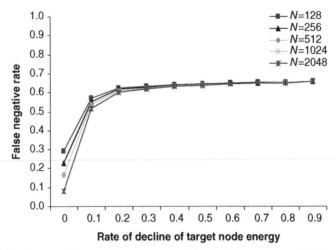

Figure 26.11   Average false negative rate vs. rate of decline of energy content in the target nodes. TI = 500. A peak false negative rate of 62% is observable for all N values when 13% of the target node's energy content is depleted.

additional attack detector nodes, to reconstitute a complete pattern vector for analysis at the base station, becomes ineffective. Therefore, even higher values of N do not affect the false alarm rates of the scheme.

26.8.5.4 Energy Decay Rate

In this section, we analyze the energy decay rates of the attack detector nodes of the network. As elaborated earlier, attack detector nodes observe and generate attack pattern vectors at the end of each time epoch Δ for communication to the base station. The total energy decay rate of the detector nodes is therefore a function of the communication cost, Cost_{comm}, and the computation cost, Cost_{comp}. With the Cost_{comp} associated with storage and generation of pattern vectors within a detector node being negligible, the energy decay rate is approximately equivalent to $f(\text{Cost}_{comm})$. The Cost_{comm} in turn is a function of the network dimensions and the node deployment densities.

The average rate of decline of energy resources, L_n, of a detector node n in the self-organizing map scheme is given by

$$\mu_{som} = \frac{\text{pkts(recv)}E_{recv} + \text{pkts(trans)}E_{trans}d^4}{t} \tag{26.10}$$

As can be seen from Table 26.3, energy exhaustion rate is higher for networks with lower densities of node deployment. This is due to the longer average distances between the detector nodes and the base station in such scenarios. On the contrary, networks with higher node deployment densities have lesser average distances for coverage by the pattern messages from the detector nodes to the base station, and therefore, smaller energy decay rates.

Table 26.3 Energy Decay Rates for the SOM-Based Centralized Detection Scheme

Node deployment density (N)	Energy decay rate (μJ/s)
128	346
256	173
512	136
1024	122
2048	90

26.8.5.5 Summary

The attack detection rates of the SOM-based scheme start reasonably high at initialization time, Figure 26.6, for all node deployment densities. For instance, the $N = 2048$ network exhibits a detection rate greater than 85% for all traffic intensities. However, the scheme fails to sustain the high detection rates for long, as can be inferred from Figure 26.7. This phenomenon occurs due to the inability of the SOM to update its trained neurons, to reflect energy decay rates of the target nodes, in scenarios where over a period of time the expected numbers of requests by the target nodes is to be reduced. Therefore, the detection scheme is totally ineffective when 90% of a detector node's lifetime is reached.

The false alarm rates show a similar trend to the detection rate, albeit inversely. For the false positive rate of the scheme for $N = 2048$, Figure 26.9 shows a steady increase from nearly 5% at network initialisation time to nearly 27% when 10% of the target node's energy content is depleted. Similarly, the false negative rates also approach close to 60% around this time. The inability of the SOM-based scheme to update pattern values and re-train the neurons to achieve higher accuracies are clearly exhibited in these figures. For schemes demanding constant updates in pattern values, there exists the need for a constant pattern update mechanism in place, accompanied with distributed pattern recognition, to achieve higher rates of success in attack detection.

26.9 CONCLUSIONS

In this chapter, we have modeled a distributed denial of service attack in wireless sensor networks and postulated a derivation of such attacks from other existing attacks in such networks. The attack is modeled as a set of threshold subpatterns from several regions of the network that need to be collated to reconstruct a holistic view of network traffic flow. A centralized approach toward the detection of such attacks based on self-organising maps is proposed. The SOM neural network is trained with patterns of network traffic, consisting of both attack and normal. Subsequently, the SOM application, running on the base station, clusters network traffic flow observations received from individual attack detector nodes of the network. The SOM-based approach is promising for detecting such attacks in environments, wherein constant network traffic is expected by individual target nodes. However, this approach does not hold in sensor networks, wherein, the activity of target nodes is expected to reduce with the passage of time, that is, diminishing energy resources.

REFERENCES

1. Z. A. Baig. A performance analysis of an application-level mechanism for preventing service flooding in the internet. Masters Thesis, University of Maryland, USA, 2003.
2. E. Catterall, K. Van Laerhoven, and M. Strohbach. Self-organization in ad hoc sensor networks: an empirical study. In *Proceedings of the Eighth International Conference on Artificial Life*, 2002.
3. H. Chan and A. Perrig. Security and privacy in sensor networks. *IEEE Comput. Mag.*, 36:103–105, 2003.
4. R. Chang. Defending against flooding-based distributed denial of service attacks: A tutorial. *IEEE Commun. Mag.*, 40:42–51, 2004.
5. T. Cormen, C. Leiserson, R. Rivest, and C. Stein. *Introduction to Algorithms*. MIT Press, 2001.
6. J. Deng, R. Han, and S. Mishra. Intrusion tolerance and anti-traffic analysis strategies for wireless sensor networks. In *IEEE Intl' Conference on Dependable Systems and Networks (DSN'04)*, 2004.
7. S. Dietrich, N. Long, and D. Dittrich. Analyzing distributed denial of service attack tools: the shaft case. In *14th Systems Administration Conference*, May 2000.
8. J. Elliot. Distributed denial of service attacks and the zombie ant effect. *IT Pro*, 2000, pp. 55–57.
9. V. D. Gligor. Guaranteeing access in spite of service-flooding attacks. In *Proceedings of Int'l Workshop on Security Protocols*, 2003.
10. J. Jung, B. Krishnamurthy, and M. Rabinovich. Flash crowds and denial of service attacks: characterization and implications for cdns and web sites. In *International World Wide Web Conference*, 2002.
11. H. G. Kayacik, A. N. Zincir-Heywood, and M. I. Heywood. A hierarchical SOM-based intrusion detection system. *Eng. Appl. Artif. Intell.*, 20, 2007.
12. K. Labib and V. R. Vemuri. Nsom: a tool to detect denial of service attacks using self-organizing maps. Technical Report, University of California, Davis.
13. S. Lee and T. Chung. Data aggregation for wireless sensor networks using self-organizing map. *Lecture Notes in Artificial Intelligence*, 2005, pp. 508–517.
14. J. Mirlovic, J. Martin, and P. Reiher. A taxonomy of ddos attacks and ddos defense mechanisms. *ACM SIGCOMM*, 34(2):39–53, 2004.
15. J. Newsome, E. Shi, D. Song, and A. Perrig. The sybil attack in sensor networks: analysis and defenses. In *Proceedings of IEEE Conference on Informantion Processing in Sensor Networks (IPSN'04)*, 2004.
16. B. Parno, A. Perrig, and V. Gligor. Distributed detection of node replication attacks in sensor networks. In *Proceedings of IEEE Symposium on Security and Privacy*, 2005.
17. T. Peng. Defending against distributed denial of service attacks. PhD Thesis, The University of Melbourne, 2004.
18. A. Perrig, R. Szewczyk, V. Wen, D. E. Culler, and J. D. Tygar. SPINS: security protocols for sensor netowrks. In *Proceedings of Mobile Computing and Networking*, 2001, pp. 189–199.
19. A. Perrig and J. D. Tygar. *Secure Broadcast Communication in Wired and Wireless Networks*. Kluwer Academic Publishers, 2002.
20. M. Ramadas. Detecting anomalous network traffic with self-organizing maps. PhD Thesis, Ohio University, 2003.
21. B. C. Rhodes, J. A. Mahaffey, and J. D. Cannady. Multiple self-organizing maps for intrusion detection. In *23rd National Information Systems Security Conference*, 2000.
22. S. Sancak, E. Cayirci, V. Coskun, and A. Levi. Sensor wars: detecting and defending against spam attacks in wireless sensor networks. In *IEEE Intl' Conference on Communications*, 2004.
23. A. Wood and J. Stankovic. Denial of service in sensor networks. *IEEE Comput. Mag.*, :54–62, 2002.

Chapter 27

Energy-Efficient Pattern Recognition for Wireless Sensor Networks

M. Baqer[1] and A. I. Khan[2]

[1] *Department of Computer Engineering, University of Bahrain, mbaqer@itc.uob.bh*
[2] *Clayton School Of Information Technology, Monash University,*
Asad.Khan@infotech.monash.edu.au

27.1 Introduction	627
27.2 Principles of Event Recognition for Sensor Networks	629
27.3 Voting Graph Neuron Approach	633
27.4 Simulations	645
27.5 Comparison	656
27.6 Conclusion and Future Work	657
References	658

27.1 INTRODUCTION

There have been several attempts to develop efficient and practical approaches to capture the different states of phenomena in the physical world using sensor networks. The most predominate model used for sensor network application involves sending sensory data to a base station for analysis [1]. Processing sensory data in the base station causes two fundamental problems. First, communicating sensory data from the physical environment to centralized servers is an expensive task in terms of energy resource consumption of sensor nodes. Second, sending streams of raw data from each sensor node to be analyzed in centralized servers may overwhelm the processing

Mobile Intelligence. Edited by Laurence T. Yang, Agustinus Borgy Waluyo, Jianhua Ma, Ling Tan, and Bala Srinivasan

capacity of these centralized servers. In both the cases, sensory data may encounter delays that could diminish its significance.

In addition to sensing the environment, sensor nodes are capable of locally processing and wirelessly communicating sensory data. The network is, therefore, capable of in-network processing sensory data to replace the need for processing sensory data in centralized locations, that is, base stations [15]. Threshold-based techniques are one of the most commonly used techniques for event detection in sensor networks. A threshold-based technique can be as simple as a single value hardcoded into the sensor network application. An event is detected when the sensory readings exceed the threshold value. This approach, however, cannot be used to detect complex events. Generally, physical world phenomena necessitate sensor node collaboration to mitigate their complexity and infer their conditions. In particular, CSIP-based, neural network-based, and pattern-based approaches exploit sensor node collaboration to detect events within sensor networks [17, 18, 24]. These approaches represent sensory data in patterns that depict unique states of events in the physical environment and then use them to detect the occurrence of events of interest by comparing them to reference patterns.

One of the main motives driving sensor network research is the ability to operate sensor networks in an *untethered* and *unattended* manner for long-term outdoor deployment. A sensor node continues to provide sensory data from the physical environment until it depletes its own on-board battery. Replacing or recharging batteries is not possible in sensor networks owing to the sheer number of deployed sensor nodes and the inaccessibility of the physical environment.

Incorporating energy-saving techniques is central to the design of sensor network applications to prolong the lifetime of the network. Significant amount of energy can be conserved during the sleep interval. For instance, Mica2 node's processor active and sleep modes consume 33.0 mW and 75.0 μW, respectively [5]. Although the idle mode allows some energy conservation, more significant energy savings may be achieved by totally switching off the sensor node instead of operating it in the receive or idle modes, for example Mica2 node receive, idle, and sleep modes consume 7.9 mA, 7.0 mA, and 1.0 μA, respectively [5].

In this chapter, we propose a pattern-recognition approach that employs collaboration among sensor nodes to detect events. Patterns are stored in a distributed manner within the network. The approach recognizes events of interest by measuring the similarity of the input sensory data with reference patterns. Moreover, the approach proposes a novel node life-cycle management technique that functions on patterns' similarity. This technique alternates sensor nodes between their active and sleep modes to dynamically control their collaboration and conserve energy.

This chapter describes the basis of template matching, offers an event-recognition approach called the VGN approach, and examines the outcomes of simulations in the following order: Section 27.2 provides a foundation to template matching. The VGN approach is presented in Section 27.3. Section 27.4 presents the simulation results and analysis. The voting graph neuron (VGN) approach is compared with the GN and Hopfield networks in Section 27.5. Section 27.6 concludes the chapter.

27.2 PRINCIPLES OF EVENT RECOGNITION FOR SENSOR NETWORKS

In this section, we provide the foundation and discuss the concepts applied in our approach for event detection using sensor nodes.

27.2.1 Event Patterns

The occurrence of events of interest is captured using patterns. A pattern may be defined as the opposite of chaos [26]. For example, severe weather conditions such as thunderstorms have weather maps or patterns that can be used to predict their occurrence. Similarly, in sensor networks, an event pattern could be a weather map, the behavior data of sea birds, or stress levels in a bridge.

Sensory data convey information about the conditions of the environment. Using patterns instead of handling individual sensory data points provides a practical method for capturing multidimensional and complex events of the physical environment. We assume that an event pattern is represented by $M \times N$ cells. Each cell is associated with one sensor node and corresponds to a specific region of the monitored area. We assume that the number of active sensor nodes is equal to the number of cells ($M \times N$). In scenarios where the number of deployed sensor nodes might be greater than $M \times N$, cell leaders may be elected to represent the cells and aggregate the sensory data of sensor nodes belonging to their cell. Consequently, a matrix of sensor nodes of size $M \times N$ is created to correspond to event patterns. The matrix, however, can be of regular or irregular shape, depending on the deployment scheme. Therefore, an event pattern may be defined as an $M \times N$ snapshot matrix of sensory data at a particular time t in a specific area A monitored by $M \times N$ sensor nodes, as illustrated in Figure 27.1.

27.2.2 Template-Matching

The objective of any event-detection system is to discover the occurrence of events of interest in the monitored environment. Since we are detecting events using event patterns, the objective of the proposed approach is to find events of interest using user-defined reference patterns. In general, pattern-recognition systems use reference patterns to establish the identity of an unknown pattern. The reference patterns may be used as examples that describe the event patterns themselves or the main classes that events of interest relate to.

Template matching is one of the earliest pattern-recognition techniques [7, 22]. The technique is commonly defined as the process of detecting whether two given patterns correlate to each other with high similarities. Template matching can be employed in data-filtering applications [2]. There are mainly two template-matching problems [14]. First, there is the problem of trying to find a match for a subpattern in a larger pattern [14, 25]. Take for example, trying to find the locations of a jigsaw piece in a large jigsaw puzzle. The system may try to sequentially match the given

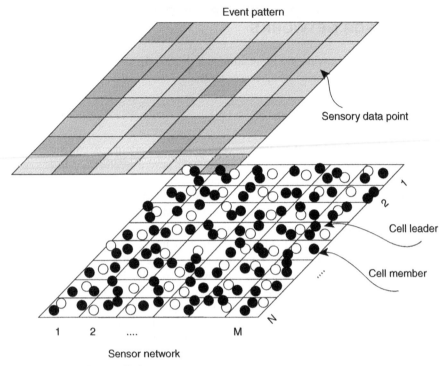

Figure 27.1   Mapping physical world events into event patterns.

jigsaw piece with all the available locations of the jigsaw puzzle. A match is declared when one of the location matches the given jigsaw piece. On the other hand, there is the problem of trying to find a match for an unknown pattern among a collection of reference patterns. In this case, the input pattern is equal in size to the reference patterns. The system tries to find a match by comparing each reference pattern one at a time to the unknown pattern. A match is found where any of the known patterns is of high similarity or equal to the unknown pattern.

In general, the advantage of using template-matching techniques is that the techniques utilize similarity-based comparisons to recognize patterns. Similarity-based approaches discover patterns by operating directly on the sensory data. In contrast to many pattern-recognition approaches, such as Maximum Likelihood and Bayesian learning [4], similarity-based approaches do not use feature vectors or probability density distribution functions [6]. Instead-patterns are recognized by measuring the similarity between patterns using raw sensory data. The geometry and structure of sensory data are defined by the similarity measurement. This is important since formulating feature vectors that represent all objects in proper classes and creating feature vectors for high-dimensional data may be hard or impossible [23]. When the dimensionality of data is reduced as a result of creating feature vectors, important sensory data may be lost, resulting in misrepresentation of objects and having different objects mapped to the same feature vector. Pattern-recognition approaches that depend on the probability density distribution to classify patterns are not useful for sensor network applications. To create a probability density distribution, one must have a very large number of samples of the object/pattern to be detected. This is not always the case in sensor network applications due to their exploratory nature.

Exploiting template-matching for recognizing events using sensor networks has numerous advantages. Template-matching functions directly on raw sensor data. Raw sensory measurements may be matched with a pattern of the set of reference measurements to discover the occurrence of events of interest with little or no data preprocessing. Finally, template-matching does not require complex computational power. The similarity among patterns can be calculated using logical or simple arithmetical operators; sensor nodes may simply match patterns by using simple comparison operators.

Traditionally, template-matching approaches have suffered from a number of limitations. The template-matching performance may degrade, for instance, as a result of using templates of various sizes or when templates are rotated [14]. However, this problem is associated with three-dimensional object recognition and character recognition. Each pair of an event pattern is associated with a designated sensor node in a specific location. For instance, the temperature of a specific region either matches a given pattern at a specific time or not. Therefore, the template-matching problem in sensor networks is generally not affected by deformation, shift, or rotation, and thereby we can focus on providing an energy-efficient pattern-recognition approach for sensor networks.

27.2.2.1 *Template-Matching Operations:*

Template-matching interprets the similarity measurement to recognize patterns. The concept behind the use of similarity measurement in template matching is derived from the compactness theorem [23]; objects belonging to the same class have to be close in their representations. Similarity-based recognition may be considered as a connection between perception and higher-level knowledge [23]. In this instance, the perception is the sensor modules of a sensor node. Higher-level knowledge is attained by sensor collaboration and exchanging information wirelessly.

The distance, $d(x, y)$, between any two patterns measures the similarity of the patterns to each other. The distance measurement may be used for establishing the membership of an unknown event pattern to one of the reference patterns' categories. To categorize an unknown pattern, the distances between the unknown pattern and each of the reference patterns need to be computed. For instance, a distance measurement called the Hamming measurement counts the locations at which patterns differ [27]. Other distance measurements include Euclidean distance, City-Block distance, and Canberra distance, which can also alternatively be used [27]. The distance $d(x, y)$ of two patterns x and y abide by the following conditions:

$$d(x, y) = 0 \tag{27.1}$$

$$d(x, y) > 0 \tag{27.2}$$

$$d(x, y) = d(y, x) \tag{27.3}$$

The distance measurement properties state that $d(x, y) = 0$, if and only if patterns x and y are identical. In this case, the unknown event pattern matches exactly one of the reference patterns. Nevertheless, it may not always be the case that patterns x and y are identical. In situations where patterns x and y are not equal, the distance between them is a nonnegative value. From the compactness theorem, patterns of the same category are of high similarities. Thus, when $0 < d(x, y) \leq \epsilon$ for a positive and small ϵ, the unknown pattern x is not equal to pattern y but belongs to its category. Finally, Eq. (27.3) shows that the distance measurement is symmetric.

Although collecting sensory data of event patterns is inherently distributed owing to the geographically distributed nature of sensor networks, the distance measurement used by template-matching approaches cannot be computed in a distributed manner. In a spatially distributed network, each sensor can be designated for a specific geographical location (subpattern) of the global event pattern matrix, and hence may at the most only be aware of its neighboring sensor nodes. It is desirable to reduce the amount of communication to enable performing the distance measurements locally within the network and then communicating the distance measurements to the base station instead of raw sensory data.

In theory, when any unknown event pattern exactly matches any reference pattern, the distance measurement of each sensor node needs to be equal to 0 ($d_i(x_i, y_i) = 0$), where i is the index of the node, x_i and y_i are the subpatterns of the unknown event and reference pattern, respectively. Consequently, the sum of local distance measurements needs to be equal to 0, as shown in formula (27.4). Furthermore, an unknown event pattern does not exactly match any reference pattern when the sum of local distances is greater than zero, as shown in formula (27.5).

$$\sum_{i=1}^{n} d_i(x_i, y_i) = 0 \qquad (27.4)$$

$$\sum_{i=1}^{n} d_i(x_i, y_i) > 0 \qquad (27.5)$$

where n is the number of active sensor nodes, that is, equal to the dimensions of the patterns $n = M \times N$.

In practice, however, the formulas in (27.4) and (27.5) cannot be generalized to attain accurate global pattern-matching results. Subpatterns may incorrectly mismatch reference patterns. A new unknown event pattern—which does not match any reference pattern—may locally be matched to reference patterns when any sensor node attains $d_i(x_i, y_i) = 0$ as a result of having the subpattern at location i exactly matching one or more reference patterns at the same location and may globally be matched when the sum of local distance measurements satisfies Eq. (27.4). Therefore, it is necessary to design a distance measurement operation that can be executed in a distributed manner to enable distributed pattern matching over the sensor network and provide accurate template matching. The local distance measurement of a sensor node needs to be equipped with sufficient criteria to produce the best possible decision without depending on any global information.

Moreover, the sensor network needs to conserve energy whenever possible. Classical pattern-recognition algorithms are generally executed on traditional computer systems, which are not constrained by energy consumption. Energy conservation is imperative to sensor networks, since the total energy consumption of the network to perform any task reduces the energy resources of the network. The energy consumption of the sensor network may be enhanced by suppressing redundant operations and communications.

In this research, the problem of event recognition will be modeled according to the second template-matching problem. Event patterns of sensory data from the physical environment will be representing the unknown pattern and compared with a collection of reference patterns that depict the various interesting conditions of the physical environment, an unknown pattern P is classified based on its similarity to n reference patterns. Event patterns will be matched using an energy-efficient and distributed template-matching approach.

27.3 VOTING GRAPH NEURON APPROACH

In this section, the VGN approach is proposed. The VGN approach recognizes events by utilizing a novel energy-efficient template-matching approach. In essence, the VGN approach uses graph neuron (GN) concepts [11–13]. The VGN represents sensory data in patterns that are used for detecting events of interest. Sensor nodes are depicted as memory cells that can store subpatterns. The VGN approach is concerned with how to infer global information from the local information of sensor nodes. Patterns are matched by means of votes. Contrary to the GN algorithm, which uses a simple binary decision, the votes of sensor nodes are the units of collaborations, list all possible reference patterns matching the unknown patterns, and are designed to allow sensor nodes to exchange decisions to reach a network decision. Additionally, node collaboration, distributed in-network processing, distributed storage, and life-cycle management are employed to enable practical and energy-efficient pattern-recognition using resource-constrained sensor nodes.

27.3.1 VGN Overview

Performing event recognition in sensor networks is a challenging problem. Events need to be recognized using geographically distributed and ad hoc infrastructures. The processing of information needs to be conducted within the network using very small, resource-constrained sensor nodes. The sensor nodes wirelessly communicate using a shared wireless channel. The channel's bandwidth and its varying quality represent an additional challenge to the recognition process.

The objective of the VGN approach is to perform template-matching in a distributed manner while conserving energy. The approach proposes a distributed template-matching approach that employs the sensor network as distributed processing system to utilize localized processing of sensory data to decrease communication by transmitting only high-level partial pattern-matching results. This is necessary because the implementation of the approach needs to fit well within the memory's footprint, executable using constrained processors, and requires a minimal amount of data communication and energy. The approach is also decentralized in a way that any sensor node can fuse information received from the network to produce global event detection results. Nevertheless, sensor nodes do not have any global information about the event pattern's dimensionality, number of sensor nodes, or topology. This provides an added feature to the approach, since more sensor nodes can be deployed to replace depleted nodes and cover a larger area or increase the resolution of the sensory data without disturbing the operations of the network. Sensor nodes dynamically team up and use internode communications to match patterns. Finally, the longevity of the sensor network is of highest caliber, since the template-matching functionality depends on sensor node availability and replacing sensor nodes is generally infeasible. Therefore, the approach conserves the energy resources of sensor nodes to provide event-recognition services for the longest possible interval. The energy conservation is attained by switching off redundant sensor nodes and suppressing redundant communications.

The VGN approach comprises three techniques, namely pattern storage, template-matching, and energy conservation. Instead of storing reference patterns in the base station, the reference patterns are spatially partitioned and mapped onto designated nodes for storage, thus allowing *in-network* pattern storage. Patterns are partitioned into subpatterns and each subpattern is stored in its designated sensor node. Consequently, the template-matching technique occurs in a distributed manner within the network. Each node matches its local sensory data with the stored reference patterns to produce a local pattern-matching decisions. Sensor node collaboration is needed since local pattern-matching decisions of individual nodes may not be able to match global event patterns of the environment.

Node collaboration can be categorized into the communication and updating of local template-matching decision. Sensor nodes collaborate by communicating their local template-matching decision in the network. Upon receiving any template-matching decision from other sensor nodes, the sensor nodes update their own local decisions. The communication and updating process continues until the network reaches a global template-matching consensus.

The node collaboration is fundamentally important in sensor networks. Generally, individual nodes cannot infer the status of the global environment with their local information. In the VGN approach, sensor nodes collaborate to match patterns by exchanging their local template-matching results. The node collaboration allows the aggregation of sensor node template-matching results to improve the network's decision and consequently attain energy conservation as a result of the reduction in the amount of communication in network. Moreover, the approach tries to conserve energy by selecting sensor nodes with significant local decisions that may support reaching a template-matching consensus; other nodes are switched off. Since patterns are matched by measuring their similarity to reference patterns, similarity among patterns also determines the significance of sensor nodes. Patterns of high similarity generally have areas where local template-matching results of sensor nodes are identical. The VGN approach takes advantage of these areas and sets as many sensor nodes as possible into the sleep mode. In addition to the obvious energy conservation gain of switched off nodes, the performance of the network as a whole is improved as a result of having less nodes competing for the network resources and communicating redundant data throughout the network.

27.3.2 VGN Approach Details

Algorithm 27.1 shows the VGN approach pattern-matching cycle. The VGN algorithm exploits the sensor network as a distributed computing environment. Each sensor node provides processing and storage capabilities necessary to match patterns. The algorithm is executed concurrently by all sensor nodes. Each sensor node lists the reference pattern labels matching its subpattern in its own vote set (v_i). Every sensor node attempts to communicate its own vote set in the network. Nevertheless, the ability to communicate vote sets depends on the network and physical layers. For instance, deploying sensor nodes within each other's communication range may reduce the amount of multihop communication, but also reduce the nodes' abilities to communicate their vote sets concurrently. When a sensor node is finally able to communicate its vote set, the sensor node enters the sleep mode. On the other hand, sensor nodes unable to communicate their vote sets remain active and listen to the communication channel for any communicated vote sets (v_c). Upon receiving any vote set, a sensor node examines its content and determines if the communicated vote set is equal to its own ($v_i = v_c$), and consequently enters the sleep mode. When $v_i \neq v_c$, the sensor node updates its v_i by listing the patterns labels common in both vote sets, this is depicted by the intersection operation between v_i and v_c. The algorithm's

---

Algorithm 27.1: VGN algorithm

Every sensor node executes the following

Initialisation
$$v_i \leftarrow \{pattern \quad labels \quad matching \quad sub - pattern\}$$

if not sleeping **then**
  repeat
    try Communicate (v_c)
        enter sleep mode
    upon Receiving(v_c)
        if $v_i \neq v_c$
            $v_i = v_i \cap v_c$
      else
            enter sleep mode
      end if
    until ($|v_c| = 1$) OR ($v_c = \phi$)
  end if

---

cycle continues until a consensus in regards to the unknown pattern is reached, which is indicated when v_c contains the matched pattern label or ϕ when no match can be found.

As mentioned earlier, the VGN approach is characterized by three important operations—the pattern storage, pattern matching, and sleep mode. In the following sections, the details of these operations are presented.

27.3.2.1 Pattern Storage

A set of user-defined reference patterns are used for detecting events in the environment. In classical approaches, the natural choice will be to store the reference patterns in a centralized manner. In sensor networks, as a result of the small storage capacity sensor nodes, a single sensor node cannot store all reference patterns, and storing the reference patterns at the base station is not desirable due to the problems associated with centralized design in sensor networks. The VGN approach overcomes this problem by storing the patterns in a distributed manner.

A reference pattern (P) is first partitioned and represented by an array of pairs (p_i), hence let $P = \{p_1, \ldots, p_i, \ldots, p_{M*N}\}$ be a $M \times N$ array of P, where p_i is the fundamental pattern element pair at index i of pattern P, $i \leq M * N$. Each pair includes information about the subpattern value, the node to which the pair should be mapped, and the pattern's unique number, the value, position, and pattern label fields, respectively. The value domain (D) of the pattern is a set of all possible values that the pair's value field may be initialized with, for instance the value domain for binary patterns is $D = \{0, 1\}$. The pair's value field $\in D$, and position is the node ID marking the location, that is, the relative location or the Euclidean coordinate for this node. Nodes are also required to store pattern labels (L_m) that are unique numbers assigned to the each of the reference pattern, where m is the number of reference patterns. Each pair p_i is mapped to its designated sensor node s_i of the sensor network $S = \{s_1, \ldots, s_i, \ldots, s_{M*N}\}$.

The implementation details of the memory and data structures used for storing the pairs in each of the individual sensor nodes are out of this research's scope. Nevertheless, the pairs should be stored in a manner that optimizes the space and time required to store and retrieve them, respectively. Moreover, the memory structure employed for storing the pairs should be designed to provide accurate retrieval of all pattern labels matching the input pair. To produce the local template-matching, the memory needs to list all pattern labels matching the input pattern. Therefore, techniques that may increase the pattern labels' retrieval performance are desirable.

For instance, partitioning the memory into various categories that depicts all possible combinations of pairs' values and storing pattern labels in their respective categories may facilitate repaid listing of all pattern labels matching the input pair. The categories of the memory depict the various combinations of pair values of the pervious sensor node p_{i-1}, its own p_i, and the successor sensor nodes p_{i+1}, respectively. For binary patterns $(D = \{0, 1\})$, the categories (p_{i-1}, p_i, p_{i+1}) needed to list all patterns are $\{000, 001, 010, 011, 100, 101, 110, 111\}$. Any pattern at the location i corresponds to one of the categories. Each category lists all pattern labels of equal values and corresponding to (p_{i-1}, p_i, p_{i+1}). Instead of searching the domain of stored patterns for pattern labels corresponding to the unknown pattern, all pattern labels of reference patterns that matches the unknown pattern at location i are retrieved by only matching the unknown pattern's pairs at location i with one of the categories.

27.3.2.2 Template Matching

The VGN approach proposes a distributed template-matching mechanism to detect events. To match patterns, the distance between the unknown pattern and the reference patterns need to be computed. Since reference patterns are stored in a distributed manner, the distance measurement needs to be conducted in a distributed manner as well.

Each sensor node performs a local template-matching by utilizing its local input sensory data to find the pattern labels of reference patterns equal to the input sensory data. Once all pattern labels of patterns that match the input sensory data at location i are found, they are listed in a *vote set* that represents the node's local pattern matching results. A vote set (v_i) is a set that lists all pattern labels that may possibly match the global event pattern. In situations where there exists no reference patterns that match input sensory data at location i, an empty vote set is assigned to the vote set $(v_i = \{ \})$. This local template-matching method corresponds to exact template-matching, thus $d(x, y) = 0$.

The global distance measurement is computed implicitly by voting instead of explicitly summing the distance measurement of each sensor node at the base station. When an unknown event pattern exactly matches any of the reference patterns, the reference pattern's label is included in the vote sets of all sensor nodes. On the other hand, a common pattern label is not included in all sensor nodes' vote sets when the unknown event pattern matches more than one reference pattern. This type of consensus has several aliases in literature, for example, common consent, unison, or unanimity voting [3, 16].

Instead of counting votes, the network performs exact pattern matching by performing a distributed search for the common pattern label included in all sensor nodes' vote sets. The search for the common pattern label is attained by applying the intersection operation on the vote sets. The intersection of two sets results in a new set containing the elements (pattern labels) that belong to both sets and remove elements that do not belongs to both vote sets. As a result, when a pattern exactly matches one of the reference patterns, the intersection of all vote sets yields the matched pattern label, or an empty set when a match cannot be not located. The similarity measurements formulas can be depicted as intersection of vote sets, as shown in formulas (27.6) and (27.7). When an unknown pattern x exactly matches the jth reference pattern (y_j), the intersection of vote sets of n nodes results in the pattern label (L_j) of the jth reference pattern.

$$d(x, y_j) = 0 \rightarrow \bigcap_{i=1}^{n} v_i = L_j \qquad (27.6)$$

$$d(x, y_j) > 0 \rightarrow \bigcap_{i=1}^{n} v_i = \phi \qquad (27.7)$$

Event patterns are matched as a result of sensor node collaboration. The network collaboration is defined by the vote set exchange among sensor nodes. Each node tries to communicate its vote set and sensor nodes update their decisions upon receiving any vote set. For this instance, the communication model is abstracted to depict a single-hop cluster. The update is attained by applying the intersection operation on the node's vote sets and the communicated vote set resulting in a new updated vote sets, as shown in Eq. (27.8). The new vote set lists pattern labels that can be found in both vote sets and removes all uncommon pattern labels. As a result, the vote sets' significance are usually improved after each update; the number of reference pattern labels in the vote sets is reduced, thus reducing the number of reference pattern label candidates of matching the unknown pattern of the physical phenomenon.

$$v_i = v_i \cap v_c \qquad (27.8)$$

A network consensus is reached when either all sensor nodes agree on a pattern label that can be found in all of their vote sets or when any of the nodes produces an empty vote set, depicting a match and reject, respectively. The following equation depicts the sensor network consensus while using the intersection operator:

$$C = \bigcap_{i=1}^{n} v_i = \begin{cases} \text{no match} & \text{if } C = \emptyset \\ \text{matching pattern } L_j & \text{otherwise} \end{cases} \qquad (27.9)$$

where C is the network consensus and n is equal to the number of nodes ($M * N$).

The global consensus depends on the node collaboration that employs internode communication to exchange local decisions and consequently match patterns. The node collaboration may be improved by adopting en route sensor network processing techniques. The global consensus can be reached by means of vote set communication and en route vote fusion while being conveyed to the base station [8, 19]. The distributed characteristic of the VGN makes it implementable with miscellaneous network topologies and multihop communication. The information processing and sleep mode techniques of the VGN approach are beneficial to reduce the amount of information communication and remove redundant information. Any node may perform template-matching operations of the VGN approach by functioning as a fusion center [10]. Since the vote sets are designed to be processed in intermediate nodes, fusion centers may process and consolidate vote sets into a single output of refined data.

27.3.2.3 Sleep Mode

In addition to conserving the network's energy, by in-network processing of vote sets, the VGN achieves further energy conservations by employing a sleep mode-based strategy. Instead of demanding that all sensor nodes be utilized to match patterns, event patterns are matched using a subset of nodes while the rest of the network is switched off. Consequently, event patterns are matched using a team of dynamically and autonomously selected sensor nodes.

The sleep mode strategy incorporated in the VGN approach switches off sensor nodes. Switched off nodes are selected event-recognition on the basis of their significance to the event-recognition process. Sensor nodes determine to participate in pattern matching or enter the sleep mode autonomously without depending on any base station. This is crucial to the approach to support needed scalability in case additional sensor nodes are deployed in the network and to preserve the decentralized properties of the VGN approach.

The VGN approach attempts to improve the energy consumption and the longevity of the network while preserving the pattern-matching performance by setting as many sensor nodes as possible in the sleep mode and not allowing the communication of redundant vote sets. Communicating identical vote sets does not enhance the pattern-matching consensus and depletes the network's energy. A subset of nodes of significant vote sets may be used to match event patterns, instead of using all sensor nodes to match patterns.

To understand the sleep mode strategy, we need to review the mechanism that sensor nodes use to recognize patterns. To match an unknown pattern, a sensor node s_i produces a vote set comprising one of the following sets: $\{\Phi\}$, $\{X\}$, $\{Y\}$, or $\{X, Y\}$; when storing two patterns X and Y. Figure 27.2 illustrates various areas of stored patterns X and Y. Vote sets are assigned one of the sets depending on the location of the sensor nodes and which of the patterns matches the unknown pattern. A vote set can possibly comprise ($\{X\}$ or $\{X, Y\}$) or ($\{Y\}$ or $\{X, Y\}$) when the unknown input pattern matches patterns X or Y, respectively. An empty set is assigned a vote set when a sensor node fails to match the unknown pattern to any of the reference patterns (X or Y).

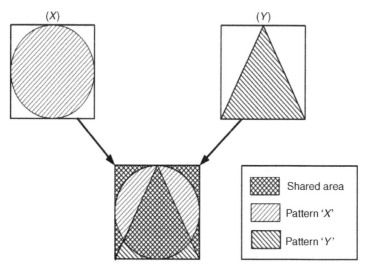

Figure 27.2   Various areas created as a result of storing pattern X and Y.

In either of the situations, vote sets comprising the set $\{X, Y\}$ are located in areas where patterns X and Y intersect. This shared area represents the spatial locations where the interference between patterns X and Y occurs. The sensor nodes located at the shared area are insignificant to the matching process, since their vote sets comprise pattern labels of all of the stored patterns, and thus cannot be used for distinguishing between which of the patterns matches the input sensory pattern. Since nodes of the shared area cannot be used to match patterns, the objective is to switch off these nodes to allow nodes located at areas unique to patterns X or Y to communicate their vote sets as soon as possible.

Any sensor node can individually decide to enter the sleep mode if it satisfies any of the following two conditions. First, a sensor node may enter the sleep mode subsequent to communicating its own vote set. Second, a sensor node may do so upon receiving an identical vote set. A sensor node identifies its vote set as identical to the communicated vote set if the two vote sets include exactly the same pattern labels, as shown in formula (27.10). The set difference operator (–) lists the elements of the first argument that are not included in the second argument. When the two vote sets are identical, they produce empty sets (ϕ), that is, zero element difference. When any node located in the shared area communicates its vote set, all nodes of the shared area enter the sleep mode. Consequently, allowing sensor nodes with more significant vote sets, that is, vote sets comprising of $\{X\}$ or $\{Y\}$, to communicate their votes that results in the network reaching a consensus.

$$|A - B| = |B - A| = 0 \qquad (27.10)$$

where sets A and B are two identical sets.

The VGN approach dynamically adapts to the similarity among patterns. Storing patterns that consist of a large shared area results in a large number of nodes of identical vote sets. Instead of communicating redundant vote sets from the shared area, as soon as a node located in the shared area communicates its vote set, all nodes in the shared area receiving the vote set enter the sleep mode. This process is repeated to set all redundant nodes of various shared areas into the sleep mode, when there are more than two reference patterns. When patterns comprising large areas of unique pattern types are stored, the VGN is capable of rapidly matching patterns. Due to the large differences of these type patterns, the vote sets are smaller in size and there are more nodes with significant vote sets that make the network reach a rapid consensus. As a result, the VGN approach attempts to concentrate the node collaboration in sensor nodes with significant vote sets (nonidentical vote sets). In addition to the energy conservation, which results from setting sensor nodes to the sleep mode and not communicating redundant vote sets, the VGN approach is capable of dynamically selecting a smaller subset of nodes to match the input event patterns.

27.3.3 Example

In this example, we will demonstrate how the VGN approach matches patterns. The example shows a sensor network comprising nine sensor nodes capable of accepting

Stored patterns

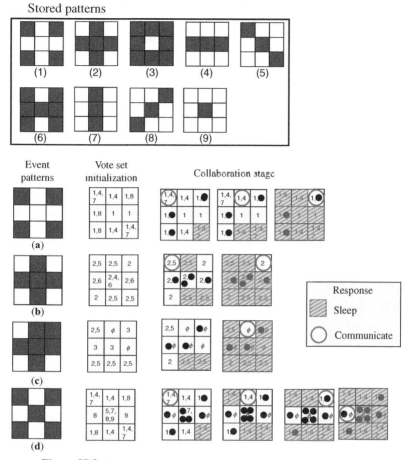

Figure 27.3   The process of reaching consensuses using the VGN approach.

input patterns of size 3×3. The network is employed to store nine binary patterns, in Figure 27.3 (top). It is assumed that a communicated vote set from any sensor node is received by all other sensor nodes in the network.

In this instance, the network's response to various unknown patterns is presented. From the perspective of the VGN approach, the network's response depends on the type of unknown event patterns and can be categorized to either matching the unknown event pattern when the event pattern exactly matches any reference pattern or otherwise rejecting the unknown event pattern. In Figure 27.3, sensory pattern inputs (a) and (b) are chosen from the stored patterns to illustrate the manner in which a sensor network matches patterns. The network starts with the vote set initialization, where each vote set is assigned a list of all pattern labels that exactly match the patterns' pairs at each of the nodes' locations. In this instance, the nodes are sequentially traversed and allowed to communicate their vote sets. When a sensor node communicates its vote set, it enters the sleep mode. Upon receiving any vote set, the nodes apply the intersection operation to update their vote sets. The updated vote sets comprise the common pattern labels of the communicated vote set and the pattern labels of the vote set of the recipient sensor node. The unknown patterns in the example are matched when any of the nodes communicate a vote vector containing one pattern label and, thereby, the remaining sensor nodes enter the sleep mode.

On the other hand, event patterns of Figure 27.3(c) and (d) show how unknown event patterns are rejected when they do not correspond to any of the reference patterns. An unknown pattern is rejected when any of the nodes communicate an empty vote set. In addition to being able to create empty vote sets when the unknown pattern does not match any reference pattern in the vote set initialization stage, an empty vote set is created when there are no common pattern labels among the received vote set and the local vote set at any sensor node. Figure 27.3(c) shows nodes creating empty vote sets during the vote set initialization stage and Figure 27.3(d) shows nodes creating empty vote sets as a result of a vote set update. In both cases, the network is able to reject the unknown event patterns when any of the empty vote set is communicated.

27.3.4 Analysis

In this section, the VGN approach is analyzed. The number of vote sets required to match event patterns is the focus of this study, since communication is one of the highest consumers of energy in sensor networks. The study provides details in respect to the worst case scenario of the VGN approach in respect of the number of vote sets required to match patterns. Communicating messages in resource-constrained sensor networks consumes a significant amount of the network's energy and influences the network's delays and congestions. In critical applications, energy consumption needs to be minimized as much as possible.

Typically, to construct a global view of the environment, information needs to be collected from all sensor nodes to create the global event pattern, that is $M \times N$ messages need to be communicated to infer the global environment status; where $M \times N$ is the dimension of patterns corresponding to the number of sensor nodes. The VGN approach, however, attempts to reduce the number of required vote sets by eliminating the redundant vote sets. Interference between reference patterns produces shared areas that yield redundancy in vote sets. The VGN approach prevents these identical vote sets from being communicated by allowing nodes receiving identical vote sets to their own vote set to be switched into the sleep mode. As a result, a subset of sensor nodes is dynamically selected to communicate their vote sets to match patterns.

Despite the occurrence of various shared areas among patterns, in the worst case scenario, the VGN approach is not able to set any sensor node into the sleep mode before communicating its vote sets, and vote sets from all sensor nodes need to be communicated to match the unknown pattern. The worst case scenario occurs when the unknown patterns are of $d(x, y) = 1$ to all stored reference patterns. In this instance, the difference between any two vote sets is one pattern label, that is almost identical to each other. Consequently, a sensor node can enter the sleep mode only when it communicates its own vote sets. The intersection among any two vote sets reduces the vote sets by one pattern label. To match the unknown event pattern of the worst case scenario, all sensor nodes need to communicate their vote sets, that is $M \times N$ vote sets need to be communicated.

Figure 27.4(a) shows an instance of the worse case while using 3×3 patterns. The network stores nine reference patterns of which eight patterns are of distance $d(x, y) = 1$ to the first pattern. Figure 27.4(b) depicts the unknown event pattern using the first pattern of Figure 27.4(a). Figure 27.4(c) shows the vote sets of the network in response to the input pattern. Although the vote sets of Figure 27.4(c) are almost identical, they are not beneficial to the sleep mode strategy of the VGN approach. It is assumed that any vote set may be selected to be communicated to all other sensor nodes. When any sensor node receives a vote set, it applies $v_i \cap v_c$, which results in a new vote set that only lists pattern labels included in both vote sets. After

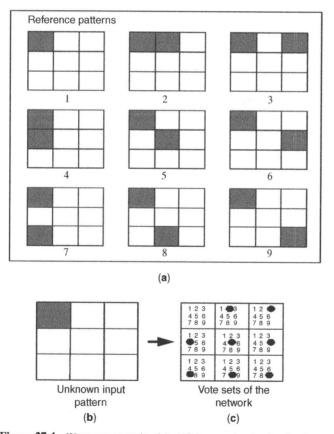

Figure 27.4   Worst case scenario of the VGN approach using 3×3 patterns.

each vote set communication, the vote sets of the rest of the network reduce in size by one pattern label only. Consequently, the unknown pattern can be matched after communicating all vote sets of all nodes.

Nonetheless, the affect of the worst case scenarios is limited by the number of reference patterns of $d(x, y) = 1$ to the unknown event pattern. The total number of binary patterns of $d(x, y) = \gamma$ can be computed using formula (27.11). The formula is a combination formula that calculates the number of patterns of a distance γ of patterns of size n.

$$(n : \gamma) = \binom{n}{\gamma} = \frac{n!}{\gamma!(n - \gamma)!} \tag{27.11}$$

The number of patterns that causes the worst case scenario $(d(x, y) = \gamma = 1)$ is very small $(n : 1) = n$ as compared to the total number of possible binary patterns (2^n). Figure 27.5 depicts the distribution of reference patterns according to their distance to a randomly selected pattern. The figure measures the distance in percentage of the value of the distance to the size of pattern to generalize the results to any pattern size. The figure shows that among all possible patterns there are less than 5% of all possible patterns of which $d(x, y) = 10\%$ of their pairs are different. The figure shows that the distributions of patterns' differences follows a normal distribution with average pattern is the center of the distribution, that is 50% difference. Therefore, when storing patterns, it is necessary to avoid storing patterns of $d(x, y) = 1$ to each other. This is possible since there are only n patterns of $d(x, y) = 1$ out of all possible patterns (2^n), hence only $n/2^n * 100\%$ of all patterns fit the worst case scenario. The worst case scenario is comparable to the situations where all data from all nodes are communicated to a centralized base station. Nevertheless, in all other situation, the VGN approach is capable of reducing the number of vote sets communication to be less than n. That is because the VGN approach is capable of taking advantage of the shared areas among stored patterns to switch as many sensor nodes as possible.

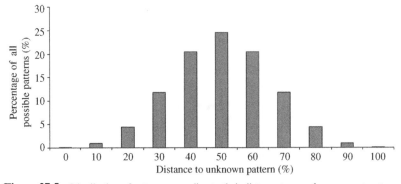

Figure 27.5   Distribution of patterns according to their distance to an unknown event pattern.

27.4 SIMULATIONS

In this section, the VGN approach is analyzed using simulations. The VGN approach provides an energy-efficient strategy to match patterns. Similarity among patterns produces shared areas that are utilized to conserve energy. The VGN approach considers these areas of redundancy important to pattern matching, as they produce identical vote sets. Instead of using all the sensor nodes to match patterns, the VGN approach is capable of dynamically selecting a subset of sensor nodes based on their significance to create a team of sensor nodes capable of collaborating to matching event patterns.

In Section 27.3.4, the VGN approach performance under the worst case scenario has been studied. The communication requirement in the worst case scenario is approximate to scenarios where all sensory data are communicated and processed at the base station. The worst case scenario is limited to patterns that are of $d(x, y) = 1$. These patterns represent a very small portion of the total possible number of patterns, thus their effect is limited. Since the occurrence of the worst case scenario is rare, the VGN approach is always capable of eliminating redundant communication and conserving energy and attaining a performance better than the worst case or base station scenarios.

Nonetheless, the VGN performance is affected by the similarity among patterns. Reference patterns of various similarity result in various VGN performances. It is, therefore, necessary to simulate the VGN approach using various types of patterns to understand the performance of the approach and specially the energy conservation gain.

In order to evaluate the performance of the approach, we modeled the VGN approach in MATLAB [20]. The model allows the construction of VGN nodes capable of performing the tasks of storing patterns, creating vote sets, collaborating, and matching patterns. The model constructs an array of nodes corresponding to the dimensionality of stored patterns. Patterns are decomposed into pairs and distributed into their designated sensor nodes. Each sensor node then stores the received pairs in the bias array according to the pattern storage mechanism of the VGN approach.

The simulations use novel metrics to evaluate the performance of the VGN approach. Sensor nodes collaborate to achieve 100% pattern matching accuracy. Therefore, traditional pattern-recognition metrics are not of importance, since they are concerned with false alarms and accuracy rates. Energy is one of the most important resources of a sensor network. The services provided by senor nodes are coupled with the availability of their on-board battery power. Sensor network applications need to be energy efficient and conserve energy whenever possible. Therefore, the performance of the sensor networks needs to be correlated with energy consumption levels of the network to perform the pattern-matching tasks and operations.

In addition to matching patterns, the VGN approach prolongs the sensor network lifetime by conserving energy by eliminating redundant communications and setting as many sensor nodes as possible into the sleep mode. The proposed performance metrics measure the energy consumption of the sensor networks while performing pattern matching in various scenarios. The first metric counts the number of vote sets required to reach a pattern-matching consensus. Since the VGN approach matches patterns by node collaboration via vote set exchanges, it is, therefore, important to determine the amount of communication associated with vote set exchanges required to match patterns. The actual amount of energy dissipated in vote set communications depends on the sensor node's hardware configurations. For instance, the energy consumed to communicate 1-bit data depends on the antenna and communication modulation. Nevertheless, assessed the communication requirement can be by counting the total number of vote set communications required to match a pattern and the amount of communication contribution of each individual sensor node. The VGN approach attempts to focus the node collaboration on nodes of highest importance in the network while setting other insignificant nodes into the sleep mode. As a result, communication contribution of sensor nodes is expected to be lower in shared areas than in the rest of the network. In addition to the overall amount of communication required to match patterns, identifying the communication contribution of each sensor node is important to locate the areas of high energy consumption. The time required to match patterns, which may affect the rate at which responses can be generated, can be inferred from the amount of communications required to match patterns. The time to match a pattern can be computed by summming all the time required to communicate all vote sets required to match a pattern. Another time factor is the amount of time sensor nodes remains in the active mode. One of the objectives of the VGN approach is to set as many sensor nodes into the sleep mode as soon as possible; the less time a sensor node remains active, the more energy is conserved. To measure the performance of the sleep mode strategy, we need to consider the wake up time interval per sensor node from the start of the voting phase until entering the sleep mode.

27.4.1   Test Patterns

Since the VGN approach operates on the similarities of patterns, the performance of the VGN approach depends on the type of stored patterns. The simulation results depict the VGN approach performance with the utilized patterns and provide a guideline on performance in terms of the VGN approach with various pattern types.

The performance was evaluated by comparing the VGN approach's performance using two types of patterns: random pattern signatures and conventional patterns in Figure 27.6 and Figure 27.7, respectively. Random pattern signatures may be defined as unique patterns that are not equal and contain only small amounts of similarities. Binary random patterns have been created by a random pattern generator built specifically for simulation purposes. The random pattern generator simply creates a random pattern of a specific dimensionality and then searches the collection of previously generated patterns to verify whether the newly generated random pattern had been previously created; if the pattern has no match, it is added to the collection of random patterns. Random pattern signatures were mainly used to evaluate the scalability of our approach with the variations in pattern dimensionality and number of stored patterns.

Figure 27.6 An example of a binary random pattern signature of size 32 × 32 pixels.

On the other hand, the conventional patterns consisted of binary images of hand-written digits of the MNIST database [21]. The MNIST database contains 60,000 examples of handwritten digit patterns of size 28 × 28 pixels. The database is categorized into numbers from 0 to 9. Each number category has a large collection of

Figure 27.7 A sample of the MNIST digits database [21].

images of handwritten digits of the same number class. It is important to note that the VGN algorithm is not a character-recognition algorithm; the patterns are shaped after digits to provide an excellent metric to evaluate the algorithm using patterns with high similarities. For simulation purposes, MNIST patterns have been converted into binary patterns.

The random pattern signatures and MNIST digits represent two different classes of patterns, patterns of approximately no shared areas and patterns that have shared areas, respectively. Figures 27.8(a–e) shows the average, minimum, and maximum distance measured for random patterns and patterns of digits with the increase in the number of stored patterns. The employed distance measurement follows the Hamming distance, which measures the similarity of the patterns by counting the number of locations (pairs) where the patterns differ. The distance was measured between each of the stored patterns to create tables that show the distance measurement for each

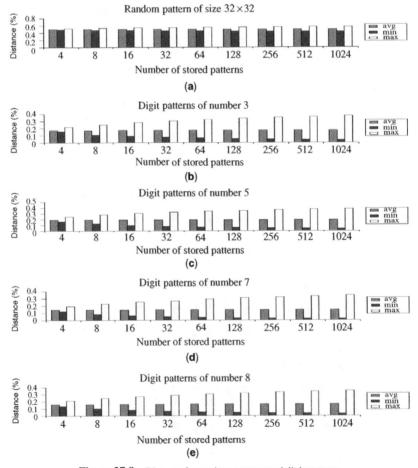

Figure 27.8   Distance for random patterns and digit patterns.

of the pattern types with the increase in the number of stored reference patterns. The figure shows the results of the created tables and depicts the distance measurement as the percentage of the number of locations where the patterns differ to the total number of locations (dimensionality of pattern). The larger the distance measurement percentage, the less similar the patterns.

The figure shows that random patterns have a larger distance measurement percentage than results of patterns of digits. Moreover, the average, minimum, and maximum distance measurement results of random patterns are approximately equal and remain constant with the increase in the number of stored patterns. Figure 27.8(a) shows that random patterns contain limited similarity; the distance measurement results between the stored patterns show that random patterns are different by approximately 50% of the patterns' locations (pairs). This concurs with the purpose of using random pattern signatures to simulate patterns of approximately no shared areas.

On the other hand, patterns of digits show high similarity. The patterns of digits 3, 5, 7, and 8 all have the same trend. The average distance measurement remains approximately constant with the increase in the number of stored patterns, as shown in Figure 27.8(b)–(e). The overall average distance measurement remains constantly under 20%, thereby confirming that patterns of digits are of high similarity. The maximum distance measurement increases with the decrease in the number of stored patterns. On the contrary, the minimum distance measurement is inversely proportional to the number of stored patterns, as the minimum Hamming distance decreases with the increase in the number of stored patterns.

27.4.2 Communication Analysis

In the first simulation, we selected the number of vote sets required to match a pattern as a metric to evaluate the performance of the algorithm. The VGN matches patterns by exploiting communication among nodes to permit node collaboration. The communication requirements of the approach were analyzed by observing the performance of the approach with the increase in pattern dimensionality and the number of stored patterns. We studied the number of communicated vote sets using pattern sizes of 256, 1024, 4096, and 16,384 mapped into sensor networks of respective sizes. Each network was simulated while storing a range from 4 to 4096 of random pattern signatures. Test patterns were randomly picked from the stored patterns and used as unknown patterns. The sensor nodes were needed to create and communicate vote sets into the network to match the patterns. Sensor nodes were assumed to be organized in a single-hop network where all sensors can receive the communication of each other, and the nodes were sequentially selected to communicate their vote sets. These assumptions allowed focusing on the characteristics of the VGN approach and mitigating various network factors, such as network latency, channel contention, and multihop communication.

Figure 27.9 shows that the number of vote sets required to reach a consensus is remarkably small and follows a logarithmic curve with the increase in the number of stored patterns. The figure also clearly shows that the total number of vote sets

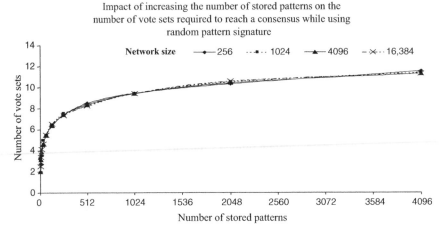

Figure 27.9 Average number of vote sets required to reach a consensus using 256, 1024, 4069, and 16,384 sensor nodes using random pattern signature.

required to match patterns remains approximately the same regardless of the network size. We may confidently conclude that, for patterns of random similarities, the process of reaching a consensus is relatively independent of the network size and the dimensionality of stored patterns. These results clearly highlight the scalability of the VGN approach and demonstrates that the communication cost for recognizing events is very small.

Furthermore, digit patterns from the MNIST database were also simulated to measure the performance of the VGN approach using patterns of high similarity. The simulation included storing a range of patterns from 4 to 1024 of randomly selected patterns of digit 3 and then the simulation was repeated using patterns of digits 5 and 8; all pattern digits are of size 28×28. Although the graph in Figure 27.10 shows a similar trend as the graph of Figure 27.9, the simulations of digit patterns show an overall increase in the number of communications required to reach a consensus with the increase in the number of stored patterns. Moreover, the different digit classes show various communication requirements; patterns of digit 5 required slightly more communication than patterns of numbers 3 and 8. Similar trends are observed in other digit patterns.

Figures 27.11 and 27.12 show the ratio of the average per node communication (ANC) for networks storing random pattern signatures and the digit patterns from the MNIST database, respectively. Each of the images in both figures represents a network of sensor nodes with the x-axis and y-axis depicting the nodes' coordinates. The bars on the side of each image show gray scales depicting the various ANC ratios of the image. The ANC measures the average number of communicated vote sets per sensor node required to recognize all of the unknown patterns. The simulations of various random pattern sizes and various digit patterns show similar trends. The results of the simulation of patterns of number 3 and random patterns of size 32×32 are selected to represent and compare the two pattern categories. Figures 27.11(a)–(i)

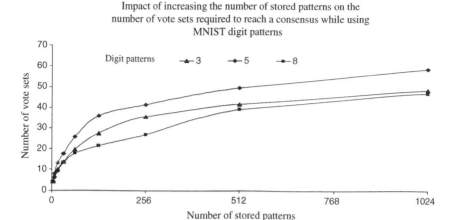

Figure 27.10   Average number of vote sets required to reach a consensus using patterns of digit 3, 5, and 8 of the MNIST database.

and 27.12(a)–(i) show ANC ratios for sensor networks storing a range from 4 to 1024 patterns.

The ANC results of both figures show that most of the nodes do not communicate their vote sets. This is a very significant result, as it complies with the objective of the VGN approach to maximize local processing and minimize communication. The ANC of nodes increases gradually with the increase in the number of stored patterns, since more vote sets are required for pattern matching.

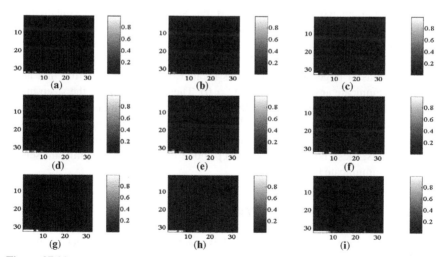

Figure 27.11   (a)–(i) ANC results of networks of size 1024 nodes storing 4, 8, 16, 32, 64, 128, 256, 512, and 1024 random pattern signatures.

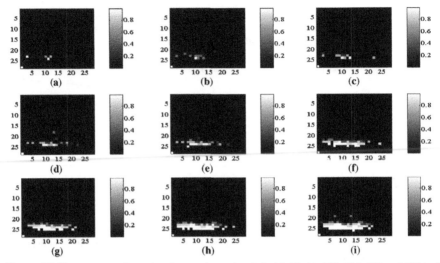

Figure 27.12 (a)–(i) ANC results of networks storing 4, 8, 16, 32, 64, 128, 256, 512, and 1024 of patterns of digit 3 of the MNIST database.

Both Figures 27.11 and 27.12 show high ANC numbers for sensor nodes located in the lower section of the patterns. The high ANC numbers for these nodes indicate that the nodes are repeatedly communicating their votes. The fact that these particular nodes show high ANC results from the node selection mechanism employed in the simulation, which sequentially traverses these nodes in each simulation. However, the resulting trend can be easily overcome by randomly selecting nodes to balance the ANC of nodes over the network.

Digit patterns outputs are different from the ANC results of random pattern signatures, as depicted in Figure 27.12. First, since digit patterns contain more similarities and require more vote set exchanges than random pattern signatures, the number of sensor nodes with high ANC is also larger than corresponding simulations in Figure 27.11. Second, although the sensor nodes were sequentially traversed in the simulations, a large number of nodes located immediately after the first node do not communicate any vote sets (ANC = 0). These nodes exploit the sleep mode strategy and enter the sleep mode after the first vote set communication. The first node and the consecutive nodes are all located in the white margins that border the number in digit patterns, thus having identical vote sets. Instead of communicating their vote sets, these nodes enter the sleep mode to allow more significant sensor nodes to communicate their vote sets.

27.4.3 Sleep Mode Analysis

One of the objectives of the VGN approach is to set as many sensor nodes as possible into the sleep mode, hence recognizing patterns using the smallest possible set of active nodes. The length of the interval for which a sensor node remains active can

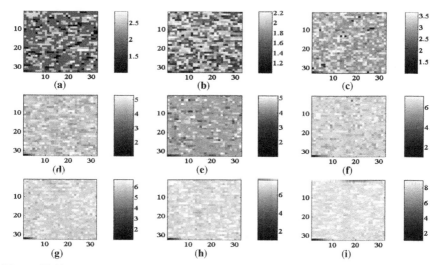

Figure 27.13   (a)–(i) AWT of networks of size 1024 nodes storing 4, 8, 16, 32, 64, 128, 256, 512, and 1024 random pattern signatures.

be used to evaluate the sleep mode strategy. To do so, we introduce a metric called the average wake up time (AWT). The AWT measures the average active time of each individual sensor node; the time scale in the simulations is measured as the time required to communicate a vote set. Sensor nodes remain active until receiving identical vote sets to their own vote sets or when the network reaches a consensus. It can also be inferred from the AWT that sensor nodes are teamed together by analyzing which nodes enter together the sleep mode as a result of identical vote sets. Nodes of each group are redundant to each other. Thus, nodes belonging to a particular group can replace or respond on behalf of lost nodes of the same group.

AWT simulations for random patterns of sizes 256, 1024, and 4096 and digit number patterns of 3, 5, and 8 were conducted. Figures 27.13 and 27.14 show examples of the AWT simulation results for networks storing 4, 8, 16, 32, 64, 128, 256, 512, and 1024 random patterns of size 32×32 and patterns of number 3, respectively; other pattern sizes and number digits show similar results. Each image in both figures represents a network of sensor nodes with the x-axis and y-axis depicting the nodes' coordinates and the bar showing the scale of the AWT values.

While using random pattern signatures and digit number patterns, the AWT of individual nodes increases with the increase in the number of stored patterns. Two factors justify the increase in AWT. First, as the number of stored patterns increases, the number of communicated vote sets increases concurrently, as shown in Figures 27.9 and 27.10. The last group of nodes has to wait until the last vote set is communicated before being able to enter the sleep mode. Second, pattern interference complexity amplifies with the increase in the number of stored patterns causing shared areas among patterns to reduce and thus decreasing the possibility of finding identical vote sets. Thus, sensor nodes need to wait longer to receive identical vote sets.

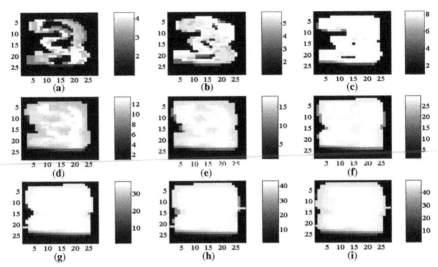

Figure 27.14   (a)–(i) AWT of networks storing 4, 8, 16, 32, 64, 128, 256, 512, and 1024 of digit 3 patterns of the MNIST database.

The performance of the VGN approach has a diverse output that is dependent on whether random patterns or patterns of digits are utilized, as shown in Figures 27.13 and 27.14, respectively. The affect of having high similarity among patterns on the AWT can be observed clearly while storing patterns of digit patterns. All digit patterns share white margins surrounding their actual digits, as shown in Figure 27.7. Nodes located in the white margin are of little significance, since the white margin creates a shared area among all patterns and thus nodes located in this area cannot distinguish between any of the patterns. Most of the nodes located in the white margin area enter the sleep mode in early stages, that is have low AWT. These nodes can conserve their energy to be utilized in other tasks, such as the relay of sensory data. The location of other groups of sensor nodes of the same AWT can be observed in the AWT results. Nodes of each color group are redundant to each other, as they have indentical vote set. This redundancy can be employed to prolong the network's lifetime by rotating the communication contribution among nodes of the same group. Consequently, the ANC of nodes will approximately be equal and the energy consumption will be uniformly distributed over the network. In random patterns, stored patterns have few similarities, thereby sensor nodes exhibit high AWT and approximately all sensor nodes have equal AWT.

Taking histograms of the AWT of each of the simulations provides additional information with regard to the number of sensor nodes entering the sleep mode after each vote set communication. The histogram's x-axis and y-axis depict the time at which sensor nodes enter the sleep mode and the number of sensor nodes entering the sleep mode in the y-axis, respectively. The histograms for the simulations of random patterns and digit number patterns are depicted in Figures 27.15 and 27.16, respectively.

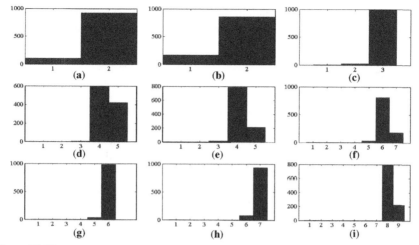

Figure 27.15 Histograms of AWT sensor nodes of networks of size 1024 nodes storing 4, 8, 16, 32, 64, 128, 256, 512, and 1024 random pattern signatures are shown in (a)–(i), respectively.

The histograms Figures for each pattern category show different trends. Random patterns are matched faster and therefore have a smaller number of categories in the *x*-axis. A common trend in all histograms of random patterns is to have a majority of sensor nodes entering the sleep mode right before reaching a network consensus as depicted in Figure 27.15. As a result, in random pattern signatures, a majority of sensor nodes remains active until late stages of collaboration. The VGN approach, however, compensates this drawback by reaching a consensus very rapidly.

To the contrary, digit patterns highly benefit from the sleep mode strategy, as shown in 27.16. A larger number of sensor nodes enter the sleep mode immediately

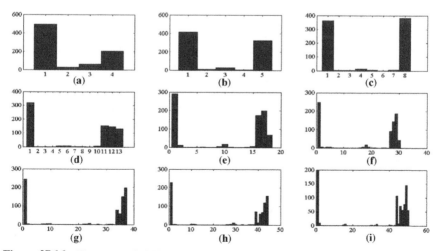

Figure 27.16 Histograms of AWT sensor nodes of networks storing 4, 8, 16, 32, 64, 128, 256, 512, and 1024 of digit 3 patterns of the MNIST database are shown in (a)–(i), respectively.

after the first vote set communication. However, similar to random patterns, a significant number of nodes enter the sleep mode in the stages close to reaching a consensus.

27.5 COMPARISON

The VGN approach can be compared to the GN algorithm. Nonetheless, the GN algorithm is a neural network-based approach in which sensor nodes are mapped into a neural network to store and recognize patterns in a distributed manner. The GN algorithm recognizes events by local node processing and communication The GN algorithm does not incorporate any sleep mode-based strategy, and thus all nodes remain active all the time. Each GN node processes subpatterns (pairs) to produce a boolean result indicating a match or no match to the unknown pattern. Although to recognize any pattern, positive recognition results need to be collected from all the nodes, the GN algorithm's recognition accuracy is very low due to pattern interference. In the existence of pattern interference, similar to shared areas discussed in Figure 27.2, boolean local results do not provide enough information to recognize patterns. To improve the GN algorithm's accuracy, the set of reference patterns are required to be selected carefully to ensure that the stored pattern set is an interference-free set—which may not be always feasible. On the contrary, the VGN approach is not affected by pattern interference, since the node collaboration and vote set exchange provide sufficient information to resolve interferences and match patterns correctly.

In regard to the communication requirement, the GN algorithm depends solely on local collaboration to recognize patterns. Each GN node exchanges its local results with its two neighboring nodes, namely the successor and predecessor nodes. Upon receiving the local results of adjacent nodes, a sensor node can produce its final decision. Unknown patterns are recognized when all nodes produce a positive boolean value that indicates that the unknown pattern matches locally stored patterns. Therefore, to detect any unknown event, each sensor node communicates two messages, thus the sensor network needs to communicate $2n$ messages to match patterns, where n is the size of the network. The number of messages can be reduced to n when proper node deployment is used to adjust the location of adjacent sensor nodes to receive each other's communications. Nevertheless, a constant number of n messages are needed to be communicated to detect any event regardless of the number of stored reference patterns. Unlike the VGN algorithm, which considers the significance of local decisions and the similarity among patterns to set nodes into the sleep mode and to reduce the amount of required communication and the number of nodes needed to detect events, the GN algorithm activates all sensor nodes to detect events. Since all nodes are required to participate to match any event pattern, the ANC and AWT are uniform and equal over the network, that is all nodes have the same ANC and AWT. All nodes need to communicate their decision causing all nodes to have an ANC = 1, and the AWT of all nodes is equal to the lifetime of the network, since no sleep mode strategy is utilized. Therefore, the performance of the AWT and ANC of the VGN approach is far superior than that of the GN algorithm.

Similar comparisons can be derived for other approaches. For example, the Hopfield network (HN) algorithm also stores patterns in a distributed manner. The HN algorithm recognizes patterns when it is able to retrieve a corresponding pattern from the distributed memory. This process demands a large number of iterations where each iteration requires that all nodes communicate their local results to each other regardless of their results' significance, thus a communication cost of $n^2 I$, where n is the number of nodes in the network and I is the number of allowed iterations. Consequently, the HN's AWT and ANC are comparable to the GN algorithm's result and higher than the AWT and ANC results obtained for the VGN approach. Moreover, at the end of the iteration, the HN may reach spurious states. Pattern inference can lead to spurious states that are defined as stable states or result patterns of the HN that do not match any of the stored patterns [9].

27.6 CONCLUSION AND FUTURE WORK

In this chapter, an energy-efficient and distributed template-matching approach to detect large-scale events using sensor nodes has been proposed. Instead of performing template-matching at the base station, the VGN approach matches patterns within the network. The proposed approach uses the information processing-based and sleep mode-based techniques to exploit in-network processing, reduce the amount of communication, and prolong the durability of the network. The event-detection problem is converted into template-matching. Patterns depicting interesting phenomena of the environment are used to detect and classify events of interest. The patterns are stored in a distributed manner within the network by partitioning these patterns into their fundamental pattern elements (pairs). The approach dynamically manages the sensor network's collaboration by concentrating the pattern-matching process within sensor nodes of higher significance. The node collaboration results in converting interesting local sensory information into meaningful global event-recognition consensus. Incorporating a sleep mode strategy facilitates on-demand processing and thus leads to better throughput utilization and prolongs the lifetime of the sensor network.

The simulation results have shown the capability of the VGN approach to match patterns within the network. We have proposed novel metrics to evaluate the performance of pattern matching for sensor networks. The metrics evaluate the parameters that effect the energy consumption in sensor networks, such as communications and sleep intervals. The approach results have shown that patterns of low similarity can be matched very fast. The results have also shown that changes in pattern dimensionality do not effect the communication overhead. On the other hand, the simulations of networks storing patterns of high similarity have shown that there is an increase in the amount of communication overhead required to reach a consensus. Nevertheless, this increase in communications is compensated by the sleep mode strategy being exploited effectively when redundant sensor node are switched into the sleep mode.

Furthermore, the simulations have demonstrated that some nodes are more significant than others. These nodes usually deplete their energy resources earlier than other nodes. By showing the processing and energy contribution of individual sensor nodes, the proposed evaluation metrics can be used to improve the network's performance and lifetime. Additional sensor nodes or sensor nodes equipped with larger batteries can be deployed in areas showing sensor nodes with high contributions to the pattern-matching process.

As part of future work, we plan to improve our model by considering other sensor network and pattern-recognition factors. The current VGN voting mechanism indirectly tolerates noise and packet loss by making use of the similarities of patterns and redundancies of identical vote vectors. We plan to further investigate the effect of noise on the detection accuracy and energy resources. Inexact pattern-matching techniques need to be considered for recognizing patterns affected by noise, rotation, and deformation. We would like to extend our study to evaluate other voting mechanisms, for example, majority voting, under constraints of the network. Finally, sensor network applications require the co-design of various layers of the application to coordinate and optimize their energy utilization. Therefore, we plan to analyze the current and planned voting mechanisms using various sensor network routing and medium access control (MAC) protocols to enhance the performance of our approach.

REFERENCES

1. I. F. Akyildiz, W. Shin Su, Y. Sankarasubramaniam, and E. Cayirci. A survey on sensor networks. *IEEE Commun. Mag.*, 19:102–114, 2002.
2. D. H. Ballard and C. M. Brown. *Computer Vision*. Prentice Hall, 1982.
3. R. Battiti and A. Colla. Democracy in neural nets: voting schemes for accuracy. *Neural Networks*, 7(4):691–707, 1994.
4. C. M. Bishop. *Neural Networks for Pattern Recognition*. Clarendon Press, Oxford, 1995.
5. Crossbow. Mica2 datasheet. [online]. (http://www.xbow.com /Products/Product_pdf _files/Wireless _pdf/MICA2_Datasheet.pdf), Last accessed 21 February 2008.
6. R. P. W. Duin and E. Pekalska. Structural inference of sensor-based measurements. In *Proceedings of the Structural, Syntactic, and Statistical Pattern Recognition (SSSPR '06)*, Vol. 4109. Springer Verlag, 2006, pp. 41–55.
7. D. A. Forsyth and J. Ponce. *Computer Vision: A Modern Approach*. Prentice Hall, 2002.
8. J. Gao, L. J. Guibas, J. Hershberger, and N. Milosavljevic. Sparse data aggregation in sensor networks. In *6th International Conference on Information Processing in Sensor Networks (ISPN'07)*, 2007, pp. 430–439.
9. S. Haykin. *Neural Networks: A Comprehensive Foundation*. Prentice Hall, 1998.
10. Z. Kamal, M. Salahuddin, A. Gupta, M. Terwilliger, V. Bhuse, and B. Beckmann. Analytical analysis of decision and data fusion in wireless sensor networks. In *Proceedings of the 2004 International Conference on Embedded Systems and Applications (ESA04)*, 2004, pp. 263–272.
11. A. I. Khan. A peer-to-peer associative memory network for intelligent information systems. In *The 13th Australasian Conference on Information Systems*, Vol. 1, Melbourne, Australia, 2002, pp. 317–326.
12. A. I. Khan, M. Isreb, and R. Spindler. A parallel distributed application of the wireless sensor network. In *Seventh International Conference on High Performance Computing and Grid in Asia Pacific Region*, Tokyo, Japan, 2004, pp. 81–88.
13. A. I. Khan and P. Mihailescu. Parallel pattern recognition computations within a wireless sensor network. In *ICPR*, Vol. 1, 2004, pp. 777–780.
14. A. Konar. *Artificial Intelligence and Soft Computing: Behavioral and Cognitive Modeling of the Human Brain*. CRC, 1999.

15. B. Krishnamachari, D. Estrin, and S. Wicker. The impact of data aggregation in wireless sensor networks. In *Proceedings of the 22nd International Conference on Distributed Computing Systems (ICDCSW '02)*, IEEE Computer Society, 2002, pp. 575–578.

16. L. I. Kuncheva. *Combining Pattern Classifiers: Methods and Algorithms*. Wiley, 2004.

17. M. Li, Y. Liu, and L. Chen. Non-threshold based event detection for 3d environment monitoring in sensor networks. In *Proceedings International Conference on Distributed Computing Systems(ICDCS '07)*, IEEE, 2007.

18. J. Liu, J. Reich, and F. Zhao. Collaborative in-network processing for target tracking. *J. Appl. Signal Process.*, 4(8):378–391, 2003.

19. S. Madden, M. J. Franklin, J. M. Hellerstein, and W. Hong. Tag: a tiny aggregation service for ad-hoc sensor networks. In *5th Annual Symposium on Operating Systems Design and Implementation (OSDI)*, 2002, pp. 131–146.

20. MATLAB. Matlab(http://www.mathworks.com/), 2007,

21. MNIST. Mnist database. [online]. (http:// yann.lecun.com/ exdb/ mnist/), Last accessed 21 February 2008.

22. M. S. Nixon and A. S. Aguado. *Feature Extraction and Image Processing*. Newnes, 2002.

23. E. Pekalska and R.P. W. Duin. *The Dissimilarity Representation for Pattern Recognition, Foundations and Applications*. World Scientific, 2005.

24. K. Römer. Distributed mining of spatio-temporal event patterns in sensor networks. In *Proceeding International Conference on Distributed Computing in Sensor Systems (DCOSS '06)*, San Francisco, USA, June 2006, pp. 103–116. IEEE.

25. J. D. Tubbs. A note on binary template matching. *Pattern Recog.*, 22(4):359–366, 1989.

26. W. Watanabe. *Pattern Recognition: Human and mechanical*. Wiley, 1985.

27. A. R. Webb. *Statistical Pattern Recognition*. Wiley, 2002.

Index

Access control
    engine (ACE), 442, 447, 457–464
    preference manager, 462
Access point (AP)
    fixed, 176
    functions of, 93
    MDAS query processing, 114
Access time, 193–195, 197, 202, 204,
    210
Accuracy, significance of, 155–157, 160,
    231, 234, 305, 392, 407, 438,
    443–445, 501, 506
Accusations, false, 359
Achievement goals, 331
ACID (atomicity, consistency, isolation,
    durability) properties in
    transaction management, 114,
    116–117
Acknowledgment (ACK), 46–47, 49, 52,
    54, 59, 159, 364–366, 370, 373
ACQUIRE (Agent-based Complex
    Querying and Information
    Retrieval Engine), 111–112, 115
Action thriller films, 535–536
Actions, 451
Active mode, 193
Active Badges, 304
Actor constraints, 234
ActorNet, 130–131
Actual attribute region, 584–586
Ad-hoc authorization, 234–235

Ad hoc on-demand distance vector (AODV)
    routing protocol, 6, 10, 13–17,
    23, 57, 95, 103, 122, 352–353, 416
Ad hoc wireless networks, 163–187
Adaptability, 27
Adaptive optimization, 115
Adaptive probing, 360
Adaptive relays, 123
Addition, AgentSpeak(L), 331–332
Address decoupling, 68
Advanced forward link trilateration
    (A-FLT), 215
Advanced Systems Format (ASF), 532
Adversary class/nodes, 604–607
Advertisements, 64, 70–71, 81
Affiliation, location-based access
    control, 451
Affinity profile, 498–499, 503
Agent-based community-oriented routing
    network (ACORN), 109–110
Agent management, Agilla sensor
    network, 129
Agent-oriented architectures
    BDI (belief-desire-intention)
        and AGENTSPEAK(L), 330–332, 340
        context-aware
            communication, 332–340
        consolidated presence information (CPI)
            322–324, 326, 332, 336, 340
    elements of
        context-aware call handling, 328–329

Mobile Intelligence. Edited by Laurence T. Yang, Agustinus Borgy Waluyo, Jianhua Ma,
Ling Tan, and Bala Srinivasan
Copyright © 2010 John Wiley & Sons, Inc.

Agent-oriented architectures (*continued*)
context information server (CIS), 324,
327–329, 335
overview of, 323–326, 340
policies and preferences manager
(PPM), 326, 328–329, 332, 335
personal communication manager
(PCM), 325, 327–328, 332,
334–335
presence information manager
(PIM), 325, 327–328
presence management strategy (PMS)
326–328
Mercury, 340
Mobile People Architecture, 340
motivation, 321–322
Personal Proxy, 340
Universal Inbox project, 340
AgentSpeak(L), 330–332, 340
AgentTalk, 330
Aggregation points, 608–609. *See also* Data
aggregation
Agilla/Agilla Engine, 129–131
AGPS/GPS, 444
Air indexing, wireless broadcast systems
basic concepts of, 194–195
cell-based distributed air index (CEDI),
see Cell-based distributed air
index
data organization, 195–196
nonspatial data
exponential index, 200
hashing-based index, 198
hybrid air index, 199–197
signature-based index, 198–199
tree-based index, 196–198
overview, 193–194, 210
for spatial data
cell-based distributed air index
(CEDI), 204–210
space filling curve-based
indexes, 201–204
space partition-based indexes, 201
tree-based index, 200
Ajax technology, 218–220
Alert-in-triangulation (AIT) test, 589–590
Alert sensors, 585, 591–592
Alignment rule, 132
All-Kth-Order PPM, 170
Alzoubi's algorithm, 36

AND operation, 199
Angle assignment, 522–526, 566
Animation films, 535–536
Anonymity-based privacy
techniques, 438–439
Anonymized data, 234
Ant-based evidence distribution
(ABED), 414
Ant colony systems (ACS), 129
ANTARCTICA (Autonomous ageNT bAsed
aRChitecture for cusTomized
mobile Computing
Assistance), 113, 115
Antenna, channel utilization and, 57. *See
also specific types of antennas*
AOL, 474
Apache Axis, 274–275
APIs, 228, 270, 309, 485
Appearance count, 169
Application layer, 90
Application service providers, 446
Application zones, 438
Approximate Taylor expansion, 183
Approximation ratio, 29–30, 34
Apriori algorithm, 248
Apriori-like group pattern (AGP)
mining, 248
ARAN, 355
Area
algorithm, 27–28, 34–36, 39–40
ID, 31
location-based access control, 455–456,
458–461
arguments, 451
ARIADNE, 304, 355
Artificial intelligence, 131
Assembly line automation, 64
Association rule mining, 247
Associative memory (AM), 381–382, 409
Associativity-based routing (ABR), 351
Asynchronous Javascript and XML, 218
Atomicity, in transaction management, 114,
116
Attacks
intrusion detection systems
(IDS), 124–125
traffic, 609
types of, 353–355, 408
Augburg Indoor Location-Tracking
Benchmarks, 305

Authentication, 345–346, 353–355, 358, 367–368
Authorization policies, 234
Authorization rule, location-based, 451
Automated planning, OC: middleware, 312
Autonomy
  information retrieval systems, 111–112, 115
  mobile agent technology, 113
Average per node communication (ANC), 650–652, 656–657
Average query delay, 99–102
Average query distance, 99, 101
Average wake up time (AWT), 653–657
AWF algorithm, 29, 33–34, 36, 38–39. *See also* Rule 1&2

Backdoors, intrusion detection systems, 125
Backoff
  counter, 46–47, 50–51
  procedure, 49, 51
  timers, 54
Backup data, 552
Bandwidth
  intrusion detection system, 125
  location tracking, 164
  mobile ad hoc networks (MANETs), 4–5, 7, 9, 22
  mobile agents, 107
  network constraints, 65
  predictive location tracking, 186
  query-processing systems, 115
  sensor networks, 128
  simulations, 99
  wireless sensor networks, 596, 602
Base station
  ANTARCTICA query processing, 113
  attacks against, 596, 599
  centralized, 607–608
  location tracking, 165
  sensor networks, 126–127
  threat detection, 407
Basic access mechanism, 47
Basic actions, AgentSpeak(L), 331–332
Basic data-centric mechanisms, 550–552, 570–571
Batteries
  ad hoc networks, 30
  burdens on, 231
  life time, 192–193

mobile ad hoc networks (MANETs), 4–5
power from, 645
Battlefield monitoring, 596
Bayesian filters, 579
Bayesian learning, 630
Bayesian networks, 305–306
Bcast, 194–195, 201–202, 208
bcast_pointer, 194
BDI (belief-desire-intention) agent architecture
  and AGENTSPEAK(L), 330–332, 340
  characteristics of, 330
  context-aware communication
    BDI mapping, 332–333
    example of, 333–335
    proof of concept, 335–340
  development of, 330
Belief atom, 331
Belief literals, 331
Bellman-Ford distance-vector algorithm, 349
Berners-Lee, Tim, 217
b-frames, 532
Bidirectional communication, 67
Bidirectional identification, 442
Binary communications, 268
Binary exponential backoff, 47
Binary messages, 316
Binary random patterns, 646
Bit rates, energy consumption effects, 538–539, 543–544
Bit string, air indexes, 205
Blackberry devices, 64
Blackhole attacks, 408
Block extensible exchange protocol (BEEP), 125
Blocking, 45
Blogs, 219
Bluetooth, 63–64, 440, 511, 514, 528
bool_value, 449
Border dominator, 31–32
Border nodes, zone routing protocol (ZRP), 17–19
Border sensors
  functions of, 591–592
  selection, 585, 590–592
Bottlenecks, 68, 124, 301
Boundary sensor clustering, 581
Boundary sensor selection, 581
Boundary tracking, 581–582

B+ tree, 178, 182
Branch-and-bound investigation, 183
Broadcast cycle, 194, 201–202
Broadcasting
  ad hoc networks, 27, 29, 31
  extra frame transmission (EFT), 52
  location services, 159
  MANET routing protocols, 9
  max degree algorithm, 31
  proactive routing, 349
  sensor networks, 129
  transaction management, 117
Broadcast query (BQ), 351
Broker network, publish/subscribe, 66–67,
    69–70, 72–73, 84
B-trees, 180, 185
Bubble function, 616
Bucket
  broadcast cycle, 194
  loss, 204
bucket_id, 194
bucket_type, 194
Buddy lists, 322
Building structure monitoring, 596
Bush fire monitoring, 596
Business-to-business (B2B)
    scenarios, 266
Busy-tone multiple access (MTMA)
    protocol, 45
Byzantine attacks, 408

Cache, *see* Caching systems, cooperative
  locality, 181
  size, simulations, 99, 101
CacheData, 92–93
Cached item IDs, 96–97
CachePath, 92–93
Caching agents (CAs), 94
Caching systems, cooperative
  characteristics of, 92–93
  using general routing protocols, 92–94
Call forwarding (CF), 338–339
Call handling, 328–329
Camouflage attacks, 302
Canberra distance, 631
Cancellation procedures, distributed
    information retrieval, 116
Cancel RTS (CRTS), 45, 56
Capability, location-based services, 232

Carrier sense multiple access
    (CSMA), 43–44
Carrier sense multiple access with collision
    avoidance (CSMA/CA), 44, 46
Carrier sensing, 44, 50
CBTC algorithm, 28
CEDAR, 29
Cell-based distributed air index (CEDI)
  applications, 194
  design goals, 204
  implicit cell data filtering, 207–208
  index structure and broadcast channel
    organization, 205–207
  query processing, 208–210
  space partition, 205
Cell-based mobile phone network, 240
Cell-Id technology, 214–215, 445
Cell/cellular phones
  agent-based architectures, 329
  characteristics of, 270, 444
  SMS capabilities, 63
  Web-enabled, 322
Cell topology, 246
Cellular network
  GSM, 214
  handover procedure, 164
  operators, 215–216
Cellular robotic systems, 131
Cellular systems, hexagonal
    architecture, 176
Cellular wireless networks, predictive
    location tracking, 163–187
Centralized attack detection
  characteristics of, 614–615
  data classification, traffic
    features, 617–618
  detection algorithm
    applications, 618–619
    map initialization, 619
    optimal time epoch, 619
  evaluation
    detection rates, 620–621
    energy decay rates, 624–625
    false negative rates, 623–624
    false positive rates, 621–623
    importance of, 620, 625
  learning phase
    neighborhood function, 616
    overview of, 615–616

update function, 616–617
Certificates, digital, 355, 361
Channel(s)
  contention, 649
  release schemes, 44
  reuse, 44, 48, 51, 53, 60
  stability, 51
  utilization, *see* Channel utilization
Channel utilization
  degradation of, 45
  enhancements for
    combination of RTS validation and
      extra frame transmission, 53–55
    compatibility with IEEE 802.11, 56
    extra frame transmission (EFT), 51–54
    modification of NAV operation, 51–52
    reverse extra frame transmission
      (R-EFT), 51, 54, 56
  medium access control (MAC) layer, 29,
    45–51
  performance evaluation
    extra frame, 58–59
    interruptions, 59–60
    packet delivery ratio (PDR), 59–60,
      429
    RTS/CTS handshaking, 56, 59–60
    RTS validation, 51, 57–59
    simulation scenario, 57
    unicast frame, 59
checkState and tracking, 292
Child finder and tracking services, 221,
    223, 227
Children node redistribution, 180
City-Block distance, 631
Client populations, growth in, 163
Client-server
  architecture, 117–118, 127
  communication, 218
Clock synchronization, 9, 354–355
Cloning, 115, 129
Cluster-based adaptive cooperative caching
    scheme
  cross-layer based COCA, overview
    of, 95, 102–103
  proposed modules
    descriptions of, 95–98
    experimental results and
      discussions, 98–102
    future trends, 102–103

Cluster-based location service protocol,
    KCLS, 143–161
Cluster-based network topology, 608, 611
Cluster-detection algorithm, 251
Cluster head (CH)-based network
    topology, 406–407
Cluster heads (CHs)
  adaptive cooperative caching, 94
  DDoS attacks, 604
  KCLS, 147–148
  tracking wireless sensor
    networks, 578–579
Cluster-hop distance, 155
Clustering
  architecture, 95–97
  medoid, 252
  moving, example of, 251
  moving object databases
    continual maintenance of moving
      clusters, 250–251
    detecting moving clusters from object
      movement histories, 251
    positional uncertainty and, 251–252
  techniques, 491
  topology control and, 27
  tracking wireless sensor networks, 575,
    578–579, 581
Cluster management, 251
Cluster state (CS) packets, 147–149, 152,
    155–157
Codec standards, 531
Cohesion rule, 132
Collaborative attacks, intrusion detection
    systems (IDS), 125
Collaborative event deteCtion and tracking
    protocol (CollECT), 583,
    585–593
Collaborative filtering systems, 473–474
Collection, mobile agent technology, 110
Collection descriptor, mobile agent
    technology, 110
Collision
  ambiguous, 359
  avoidance of, 48
  backoff times and, 47
  DATA packet, 45
  detection, 47, 49
  hash function, 198
  routing protocols and, 4

Collision (*continued*)
  RTS frame and, 54
  sources of, 52
  wireless sensor networks, 588
Collusion, 359, 362, 370–371, 414, 423
Comb-needle model, 559
Common ancestor, 75
Communication
  abstraction, 65–66, 82
  channel, 164
  cross-layer design framework, 88
  flow, 415
  graph, 27
  infrastructure, 65–66
  intrusion detection systems, 125
Communication layer, location
  middleware, 462
Communication phase, detection
  algorithm, 619
Community services, 219
Compactness theorem, 632
Compass routing algorithm, 347–348
Compatibility problems, 44–45, 60
Complete/full binary tree, 172–173
Complicated attacks, intrusion detection
  systems (IDS), 124
Compromised sensor nodes, 604
Conceptual partitioning method
  (CPM), 183–184
Conditional probability, 169–170
Cone selection, 521
CONFIDANT (Cooperation of Nodes:
  Fairness In Dynamic Ad-hoc
  NeTworks), 359, 414
Confidence intervals, 152
Conflict resolution, 334–335
Connected dominating set (CDS), ad hoc
  networks
  area-based algorithm
    max degree algorithm, 31–34
    mobility issues, 35–36
    overview, 30
    performance analysis, 34–35
  characteristics of, 27–30
  experiential simulations, 36–39
  future directions for, 39–40
Connectivity
  ad hoc networks, 27
  KCLS, 147, 151

MANET routing protocols, 16
mobile agent based query-processing
  systems, 115
mobility management and, 164–165
sensor networks, 130
topology control, 35
transaction management, 118
wireless network, 64–65
Connectors, ad hoc networks, 29–30
Consolidated presence information
  (CPI), 322–324, 326, 332, 336,
  340
Constant bit rate (CBR), 57, 372, 375, 427,
  429–431
Constraints, location-based services, 234
Consumer, in service-oriented
  architecture, 267, 269
Content-based publish/subscribe system, 68
Contention window (CW), 47, 52
Content provider, 224, 226
Context, defined, 167
Context awareness
  action systems
    actions, 282–283
    building blocks, 283
    composition of, 284
    refinement of, 284–285, 287–290
  background to, 280
  case study, 290–292
  context models, 285–287
  data propagation/composition, 286
  formalizing, and context
    dependency, 281–285
  related work, 280–281
  wireless sensor networks, 281
Context-aware systems
  call handling, 328–329
  characteristics of, 321
  communication, 332–340
Context information server (CIS), 324,
  327–329, 335
Context manager, Agilla sensor
  network, 129
Context refinement
  action systems, 284–285
  compositional, 288–290
  context-utilizer, 287–288
  context variable, 288
  parallel composition rule, 287

Context tree weighting (CTW)
    scheme, 172–173
Context-utilizer, 287–288
Continuous query, 177
Contribution potential, 564
Control center, mobile agent
    technology, 110
Control messages, 5, 9
Control packet size, 45
Conventional patterns, 646
Convergence, 249
Convergence phase, detection
    algorithm, 618
Convex hull property, 183
Convoy tree, 576–577
Cooperative caching (COCA)
    cluster-based adaptive schemes, 95–98
    overview of, 95
Cooperative caching without prefetching
    (CCNP), 99–103
Cache management, COCA schemes, 95,
    97–98
Cooperative caching with cross-layer design
    (CCCL), 99, 101–102
Cooperative caching without cross-layer
    design (CCNCL), 99, 101
Cooperative caching with prefetching
    (CCPF), 99–102
CORBA services, 268, 309, 475
Core, ad hoc networks, 29
Corporate Memory Management through
    Agents (CoMMA) project
    architecture of, 109
    mobile agent technology, 109
    prototype, 109
    resource management, 109
Corruption, 204, 408
Coverage area, mobile agent
    technology, 113
Covering relations, publish/subscribe
    systems, 71–72
C++ programming language, 24, 503
CP2U, 226
Credit clearance service (CCS), 358–359
Cross-layer design, wireless networks
    benefits of, 88–89
    characteristics of, 88, 90–91
    cooperative caching, 97–102
    implementation framework, 89

proposals, 89
    stack-wide, 91
Cross-layer information exchange,
    96–97
Cross-layer signalling shortcuts
    (CLASS), 90
Cross-referencing LBS, 221, 223
Cross-talk, 384–385
Cryptlib crypto toolkit, 371
Cryptography, 354
CTS frame, 49
Cyberspace, 214
Cyclic overlays, 73

Data accessibility
    delay, 88
    ratio, 98–100
Data aggregation (DA)
    location services, 160
    threat detection, 406
    topology, 608–609, 611–612
    tracking in wireless sensor
        networks, 579, 604, 608–609
Database management, 117
Data center (DC), 95–96
Data-centric (DC) storage mechanisms
    basic, 550–552, 570–571
    characteristics of, 549–550, 571
    future directions for, 568–570
    multiple sink nodes, 569–571
Data-cleaning agents, 125
Data collection service (DCS), 272
Data compression, 166
Data delivery model, 608
Data frame size, 57
DATA frame transmission, 45
Data fusion, 127–128
Data heterogeneity, query-processing
    systems, 115
Data item size, simulations, 99
Data link layer, 96
Data listening time, 194
Data mining
    affinity-based, 494
    algorithms, 125
    multimodal, 496
    service (DMS), 272
    techniques, 111
    video event detection, 491

Data packet
  delivery ratio, 372–373
  length, 90
  location service, 157
  wireless sensor networks, 551–552
Data providers, 271
Data push, 69
Data rate, 90
Data refinement, 285
Data-replica mechanism
  characteristics of, 550–553, 570–571
  ring-based, 553–555, 571
  tree-based, 555–557, 571
  zone-based, 557–560, 571
Data request, 16
Data selection tasks, 265
DATA-send (DS) packet, 45
Data source (DS), 94
Data traffic, 369
Data transmission, mobile agent
        technology, 122
Data volume, 65
Data warehouse technologies, 125
DBMS-Aglet framework, transaction
        management, 118
DCF interframe space (DIFS)
        interval, 46–47
DCOM, 268
DCTC framework, 576–577
Deadlock state, 44
Dead reckoning algorithm, 231
Decentralization, 381
Decision maker, 46
Decoding, 534
Decomposed CTW (DeCTW), 173–174
Decryption, 356–357
Delaunay triangulation, 28, 578
Delayed algorithm, 80–82
Delays, 231, 628
Delay-tolerant networks, 165
Deletion
  AgentSpeak(L), 331
  and re-insertion, 186
Demand-based routing, 346
Denial of service (DOS) attacks
  characteristics of, 596
  routing security, 353–354, 362
  threat detection, 408
  wireless sensor networks, 600–602

Density-based clustering, 248, 252
Dense regions
  data mining techniques, 247
  detection of, 253
Density of network, impact of, 36
De-obfuscation, 439, 464
Dependent set (DS), ad hoc
        networks, 29–30
Destination address, 363
Destination sequenced distance vector
        (DSDV) routing protocol, 6–9,
        14, 17
Detection-reaction approach, 413, 415,
        420–421. *See also* SMRT1
        (Secure Mobile Ad Hoc Network
        Routing with Trust Intrigue)
Deviation-avoidance tree (DAT), 578
Dictation (Careflow), 475
Diffie-Hellman key exchange protocol, 362,
        596
Digital Rights Management, 531
Digital search trees, 169
Digital signature, 354, 356, 359, 367–368,
        373
Digit patterns, 649–650, 652, 654–655
Dijkstra's shortest path algorithm, 7–8, 147
Direct attacks, 600–601
Directed acyclic graph (DAG), 6, 19–22,
        352
Directories, 475
Direct reputation, 417, 421–422, 431–433
Disconnection(s)
  frequency of, 192
  MANET routing protocols, 17
  MDAS query processing, 114
  mobile agent technology, 107, 132
  publish/subscribe system, 69, 73–74, 77,
        84
Discrete event simulator, 23
Discrete sequence prediction
        problem, 167–168, 175
Distance-based updating, 231–233
Distance routing effect algorithm for
        mobility (DREAM), 144
Distance-vector algorithms, 349
Distorted patterns, 393–395, 397–404
Distributed access protocol, 46
Distributed coordination frame
        (DCF), 46–47

Distributed coordination function
    (DCF), 365
Distributed datastore, OC:
    middleware, 312–315
Distributed denial-of-service (DDoS) attacks
  characteristics of, 595–596
  defined, 596
  models of
    attacks against base station, 599
    identity attacks, 597–599
    physical layer attacks, 599
  threat detection, 381, 405–408
  traffic and, 602
  wireless sensor networks
    adversary model, 604–607
    centralized attack detection, 614–625
    derivation of, 601
    detection requirements, 603–604
    direct attacks, 600–601
    flooding-based attacks, 601–602
    network model, 607–609
    probability of attacks, 602
    reflector attacks, 601
    threshold pattern modeling, 603,
      609–612
    traffic flow observation table, 603,
      613–614
Distributed document clustering, 131
Distributed environment sensing, 64
Distributed flocking algorithm, 132
Distributed hierarchical graph neuron
    (DHGN)
  characteristics of, 381, 388–389
  pattern recognition case study
    accuracy in, 392
    simulator design, 389–391
    standard-form, 397–405
    variable-form, 391–397
Distributed index for multidimensional data
    (DIM) mechanisms, 560–562
Distributed indexing
  scheme, 196
  table, 203
Distributed information retrieval
  characteristics of, 108, 111
  distributed query processing and
    information retrieval, 109–114
  global transactions, 116–117
  transaction management, 114, 116–120

Distributed intrusion detection system
    system using mobile agents
    (DIDMA), 125
Distributed matching, 68
Distributed model predictive control
    (DMPC), 123
Distributed predictive tracking (DPT),
    wireless sensor
    networks, 578–579
Distributed query processing, information
    retrieval systems
    mobile data access system
      (MDAS), 112–114
    wired networks, 111–112
Distributed routing protocols, 144
Distributed virtual backbone mobility
    management scheme, 144
DivX, 531–532, 535–538, 541–542, 544
DMS Stub, 273–274
Dominatees, ad hoc networks, 29, 31–33
Dominators, ad hoc networks, 29–30
Doubling circles scheme, location-based
    mobile information services, 144
Downlink channel, 191–192
Downloads/downloading, 193–195, 199,
    208, 226, 265
Downstream packet, 425
Doze mode, 193
Drain-and-balance (DAB), 577
DREAM location service (DLS), 144
Drill-down, 244
DSI, 201–203, 210
D-tree, 201
Duality-transform, 254
Duplicate events, 75
Dynamic backup routes routing protocol
    (DBR^{2}P), 94
Dynamic destination-sequenced
    distance-vector routing
    (DSDV), 349
Dynamic loading, 123
Dynamic networks
  location-based mobile information
    services, 144
  swarming intelligence, 131
  topology
    mobile agent technology, 131
    query-processing systems, 115
Dynamic programming, 169

Dynamic source route (DSR) protocol
  KCLS, 157
  routing security, 350, 356, 372, 375
  SMRTI, 414, 417–418, 422, 427–428,
    431–432
  state-of-the-art, 6, 10–13, 15, 17, 23
Dynamic time warping (DTW), 257–259

EASE, 566–567
Eavesdropping, 408
E-commerce, 473
Edit distance, 169
Edit distance on real sequence (EDR), 259
Electric grids, cascading failure, 123–124
E-mail
  retrieval, 299–300
  sending, 63–65
Embedded Sensor Board ESB 430, 304
Emerging mobile applications, 62
Encapsulation effect, 393, 396–397, 401
Encoding, 534
Encounter, 249
Encryption, 356, 367
End-to-end communication, 130
Energy conservation, 175
Energy conservation
  mobile computing
    environments, 193–194, 204, 210
  wireless sensor networks, 633
Energy consumption
  analysis, 531
  experiments, 534–535
  mobile agent-oriented applications, 128
  mobile computing
    environments, 193–194
  wireless sensor networks, 633, 639
Energy decay rate, 624–625
Energy efficiency
  components of, 543–544
  energy consumption effects, 538–542
  experiments
    results, 535–543
    setting, 532–534
  pattern recognition, *see* Energy-efficient
    pattern recognition, wireless
    sensor networks
  in threat detection schemes, 405, 407,
    409
  video codec

DivX, 531–532, 535–538, 541–542,
    544
  H.263, 531, 536–538, 544
  MPEG-4, 531–532, 535–538,
    541–542, 544
  WMV, 532, 535–538, 541–542, 544
  XviD, 532, 535–538, 541–542, 544
Energy-efficient pattern recognition,
    wireless sensor networks
  benefits of, 628
  event recognition principles
    event patterns, 629
    template-matching, 629–633
  future directions for, 657–658
  simulations
    applications, 645–646
    communication analysis, 649–652
    sleep mode analysis, 652–656
    test patterns, 646–649
  voting graph neuron (VGN) approach
    analysis, 642–644
    details of, 635–640
    example of, 640–642
    graph neuron algorithm compared
      with, 633, 656–657
    overview of, 633–635
Energy-efficient query processing, 195
Enhanced 911 (E911), 215–216
Enhanced observed time difference
    (E-OTD), 215
Enterprise databases, 475
Entertainment services, 62
Entity-relationship data model, 336
E-OTD, 445
Epidemic routing, 35
Error messages, 361
Error-prone wireless
  channels, 204
  transmission environments, 210
Error resilience, air indexes, 204
Estimated attribute region, 584
Estimated event region, 584
Euclidean distance, 257, 259, 611, 613,
    615–616
Euclidean metrics, 183
Event boundary, tracking wireless sensor
    networks, 579–580
Event-condition-action (ECA) rules, MDAS
    query processing, 113

Event determination, 585, 587–590
Event Discovery (EVT) packet,
        589–590
Event Dispatcher, OC:
        middleware, 307–310
EventMessage, OC: middleware, 310
Event recognition
    event patterns, 629
    template-matching, 629–634
Events, AgentSpeak(L), 331
Event tracking, 592
Exclusion links, 368 370
Explorer agent, mobile agent
        technology, 132
Exponential air index, 200
External storage (ES) mechanism, 550
Extra frame transmission (EFT), 44, 51–54,
        58–59

Failure(s)
    detection, 116, 312–313
    location-based services, 231
False blocking, 44–45, 49–51
False location information, 455
False recall, 384
Fault detection
    and isolation, 365
    power system management, 123
Fault-tolerance
    capability, 550, 553
    protocol
        sensor networks, 129–130
        transaction management, 117
FAult-TOlerant Mobile Agent System
        (FATOMAS), 116
Feature extraction, 520
Feature profile, 499–500
Feature weight profile, 500–501
Federal Communications Commission
        (FCC), 215
Feedback process, 491–492, 497, 500
File integrity checker agent, intrusion
        detection systems, 125
File sharing, 514
File transfer, 300
Film resolution, energy consumption
        effects, 540–544
Filtering
    air indexes, 200, 204, 206

capabilities, mobile applications, 62–63,
        65, 84
cell-based distributed air index, 204, 206
predictive location indexing, 181
in SMRTI, 473–474
wireless data broadcast system, 192
Final relevance, privacy management, 447
Financial data storage, 64
Finder services, 216–217, 222
Fingerprinting, 227, 229
Finger tree, 556
Fire tracking applications, sensor
        networks, 130
First-order Markov chain, 244
Fitts' law, 523, 525
Fixed length Markov chains/predictors, 167
Fixed network servers, 176
Flash crowd, 601
Flash players, 218
Flat addressing, 68
Flat topology, 406, 608, 610
Flexible office organization, 295–295
Flickr, 219
Flock, 248
Flooding
    limited, 366, 368, 370–371
    location services, 159
    MANETs
        cooperative caching, 93–94
        routing protocols, 19
    publisher mobility system, 81
    SYN, 408
Flooding-based attacks, 599, 366, 368, 408,
        415, 427, 601–602, 604–606
Flood subscriptions, 81
Forerunner (FR) packet, 365–366, 369, 373,
        376
Forwarding node, MANET routing
        protocols, 5
Forwarding packets, location services, 158
Forward path, ad hoc on-demand distance
        vector (AODV) protocol, 14–16
Four-ary position, 450
FP-growth, 248
Frame exchange protocol. See RTS/CTS
        handshaking
Frame rates, energy consumption
        effects, 539–540, 544
Free-riding, 415, 424, 435

Frequency, cooperative caching, 103
Front end, defined, 224
Full reversal, Gafni-Berksekas
     protocol, 21–22
Fuzzy associated retrieval
   characteristics of, 490, 492, 501, 507
   fuzzy-associated recommendation, 503,
     505
   generalized recommendation, 501–502
   personalized recommendation, 501–503
Fuzzy logic, 491, 501

Gafni-Bertsekas routing protocol, 19–22
Galileo, 225
Games, 62
Gateway mobile location center
     (GMLC), 217
Gateway selection process, KCLS, 149
Gaussian function, 616
Gaussian noise, 252
GEM, 565–566
Generic conditions, 451
Genetic encoding, 128
Geocast, 159
Geographical leashes, 355
Geographic hash table (GHT), 551–552,
     557, 562, 571
Geographic information system
     (GIS), 131–133, 214
Gesture recognition device, 519
Get(), 551–552
getPoS, 292
GLIDER routing mechanism, 556–557
Global association rule mining, 265
Global graph property, 27–28
Global intercluster link state, 155
Global Motion Compensation (GMC), 532
Global positioning system (GPS)
   access control systems (ACS), 440, 444
   air index scheme and, 192
   cluster-based location service
     protocol, 144
   intelligent mobile services, 472
   publish/subscribe systems, 63–64
   routing security, 347
   topology control, 23
   Web 2.0 and, 214–215, 219, 225–227,
     235
   wireless sensor networks, 584

Global village, 214
Goals, AgentSpeak(L), 331, 334
Google, 218, 227, 474
GPRS, 64, 214, 217, 219, 231, 448
GPS/ant-like routing algorithm
     (GPSAL), 144
GPSR routing protocols, 550–552, 558,
     561, 569
Graph neuron (GN)
   associative memory concept, 382
   cross-talk phenomenon, 384–385
   defined, 381
   distributed denial-of-service
     (DDoS), 405–408, 410
   energy efficient, 405, 407
   hierarchical (HGN)
     characteristics of, 384–386
     communications, 386–387
     distributed, 388–405, 408–410
     node requirements, 387
     pattern recognition, 385
   simple graph neuron approach
     bias array update, 383–384
     overview of, 382
     pattern input phase, 383
     synchronization phase, 383
     storage requirements, 383–384
   threat detection in WSN case
     study, 405–408
   wireless sensor networks
     (WSN), 405–406, 410, 633,
     656–657
Greedy routing algorithm, 347
Grid location service (GLS), 144
Grid partition index, 201
Grouping, OC: middleware, 312
Group keys, 414
GSM network, 214, 327, 444–445, 448, 488

Hamiltonian distance, 615–616
Hamming distance, 648–649
Handheld devices, 489
Handoff management, 165
Hands-free speech, 475
Handshake timeout, 53–54, 56
Handshaking schemes, four-way and
     two-way, 46
Hash function, 198, 354, 551, 558–559,
     571, 599

Hashing-based air index, 198
Height, temporarily ordered routing
     algorithm (TORA), 22–23
HELLO messages, 29
Hidden Markov models, 305–306
Hiearchical Markov model mediator
     (HMMM) database, 490, 492,
     494–504, 506–507
Hierarchical addressing, 68
Hierarchical clustering methods, 252
Hierarchical management mechanisms
   characteristics of, 551, 563, 570–571
   cluster-based, 551, 564–568, 571
   zone-based, 551, 563–564, 571
Hierarchical state routing (HSR), 145
Hierarchical topology, 27
High-security classification, 356
High-speed wireless networks, 192
High throughput, MANET routing
     protocols, 18
High-traffic region, data mining
     techniques, 247
Hilbert curve index (HCI), 201, 210
Histograms
   density, 253
   historical synopses, 254
   mobility, based on Markov chains
      logical level: data cubes, 255–256
   mobility prediction, 244
moving queries on moving objects, 254
   pattern recognition, 655
   physical, 256–257
   predictive location indexing
      techniques, 181
   spatio temporal, 254
   static queries on moving objects, 254
History Calls, 475
Hit probability, KCLS, 155–157, 160
HMAC, 354
Home-broker algorithm, 76, 78
Home region scheme, location database
     systems, 144
Honest-elicitation problem, 413–414, 424,
     435
Hopfield network (HN) algorithm, 382,
     628, 657
Hop limit, 93
Hops, *see* One-hop; Two-hop
   multihop, 43, 143–145, 635, 649

on-demand routing, 350–351
   proactive routing, 349
HORS, 354
Hotel registration system, 327
HTML Web pages, 218, 224
HTTP, 218, 266–267
H.263, 531, 536–538, 544
Hybrid air index, 199–197
Hybrid broadcast, 193
HybridCache, 92
Hybrid reference model, 88

ICMP router discovery message, 353
ID
   cluster, 147–148, 157, 160
   MANET environment, 95–96
   sensor networks, 128
Identity attacks
   characteristics of, 599
   node replication attack, 599–602
   Sybil attack, 599, 601–602
Idle medium, 46–47
IEEE 802.11
   AGPS/GPS, 444
   MAC protocol, 365, 369, 408
   physical layer, 46, 408
   standards, 12, 44–45, 56, 60, 361
   WiFi, 444
   WLAN, 515
IEEE 802.15.4, 515
Image processing mechanisms, tracking
     WSNs, 580
Immediate updating, 230
Impersonation attacks, 353
Implicit cell data filtering (IDF), 207–208
Imprecision concept, 439
Incentive-based routing schemes, 358–359,
     376
Index listening time, 194
index_pointer, 194–195
Indoor LBS, 222–223
Information filtering, 473–474
Information Pools, OC:
     middleware, 308
Information retrieval systems, 111, 492
Information search, cooperative caching
     system, 95, 97
Information service providers, 474
Information sharing, 440

Information-theoretic data
    compression, 166
Infrared interfaces, 63
Initialization, Gafni-Bertsekas
    protocol, 19–20
Initial relevance, privacy management, 446,
    456
Injected sensor nodes, 604–605
Inproxy nodes, wireless sensor
    networks, 567
Insertion path selection, 180
In-session mobility management, 165
Instant messaging, 64, 299, 322, 329
Instead-patterns, 630
Instruction manager, Agilla sensor
    network, 129–130
Integrity checks, 367
Intelligent analyzing agent, power system
    management, 123
Intelligent network, wireless sensor
    networks, 549–658
Intelligent software, 107
Intel Xscale, 513
Intentions, AgentSpeak(L), 331, 333
Interaction
    location-based predicates, 449–450,
        458–459
    patterns, 62–63, 65, 83
Interceptions, 408
Intercluster
    communication, 434
    link failure, 157
    location, 151–152
Interference range, 45
Interleaving, 175, 198
Interlink failures, location services,
    160
Intermittent connections, 84
Internet
    benefits o f, 63, 88, 214
    protocols, 267–268
    technological impact, 471
Internet control message protocol
    (ICMP), 90
Internet Engineering Task Force (IETF)
    functions of, 4
    MANET Working Group, 346
Interoperability, location-based control
    systems, 443–444

Inter-router authentication schemes, 353
Interruption detection, 44, 54, 59–60
Intersection
    operations, 637–638
    relations, publish/subscribe
        systems, 70–71
Interval skip graph, 568
Intranet, 123
IntraR Table, 147
Intrusion detection schemes
    (IDS), 123–124, 360, 381, 409
Intrusive messages, 316
INVITE message, 335
I/O resources, limitation of, 517
IP address, 14, 353, 361
IPv6 address, 360
ISM band, 514, 528
Isolation mechanisms, 359–360, 376
I-throw, 516, 520, 528
ITU, 532
ITU-T, 531

Jack Intelligent Agents, 330
Jam Agents, 330
Jamming attacks, 408, 415, 599, 601
Jason, 341
Java
    Database Connectivity (JDBC), 118–119
    Native Interface (JNI), 514
    2 Micro Edition (J2ME), 224, 270, 273,
        504
    Wireless Toolkit, 274–275, 504
JINI, 309
Jitter, 305
JMX, 309
Join request, 16
Joint pdf, 445, 452, 454
JSR-172 library, 270, 273
JXTA, 301, 307

KAIST Ubiquitous Service Platform
    (KUSP), 519
Kalman filtering, 168
K-center optimization, 252
K-clustering, 252–253
k-d trees, 200, 256
Key exchange protocols, 362, 367, 596
K-hop cluster-based location service (KCLS)
    location service and applications

cluster-level routing, 157–159
Geocast, 159
sensor data aggregation, 160
performance analysis
accuracy in location service, 155–157
cost in location maintenance
stage, 151–155
overheads in initial stage, 150–151
significance of, 149–150
protocol
characteristics of, 145–146, 161
intercluster location update, 147–149
location management, 146–147
K-hop compound metric-based clustering
(KCMBC), 145, 150
K-means algorithm, 110
K-nearest static neighbors (kNN)
queries, 182–184, 193, 200–201
Krichevsky-Trofimov estimator, 172–173

LabVIEW, 534
Language(s)
actor, 130
C++, 24, 503
P3P, 441
representation, 70
Laptop-class nodes, 604
Latency
average data packet, 373, 375
average end-to-end, 372
cooperative caching, 103
KCLS, 152
location services, 157
MANET routing protocols, 5–6,
17–18
mobile agent technology, 122
network, 65, 649
sensor networks, 128
Latency_opt, 195, 202
Lawful interception, 233
LCS data-replica mechanism, 554–555
Leader
hierarchical management
mechanisms, 563, 565
node, MANET routing protocols, 17
Leadership, 248
Leaf-end sensors, 612
Leaf nodes, 600
Learning automata, 168

Least cluster change (LCC) clustering
algorithm, 95–96
Least common subsequence (LCSS)
distance, 258–259
Lempel-Ziv-78 (LZ78) Markov
predictor, 171, 174–176
Length, Markov chain/predictors, 167–168
LESS, 329
Level-P partitioning, 243
LeZi-Update, 242
Linearization process, predictive location
tracking, 185
Link breakage
location services, 159
MANET routing protocols, 12–13,
15–17, 23
Link-error probability, 195
Link failures
location services, 158
sensor networks, 127
Link layer
channel utilization and, 57
error message, 361, 364
Link reversal routing protocols
Gafni-Bertsekas protocol, 19–22
temporarily ordered routing algorithm
(TORA), 22–23
Link state
protocol, location services, 160–161
routing (LSR), KCLS, 155
Linux Watch (IBM), 527
Load-balance, publish/subscribe system, 69
Load-balanced mechanisms, 550–551,
559–563, 570–571
Local connection (LC), 146
Local dependent query (LDQ), 112–113
Local link change (LLC) packets, 146–147,
151–152
Local storage (LS) mechanism, 550,
566–567
Local topology, mobile agent
technology, 122
Location-aware traffic report
services, 64–66
Location-based access control (LBAC)
access request, 442
accuracy, 443–445
architecture, 441–443
healthcare scenario, 452–453

Location-based access (*continued*)
  information negotiation, 442
  initialization, 442
  location accuracy, 438, 445
  location-based predicates, 448–451
  location measurements, 443–445
  policies, 438, 451–452
  privacy-aware
    architecture, 443
    middleware, 443, 461–464
    predicates evaluation, 457–461
  privacy protection, 443
  privacy requirements, 438
  relevance, 438–439, 446–447, 464
  service provision, 443
  terminology, 441–442
  user, 441–442
Location-based access control systems,
      privacy management
  importance of, 438
  obfuscation techniques for user-privacy
    enlarging the radius, 452, 454
    reducing the radius, 454–455
    research studies, 438–439
    shifting the center, 455–457
  privacy preferences, 439–440
  types of privacy techniques, 438–439
Location-based conditions, 45
Location-based mobile information
      services
  accuracy of, 144, 155–157, 160
  efficient air index scheme for spatial data
      dissemination, 191–210
  KCLS, 143–161
  merging positioning and Web
      2.0, 213–235
  predictive location tracking, 163–187
  reactive, 145
  up-to-date, 144
Location-based semantics, 84
Location-based service environment
  ubiquitous fashionable computing (UFC)
    adaptive angle assignment of resolving
      the difficulty in target selection
      procedure, 522–526
    characteristics of, 515–516
    recognition of gesture, 520–521
    target-detection algorithm,
      521–522

  user-friendly interaction using
      i-throw, 516, 520
Location-based services (LBSs)
  classification, 220–223
  commercial, 216
  cross-referencing, 221
  defined, 214
  first-generation, 215–217
  front end, 228
  location middleware (LM), 442, 457–461
  location privacy, 233–235
  location services (LS), 227–233,
      441–442
  next-generation, 217–218
  privacy issues, 440
  provider, 224, 226
  publish/unsubscribe systems, 62, 64, 68,
      84
  user, 221, 224
  Web 2.0
    characteristics of, 217–220, 235
    supply chain, 223–227, 235
Location constraints, 234
Location-dependent information services
      (LDIS), 192–194
Location-dependent mobile devices, 65–66,
      84
Location disclosure attacks, 408
Location enquiry packet, 151–152, 155–156
Location gateway, 440
Location-independent mobile devices, 66
Location maintenance, 151–155, 160
Location management
  defined, 165
  proactive, 165
  reactive, 165
  strategies for, 146–147
Location obfuscation, 438–440, 462
Location-of-target LBS, 222
Location prediction, 164, 305–306, 317
Location provider, 224
Location registration, defined, 165
Location-sensing technologies, 446
Location server, 519
Location service protocol, simulation
      parameters, 150
Location tracking
  assumptions, 177
  benefits of, 186–187

characteristics of, 164–165
defined, 165
device, UWB, 516–517
indexing structures
    functions of, 176–177
    moving objects, 177, 179–182
    for nearest neighbor (NN)
        queries, 177, 182–185
    predictive location indexing
        techniques, 176–178, 184–185
    for windows, 177, 179–180, 184–185
predictive techniques
    comparison of, 173–176
    design challenges, 166
    discrete sequence prediction
        problem, 167–168, 175
    Markov predictors, 168–173
    mobility model, 166
    preliminaries of, 164–166
    proactive, 186
    reactive, 186
    significance of, 187
    strategies for, 164, 304–305, 317
    success factors, 166
    technologies, 64
Logging algorithm, 76–78
Logical clocks, 9, 14
Logical neighborhood/neighbors, 584
Logical triangle (LT), 584, 586, 588–591
Loop-free routing, 346–347, 349
Loose coupling, 62
Low-energy adaptive clustering hierarchy
        (LEACH), 27
Lowest ID clustering (LIC), 93–96
LP2U, 225–226
LRU algorithm, 94–95, 98
LRU-MIN algorithm, 97–98
Lumi masking, 532

MACAW protocol, 45
Malicious agents, 356, 358
Malicious attacks. *See* Attacks
Malicious nodes, 427–428, 430
Malicious services, 315
MAMDAS (mobile agent-based mobile data
        access systems), 110, 115
Man-in-the-middle, 408
Many-to-many communication, 62, 65
Mapping

BDI, 332–333
event patterns, 629–630
wireless sensor networks (WSN), 558,
        563
Markov chain, mobility prediction, 243–244
Markov model mediator (MMM), 494–495
Markov patterns, 255–257
Markov predictors
    discrete sequence prediction
        problem, 167–168
    domain-independent
        characteristics of, 241
        string compression technique
            applications, 241–242
    families of
        characteristics of, 169–170
        context tree weighting (CTW)
            scheme, 172–173, 175
        Lempel-Ziv-78 (LZ78) scheme, 171,
            174–176
        prediction by partial match (PMM)
            scheme, 170–172, 174–176
        probabilistic suffix tree (PST)
            scheme, 171–172, 174–175
    location tracking, 167
    Markov chain, 243–244
    models of, 305–306
    power of, 168–169
    stationarity factor, 176
MAS-SOC, 340
Mashups, 215, 218–219, 235
Masked node problem, 45
Mate virtual machine, 128–129
MATLAB, 645
Max degree algorithm
    area connection, 32
    area formation, 31
    components of, 31
    example of, 32–34
    topology control, 30, 34, 36–39
Maximal independent set (MIS), ad hoc
        networks, 29–30, 33, 35
Maximum-Likelihood, 630
Max-min heuristic, 145
Mean, TTL, 99
Mediator, intrusion detection systems, 125
Medium access control (MAC) layer
    access control, 46–47
    characteristics of, 29

Medium access control (*continued*)
  data rate, 90
  distributed coordination function
      (DCF), 427
  false blocking, RTS/CTS-induced, 49–51
  IEEE 802.11 functional areas, 45
  reliable data delivery, 46
  research on, 44
  RTS/CTS handshaking mechanism, 49
  sensor network routing, 658
  sublayer, 46
  terminal problems, hidden and
      exposed, 47–49
  threat detection, 353, 361
  throughput, 57–58
  virtual carrier sensing, 49–50
Medoid clustering, 252
Memory
  associative, 380–382, 409
  bottleneck, 68
  Markov chain/predictors, 167
  pattern recognition and, 636–637
  short, 169
Mercury, 340
Message(s)
  complexity, 30, 34–35, 83
  flooding, 553
  injection attacks, 605
  integrity codes (MICs), 354
  passing interface (MPI), 389
Message-origin authenticity, 599
Messaging, typed, 306, 309–310
Metadatabase, 117
MIAMI project, 121
Mica2Dot, 129
MICA2 motes, 130
Microclusters, 250–251
Microsoft, 531, 534
Middleware
  adaptive cooperative caching, 90,
      96–97
  architecture, 439
  COCA, 95–96
  location, 441
  mobile agent-oriented applications, 118
  publish/unsubscribe systems, 66
  quality of service (QoS), 441
  smart office project
    experiences, 297

  organic computing for middleware for
      ubiquitous
      environments, 306–317
  UFC platform, 514
  video semantic summarization, 491
MIDlet, 273–275
Minimum bounding rectangle
      (MBR), 177–179
Minimum CDS (MCDS), 27–28, 30
Minimum distance, 439
Minimum dominating set, 29
Minimum spanning tree, 28
Minimum traffic threshold, 247
Min ID algorithm, topology control, 30,
      33–34, 36–39
Mining servers, 271–272. *See also* Data
      miningMin-max property, 245
Min-support, 245
MITthril, 526
Mix zones, 438–439
MNIST database, 647–648, 650, 652
Mobile ad hoc networks (MANETs)
  characteristics of, 3–4, 87, 164
  design of, 4
  Internet-based, 93
  mobile agent technology, 122
  mobile devices, 87
  network topology, 92
  routing protocols
    link reversal, 19–23
    proactive, 5–10
    reactive, 5–6, 10–17
    taxonomy of, 5–6
    zone, 6, 17–19
  routing security, 345–376
Mobile agent (MA), intrusion detection
      system, 125
Mobile agent-based query-processing
      systems, 114–115
Mobile agent-oriented applications
  distributed information retrieval
    distributed query processing and
        information retrieval, 109–114
    transaction management, 114,
        116–120
  future trends, 136–137
  overview of, 106–107
  pervasive computing and education case
      study

overview, 133
pervasive, continuous
 continuum, 134–136
static teaching environment, 134
sensor networks
 characterized, 126–127
 data aggregation, 127–128
 reconfigurable, 126, 128–131
 swarming intelligence, 107, 131–133
system administration
 resource management, 121–124
 system security, 123–125
Mobile agent technology
Agent-based community-oriented
 routing network (ACORN),
 109–110
applications, *see* Mobile agent-oriented
 applications
benefits of, 137
characterized, 106–107, 111, 137
system management, 110
traditional networking environment, 108
Mobile applications
example, 63–64
device and application
 characteristics, 64–65
infrastructure requirements, 65–66
Mobile clients, 271, 273
Mobile commerce, 62. *See also*
 E-commerce
Mobile computing
air indexing, 191–210
applications, 192
growth of, 191
on-demand approach, 192
wireless data broadcast
 approach, 192–194
Mobile context-aware
agent-based architecture for providing
 enhanced communication
 services, 320–340
formal foundation, 279–292
MANET routing security, 345–376
online scheme for threat detection
 within mobile ad hoc networks,
 380–410
privacy management in location-based
 access control systems, 437–464
smart office project, 294–317

SMRTI (secure mobile ad hoc network
 routing with trust
 intrigue), 412–435
Mobile data access system
 (MDAS), 112–114
Mobile devices, *see specific types of mobile
 devices*
applications on, 64–65, 530–531
battery-powered, 530
emerging market, 217
Mobile host dispatcher (MAD), 125
Mobile hosts (MHs), 94–95
MobileIP, 77
Mobile mining, data mining
for moving object databases,
 239–259
on small devices through web services
 data mining, 264–266, 274
 mobile web services, 266–270
 system design and
 implementation, 270–275
Mobile multimedia
energy efficiency for mobile multimedia
 replay, 530–544
ubiquitous computing system
 environment with an i-throw
 device on LBS
 environment, 510–528
user adaptive video retrieval on mobile
 devices, 488–508
VoiceXML-enabled intelligent mobile
 services, 471–485
Mobile People Architecture,
 340
Mobile phones, 175, 265. *See also*
 Cell/cellular phone
MobileSpaces, 121
Mobile switching center, 165
Mobile web services, 267
Mobility
defined, 214
history, 166
model, 166
patterns
 characteristics of, 249–250
 frequent mining, 245–246
rules, 246
MobiMine, 265–266
Modification attacks, 423, 435

Monitoring, passive, 413, 423
Monitor Queues, OC: middleware, 309
MonitorTracker agent, ANTARCTICA
    query processing, 113
Most forwarded within radius (MBR), 348
Most-valued nodes, 30
Motion
  matrices, 181
  vectors, 176
Motivation, UFC platform, 517–519
Movein operations, 73, 80
Movement
  approximation, 181
  histories, 241–242, 251
  location-based predicates, 449–450, 458
Moveout operations, 73, 80
Moving clusters
  continual maintenance of, 250–251
  detection of, 251
Moving microclusters, 250
Moving object databases
  data mining technology
    applications, 240
    clustering moving objects, 250–253
    complex patterns, 249–250
    dense regions, 253
    group patterns, 248
    Markov patterns, 255–257
    moving object trajectories, 257–259
    periodic movement patterns, 248
    REMO (Relative Motion)
      framework, 248–249
    selectivity estimation, 253–254
    sequential pattern mining, 244–246
    spatio temporal association rules, 247,
      254
  defined, 239
  mobility prediction using movement
    histories
    domain-independent Markov
      predictors, 240–242
    domain-specific methods, 240
    Markov chain model over spatiakl grid
      cells, 243–244
    overview of, 240–241
Moving object trajectories, comparison of
  distance between trajectories, 257
  dynamic time warping (DTW), 257–258
  robust distance measures, 258–259

MoVR (mobile-based video retrieval)
  fuzzy associated retrieval, 490, 501–503
  overall framework of, 496–497
  user feedbacks and profiles, 498–501
MPEG, 531
MPEG-1/2 video transcoders, 491
MPEG-4 video codec, 531–532, 535–538,
    541–542, 544
MPEG-7, 491
MSN, 474
Multiagent-based social simulation, 340
Multibase station, 599
Multicast activation, 17
Multicast group, MANET routing
    protocols, 14, 16–17
Multicasting, 68–69, 149, 159, 349
Multicast/multicasting tree, 16–17, 73–76,
    79–81
Multidatabase systems (MDBAS), 117,
    119–120
Multidimesional trie, 256–257
Multiforwarded data packets, 99
Multihop communication, 43, 635, 649
Multihop networks
  characteristics of, 45
  mobile, 143–145
  wireless network, 144–145, 159–161
Multihop wireless topology, 57
Multilayer perceptron, 306
Multilevel topology, location-based mobile
    information services, 145
Multimedia communications, 27
Multimedia messaging service, 64
Multimedia mobile services, 489
Multimodal Web browsers, 475
Multipath AODV (AOMDV), 15–16
Multiple access with collision avoidance
    (MACA) protocol, 45
Multiple path traversals, 181
Multiple privacy preferences, 462
Multipoint relays, MANET routing
    protocols, 9–10
Multitarget LBS, 221
μ-ware, 516, 518–519

NAV (network allocation vector)
  duration, 53
  operations, modifications of, 44–45,
    51–52

RTS/CTS handshaking mechanism, 49
updating schemes, 53, 58–59
Navigation systems. *See* Global positioning
system (GPS)
Navigator agents, 121–122
Nearest Neighbour in Signal-strength Space
(NSSS), 304–305
Nearest neighbor (NN) queries, 177,
182–185, 201
Nearest surrounder (NS) queries, 184
Negative ACKs, 370–371
Negative selection mechanism, 315
Negotiation manager, 462
Neighborhood, centralized attack
detection, 616
Neighbors, MANET routing
protocols, 7–10, 15
Network(s)
allocation vector. *See* NAV (network
allocation vector)
assumptions, 361–362
bottleneck, 68
capacity, 27
consensus, 638, 640–641, 653
constraints, 65
hierarchy, 75
intrusion, 602
longevity, 27
monitors, intrusion detection systems
(IDS), 124–125
routers, 125
simulator, 23
stability, location-based mobile
information services, 145
topology
ad hoc wireless networks, 164
distributed denial-of-service (DDoS)
detection, 406–408, 607–609
flat, 406, 610
MANET routing protocols, 5, 7, 92
mobile agent technology, 110, 122
tracking wireless sensor networks, 577
traffic, 96–97
Network-capable applications, 64
Neural networks, 305–306, 381
News kiosk, 521–522
Node(s)
collaboration, 634, 645–646, 657
compromise, 601–602

degree, 30
detectors, intrusion detection systems
(IDS), 124–125
identification, 612
implant attack, 601–602
replication attack, 598–600, 602
speed, simulations, 99
Noise
mobility rules, 246
symbols, 175
Nomadic users, 264
Nonblocking agent-based transaction
model, 117
Nonseed dominator, 31
Nonselfs, defined, 315
Non-tree nodes, MANET routing
protocols, 16
Normalized appearance count,
169–170
NP-complete
defined, 27
disconnection, 132
NS-2 simulator, 13, 23–24, 57, 98, 428

Obfuscation-based privacy
techniques, 438–440
Obfuscation techniques, privacy
management, 447
object_expression, 451
Object movement histories, 251
Object profile, 451
Object servers, in transaction
management, 117–118, 120
Object trajectories, 259
Object transaction service (OTS), 118
Observation-based Cooperation
Enforcement in Ad hoc Networks
(OCEAN), 415
Observation phase, detection algorithm,
618
Observed reputation, 417, 422–424,
432–433
Observed time difference (OTD), 445
Observed time difference of arrival
(OTDOA), 445
Offered traffic load, 59–60
Ohm's Law, 534
Omnidirectional antenna, 57
On-chip processing, 596

On-demand
  broadcast, 193
  protocols, 346, 350–353
  update, topology control, 36
One-dimensional index, 180–181
One-hop
  connector, 29, 35
  dominator, 32
  neighbors, 9–10, 29, 38–39, 95, 97, 368
  nodes, 414–415
1,m indexing, 195–197
One-to-many communication, 65
One-to-one communications, 65, 192
On-Line Analytical Processing
       (OLAP), 244
Online Internet-enabled services, 472
Optimized complete reconfiguration
       (OCR), 577
Optimized interception reconfiguration
       (OIR), 577
Optimized link state routing (OLSR)
       protocol, 6, 9–10
Order, Markov chain/predictors, 167
Ordinary sensor, 585
Organic Manager, OC: middleware, 308
Originating call screening (OCS), 338–339
Origin ID (oid), 23
Outdoor LBS, 222–223
Out-of-bind signaling shortcut, 90
Out-of-order transmission, 53
Out-of-session mobility management, 165
Overhead
  data-replica mechanisms, 552
  graph neurons, 382
  location-based mobile information
       services, 144, 150–151
  mobile agent technology, 128
  routing, 372, 374–375
Overhearing
  cancel RTS (CRTS), 45
  false blocking, RTS/CTS-induced, 53
  RTS/DATA frame, 52
  traffic, 5
Overlap
  air indexes, 203
  location services, 160

Packet(s)
  buffer, 416–417, 421

cluster state, 148
delivery ratio (PDR), 59–60, 372, 429,
       435
drop attacks, 415, 427
dropped, 103, 367
headers, 90
leashes, 355
MANET routing procotols and, 4–6, 8–9,
       12–13, 19
salvaging, 13
-sensing range, 45
Pagers/paging, 165, 329, 475
Palm-OS devices, 491
Palmtops, 448
Parallelism, query-processing
       systems, 115–116
Parsing, 270
Partial dropping, 359, 366
Partial reversal, Gafni-Bersekas
       protocol, 21–23
Partitioning
  air indexing, 194
  clustering moving objects,
       252–253
  data-centric storage, 551
  k-hop cluster-based location service
       (KCLS), 145
  location services, 159
  mobility prediction, 243–244, 254
  moving clusters, 250
  moving object databases, 254
  pattern recognition, 636
  predictive location indexing, 183, 185
  tracking wireless sensor networks, 580,
       582
  wireless sensor networks
       (WSN), 556–558, 565
Partitions
  ad hoc networks, 27
  MANET routing protocols, 22–23
  max degree algorithm, 31
  topology control, 35–36, 39–40
Partition trees, 178, 180
Path loss, sensor networks, 128
Path privacy, 439
Pathrater, 359
Path-sharing mechanisms, 569
Patrol agent, power system
       management, 123

Pattern inference, 657
Pattern interference, 656
Pattern matching, 168
Pattern mining, 246
Pattern recognition
    accuracy in, 407
    attack detection, 625
    distributed hierarchical graph neuron case
        study
        simulator design, 389–391
        standard-form DHGN, 397–405
        variable-form DHGN, 391–397
    energy-efficient, 627–658
    tracking wireless sensor networks, 580
Pause time, 57, 99, 433–434
Payment services, mobile, 64
PDAs (personal digital assistants), 63–64,
        66, 68, 175, 265–266, 270, 276,
        334, 336, 474, 491, 526, 534,
        537–538
PE2T, 225
Peer-to-peer (P2P)
    Decentralized Information Ecosystem
        Technologies
        (P2P-DIET), 114–115
    network, 268
    topology, 113–114
Peer-to-peer networks, 73, 414
Performance analysis, 34–35
Performance degradation, 44
Performance element agent (PEA), 121
Performance metrics
    cooperative caching system, 98
    robust source routing (RSR), 372
Performance monitor agent (PMA), 121
Performance negotiation agent (PNA), 121
Period density query, 253
Periodic pattern mining, 248
Periodic updating, 230–231
Personal communication manager
        (PCM), 325, 327–328, 332,
        334–335
Personal information management (PIM)
        utilities, 63, 327
Personal Proxy, 340
Pervasive computing, mobile agent
        technology case study
    curriculum, 134–136
    overview of, 133

static teaching environment, 134
Physical carrier sensing, 44, 49, 59
Physical histograms, 256–257
Physical layer attacks, 599
Piggy-backing, 312
Pivot queries, 182
Plan failure recovery, agent-oriented
        architectures, 334–335
Plans, AgentSpeak(L), 331, 334–335
Platform decoupling, 68
Platform-dependent system activity
        agents, 125
Plausible deniability, 234
PoBox/PoBoxAdder, 302–303
Point coordination function (PCF), 46
Pointer table, air indexes, 206
Poisson distribution, 150
Policies and preferences manager
        (PPM), 326, 328–329, 332, 335
Policy-based privacy techniques, 438, 440
Policy Server (PS), agent-oriented
        architectures, 325, 333
Polling, 227–228, 231
Port scanner agent, intrusion detection
        systems, 125
Position, *see* Positioning
    location-based predicates, 448–449, 458
    management, 231
Positional uncertainty, 251–252
Position-based routing protocols
    compass algorithm, 347–348
    greedy algorithm, 347
    most forwarded within radius
        (MFR), 348
    randomized compass algorithm, 348
    topology control, 23
Positioning
    clustering moving objects, 252
    daemon, 227–228
    enabler, 224
    handovers, 229
    technologies, 227
    transparency, 225
Positive acknowledgment scheme, 45
Positive selection mechanism, 315–316
PostgreSQL, 503
Power supply, 489
Power system management, 121,
        123–124

Prediction of partial match (PMM)
    scheme, 170–172, 174–176
Predictive indexings, evaluation
    of, 185–186
Predictor, defined, 167
Prefetch-delayed algorithm, 78, 80–82
Prefetching
    algorithm, 75–80, 92
    COCA schemes, 95–96, 98, 102
    transaction management and, 120
Presence information manager (PIM), 325,
    327–328
Presence user agents (PUAs), 327
Priority queue, 180
Privacy
    ACORN technology, 110
    location, 233–235
    manager, 462
Proactive LBSs, 221–222
Proactive protocols, 349
Proactive routing protocols, MANETs
    characteristics of, 5–6, 346
    destination sequenced distance vector
        (DSDV) protocol, 6–9, 14
    optimized link state routing (OLSR)
        protocol, 6, 9–10
Probabilistic suffix tree (PST)
    scheme, 171–172, 174
Probability density distribution
    functions, 630–631
Probability density function (PDF), 445,
    454
Probe wait, 194
Probing packets, 360
Processing
    bottleneck, 68
    elements (PEs), sensor networks, 127
    speed, 489
Propagation
    delay, 45
    multicast *vs.* unicast, 69
    radio, 426
    subscription, 71
Protocol layer, 91
Provider, service-oriented architecture, 267,
    269
Proximity
    detection, 232–233
    queries, 177

Proxy
    algorithm, 80–82
    nodes, wireless sensor
        networks, 567–568
Pruning rules, ad hoc networks, 28–29
Pseudodeadlock, 51
P3P language, 441
P2P-ECA, 115
P2U, 225
Publication message, 67
Public Safety Answering Point (PSAP),
    215
Publisher
    defined, 66
    mobility
        delayed algorithm, 80–82
        message cost, 81
        prefetch-delayed algorithm, 78, 80–82
        prefetching algorithm, 78–80
        proxy algorithm, 80–82
        standard algorithm, 78–81
Publish/subscribe systems
    benefits of, 68–69
    broker networks, 66–67, 69–70, 72–73,
        84
    client mobility
        publisher, 78–82
        subscriber, 74–78
    content-based, 67, 84
    complex interactions, 83
    decoupling, 68, 83
    distributed, 73
    example application, 82–84
    impact of, 62
    model of, 66–68
    routers, 69–72
Push-based broadcast, 193
Push-pull strategies, 559–560
Put(), 551–552

QBIC, 527–528
QoSDREAM, 441
QPEL, 532
QRY packet, 20, 352
Quality of service (QoS)
    location-based privacy, 440–441,
        446–447
    MANET routing protocols, 5, 88
    mobile agent technology, 131–132

smart office project, 310–311
SMRTI, 432
Query by example (QBE), 491
Query delay, 88, 93, 99–102
Query execution module, ACQUIRE
    system, 112
Query optimization module, ACQUIRE
    system, 112
Query packet, wireless sensor
    networks, 553–554, 558
Query planning module, ACQUIRE
    system, 112
Query processing, 111

R, 449, 454, 456
RADAR system, 304
Radio
    frequency identification (RFID), 229
    propagation, 426
    signal strength, 305
    transmission range, 150
Radius, routing zone, 17, 19
Random access schemes, 44
Random pattern signature, 646, 648–649,
    651, 653, 655
Random walk, 169, 250
Random waypoint mobility, 24, 57, 432
Range queries
    circular static, 181
    defined, 177
    range aggregate, 177, 181–183
    range reporting, 177–181
    TPR-trees and, 179–180
Rate adaptation, 90
Reachability matrix, 606
Reactive LBSs, 221–223, 226
Reactive routing protocols
    ad hoc on-demand distance vector
        (AODV), 6, 10, 13–17
    design of, 5–6, 10–11
    dynamic source (DSR), 6, 10–13, 15
    zone routing protocol (ZRP), 17
Ready-to-receive gesture, 521
RECALL command, 390–391
Recall rates, 393–395, 397–404
Received Signal Strength Indicator
    (RSSI), 515
Receive$_{EVT}$ algorithm, 589–590
ReceiveVTREQ packets, 586–587

Recommendations, dissemination
    of, 414–415
Recommended reputation
    convention recommendations,
        processing, 424–425
    deriving recommendations, proposed
        approach for, 425–426
    significance of, 424, 432
Recommender systems, 473–474
Reconfigurable sensor networks, 126,
    128–131
Reconnection, publish/subscribe
    system, 69, 73–74, 76, 79
Reconstruction
    multicast trees, 81
    topology control, 36
Record management system (RMS), 273
Recursive dynamic partition (RDP), 582
Redundancy
    location-based services, 231
    mobile agent technology, 135–136
    sensor networks, 127
    spatial, 532
    VGN performance, 642, 645, 653–654
Redundant data, air indexes, 202–203
Redundant nodes, ad hoc networks, 27
Reference points, 224
Reflector attacks, 601
REGISTER message, 335
Registry, service-oriented architecture, 267,
    269
Regression models, 252
Re-insertion, node, 180
Reject list, 416–417, 421
Relational databases, 335
Relative neighborhood graph, 28
Relevance
    location-based access control, 460
    privacy management, 446–447
Reliability, publish/subscribe system, 69
Remote control, 266
Remote denial of service, 601
Remote location server, 228–229
Rendezvous regions (RRs), 557
ReOrganizeLT algorithm, 588
Repetitive user routes, 176
Replay, 408
Replication path, 557
REPLY packet (REP), 351, 355

Representation (semantic) decoupling, 69
Representation language, 70
Repudiation, 408
Reputation
    capture, 416, 421, 433
    evaluation, 416, 421
    ratings, 417, 428
    systems, 359
    table, 416–417, 421
Request id, 363–364
Request intervals, simulations, 99
Request packet, cluster-based adaptive
            cooperative caching
            scheme, 96–98
Required relevance, privacy
            management, 447
Rerouting process, 158
Resource allocation, 132, 164
Resource consumption attacks, 408
Resource management
    algorithms, 164
    mobile agent technology, 109, 121–124
    network management and
            routing, 121–122
    power systems, 121, 123
Retransmission, 47, 52
Retrieval path, 557
R_{EVAL}, calculation of, 457–461
Reverse channel, 165
Reverse extra frame transmission
            (R-EFT), 44, 51, 54, 56, 59
Reverse path, MANET routing
            protocols, 16
Reverse route, ad hoc on-demand distance
            vector (AODV) protocol, 14–15
Reversed nearest neighbor (RNN)
            searching, 177, 182–183
R^{EXP}-trees, 180
RMI, 268
Roaming, 187
Robustness
    information retrieval systems, 110–111
    KCLS, 147
    moving object, 258–259
Robust source routing (RSR)
    details of, 362–366, 376
    malicious behaviors and mitigation
            strategies, 369–371
    problem definition and model, 361–362

robust source routing, 367–368
route discovery, 360–361
route maintenance, 361
route utilization, 361
simulation evaluation, 371–375
Role-based access control (RBAC), 440
Roll-up, 244
Romance films, 535–537
Root replacement, convoy tree, 576–577

Route discovery, 11, 14, 16–17, 19,
            360–361
Route discovery packet (RDP), 355
Route error (RERR), MANET routing
            protocols, 12–13, 15–16
Route failure, 51
Route maintenance, 11, 14, 19, 21
Route reply (RREP) packet
    ad hoc on-demand distance vector
            (AODV) protocol, 14–17, 353
    basic routing security schemes, 354
    dynamic source routing (DSR)
            protocol, 11–13
    route discovery, 364
    simulation, 371
    SMRT1, 415
    trust-based routing schemes, 356–357
    zone routing protocol (ZRP), 18
Route request (RREQ) packet
    ad hoc on-demand distance vector
            (AODV) protocol, 14–17,
            352–353
    dynamic source routing (DSR)
            protocol, 11–13, 350, 356
    forward, 362
    robust source routing (RSR), 370
    route discovery, 363–364
    route utilization and maintenance, 365
    simulations, 371
    SMRT1, 415
    trust-based routing schemes, 356–357,
            415
    zone routing protocol, 17–19
Routers, publish/subcribe
    forwarding of advertisements, 70–71
    forwarding publications, 72
    forwarding subscriptions, 71–72
    operations, overview of, 70
Route table

ad hoc on-demand distance vector
(AODV) protocol, 15
dynamic source route (DSR) protocol, 12
zone routing protocol (ZRP), 17–18
Routing
capabilities, 62–63
cluster-level, 158–160
content-based, 68
control messages, replay of, 353
loops, 16
protocols, location-based mobile
information service, 144
security, *see* Routing security
sensor networks, 128
sources of problems, 164
Routing security
cryptographic tools, selection
of, 367–368
exclusion links, 368–370
forerunner packets mechanism, 369
importance of, 345–346
overview, 346–347
position-based protocols
compass algorithm, 347–348
greedy algorithm, 347
most forwarded within radius
(MFR), 348
randomized compass algorithm, 348
robust source routing (RSR)
details of, 362–366
malicious behaviors and mitigation
strategies, 369–371
problem definition and
model, 361–362
route discovery, 360–361, 369
route maintenance, 361
route utilization, 361
simulation evaluation, 371–375
secure routing proposals
basic routing security
systems, 353–355
incentive-based schemes, 358–359,
376
schemes that employ detection and
isolation mechanisms, 359–360,
376
trust-based routing schemes, 355–358
tabu list, 364, 368–369
topology-based protocols

on-demand, 350–353
proactive, 349
Routing table
ad hoc on-demand distance vector
(AODV) protocol, 16
destination sequenced distance vector
(DSDV) protocol, 7–9
KCLS, 147
link reversal routing protocols, 19–20
location service (LS), 158
mobile agent technology, 122
publish/subscribe broker
network, 72
single stability-based adaptive
routing, 350
Routing zone, 17
Row table, air indexes, 206–208, 210
RPC communication, 270
RPY packets, 20
RSS feeds, 219
R-trees, 178, 182–184, 200
R*-trees, 178–179, 200
RTS
frame, 49
validation scheme, 44–45, 51, 54, 56,
58–59
RTS/CTS
failure, 44
handshaking, 43–45, 49, 52, 56, 59–60
mechanisms, 45
Rule-based policies, 440
Rule 1&2, 36, 38–39
Runtime
APIs, 270
intrusion detection systems, 125

Satellite technology, 488
Scalability, 69, 145,155, 61, 192–193
Scalable tracking using networked sensors
(STUN), 577
Scheduling, OC: middleware, 312
Search engine, mobile agent
technology, 110
SecAODV, 360
Second-order Markov transition
probability, 243
Secret keys, 367
Secure AODV (SAODV), 354–355
Secure-aware ad hoc routing (SAR), 356

Secure distributed anonymous routing
    protocol (SDAR), 357
Secure efficient ad hoc distance vector
    routing protocol (SEAD), 354
Secure routing protocol (SRP), 354
Security
  MANETs, 5
  publish/subscribe system, 69
  strategies for, 123–125
Seed denominator, 31–32
Selection functions, AgentSpeak(L), 332
Selective listening, air indexes, 194–195
Self-configuration, 306, 310–311, 317
Self-healing, 306, 312–315, 317
Self-optimization, 306, 311–312, 317
Self-organized environments, 131
Self-organizing map (SOM), intrusion
    detection schemes, 614–622,
    624–626
Self-protection, 306, 315–317
Self-referencing LBS, 221, 223, 226, 228
Selfs, defined, 315
Sense-predict-communicate-sense, 578
Sensing technology
  physical attributes of, 440
  privacy management, 446–447
Sensor networks
  applications, 64
  characterized, 65, 126–127
  data aggregation, 127–128
  reconfigurable, 126, 128–131
Sensornet network, 282
Sensor nodes, defined, 126
Sensor selection algorithm, 578–579
Separation rule, 132
Sequence numbers
  ad hoc on-demand distance vector
    (AODV) protocol, 14–15
  MANET routing protocols, 7–9
Sequential pattern mining
  characteristics of, 244–245
  frequent mobility pattern mining from one
    long trajectory, 245–246
  mobility rules, consideration of cell
    topologies and noises, 246
  support threshold, 246
  TrajPattern, 245
Server addition request, 558
Service discovery platform, 519

Service level agreements (SLAs), 224, 440
Service-oriented architecture
    (SOA), 266–267
Service Proxy, OC: middleware, 308
Services, OC: middleware, 308
Session hijacking, 408
Set-cover problem, KCLS, 149
Setting agent, power system
    management, 123
Share & Care, 227
Shared keys, 362, 367
Short interframe space (SIFS) interval, 47
Short memory principle, 169
Short message service (SMS)
  location-based services, 227
  mobile devices, 63–64, 66
  VoiceXML, 485
Shortest path
  clustering moving objects, 252
  MANET routing protocols, 7–8, 13
  predictive location tracking, 184
  proactive routing protocols, 349
  SMRTI, 432
  tracking wireless sensor networks, 578
Signal(s)
  detection, 64
  interference, 204
Signature-based air index, 198–199
Signatures, verification of, 364, 367–368.
    See also Digital signatures
Similarity matching, 168–169
Similarity rule, 132
Simple caching (SC), 99
Simple location service (SLS), 144
Simple network management protocol
    (SNMP), 121, 309
Simple object access protocol (SOAP), 218,
    268–269
Simulation(s)
  agent-based architecture, 335, 340
  channel utilization, 57
  OC: middleware, 311–312, 315–316
  robust source routing, 371–375
  sensor networks, 128–129
  SMRT1, 426–435
  topology control, 36–39
  wireless sensor networks
    applications, 645–646, 657
    communication analysis, 649–652

sleep mode analysis, 652–656
  test patterns, 646–649
Single-level *k*-hop clustering structure, 145
Single stability-based adaptive routing
      (SSA), 350–351
Single-target LBS, 221, 223, 226
Singular value decomposition
      (SVD), 490–491
Sink nodes
  sensor networks, 126–128, 160
  wireless sensor networks, 567–569
SIP-based architecture, 335, 340
Sketch, 254
Sleep mode analysis, 638–640, 652–656
"Sleep" period, 346
Smart client, 270
Smart homes, 176
Smart network, 166
Smart office
  benefits of, 295–296
  development of, 294–295
  location
    prediction, 305–307, 317
    tracking, 295, 304–305, 307, 317
  organic computing (OC:) middleware for
      ubiquitous environments
    components of, 306–308, 317
    monitoring, 309
    self-configuration, 306, 310–311, 317
    self-healing, 306, 312–315, 317
    self-optimization, 306, 311–312, 317
    self-protection, 306, 315–317
    typed messaging, 306, 309–310
  reflective mobile agents within
    location-based services, 300–301
    UbiMAS, 296, 301–304
  smart doorplate
    email retrieval, 299–300
    file transfer, 300
    as signpost, 297–298
  visitors
    arrives in absence of office
      owner, 298–299
    visitor in front of the office and office
      owner present, 297–298
Smart phones, 474
Smart techniques, 164
SMRT1 (Secure Mobile Ad Hoc Network
    Routing with Trust Intrigue)

architecture
  detection component, 421–426
  overview, 416–417
  reaction component, 420–421
  system operation, 417–420
  assumptions, 415–416
  bias, 413, 415, 424
  characteristics of, 413–414, 435
  honest-elicitation problem, 413–414,
    424, 435
  related research work, 414–415
  simulation results
    malicious nodes and, 427–428,
      430–434
    performance analysis, 429–435
    scenarios and performance
      metrics, 428–429
    significance of, 426–427
  terminology, 415–416
  trust evaluation, 420
  trust over reputation, 421
Social networking, 219
Source address, 363–364
Source node, MANET routing
    protocols, 13–14
Source route, 12, 363
Source-sink communication path, 407, 608
Space decoupling, 69
Space filling curve-based air
    indexes, 201–204
Space partition-based air indexes, 201
Spam attacks, 600
Sparse network, 36
Spatial bounding rectangle, 254
Spatial grid cells, 243–244
Spatial network, 252
Spatial queries, 193
Spatio temporal
  association rules, 247, 254
  database, 259
  pattern queries, 249
Special forward control channel, 165
Speech-based communication, 472, 475
Spoofing attacks, 599
Spread spectrum, 415
Sprint, 474
Sprite, 358–359
Stack profile, 95–97
Standby mode, 165

STAT, 304–305
State predictor methods, 306
Static agents (SAs), intrusion detection
    system, 125
Static wireless network, 20–21
Stationary Markov chain, 167
Statistically unique and cryptographically
    verifiable (SUCV) identifiers, 360
Stochastics, 579
Stock market data, 265–266
Stock quotes, 66
Storage sensor, hierarchical management
    mechanisms, 563
Storage space, 489
STP-tree, 181
Streaming video, 164
String alignment, 246
String compression techniques, 241–242
STRIPE index, 180–181, 185
Strong authentication, 354
subject_expression, 451
Subnet monitors, intrusion detection
    systems (IDS), 124–125
Subnetworks, mobile agent technology, 122
Subqueries, information retrieval
    systems, 111–112
Subscriber mobility
    home-broker algorithm, 76, 78
    logging algorithm, 76–78
    prefetching algorithm, 75–78, 92
    standard algorithm, 74–77
    subscriptions-on-device, 76
Subscription
    forwarding, 71–72
    language, 67
    message, 66
    propagation, 71
Substations, power system
    management, 123
Summary schemas model (SSM), 136
SUN grid system, 393
SUPER©, 533
Super-peer network, MDAS query
    processing, 114
Supply chain, location-based
    services, 223–227, 233, 235
Support threshold, 246
Suspicious events, intrusion detection
    systems (IDS), 124

Swarm intelligence, 107, 131–133, 414
Sybil attack, 599, 601–602
SYN-ACK response, 601
System log routers, 125
System Monitor, OC: middleware, 309
System profile, 90

Tabu list, 364, 368–369
Tampering attacks, 605
Target descriptors (TDs), 578–579
Target-detection algorithm
    adaptive angle assignment in target
        selection procedure, 522–526
    characteristics of, 521–522
Targets-at-location LBS, 222
Taxonomies, mobile ad hoc networks
    (MANETs), 5–6
TCP
    characteristics of, 46, 504, 601
    connections, 75
    SYN packets, 601
TCPMP (The Core Pocket Media
    Player), 533
Telecom providers, 64
Telecommunication management
    network with Q3 interface
    (TMN Q3), 121
Template matching
    applications, 629–631
    operations, 631–633
    pattern recognition, 637–638
Temporal leashes, 355
Temporally-ordered routing algorithm
    (TORA), 22–23, 351–352
Terminal problems
    exposed, 44, 48
    hidden, 43–45, 47–49
TESLA, 354
Test goals, 331
Testing, 520, 589–590, 646–649
Text messages, 485. *See also* Messaging
Text-to-speech translator, 480
Thin client, 270
Third-party service providers, 214, 217
Threat detection
    graph neuron
        associative memory concept, 382
        cross-talk phenomenon, 384
        hierarchical, 384–389

simple graph neuron
    approach, 382–384
overview of, 380–381
pattern-recognition approach, 380
single-cycle associative memory
    algorithm, 380
Threat model, 362
3G technology, 444, 448, 488
Three-hop dominator, 32–33
Threshold pattern modeling, 603, 609–612
Ticks, associativity, 351
TIK, 354–355
Time constraints, 234
Time decoupling, 69
Timelines, mobile subscriber, 74
Timeout, 449
Time-parameterized bounding rectangles
    (TPBRs), 179–180
Time-parameterized (TP) queries, 184
Time-slice queries, 179–180
Time-synchronization hardware, 354–355
Time-to-live (TTL)
    cooperative caching schemes, 92, 94,
        97–101
    wireless sensor networks, 555, 564
Timing, RTS/CTS sequence, 53
TinyOS, 129–130
T-Mobile, 474
Topic-based publish/subscribe system, 68
Topology, *see* Network topology
    KCLS protocol, 146
    location-based mobile information
        services, 145
Topology-based routing protocols
    on-demand, 350–353
    proactive, 349
Topology control, ad hoc networks
    clustering, 27, 29
    connected dominating set (CDS), 27–37
    network assumptions and
        preliminaries, 29–30
    significance of, 26–27
    transmission range, 27–28, 35–37, 39–40
TPR-trees, 178–180, 183, 185–186
TPR*-trees, 180–181, 185
Tracker agent, ANTARCTICA query
    processing, 113
Tracking, wireless sensor networks (WSNs)
    classification of scenarios, 573–574, 592

importance of, 573–574
illustrations of, 574–575
scenario characteristics of, 574–575
techniques
    event boundary determination, 575,
        579–582, 593
    target tracking, 574–579, 592
Traffic
    ad hoc networks, 27
    advisories, 64, 83
    analysis attacks, 599
    bursty, 46
    congestion, 103
    constant bit rate (CBR), 57
    cooperative caching system, 98
    drain-and-balance structures, 577
    dropped, 366
    flow observation table, 603, 613–614
    flow rate, 369
    increased load, 53
    intensity (TI), 610, 620–621
    interference with, 48
    load, 164
    location-based services and, 233
    MANET routing protocols, 5, 13
    mobile agent technology, 131–132
    mobile web services, 269
    offered load, 59–60
    overhearing, 5
    types of, 609
Trajectory clustering, 252
TrajPattern, 245
Transactional agents, 116–117
Transactional client movement, 69
Transaction management, distributed
        information retrieval
    database management, 117
    distributed objects, 117–118
    multidatabase systems, 119–120
    transactional agents, 116–117
    web-based distributed database, 118–120
Transfer Strategies, OC: middleware, 312
Transmission
    error detection service, 362
    failure, 47
    loss, 47
    MANET routing protocols, 9, 15, 19
    range, *see* Transmission range
    rate, 57

Transmission range
  ad hoc networks
    characteristics of, 28
    implications of, 27
    topology control, 35–37, 39–40
  MANET routing security, 361
  simulations, 99
  topology control, 36–37, 39–40
TransportConnectors, 307
Transport Interface, OC: middleware,
    309
Transport layer
  communication primitives, 62
  OC: middleware, 309
TraX, 227
Tree-based air index, 196–198, 200
Tree construction, target tracking
    WSNs, 575
Trellis quantization, 532
Triangulation, 585–588
Triggering/triggering event,
    AgentSpeak(L), 331
Trojan ports, 125
Trust
  evaluation, 417
  models, 415
  over reputation, 417, 421
  query messages, 358, 376
  relationship, 414–415
  table, 358
Trust-based routing schemes, 355–358
Trust model, 414
Trustworthy route, 414, 417, 419–421,
    425–426, 431–433
T2LP, 225, 227
Tune_opt, 195
Tuning time, 193–194, 202
Tuple space(s)
  manager, Agilla sensor network, 129–130
  sensor networks, 129
Two-dimensional index, 180–181
Two-hop
  connector, 29, 38
  dominator, 32, 34–35
  neighbors, 9–10, 17, 29, 32
  traffic flow path, 608
Two-phase commit (2PC) protocol, 117
Two-tier Sensor storage Architecture
    (TSAR), 567–568

Two-way ground reflection model, 426
Type-based publish/subscribe
    system, 68

Ubiquitous fashionable computer (UFC)
  characteristics of, 511, 519, 528
  design philosophy, 512
  gesture interface, 519–520
  operation sets, 526–527
  platform, 513–514, 521
Ubiquitous Mobile Agent System
    (UbiMAS)
  framework, 301–302
  host service, 302–303
  mobile agents, 303–304
Ubiquitous service discovery
    (USD), 519
u-display, 522
UDP, 504
UMTS, 214, 217, 219, 231
Unauthenticated packets, 368
Unicast frame, 59
Unicasting, 14, 69, 129
Unicast messages, 75–76
Unicast routing packets, 362–363
Unit disk graph (UDG), 27, 30, 35
Universal description discovery and
    integration (UDDI), 268
Universal Inbox project, 340
Unreliable connections, 65
UPD packet, 352
Update
  interval, 230
  messages, 354
  mode, 165
Updater agent, ANTARTICA query
    processing, 113
Updating, 227–230
Uplink channel, 191, 193
u-printer, 522
u-projector, 522
Upstream packet, 425
Urgent sensors, 585, 591
User adaptive video retrieval
  challenges of, 489
  content management, 489
  development of, 488–490
  implementation and
    experiments, 503–508

modeling
   hiearchical Markov model mediator
      (HMMM), 490, 492, 494–500,
      507
   Markov model mediator
      (MMM), 494–495
   MoVR (mobile-based video retrieval)
   accuracy of, 501, 506
   fuzzy associated retrieval, 490,
      501–503, 505
   overall framework of, 496–497
   scope, 506
   soccer retrieval system
      illustration, 503–507
   user feedbacks and profiles, 490, 492,
      497–501
   system architecture, 492–494
User-centricity, 217, 219, 235
User-generated
   content (UGC), 215, 219
   services, 219
User identifiers (UIDs), 448–449
User mobility patterns, 246
User profiles, 451, 490, 492, 498–501, 507
User session state/context manager
      (USSCM), 477, 479
U-speaker, 521–522
u-trash, 522

Valid group-growth (VG-Growth), 248
VAP, 214
Variable length Markov
      chains/predictors, 168, 170
Variable-order Markov predictors, 175
VEhicle DAta Stream mining (VEDAS)
      system, 266
Velocity
   bounding rectangle, 254
   KCLS, 150
   SMRTI nodes, 433–434
   vector, 179
Venn sampling, 182, 186
Verdict, 619
Verizon, 474
Vertical calibration, 90
Vertical layer, 91
Vicinity triangulation, 585–588
Victim host list (VHL), 125
Victim nodes, 603, 606–607

Video
   bit rates, 538–539, 543–544
   codecs, 531–532, 535–538, 541–544
   energy-efficient, 530–531
   file format, 544
   film resolution, 540–543
   frame rates, 539–540
   on mobile devices. *See* User adaptive
      video retrieval
Virage, 491
Virtual carrier sensing, 44
Virtual clients, 493–494, 497, 504, 507
Virtual Reality Model Lnaguage
      (VRML), 531
Virtual sensing, 49–50
Vocera communication system, 475
Voice Law Enforcement Tactical System
      (VoiceLETS), 475
Voice mail, 475
Voice operated interchangeable control
      environment (VOICE), 472. *See*
      also VoiceXML-enabled
      intelligent mobile services
Voice-recognition technology, 475
VoiceXML-enabled intelligent mobile
      services
   application variables, 477, 481–482
   architecture
      context information database, 479–480
      flow control module, 477, 479
      future directions for, 485
      illustration of, 478
      user session state/context manager
         (USSCM), 477, 479
      VoiceXML converter, 477, 479–480
      XML applicaiton schema, 480
   characteristics of, 472–473
   context-aware applications, 474–475
   future directions for, 485
   recommender systems, 473–474
   Restaurant Search Guide case
      illustration, 476–477, 479,
      481–482, 484–485
   speech-enabled applications, 474–475
   usability experiments, 484–485
   user session states
      available, 482–484
      Data Source adapter, 477, 483–485
      overview of, 480–481

VoiceXML-enabled (*continued*)
Query Designer, 477, 483
Query Manager, 477, 483
Query Parser, 477, 483–484
Volume, MANET routing protocols, 9
Voronoi Diagram (VD), 201, 612
Voronoi neighbors, 556
Vote sets, 636–639, 642–643, 649–650,
652–653, 656
Voting graph neuron (VGN), wireless sensor
networks
analysis, 642–644
details of
overview of, 635–636
pattern storage, 636–637, 657
sleep mode, 638–640
template matching, 637–638
example of, 640–642
graph neuron algorithm compared
with, 656–657
overview of, 633–635
pattern recognition, 634
VTREQ packet, 586–590

WAN, transaction management, 118
WAP browser, 228
Wasted channels, 51, 53, 60
Watchdog, 357, 359
WaveLAN, 440
Wearable computing system, 510–511,
526–527
WearARM, 527
Web
access, 63
browsers, 475
browsing, 65
caching systems, 91–92
searches, mobile agent technology, 111
service agent, 269
services
benefits of, 266–268
description (WSDL), 268
in mobile environments, 268–270
2.0, 214–215, 217–220, 223–227, 237
Web-based distributed database, 118–120
Weight, ad hoc networks, 27
Well-sited stations, 48
While-loop, 209

Widehand Code Division Multiple Access
(WCDMA) networks, 445
WiFi
cards, 448
interfaces, 63–64
network, 222, 225, 227
Window queries, 179–180, 184–185, 193,
202, 208–210
Wireless access medium, 373
Wireless ad hoc networks, 186
Wireless cellular networks, 113
Wireless data broadcast system,
192–193
Wireless extended Internet
architecture, 268–269
Wireless homogenous sensor networks
(WHSNs)
future research directions, 593
illustration of, 583
tracking in
collaborative event deteCtion and
tracking protocol (CollECT), 583,
585–593
network model, 584–585
problem statement, 582–583
Wireless local area networks
(WLANs), 45–46, 113, 176, 488,
511, 514, 528
Wireless sensor networks (WSNs)
applications of, 129–131, 136, 549
characteristics of, 549–550
DDoS attack modeling and detection
in, 595–626
efficient data-centric storage
mechanisms, 549–571
energy-efficient pattern
recognition, 627–658
homogenous, *see* Wireless homogenous
sensor networks (WHSNs)
infrastructure, 288–289
location-based services, 160
publish/subscribe systems, 73
threat detection, 381
topology, 27
tracking in, 573–593
WMV, 532, 535–538, 541–542, 544
World Wide Web (WWW), 164, 215. *See
also* Internet; Web

World Wide Web Consortium, 472
Wormhole attacks, 355, 362, 408, 602

X-TRA, 118
XML:
   Application Schema, 477, 479–481
   characteristics of, 266–270
   data, 68
XviD, 532, 535–538, 541–542, 544
Xybernaut, 527–528

ZigBee communication, 511, 514–515, 520,
   528
Zipf parameter, simulations, 99
Zombies, 600
Zone-based data-replica mechanism, 559
Zone-based updating, 231
Zone routing protocol (ZRP), 6, 17–19
Z-ordering
   method, 243
   numbering, 256

**WILEY SERIES ON PARALLEL
AND DISTRIBUTED COMPUTING**

Editor: Albert Y. Zomaya

A complete list of titles in this series appears at the end of this volume.

WILEY SERIES ON PARALLEL AND DISTRIBUTED COMPUTING
Series Editor: Albert Y. Zomaya

Parallel and Distributed Simulation Systems / Richard Fujimoto

Mobile Processing in Distributed and Open Environments / Peter Sapaty

Introduction to Parallel Algorithms / C. Xavier and S. S. Iyengar

Solutions to Parallel and Distributed Computing Problems: Lessons from Biological Sciences / Albert Y. Zomaya, Fikret Ercal, and Stephan Olariu (Editors)

Parallel and Distributed Computing: A Survey of Models, Paradigms, and Approaches / Claudia Leopold

Fundamentals of Distributed Object Systems: A CORBA Perspective / Zahir Tari and Omran Bukhres

Pipelined Processor Farms: Structured Design for Embedded Parallel Systems / Martin Fleury and Andrew Downton

Handbook of Wireless Networks and Mobile Computing / Ivan Stojmenović (Editor)

Internet-Based Workflow Management: Toward a Semantic Web / Dan C. Marinescu

Parallel Computing on Heterogeneous Networks / Alexey L. Lastovetsky

Performance Evaluation and Characteization of Parallel and Distributed Computing Tools / Salim Hariri and Manish Parashar

Distributed Computing: Fundamentals, Simulations and Advanced Topics, *Second Edition* / Hagit Attiya and Jennifer Welch

Smart Environments: Technology, Protocols, and Applications / Diane Cook and Sajal Das

Fundamentals of Computer Organization and Architecture / Mostafa Abd-El-Barr and Hesham El-Rewini

Advanced Computer Architecture and Parallel Processing / Hesham El-Rewini and Mostafa Abd-El-Barr

UPC: Distributed Shared Memory Programming / Tarek El-Ghazawi, William Carlson, Thomas Sterling, and Katherine Yelick

Handbook of Sensor Networks: Algorithms and Architectures / Ivan Stojmenović (Editor)

Parallel Metaheuristics: A New Class of Algorithms / Enrique Alba (Editor)

Design and Analysis of Distributed Algorithms / Nicola Santoro

Task Scheduling for Parallel Systems / Oliver Sinnen

Computing for Numerical Methods Using Visual C++ / Shaharuddin Salleh, Albert Y. Zomaya, and Sakhinah A. Bakar

Architecture-Independent Programming for Wireless Sensor Networks / Amol B. Bakshi and Viktor K. Prasanna

High-Performance Parallel Database Processing and Grid Databases / David Taniar, Clement Leung, Wenny Rahayu, and Sushant Goel

Algorithms and Protocols for Wireless and Mobile Ad Hoc Networks / Azzedine Boukerche (*Editor*)

Algorithms and Protocols for Wireless Sensor Networks / Azzedine Boukerche (*Editor*)

Optimization Techniques for Solving Complex Problems / Enrique Alba, Christian Blum, Pedro Isasi, Coromoto León, and Juan Antonio Gómez (*Editors*)

Emerging Wireless LANs, Wireless PANs, and Wireless MANs: IEEE 802.11, IEEE 802.15, IEEE 802.16 Wireless Standard Family / Yang Xiao and Yi Pan (*Editors*)

High-Performance Heterogeneous Computing / Alexey L. Lastovetsky and Jack Dongarra

Mobile Intelligence / Laurence T. Yang, Augustinus Borgy Waluyo, Jianhua Ma, Ling Tan, and Bala Srinivasan (*Editors*)

Advanced Computational Infrastructures for Parallel and Distributed Adaptive Applicatons / Manish Parashar and Xiaolin Li (*Editors*)

Market-Oriented Grid and Utility Computing / Rajkumar Buyya and Kris Bubendorfer (*Editors*)

Printed and bound by CPI Group (UK) Ltd, Croydon, CR0 4YY

27/10/2024

14580254-0005